McGRAW-HILL SERIES IN NUCLEAR ENGINEERING

WALTER H. ZINN, *Consulting Editor*

NUCLEAR ENGINEERING

NUCLEAR ENGINEERING

Theodore Baumeister

Charles F. Bonilla

John R. Dunning

Gioacchino Failla

Alfred M. Freudenthal

William W. Havens, Jr.

John W. Hoopes, Jr.

R. Wayne Houston

George L. Kehl

John W. Landis

John G. Palfrey

David C. Peaslee

Chien-Shiung Wu

Edited by

CHARLES F. BONILLA

Professor of Chemical Engineering, and Chairman, Nuclear Engineering Committee, Columbia University; Consultant, U.S. Atomic Energy Commission, Brookhaven National Laboratory, E. I. du Pont de Nemours & Co., Inc., Westinghouse Atomic Power Division, etc.

1957

McGraw-Hill Book Company, Inc.

NEW YORK   TORONTO   LONDON

NUCLEAR ENGINEERING

Library of Congress Catalog Card Number 56-8167

THE MAPLE PRESS COMPANY, YORK, PA.

THE AUTHORS

Theodore Baumeister

Stevens Professor of Mechanical Engineering, Columbia University; Consulting Engineer: General Public Utilities Corp., South Carolina Electric and Gas Co., Central Hudson Gas and Electric Corp., The Babcock and Wilcox Co., Brookhaven National Laboratory, Worthington Corp. (CHAPTER 13)

Charles F. Bonilla

Professor of Chemical Engineering, and Chairman, Nuclear Engineering Committee, Columbia University; Consultant, U.S. Atomic Energy Commission, Brookhaven National Laboratory, E. I. du Pont de Nemours & Co., Inc., Westinghouse Atomic Power Division, Avco Advanced Development Division, Worthington Corp., etc., Formerly Knolls Atomic Power Laboratory, Pratt & Whitney Aircraft. (CHAPTERS 8 AND 9)

John R. Dunning

Dean of Engineering and Professor of Physics, Columbia University; Director, The Oak Ridge Institute of Nuclear Studies, The Vitro Corporation of America, The Atomic Energy Corporation, Fund for Peaceful Atomic Development, etc. (CHAPTER 1)

Gioacchino Failla

Professor of Radiology and Director of the Radiological Research Laboratory, Columbia University; Chairman, Advisory Committee on Biology and Medicine, U.S. Atomic Energy Commission. (CHAPTER 5)

Alfred M. Freudenthal

Professor of Civil Engineering, Columbia University; Author, "The Inelastic Behavior of Engineering Materials and Structures." (CHAPTER 11)

William W. Havens, Jr.

Professor of Physics and Director of the Nuclear Cross-Sections Laboratory, Columbia University; Member, Nuclear Cross-Sections Advisory Group, U.S.

Atomic Energy Commission; Member, Physics Staff, Brookhaven National Laboratory.   (CHAPTERS 2 AND 3)

John W. Hoopes, Jr.

Senior Chemical Engineer, Atlas Powder Co.; Formerly Assistant Professor of Chemical Engineering and Senior Staff Engineer, Heat Transfer Research Facility, Columbia University.   (CHAPTER 12)

R. Wayne Houston

Associate Professor of Chemical Engineering, University of Pennsylvania; Consulting Engineer, Esso Research and Engineering Co.; Formerly Assistant Professor of Chemical Engineering, Columbia University.   (CHAPTERS 1 AND 6)

George L. Kehl

Associate Professor of Metallurgy, Columbia University; Member, Department of Metallurgy, School of Nuclear Science and Engineering, Argonne National Laboratory; Author, "The Principles of Metallographic Laboratory Practice." (CHAPTER 10)

John W. Landis

Assistant Manager, Atomic Energy Division, The Babcock & Wilcox Co.; Formerly of the Division of Reactor Development, U.S. Atomic Energy Commission.   (CHAPTER 14)

John G. Palfrey

Professor of Law, Columbia University; Associate, Special Committee on Atomic Energy, Association of the Bar of the City of New York; Formerly Member, Office of the General Counsel, U.S. Atomic Energy Commission.   (CHAPTER 15)

David C. Peaslee

Associate Professor of Physics, Purdue University; Consultant, Westinghouse Atomic Power Division; Formerly Associate Physicist, Kellex Corporation, AEC Associate in Physics, Columbia University; Author, "Elements of Atomic Physics."   (CHAPTER 7)

Chien-Shiung Wu

Associate Professor of Physics, Columbia University; Consultant on Radioactivity Application, U.S. Atomic Energy Commission; Author, "Beta Decay." (CHAPTER 4)

. . . the United States pledges . . . to devote its entire heart and mind to find the way by which the miraculous inventiveness of man shall not be dedicated to his death but consecrated to his life.

President Eisenhower addressing
the United Nations General Assembly
December 8, 1953

PREFACE

The first production and research reactors have been operating for well over a decade. Mobile marine reactors have already proved themselves. A nuclear power plant for a central station is being built, and larger ones are on the drafting boards. However, only a beginning has been made on the myriad technical problems in the field.

The first step in the building of a nuclear reactor is the decision as to the specific purpose, limitations on size and weight, desirable operating characteristics, and safety features. Next, one of the various reactor types is selected and the critical size, the concentration of the fuel, the relative amounts and spacing of the fuel, moderator, and coolant, the reflector or breeding "blanket," control elements, and any other similar nuclear factors are determined. The possible materials of construction must be evaluated on the basis of strength, corrosion, radiation damage, and effects on the neutron economy. The reactor must be designed on a calculated margin of safety, with allowance for thermal stress, creep, phase changes, corrosion, and shielding. Alternative fluids, temperatures, and velocities for removal of the heat must be considered on the basis of feasibility, economy, dependability, safety, and any other pertinent factors. Methods of coolant circulation during normal and emergency operation must be fixed, and such an apparently minor matter as a satisfactory valve-stem packing material for contact with a molten metal may reveal itself as a vexing problem. Operating, warning, and emergency control methods, circuits, and apparatus must be selected in light of their speed and reliability. And at all times the characteristics of the power-generating cycle and equipment and of the connected load must be kept in mind.

It is evident that the planning of the reactor and associated cycles and equipment requires development engineers of the broadest background, the most fertile originality, and the surest mastery of engineering fundamentals. This book is offered to aid in the training of such engineers. Its particular goal is to carry the engineer into the details of the application of his field to nuclear-power-plant design. For more specialized applications and later developments, it will be necessary to refer to subsequent literature, particularly through *Nuclear Science Abstracts*. Most such work is reported in nomenclature generally accepted in each field. Accordingly, such nomenclature has on the whole been retained, though in minor respects it is inconsistent between various fields. Since most engineers will have major responsibilities in one of the fields only, the interrelation among the various fields has not been emphasized over the thorough development of the application of each field in nuclear power. All significant interconnections have, however, been brought out in the pertinent chapters, and study plus experience in the whole subject will develop confidence in this respect.

"Nuclear Engineering" is intended to be suitable as a text for a graduate course for engineers and physics majors. It is, however, not beyond the capabilities of interested superior seniors in these lines, or of graduate engineers who wish to prepare themselves in this field by self-instruction. It is also intended to serve as a single compact reference volume to those employed in this field on the detailed work or supervisory levels.

Engineering problems also abound in the preparation of the minerals, their reduction to the fuel metals or compounds, the processing of the spent fuel elements, the recovery of the fissionable materials, and the concentration and disposal of the radioactive fission products. This group of processing problems is highly important from the standpoint of a reactor system as an integrated operation and is crucial in the over-all picture for economic nuclear power in the future. However, these processing problems are usually not handled by the same team that designs the reactor and associated equipment and, if considered at all, are not so critical from the standpoint of timing, space, weight, and speed. These phases have thus not been covered in the present volume. Construction of the nuclear reactor and power plant is of course another engineering operation. Here, however, each technician and engineer works in his own standard field from complete instructions, and a well-rounded preparation in nuclear applications of engineering is not so necessary.

To cover the pertinent fundamentals of so many fields, a concise or handbook style has been used on all the topics that are normally included in the student's undergraduate training or that are discussed at length in readily available references. On the other hand, topics not generally referred to, or not adequately organized and correlated elsewhere, have been discussed in greater detail. In both cases extensive bibliographies are provided.

It is impossible to list the many workers in the government and contractors' laboratories and in other universities who have been of assistance to the authors in the preparation of this book; it must suffice to recognize that without them it could not have been written.

Undoubtedly some errors have slipped through in spite of best efforts to prevent it. The authors and editor will greatly appreciate having any such errors brought to their attention.

CHARLES F. BONILLA, *Editor*

CONTENTS

CHAPTER 1

INTRODUCTION

By John R. Dunning and R. Wayne Houston

From our experience in other fields it is obvious that the atomic age could not have come into being suddenly. Nuclear engineering and technology constitute only one development in one of the fields opened up by the patient efforts and the occasional brilliant discoveries in the last 50 years by hundreds of scientists the world over.

It is quite true, however, that in only 15 years unprecedented progress has been made in transforming laboratory discoveries into the enormous ventures which now support over 100,000 people in this country alone. This has been accomplished through a unique cooperation between scientists, engineers, and production men on the one hand, and government and the taxpayer on the other. Although continuing progress in large-scale nuclear power still must rest to a considerable extent with government, it is significant that the transition to a privately financed program is now well under way.

It is well to reflect briefly upon the course of events that has brought us to the threshold of large-scale nuclear power.

1-1. Historical Basis of Atomic Physics. Our progress in exploring the nature of matter and energy naturally has roots far in the past. Of course, if one wishes, early atomic ideas can be traced back to Democritus and the ancient Greeks. Certainly one must at least recognize the landmarks in the development of the atomic concepts achieved by John Dalton, Robert Boyle, Avogadro, Mendeleev, and others. The ideas represented by the kinetic theory of gases and radiation and statistical concepts of the atoms by Maxwell, Boltzmann, Planck, and others in the last century laid a solid foundation.

In the period just before the turn of the century, about 1895–1900, several remarkable discoveries opened up new avenues leading down to the present day. The work on electromagnetic radiation by Hertz and Maxwell was being put to practical use for communication by Marconi and others. In Germany, Roentgen discovered X rays and thus provided a new tool for medical diagnosis and therapy. To the physicist this tool became a new method for exploring the structure of matter and the interior of atoms. In England J. J. Thomson and others were beginning to take apart the atom, and they established that a new particle, the electron, was a fundamental building unit of all atoms. In that same period in France Becquerel discovered the natural radioactivity of uranium minerals and, followed by the Curies and many others, opened up a new field which was to become the study of the atomic nucleus—or nuclear physics.

Rutherford made a major step forward when he fired α particles at thin gold foils and discovered from the way the α particles were scattered that virtually all the mass of an atom is concentrated in a tiny central core less than one ten-thousandth the diameter of the whole atom.

The modern concept of the nuclear atom again made great progress when Bohr first combined Newton's mechanics with Planck's quantum hypothesis and gave us what may crudely be called the solar-system atom. The Rutherford-Bohr atom postulated a tiny positively charged central sun, or nucleus, surrounded by one or more negative electrons which, roughly speaking, revolve much like planets around the sun. Energy changes, however, could occur only in definite amounts. Bohr's quantum mechanics was remarkably successful in explaining the discrete sharp spectral lines emitted by the hydrogen atom when it is excited in an electrical discharge.

This early atom model has been extended through the work of Heisenberg, Schrödinger, Pauli, and others on the more mathematical bases of matrix mechanics and wave mechanics. It is probably safe to say that, where present-day quantum mechanics has been applied carefully, all phenomena in nature outside the atomic nucleus—i.e., extranuclear phenomena in physics and chemistry, such as atomic and molecular spectra, radiation processes, electricity and magnetism, molecular phenomena, and chemical reactions—appear to be satisfactorily "explained." It must be emphasized that, as Bohr has continually pointed out, physicists have no illusions that our "explanations" represent ultimate reality. Rather they represent simply consistent logical quantitative descriptions in terms of the special language we have adopted.

1-2. Advent of Nuclear Physics Research. Our interest, however, lies deep in the center of the atom, in the tiny atomic nucleus, for here is where over 99.9 per cent of the mass of the atom is concentrated and therefore, in terms of the equivalence of mass and energy, where virtually all the energy of the universe is concentrated. The "fundamental" building stones of the nucleus appear to be the proton and the neutron, often called "nucleons." The proton is the nucleus of the hydrogen atom, the hydrogen ion to the chemists, with a unit positive charge and 1.007600 mass units (in the physicists' mass scale). The neutron, finally conclusively identified by Chadwick in 1932, has no electrical charge, which gives it very unique properties, and it has a mass only slightly more than that of the proton, namely, 1.008982.

Just how can the physicist study this exceedingly small atomic nucleus —so small that even in a tiny atom it is like a pea in the center of an auditorium? The problem is exceedingly difficult, and obviously one must use probes of the same order of size as the nucleus itself in order to gain useful information. One cannot effectively use an elephant's foot as a probe to study the anatomy of a fly. The essential methods used follow the footsteps of Rutherford, who began the use of α particles (helium nuclei) as probes or projectiles to study other nuclei. Ever since 1919, when Rutherford directed high-speed α particles at nitrogen and produced the first nuclear transmutation by observing that even more energetic protons were shot out, the physicists have been busy at an ever-increasing pace finding newer and better ways to use the fundamental particles as high-speed probes to investigate atomic nuclei.

Clearly, positively charged particles must have very high speeds if they are to penetrate through the outer electron cloud, and even then most of these subatomic particles will go through the vast open space inside the atoms, missing the nucleus. A few will be directed by pure chance at the almost infinitesimal target represented by the nucleus, but they must still have very high energy if they are to overcome the enormous and rapidly increasing repulsive forces as they approach the positively charged nucleus in order actually to hit it.

Artificial methods of accelerating particles such as protons or deuterons were gradually developed. First came direct high-voltage devices, such as rectifiers and huge electrostatic generators which developed millions of volts. These high voltages could be used to speed up particles in a high-vacuum tube in a single large step so they would have the required energy. Most successful of all for reaching very high voltages have been the cyclic devices, such as the cyclotron first developed as a practical device by Lawrence and his co-workers at Berkeley, Calif. Many new devices and variations have come along in recent years—the linear accelerator, the synchrocyclotron, the synchrotron, and others. All follow essentially the same idea, using a comparatively small accelerating voltage on a particle many successive times in proper synchronism. Thus a small push used many times achieves the same effect as one big push and avoids the as yet insuperable difficulties of handling tens or hundreds of millions of volts in one step.

The game of "atom smashing" which the physicists and chemists have been playing for the past 20 years or so has been, in essence, to place every element and isotope obtainable into such beams and see what happens. What does happen for nuclear machines in the range of about

10 million volts may be summarized in the following way. Most of the high-speed particles go through the wide-open spaces in the atoms and cause ionization excitation of the outer electrons only. A very few by pure chance will be directed at or near the nucleus. Those which only pass near are deflected or scattered by the nuclear force fields and thus provide a tool for exploring these fields. Those which are directed virtually at the nucleus and have enough energy to overcome repulsive forces, actually hit the nucleus and are normally *captured* to form at least momentarily an excited *compound* nucleus. Sometimes the new nucleus achieves stability by simply emitting the excess energy as γ-ray photons. Usually the compound nucleus is so excited that one or more protons or neutrons or even α particles are ejected almost instantaneously.

Thus, by using the projectiles at our disposal—protons, deuterons, and α particles—and the secondary particles such as neutrons, it has been possible to transmute almost any nucleus into a new nucleus one or two steps up the periodic table or one or two steps down by adding particles to the target nucleus or "chipping off" particles from it. Whereas nature has some 276 stable and 50 radioactive isotopes for all the elements, now with the atom smashers, and with the large number of new isotopes formed in the fission processes, some 850 new man-made radioactive isotopes have been formed.

Nuclear transmutation processes form essentially a nuclear chemistry whose reactions can be described in the same general way as those of ordinary chemistry. Most of the synthetic nuclei formed are artificially radioactive isotopes and decay in an exponential manner, with half-lives from fractions of a second to thousands of years, by emitting a high-speed electron or positron before finally becoming stable.

The energy changes in nuclear reactions are particularly interesting to the physicists. In ordinary chemical reactions we have long been accustomed to exothermic and endothermic processes where, in atomic rearrangements, energy is given out or absorbed.   If we take, for example, a process such as the burning of coal, which is largely carbon, we write

$$C + O_2 \rightarrow CO_2 + 14,000 \text{ Btu/lb}$$
$$\rightarrow CO_2 + 1.56 \times 10^{-19} \text{ cal per molecule}$$
$$\rightarrow CO_2 + 4.1 \text{ ev per molecule}$$

If you are a power-plant engineer feeding coal to the combustion chamber, you will write the upper equation, which indicates an exothermic reaction liberating around 14,000 Btu/lb, depending on the grade of coal. A chemist would write the reaction as giving 1.56×10^{-19} cal per molecule of CO_2 formed. Physicists and many chemists as well would normally write the energy release as 4.1 ev per molecule of CO_2. When you introduce food into your stomach by eating, you employ the same type of reactions (including, of course, the hydrogen contained), since the human power plant operates for internal heating and supplies energy for physical processes by combustion with oxygen in the same manner—incidentally with over-all efficiencies in the same range as modern power plants, 20 to 40 per cent.

In terms of the equivalence of mass and energy, $E = mc^2$, this energy released by atomic rearrangement means that the C and O_2 entering the reaction actually weigh somewhat more than the CO_2 formed as a reaction product. The difference expressed in weight is only 10^{-10} lb for each pound of CO_2 formed; this is too small to be detected as a weight change by present available methods, but nevertheless such a change does occur.

In nuclear reactions we use essentially the same language. When we combine $_5B^{10}$ with slow neutrons we write the reactions as in Fig. 1.1.

Nuclear reaction

$_0n^1 + {}_5B^{10} \rightarrow {}_5B^{11} \rightarrow {}_2He^4 + {}_3Li^7 + W$ (energy)

Mass plus energy balance in physical mass units (pmu)

Mass:　　　　$_0n^1 =$　1.008982　Mass:　　　　　　　$_2He^4 =$　　4.003873

　　　　　　$_5B^{10} =$　10.016114　　　　　　　　　　$_3Li^7 =$　　7.018223

　　　　　　　　　　　　11.025096　　　　　　　　　　　　　　　　　11.022096

Energy:　　　$_0n^1 =$　0.000000

Mass + energy = 11.025096　Mass + energy　　　　　=　11.025096

　　　　　　　　　　　　　　Energy (by difference)　=　+ 0.003000 pmu

　　　　　　　　　　　　　　Energy (by difference)　=　+ 2.794 Mev

　　　　　　　　　　　　　　Energy (experimental re-

　　　　　　　　　　　　　　sults):　　(1)　　　　　=　+ 2.788 Mev ± 0.010

　　　　　　　　　　　　　　　　　　(2)　　　　　　+ 2.793 Mev ± 0.027

(exothermic)

FIG. 1-1. Slow-neutron and boron-10 nuclear reaction. [*Masses from E. Segrè (ed.),* "*Experimental Nuclear Physics,*" *vol. 1, John Wiley & Sons, Inc., New York, 1953.*]

The intermediate state has a very short lifetime, generally less than 10^{-11} sec. But there is an enormous difference from chemical reactions. When B^{10} combines with neutrons in a nuclear reaction, we get about 1 million times as much energy as when C and O_2 are combined in an ordinary chemical reaction, where only the low-energy valence electrons are involved. This is characteristic of nuclear reactions, in which the energy changes usually involved are millions of times greater than in the ordinary physical-chemical processes we have known heretofore.

In nuclear reactions the energy changes are so large that the mass changes are appreciable. To make our books balance we must consider that (mass + energy) is conserved:

Mass + energy of initial reacting particles

　　　　　　　　　　　= mass + energy of final reaction products

Mass + energy checks have been made for a large number of nuclear reactions, and the balance for the initial and final states is within experimental error.

Nuclear transformations of the type described, it should be carefully noted, involve mass conversions into energy generally less than 1 per cent. In the uranium-fission case, the mass change is only about 0.1 per cent. Nuclear processes observed up to now do not involve the destruction or conversion into energy of the fundamental particles themselves.* They are really simply *rearrangements* of the nucleons in which the energy changes, positive or negative, are the energy differences between different arrangements. These are naturally only small fractions of the masses of the particles involved, for the binding energy of nucleons in atomic nuclei is normally only 6 to 8 Mev per nucleon. This is quite analogous to ordinary chemical reactions in which the atoms in molecules are simply rearranged, and the "arrangement energy" changes may be either positive or negative.

What is behind these rearrangements leading to large energy effects may be paraphrased in this way. From the well-known Coulomb law in elementary physics it is known that like electrical charges repel each other with an inverse-square-law force given by

$$F = k \frac{Q_1 Q_2}{d^2}$$

where k = constant, depending on the unit used
Q_1 and Q_2 = electrical charges on the particles
      d = distance between the particles

Now at the almost infinitesimal distances between nuclear particles, of the order of only 10^{-13} cm, where so far as we can see the Coulomb electrical forces still operate, the denominator d^2 is of the order of 10^{-26} cm^2. Thus, owing entirely to the exceedingly small nuclear distances, enormous electrical repulsion forces exist between protons in an atomic nucleus. On this basis the nucleus should explode violently. Actually, however, most atomic nuclei in the world are obviously quite stable. Clearly there must be a new type of *attractive* force operating in the nucleus whose effect at the tiny nuclear distances must *more* than counterbalance the enormous Coulomb electrical repulsive forces.

The nature and origin of this new nuclear force constitute one of the major problems which nuclear physicists are attacking today. Evidently it must be a short-range force, which shows "saturation," and is probably of the nature of an "exchange force" somewhat similar to that by which the H_2 molecule is stable because of electron exchange between the two H atoms.

* This does occur, however, in the "annihilation" reaction between an electron and a positron, in which only a γ ray remains (Sec. 2-17), and in the proton-antiproton reaction.

This has stimulated numerous theoretical attempts, beginning with that of the Japanese physicist Yukawa, to explain the neutron-proton attractive forces as arising from exchange of mesons, the newest of the fundamental particles to be discovered. Indeed, physics is now almost embarrassed by the variety of mesons and other particles that have been discovered.

1-3. Uranium Fission. Production of nuclear reactions with charged particles as in a cyclotron is inefficient in an over-all sense. Charged particles must have high energy to penetrate the atomic nucleus, and even with 400-Mev particles, only about one-third of the reactions are nuclear. Most of the energy of bombarding charged particles is frittered away ionizing electrons in the vast wide-open spaces away from the nucleus.

With the discovery of the uncharged neutron in 1932, the situation changed. Particularly with the opening of the slow-neutron field in 1934–1935, completely new possibilities became clear. In the first place, neutrons interact only with atomic nuclei, and as they bounce around in matter from nucleus to nucleus they lose energy by nuclear collisions. Even though finally slowed down to ordinary thermal molecular equilibrium speeds, eventually they are captured by atomic nuclei. In the slow-neutron-capture cases, the nuclear processes are always exothermic.

It is well to emphasize a very fundamental point. There are many kinds of exothermic nuclear reactions known, and there are many kinds of high-energy processes observed, such as cosmic-ray reactions, which may involve energies a million times greater—10^{12} ev and more. Unfortunately it is common to speak of these loosely as "releasing" energy. Such processes are very interesting from a scientific standpoint, but in general all these energy releases are completely meaningless from the standpoint of practical energy unless there is some mechanism for making the reactions self-perpetuating so there is a continuous self-maintained energy release involving very large numbers of atoms rather than single atoms.

It is known that if one takes a piece of paper such as this page, which is in part made of carbon and hydrogen atoms, combining the hydrogen and carbon atoms in it with more oxygen will result in exothermic processes. If the flame of a lighted match is brought near one corner of the paper, and the temperature raised to some threshold level (the "kindling temperature," loosely speaking), then the C and H atoms in the paper start reacting with O atoms in the air. Now the important thing is not alone that the reaction is exothermic; the important thing is that, as C and H atoms combine with O and produce $CO_2 + H_2O$ + energy, they simultaneously "trigger" off their nearest neighbors. Thus the reaction, once started, proceeds from molecule to molecule and is propagated as a "chain reaction" throughout the system.   The method of propagation here is essentially that the photons (heat energy) liberated in the reaction trigger off reactions with neighboring atoms.

The uranium-fission reaction gave the necessary key to a practical energy release.   The reaction of uranium with neutrons had long before 1939 been a subject of much work and speculation.   The first measured neutron-uranium cross sections (interaction probabilities) at Columbia in 1935 appeared anomalously high.   Fermi and his co-workers separated out chemically various artificial radioactivities from the neutrons-plus-uranium reaction end products and concluded that "transuranic" elements above uranium were formed.   The Joliot-Curies in Paris elaborated on the large number of radioisotope periods, suggesting series transformations.   Meitner in Berlin in 1936–1938 carried out the most complete investigations of neutron-induced radioactivities in uranium and postulated a number of series of radioactive decay chains of transuranic elements, among which was an isotope similar to radium.   After Meitner left Berlin, Hahn and Strassmann continued to carry forward the uranium work.   Hahn, a pioneer from the early days of radium natural radioactivity, carried out a series of elegant but simple tracer "carrier" fractional crystallization separation methods in which he showed conclusively that the radium-like artificial isotope, supposedly heavier than uranium, was in fact identical with ordinary barium.   This exciting discovery immediately led physicists to look for the enormous energy release that would have to accompany a process of fissioning uranium into two or more smaller fragments.   It was obvious that if element 92 splits in half, each half having, say, 46 positive charges, at the instant of separation the repulsive force between the two halves would be enormous.   If one simply integrates the Coulomb force law from a typical nucleon distance out to infinity, something like 200 Mev of energy should be released.   This is about 50,000 times the energy from burning a single atom of carbon with oxygen.   That such was the case was quickly confirmed abroad by Frisch and Meitner and in several laboratories in this country.

The enormous energy release in uranium fission by itself would be quite useless from a practical standpoint were it not for the fact that the splitting is so catastrophic that several extra neutrons are flung out or "evaporated" in the process.   These secondary neutrons, which were quickly observed in various laboratories, are the key to a sustained chain reaction.   The average number of neutrons released per fission is usually between 2.5 and 3.0, depending on the fuel isotope.   Only one of these is needed to cause another fission and thereby sustain the chain reaction.   However, if the number of neutrons released per fission had been 2, there could practically not be a major atomic age.

1-4. Advent of the Nuclear Chain Reactor. Although a fission chain reaction is simple enough in principle, the vagaries of neutron interactions with atomic nuclei set strict requirements on its achievement. Comparatively few of the fast, or high-energy, neutrons born in fission will cause another fission when striking uranium nuclei directly. Most such impacts result in inelastic scattering or nonfission capture processes. In addition, fast neutrons move through distances of the order of inches before striking any nuclei at all and therefore have an opportunity to escape completely from the system without undergoing any reaction. This gives rise to the concept of a "critical" amount required to sustain a chain reaction. With natural uranium, the nonfission capture processes predominate to such an extent that no matter how much of it is brought together a chain reaction cannot result. If uranium is enriched in its isotopic content of U^{235}, there is a "critical" concentration for which an infinite mass will just be able to sustain a chain reaction. For greater enrichment, neutrons in excess of those required to sustain the reaction must be allowed to escape from the system to prevent an "explosion" or runaway. There exists a critical mass corresponding to each degree of enrichment.

Contrary to the case with charged particles, neutrons are usually much more effective in producing nuclear transformations when they are slow than when they are fast. Thus another avenue of approach is to "moderate" or slow down the fast neutrons before bringing them into contact with additional uranium nuclei. If a mutual force of attraction exists between a neutron and a nucleus, the longer the time allowed for the force to operate, the greater the probability of interaction. Experimentally it is found that many neutron reactions occur at rates strictly proportional to the inverse of the neutron velocity.

In uranium the question is, is there a strong capture tendency which will give a much increased fission rate with slow neutrons? Water is a good moderator because H nuclei have the most efficient mass of all nuclei for slowing down neutrons, since an equal mass removes, on the average, half the energy of the impinging particle—the maximum possible. The process becomes one of statistical neutron diffusion. So if one surrounds a neutron source and uranium with water, the fast neutrons will travel a few centimeters, make a collision with an H nucleus, lose energy, and bounce to collide with some other H (or O) nucleus. In a sufficient series of collisions, the neutrons will come down to *thermal equilibrium* with the moderator molecules and have ordinary molecular kinetic energies appropriate to the temperature of the medium. Such equilibrium speeds are only a few thousand feet per second instead of many tens of thousands of miles per second.

With normal uranium and ordinary water, it appears that no matter how much uranium and how much water is brought together, critical chain-reacting conditions cannot be achieved.* The major reasons are two: (1) ordinary H captures neutrons, in the parasitic reaction

$$_1H^1 + _0n^1 \rightarrow 1D^2 + 2.2 \text{ Mev } (\gamma \text{ ray})$$

and (2) uranium itself also has a considerable amount of parasitic non-fission resonance capture processes, where neutrons are lost by capture without producing fission, thus inhibiting a chain reaction. In addition, in any system of finite size there is always some small loss of neutrons by "leakage" through the outer boundary.

One line of approach was to minimize parasitic capture by employing a different moderator. Carbon was used by Fermi, Szilard, and Pegram in their early work at Columbia in 1939–1941. Carbon, atomic weight 12, serves to transfer only about one-third as much energy on the average per neutron collision as the hydrogen in H_2O, but the capture cross section of carbon is so much lower that, even though many times as much carbon moderator, and therefore also uranium, is required, nevertheless carbon is feasible for a successful chain reactor with normal uranium while H_2O is not. Similarly D_2O in large enough quantities is feasible. Producing graphite and uranium in large quantities in pure enough form, free of additional impurities like boron which capture neutrons parasitically, proved to be a serious production problem. The first graphite-uranium lattice test piles, 8- to 10-foot cubes set up at Columbia University, gave the basic physical data to proceed with full chain-reacting pile design. After Fermi and his group had moved to the University of Chicago, where The Metallurgical Laboratory was established under the direction of Dr. Arthur Compton, the first sustained reaction was achieved Dec. 2, 1942, at West Stands–Stagg Field.

Next came the large air-cooled graphite-uranium lattice pile at Oak Ridge, Tenn., which was built as a pilot plant by the du Pont Company. This was the first reactor to involve real engineering problems of heat transfer, coolant flow, fuel-element fabrication and handling, and instrumentation and control. Later came the huge Hanford Engineer Works construction at Richland, Wash. Here the du Pont Company constructed several large graphite-uranium piles, using the Columbia River water as a coolant. While energy is liberated at the rate of hundreds of thousands of kilowatts in the Hanford piles, the piles were not designed to produce the high temperatures required for effective practical use of heat energy. The major purpose of this huge plant was to use the extra by-product neutrons from the chain reaction to produce plutonium by the reaction

$$U^{238} + n \rightarrow U^{239} \xrightarrow{\beta} Np^{239} \xrightarrow{\beta} Pu^{239}$$

* Recent measurements indicate that this may be possible at high temperatures.

Since only one of the secondary neutrons emitted in each fission is required to maintain a chain reaction, the remainder is available to allow for losses due to parasitic neutron capture, leakage of neutrons from the pile, and by-product processes. Typical of such processes is the neutron capture by the abundant uranium 238 isotope to form U^{239} (23-min half-life), which in turn in two steps of disintegration emitting β particles becomes first neptunium 239, element 93 (2.3-day half-life); and then plutonium 239, element 94. $_{94}Pu^{239}$ is an α emitter with a half-life of about 25,000 years. More important, like U^{235}, it is fissionable with slow neutrons. Plutonium 239, a synthetic man-made isotope, is thus a substitute for U^{235} for both power and bomb purposes. It has some good properties and some undesirable properties compared with U^{235}. Thus the final success in making a chain reaction work with natural uranium culminated in the production of a synthetic isotope which could be separated from irradiated uranium in concentrated form by chemical methods. The chemical processing plants represent a considerable part of the total capital investments at Hanford, and developing improved designs is a continuing high-priority part of the Atomic Energy Commission's present program. It is now clear that one of the major factors in achieving low-cost atomic power in peacetime application is the achievement of reprocessing systems with low capital and operating costs.

1-5. Reactor Development for Nuclear Power. While some interesting long-range possibilities exist for conversion of fission fragment energies in part directly into electrical energy, all practical methods for nuclear-energy utilization considered feasible in the immediate future involve degradation of nuclear energy into heat and transfer of the heat to some working fluid in one or more steps, either (1) to some process heat application, or (2) to some more or less conventional prime mover such as a steam or gas turbine. While many people have unfortunately spoken loosely about using nuclear energy in low-temperature form, it is quite clear that, in general, nuclear energy must produce heat available for use at least in the temperature ranges commonly used in industrial and power plant practice, or else it cannot be competitive with other methods of producing heat energy (possibly with the exception of a few special purposes). If steam is the final working fluid, temperatures for most purposes must be in the range of 700 to 1000°F to be really interesting, which means that the nuclear system itself must be appreciably hotter to allow for the necessary heat-transfer temperature differentials.

It is now generally agreed that high-level nuclear reactors feasible for producing practical operating temperatures and over-all economy should preferably not be based on natural uranium. Concentrated nuclear fuels are desirable, such as U^{235} concentrated at least in some degree. It was this strong conviction rather than possible bomb applications which led some physicists in 1940 to press the development of large-scale U^{235} separation, since it was believed firmly that no nuclear-power industry could be developed without having U^{235} separation plants as a base or steppingstone.

There is one other strong premise for a large-scale nuclear-power industry based on a uranium-type chain reaction. Uranium 235 is surprisingly cheap by the automatic production processes employed— some \$7,000 to \$10,000 per lb on various published unofficial cost estimates,* roughly comparable to \$7 to \$10 per ton of coal for the same energy equivalent. Nevertheless, in view of the comparatively larger capital investment costs envisaged in nuclear plants for the immediate future on the one hand, and on the other hand the obvious limitations of a U^{235} nuclear fuel which is only 1 part in 140 of normal uranium, the use of U^{235} alone in power units does not appear likely for more than a relatively limited number of special power applications.

1-6. Breeding.† Large-scale utilization of nuclear-fission power in our national economy does not seem possible for the long-term future unless "net-gain breeding" of new fissionable material, such as $_{94}Pu^{239}$ or $_{92}U^{233}$, is employed. The possibility of net-gain breeding arises because an excess number of neutrons are emitted per fission under some conditions. One neutron per fission is required to maintain a chain reaction. If the over-all "neutron economy" is such that, after allowing for neutron leakage from the nuclear reactor and true parasitic neutron capture processes, more than one net neutron is still available to react with "fertile materials" like U^{238} to form Pu^{239}, or to react with Th^{232} to form U^{233} in a similar process, then it is possible to produce more new fissionable fuel materials, Pu^{239} or U^{233}, by "conversion" than we burn.

If a reactor can be designed from both an engineering and a heat-transfer viewpoint with enough margin to produce a net gain of Pu^{239} or U^{233} and allow for recycling and reprocessing losses, then in principle, by recycling nuclear fuel and fertile materials, we can look forward eventually to converting an appreciable fraction of the U^{238} or Th^{232} in the world into high-grade fissionable fuels. The consequent availability of hundreds of times the amount of nuclear fuel as compared with U^{235} alone and the effect of this upon over-all nuclear power economics are obvious. A practical "net-gain power breeder reactor" would be a tremendous step forward, even though one must recognize that a net-gain breeder acts like compound interest and that the time required to double the amount of fissionable material will be long, even with net gains of, say, 25 per cent per cycle. The success of the U.S. Atomic

* Also see App. K.

† This term is often used in the more restricted sense of what is here called "net-gain breeding."

Energy Commission's Experimental Breeder Reactor (EBR) at Arco, Idaho, has been the first step in this direction.   Breeding gains in the range of 0.5 to 1.0 of course also markedly reduce "apparent" fuel costs, and it is not necessary to achieve net-gain breeding above 1.0 to effect substantial economies.

1-7. Reactor Types.   The development of nuclear reactors has taken place along several basic lines of approach.   In order to understand these developments better, it is necessary to consider the classifications of different reactors.

Broadly speaking, reactor types may be characterized in several ways. In terms of effective "neutron spectrum," at one extreme the "thermal reactor" involves the fissionable material with a considerable amount of moderator, so that the neutrons are largely near thermal velocities. Conversely, at the other extreme the fissionable material may have very little or no associated moderator, so that the neutrons are not slowed appreciably and we have therefore a "fast" reactor.   With intermediate amounts of moderator, the neutron spectrum will be intermediate between thermal and fast, and one has an "intermediate" type of reactor.   The fissionable fuel may be U^{235}, Pu^{239}, U^{233}, or mixtures.

The reactor fuel may be in many forms—metals, compounds, alloys, mixtures, etc.   For moderators there is a considerable choice—ordinary water, heavy water, carbon, beryllium, etc.   Should the reactor be a breeder, U^{238} or thorium may be used for fertile material.   The reactor may be "homogeneous" with working materials mixed, or "heterogeneous," i.e., a lattice type with components segregated.   To remove the heat for useful purposes a heat-transfer medium is necessary; it may be a gas like helium, a liquid like water, or molten metals of low melting point like sodium.

Clearly, a nuclear reactor involves a complex mixture of nuclear physics and process engineering.   Underlying all possible designs is the all-important question of materials, for extremely severe limitations are placed on the materials of construction.   Not only must they stand the operating temperatures and pressures and corrosion conditions, but their nuclear properties must be such that their parasitic neutron capture is not excessive, and in addition they must have high "radiation stability." The enormous radiation flux, in which the materials are bombarded with neutrons and γ rays, is serious, since many materials suffer marked deterioration of their physical properties.   Fast neutrons knock atoms out of their lattice positions and produce changes in structure.   Actual transmutation of the atoms into other types of atoms further changes the character of materials.   Some of the best high-temperature alloys developed for ordinary power-plant use may be useless.   The testing of existing materials and the development of new materials, metals, alloys, ceramics, etc., to meet the special problems of nuclear reactors are matters of the utmost importance. Materials formerly little used and expensive find here new applications which will bring their production costs down. Beryllium and, more recently, zirconium are examples.

The second decade of nuclear-reactor development is marked perhaps most notably by the attention to economics, a sure sign of the transition from science to engineering. Reactor designs undergoing current development appear to hold the most promise for nuclear-power generation competitive with conventional central power stations. These draw heavily on experience gained from the Hanford production reactors (natural-uranium–graphite, water-cooled), which have now operated for over 10 years; the various research reactors, notably the Materials Testing Reactor (MTR, enriched uranium, light-water-moderated and -cooled) and its predecessors; the Experimental Breeder Reactor (EBR, enriched-uranium liquid-metal-cooled fast reactor); and the submarine reactors. In the latter, the U.S.S. *Nautilus* is powered by a thermal enriched-uranium pressurized-light-water-cooled reactor (STR), while the U.S.S. *Sea Wolf* has an intermediate enriched-uranium liquid-sodium-cooled reactor (SIR).

The current "five-year" reactor development program of the Atomic Energy Commission is outlined in Table 13-15. The pressurized-water reactor may be regarded as stemming partially from MTR experience and more directly from the submarine thermal reactors. A chief advantage, the established technology, is temporal. Its great disadvantage is the high pressure, about 2000 psi, which the reactor vessel must withstand with only very limited inspection possible. The intermediate-size breeder reactor, a fast reactor, stems from EBR but also has a much older generic ancestor in Los Alamos' "Clementine," an immediate postwar mercury-cooled, plutonium-fueled reactor of small size. Aside from the long-range importance of breeding, liquid-metal-cooled reactors enjoy advantages of higher temperatures and lower pressures, and with sodium coolant, surprisingly perhaps, the corrosion problems of fuel elements are less severe than with water cooling. Advantage of this fact is also taken in the sodium-cooled–graphite experiment (SRE).

One of the most intriguing possibilities is the use of the reactor core as the boiler itself. The above-mentioned reactors require an external heat exchanger to produce steam for driving turbines. Boiling-water reactors offer potential economic savings in the form of lower pressures, lower pumping costs, and lesser external shielding requirements. Nuclear stability due to fluctuations in moderator density does not appear to be as great a problem as originally anticipated. Corrosion and erosion could be problems, since deposition of radioactive material in the turbine would hinder maintenance, but current tests indicate these difficulties not to be serious.

All heterogeneous reactors suffer the serious disadvantage of requiring periodic removal of fuel elements and fabrication of new ones. Long-range planning will not permit once-through operation of uranium in solid fuel elements, since it does not appear feasible to extract more than a few per cent of the available energy because of physical breakdown of the fuel elements. Reuse of fuel requires removal of the highly radioactive fission products before refabrication into new fuel elements because of the health hazard as well as neutron losses. In addition, the strong α activity of the fuels themselves, particularly plutonium, requires expensive, complicated fabrication facilities to eliminate the ingestion hazard. It is thus clear that fuel-element design is highly important from the point of view of life and manufacture as well as use.

This problem is avoided by using a homogeneous reactor, for which the handling of fluid fuels is merely a matter of pumping through pipes and necessary equipment. The homogeneous water reactors under development at Oak Ridge may be regarded as a direct outgrowth of experience with the so-called "water boilers" originating at Los Alamos. A 1000-kw reactor (HRE, Homogeneous Reactor Experiment) has operated successfully and is currently being scaled up by a factor of 3. The disadvantages of such a reactor relate to the handling of the very highly radioactive fuel solution, the extreme importance of avoiding leaks, and the production of large quantities of radiolytic gas. The latter is principally hydrogen and oxygen due to decomposition of water in the reactor, as well as volatile fission products. The evolution of this gas tends to produce nuclear instability by density variation, and the mixture is also highly explosive. For reasonable efficiency the vessel must, of course, be under pressure, to operate at elevated temperatures. Not listed, but also of great potential importance, is the Homogeneous Thorium Reactor (HTR), also under development at Oak Ridge. This reactor will take advantage of the fact that breeding of Th^{232} to U^{233} is best accomplished with thermal neutrons.

A longer-range development in the homogeneous field uses fuel dissolved in liquid metal, e.g., bismuth. Chemical processing to remove parasitic fission-product poisons poses many difficult problems by comparison with an aqueous system.

Although there are several natural-uranium heavy-water-moderated reactors in existence, high fixed charges for a large inventory of heavy water constituted in the past a considerable barrier for economic power production. Decreased costs (App. K) and technology gained from the operation of the large production reactors at Savannah River may, however, alter this picture.

1-8. Future of Nuclear Power Plants. The nuclear-energy field is growing with startling rapidity. There is no longer any real doubt that electrical power generated from nuclear energy will take its place alongside conventional sources of energy. In many energy-starved areas of the world nuclear power will soon dominate and will raise productivity to hitherto undreamed-of levels. Technical information required for the design of nuclear power plants is becoming available at ever-increasing rates.

One of the most important factors involved in the future of nuclear power plants is location. Economics dictates that power must be generated relatively near the area of utilization, otherwise distribution costs become excessive. Although nuclear reactors can be built with inherent safety from the nuclear-stability point of view, they must also be safe from any other malfunction or accident that could cause the spread of the dangerous radioactive contents over their immediate surroundings. This problem can be minimized by containment of the entire reactor structure in an essentially leakproof vessel, such as the 225-ft 3800-ton steel sphere at West Milton, N.Y., for the first model of the SIR. Whether or not it will be possible to locate a large reactor, with or without such containment, in the heart of a heavily populated area, can be determined only by operating experience. This factor may well be a deterrent to rapid growth of nuclear power in the United States, but current reactor experience reveals a remarkable safety record.

Aside from the application to large central-station power plants, it is worthwhile to mention the existing applications to marine and aircraft propulsion. Great strides have been made in both areas, particularly the former, where heavy shielding requirements are less restrictive. It is also of interest to note that more energy is produced and used in this country in the form of direct heat than as electrical power, and one can reasonably expect developments in this direction, although at a less rapid rate. Longer-range possibilities include the carrying out of new or difficult chemical reactions utilizing the purely chemical effects of neutron or γ irradiation within a reactor, or possibly even the ionizing power of fission fragments directly.

CHAPTER 2

NUCLEAR PARTICLES

By William W. Havens, Jr.

2-1. The Atomic Structure of Matter. The atomic theory of matter attempts to describe the behavior of matter in terms of the characteristic properties of the elementary particles which constitute all matter and the interaction between these elementary particles.

These properties are mass, charge, spin, magnetic moment, and statistics. The statistics of a particle are the rules to which a particular particle must conform when a large number of the particles are involved. These characteristics are capable of being measured either directly or indirectly by an experiment which can be performed in the laboratory. When all these properties of a certain particle are known, then this particle is uniquely determined.

Although the fundamental particles which constitute all matter are submicroscopic, the characteristics which uniquely describe these particles are for the most part properties which might characterize a macroscopic particle such as a billiard ball. The difference lies in the magnitude of the quantity used to describe the particle. In the macroscopic or classical theory of matter the characteristics of a particle can have any possible value, whereas in the submicroscopic system, since we are dealing with a single particle or a few particles, the properties of the system must be made up of units of this same property of the elementary particles. The units of quantization are so small that in the macroscopic world one unit difference in the property studied would not be measurable, but in the submicroscopic world the units are of the same order of magnitude as the total quantity we are dealing with, and therefore this quantization must be treated in detail. Thus the method of handling systems in classical macroscopic physics and in submicroscopic physics must necessarily be different.

A simple example easily illustrates this point. Consider the mass of a 1-mm sphere of water. This drop weighs 4.1888×10^{-3} g at 14.50°C. However, it contains 2.3251×10^{18} molecules of water. If we add or subtract one molecule of water, the change in weight would be beyond

17

the experimental error possible in even the most accurate weighing. Therefore we can assume when dealing with macroscopic properties that these quantities are continuous and can have any value whatsoever. Suppose we try to compare the mass of one atom of plutonium of mass 240 with the mass of the hydrocarbon molecule which also has mass number 240. This evidently is $C_{17}H_{36}$ and not $C_{17}H_{35}$, which obviously has the wrong mass. These masses are different by a definite amount, not by an amount that could have any value. It is not possible to lose one-half the mass of the hydrogen atom. The mass of the molecule must be an integer in units of the elementary particles constituting this molecule.

The above example may seem trivial, as we are accustomed to thinking in terms of a quantized mass for molecules. We have accepted the concept of the atomic theory of matter. However, we should realize that the theory was not always accepted.

2-2. Quantum Mechanics. The modern phase of the atomic theory had its inception in the early nineteenth century when Dalton proposed his law of definite proportions. However, it was not until the end of the nineteenth century that the atomic theory and therefore the idea of quantization of matter was generally accepted.

In order to describe adequately the radiation emitted by a hot body, the line spectra of atoms, the photoelectric effect, the Compton effect, diffraction of electrons, and many other phenomena of modern physics, it was necessary to extend the idea of quantization to quantities other than mass.

In 1896 J. J. Thomson recognized that electrons were distinct particles with a definite ratio of charge to mass. He thereby introduced the idea of quantization of charge. Up until that time charge was always thought to be continuous and could therefore have any value. In fact the electromagnetic theory, which has been so successful in describing even such a modern invention as radar, uses the concept of a continuous charge. The charge on a single electron was measured in 1907 by R. A. Milliken, and today the electrons are among the best understood and most universally accepted of the elementary particles.

In 1898 Max Planck quantized the energy of a system of oscillators in order to explain black-body radiation. He had to assume that the energy of each individual oscillator could not be arbitrary but had to be in units of a quantum of action which he called h. Thus the energy E of an oscillator would have to be $E = nh\nu$ where n was an integral number, h the quantum of action, and ν the frequency of the oscillator. The magnitude of this quantum of action is extremely small,

$$h = 6.624 \times 10^{-27} \text{ erg-sec}$$

The minuteness of this quantity explained why it was never of importance in macroscopic physics. However, in submicroscopic physics this constant is one of the most important fundamental quantities.

In 1911 Bohr quantized the angular momentum of a particle in order to explain the emission of bright-line spectra. He introduced this quantization of angular momentum in a rather arbitrary manner, but by using this postulate he was so successful in explaining the interrelation of a large quantity of spectroscopic data that a great deal of progress was made using this theory of the atom in the next few years.

Another principle of submicroscopic physics was introduced arbitrarily into atomic physics by Wolfgang Pauli in 1925. This principle stated that two particles which were exactly alike in every detail could not occupy the same state in one atom at the same time. This principle, called the "Pauli exclusion principle," was very successful in explaining the periodic table of the elements and is now extremely important in all branches of modern physics.

In order to explain the photoelectric effect, Einstein in 1905 introduced the quantum of action into the theory of light. He assumed that a quantum of energy $E = h\nu$ was associated with a light wave of frequency ν. This quantum of energy, a photon, acts like a particle with an energy $h\nu$ and a momentum $p = h\nu/c$, where c is the velocity of light. Since the light wave or photon is traveling with a velocity c, the relationship between energy and momentum is the same for photons as it is for particles in ordinary relativistic mechanics.

In 1924 Louis de Broglie introduced the idea that a wave should be associated with a particle of mass m traveling with a velocity v. The wavelength is given by the formula $\lambda = h/mv = h/p$. He had no particular basis for this idea other than that he believed the universe to be symmetrical. Since a photon was associated with a light wave, then a wave should be associated with a particle.

This concept quantitatively explained the wave character of a beam of particles such as the diffraction of electrons by crystals and completed the set of ideas necessary to the logical formulation of the quantum theory of modern physics.

Several of the concepts mentioned above were arbitrarily introduced into the theory to give a good explanation of particular phenomena. Although in many cases a good explanation could be given for separate phenomena, there was no logical structure correlating the science of atomic physics until Schrödinger and Heisenberg introduced wave and matrix mechanics in 1926. About a year later Born showed that the two systems were mathematically equivalent. From that point on, quantum mechanics, as it is now called, was on a sound logical basis comparable to that of classical mechanics. It was now possible by assuming certain postulates to describe how a given system would act in a specific set of circumstances. In this respect quantum mechanics is no different from classical mechanics. However, the postulates are a little more difficult to understand in that they are not directly associated with macroscopic physical phenomena, as is the case in Newtonian mechanics.

However, there is a fundamental difference between quantum mechanics and classical mechanics when applied to the measurement of, e.g., the position and momentum of a specific particle. In classical mechanics both the position and momentum of a particle could in principle be determined with infinite accuracy if the measuring instruments were good enough. In quantum mechanics both these quantities cannot be measured simultaneously with infinite accuracy. The Heisenberg uncertainty principle of quantum mechanics says that in the simultaneous determination of both the position and the momentum the uncertainty in the position Δx multiplied by the uncertainty of the momentum Δp_x must be equal to or greater than the fundamental quantum of action h, or

$$\Delta x \, \Delta p_x \geqq h \tag{2-1}$$

This is not a practical limitation on the accuracy obtainable but a theoretical limit; no matter how accurate the instruments can be made, this relationship must hold.

Since h is an extremely small quantity, the theoretical limitation is well beyond the practical limit of measurement. However, examination of an idealized experiment will show how the wave associated with a particle places the fundamental limitation on the accuracy obtainable. Bohr gave an excellent example of such an idealized experiment. Suppose the position of an electron is to be determined with a microscope. The resolving power of a microscope can be shown to be

$$\Delta x = \frac{\lambda}{\sin \alpha} \tag{2-2}$$

where Δx = distance between two points which are just resolved
λ = wavelength of light used
$\alpha = \frac{1}{2}$ angle of cone of light coming from illuminated object to objective of microscope (Fig. 2-1)

If Δx is the uncertainty in position of the electron, Δx can be made small by using light of very short wavelength such as X rays or γ rays. However, according to Einstein, in the quantization of light the minimum amount of light that can be used is one photon, which has the energy of $h\nu$. When the electron scatters this photon into the microscope, the electron will receive some momentum from the photon. The photon can enter the microscope anywhere within the angle α. The original momentum of the photon in the x direction was $h\nu/c = h/\lambda$. Its final momentum in the positive x direction must be between 0 and $h/\lambda \sin \alpha$ in order for it to get into the microscope. Therefore the uncertainty in the x component of the momentum of the electron is

$$\Delta p_x = \frac{h}{\lambda} \sin \alpha \qquad (2\text{-}3)$$

Thus the product of the Δx and Δp in the simultaneous determination of both the position and momentum of the particle is

$$\Delta x\, \Delta p_x = \frac{\lambda}{\sin \alpha} \frac{h}{\lambda} \sin \alpha = h \qquad (2\text{-}4)$$

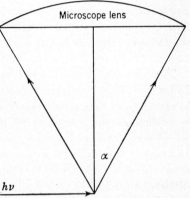

Many other idealized experiments have been conceived which attempt to contradict the uncertainty principle, but none has been successful. The uncertainty principle is inherent in the formulation of quantum mechanics.

The difference between allowable accuracy in quantum theory and that in classical mechanics comes from the fact that the magnitude of the

Fig. 2-1. Idealized experiment to determine the position of an electron by using an optical microscope.

quantities used for measurement in classical physics are extremely small compared to the quantity being measured. Thus in determining the position of a truck the momentum of the photon which is scattered by the truck does not affect the measured momentum of the truck. However, in quantum mechanics one is dealing with quantities which are of the same order of magnitude as the quantities used for measurement, and their interactions cannot be ignored. At the present time quantum mechanics seems to be the proper system to use for a description of the submicroscopic world.

2-3. Fundamental Properties of Elementary Particles—Mass. The mass of an elementary particle must be measured in terms of the standard kilogram, which is by international agreement the unit of mass. The usual means of determining the mass of an atom on a macroscopic basis is to measure the mass of 1 mole of the substance and then divide by Avogadro's number [1].* Thus 1 mole of hydrogen weighs 2.0178 g. Since this is a diatomic gas H_2, one atom of hydrogen weighs

$$\frac{2.0178}{2} \times \frac{1}{6.023 \times 10^{23}} = 1.67 \times 10^{-24} \text{ g}$$

* Bracketed numbers are keyed to the References at the end of the chapter.

Since the mass of an electron can be determined from its charge and a measurement of the ratio of e/m, the mass of the hydrogen nucleus, the proton, can be determined.

This method of determining the weight is satisfactory from the chemical point of view but is not satisfactory from the physical point of view because of the existence of isotopes. The chemical properties of an atom are determined only by the charge on the nucleus. However, two nuclei that have the same charge do not necessarily have the same mass. For instance, in nature there are two stable kinds, or isotopes, of hydrogen. One type has an atomic weight of 1.008142 and the other has an atomic weight of 2.014735. Although there are only 92 elements found in nature, there are 276 stable isotopes known at the present time.

The physicist uses a mass spectrometer for measuring the mass of isotopes. This instrument depends on the motion of a charged particle in electric and magnetic fields. It is not possible to determine by the mass spectrometer the masses of all particles, such as the neutron, which has no charge, and radioactive atoms which are too short-lived. Thus the masses are sometimes determined by other methods, but all methods refer to mass-spectrometer measurements. This leads to a slightly different unit for the physical and chemical scales. The "chemical scale" is defined in terms of the atomic weight of oxygen. Oxygen is found by a mass spectrometer to have one abundant isotope and two heavier ones, in the mass ratio of $16:17:18$, respectively. The "physical scale" has been created for convenience in discussing isotopes. It takes the atomic weight of the abundant isotope of oxygen as exactly 16. Since oxygen contains 0.03997 per cent O^{17} and 0.20470 per cent O^{18}, one unit on the chemical scale is 0.0281 per cent heavier than one unit on the physical scale.

2-4. Charge. The charge on a particle is measured in terms of the unit of charge as defined by Coulomb's electrostatic law of repulsion. One electrostatic unit of charge is defined as that charge which, when placed 1 cm from a like charge in a vacuum, will experience an electrostatic force of 1 dyne. The charge is defined as positive if it is the same type of charge as that acquired by glass when rubbed with silk, and negative when it is the same type of charge as that acquired by hard rubber when rubbed with cat's fur.

The original measurement by Milliken of the charge on the electron compared the electrical force on a charged oil drop in an electric field with the force of gravity on the same oil drop. He found the charge on the electron to be negative and found a value for it almost equal to that accepted today. The most precise value of the charge on the electron is obtained by determining the value of the Faraday constant and the value of Avogadro's number determined from the density, molecular weight, and lattice spacing of a crystal. Since 1 faraday of charge is deposited by 6.023×10^{23} ions carrying one electric charge each, the value of the unit of the electric charge is 1.601×10^{-19} coulomb.

The charge on any particle is made up of an integral number of these electronic charges.

2-5. Spin. Each particle has an intrinsic angular momentum M, which is quantized in units of $h/2\pi$. The angular momentum of a particle is due to its rotation about its axis of gyration, measured in units of its angular-momentum quantum number, or "spin," S. Classically $M = Sh/2\pi$. However, quantum mechanics gives the result that, if a particle has a spin of S, the value of the angular momentum is $\sqrt{S(S + 1)}\ (h/2\pi)$. Rather than use the more complicated expression for the actual angular momentum of the particle, the spin of the particle is usually used to designate its angular momentum. The spin of an electron, a proton, or a neutron is $\frac{1}{2}$. This means that an angular momentum of

$$M = \sqrt{\frac{1}{2}\frac{3}{2}}\frac{h}{2\pi} = \frac{\sqrt{3}}{4\pi}h \tag{2-5}$$

is used in any calculation for the angular momentum of these particles. However, the spin of $\frac{1}{2}$ is usually used to designate the angular momentum of the particle.

The value of the spin of a particle is never measured directly but is inferred from the values of a series of data which are influenced by the angular momentum levels of the particle. Thus the spin of the electron is determined from the number of lines observed or inferred in the fine structure of hydrogen or hydrogen-like atoms. The spin of the proton is determined from the hyperfine structure of atomic spectra of the hydrogen atoms, from the alternation in intensities in the band spectra of the hydrogen molecule, or from molecular beam resonance experiments.

2-6. Magnetic Moment. A charged particle, spinning about its axis, constitutes a circulating current and creates a magnetic field. If the charge distribution and the speed of all parts of the particle were known, then its magnetic moment could be calculated. However, these quantities are usually not measurable, so the magnetic moment becomes an experimental characteristic of the particle. One of the greatest challenges to the theoretical physicist of the present time is to obtain a consistent theory which will give the magnetic moment of the proton and neutron.

The magnetic moment of a particle is determined by the interaction of the particle with the internal magnetic field of an atom or with an applied magnetic field. Thus the magnetic moment of the electron, due to its rotation about its axis, interacts with the magnetic field created by the same electron rotating about the nucleus. By measuring this interaction the magnetic moment of the electron can be determined. That is, the separation of two lines in the fine structure of the hydrogen atom will give a measure of the magnetic moment of the electron.

The magnetic moment of an atomic nucleus is usually determined by measuring the energy difference between the two lines when the atom is placed in a magnetic field.

2-7. Statistics. In submicroscopic physics it is not possible to measure the properties of an individual particle such as an electron or a proton. The observable phenomena are caused by the actions of large numbers of particles. Therefore it is necessary to study the "statistics" of the particles, i.e., the rules which govern their behavior.

It has been found that every particle obeys either the Fermi-Dirac statistics or the Bose-Einstein statistics. If the particles obey the Pauli principle, which states that no two particles can occupy the same quantum state, then these particles obey the Fermi statistics. If it is possible to put two or more particles in the same quantum state, then the particles obey the Bose statistics.* The statistics of a particle is an extremely important property, for it is from measurements of the statistics and spins of the various nuclei that it is believed the nucleus is composed of protons and neutrons rather than protons and electrons.

There seems to be a connection between spin and statistics for material particles, in that all particles which have half-integral spin have Fermi statistics and all particles with 0 or integral spin have Bose statistics. However, virtual particles such as photons do not have a spin but still have a particular type of statistics. It is not believed necessary for a particle, real or virtual, with spin $\frac{1}{2}$ to have Fermi statistics or for a particle with spin 1 to have Bose statistics, but until now no particles have been discovered which do not follow this rule.

Table 2-1 shows the characteristics of the particles which are of interest in nuclear physics. The electron, positron, and neutrino are the light elementary particles and the proton and neutron the heavy elementary particles. The deuteron is known to be a neutron and a proton bound together, for a deuteron can be broken up into these two components. The α particle, or nucleus of He^4, is known to be composed of two neutrons and two protons. The role of the mesons in the structure of the nucleus is not well understood at the present time. The π mesons are produced in high-energy nuclear collisions and are thought to have some role in holding the nucleus together, serving as a "glue" between the protons and neutrons. The π mesons decay into μ mesons, and μ mesons decay into either electrons or positrons depending on the charge of the

---

* The best mathematical formulation of statistics is in terms of the wave function which describes the state of a system of particles. Letting $\psi(P_1P_2)$ be the wave function describing two particles, then if $\psi(P_1P_2) = +\psi(P_2P_1)$ the particles obey Bose statistics, and if $\psi(P_1P_2) = -\psi(P_2P_1)$ the particles obey Fermi statistics.

original π. Although mesons were postulated by Yukawa in 1934 it was not until 1948 that the π meson was distinguished from the μ meson. Since then rapid strides have been made in understanding the properties of the π meson, but the field is still in the first stages of development.

TABLE 2-1. CHARACTERISTICS OF THE FUNDAMENTAL PARTICLES

Particle	Mass, amu*	Charge, e†	Magnetic moment, magnetons‡	Spin, $\dfrac{h}{2\pi}$	Statistics
Electron...	$5.486^2 \times 10^{-4}$	-1	2	½	Fermi
Positron...	$5.486^2 \times 10^{-4}$	$+1$	2	½	Fermi
Proton....	1.00758	$+1$	$+2.7896$	½	Fermi
Antiproton	1.00758	-1	-2.7896	½	Fermi
Neutron...	1.00893	0	-1.9103	½	Fermi
Antineutron	1.00893	0	$+1.9103$	½	Fermi
Deuteron..	2.01417	$+1$	$+0.8565$	1	Bose
α particle..	4.00280	$+2$	~ 0	0	Bose
Neutrino..	0	0	Presumed 0	½	Fermi
π^+ meson..	0.149 (273 me)	$+1$	Presumed 0	0	Bose
π^- meson..	0.149 (273 me)	-1	Presumed 0	0	Bose
π^0 meson..	0.144 (264 me)	0	Presumed 0	0	Bose
μ^-........	0.113 (207 me)	-1	?	½	Presumed Fermi
μ^0........	Presumed to exist from symmetry	0	Presumed 0	Presumed ½	Presumed Fermi
μ^+........	0.113	$+1$	?	½	Presumed Fermi

* 1 amu $= 1.66 \times 10^{-24}$ g.
† $e = 4.8 \times 10^{-10}$ esu.
‡ 1 magneton $= eh/4\pi mc$.

It is risky to call a particle elementary, for physics is full of illustrations of a particle which was considered elementary but was later broken up into many component parts. It seems that this may also be true of the neutron. It is known that the neutron decays into a proton and an electron with a half-life of about 12 min. Therefore the neutron may have structure in the same way that the hydrogen atom has structure. At great distances the hydrogen atom is neutral just as the neutron is neutral. It is only when we are able to study the inside of the hydrogen atom that we find it has structure. At present, studying the inside of the neutron is on the fringe of possible measurements, but it may eventually be possible to study interaction at distances short enough to uncover the structure of the neutron. This requires particles of higher energy (shorter wavelengths), since if the uncertainty in the position is to be made smaller the momentum of the particle must be much higher. How many more mysteries will be uncovered and later solved is impossible to predict.

2-8. Interaction of Nuclear Radiations with Matter. In order to utilize the energy given off by the nucleus and to study the nucleus itself, it is necessary to know how nuclear radiations will affect the matter through which they pass. In their interaction with matter the radiations divide naturally into three categories: (1) charged particles, (2) neutral particles, (3) electromagnetic radiation or photons. Charged particles and photons lose most of their energy by interaction with the atom as a unit or, if the energy is high enough, with the separate parts of the atom. Collisions with nuclei are an extremely small fraction of the collisions and can usually be neglected. Neutrons interact primarily with nuclei. Since they carry no charge, electrical interactions with the atom are impossible. The neutron can give very little energy to an electron in a direct collision because the mass of the neutron is so much larger. There- fore only the collisions between neutrons and nuclei are important. These divide into two categories. In one the neutron and nucleus have an elastic collision like billiard balls. In the other, which is the subject of the next chapter, the neutron is absorbed by the nucleus.

2-9. Heavy Charged Particles. There are many experimental data on the interaction of heavy charged particles with matter. The ioniza- tion characteristics and range-energy relations of the α particles emitted by radioactive nuclei were studied in great detail by early investigators. Many empirical formulas are available from these studies [2]. However, many years elapsed before theoretical physics was sufficiently well advanced to give a reasonably good explanation of the observed facts. More recently, charged particles have also been accelerated to high energy and their properties studied directly to determine the character- istics of a charged particle in terms of its energy, mass, and charge [3].

In most cases it is much more convenient and correct to use the experimental relations that have been evolved than to attempt to calculate the range or the energy loss of the particle from the theoretical formulas.

One of the first observations concerning α particles was that their number remains approximately constant up to a definite distance R from a source and then drops suddenly to zero. This is a characteristic of all heavy particles traveling through matter. They have a definite range which depends on their initial velocity and the material through which they travel. Heavy particles lose energy by colliding with elec- trons as they pass through matter. A heavy particle cannot lose much energy to each electron; in fact, by classical mechanics, if momentum and energy are conserved in a collision between an α particle and an electron, the α particle will lose about $\frac{1}{2000}$ of its energy per collision. This means a large number of collisions are necessary to cause the α particle to lose all its energy. Since a collision with an electron cannot change the momentum of the α particle appreciably, the path of the α particle will be approximately straight.

Under these conditions the range of all α particles emitted with the same velocity will be approximately equal but not exactly equal, because each α particle will not collide with exactly the same number of electrons in losing all its energy. This small variation in the range of a heavy particle is called "straggling." It comes from the statistical nature

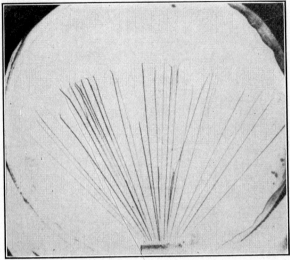

FIG. 2-2. Cloud-chamber tracks of α particles from polonium. (*From F. Rasetti, "Elements of Nuclear Physics," Prentice-Hall, Inc., New York*, 1936.)

of the stopping process and follows the laws of probability. The cloud-chamber photographs of the 5.303-Mev α particles emitted by Po^{210} with an average range in air of 3.842 cm, shown in Fig. 2-2, illustrate these points very well.

By colliding with the electrons in traveling through matter, the heavy particle will ionize some of the atoms. Thus the energy-loss phenomenon is accompanied by ionization. The "specific ionization" is defined as the number of ion pairs produced per millimeter of track of a heavy charged particle passing through matter. The "total ionization" is the total number of ion pairs produced over the complete track length. Both these quantities are extremely important in the design of instruments for the detection of heavy particles.

Another quantity which is extremely useful and which is directly proportional to the specific ionization is the "stopping power." This is defined as the amount of energy lost by a particle per unit length $-(dE/dx)$. Thus $-(dE/dx) = wI_s$, where w is the "average energy loss per ion pair produced" and I_s is the "specific ionization." The quantity w is almost constant over a wide range of stopping powers for a particular substance. However, w depends very strongly on the material through which the particle passes. A few typical values of w for different gases for polonium α particles are given in Table 2-2. Although w is the average energy required for ion pairs, it is considerably above the ionization potential of the atom, showing that energy is lost by other atomic excitation processes besides ionization.

TABLE 2-2

Gas	Z	w, ev	Ionization potential, ev
Hydrogen...........	1	35.1	13.527
Helium.............	2	30.2	24.46
Nitrogen............	7	36.3	14.48
Oxygen.............	8	32.3	13.55
Neon...............	10	28	21.47
Argon.............	18	26	15.68
Krypton............	54	22	13.93
Air.................	...	34.7	
CH_4...............	...	29	
C_2H_4..............	...	26.9	
CCl_4..............	...	26.8	
CO.................	...	34	
CO_2................	...	34	

The range of the particles is obtained by integrating the distance per unit energy loss over the total energy of the particle

$$R = \int_0^E \frac{dE}{dE/dx} \tag{2-6}$$

The range can be determined experimentally by interposing between the source and the detector the material in which it is desired to determine the range. If the range in air is desired, the detector can be moved away from the source until no more particles are detected.

Stopping power can be determined by observing the energy of a particle before and after it goes through the absorbing material, by using a magnetic deflection method, or by determining the specific ionization, which is directly proportional to the stopping power. The specific ionization may be determined by using a very small ionization chamber and observing the number of ion pairs produced in this small chamber as it is moved along the path of the particle. The specific ionization plotted against the distance away from the source is called a

"Bragg curve" and is shown in Fig. 2-3. The specific ionization increases with decreasing velocity, reaches a maximum, and then falls off very rapidly. For a single particle the falloff would be much more rapid; however, since the data shown represent the average ionization at a particular point for many particles, the straggling of the particles smooths out this rapid cutoff.

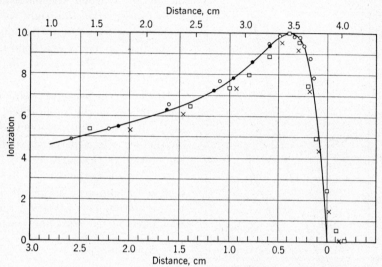

FIG. 2-3. The specific ionization for a single α-particle track, measured from the end of the track (bottom scale). The small solid circles represent the experimental work of Holloway and Livingston. The other symbols represent the results of other workers. The top scale is the distance of the detector from the source. (*From D. Halliday, "Introductory Nuclear Physics," John Wiley & Sons, Inc., New York, 1950.*)

The ranges of protons, deuterons, and α particles have been measured in standard air by several experiments. Since no single equation relates the range of the particle to the initial energy, the range-energy curves are given in Figs. 2-4 and 2-5.

2-10. Derivation of a Stopping-power Formula. Consider a heavy charged particle of charge Ze, mass m, and velocity v, passing at a distance R from an electron at rest (Fig. 2-6).

When the particle is distance x from the distance of closest approach R, the Coulomb force between the particle and the electron is

$$F = \frac{Ze^2}{r^2} \tag{2-7}$$

and acts along the line joining the particle and the electron. The component parallel to the motion is

$$F_{\parallel} = \frac{Ze^2}{r^2} \sin \beta \tag{2-8}$$

and the perpendicular component is

$$F_\perp = \frac{Ze^2}{r^2} \cos \beta \qquad (2\text{-}9)$$

When the particle is distance x to the left of 0, the component parallel to the motion is exactly equal and opposite to the parallel component of the force when the particle is a distance x to the right of 0. Since for every point to the left of 0 there is an equivalent point to the right of 0, the parallel components cancel.

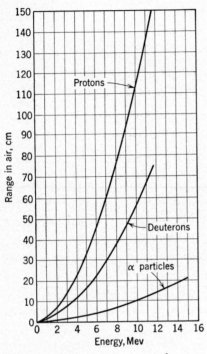

However, the perpendicular components are in the same direction and will add. Thus, the net force on the particle is perpendicular to the motion of the particle.

In distance dx the particle Ze will receive an impulse perpendicular to the motion $dI_\perp = F_\perp \, dt$. Since the particle is traveling with a velocity v, $dt = dx/v$. So

$$dI_\perp = \frac{Ze^2}{r^2} \cos \beta \, \frac{dx}{v} \qquad (2\text{-}10)$$

Since $x = R \tan \beta$,

$$dx = R \sec^2 \beta \, d\beta = \frac{R \, d\beta}{\cos^2 \beta}$$

and $r \cos \beta = R$,

$$dI_\perp = \frac{Ze^2}{Rv} \cos \beta \, d\beta \qquad (2\text{-}11)$$

Fig. 2-4. Range-energy curves for protons, α particles, and deuterons in air at 15°C and 760 mm.

Assuming that the velocity of the particle does not change appreciably (small angle of deflection), the total perpendicular impulse given to the particle by the electron is obtained by integrating over the total path.

$$I_{\text{total}} = \int_{\beta=-\pi/2}^{\beta=\pi/2} \frac{Ze^2}{Rv} \cos \beta \, d\beta = 2\frac{Ze^2}{Rv} \qquad (2\text{-}12)$$

By Newton's law of action and reaction, this must also be the impulse given to the electron by the charged particle. The impulse is equal to the change in momentum, and since the electron started off at rest, the momentum of the electron p is

$$p = \frac{2Ze^2}{Rv} \qquad (2\text{-}13)$$

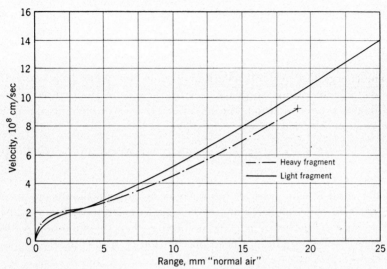

FIG. 2-5. Range of fission particles. [*From experiments by Bøggild, Brostrøm, and Lauritsen, Phys. Rev.*, **59**:273 (1941); *from E. Segrè (ed.), "Experimental Nuclear Physics," John Wiley & Sons, Inc., New York, 1953.*]

and the energy of the electron is

$$E_e = \frac{p^2}{2m_e} = \frac{2Ze}{m_e R^2 v^2} \qquad (2\text{-}14)$$

The stopping power, which is the energy loss per unit length, is obtained by integrating the energy loss per electron over the number of electrons per unit length. If there are N electrons per unit volume in a substance,

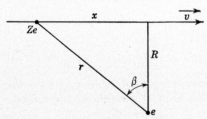

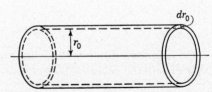

FIG. 2-6. A heavy charged particle Ze of mass m and velocity v passes at a distance R from an electron at rest.

FIG. 2-7. A cylindrical shell of thickness dr_0 and unit length at a distance r_0 from the path of the heavy charged particle.

then the number of electrons in a cylindrical shell at a distance r_0 from the path of the particle, of thickness dr_0 and of unit length, is the number of electrons per unit volume times the volume of this cylindrical shell. As shown in Fig. 2-7, therefore, $n = 2\pi r_0 \, dr_0(N)$ [1]. Each one of these electrons will receive the energy

$$E_e = \frac{2(Ze^2)^2}{m_e r_0{}^2 v^2} \qquad (2\text{-}15)$$

The energy loss per unit length is obtained by integrating over r_0 from the distance of closest approach (set by the size of the two particles) to the distance beyond which the heavy particles will have no appreciable effect on the electron.

$$-\frac{dE}{dx} = \int_{r_{\min}}^{r_{\max}} 2\pi r_0 \, dr_0 N \, \frac{2(Ze^2)^2}{m_e r_0{}^2 v^2} = \frac{4\pi N e^4}{m_e} \left(\frac{Z}{v}\right)^2 \ln \frac{r_{\max}}{r_{\min}} \qquad (2\text{-}16)$$

Bohr used a semiclassical approach in setting the limits on r_0. For $r_{\min}$ he used the classical limit of energy transfer in a direct collision. If the mass of the heavy particle is infinite, then the maximum velocity the electron can obtain in a head-on collision is $2v$. Thus,

$$E_{\max} = \frac{1}{2} m_e (2v)^2 = 2 m_e v^2 = \frac{2(Ze^2)^2}{m_e v^2 r_{\min}^2} \qquad (2\text{-}17)$$

or
$$r_{\min} = \frac{Ze^2}{m_e v^2}$$

He took the minimum energy that the particle could give to the electron as the effective binding energy I_b of an electron in an atom, so

$$E_{\min} = I_b = \frac{2(Ze^2)^2}{m_e v^2 r_{\max}^2} \qquad (2\text{-}18)$$

$$r_{\max} = \sqrt{\frac{2}{m_e I_b}} \, \frac{Ze^2}{v} \qquad (2\text{-}19)$$

Putting in these limits, we obtain

$$-\frac{dE}{dx} = \frac{2\pi N}{m_e} \left(\frac{Ze^2}{v}\right)^2 \ln \frac{2 m_e v^2}{I_b} \qquad (2\text{-}20)$$

This approximate result is correct except for a factor of 2. Bethe has worked out the relativistic quantum-mechanical formula and obtained

$$-\frac{dE}{dx} = \frac{4\pi N}{m_e} \left(\frac{Ze^2}{v}\right)^2 \left[\ln^2 \frac{m_e v^2}{I_b} - \ln\left(1 - \frac{v^2}{c^2}\right) - \frac{v^2}{c^2} \right] \qquad (2\text{-}21)$$

which for $v/c \ll 1$ is twice Eq. (2-20).

Over a large range of energies Block showed that $I_b = kZ$, and on the basis of measurements by Wilson, Wheeler showed that $k = 11.52$, which gives fairly good degrees of approximation for a wide energy region. A plot of dE/dx vs. E is shown in Fig. 2-8.

Until the particle velocity is near that of light, the energy-dependence of the terms in the parenthesis is relatively unimportant, so that the energy-dependence is given principally by Z^2/v^2.

Thus, the stopping power or the specific ionization increases as Z^2 and decreases as v^2. An α particle which has charge $Z = 2$ will have four times the specific ionization of a proton of the same velocity. For the same energy, the α particle has one-half the velocity of the proton, and consequently the specific ionization of the α particle is sixteen times that of the proton.

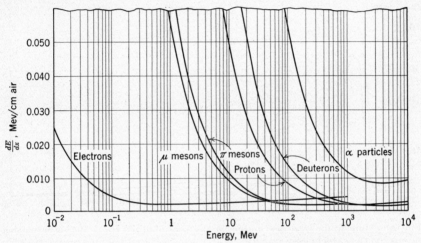

FIG. 2-8. A plot of dE/dx vs. E for several charged particles.

For particles of unit charge the formula holds fairly well and gives the energy loss of the particle over a wide range of energies. However, when particles of higher Z, such as α particles and fission fragments, pass through matter, they tend to pick up or lose charge. A rough rule for Z is that, on the average, all electron shells will be filled for which the velocity of the particle is less than the velocity of the electron in that shell. The velocity of high-energy particles exceeds the speed of any of the electrons, and the formula holds well. The range of the particles cannot be calculated exactly because of uncertain behavior at low energies, but qualitative results can be useful.

The range of fission fragments in uranium is 0.00025 in., and in aluminum it is 0.0005 in. Most of the ionization takes place at the beginning of the track before the fission fragment has picked up electrons from the surrounding material. Thus most of the energy is liberated in the fissionable material itself, which requires that it be effectively cooled (see Chap. 9).

The energy loss of a charged particle passing through a substance can usually be computed satisfactorily from the energy-loss formula. To design an instrument to measure the fast-neutron flux inside a reactor, the energy of the recoil protons would be calculated and thence the energy loss of the protons in the ionization chamber, and the detection instrument selected.

2-11. Electrons. The passage of electrons through matter is considerably more complicated than that of heavy charged particles, except for slower electrons than are usually important in nuclear physics. For

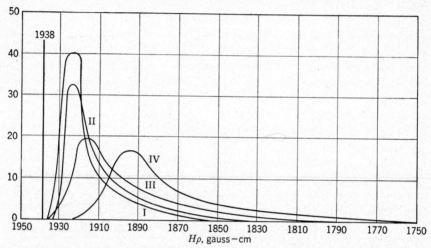

Fig. 2-9. Momentum $(H\rho)$ distribution of electrons of initial $H\rho = 1938$ after passing through various thicknesses of mica. Curve I, 2.25 mg/cm² of mica; curve II, 2.65; curve III, 3.95; curve IV, 5.72. The ordinate is the relative frequency. [*From White and Millington, Proc. Roy. Soc. (London),* **A120**:708 (1928); *from E. Segrè (ed.),* "*Experimental Nuclear Physics*," *John Wiley & Sons, Inc., New York,* 1953.]

fast electrons the energy loss is principally due to the emission of electromagnetic radiation by the electron when it is accelerated in the electric fields of the nuclei which compose the stopping material. Classically the accelerated charge emits radiation at a rate $S = \frac{2}{3}(e^2/c^2)a^2$, where e is the charge of the electron, c is the velocity of light, and a is the acceleration of the electron. An electron can receive a large acceleration in the Coulomb field of the nucleus because of its small mass. Therefore the electron emits a large amount of radiation (usually called "bremsstrahlung," or "braking radiation"). The rate of emission increases approximately as the square of the charge on the nucleus in the stopping material; consequently, higher-Z materials are much more effective in stopping electrons.* Because of the large acceleration electrons undergo, the path will be tortuous. Another difficulty encountered in the derivation

---

* It is interesting to note that the X rays produced in all X-ray tubes are the radiation of the electron accelerated down the tube and then suddenly decelerated in the target. This radiation has a continuous distribution, with the intensity in each logarithmic energy interval approximately constant.

of an energy-loss formula for electrons is the quantum-mechanical phenomenon of resonance, which has no classical analogy. It stems from the fact that, after two electrons collide, it is impossible to determine which was the incident electron and which the struck electron, because the two electrons are identical particles. The scattering is not the same as would be expected classically. A final difficulty is due to the fact that the electrons dealt with in nuclear physics have in many cases energies above the self-energy of the electron $m_e c^2$. Therefore, the mechanics used must be entirely relativistic. Consequently, it is not possible to derive easily an expression for the energy loss of electrons passing through matter.

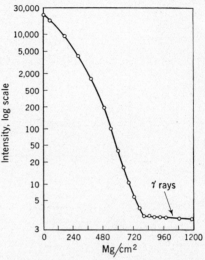

Electrons do not have a definite range. If a monoenergetic beam of electrons passes through a thin foil, the energy distribution will extend from almost the initial energy of the beam to a very low energy, as shown in Fig. 2-9. The reason is that some of the electrons went through the foil without colliding with another electron, while some of the electrons were scattered many times and suffered large energy losses.

FIG. 2-10. A semilogarithmic absorption plot for β rays from a 5.0-mc P^{32} source. The detector was an ionization chamber. The end point is 780 mg/cm² of aluminum. (*From D. Halliday, "Introductory Nuclear Physics," John Wiley & Sons, Inc., New York, 1950.*)

There are many experimental data available on the absorption of electrons as they pass through matter. Figure 2-10 shows the fractional intensity of the β rays (which are electrons emitted by a nucleus) emitted by the radioactive nucleus P^{32} as a function of the absorber thickness. The decrease in intensity is almost exponential. This is a coincidence and is due to the fact that initially the electrons are not monoenergetic. As shown by the curve, there is a maximum thickness beyond which one will not detect any electrons. This might be called the "range," although it is not a range in the true sense of the word. This "range" R is roughly related to the maximum energy of the emitted electron by the empirical formula

$$t\rho = R = 0.546 E_{max} - 0.094 \qquad (2\text{-}22)$$

where t = thickness of absorbing material, cm

ρ = density of material, g/cm³

E_{max} = maximum energy of electron, Mev

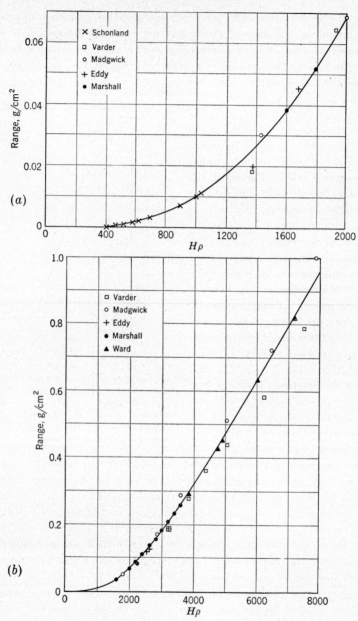

FIG. 2-11. Extrapolated range of electrons in aluminum vs. momentum $H\rho$ in gauss-centimeters. [*From Marshall and Ward, Can. J. Research*, **A15**:40, 41 (1937); *from E. Segrè (ed.), "Experimental Nuclear Physics," John Wiley & Sons, Inc., New York, 1953.*]

Although the above formula gives a reasonable answer for the range of β rays of above 0.6 Mev in matter, if more accurate data are required the range-energy or range-momentum curves should be used. The range-momentum curve for aluminum is shown in Fig. 2-11.

2-12. Cerenkov Radiation. Charged particles traveling through matter emit electromagnetic radiation when the velocity v_p of the particle

FIG. 2-12. A photograph taken by the Cerenkov radiation due to fast electrons traveling in the water surrounding the uranium. This is the Swimming Pool Reactor, SPR (or Bulk Shielding Facility, BSF), at ORNL, operated by Union Carbide and Carbon Corporation for the AEC. The fuel elements are surrounded by (beneath) the glow and are not visible here. The five large cylinders in front are waterproof holders for ionization and neutron chambers. The three smaller cylinders near the center are lifting-magnet guides for the control rod (center) and the safety rods (both sides). The long narrow cylinder to the right contains a fission chamber (see Chap. 4).

in the medium is greater than the velocity of light v_l in the medium in which the particle is traveling. (This is not a contradiction of the principle that no particle can have velocity greater than the velocity of light in a vacuum.) This radiation is called "Cerenkov radiation" and is not important until the velocity of the particle is very close to the velocity of light. The radiation is emitted coherently such that the ray makes an angle θ with the direction of motion of the particle in accordance with the relation

$$\cos \theta = \frac{v_l}{v_p} = \frac{c}{v_p} \frac{1}{\mu'} \qquad (2\text{-}23)$$

where μ' is the index of refraction of the medium. The picture of the water-cooled uranium reactor at Oak Ridge shown in Fig. 2-12 was taken with the Cerenkov radiation emitted by fast electrons traveling in the water surrounding the uranium.

2-13. Interaction of γ Rays with Matter. Gamma rays were found in the first studies of radioactivity. This radiation was the most penetrating of the three radiations emitted and therefore received the name of the third letter in the Greek alphabet. Gamma radiation is undeflected by a magnetic field, and the conclusion was reached that this emanation was electromagnetic. This radiation was much more penetrating than the X radiation that could be produced at the time and was therefore called by a different name. However, modern high-energy machines make X rays with higher energy and penetrating power than the γ radiations associated with radioactivity, and there is absolutely no difference provided the two radiations have the same energy. Currently the term "γ ray" is reserved for rays emitted by a nucleus in a nuclear transition, which are monochromatic. The radiation from the decelerations of the β rays emitted by a radioactive substance is "bremsstrahlung," and rays produced by artificially decelerated electrons are called "X rays." All these rays are electromagnetic radiation or photons.

The interaction of radiation with matter is extremely important in physics; a large fraction of atomic and molecular physics concerns the absorption and emission of photons. Only photons of energy above 1 kev need be considered here.

If a collimated beam of photons hits a sample of absorbing material, the beam is attenuated in a characteristic exponential manner (see Chap. 7). This type of absorption is characteristic of all electromagnetic radiation and is due to the fact that a single interaction removes a photon, the probability of this occurrence being constant along any portion of the path. Thus the number of photons interacting in linear thickness dx is proportional to the total area of all the atoms in the thickness dx and the number of photons in the beam at that location. If n is the number of atoms in a cubic centimeter, the change in intensity in passing through the thickness dx is

$$dI = -In\sigma \, dx \qquad (2\text{-}24)$$

Integrating, we obtain

$$I = I_0 e^{-n\sigma x} \qquad (2\text{-}25)$$

where I is the intensity of the radiation after passing through thickness x, and I_0 is the intensity incident on the surface of the material. Tables give the "microscopic" cross section σ, the "macroscopic" cross section $\Sigma = n\sigma$, or the mass-absorption coefficient $\mu = n\sigma/\rho$, where ρ is the density of the material. The mass-absorption coefficient vs. energy is plotted for several materials in Fig. 2-13.

In the energy region in which we are interested, three processes account for the interaction of radiation with matter: (1) the photoelectric effect,

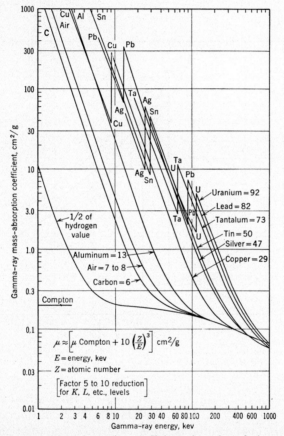

FIG. 2-13. Gamma-ray mass-absorption coefficients for selected elements. [*Data for energies above K absorption limit from J. A. Victoreen, J. Appl. Phys., 14:95 (1943). Data for lower energies from C. D. Hodgman (ed.), "Handbook of Chemistry and Physics," 37th ed., pp. 2405–2412, Chemical Rubber Publishing Company, Cleveland, 1955.*]

(2) the Compton effect, and (3) pair production. The total cross section for γ radiation is the sum of the three cross sections

$$\sigma_{\text{total}} = \sigma_{PE} + Z\sigma_{\text{Compt}} + \sigma_{\text{pair}} \tag{2-26}$$

For low energies σ_{PE} predominates and varies as Z^5. In the intermediate-energy region the Compton effect, which varies as Z, predominates, and at high energies the pair-production cross section, which varies as Z^2, is most effective. The γ-ray absorption coefficients have been checked experimentally from 50 kev to 100 Mev, and the agreement between theory and experiment is fairly good.

2-14. The Photoelectric Effect. In the photoelectric effect the photon is absorbed and the energy is transferred to one of the electrons in the atom. For low-energy radiation almost all the interaction in heavy elements is due to the photoelectric effect. If an electron is bound, say,

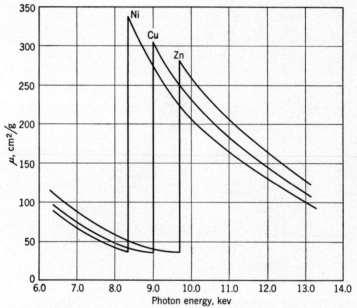

FIG. 2-14. Mass-absorption coefficients for photons of various energies in nickel, copper, and zinc. (*From D. Halliday, "Introductory Nuclear Physics," John Wiley & Sons, Inc., New York, 1950.*)

in the K shell of the atom, then the energy of the emitted electron will be given by $T = h\nu - I_K$ where ν is the frequency of the radiation and I_K is the ionization potential of the K electron. The total photoelectric-absorption coefficient is the sum of the absorption coefficients for all the electrons. If the energy of the photon is less than the ionization energy of a particular electronic shell, then the photoelectric-absorption coefficient for that shell is zero. For frequencies where the energy of the photon is above the ionization limit, the photoelectric-absorption coefficient falls off as ν^{-3}. The photoelectric-absorption cross section increases markedly as the energy of the photons becomes higher than the ionization limit, as shown in Fig. 2-14. Actually the contribution of the L shell to the photoelectric cross section above the K absorption limit is only about one-fifth that due to the K shell. Above the K absorption limit but below highly relativistic energies such that $h\nu \gg m_0c^2$ (i.e., 0.5 Mev), the photoelectric cross section is well represented by the formula

$$\sigma = 47.5\lambda^{7/2}Z^5 \qquad cm^2 \tag{2-27}$$

for λ in centimeters.

Above 0.5 Mev the photoelectric cross section for the K shell is

$$\sigma = 1.16 \times 10^{-23}\lambda Z^5 \qquad cm^2 \tag{2-28}$$

For low energies it can be seen that the cross section decreases very rapidly with increasing energy, whereas at high energies the cross section

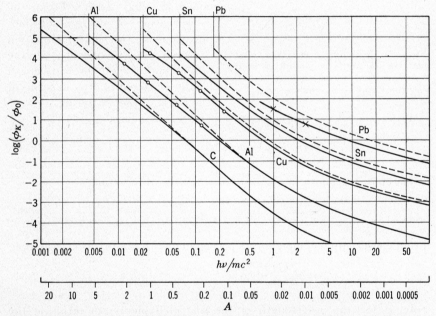

FIG. 2-15. Photoelectric cross section as a function of γ-ray energy for various atoms. The lower abscissa scale gives the wavelength in angstroms. Solid curves, exact calculation; dotted curves, Born approximation. [*From E. Segrè (ed.), "Experimental Nuclear Physics," John Wiley & Sons, Inc., New York, 1953.*]

falls off only as $1/\nu$. The calculated cross sections using the exact formulas which are derived in Heitler [4] are shown in Fig. 2-15.

In both cases the photoelectric cross section is proportional to Z^5 and therefore is important for heavy elements even at very high energies.

A quantitative idea of the energy at which the photoelectric cross section is no longer predominant can be obtained from the following table showing the energy at which it is approximately equal to the Compton cross section for several elements:

Element	H	C	Al	Cu	Pb
Z	1	6	13	29	82
Kev	4	24	48	130	420

2-15. Compton Scattering. In Compton scattering the photon acts like a particle of energy $E = h\nu$ and momentum $M = h\nu/c$. If the electron recoils at an angle ϕ with the path of the incident photon, the photon will be scattered at an angle θ (Fig. 2-16). From the conservation of energy and momentum one can obtain a relation between the wavelength of the initial photon and the wavelength of the scattered photon in terms of the angle of the scattering.

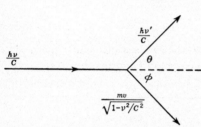

FIG. 2-16. A photon of energy h collides with an electron. A photon of energy $h\nu'$ is emitted at an angle θ, and the electron recoils at an angle ϕ.

From conservation of energy,

$$h\nu = h\nu' + mc^2 \left(\frac{1}{\sqrt{1 - v^2/c^2}} - 1 \right) \qquad (2\text{-}29)$$

From conservation of momentum in the forward direction,

$$\frac{h\nu}{c} = \frac{h\nu'}{c} \cos \phi + \frac{mv}{\sqrt{1 - v^2/c^2}} \cos \theta \qquad (2\text{-}30)$$

From conservation of momentum perpendicular to the direction of motion,

$$0 = \frac{h\nu'}{c} \sin \phi + \frac{mv}{\sqrt{1 - v^2/c^2}} \sin \theta \qquad (2\text{-}31)$$

Solving for the wavelength of the scattered photon in terms of the wavelength of the initial photon and the angle of scattering,

$$\lambda' - \lambda = \frac{h}{mc} (1 - \cos \theta) = 24.17(1 - \cos \theta) \times 10^{-11} \qquad \text{cm} \qquad (2\text{-}32)$$

Thus the change in wavelength is independent of the initial wavelength of the photon and depends only on the scattering angle. The universal constant h/mc is called the "Compton wavelength." The wavelength shift predicted by this formula has been verified experimentally many times and serves as one of the best examples of evidence for the particle properties of the photon.

The Compton scattering cross section of a free electron for electromagnetic radiation can be calculated rather simply from classical theory, and although the results obtained are not accurate at high energy, they do give a good approximation for the low-energy limit of the Compton effect.

Consider a free electron in a varying electromagnetic field in which the electric field is polarized in the x direction and is given by

$$E = E_0 \sin \omega t \tag{2-33}$$

The force on the electron will be

$$F = m\ddot{x} = eE_0 \sin \omega t \tag{2-34}$$

Classically the average power P radiated by an accelerated charge is

$$P = \frac{2}{3}\frac{e^2}{c^3}\ddot{x}^2 = \frac{2e^4E_0^2}{3c^3m^2}\sin^2\omega t \tag{2-35}$$

Averaging this over time gives

$$P_s = \frac{1}{3c^3}\frac{e^4E_0^2}{m^2} \tag{2-36}$$

The scattering cross section is given by the ratio of the power scattered by the electron to the power incident on the electron. Thus

$$\sigma = \frac{P_s}{P_{inc}} \tag{2-37}$$

The energy contained in an electric field is $E^2/8\pi$, and since the wave is traveling with a velocity c, the power incident on the electrons will be

$$P_{inc} = \frac{cE^2}{8\pi} \tag{2-38}$$

Thus $\quad \sigma = \dfrac{(1/3c^3)e^4E_0^2/m^2}{cE_0^2/8\pi} = \dfrac{8\pi}{3}\left(\dfrac{e^2}{mc^2}\right)^2 = 6.64 \times 10^{-25} \text{ cm}^2 \tag{2-39}$

This classical expression was originally derived by J. J. Thomson and predicts the scattering cross section to be independent of energy.

In the classical Thomson theory the scattered radiation is of the same frequency as the incident radiation. Thus this theory is a reasonably good approximation when the wavelength shift of the Compton scattering is rather small, as occurs for low-energy photons. Here the cross section is approximately constant.

A satisfactory quantum-mechanical theory of Compton scattering has been worked out by Klein and Nishina [4]. The result is

$$\sigma_s = \frac{2\pi e^4}{m^2c^4}\left[\frac{1+\omega}{\omega^2}\frac{2(1+\omega)}{1+2\omega}\frac{1}{\omega}\ln(1+2\omega) + \frac{1}{2\omega}\ln(1+2\omega)\right. \\ \left. -\frac{1+3\omega}{(1+2\omega)^2}\right] \tag{2-40}$$

where $\omega = h\nu/mc^2$. In the limiting case of small energies, $\omega \to 0$ and this formula reduces to the Thomson formula. For the cases of very high energy, such that $h\nu \gg mc^2$,

$$\sigma_s = \frac{\pi c^4}{m^2 c^4}\frac{1}{2\omega} + \frac{1}{\omega}\ln 2\omega \qquad (2\text{-}41)$$

and the scattering cross section is approximately proportional to $1/\nu$. In Fig. 2-17 the cross section for photoelectric scattering and Compton scattering is plotted in terms of the classical Thomson cross section.

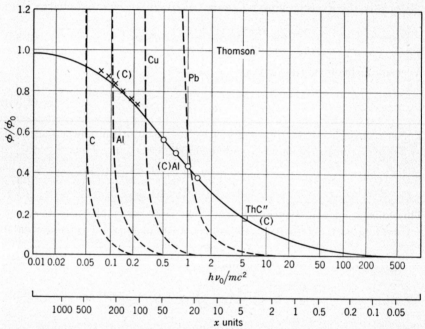

Fig. 2-17. Solid curve, cross section for the Compton effect, in units of $\phi_0 = (8\pi/3)r_0^2$, as a function of the incident γ-ray energy in units of mc^2 (lower scale gives wavelength). The dotted curves give the photoelectric cross section for various elements in the same units. Crosses, circles, and square represent experimental points. [*From E. Segrè (ed.), "Experimental Nuclear Physics," John Wiley & Sons, Inc., New York, 1953.*]

2-16. Pair Production. At yet higher energies the phenomenon of pair creation occurs, which has no classical analogy. In this process the photon disappears and a positron and electron appear. This process must take place in the vicinity of a nucleus in order to conserve both momentum and energy. A pair of particles of equal and opposite charge must be produced because charge must also be conserved. The energy of the photon is transformed into the mass of the positron and electron and the kinetic energy of the two particles such that

$$E = 2m_0c^2 + E^+ + E^-$$

The energy E^+ and E^- may range from zero to the initial energy of the photon minus $2mc^2$. Figure 2-18 shows the probability distribution of the energy of the particle for a photon of energy $E_{\max} + 2m_0c^2$.

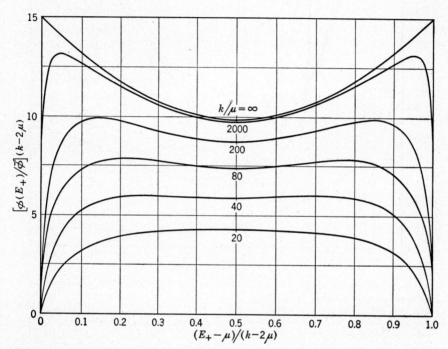

FIG. 2-18. Energy distribution of the electrons in an electron pair. Abscissa, kinetic energy of one electron, divided by available energy, $h\nu - 2mc^2$. The scale of the ordinate is so chosen that the area under each curve gives the total cross section in units of $\overline{\phi} = Z(Z + 1)r_0^2/137$. The number on each curve indicates the γ-ray energy in units of $mc^2 = \mu$. [From E. Segrè (ed.), "Experimental Nuclear Physics," John Wiley & Sons, Inc., New York, 1953.]

Since the process of pair production requires the minimum energy of the photon to be equal to twice the rest mass of the electron, this process cannot take place for energies less than $2m_ec^2 = 1.02$ Mev.

The complete theory of pair production has been worked out by Bethe and Heitler, who find that the pair-production cross section increases approximately as Z^2. However, no closed simple form for the energy-dependence can be given.

For extremely high energies the pair-production cross section is independent of energy, but for lower energies the results are best given

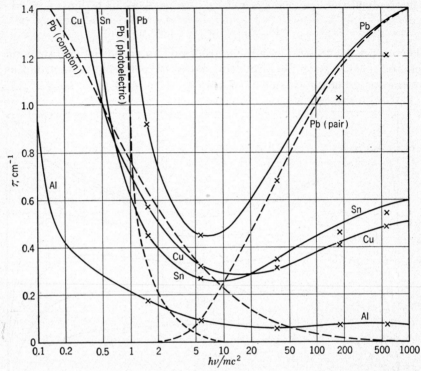

FIG. 2-19. Absorption coefficient τ for γ rays in Pb, Sn, Cu, and Al as a function of the frequency, in a logarithmic scale. The dotted curves show the three components of τ for lead. The measurements are for $h\nu/mc^2 = 1.635, 5.40, 34.4, 172,$ and $550.$ (*From W. Heitler, "The Quantum Theory of Radiation," 3d ed., Oxford University Press, New York, 1954.*)

graphically. In Fig. 2-19 the photoelectric, Compton, and pair-production cross sections and the total cross section for lead by all three methods of photon disappearance are plotted against energy. The form of this curve is the same for all elements, with the minimum occurring for typical elements at the following energies:

	Element		
	Al	Cu	Pb
Energy of minimum, Mev..........	35	10	3½
Half-thickness, g/cm²..............	47	31	25

2-17. Annihilation Radiation. The inverse of pair production is annihilation of a pair. When a positron is slowed down in matter, it can combine with an electron to form two γ rays. Two γ rays, which travel in opposite directions in the center of mass system, must be produced in order to conserve momentum. Thus, when a positron emitter is present in a radioactive substance, the 0.51-Mev ray due to annihilation radiation will always be present. This is a very useful spectral line with which to calibrate γ-ray instruments.

2-18. Slowing Down of Neutrons. Excluding nuclear reactions (Chap. 3), the principal interaction between a neutron and the matter which it traverses is elastic scattering. Classical mechanics applies, and both the neutron and nucleus may be assumed impenetrable spheres.

As in similar problems in classical mechanics, it is convenient to use

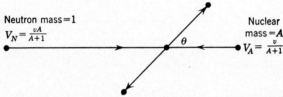

FIG. 2-20. A neutron of mass 1 collides with a nucleon of mass A in the center-of-mass system.

two simply related reference systems: (1) the "laboratory reference" system, which is determined by considering the target at rest before collision; and (2) the "center-of-mass" system, in which the center of mass of both the target nucleus and the neutron remains fixed.

Consider a neutron of mass $m = 1$, traveling with a velocity v in the laboratory system and colliding with a nucleus of mass A which is stationary in this system. The total mass of the system is $A + 1$, its momentum is mv, and the velocity of the center of mass is $v/(A + 1)$.

In the center-of-mass system the nucleus is moving with a velocity $v/(A + 1)$ and the neutron is moving with a velocity

$$v - \frac{v}{A + 1} = \frac{Av}{A + 1}$$

These velocities must be in opposite directions both before and after collision, since in the center-of-mass system the initial momentum by definition is zero. After collision the magnitudes of the velocities must be unchanged because of the simultaneous applicability of the laws of conservation of energy and of momentum. However, the particles may come off in a different direction (see Fig. 2-20).

Let the angle between this initial direction of the neutron and the direction which it recoils in the center-of-mass system be θ. Then we can easily transform from the center-of-mass system to the laboratory system by adding vectorially the velocities of the center-of-mass system and the velocity of the neutron (see Fig. 2-21).

Adding these gives the velocity of the scattered neutrons in the laboratory system. From the diagram, the angle between the initial direction of the neutron and the direction of the scattered neutron must be smaller

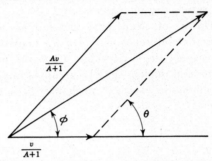

FIG. 2-21. The vertical relationship between the velocity of the neutron in the center-of-mass system, the velocity of the center of mass, and the velocity of the neutron in the room system.

in the laboratory system than it is in the center-of-mass system. The relation between θ and ϕ comes from the sine law

$$\frac{\sin (\theta - \phi)}{\sin \phi} = \frac{v/(A + 1)}{Av/(A + 1)} = \frac{1}{A}$$

$$\tan \phi = \frac{A \sin \theta}{1 + A \cos \theta}$$

(2-42)

Since we are usually interested in the energy of the neutron in the laboratory system both before and after collision, we want an expression relating these energies in terms of the angle ϕ or θ.

The energy of the neutron E before collision is $\frac{1}{2}mv^2$, and the energy E' after collision is $\frac{1}{2}mv'^2$. Thus the ratio of the two energies is v'^2/v^2. From Fig. 2-21 and the cosine law,

$$v'^2 = \left(\frac{v}{A + 1}\right)^2 + \left(\frac{vA}{A + 1}\right)^2 + \frac{2Av^2}{(A + 1)^2} \cos \theta$$

or

$$\frac{E'}{E} = \frac{A^2 + 1 + 2A \cos \theta}{(A + 1)^2}$$

(2-43)

Combining Eqs. (2-42) and (2-43), the energy of the neutron after collision can be obtained in terms of the angle through which the neutron is scattered in the laboratory system.

We are usually not interested in the behavior of one neutron in a collision but in the statistical behavior of a large number of neutrons. It can be shown that the number of particles scattered into any solid angle in the collision between a point particle and a sphere is independent of the angle.

For isotropic scattering in the center-of-mass system, all energies between the maximum energy $E'_{max} = E$ for $\theta = 0$ and the minimum energy $E'_{min} = [(A - 1)/(A + 1)]^2$, where $\theta = \pi$ for a head-on collision, are shown below to be equally probable.

The number of neutrons dN that will have the energy E' after collision will be equal to the number of the neutrons scattered between the angle θ

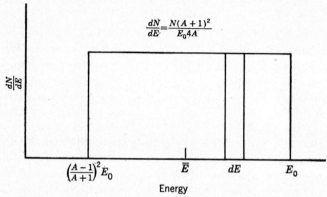

FIG. 2-22. The probability-distribution function of the energy loss of a neutron of energy E_0 in an elastic collision with a nucleon of mass A.

and $\theta + d\theta$, which equals the area between θ and $\theta + d\theta$ divided by the total area of the sphere, thus,

$$dN = N \frac{2\pi r \sin \theta r \, d\theta}{4\pi r^2} = \frac{N}{2} \sin \theta \, d\theta \qquad (2\text{-}44)$$

But $\qquad dE' = -E \frac{2A}{(A + 1)^2} \sin \theta \, d\theta = -E \frac{4A}{(A + 1)^2} dN$

The negative sign means that, on increasing θ, E' decreases. The probability that the final energy is between E' and dE' is

$$\frac{dN}{dE'} = N \frac{(A + 1)^2}{4A} \frac{1}{E} \qquad (2\text{-}45)$$

and is independent of E'. Thus all energies E' are equally probable. A graph of this energy distribution of neutrons after collision is shown in Fig. 2-22.

The average energy of the neutron is $E_0(A^2 + 1)/(A + 1)^2$. The average energy loss per collision is $2E_0A/(A + 1)^2$. The average energy and the average energy loss for more than one collision can be computed, but they become inconvenient to use after a few collisions. Since the fractional energy loss is on the average the same for successive collisions, $\log E$ changes on the average by a fixed amount. Thus we define a new quantity ξ, such that

$$\xi = \left(\ln \frac{E}{E'}\right)_{\mathrm{av}} = \int_{E_{\min}}^{E_{\max}'} \ln \frac{E}{E'} \, dN$$

$$= \int_{\frac{(A-1)^2}{(A+1)^2}E}^{E} \ln \frac{E}{E'} \frac{(A+1)^2}{4A} \frac{dE'}{E} \tag{2-46}$$

which for $A \gg 1$ integrates to

$$\xi = 1 + \frac{(A-1)^2}{2A} \ln \frac{A-1}{A+1} \tag{2-47}$$

The approximate formula $\xi \approx 2/(A + \frac{2}{3})$ is within 1 per cent for $A > 10$.

The quantity ξ is important for neutron moderation (Chap. 6). The average number of collisions n' required to moderate a neutron from an initial energy E_i to a final energy E_f can be calculated from ξ by

$$n'\xi = \ln E_i - \overline{\ln E_f} \tag{2-48}$$

The value of ξ for various moderating materials and the number of collisions required to reduce the energy of a 2-Mev neutron, which is the average energy of a fission neutron, to 0.025 ev (thermal energy) is given in the following table for the light elements which have been used as moderators or reflectors and for uranium and lead.

Material	A	ξ	n'
H...............	1	1.000	18
D (H^2)...........	2	0.725	25
Be...............	9	0.209	87
C...............	12	0.158	115
Pb...............	208	0.0096	1909
U................	238	0.0084	2180

The advantage of using light elements as neutron moderators is obvious from this table. Of course, moderators must also have a very low neutron-capture cross section to be useful (Chap. 6).

REFERENCES

1. Richtmyer, F. K., and E. H. Kennard: "Introduction to Modern Physics," 5th ed., McGraw-Hill Book Company, Inc., New York, 1955.
2. Rutherford, Sir Ernest, James Chadwick, and C. D. Ellis: "Radiations from Radioactive Substances," Cambridge University Press, New York, 1951.
3. Segrè, E. (ed.): "Experimental Nuclear Physics," vol. 1, John Wiley & Sons, Inc., New York, 1953.
4. Heitler, W.: "The Quantum Theory of Radiation," Oxford University Press, New York, 1954.
5. Halliday, D.: "Introductory Nuclear Physics," John Wiley & Sons, Inc., New York, 1950.
6. Rasetti, F.: "Elements of Nuclear Physics," Prentice-Hall, Inc., New York, 1936.

CHAPTER 3

NUCLEAR PHYSICS

By William W. Havens, Jr.

3-1. Introduction. Although the science of nuclear physics began in 1896 with the discovery of radioactivity by Henri Becquerel, it was not until 1911 that Rutherford introduced the nuclear theory of the atom and that radioactivity was associated with the nucleus. For years after the existence of the nucleus was recognized there was very little interest in its structure. Physicists were mainly concerned with unraveling the mysteries of the atom, and the structure of the nucleus was relatively unimportant in the study of the atom. It was sufficient for the atomic physicist to assume the nucleus a point particle with a charge Z and mass M. It was not until 1924 that the angular momentum or spin of the nucleus was introduced to explain the hyperfine structure observed in atomic spectral lines.

In 1932 Chadwick discovered the neutron, Lawrence built his first cyclotron, and modern nuclear physics began to make real progress. Since then the field has grown so rapidly that it is now almost impossible to keep up with the latest developments.

Nuclear physics is now in a transitional state. We have a large amount of information about many nuclei, and in certain categories we have theories which will predict what will happen from the results of previous experiments. However, there is no internally consistent, comprehensive nuclear theory from which we can predict what will happen to *any* specific nucleus in *any* particular experiment.

We believe that quantum mechanics is the correct system with which to describe nuclear phenomena. There are innumerable examples in which quantum mechanics gives a description in agreement with experiment. We also know the composition of the nucleus; there is good evidence to show that the nucleus is made up of neutrons and protons. The missing piece of information is the law of the force that holds the particles in the nucleus together. This force is different from any heretofore used to describe matter. Neither the gravitational force, which explains the macroscopic behavior of matter, nor electromagnetic forces, which describe the behavior of the atoms and molecules, explain why nuclei are stable. Since we do not know the fundamental law of force which holds the nuclear particles together, it is impossible to give a comprehensive description of the behavior of nuclear matter.

There are two methods of attack which have been used with some success to solve the problem of the nucleus. One method, so successful in atomic physics, is to look at the details of the interaction between the elementary particles, and from this attempt to explain the stability of the nucleus as the result of the interaction between the elementary particles. The other method is to assume some general properties of a system of particles and then see how these assumptions must be changed in order to fit the predictions of the theory to the experimental results. The first method is usually more soul-satisfying for the physicist, but the second very often also produces useful results.

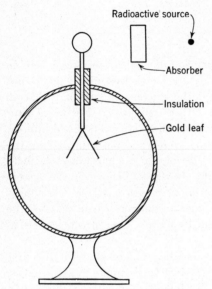

FIG. 3-1. A simple electroscope.

3-2. Radioactivity. It was radioactivity which originally drew attention to the existence of the nucleus, and most of the information which is available about the nucleus has come from studies of both natural and artificial radioactivity. Naturally radioactive substances emit three types of radiation. These are called α, β, and γ radiation after the first three letters of the Greek alphabet. These radiations can be distinguished very easily by placing the radioactive substance a few centimeters away from the terminal of an electroscope, as shown in Fig. 3-1, and placing absorbing material in between. If all three types of radiation are present and the rate of discharge of the electroscope is plotted against the thickness of the absorber, as shown in Fig. 3-2, three distinct groups are observed. The rate of discharge of the electroscope will decrease rapidly when something as thin as a sheet of paper is placed between the radioactive substance and the electroscope. The radiation stopped by the paper is called "α radiation." The rate of discharge will decrease slowly until filtering material equivalent to about $\frac{1}{4}$ in. of aluminum is placed between the radioactive substance and the electroscope, at which point the rate of discharge will decrease considerably. The radiation which will pass through a thin sheet of paper and will be stopped by the

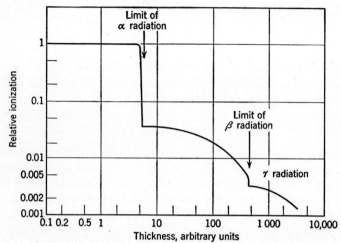

FIG. 3-2. Absorption curves of α, β, and γ radiations.

aluminum is called "β radiation." The residual radiation is found to penetrate several inches of lead, with the rate of discharge of the electroscope decreasing exponentially as the thickness of the absorber is increased. This residual radiation is called "γ radiation." Many experiments have shown that the α radiations are identical to nuclei of helium atoms, β radiations are electrons, and γ radiations are high-energy photons.

Soon after radioactivity was discovered, the radioactivity of a pure substance was found to decrease with time. When a pure radioactive substance such as P^{32} is isolated and the radioactivity is plotted as a function of time, the graph shown in Fig. 3-3 is obtained. The radioactivity is plotted on semilog paper to show that the curve is linear on this plot. This indicates that the radioactive decay is logarithmic.

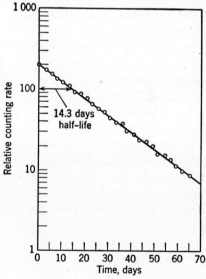

FIG. 3-3. The decay of P^{32}.

If we assume that the radioactive decay of a single nucleus is independent of that of any other nucleus, then the number of particles dN which will disintegrate in the time dt will be proportional to the number N of radioactive atoms present and to *dt.* Analytically this is

$$dN = -\lambda N \, dt \tag{3-1}$$

where λ is the constant of proportionality called the "decay constant." The negative sign is introduced in this equation because the number of atoms decreases. Integrating gives

$$N = N_0 e^{-\lambda t} \tag{3-2}$$

This equation shows that the number of particles which exist at any time is equal to the product of the initial number of radioactive atoms and a negative exponential of the time, yielding the results plotted in Fig. 3-3.

One of the characteristics of a radioactive decay process is the half-life T. The half-life is defined as the time required for one-half the atoms originally present to decay. If we put $N = 0.5N_0$, we find that

$$\lambda T = \ln 2 = 0.69 \qquad \text{or} \qquad T = \frac{0.69}{\lambda} \tag{3-3}$$

which is the relation between the decay constant and the half-life. Another quantity which is sometimes used is the mean life; by integration of Eq. (3-2) with respect to time this is seen to be the time required for the number of atoms to reach $1/e$ of the original number present. This mean life = $1/\lambda$.

The curie is the unit of radioactivity and is defined as the number of disintegrations which take place per second in 1 g of radium. Experimentally the half-life of radium is 1600 years. This gives a disintegration constant of 13.8×10^{-12} sec^{-1}. Since the mass of radium is 226, and Avogadro's number is 6.023×10^{23}, for 1 g of radium $dN/dt = 3.7 \times 10^{10}$ disintegrations per second.

In many cases in radioactivity, the nucleus to which the original radioactive nucleus disintegrates is also radioactive. In fact, uranium undergoes many disintegrations before it finally becomes lead. This series of disintegrations is called a "radioactive chain." For the first disintegration we have, as before,

$$dN_1 = -N_1 \lambda_1 \, dt \tag{3-4}$$

For the second disintegration we have

$$dN_2 = -\lambda_2 N_2 \, dt + \lambda_1 N_1 \, dt \tag{3-5}$$

As before, the first equation yields

$$N_1 = N_{1,0} e^{-\lambda_1 t} \tag{3-6}$$

Eliminating N_1 by Eq. (3-6), Eq. (3-5) yields

$$N_2 = \frac{\lambda_1}{\lambda_2 - \lambda_1} N_{1,0}(e^{-\lambda_1 t} - e^{-\lambda_2 t}) + N_{2,0}e^{-\lambda_2 t} \qquad (3-7)$$

For subsequent disintegration products the expressions become successively more complicated. The expressions for an n-component chain have been worked out and are given in some of the references at the end of the chapter.

If isotope 1 is very short-lived compared with isotope 2, we have

$$N_2 = N_{1,0}e^{-\lambda_2 t} \qquad (3-8)$$

if $N_{2,0} = 0$. Another case of special importance is that in which the half-life of isotope 1 is much greater than the half-life of isotope 2. This means that λ_2 is much greater than λ_1. In this case we can neglect the term $e^{-\lambda_2 t}$ in Eq. (3-7), and we have, after a long period of time,

$$N_2 = N_1 \frac{\lambda_1}{\lambda_2} \qquad (3-9)$$

or

$$\lambda_1 N_1 = \lambda_2 N_2 \qquad (3-10)$$

This is called "secular equilibrium." For a radioactive decay chain from a long-lived starting nucleus, all the members will be in secular equilibrium and we have

$$N_1\lambda_1 = N_2\lambda_2 = N_3\lambda_3 = \cdots \qquad (3-11)$$

This is an extremely important equation in dealing with natural radioactive substances, since we can obtain the activity of any member of the chain by knowing the activity of a single member, or we can obtain the quantity of a single isotope in the chain if the activity and decay constant of that isotope are known.

Artificial radioactivity is no different from natural radioactivity in that both obey the same laws of disintegration, but the half-lives involved are usually much shorter. The detection of short-lived artificially radioactive isotopes improves with the advancement of our measurement techniques. Before 1940 anything which had a half-life of less than 10^{-3} sec might be considered to have disintegrated immediately, whereas now half-lives down to 10^{-10} sec are being measured, and in some special cases half-lives as low as 10^{-13} sec have been measured. Artificially radioactive nuclei can also be said to emit protons, deuterons, and neutrons with a finite half-life, but since this half-life is usually less than 10^{-14} sec, the reaction is considered to go instantaneously. Most of the artificially radioactive materials whose half-lives are measurable emit either positive or negative electrons or γ rays or both. Although neutrons are emitted instantaneously, a neutron emitter may be in secular equilibrium with its longer-lived β active predecessor. Then the neutron emission proceeds as if it were characterized by a finite half-life. This phenomenon is extremely important from the point of view of reactor technology, because it is the delayed neutrons which are emitted with a finite half-life which enable a reactor to be controlled with ease and make it safe to operate.

Artificial radioactivity is usually produced by bombarding a sample with a constant beam of particles. In this case the activity will increase

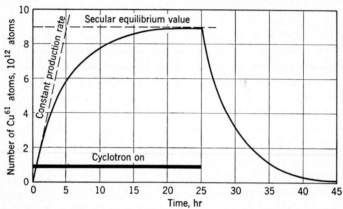

FIG. 3-4. The activity of Cu⁶¹ present in a nickel target during and after a cyclotron bombardment.

while the sample is being bombarded, but decay will also occur during this time. After the sample is removed from the beam, it will decay with its natural half-life according to Eq. (3-2) given above.

During the bombardment the rate of growth of the activity N_{ac} is given by the difference between the rate of formation of the active nuclei, which is assumed a constant C, and their rate of decay:

$$\frac{dN_{ac}}{dt} = C - \lambda N_{ac} \qquad (3\text{-}12)$$

Integrating, we have

$$N_{ac} = \frac{C}{\lambda}(1 - e^{-\lambda t}) \qquad (3\text{-}13)$$

If we irradiate for a time t_1 and measure the activity at a time t_2 after the irradiation has stopped, then

$$N = \frac{C}{\lambda}(1 - e^{-\lambda t_1})e^{-\lambda t_2} \qquad (3\text{-}14)$$

a graph of which is plotted in Fig. 3-4 for the production of Cu⁶¹ from nickel by cyclotron bombardment.

3-3. Composition of the Nucleus. The most important property of the nucleus is its charge. It is this quantity which determines the number of electrons which surround the nucleus and therefore determines the chemical properties of the atom of which the nucleus is a part. The charge on a nucleus has been found to be an integral multiple of the charge on the electron, and to be of opposite sign. The number of electronic charges on a nucleus is designated its atomic number Z. Although the examination of the X-ray spectrum of the atom is the best method of determining the nuclear charge, simpler methods usually will suffice.

All elements with Z from 1 to 102 have been found. This number increases as elements are created artificially in nuclear explosions, reactors, and electronuclear machines. The highest Z found in nature is 92, which is the charge on the nucleus of the element uranium.

Another nuclear quantity which has been of utmost importance in the study of atomic structure is the atomic mass. In fact, it was the placing of the elements in the order of their atomic mass which first led to the periodic grouping of the elements. It is indeed fortunate that there was some relationship between the mass and the charge, since the periodic nature of the elements does not depend on the mass but on the charge.

The regular increase in atomic masses in terms of a subunit is very striking. As early as 1813 Prout suggested that all atomic masses were integral multiples of that of the hydrogen atom. As pointed out previously, there were some notable exceptions to this rule, but with the discovery of the three isotopes of neon by J. J. Thomson and the extensive investigation of the masses of a large number of isotopes by Aston, it soon became clear that Prout's hypothesis was very close to the experimental situation.

Atomic masses have been studied in great detail. In fact, the latest tables give the abundance of some 325 isotopes found in nature and list about 1175 isotopes in all. This number is always increasing as measurement techniques improve. It is by the study of the masses of nuclei that we obtain some of our best information about the composition of the nucleus and the nature of nuclear forces.

Before nuclear masses can be studied in detail, the units and conventions used must be discussed. Masses in nuclear physics are usually given in atomic mass units, amu [1 amu = $1/(6.023 \times 10^{23})$ g = 1.66×10^{-24} g]. The mass used is the mass of the atom, not the mass of the nucleus. Thus the mass of the U^{235} atom, which consists of a nucleus plus 92 electrons, is 235.11704 amu. This convention may seem strange in discussion of the nucleus, but it is convenient in most cases. Suppose one proton is added to a nucleus; then this nucleus immediately picks up one electron, and the mass of the atom is increased by the mass of the hydrogen atom—not just the mass of the proton. Also, since the masses of singly ionized atoms and not of stripped nuclei are usually measured, the tables of atomic masses are in effect tables of experimental results rather than tables of interpreted or corrected nuclear masses.

Another quantity which is commonly used in nuclear physics is the mass number A. This number is the integral number nearest to the mass of the isotope. The mass of one of the uranium isotopes is 235.11704 amu per atom. This is designated as U^{235} with mass number $A = 235$. Another isotope has $M = 238.12493$ and is designated as U^{238}. The chemical symbol designates the number of positive charges in the nucleus, but since many people do not associate the number of charges on the nucleus with the chemical symbol, the nuclear charge Z is sometimes used as a subscript before the chemical symbol, although this is admittedly redundant. Thus $_{92}U^{325}$ and $_{92}U^{238}$ are sometimes used to designate these isotopes.

Although Prout's hypothesis was very attractive, one fact that could not be discounted was that the masses of the isotopes were almost integral multiples of the mass of the hydrogen atom, but not quite. For instance, the isotope O^{18} weighs 18.004874 amu and F^{19} weighs 19.004456 amu. These differ in mass by 0.999582 amu, 0.9 per cent different from the mass of the hydrogen atom, which is 1.008142 amu. An explanation for this discrepancy in mass was advanced from the Einstein relation between mass m and energy E:

$$E = mc^2 \qquad\qquad (3\text{-}15)$$

where c is the velocity of light. The nucleus is said to be "bound" by a certain energy which is equivalent to the loss in mass. Thus, in the case of the last proton in F^{19} the binding energy is

$$1.008142 - 0.999582 \text{ amu} = 0.008560 \text{ amu}$$

From the Einstein equation of the equivalence between mass and energy,

$$1 \text{ amu} = 931.376 \text{ Mev}$$

where 1 ev is 1.60203×10^{-12} erg. The electron volt is the unit of energy commonly used in nuclear physics, and the conversion factors given above should be committed to memory. Thus the difference between the mass of $O^{18} + H^1$ and the mass of F^{19} is interpreted as the binding energy of the last proton in F^{19} and is usually given as 7.97 Mev.

Since a good explanation which has been experimentally verified can be given for the difference between the mass of the nucleus and the sum of the masses of its component parts, the nucleus can be accepted as being made up of protons and neutrons. One possible flaw in this argument is the fact that radioactive nuclei emit electrons.

There are several good reasons for believing that electrons as such do not exist in the nucleus. The measured angular momentum of the deuteron, H^2, is 1 unit. If the deuteron were composed of two protons and one electron, then the total number of particles would be three. Since the total angular momentum must be the sum of the angular momenta of all the particles and the spin of each particle is $\frac{1}{2}$ unit, it is impossible to obtain a spin of 1 by adding the possible momenta of the three particles. If we assume the deuteron is composed of a neutron and a proton, then the total spin of 1 can be obtained by having the spin of the two particles parallel. There are several other examples to support this argument, such as N^{14}, which would have 21 particles if composed of electrons and protons but 14 if composed of neutrons and protons. The spin of N^{14} has been measured as 1, which supports the argument that the nucleus is composed of neutrons and protons.

Another strong argument for the existence of only heavy particles in the nucleus comes from the measured magnetic moments of nuclei. From electromagnetic theory, the magnetic moment μ_0 of a particle should be

$$\mu_0 = \frac{eh}{4\pi mc} \tag{3-16}$$

where μ_0 = unit magnetic moment ("magneton")
    e = charge of particle
    h = Planck's constant
    m = mass of particle

Thus we can see that the ratio of the magnetic moment of the proton to that of the electron should be the inverse ratio of their masses, which is $\frac{1}{1837}$. If there were electrons in the nucleus, the magnetic moments of nuclei should be observed to be of the order of electron magnetons rather than heavy-particle magnetons. Since all the magnetic moments of nuclei which have been measured are of the order of magnitude of heavy-particle magnetons, electrons cannot exist as such in the nucleus.

There are many other independent arguments to support the hypothesis that there are no electrons in the nucleus. But how, then, one might ask, does a nucleus emit electrons if there are no electrons in the nucleus to emit? In 1934 Fermi gave the first reasonable explanation for the emission of electrons by a nucleus which contains no electrons. He postulated that the electrons are created when a neutron changes into a proton in the same way that photons or light quanta are created when an electron jumps from one orbit in an atom to another orbit. One does not postulate that the photon exists in the atom and is emitted when the electron jumps to another orbit; in the same way, one does not have to assume that electrons exist in the nucleus.

Many other difficulties have been encountered with the β disintegration of nuclei which were of fundamental importance because the laws of conservation of energy and momentum seemed to be disobeyed. In the study of natural radioactivity the α rays and γ rays had definite energies, which would be expected from a quantized system. However, β rays are given off with any energy between the maximum allowed by the energy available from the mass difference between the two nuclei and zero energy. The average energy given off in the β disintegration is also much less than the mass difference energy. Many experiments were performed to determine how this energy disappeared, but all were unsuccessful. The question arose as to whether or not energy was actually conserved.

Another difficulty encountered in β disintegration is the lack of conservation of angular momentum. To illustrate this we use the decay of the neutron into an electron and a proton. The neutron has a half-life of about 12 min, and it emits electrons with energies ranging from zero to 0.781 Mev, which is the mass energy difference between the neutron and the hydrogen atom. The spin of each of the three particles—neutron, proton, and electron—is $\frac{1}{2}$. How then can the particle with a spin $\frac{1}{2}$ disintegrate into two particles each of spin $\frac{1}{2}$ and still conserve angular momentum? Here again the question was asked, Is angular momentum always conserved?

Pauli suggested a solution to the problem by inventing a particle called a "neutrino." The neutrino had no charge, a spin of $\frac{1}{2}$, and no rest mass. If a neutrino is also given off in a β disintegration, then both energy and momentum can still be conserved. The neutrino does not interact strongly with matter, and it was not detected directly until 1956.

In reactors there are many β disintegrations after fission, and a considerable amount of energy is released by this process. The neutrinos which are emitted are lost, since they pass through the reactor shield—and, for that matter, even the earth—without colliding with another particle. This energy cannot be utilized because the mean free path of the neutrino, even in the center of the sun, where matter is extremely dense, would still be a few thousand miles.

In the Fermi theory of β decay, electrons as such play no part in the structure of the nucleus. When a positive electron is emitted, a proton is converted into a neutron and a neutrino is emitted; and when a regular electron is emitted, a neutron is converted into a proton and a neutrino is also emitted. These reactions can be written

$$n \rightarrow p^+ + e^- + \nu \qquad (3\text{-}17)$$
$$p \rightarrow n + e^+ + \nu \qquad (3\text{-}18)$$

The first reaction is exothermic, and it can proceed spontaneously. The second reaction is endothermic, and energy must come from the remainder of the nucleus in order for the reaction to proceed.

With the substantial evidence which shows that no electrons exist in the nucleus, we can proceed with our original hypothesis that the nucleus consists of protons and neutrons. If we have a nucleus of atomic number A which has Z protons, then there are $N = (A - Z)$ neutrons in the nucleus. Nuclei which have the same Z but different A are called "isotopes"; nuclei with the same A but different Z and N are called "isobars"; and nuclei with the same N but different A and Z are called "isotones."

3-4. Binding Energy. Since a nucleus is made up of neutrons and protons, the total binding energy (TBE) of the nucleus is the difference between the mass of particles which constitute the atom and the mass M of the atom as determined by a mass spectrograph, i.e.,

$$\text{TBE} = (Zm_H + Nm_n) - M \tag{3-19}$$

In order for a nucleus to be stable, the binding energy must be a positive quantity. The above equation also can be used to determine the binding energy of the last neutron or the last proton in a nucleus, as was illustrated in the last section for the binding energy of the last proton in F_{19}.

Although the criterion that the total binding energy be positive is a necessary condition for the stability of a nucleus, it is not a sufficient condition. The sufficient condition on the energy for stability is that the mass of the nucleus must be less than the sum of the masses of any combination of particles from which the nucleus can be constructed. Of course, nuclei whose mass is greater than two possible components may be stable for other reasons, but if these nuclei could be broken up, then energy would be released.

Let us use the nuclear reactions in the very light nuclei or the so-called "thermonuclear reaction" as an example of how this works. These reactions revolve about the first three elements, hydrogen, helium, and lithium. Lithium 6 is a stable isotope and therefore must be lighter than any combination of its component parts. The total binding energy of Li^6 is

$$\text{TBE} = (3 \times 1.008142 + 3 \times 1.008982) - 6.017021$$
$$= 0.034351 \text{ amu} = 31.98 \text{ Mev}$$

Reversing this means that, if we can fuse three protons and three neutrons into a Li^6 nucleus, the reaction will give up 31.98 Mev of energy per nucleus formed. It would be very improbable, even in a very dense medium, that six particles would collide simultaneously to form Li^6; therefore, for energy production this reaction need not be considered. However, this nucleus can be built up in other ways. Lithium 6 must be stable against disintegration into H^3 and He^3 or into H^2 and He^4. The energy released when Li^6 is made from H^3 and He^3 would be

$$
\begin{aligned}
E &= M_{H^3} + M_{He^3} - M_{Li^6} \\
&= 3.016997 + 3.016977 - 6.017021 \\
&= 0.016953 \text{ amu} = 15.8 \text{ Mev}
\end{aligned}
$$

and from H^2 and He^4 it would be

$$
\begin{aligned}
E &= 2.014735 + 4.003873 - 6.017021 \\
&= 0.001687 \text{ amu} = 1.478 \text{ Mev}
\end{aligned}
$$

Thus we see that Li^6 is very stable against disintegration into three neutrons and three protons, is quite stable against disintegration into H^3 and He^3, and is only moderately stable against disintegration into H^2 and He^4. Conversely, the amounts of energy that would be released if three neutrons and three protons were combined would be greater than the energy released when H^3 and He^3 are combined to form Li^6, which in turn is greater than the energy released when H^2 and He^4 are combined to form Li^6. This illustrates the origin of nuclear energy, since whenever a more stable nuclear configuration is formed matter is converted into energy.

Actually, none of the above reactions will probably be used for the propagation of a controlled thermonuclear reaction, because like charged particles repel each other. The velocity of the particles would have to be very high in order to get within the range of action of the nuclear forces. We can calculate the energy required to bring two particles of charge $Z = 1$ and charge $Z = 2$ to a distance r of 1.4×10^{-13} cm, which is of the order of distances between neutrons and protons in a nucleus. The electrostatic energy of two charges q_1 and q_2 is given as

$$
W = \frac{q_1 q_2}{r} = \frac{4.8 \times 10^{-10} \text{ esu} \times 9.6 \times 10^{-10} \text{ esu}}{1.4 \times 10^{-13} \text{ cm}} \tag{3-20}
$$
$$
= 3.3 \times 10^{-6} \text{ erg} = 2.06 \times 10^6 \text{ ev}
$$

or about 2 Mev. Since 0.0253 ev corresponds to the average energy of an atom in a monatomic gas at room temperature, which is about $300°K$ on the absolute temperature scale, the temperature required to have the average energy of the atom at 2 Mev is

$$
300 \times \frac{2 \times 10^6}{2.53 \times 10^{-2}} = 2.4 \times 10^{10}
$$

Actually, this reaction will take place at a much lower temperature because there is a distribution of velocities of atoms in a gas and some atoms have velocities considerably higher than the average, but even so the temperatures required are fantastic.

The reaction which releases the most energy at the lowest temperature is the reaction between H^2 and H^3 where

$$H^2 + H^3 \rightarrow He^4 + n \qquad (3\text{-}21)$$

or, in the shorthand of nuclear reactions,

$$H^2(H^3,n)He^4$$

The neutron which is released is then available to be absorbed by the H^2 to form H^3. Consequently, this reaction can proceed starting with

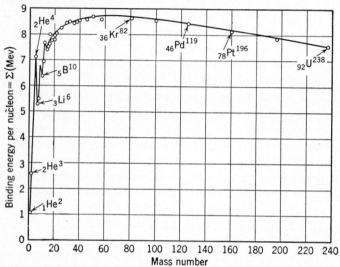

FIG. 3-5. The binding energy of nuclei as a function of mass number.

mostly H^2 to which a small quantity of H^3 has been added, provided the temperature is high enough, which temperature has been calculated to be about 10 million°C. The necessary temperature is evidently achieved in atomic bomb tests, for thermonuclear devices are known to work.

Although the total binding energy of a nucleus is of interest for the very light nuclei, it is usually not of interest for intermediate and heavy nuclei. In a nuclear reaction usually one or at most a few particles are involved; consequently, a quantity which is more useful is the binding energy per particle. This is usually just designated the binding energy, or BE. The BE is plotted vs. the mass number A in Fig. 3-5. Another quantity which is related to the binding energy per particle is the packing fraction, which is defined as $P = (M - A)/A$. The packing fraction is very often used in textbooks of nuclear physics and can be used instead of the binding energy; it is not so significant physically, although it is sometimes more convenient.

The BE curve rises fairly rapidly as A increases, with some remarkably large fluctuations, and then levels off at about $A = 20$. It remains almost constant until A is about 200, then decreases again as A increases. It is this curve that tells us how nuclear energy can be obtained. By fusing light elements together to form an intermediate element, energy will be released because the binding energy per particle is larger in the intermediate elements than in the very light elements. Also, of more practical significance is the fact that the binding energy per particle in the very heavy elements is less than in the intermediate elements. This means that by fissioning one of the very heavy elements into two inter-mediate elements, energy will be released. The amount of energy released can be calculated from Fig. 3-5. From the curve we find that the average binding energy per particle in U^{238} is about 7.6 Mev, whereas the average binding energy per particle for Sn^{120} is about 8.5 Mev. This means that, if we could split U^{238} into two equal nuclei, then we would get out about 0.9 Mev per particle, or about 210 Mev. Actually, the process is much more complicated than this, because the number of neutrons per proton in the nucleus increases with increasing A faster than the number of protons. Thus when a heavy nucleus splits up there are many more neutrons than can be contained in two nuclei with half the atomic number of the heavy nucleus. The fact that energy was available in this form was known many years prior to 1939, when nuclear fission was first discovered, but the method by which it could be released in quantity awaited the discovery of slow-neutron-induced fission.

Another very significant fact, from the point of view of obtaining a model of the nucleus, comes from this BE curve. In the intermediate nuclei the BE is approximately constant at about 8.5 Mev per particle. This indicates that the forces which bind nuclei together are not like the Coulomb forces which hold the atom together. With Coulomb forces and gravitational forces, every particle interacts with every other particle. Nuclear forces are more like molecular forces which saturate. In a molecule such as CH_4 there are four bonds associated with the carbon, and when all these bonds are filled there is no longer any force between the carbon atom and a fifth hydrogen atom. There are many examples of saturated systems in molecular physics. The molecule H_2 is the simplest example. The two electrons are shared by the two protons, giving very strong molecular binding. There is no room for an extra electron between the two particles, and the system is therefore saturated, since both bonds are filled. There can be no neutral H_3. Forces between two particles which exist only because of the presence of a third particle (an electron, in the case of H_2) are called "exchange forces." The forces holding the hydrogen molecule together are exchange forces, and since nuclear forces are similar to this, they are also said to be at least partially an exchange force.

If we are guided by the chemical analogy to the hydrogen molecule to describe nuclear binding, then a saturated system should be one in which two protons or two neutrons are paired. From this argument alone we would expect to find that the dineutron (i.e., two neutrons combined) or He^2 would be stable. Neither of these two nuclei is found in nature. The situation is evidently much more complicated than can be described here.

3-5. Nuclear Size. The first indication of the size of the nucleus was obtained by Rutherford in his famous α-particle scattering experiment, which led to his nuclear theory of the atom. He found that α particles, which have a positive charge, were deflected more than could be accounted for by an electrical charge the size of the atom. He concluded from these experiments that the positive charge of the atom must be concentrated in a sphere of radius less than 10^{-12} cm. Thus the nucleus must be much smaller than the atom, which was known from many experiments to be about 10^{-8} cm in diameter.

Actually, α-particle scattering can still be used to determine the nuclear radius, but since the observed effects are due mostly to electromagnetic forces, purely nuclear effects must be deduced by subtracting the electromegnetic interaction from the experimental results. The results obtained this way are not very accurate, so othar methods are usually used to determine nuclear size.

The scattering of high-energy neutrons by nuclei is probably the best way of determining the nuclear size, since no electromagnetic interaction can take place because the neutron has no charge. The relationship between the energy of an α particle emitted by a heavy nucleus and the half-life of the disintegration is a rapidly varying function of the radius of the nucleus. The determination of the nuclear radius from the energy and half-life of the emitted α particle leads to a value of the nuclear radius which is in good agreement with other methods. Another method for determining the size of the nucleus is to compare the binding energy of two nuclei A and B which contain the same number of particles but in which A has N neutrons and Z protons and B has $N - 1$ neutrons and $Z + 1$ protons. If the nuclear configuration is the same, the nucleus with more charge should have less binding energy because of the electrostatic repulsion of the extra proton. If we assume that the nucleus is a uniform spherical charge, then the radius of this charge can be calculated from the difference in binding energy of the two nuclei. A more recent method of determining the size of the nucleus depends on making atoms in which an electron is replaced by a μ meson. By observing the energy of the spectral line emitted by this atom, the nuclear radius can be determined. This occurs because the radius of the orbit of the meson about the nucleus is $\frac{1}{200}$ the radius of the orbit of the electron; consequently the size of the nucleus will have a much greater effect on the meson orbit than on the electron orbit. The diffraction pattern obtained when extremely high-energy electrons are scattered by a nucleus has also recently been used to determine the nuclear size.

All the methods given above for determining nuclear size are independent, and one might fear that different results would be obtained. Actually, all the experiments give approximately the same results. The radius of the nucleus can be closely represented by the equation

$$r = r_0 A^{1/3} \tag{3-22}$$

where r = radius of the particular nucleus

r_0 = constant

A = total number of particles in nucleus

The constant r_0 is the subject of debate at the present time. All experiments with neutrons, α particles, and protons showed r_0 to be 1.4×10^{-13} cm. However, recent experiments on the interaction of 500-Mev electrons and of π and μ mesons with matter (1953) indicate that r_0 should be 1.2×10^{-13} cm. It should be pointed out that in the case of the μ meson atom and electron diffraction, the effect of the charge is being measured, whereas in the other experiments the nuclear boundary plus the size of the particle used as a probe is measured. It seems probable that the two sets of experimental data will be made compatible very shortly.

One extremely important conclusion which can be drawn from the experimental evidence is that the volume $V = \frac{4}{3}\pi r^3 = \frac{4}{3}\pi r_0^3 A$ is directly proportional to the number A of particles in the nucleus. This means that nuclear density is almost always constant, unlike the density of atomic matter. Another conclusion that can be drawn is that the nuclear forces must be short-range.

The main effect the nucleus has on the atom is due to its mass and charge; the purely nuclear forces are not observable much beyond the distance of the nucear radius. This shows that the nuclear forces, unlike Coulomb forces and gravitational forces, cannot have a range much greater than the size of the nucleus. In fact, the range of the nuclear forces can be shown to be much less than the nuclear size.

3-6. Nuclear Systematics. A study of all the stable isotopes reveals many facts which must be explained by any theory of nuclear structure. It was very early in the study of nuclear masses that the correlation between the even-odd properties of the number of neutrons and the number of protons in the nucleus was noted. The facts known at present are as follows: the number of stable nuclei which have (1) an even number of protons and an even number of neutrons is very much larger than the number of nuclei with an odd number of protons or neutrons. The number of nuclei which have (2) an odd number of neutrons and an even number of protons is not significantly different from the number of nuclei which have (3) an odd number of protons and an even number of neutrons. There are only eight nuclei which are stable that have (4) both an odd number of neutrons and an odd number of protons. Only three of these odd-odd nuclei have an appreciable isotopic abundance. The other five are only a small fraction of the total amount of the elements found. The three with significant abundance are Li^6, B^{10}, and N^{14}. All these are extremely light nuclei which might be expected to be governed by different rules of stability from those for nuclei with a large number of particles. The numbers of isotopes in the four categories listed above are shown in Table 3-1.

TABLE 3-1

N	Z	
	Even	Odd
Even............	167	52
Odd.............	56	8

A plot of the number of neutrons in a nucleus vs. the number of protons in the same nucleus for both stable and radioactive nuclei is shown in Fig. 3-6. On this type of plot the isotopes (same Z) are on the same vertical line, isotones (same N) on the same horizontal line, and isobars (same A) are on the same diagonal.

The stable isotopes lie in a narrow region near the center of all the known isotopes. Radioactive isotopes which are below the stability line are found to emit positive particles, converting protons to neutrons and thereby getting back to the stable region. Those radioactive isotopes which are above the stability line emit negative particles, converting neutrons to protons and thereby coming back to the stability region. If the radioactive isotope is far from the stability line, several disintegrations may take place before the nucleus reaches the stability region. This series of disintegrations is called a "radioactive chain."

In the light nuclei the stability region is centered around $N = Z$, but for heavier elements the stability region deviates toward a higher number of neutrons than protons. Both these facts are extremely important in the development of a nuclear theory.

Although there is no strong periodic structure to the chart of the isotopes as there is when the masses of the elements are plotted vs. the chemical valence, there is some significant structure which has received considerable attention recently. The number of stable isotopes with $Z = 20$ and 50 is much greater than for any Z close by. The number of isotopes with the neutron number $N = 20$, 50, and 82 is significantly

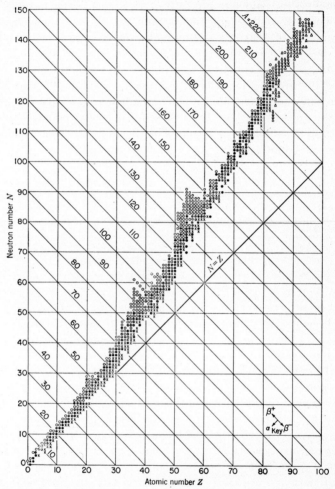

FIG. 3-6. A chart of the known isotopes. ●, stable isotopes; △, naturally radioactive alpha emitters; ✕, radioactive positron emitters or electron absorbers; ○, radioactive electron emitters.

higher than for neighboring N's. Also, if one looks at the binding energy of the isotopes with these numbers of particles, it is found to be higher than that of any of the neighboring isotopes. A systematic study of all the isotopes reveals special stability properties when N or $Z = 2$, 8, 20, 50, 82 and for $N = 126$. This, together with other evidence obtained from nuclear reactions, leads to a theory in which the nucleus consists of shells of particles in much the same way as the atom consists of shells of electrons. Although the shell-structure theory is not as obvious as in the case of atomic structure, a close study of Fig. 3-6 will demonstrate some of these properties.

3-7. Semiempirical Mass Formula. A semiempirical formula for the masses of stable nuclei has been derived which is extremely useful, not only for the determination of these masses but for the discussion of nuclear stability in terms of the types of particles in the nucleus. Although at present there is no good theory of the nucleus, for lack of exact knowledge of nuclear forces, we can predict the general behavior of nuclei from the following empirical knowledge:

1. Nuclear matter has a constant density, since the nuclear volume is directly proportional to the number of particles in the nucleus.

2. The total binding energy of nuclei is approximately proportional to the mass of the nucleus and therefore is proportional to its volume. Thus nuclear forces show saturation. Nuclear forces also have a very short range. The relatively short range of nuclear forces can best be seen from the fact that the nuclear forces have no effect on nuclear particles unless they are closer than 10^{-12} cm. Some experimental evidence points to the fact that the range of nuclear forces must be about 2 or 3×10^{-13} cm. This means that each particle interacts most strongly with its nearest neighbor but not very strongly with other particles.

3. The protons in the nucleus repel each other because of electrostatic forces. Since these forces tend to disrupt the nucleus, the effect decreases the binding energy of the nucleus.

4. In light nuclei the number of neutrons is approximately equal to the number of protons.

5. Nuclei with both Z and N even are the most stable. Nuclei with odd N, even Z, and with odd Z, even N, are of about equal stability. Nuclei with odd N and odd Z are unstable except for the very light nuclei.

Suppose we assume that the nucleus behaves like a liquid drop. Then from (1) and (2) we assume a term in the mass formula which is directly proportional to the volume of the nucleus:

$$E_{vol} = -a_1 A \qquad (3\text{-}23)$$

which is known as the "volume-energy term." This is negative, since an increase in binding represents a decrease in mass.

By taking the binding proportional to the volume, we have overestimated the energy, since a nuclear particle at the surface is not in contact with as many particles as a particle on the inside of the nucleus. The number of particles on the surface would be proportional to the surface area; consequently we take a term in the mass formula which is proportional to the surface area of the nucleus:

$$A_s = 4\pi r^2 = CA^{\frac{2}{3}} \qquad E_s = a_2 A^{\frac{2}{3}} \tag{3-24}$$

which is known as the "surface-energy term." This is a positive term, since it represents a decrease in binding.

The Coulomb energy can be determined by finding the energy of Z protons distributed throughout the volume of a sphere of radius r. In electrostatics this energy is

$$E = \frac{3}{5} \frac{Z^2 e^2}{r} \tag{3-25}$$

Thus for the Coulomb energy we write

$$E_{Coul} = \frac{a_3 Z^2}{A^{\frac{1}{3}}} \tag{3-26}$$

The term in the semiempirical mass formula which takes account of the fact that in the light nuclei the number of protons is equal to the number of neutrons can be determined by plotting the binding energy of different isobars against Z after subtracting the Coulomb energy. A plot of the quantity $M - A$ for the isobar $A = 125$ is given in Fig. 3-7. This curve is almost symmetric about the minimum. If it were perfectly symmetric the curve would be an even function of A. Thus we take a parabolic dependence as the simplest even function. However, this is not the whole story, since A has been kept constant. What we have been looking at is the effect of the asymmetry on the binding of the last particle in the nucleus. We must also determine how this term depends on A. Suppose we have two nuclei with the same Z/A but one having twice the A of the other. If we associate a certain amount of energy with each asymmetrical particle, the total amount of energy due to the asymmetry should be twice as much in the nucleus with twice as many particles. Thus the total asymmetry energy is proportional to A. Therefore the asymmetry energy is

$$E_{asy} = a_4 A \left(\frac{1}{2} - \frac{Z}{A} \right)^2 = \frac{a_4 (A/2 - Z)^2}{A} \tag{3-27}$$

Our last term gives a binding energy associated with an odd or even number of protons in the nucleus, which is just an empirical fact. This term is $E = \delta$, where

$$\begin{aligned} \delta &= 0 & \text{for } A \text{ odd} \\ &= \frac{-0.036}{A^{\frac{3}{4}}} & \text{for } N \text{ even, } Z \text{ even} \\ &= \frac{+0.036}{A^{\frac{3}{4}}} & \text{for } N \text{ odd, } Z \text{ odd} \end{aligned} \tag{3-28}$$

Combining all these terms, we have the total binding energy as

$$\text{TBE} = -a_1 A + a_2 A^{2/3} + a_3 \frac{Z^2}{A^{1/3}} + a_4 \frac{(A/2 - Z)^2}{A} + \delta \quad (3\text{-}29)$$

and it remains to determine the magnitude of these coefficients. These

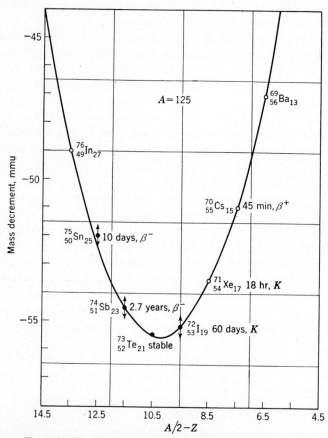

Fig. 3-7. The relative energies of the isobars of $A = 125$.

can be determined in many ways, since there are many more stable nuclei than constants in this formula. A recent formula [1] is

$$M(A,Z) = 0.99395A - 0.00084Z + 0.0141A^{2/3} + 0.000627 \frac{Z^2}{A^{1/3}}$$
$$+ 0.083 \frac{(A/2 - Z)^2}{A} + \delta \quad (3\text{-}30)$$

This formula gives a good approximation for all the nuclear masses and can be used to determine the stability of one nucleus when compared to another. Exercises in the use of this formula can be found at the end of this chapter.

Using this formula, the mass of Ni^{60} has been calculated as 59.9503, and its measured mass is 59.94977. The calculated mass is only 0.02 per cent different from the measured mass. This indicates the accuracy of Eq. (3-30), which should not be surprising since the semiempirical mass formula was determined from known masses.

3-8. Nuclear Forces and the Deuteron. The simplest nucleus aside from the proton is that of heavy hydrogen, the deuteron. Since it is easiest to work with the simplest system, we might expect to find out most about the structure of nuclei from the study of the deuteron. After all, it was by studying the spectrum of the hydrogen atom that Balmer first determined the recurrence formula which was so successful in describing the position of the spectral lines of the hydrogen atom. Bohr's theory of the atom was first applied to hydrogen because it was the simplest system to describe, and greatest success has been achieved when describing hydrogen-like atoms.

Unfortunately, the situation in studying the nucleus is not the same as in the study of the atom. There are no excited states of the deuteron in the same way that there are excited states of the hydrogen atom, and consequently we cannot build up a theory of nuclear structure in the same way that a theory of atomic structure was built up. However, it is worthwhile to look at our present concept of the deuteron, since it gives a great deal of information about nuclear forces and the same general scheme is used to describe more complex nuclei even though the justification for using the same two-body system to describe a more complex nucleus is not too clear.

We first assume the deuteron is made up of a neutron and a proton. The fundamental properties of the proton, neutron, and deuteron are given in Table 3-2.

TABLE 3-2

Particle	Symbol	Charge (relative)	Approximate rest mass (relative)	Rest mass μ, amu	Spin	Magnetic moment, nuclear magnetons
Neutron......	n	0	1	1.008937	$\frac{1}{2}$	−1.91354
Proton........	H^1, p	1	1	1.008130	$\frac{1}{2}$	2.79353
Deuteron......	H^2, d	1	2	2.01472	1	0.857648

From the masses of the three particles we find that the binding energy of the deuteron is 2.23 Mev. Since no bound excited state of the deuteron has been found, we must assume the ground or stable state of the deuteron is at $E = -2.23$ Mev.

If we add the magnetic moment of the neutron and proton, we find the sum is almost equal to the magnetic moment of the deuteron. The spin of the deuteron is 1 and that of the neutron and proton each $\frac{1}{2}$. Since the magnetic moments are almost equal and the spins do add up, it is simplest to assume that the deuteron consists of the neutron and proton extremely close together and that each particle is spinning on its own axis with the direction of the spins parallel. We also assume that there is no angular momentum due to the particles rotating about their mutual center of mass. We could also assume that the directions of spins are antiparallel and the angular momentum of the two particles about their mutual center of gravity is 1 unit, to give a spin of 1 for the deuteron. However, the magnetic moment calculated from this assumption is found to be completely different from the observed magnetic moment of the deuteron. We therefore assume the simpler model is approximately true.

We do not know the exact nature of nuclear forces, so we must assume something about them before we can go any further. Since the experimental evidence shows that the forces are extremely short-range forces, we can assume for simplicity that the forces depend only on the distance between the two particles. We also make the very simple assumption that the forces between the particles are zero until the distance between them is extremely small and then the force suddenly becomes very large. Since systems of forces are difficult to work with because of their directional properties, the potential function which represents the energy between the two particles is usually assumed and the properties of the system derived from it. This is common practice in physics and is used in electromagnetic theory, gravitational theory, and atomic theory as well as in nuclear theory. The simplest potential between the neutron and proton that can be assumed is shown in Fig. 3-8. This is usually called a "potential well" because the force is attractive and it requires energy to remove the particle.

Of course, this is not the physical potential in which the particles actually move, since physical functions do not have abrupt changes, but it is the easiest with which to work. The position of the ground state of the deuteron on this potential-energy diagram is known to be 2.23 Mev below zero energy and is one of the important experimental quantities used in obtaining the theory of the deuteron.

Using this simple model, many properties of the deuteron can be calculated; and fortunately from the point of view of simplicity, but unfortunately from the point of view of finding more about the structure of the nucleus, the results obtained from this simple model agree fairly well with the experimental results. In fact, many different potential-well functions have been tried, but all give substantially the same result.

As one might expect, this simple model does not explain some properties very well. It was noted at the beginning of this section that the magnetic moment was almost, but not quite, equal to the sum of the magnetic moments of the neutron and the proton. The difference is about 3.8 per cent, which has been explained by assuming the forces between the proton and the neutron have a more complicated dependence than only on the distance between them. This more complicated dependence is

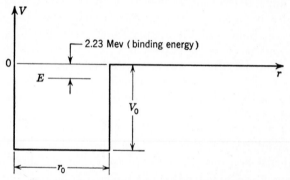

Fig. 3-8. A possible neutron-proton potential.

said to be caused by a tensor force similar in properties to the restoring force in a crystal when a force is applied in some direction other than along the direction of the crystal axes. In this case the displacement is not in the direction of the force but in some other direction depending on the directional properties of the particular crystal and the point on the crystal at which the force is applied. With the introduction of this tensor force, the calculated magnetic moment of the deuteron can be made to agree with the measured magnetic moment.

This simple model of the nucleus could not explain quantitatively how neutrons are scattered by protons. The number scattered was about four times the number that would be expected. In order to account for this discrepancy, the forces between the neutron and proton were assumed to be dependent on their relative spin orientation. This assumption leads to two different potential functions between the neutron and proton, one when the spins are parallel and the other when the spins are antiparallel. The corresponding states are called the "triplet state" (spins parallel) and the "singlet state" (spins antiparallel). If the nuclear forces are spin-dependent the results of certain experiments would be entirely different from what they would be if the forces are not spin-dependent. In particular, there are two types of hydrogen molecules: orthohydrogen, in which the spins of the protons in the molecule are parallel, and parahydrogen, in which the spins are antiparallel. If the force when the spin of the proton is parallel to the spin of the neutron is stronger than when the spins are antiparallel, the number of neutrons scattered from orthohydrogen would be expected to be much larger than the number scattered from parahydrogen. The experiment to test this hypothesis has been performed many times, with the most recent results showing

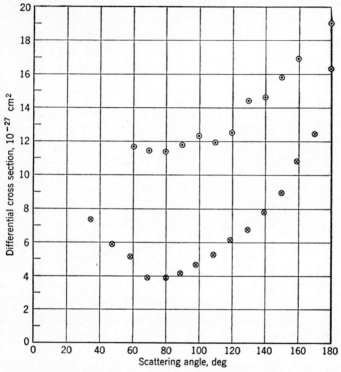

FIG. 3-9. The angular dependence of (n,p) scattering on differential cross section. The upper curve is for 40 Mev, and the lower curve is for 90 Mev.

the scattering from orthohydrogen to be approximately thirty-six times the scattering from parahydrogen. This is conclusive evidence that nuclear forces are spin-dependent.

Another property of nuclear forces has been obtained by studying the neutron-proton scattering at very high energies. Using a model similar to the simple model given above, the angular distribution of neutrons scattered by protons can be calculated and is found to give a strong peak in the forward direction. Now suppose that on collision a neutron and proton exchange the charge so that the neutron becomes the proton and the proton becomes the neutron. In this case, since the particles are reversed and since for conservation of momentum the proton must move in the opposite direction to the neutron in the center of mass system, the neutron distribution will show a strong peak in the backward direction. (This force which changes the proton into a neutron is called an "exchange force.") The experimental results for (n,p) scattering at 40 and 90 Mev are shown in Fig. 3-9. These curves show a peak in both the forward and backward direction, illustrating that the forces are not of a simple attractive type, or only of an exchange type. Actually a mixture of about 50 per cent of each type of force fits the experimental data fairly

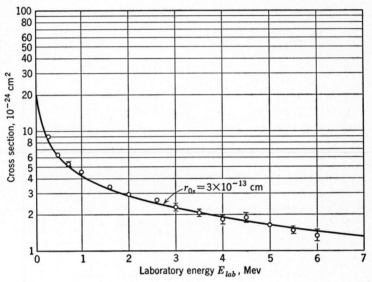

FIG. 3-10. Total cross section for (n,p) scattering. The curve is theoretical, based on a radius of 3×10^{-13} cm of the well for the singlet state. Φ represents the experimental points and their probable error.

well, but it will then not account for the saturation properties of nuclear forces. However, this experimental evidence is the best experimental verification of the exchange character of nuclear forces.

In order to indicate how good are the results obtained from this simple potential-well theory, the neutron-proton scattering cross sections calculated from this simple model are shown with the experimental results in Fig. 3-10. Unfortunately the shape of the well, and thus the specific character of the nuclear forces, has very little effect on the shape of the curve.

Another important property of nuclear forces can be inferred from a comparison of the proton-proton and the neutron-proton collision process. After the electromagnetic effects are subtracted from the (p,p) scattering, the interaction between two protons is the same as the interaction between a neutron and a proton in the singlet state, to within the accuracy of present experimental results. It is not possible to have a triplet interaction in the lowest-energy state for two protons because of the Pauli exclusion principle; consequently the only interaction that can be compared is that in the singlet state. Since the (n,p) and (p,p) interactions are the same, the assumption has been made that nuclear forces are not charge-dependent. The best evidence for the charge-independence of nuclear forces comes from studying the three nuclei C^{14}, N^{14}, and O^{14}. After the Coulomb energy and the neutron-proton mass difference are corrected for, the ground states of these nuclei occur at the same energy. Since the number of (nn), (np), and (pp) bonds is different for the three nuclei, the nuclear forces do not depend on whether the particle in the nucleus is a neutron or a proton. Thus all particles in the nucleus can be assumed to be exactly alike and are called "nucleons." By exchanging an electron, a particle can change from a neutron to a proton or vice versa.

Although we get little help from the structure of the deuteron in building up a detailed theory of the nucleus, it does give us the characteristics of nuclear forces which are essential to our qualitative understanding of the nucleus.

Summing up, we have the following properties of nuclear forces:

1. Short range ($r_0 \approx 2.6 \times 10^{-13}$ cm)
2. Noncentral (tensor forces needed)
3. Spin-dependent
4. Exchange in character
5. Charge-independent

3-9. Nuclear Cross Sections. In the study of nuclear physics and nuclear reactions, the term "cross section" is one of the most important terms used. The cross section of a nucleus may be visualized as the area presented by the nucleus for a particular process. Suppose we look at the classical picture of the cross section. Assume we have a geometrical arrangement of billiard balls of radius r as shown in Fig. 3-11 such that the projected area, πr^2, of any billiard ball is very small compared to the area a^2, where a is the distance between billiard balls. Suppose another ball, which is small in radius compared to the billiard balls in the array, is projected in the x direction at the array of billiard balls, which has a height and width of 1 cm.

If the position of the projectile is random with respect to the geometrical array, the probability of the projectile having a collision with the first plane of billiard balls is just the total area of all the billiard balls divided by the total area, which is 1 cm². Since the billiard balls are a cm apart, there are $1/a^2 = m$ billiard balls in 1 cm². Thus the prob-

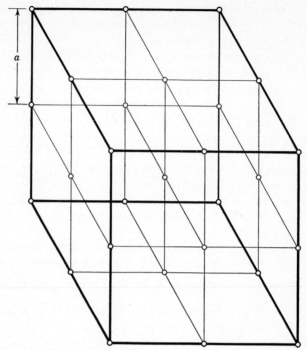

FIG. 3-11. A representation of matter for the purpose of developing the classical concept of cross section.

ability P of having a collision in the first plane is

$$P_1 = \frac{m\pi r^2}{1} \tag{3-31}$$

The probability of not having a collision in the first plane is

$$P_1' = 1 - m\pi r^2 \tag{3-32}$$

The probability of not having a collision in the second plane is

$$P_2' = (1 - m\pi r^2)^2 \tag{3-33}$$

If there are n planes of billiard balls in 1 cm, the probability of *not* having a collision in x cm is

$$P_n' = (1 - m\pi r^2)^{nx} \tag{3-34}$$

If N is the number of billiard balls for 1 cm³, then

$$nmx = Nx$$

and we have

$$P'_n = \left(1 - \frac{N\pi r^2 x}{nx}\right)^{nx} \tag{3-35}$$

If nx is a very large number, $P'_n = e^{-N\pi r^2 x}$, since

$$\lim_{n \to \infty} \left(1 - \frac{x}{n}\right)^n = e^{-x} \tag{3-36}$$

Of I_0 incident particles, the number which will travel the distance x is

$$I = I_0 e^{-N\pi r^2 x} = I_0 e^{-N\sigma x} \tag{3-37}$$

where σ is defined as the total cross section for the process. This is the total cross section because we have assumed that any collision would deflect the particles from their original path. Thus, if a collision occurs the particles are no longer in the beam. This indicates how the total cross section can be measured. If a highly collimated beam of particles is incident on a sample, by measuring the number of the particles which are traveling in the initial direction after passing through thickness t which has N nuclei per cubic centimeter, the total cross section can be obtained from

$$\sigma_T = \frac{1}{Nt} \ln \frac{I_0}{I} \tag{3-38}$$

From our classical argument the cross section would always be a constant πr^2, but since quantum mechanics holds in the nucleus, there are resonance phenomena and the cross section of a particular nucleus can vary by large amounts in a very small energy interval.

There are many other cross sections besides the total cross section. In a nuclear process a particle may be scattered by the nucleus without net energy loss. The cross section for this process is called the "elastic scattering cross section." A particle may be absorbed by the nucleus; the cross section for this process is called the "absorption cross section." Sometimes we are interested in the angular distribution of the particles which collide with the nucleus, in which case we have the "differential scattering cross section" $\sigma(\theta)$. The integral over all angles of the differential scattering cross section equals the total scattering cross section. There are particle cross sections for a particular process which are associated with the particular process and vary widely with energy and target nucleus.

3-10. The Compound Nucleus. In the study of nuclear reactions, one of the generally adopted concepts is that of the compound nucleus, introduced by Bohr in 1936. In this model any nuclear reaction is a three-step process:

$$\overset{\text{I}}{\text{Incident particle}} + \text{initial nucleus} \rightarrow \overset{\text{II}}{\text{compound nucleus}}$$

$$\rightarrow \overset{\text{III}}{\text{final nucleus}} + \text{outgoing particle}$$

Since the distances between particles in the nucleus are of the same order of magnitude as the range of nuclear forces, it is not very probable that a particle would pass through a nucleus without interacting with the nucleus. In the case of the atom, where quantum theory has worked so well, there is small probability of an incident particle's even striking an electron, because the atom is mostly empty space. When a particle in the atom is struck by a projectile, it has a high probability of leaving the atom if there is sufficient energy. However, when a particle hits a nucleus it does not necessarily knock out the particle it hits. Because the struck nucleon is so tightly bound to the adjacent nucleons, the energy of the struck nucleon will be shared with the other nucleons in its vicinity. These nucleons will in turn give some of their energy to their neighbors, etc. By this collision process the energy of the incident particle will be distributed among all the particles in the nucleus. This leaves the complete nucleus with an excess of energy rather than with the excess energy concentrated in one particle, as is the case with an atomic collision. The nucleus is then said to be in an "excited" state. This excited nucleus is called the "compound nucleus." Because the incident nucleon has lost some of its energy to other nucleons, it no longer has sufficient kinetic energy to escape and is now bound in the compound nucleus.

The compound nucleus lives a relatively long time compared to the time involved in a nuclear collision. The natural nuclear time with which to compare the lifetime of the compound state is the time required for a nucleon to cross a nucleus. The binding energy of the last nucleon in the nucleus is about 8 Mev, which means it has a velocity in the nucleus of about 4×10^9 cm/sec. The diameter of a heavy nucleus is about 10^{-12} cm, so the time for the neutron to cross the nucleus if no collision occurs would be about $10^{-12}/(4 \times 10^9) = 2.5 \times 10^{-22}$ sec. If a compound nucleus lives 2.5×10^{-18} sec, it lives 10^4 times the natural nuclear time, which is considered long even though the time the nucleus lives is much too short to measure. The time of existence of the compound nucleus can be determined from the uncertainty principle, which in another form than that given in Chap. 2 can be stated as

$$\Delta E \, \Delta t \approx h \qquad (3\text{-}39)$$

The ΔE observed in low-energy neutron resonances are as low as 0.03 ev; therefore $\Delta t = (6.63 \times 10^{-27})/(0.03 \times 1.6 \times 10^{-12}) = 1.4 \times 10^{-13}$ sec.

This means that there can be 10^9 nuclear collisions before a compound nucleus disintegrates, which is good justification for the concept of the compound nucleus.

Since a compound nucleus lasts a long time, it has no memory of its formation and decays in various ways according to its excitation energy independent of the way in which it was formed. To illustrate this, consider the compound nucleus N^{15}. This can be formed by the reactions

$$B^{11} + He^4 \rightarrow N^{15}$$
$$C^{12} + H^3 \rightarrow N^{15}$$
$$C^{13} + H^2 \rightarrow N^{15} \qquad (3\text{-}40)$$
$$C^{14} + H^1 \rightarrow N^{15}$$
$$N^{14} + n \rightarrow N^{15}$$

The compound nucleus can then decay by reversing the reactions given above and disintegrate into any of the nuclei from which it can be formed,

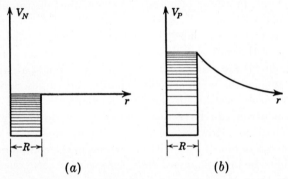

(a)   (b)

FIG. 3-12. (a) Interaction potential between a neutron and a nucleus.   (b) Interaction potential between a proton and a nucleus.

but the way it disintegrates does not depend on the way it was formed. Using this idea of the compound nucleus, the study of nuclear dynamics is the same as the study of nuclear statics. In a nuclear reaction we are studying the metastable states of the compound nucleus.

Suppose that the nucleus is a region of space in which a nucleon experiences a force that can be derived from a potential. At large distances we know the potential energy of a neutron in the field of the nucleus is zero, because neutrons do not react with nuclei unless there is a direct collision; therefore, we assume that the potential is zero at large distances. We also know that within the nuclear radius the force holding the neutrons in the nucleus is very strong and thus the potential is very large. The simplest potential that can be assumed for the interaction between the neutron and the nucleus is shown in Fig. 3-12a.

The potential for the proton is somewhat different because of its charge.

At large distances we know the potential energy of the proton is just the Coulomb energy. Therefore, the proton moves in the Coulomb potential $V = Ze/r$ at large distances. At short distances the nuclear force on the proton overcomes the Coulomb repulsive force, and we assume that the proton also moves in a potential well. The complete potential for the proton is shown in Fig. 3-12b.

When a neutron moves in a potential well of the type shown in Figure 3-12a, its quantum mechanical behavior is considerably different from that expected in classical mechanics. When the particle is inside the range of the nuclear forces and is bound in the nucleus, its energy is below zero. In this case the system has specific energy levels. The energy levels have a definite position depending on the depth of the well and the radius of the well (nuclear radius). Near the bottom of the well the energy levels are quite widely spaced. The spacing decreases as the energy between a particular energy level and the lowest energy level increases. A schematic set of energy levels for the neutron and proton is also shown in Fig. 3-12.

From the lowest energy level in the well—which, incidentally, is not at the bottom of the well, as it would be classically—to zero energy on the diagram, the levels are said to be bound levels. When a particle is in a bound level it must have a particular energy and cannot have any arbitrary energy, as is possible in classical mechanics. Above zero kinetic energy, the neutron can have any arbitrary kinetic energy, just as in classical mechanics. However, when the kinetic energy of the particle equals the energy of one of the "virtual" levels in the compound state, the system is said to be in resonance. The cross section at a resonance can be very different from the cross section of πr^2 that would be expected from classical analogy.

In Fig. 3-13 the neutron cross section of silver is plotted vs. the kinetic energy of the neutron. The first resonance occurs at 5.12 ev. The maximum cross section at the peak of the energy level is about

$$3.9 \times 10^{-20} \text{ cm}^2 = 3.9 \times 10^4 \text{ barns}$$

For cross-section measurements, the unit used is the barn, 10^{-24} cm². The geometrical cross section πr^2 for this nucleus is 1.4 barns, which illustrates how widely different the peak resonance cross section can be from the so-called "geometrical cross section." The level in silver at 5.12 ev occurs when a neutron is incident on Ag^{109}. This level corresponds to an energy level in the compound nucleus (Ag^{110}). No theory exists at the present time which can predict the positions of levels in a nucleus as complicated as Ag^{110}. The positions, the spacings, and other characteristics of the energy levels have been obtained from experiment.

Using this simple potential-well model of the nucleus, many aspects of both nuclear statics and nuclear dynamics can be explained.   If nuclear forces are assumed to be charge-independent, no distinction can be made in the nucleus between the neutrons and the protons.   The energy levels

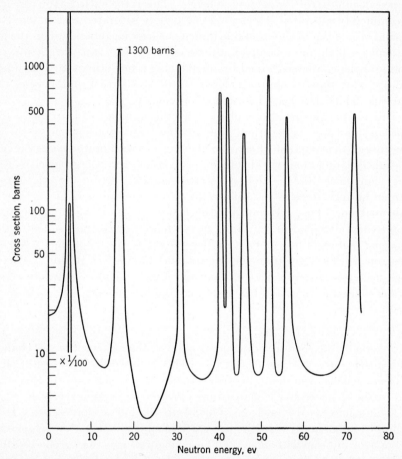

FIG. 3-13. The total neutron cross section of silver as a function of neutron energy.

which exist are filled with nucleons.   The stability of the deuteron can be explained from the spin-dependence of nuclear forces.   The state with spins parallel has much lower energy than that with spins antiparallel. Because of the Pauli exclusion principle, it is impossible to place either two protons or two neutrons in the lowest level with their spins parallel. Thus the most stable configuration is one in which a neutron and proton are in the lowest state with parallel spins.   Once the lowest level has both a neutron and a proton, one can add either a neutron or a proton to the second level.   This theory would predict that H^3 and He^3 should have approximately the same energy, and experimentally this is found to be the case.   The next nucleus would be composed of four particles, with two neutrons of opposite spins in the lowest neutron level and two protons of opposed spins in the lowest proton level.   This is the nucleus He^4, which is a completely saturated system.   A completely saturated system like He^4 should be and is much more stable than its nearest neighbors.   This theory becomes a little more complex after one passes He^4, but the stability of many nuclei can be explained by introducing a few other simple assumptions.   This type of approach leads to the shell-structure theory of the nucleus, which has been mentioned previously.

This simple theory also explains why the number of neutrons in the nucleus increases faster than the number of protons in the nucleus. Because of the Coulomb effect, the distance between proton energy levels is slightly larger than the distance between neutron energy levels.   For very light nuclei the position of a proton energy level for the nth proton cannot be very much different from the position of the energy level for the nth neutron.   This accounts for the fact that in the light nuclei N is approximately equal to Z.   However, as more particles are added to the nucleus, there will be a point at which the energy level for the $(n + 1)$st neutron will be lower than the energy level for the nth proton.   Therefore, it is easier to add a neutron than a proton.   If a proton already exists in the excited nucleus and has more than 0.5 Mev energy

$$(E = m_0 c^2 = 0.5 \text{ Mev for an electron})$$

excess above the corresponding neutron level, then the proton will change into a neutron, emitting a positron.   This is called "β^+ disintegration." If a nucleus has too many neutrons for its proper point on the stability curve, then a neutron will change into a proton and emit an electron.

In any nucleus the nucleons will fill up the lowest states first.   It is found experimentally that the last state filled is about 8 Mev below zero energy for intermediate nuclei.   This configuration is the "ground state" of the nucleus.   There are many allowable states for both neutrons and protons between the last filled state and zero energy.   A nucleus can be excited in many different ways.   One proton or neutron can be in a state which is several Mev above the ground state, or several particles can have a small amount of energy in states not far above the ground states. If one nucleon is in a highly excited state and falls to the ground state, it will emit a high-energy γ ray.   If several particles are slightly above the ground state and fall back to the ground state, several γ rays of low energy will be emitted.

To explain α-particle emission, a concept completely foreign to classical mechanics occurs in quantum mechanics. Suppose that the kinetic energy of an α particle inside the nucleus is greater than zero energy, but not high enough to be above the Coulomb energy of the α particle at the nuclear radius. Classically, it is impossible for the particle to go over the Coulomb barrier and appear on the outside. Quantum-mechanically, there is a finite probability that the particle will be found on the outside. The particle that appears on the outside can be said to have leaked through the barrier. The higher the energy of the particle, the higher the probability that the particle will leak through the barrier. It is this potential barrier leakage which explains natural α radioactivity. The α particle in the nucleus collides with the nuclear barrier a great many times before it escapes. In the case of U^{238} the energy of the α particle is found to be 4.2 Mev or $v = 1.4 \times 10^9$ cm/sec. For the uranium nucleus $r = 1.4 \times (238)^{1/3} \times 10^{-13}$ cm $= 8.7 \times 10^{-13}$ cm. This means it takes the α particle about $(8.7 \times 10^{-13} \times 2)/(1.4 \times 10^9) = 1.2 \times 10^{-21}$ sec to travel all the way across the nucleus. The half-life of U^{238} is

$$4.51 \times 10^9 \text{ years} = 1.42 \times 10^{17} \text{ sec}$$

Thus an α particle has $(1.42 \times 10^{17})(1.2 \times 10^{-21}) = 1.2 \times 10^{38}$ collisions with the nuclear boundary before it has a probability of $\frac{1}{2}$ of escaping. It is no wonder that this phenomenon is not observed in classical physics.

3-11. Neutron Reactions. In reactor physics we are particularly interested in the way neutrons react with various materials. A summary of a number of the important types of neutron interactions is given in Table 3-3. Activation data for the principal natural elements useful in reactor construction are summarized in Table 3-4.

TABLE 3-3. TYPES OF NEUTRON INTERACTIONS

$$\sigma_t \quad = \quad \sigma_a \quad + \quad \sigma_s$$

Total =	Absorption processes	+	Scattering processes
	(n,γ)		Elastic nuclear scattering
	(n,p)		Inelastic nuclear scattering
	(n,α)		Resonant nuclear scattering
	(n,d)		Coherent crystal (diffraction)
	Also		Ferromagnetic scattering
	(n,2n)		Paramagnetic scattering
	(n,fission)		Inelastic molecular scattering
			Neutron-electron scattering

Absorption. In absorption reactions, the excited compound nucleus changes to a more stable state by retaining the incident neutron and emitting either a photon (γ) or one of the other particles listed. Because charged particles must penetrate the potential barrier of the nucleus before they can escape, the (n,p), (n,d), and (n,α) reactions occur primarily with high-energy neutrons and with elements of low atomic number. The

TABLE 3-4. NEUTRON CROSS SECTIONS OF SOME COMMON MATERIALS*

Element	Thermal σ_a, barns	Thermal σ_s, barns	Activities formed				Resonances, ev <1000 ev
			Half-life	Radiation	Energy, Mev	σ	
$_1$H	0.33	38.4	....	None		....	None
$_1$H^2	0.0005	7.0	12.5 years	β^-	0.0189	0.5 mb	None
$_5$B	755	4	<10^{-14} sec	α	2.792	755 b	None
$_6$C	0.003	4.8	5570 years	β^-	0.15	9 μb	None
$_7$N	1.88	10	7.4 sec	β^-	4.3	0.2 μb	None
				γ	6.2		
$_8$O	<0.0002	4.2	5570 years	β^-	0.15	1.75 b	None
			5570 years	β^-	0.15	200 μb	
			29 sec	β^-	4.5	0.1 μb	
				γ	1.6		
$_{11}$Na	0.56	4	15 hr	β^-	1.4	0.56 b	None
				γ	2.76, 1.38		
$_{12}$Mg	0.063	3.6	9.5 min	β^-	1.8	5.6 mb	None
				γ	1.0, 0.84		
$_{13}$Al	0.21	1.4	2.3 min	β^-	3.0	0.21 b	None
				γ	1.8		
$_{14}$Si	0.13	1.7	2.6 hr	β^-	1.8	3.3 mb	None
$_{18}$A	0.62	1.5	110 min	β^-	1.18, 255	0.53 b	None
				γ	1.37		
$_{19}$K	1.97	1.5	1.3×10^9 years	β^-	1.7, 1.41	2 b	None
				γ	1.54		
$_{22}$Ti	5.6	1.4	5.8 min	β^-	1.6	7 mb	None
$_{23}$V	5.1	5	3.8 min	β^-	2.7	4.5 b	None
				γ	1.4		
$_{24}$Cr	2.9	3.0	27.8 days	γ	0.32, 0.27	0.65 b	None
			3.6 min	....	....	9 mb	
$_{25}$Mn	13.4	2.3	2.6 hr	β^-	2.8	13.4 b	337
				γ	2.1, 1.8, 0.85		
$_{26}$Fe	2.5	11	2.9 years	γ	....	0.13 b	None
			46 days	β^-	0.46	3 mb	
				γ	1.3, 1.1		
$_{27}$Co	37	7	10.4 min	β^-	1.6	16 b	132
				γ	1.33, 1.17, 0.06		
			5.2 years	β^-	0.32	20 b	
				γ	1.33, 1.17		
$_{28}$Ni	4.6	17.5	2.6 hr	β^-	2.1	30 mb	None
				γ	1.5, 1.1, 0.4		
$_{30}$Zn	1.1	3.6	250 days	γ	1.1	0.25 b	225, 455, 530
			13.8 hr	β^-	0.86	18 mb	
				γ	0.44		
			52 min	β^-	0.86	180 mb	
			2.2 min	β^-	2.1	0.5 mb	
				γ	0.8		
$_{40}$Zr	0.180	8	17 hr	β^-	2.2	0.02 b	None
				γ	0.8		
			65 days	β^-	1.0, 0.4		
				γ	0.73, 0.23, 0.92		
$_{41}$Nb	1.1	5	6.6 min	β^-	1.3	1.1 b	None
$_{48}$Cd	2550	7	Many	Many		~0.5 b	Many
$_{50}$Sn	0.6	4	112 days	γ	0.09	12 mb	21 known resonances between 46 and 460 ev
			14.5 days	γ	0.17	840 μb	
			250 days	γ	0.07	2.4 mb	
			> 400 days	β^-	0.4	0.3 mb	
			27.5 hr	β^-	0.4	0.05 b	
			40 min	β^-	1.3	0.08 b	
				γ	0.16		
			130 days	β^-	1.4	40 μb	
				γ	0.4		
			10 min	β^-	2	12 mb	
				γ	1.9, 0.3		
			10 days	β^-	2.4	25 μb	
$_{80}$Hg	380	20	5.5 min	β^-	1.62	~1.3 b	Many
			47 days	β^-	0.21, 0.11		
				γ	0.286		
$_{82}$Pb	0.170	11	3.32 hr	β^-	0.70	?	None
$_{83}$Bi	0.032	9	5 days	β^-	1.17	0.019 b	2

* Compiled by L. Lidofsky. See also App. Q.
NOTE: For energies other than thermal, σ_s is relatively constant and, in general,

$$\sigma_a(E) \approx \sigma_{a,th} \left(\frac{0.025}{E,\text{ ev}}\right)^{1/2}$$

(n,γ) reaction is by far the most common absorption process, particularly for slow neutrons. This reaction is frequently called "radiative capture," because the γ ray is emitted from the compound nucleus very shortly after it is formed ($\sim 10^{-14}$ sec). The atomic number of the nucleus does not change in this process, but the mass number increases by one unit. If the isotope formed does not normally exist in nature, it will be radioactive. Over 150 radioactive isotopes have been formed by the (n,γ) reaction.

Scattering. When a neutron rather than another particle is emitted from the compound nucleus, scattering results and the process is called an "(n,n) reaction." The neutrons are emitted nearly isotropically from the compound nucleus, and consequently very few of them reach the detector if good geometry is maintained. For good geometry, the incident neutron beam should be essentially parallel and the detector should be sufficiently distant from the sample that it subtends only a negligible solid angle.

Inelastic Scattering. If the nucleus is left in an excited state after the compound nucleus emits a neutron, the reaction is said to be an "inelastic nuclear collision," because a part of the kinetic energy of the neutron is used to excite the nucleus. This excited nucleus will then return to a more stable state by emission of a γ ray. The probability of such collisions increases with increasing neutron energy; it becomes significant for the heavy elements only above 0.1 Mev and for the lighter elements above 1 Mev. Consequently, the slowing down of neutrons by this process is appreciable only for fast neutrons. Slow neutrons are involved in inelastic molecular collisions when they collide with a molecule and a part of their kinetic energy is used up in breaking chemical bonds or in exciting higher vibrational and rotational energy states of a molecule. On the other hand, subthermal neutrons may receive energy from the molecular motions of molecules.

Elastic Scattering. When the total kinetic energy of the neutron and the nucleus remains constant, the collision is called an "elastic" one. In such collisions with heavy nuclei, the neutron leaves the compound nucleus with the same energy that it had prior to the collision. With light nuclei, however, the energy is distributed between the scattered neutron and the nucleus. In a head-on collision with hydrogen, where the masses are nearly equal, the laws of conservation of energy and momentum require that practically all the energy be transferred to the hydrogen atom. At higher energies some protons are produced, and their maximum range can be measured to give the energy of the incident neutrons. If the collision is not head-on, the energy transfer will not be so great. On the average, however, about one-half the energy of fast neutorns will be transferred to the proton at each collision.

Such elastic collisions with light nuclei are the main process by which fast neutrons are slowed down to thermal velocities. Neutrons with an energy of 2 Mev, for example, are slowed down on the average to thermal energies (0.026 ev) by about 18 collisions with H, 40 with He, 80 with Be, 110 with C, and 2100 with U. A thickness of about 5 cm of paraffin would be required for this slowing-down process. If the neutrons do not escape from the paraffin, they are ultimately absorbed by the hydrogen and to a lesser extent by carbon. The mean free path of thermal neutrons in paraffin is about 0.3 cm, and their average life is about 200 μsec.

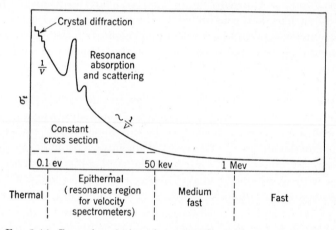

FIG. 3-14. General variation of cross section with neutron energy.

Variations in Cross Sections with Neutron Velocity. Marked variations exist in the cross sections for neutron interactions as the velocity of the neutron is changed. At high energies (1 Mev) where the wavelength of the neutron is of the same order of magnitude as the radius of the nucleus ($\lambda \approx r$), the effective total cross section is roughly twice the geometric size of the nucleus as measured by other methods, or $2\pi r^2$, i.e., from 1 to 10×10^{-24} cm^2 (1 to 10 barns). In this high-energy region, the principal processes involve particle emissions, primarily scattering [(n,n) reactions]. However, there are some (n,p) and (n,α) reactions. As the energy of the neutrons is decreased, the cross section increases approximately as $1/v$, as illustrated in Fig. 3-14.

Resonance Absorption and Scattering. Many of the elements, particularly those with atomic masses above 100, show regions of resonance absorption and scattering where a marked increase in cross section occurs at definite neutron energies (Fig. 3-14). Many resonances have been observed experimentally in a large number of nuclei.

The shape of the resonance curve is given by the Breit-Wigner formula. The cross section at an energy E (electron volts) in the region of a resonance for an (n,γ) reaction is

$$\sigma_{(n,\gamma)} = \frac{\lambda^2}{4\pi} g \frac{\Gamma_n \Gamma_\gamma}{(E - E_0)^2 + (\Gamma/2)^2} \tag{3-41}$$

where λ = wavelength ($\lambda = h/mv$) of incident neutron

g = statistical weight factor for spins of the target nucleus, compound nucleus, and neutron

Γ_n = partial level width (proportional to reaction probability) of (n,n) process (it is sometimes convenient to include the value of g in this neutron width)

Γ_γ = partial level width with respect to (n,γ) process

$\Gamma = \Gamma_n + \Gamma_\gamma$ = full width of resonance peak at half its maximum value σ_0

E = energy of incident neutron

E_0 = resonance energy

For low-velocity neutrons, Γ_γ is nearly independent of velocity, while Γ_n is proportional to v. The maximum possible cross section for the (n,γ) process would be

$$\sigma_{0(n,\gamma)} = \frac{\lambda^2}{\pi} \frac{\Gamma_n \Gamma_\gamma}{\Gamma^2} \tag{3-42}$$

or $\lambda^2/4\pi$ at $E = E_0$ if $\Gamma_n = \Gamma_\gamma$. This yields a possible total cross section of $\lambda^2/2\pi$ for the sum of both $\sigma_{0(n,\gamma)}$ and $\sigma_{0(n,n)}$. A few nuclei such as Pd^{108}, Sm^{152}, and W^{186} appear to have levels in which $\Gamma_\gamma \approx \Gamma_n \approx 0.1$ ev.

Resonance Capture. In most cases, however, $\Gamma_n < \Gamma_\gamma$ by a factor of 10 to 100, so that $\Gamma_\gamma \approx \Gamma$, and the maximum possible cross section for resonance absorption is $\sigma_0 = \lambda^2 \Gamma_n/\pi\Gamma$. Since Γ_n varies as the velocity v or as $\sqrt{E}$, the cross section $\sigma(E)$ for an energy E other than the resonance energy can be obtained by bringing Eq. (3-41) into the following form:

$$\sigma(E) = \left(\frac{E_0}{E}\right)^{1/2} \frac{\sigma_0 \Gamma^2}{4(E - E_0)^2 + \Gamma^2} \tag{3-43}$$

At energies low compared to E_0, Eq. (3-43) shows that $\sigma(E)$ is proportional to $1/\sqrt{E}$ or to $1/v$. This is the well-known $1/v$ law. Examples of the above are Rh^{103}, In^{115}, and Au^{179} in which Γ_γ is about 0.1 ev compared to 0.001 ev for Γ_n.

Resonance Scattering. The cross section $\sigma_{(n,n)}$ for elastic scattering of low-energy neutrons in the region of a resonance is

$$\sigma_{(n,n)} = \frac{\lambda^2}{\pi} \frac{\Gamma_n^2}{4(E - E_0)^2 + \Gamma^2} \tag{3-44}$$

which, at the resonance energy E_0 gives a maximum cross section of λ^2/π. There are cases of levels in which scattering is the more favored process, such as in Mn and Co. For Co, $\Gamma_n \approx 5$ ev and $\Gamma_\gamma \approx 0.1$ ev.

The $1/v$ Region. At low energies, the cross section for scattering varies as $(\lambda^2/\pi)(\Gamma_n^2/4E_0^2)$, and since λ varies as $1/v$ and Γ_n varies as v, the scattering cross section should be fairly constant. Hence, the total cross section

$$\sigma_{total} = \sigma_{a(n,\gamma)} + \sigma_{s(n,n)} \tag{3-45}$$

in the $1/v$ region is made up of a constant scattering term plus a capture term that varies as $1/v$ or as $1/\sqrt{E}$. Thus, the cross section in this region can be expressed by an equation of the form

$$\sigma(E) = a + \frac{b}{v} \tag{3-46}$$

The magnitudes of the absorption cross sections in this region, as well as the slopes of the $1/v$ curves, differ for a number of the elements.

Molecular Effects. In the case of the lighter elements, such as hydrogen and deuterium, chemical binding effects become significant below energies of 1 ev. Actually the scattering is proportional to the square of the reduced mass of the neutron and the scatterer. If the hydrogen atom is free, the reduced mass $\mu \approx 0.5$. However, when hydrogen is bound in heavy molecules such as paraffin, the reduced mass will be $\mu \approx 1$. The velocity of the neutron in the center-of-mass system for a collision with a heavy molecule will be twice that for collision with a proton, for the same velocity of neutron in the laboratory system. Thus the ratio of $\sigma_{free}/\sigma_{bound} = 1/4$. In general, the cross section σ_{bound} for a rigidly bound nucleus of mass A relative to the cross section σ_{free} of the same nucleus in a free state will be given by

$$\sigma_{bound} = \sigma_{free} \left(\frac{A+1}{A} \right)^2 \tag{3-47}$$

When the energy of the neutron is considerably higher than the vibrational energy of the C—H bond in paraffin (i.e., $E \gg h\nu \approx 0.4$ ev), H may be separated from the molecule, and the cross section for hydrogen is what would be expected for free hydrogen atoms. At energies of the order of $h\nu$, however, the neutron can lose energy to the vibration of the hydrogen atom or other groups in the molecule. In this case the hydrogen atom is essentially bound, and the cross section increases to about four times that at the higher energies. Experiments show, in fact, that the cross section for hydrogen bound in a hydrocarbon molecule does increase from about 20 barns in the region of 10 to 1 ev to over 80 barns at energies of about 0.0005 ev, as shown in Fig. 3-15.

Diffraction.   As the energy of the neutrons is decreased to the thermal region, where the associated wavelength is of the order of the distance

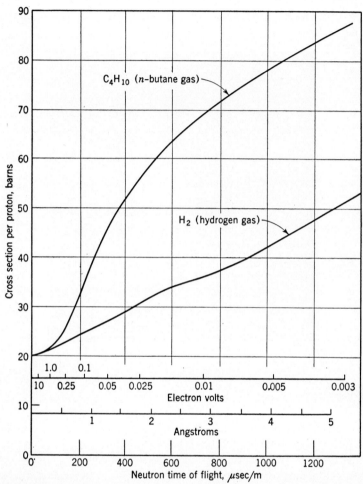

FIG. 3-15. Graph of neutron cross section of *n*-butane and hydrogen as a function of neutron energy.

between the atoms in crystals, diffraction effects are observed in accordance with the Bragg relation,

$$n\lambda = 2d \sin \theta \tag{3-48}$$

When a heterogeneous beam of thermal neutrons is passed through a single crystal, a Laue pattern can be obtained.   Since neutrons do not affect a photographic emulsion sufficiently for direct photographic detection, it is necessary to use a neutron-absorbing material in the emulsion or to place a thin sheet of substance such as indium in front of the X-ray film. Radiations resulting from the neutron capture then produce the Laue spots on the film, as illustrated in Fig. 3-16. Monoenergetic neutrons may be used with powdered crystalline material to obtain diffraction patterns.

FIG. 3-16. Neutron diffraction Laue photograph for NaCl.

From the preceding discussion it follows that the total cross section for transmission by crystals of slow neutrons is

$$\sigma_{\text{total}} = \sigma_a + \sigma_{el} + \sigma_{inel} + \sigma_{coh} + \sigma_d + \sigma_m + \cdots \qquad (3\text{-}49)$$

in which the terms refer respectively to cross sections arising from absorption, elastic, inelastic, coherent Bragg, diffuse or incoherent effects already discussed, and magnetic effects. If the absorption is large it masks the other effects, and neutron diffraction for studying structure is not very useful except in special cases.

Since neutrons of all energies are found in reactors, the details of the way the neutron cross section varies with energy are extremely important. If, for instance, a reactor is designed so that the maximum number of neutrons occur at an energy E_1, and one of the construction materials of the reactor has absorption resonance at the energy E_1, many neutrons will be removed from the reactor and will not be able to cause fission. Not only will the efficiency of the reactor be decreased, but the structural material will become extremely radioactive and also suffer radiation damage, which may impair its structural properties. Another point of importance is the side of a resonance at which a reactor is operating. Suppose, for illustration, all the neutrons in a U^{235}-fueled reactor have an effective energy of 0.075 ev. At 0.075 ev the fission cross section of U^{235} is decreasing with energy, as shown in App. P. As the temperature of the reactor increases, the average energy of the neutrons increases and the fission rate decreases, thus reducing the power and bringing the reactor back to stable equilibrium. Suppose, however, the reactor were operating such that the effective energy of the neutrons was at 0.25 ev rather than 0.075 ev. The figure shows that at this energy the fission cross section is increasing with energy. If the temperature increases, the fission rate also increases, introducing more power and further increasing the temperature. This would go on until the reactor expanded sufficiently so that the reaction became steady again or was stopped by the control mechanism or some catastrophe occurred. Of course, such a simple case is not realized in practice, since the distribution of neutrons in a reactor is very broad rather than strongly peaked at a particular energy, and many other factors also affect the temperature coefficient.

Most of the reactors which have already been built have operated in the thermal region, where there are a considerable number of very slow neutrons. When the neutrons have a long enough wavelength, the cross section for scattering is the incoherent cross section and not the total cross section. In some cases for reactor moderators such as beryllium and carbon, the incoherent cross section is quite low. Neutrons will then leak out the sides of the reactors and be lost to the fission process. Such variations in the neutron cross section must be taken into consideration when a reactor is designed. The available neutron cross sections have been summarized in BNL-325 [2].

3-12. The Fission Process. From the binding energy of a nucleon, one sees that for A greater than about 100, energy will be released in the fission process. Why, then, one asks, is fission such a rare process? This can be explained on the basis of the potential-energy diagram for the proton (Fig. 3-12b). Consider a nucleus which breaks into two fission fragments. If the energy is plotted as a function of distance between the two parts, a curve such as Fig. 3-17 is obtained. At infinite separation the energy is arbitrarily taken to be zero. When the fragments are combined, we know from measurements of the binding energies of nuclei that the total energy of binding is 200 Mev or greater. We must look at points in between to determine the stability of a nucleus against fission. Up to the distances of the order of the diameter of the fission fragments it is the Coulomb energy $(Ze/2)^2/r$ which contributes to the energy between the particles. When r is less than the diameter of the fragments the energy must change in such a way that it becomes the fission energy at $r = 0$.   If $(Ze/2)^2/r$ is smaller than, equal to, or greater than the fission energy at $r = $ the diameter of the fragments, then the three curves illustrated in Fig. 3-17 can be drawn to connect the Coulomb potential to the known energy at $r = 0$.   Stable nuclei with A greater than 100 are represented by a curve of type I.   Presumably uranium would be like curve II where the barrier is about 6 Mev.   Substances whose energy curve is like III would not exist for long, for they would spontaneously fission in a very short time (i.e., the new synthetic elements of $Z > 92$).

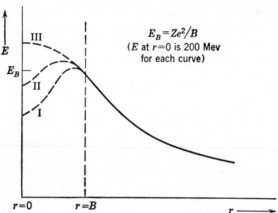

Fig. 3-17. Potential energy of nuclear fragments as a function of separation.

Consider $r = B$ to be of the order of the diameter of a fission fragment. Then, from the equation

$$r = 1.4A^{\frac{1}{3}} \times 10^{-13} \qquad cm$$

we can calculate B.   Using this value for B, we can plot E_B (the Coulomb potential at $r = B$) as a function of mass number A (Fig. 3-18).   Similarly we can draw a curve E_A, the excess of mass of a parent nucleus of mass number A over that of its two fragments (i.e., the fission energy). This latter curve becomes negative below $A = 85$ and crosses the curve for E_B at about $A = 250$.   From such a graph, one can get $E_B - E_A$ for any A.   The quantity $E_B - E_A$ is a measure of the height of the energy barrier against fission.

It is possible to investigate more precisely the shape of the energy vs. fragment-separation curve near $r = 0$ if some specific model is assumed. Considering the Bohr liquid-drop model, we assume that the original nucleus is a sphere and then calculate the change in energy for a small deformation.   Let us further assume that the sphere, in beginning to split, deforms in a very simple manner, namely, it stretches slightly in one direction and flattens out perpendicularly to this direction, thus becoming an ellipsoid. If we assume that the sphere does not change its volume on becoming an ellipsoid—and this is reasonable in view of the fact that all nuclei tend to maintain the same density of nuclear particles—the change in the energy of the nucleus upon deformation will be due to only two of the factors discussed in Sec. 3-7. First, the surface energy will tend to increase with deformation because more surface will be exposed. Second, the electrostatic energy will decrease because the repelling charges will be effectively separated to some extent. Thus we have at least two energies

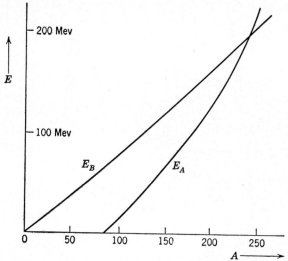

Fig. 3-18. The fission energy and the Coulomb energy of a nucleus as a function of mass number.

changing in opposite ways with deformation of a spherical nucleus. The surface energy is proportional to the surface area or $A^{2/3}$, and the electrostatic energy is proportional to $Z^2 A^{-1/3}$, which is $\sim A^{5/3}$. The latter energy becomes more important for heavy nuclei, so that for heavy nuclei it is likely that the energy of a nucleus tends to decrease with deformation, making a spherical nucleus unstable. The opposite is true for light nuclei. From this picture, it is in heavy nuclei that we would expect fission.

Our principal interest is not in spontaneous fission but in fission brought about by neutrons. Neutrons can cause fission by contributing their kinetic energy and their binding energy to the nucleus. This energy is at least 5 or 6 Mev (the binding energy of the neutron) and may raise the energy of the nucleus high enough within the barrier for fission to take place before the excess energy is lost by γ radiation. Because the binding energy of a neutron to a nucleus with an odd number of neutrons is larger than it is to one with an even number of neutrons [see Eq. (3-28)], it is reasonable to expect fission for thermal neutrons to be more prevalent for those nuclei with an odd number of neutrons. This is confirmed by experiment.• Uranium 238 is not fissioned by thermal neutrons, whereas U^{235} is. Moreover the other practical "fissionable" materials, U^{233} and Pu^{239}, both have an odd number of neutrons. From facts such as these

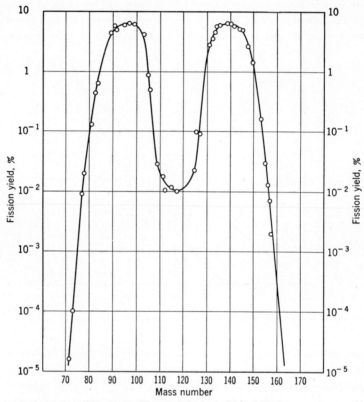

FIG. 3-19. Thermal-fission yield of U^{235}.

and photofission thresholds, one can estimate that for uranium the height of the fission barrier is of the order of 5 Mev.

When a nucleus fissions, there are many modes of decay possible. This is well illustrated by Fig. 3-19, which is a plot of the fission yield observed in the fission of U^{235} vs. the mass number of the fission fragments. (The fission yield is defined as the percentage of the total nuclear fission products of a given mass number and adds up to 200 per cent as the total fission yield, since two fragments are formed per fission.) Since at least 60 primary products of the fission of U^{235} have been observed, there must be at least 30 major different modes of fission. The fission products range in mass number from 72 to 158.

Since the number of neutrons in a nucleus increases faster than the number of protons, there will be an excess of neutrons in the fission fragments. Most of these neutrons are emitted immediately by the fission fragments, but all of these neutrons assist the spontaneous chain reaction to proceed. The number of neutrons emitted in fission has been measured many times because of its importance in nuclear reactors. This number is usually designated by ν. For U^{235} $\nu = 2.46$, for U^{233} $\nu = 2.54$, and for Pu^{239} $\nu = 2.88$ at thermal energies. The variation of ν with energy is the subject of much experimental interest at the present time (1955). The value of ν for fast neutrons seems to be slightly larger than the value of ν for slow neutrons [3]. However, this quantity ν should not change significantly with energy, because the neutrons appear to be emitted by the fission fragments, and the mode of fission of the nucleus should be relatively insensitive to the energy of an incident neutron, since the excitation energy of the compound nucleus is so large compared to the energy of the incident neutron.

The energy distribution of the neutrons emitted in fission varies over a considerable energy range. In the range from 0.1 to 10 Mev, the energy distribution is given by $n(E) = 0.484e^{-E} \sin h \sqrt{2E}$. Some of the neutrons emitted in the fission process are delayed. These neutrons are emitted with a definite half-life which is the half-life of the parent isotope. For example, the delayed neutrons which have a half-life of 55.6 sec come from the decay of Br^{87}, which decays by β emission to Kr^{87}, with a half-life of 55.6 sec. The Kr^{87} decays immediately $<10^{-14}$ sec to Kr^{86} by the emission of a neutron. The properties of the delayed neutrons in the slow-neutron fission of U^{235} are given in Table 3-5 (see also Table 6-6). It is the emission of these delayed neutrons with a definite half-life which allows a power reactor to be so easily controlled.

TABLE 3-5. PROPERTIES OF THE DELAYED NEUTRONS IN THE SLOW-NEUTRON FISSION OF U^{235}

Half-life, sec	Fraction of fission neutrons delayed	Energy, Mev
0.43	8.5×10^{-4}	0.40
1.52	24.1×10^{-4}	0.67
4.51	21.3×10^{-4}	0.41
22.0	16.6×10^{-4}	0.57
55.6	2.5×10^{-4}	0.25
Total.....	0.0073	

It is to be kept in mind that, in the consideration of the competition of fission with other processes, it is not sufficient to consider energies alone, as we have done.   For fission, one must not only have the energy rise to the top of the barrier, but it is also necessary that this energy be concentrated in the proper modes of motion for fission.   This may take some time, so that competing processes may occur at the expense of fission.   Since the number of modes, and hence of useless nonfission modes, increases with excitation energy, it may be very likely that the reason photofission with $\sim$100-Mev γ rays on lower-Z nuclei has not been observed is that the energy is not concentrated in a proper mode before it is lost in some way other than fission.

Nonfission capture by the three fissionable isotopes is an appreciable portion of the total neutron absorption cross section at 0.025 ev.   The ratio of the nonfission-capture cross section σ_c to the fission cross section σ_f is denoted in reactor terminology as α.   The quantity α definitely varies with energy in a small energy interval.   It has been shown that α for the 2.04-ev resonance in U^{235} is definitely different from the value of α at 0.0253 ev.   The variation of α with energy is now being investigated in many laboratories.

Although both ν and α are individually important, it is the effective number of neutrons emitted per neutron absorbed by a fissionable nucleus, designated by η, which is most important from the point of view of reactor design. $\eta = \nu(\sigma_f/\sigma_a) = \nu/(1 + \alpha)$.   The most recent values of σ_a, σ_f, ν, α, and η for the fissionable isotopes are given in Table 3-6.   Appendix P gives σ_f for U^{235} as a function of energy in the thermal and near-epithermal regions.

TABLE 3-6. PROPERTIES OF FISSIONABLE MATERIALS AT 0.025 Ev

Material	σ_a	σ_f	ν	α	η	σ_s
U^{233}.........	585	533	2.54	0.098	2.31	
U^{235}.........	687	580	2.46	0.184	2.08	
Pu^{239}........	1065	750	2.88	0.42	2.03	11

Recent experiments indicate that both η and α vary rapidly with energy and that ν remains constant.   Further information on these quantities should improve our understanding of the reactors presently operating and thus improve the design of future reactors.

Although a detailed satisfactory theory of the fission process is not as yet available, the enormous amount of experimental data available on the fissionable isotopes enables the properties of the fissionable nucleus in a reactor to be calculated with a reasonable degree of exactness.

Prob. 3-1. The absorption coefficient of a certain concrete is 0.12 per in. for 1-Mev γ radiation.   What is the thickness required to reduce the radiation intensity from a given source to 10^{-2}, 10^{-3}, 10^{-6}?

Prob. 3-2. Bismuth 210 decays with a half-life of five days to Po^{210}, which in turn decays with a half-life of 138 days into stable Pb^{206}. If one has pure Bi^{210} initially, at what time afterward will there be a maximum amount of Po^{210} present? At that time what will be the ratio of Bi^{210} to Po^{210} to Pb^{206} present?

Prob. 3-3. Gold 198 (half-life 2.7 days) is produced at a rate of 10^{11} atoms/g of Au^{197} per second in a certain pile. If the gold is bombarded for 10 days, what is the activity 1 day, 5 days, 10 days, 15 days after the start of bombardment?

Prob. 3-4. A deuteron has a binding energy of 2.23 Mev. What is the difference in mass between a (neutron + proton) and a deuteron?

Prob. 3-5. Compute the mass of a Au^{197} atom. Compare this with the measured value 197.04. What is the BE of the 79th proton in gold? Of the 118th neutron?

Prob. 3-6. What is the most stable grouping of 200 neutrons and 200 protons? How does the energy of this grouping compare with that of the free particles?

Prob. 3-7. What is the density of a uranium nucleus?

Prob. 3-8. On a classical basis, what is the minimum-energy α particle which will penetrate a U^{238} nucleus? Compare this to the energy of the α emitted by U^{238}.

Prob. 3-9. The thermal cross section for boron is ~ 750 barns. Plot the absorption of a layer of boron 1 g/cm² thick as a function of neutron energy of $0.001 < E_n < 1000$. Assume that the cross section has a $1/v$ dependence.

Prob. 3-10. The fission energy of uranium is ~ 200 Mev per fission. Calculate the number of calories per gram and Btu per pound released if there is total fission.

REFERENCES

1. Green, A. E. S., and N. A. Engler: *Phys. Rev.*, **91**:40 (1953).
2. Hughes, D. J., and J. A. Harvey: "Neutron Cross Sections," McGraw-Hill Book Company, Inc., New York, 1955.
   BNL-325, Government Printing Office, 1955.
3. Leachman, R. B.: Paper 592, UN Conference on Atomic Energy, Geneva, 1955.

CHAPTER 4

PARTICLE DETECTION

By Chien-shiung Wu

4-1. Introduction [1–5]. The detection of radioactivity is made possible through the interaction of radiation with matter. In interacting with matter, the various radiations directly or indirectly produce ionization. It is the detection of this ionization which constitutes the basic principle of all detection devices. The mechanisms of interaction of various radiations with matter were discussed in detail in Chap. 3. These mechanisms can be summarized by saying that fast-moving charged particles such as α particles, protons, β particles, or charged mesons passing through matter lose their kinetic energy by electromagnetic interactions with the atomic electrons of the traversed matter. The interaction results in the excitation or ionization of atoms of the matter. Therefore the ionization is directly produced in the slowing down and stopping of charged particles.

The γ ray is an electromagnetic radiation which does not produce ionization in matter directly. However, a γ ray interacts with matter by producing either a photoelectron (photoelectric effect), a recoil electron in the Compton effect, or an electron-positron pair in the process of pair creation. The photoelectron, Compton-recoil electron, or electron-positron pair thus produced are fast-moving charged particles and therefore can be used in the detection of the γ ray.

The neutron is a neutral particle and does not interact with matter electromagnetically. In other words, one could not expect to detect neutrons through direct ionization. In order to detect fast neutrons, one places a hydrogenous material in their path and detects the recoil protons which were knocked out through the elastic collision between the fast neutrons and the hydrogen nuclei. Slow neutrons, being neutral and traversing matter slowly, have very large probabilities of capture by nuclei. The products resulting from their capture are either heavy ionizing particles (such as α particles, protons, or fission fragments) or γ radiations. The detection of these products constitutes the essential means of detecting slow neutrons.

4-2. Ionization-chamber Method. An ionization chamber consists mainly of a gas volume bounded by two or more electrodes maintained at different potentials, usually a central electrode in a cylinder, or parallel plates. Ionizing particles traversing the chamber produce ion pairs. The electrostatic field attracts the positive ions and electrons to the electrode of opposite polarity. A characteristic curve of an ionization current vs. collection potential generally exhibits saturation (Fig. 4-1),

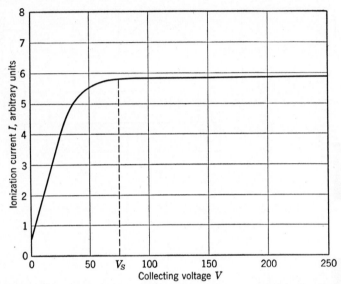

Fig. 4-1. Ionization current vs. collecting voltage V in a typical ionization chamber. V_s indicates the saturation voltage.

which requires several hundred volts per centimeter in standard air. It is desirable to work above this voltage for the strongest radiation used.

The ionization current to be measured is generally small.* Besides the electroscope and electrometer, special vacuum tubes have been constructed and electrometer circuits designed to measure ionization currents much lower than 10^{-15} amp.

4-3. Electroscopes and Electrometers. An electroscope consists of an ionization chamber of cylindrical geometry with the outside cylinder grounded. The collecting electrode consists of two small flexible metal leaves, or one leaf and a rigid element. The electroscope is electrically charged, then observed with a microscope. The flexible element returns

---

* Consider a beam of α particles of 5 Mev entering an ionization chamber at a constant rate of 10 per second. Since each α particle will produce approximately $(5 \times 10^6)/30 \approx 1.7 \times 10^5$ ion pairs and each ion carries a charge of 1.6×10^{-19} coulomb, the total current flowing will be $1.7 \times 10^5 \times 1.6 \times 10^{-19} \times 10 = 2.7 \times 10^{-13}$ coulomb/sec $= 2.7 \times 10^{-13}$ amp.

toward its uncharged position at a rate proportional to that at which ions are collected and therefore to the intensity of ionizing radiation over a large part of its range.

Electrometers contain three conductors: two similar and fixed, and a third which is flexible.

4-4. Vacuum-tube Electrometers. Electroscopes and electrometers have now been largely replaced by the vacuum-tube electrometer, which

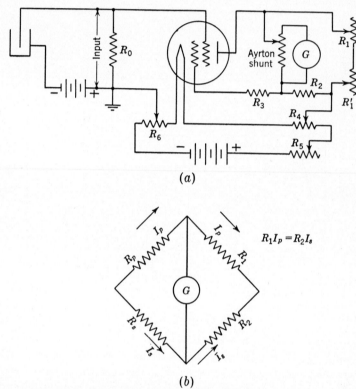

FIG. 4-2. (a) DuBridge and Brown balanced electrometer-tube circuit for d-c amplifier. (b) Equivalent Wheatstone bridge.

is as sensitive and more convenient and flexible. The FP-54 (General Electric) and D-96475 (Western Electric) electrometer tubes have been especially developed, requiring grid currents as low as 10^{-17} amp. Commercial radio tubes (RCA 38,22,954 and 38,22,959, and Western Electric 259-B) can also be used if operated at potentials low enough to reduce the grid current to $\sim 10^{-13}$ amp. The peanut-size Victoreen (Vx41) can measure down to 10^{-15} amp.

The single-tube circuit originated by DuBridge and Brown as shown in Fig. 4-2a has been widely used. Its connection may be schematically represented by a Wheatstone bridge as in Fig. 4-2b, which can measure intensities varying by more than a factor of 10^8 by changing input grid resistors.

Direct coupling, as employed in multistage d-c amplifiers, can be avoided by converting the direct into an alternating voltage. Vibrating-reed electrometers are commercially available for this purpose with minimum ranges of 10 ± 0.1 mv. The conversion is effected by applying the direct current across a condenser having one of its plates vibrating.

4-5. Ionization Chambers as Pulse Instruments. When an ionization chamber is used as described above to measure the integrated effects

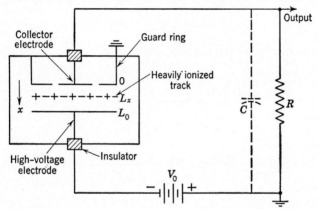

FIG. 4-3. An ionization chamber connected as a pulse instrument. R is the leakage resistor inserted from the input of an amplifier to ground. The product RC represents the time constant of the input circuit.

of a large number of ionization particles, it is considered an integrating instrument. However, when provided with a recharging resistor from the collecting electrode to ground and connected directly to a high-gain pulse amplifier (Fig. 4-3), it is a pulse instrument. When ions are produced inside the chamber by a passing ionizing agent, the potential on the collecting electrode starts to vary from ground potential according to the distribution and drift velocities of the ions.

The variation of the potential of the collecting electrode vs. time is known as the "pulse shape." A pulse shape generally consists of two distinct parts (Fig. 4-4). There is a very sharp rise of short duration $\sim 10^{-6}$ sec caused by the rapid motion of the electrons toward the collector. It is followed by a gradual rise of longer duration $\sim 10^{-4}$ sec, due to the slower drift of the positive ions away from the collecting electrode. If the discharging resistor chosen provides a time constant RC (Fig. 4-3) large compared with the collection time of electrons but small compared with the collection time of positive ions, the collecting electrode will follow the fast electron pulse approximately but will discharge through the resistor R to ground for the slow ion pulse.   The lower curve in Fig. 4-4 represents the pulse shape for a purified inert gas.   An electronegative gas such as O_2 or Cl_2 gives a distorted pulse.

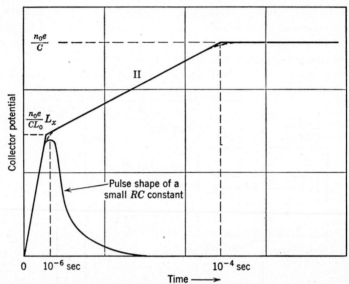

FIG. 4-4. The variation of voltage at the collecting electrode of an ionization chamber vs. time.   The upper curve represents an insulated collecting electrode.   The lower curve shows the pulse that results from employing an appropriate leakage resistor R.

4-6. Applications of Ionization Chambers as Pulse Instruments. The pulse height is proportional to the number of ions collected, and the number of ions formed by an ionizing particle terminating inside a chamber is equal to the kinetic energy of the particle divided by the energy required to produce a pair of ions in the gas.   It is therefore possible to calibrate a chamber for pulse height against particle energy.

For instance, a polonium α particle of 5.4 Mev would produce about $n_0 = (5.4/30) \times 10^6 = 1.8 \times 10^5$ ion pairs in a gas.   If the capacitance of the collector system (C) is about 30 $\mu\mu$f, the pulse height would be

$$\frac{n_0 e}{C} = \frac{1.8 \times 10^5 \times 1.6 \times 10^{-19}}{30 \times 10^{-12}} = 10^{-3} \text{ volt}$$

For the fission fragments from U^{235} exposed to slow neutrons, the total energy is around 200 Mev.   A pulse of a height forty times that of the α particles should be observed.   The heights of these pulses are much above the background of β, γ, or even α radiations and can be easily distinguished. The fission ionization chamber has become a very useful slow-neutron monitor in pile buildings.

A boron- or lithium-lined ionization chamber or one filled with BF_3 gas is also useful as a slow-neutron detector. The slow neutrons are absorbed by the B or Li nuclei by the reaction

$$_5B^{10}(n,\alpha)_3Li^7 \qquad \text{or} \qquad _3Li^6(n,\alpha)_1H^3$$

in which the heavy ionization produced by the disintegration particles (α particles in these cases) and the recoil nuclei is observed.

The ionization-chamber method is also used to count fast neutrons through the effects of recoil nuclei produced during collisions. Pure

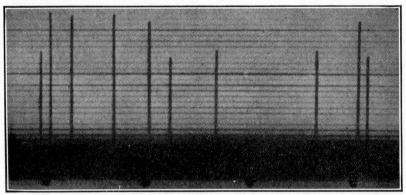

FIG. 4-5. Oscilloscope picture showing huge fission pulses above an intense background of α-particle pulses.

hydrogen at a pressure up to 90 atm has been used with satisfactory results.

4-7. Proportional Counters. Because of the ion-multiplication processes which set in at a high electric field, the proportional counter and the Geiger-Müller counter can give much larger pulses than that of the ionization chamber. To establish a sufficiently high field region without using unreasonably high potentials, the cylindrical type of chamber must be used. The field distribution of a cylindrical counter with outside cathode radius b and radius of the central wire a is given by

$$E(r) = \frac{V_0}{r[\ln (b/a)]} \tag{4-1}$$

where V_0 is the potential between the electrodes and r is the radial distance from the central wire. The distribution of the field for a commonly used counter size of $b = 1$ cm, $a = 0.005$ cm, and $V_0 = 1000$ volts is illustrated in Fig. 4-6. In the immediate neighborhood of the central wire, the field is as high as 10^4 to 10^5 volts/cm.

When an electron is accelerated in a weak electric field, the kinetic energy it gains between collisions is lost as heat in the collisions. As soon as the electron enters a field where the field gradient is strong enough so that the energy gained between collisions will excite or ionize the atoms or molecules, ion multiplication by collision begins. The secondary electrons so produced in turn produce more electrons. Eventually a

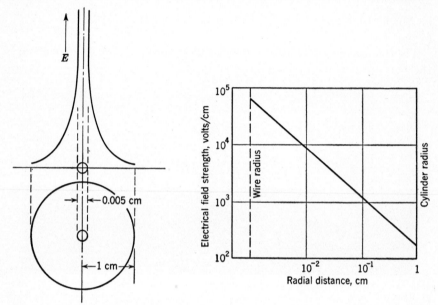

Fig. 4-6. Electric field in a cylindrical counter and its logarithmic field strength at various points along the radius (between the surface of the central wire and the inner surface of a coaxial cylinder).

"Townsend avalanche" of n electrons from each primary electron produced by the ionizing agent reaches the collecting electrode. The ion-multiplication region in gas will expand outward from the central wire, and the size of the avalanche will grow larger as the voltage across the chamber is increased.

In addition, photons are emitted during the deexcitation processes, which are capable of photoionizing the gas or releasing photoelectrons from the metal cathode and starting new avalanches. Let us define γ as the number of photoelectrons ejected per ion pair formed in the gas and n as the number of ion pairs per Townsend avalanche. The multiplication factor m, the total number of electrons produced in the gas due to every primary electron, is given by the infinite series

$$m = n(1 + \gamma n + \gamma^2 n^2 + \cdots) \tag{4-2}$$

When the voltage is not too high the photoelectric effect is small and

$\gamma n < 1$, giving

$$m = \frac{n}{1 - \gamma n} \tag{4-3}$$

A counter operated with m completely independent of the initial ionization is called a "proportional counter" and can have an m of several thousand.

The higher the m, the smaller the voltage amplification required of the pulse amplifier. However, it is advisable not to use a multiplication factor above a few hundred if strict proportionality is required, because of the large space charge.

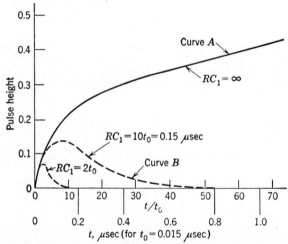

FIG. 4-7. The pulse shape from a proportional counter for three time constants: infinity, 0.15 μsec, and 0.03 μsec. (*From R. R. Wilson, D. R. Corson, and C. P. Baker, "Particle and Quantum Detectors," Preliminary Report No. 7, Nuclear Science Series, National Research Council, Washington, 1950.*)

Ion multiplication is confined to the immediate neighborhood of the central wire, and the pulse begins to rise appreciably only when the positive ions start to drift away. A fast initial rise is followed by a more gradual rise as the ions move out to the region of lower field. The completion of the collection of positive ions may take as long as 10^{-3} sec (see Fig. 4-7, curve A). A proportional counter is usually connected to ground through a leak resistor which is adjusted to give a time constant between 10^{-6} and 10^{-5} sec. The pulse thus shaped is shown in Fig. 4-7, curve B. Although the pulse heights attained with a short time constant are much reduced, they are still proportional to the amount of initial ionization.

4-8. Application of Proportional Counters. Proportional counters are most useful in detecting and measuring heavily ionizing α particles or protons against a strong background of β or γ radiations, which can be discriminated against electronically.

A valuable application is in the indication of slow-neutron flux in reactors. The proportional counter is filled with BF_3 or is boron-lined. The isotope $_5B^{10}$ has a very large cross section* for the (n,α) reaction

$$_5B^{10} + n \rightarrow {}_3Li^7 + {}_2He^4 \qquad (4\text{-}4)$$

This reaction has a cross section of 3830 barns for the B^{10} nucleus at a neutron energy of $\frac{1}{40}$ ev (2200 m/sec), and is inversely proportional to the neutron velocity. Because of this $1/v$ relation, any boron detector

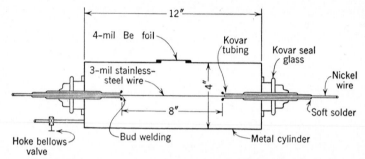

FIG. 4-8. The construction of a conventional proportional counter.

will read directly the neutron flux in a beam if the absorption of neutrons due to the presence of boron is small.

Another valuable use of proportional counters is to investigate β or γ spectra in the low-energy region. It is also possible to introduce the radioactive substance directly into the counter as a gas or as a thin-layer deposit to obtain 100 per cent geometrical efficiency and avoid the window absorption (Fig. 4-8).

4-9. Geiger-Müller Counter. In the proportional counter, the photoelectric effect is small and the dominating phenomenon is ion multiplication by collision. The avalanche will therefore terminate as it approaches closer and closer to the central wire. On increasing the potential applied to the counter, the proportionality feature will slowly disappear until all pulses have the same size no matter what the initial ionization is. Under that condition, the counter is called a "Geiger-Müller counter."

The multiplication factor [Eq. (4-2)] increases gradually with the voltage across the tube and diverges as γn approaches 1. $\gamma n = 1$ is the condition for each avalanche of n ion pairs to generate a new Townsend avalanche by the photoelectric effect in the gas or at the cathode. Therefore the multiplication process is not limited to collisions, and the ioniza-

---

* Enriched $_5B^{10}$ isotope with abundance up to 96 per cent can be obtained through Stable Isotope Division, Oak Ridge National Laboratory, Oak Ridge, Tenn.

tion spreads throughout the counter volume or along the central wire instead of being localized, and a discharge finally takes place. If γn is not too much larger than 1, the discharge will extinguish itself and the counter will register an individual pulse. The pulses from a Geiger-Müller counter can be as large as 20 to 30 volts. Therefore little or no amplification is needed for recording. In the Geiger-Müller counter the discharge spreads relatively slowly along the axis within a narrow sheath where the field strength is high. Thus, it may take a few tenths of a microsecond to reach one-tenth of the final pulse height as compared

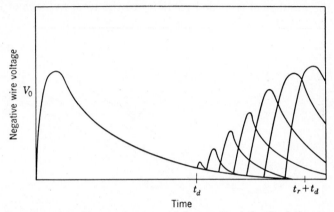

Fig. 4-9. Drawing of oscilloscope pattern of the result of the dead-time experiment. t_d and t_r are the dead time and the recovery time. They can be measured from the oscilloscope pattern if the speed of the sweep is known. [*From H. G. Stever, Phys. Rev.*, **61**:38 (1942).]

with a few hundredths of a microsecond in a proportional counter. Furthermore, the delay of the appearance of the pulse usually varies from a fraction of a microsecond to several microseconds after the occurrence of the initial ionization, according to the time taken by the electrons to drift to the multiplication region. This delay is called the "intrinsic delay" and becomes unreasonably long if negative ions formed by electron attachments are present in the gas.

The positive ion sheath surrounding the central wire weakens the electric field between the sheath and the central wire to below the threshold for multiplication by collision, and thus pulse formation. As the sheath moves out radially, the field near the wire gradually recovers. The period following a pulse in which the counter is insensitive to further ionizing particles is called the "dead time." When the ions reach the cathode and the field inside of the counter returns to normal, full-size pulses can reappear. The interval from the end of the dead time to the time when pulses of full size are resumed is called the "recovery time." Both time intervals are about 1 to 2×10^{-4} sec (Fig. 4-9), which greatly limits the maximum counting rate of a counter. In addition, it is necessary to avoid rekindling the discharge from electrons emitted when the + ions reach the cathode cylinder. This is acomplished by the presence of the vapor of a polyatomic compound such as alcohol in the gas (self-quenching) or by an external electronic circuit.

Dissociation of the polyatomic molecules in a self-quenching counter limits the lifetime to about 10^9 to 10^{10} counts. Counters in which a halogen gas replaces the polyatomic vapors show an unlimited life.

4-10. Efficiency. The efficiency of a counter may be defined as the ratio of the number of recorded particles to the total number of ionizing

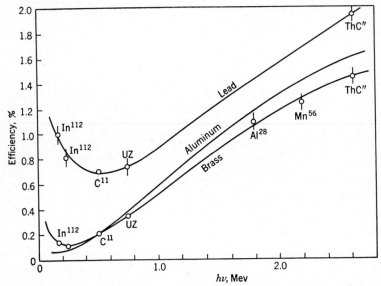

Fig. 4-10. Efficiency curves of Geiger-Müller counters with various wall materials (efficiency vs. γ-ray energies). The chemical symbols represent γ-ray sources. [*From H. Bradt et al., Helv. Phys. Acta*, **19**:77 (1946).]

particles passing through the sensitive volume of the counter. The ionizing agent may escape detection either because no ion was produced in its passage through the counter or because the passage took place during the dead time following a previous count.

The counter efficiency for charged particles is generally high. If n is the specific ionization of the ionizing particle, the average number N of primary pairs produced in a path of length d cm in a counter filled with gas to a pressure of p atmospheres is npd. The probability that no ions are produced along this path is e^{-N}, and the efficiency is thus

$$E = 1 - e^{-N} = 1 - e^{-npd} \tag{4-5}$$

For a cosmic ray crossing an ordinary argon counter where $p = \frac{1}{10}$ atm,

$d = 2$ cm, and $n = 29$,

$$E = 1 - e^{-6} = 99.8 \text{ per cent}$$

The counter efficiency for γ radiation is generally less than 1 per cent (Fig. 4-10). In order to increase the efficiency, various multiple electrodes or multiple cells have been developed to increase the production of photoelectrons and Compton electrons (Fig. 4-11).

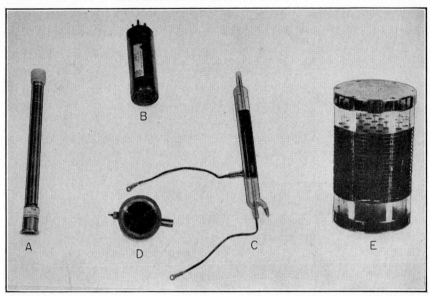

Fig. 4-11. Various types of Geiger-Müller counters: (A) thin-wall stainless-steel counter; (B) thin-mica-end-window counter, 1 to 5 mg/cm²; (C) silvered-thin-glass-wall counter, ~30 mg/cm²; (D) pancake-shaped counter with thin mica window; (E) multiple-cell γ-radiation counter.

4-11. Construction and Testing of Counters. Mass production of stable and reliable counters is now possible. It is important to have the cathode smoothed and cleaned. The central wire should be uniform and free from irregularities and sharp points. The assembled counter should be outgassed before filling.

A gas filling of 10 cm Hg of argon and 1 cm of ethyl alcohol is as good as any. Other polyatomic vapors such as ethyl acetate, lead tetramethyl, methane, acetone, carbon tetrachloride, etc., all give satisfactory results.

The performance of a counter is usually represented by a curve which plots the counting rate vs. the counter voltage, for a given source of radiation. Figure 4-12 shows a typical "plateau" and the threshold voltage. The threshold voltage should not be too high, and the plateau should be nearly flat and extended so that slight instability of the high voltage supply will not change the counting rate. The counter also should not be temperature-dependent and should have a long life.

Ionizing particles arriving during the dead-time period of the counters are not counted. Let the input counts which would be obtained if there were no dead time be N. If the output counting rate observed is n and the dead time is τ, then the average number of input counts missed

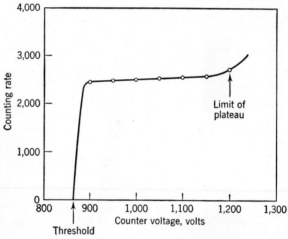

FIG. 4-12. Plateau of a Geiger-Müller counter. Counting rate, in counts per minute, vs. collecting voltage; 250-volt plateau; 2 per cent rise in 100 volts.

during the insensitive interval is $Nn\tau$. So the loss is

$$N - n = Nn\tau \tag{4-6}$$

For a dead time $\tau = 2 \times 10^{-4}$ sec an output rate of 50 counts per second gives a 1 per cent counting loss.

4-12. Scintillation Counters [6]. The detection under a microscope of individual α particles by their scintillation flash on a phosphor screen was used by Rutherford in his α-particle scattering work that established the present-day nuclear model. After the development of the photomultiplier tube and the discovery of new phosphors, scintillation counting again resumed importance as a particle or quantum detection method.

A scintillating phosphor is a substance which fluoresces on passage of ionizing particles. After the fluorescent levels have been excited by the ionizing particle, they decay by emitting light quanta according to the exponential law

$$n(t) = \text{const} \, (1 - e^{-t/\tau}) \tag{4-7}$$

where n is the total number of photons emitted during a time t after the passage of the particle. τ is the time necessary for the number of excited levels to decay to $1/e$ times the initial number.

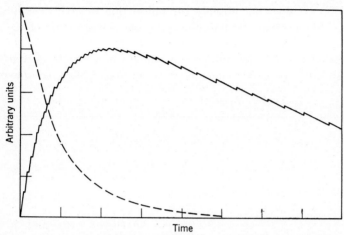

FIG. 4-13. Logarithmic build-up of a scintillation pulse.   The dashed curve represents the decay of the excited states in the phosphor.

TABLE 4-1. A LIST OF COMMON PHOSPHORS*

Phosphors†	Density, g/cm³	Refractive index	Spectrum, A	Decay time, sec	Efficiency
Inorganic:					
Calcium tungstate..................	6.06	1.934	4300	$>10^{-8}$	1
Sodium iodide, Tl (hygroscopic)......	3.67	1.7745	4100	2.5×10^{-7}	2
Cadmium tungstate................	7.90	2.2–2.3	5200	7×10^{-6}	2
Lithium iodide (Eu) (very hygroscopic)	4.06	1.955	Blue, green	2×10^{-6}	0.7
ZnS (Ag)(only powder or small crystal)	4.09	2.368	Blue	$\sim 10^{-5}$	~ 2
Potassium iodide..................	3.13	1.677	4100	$>10^{-6}$	0.5
Cesium iodide (Tl) (not hygroscopic).	4.51	1.7876	Blue	$\sim 10^{-6}$	~ 0.6
Cesium iodide (77°K) (not hygroscopic)	4.51	1.7876	Blue	$\sim 5 \times 10^{-7}$	
Organic:					
Anthracene.......................	1.25	1.595	4400	$\sim 2 \times 10^{-8}$	1
1,2-Diphenylacetylene..............	.....	....	....	0.7×10^{-8}	~ 0.5
Terphenyl........................	1.23	....	~ 4000	$<10^{-8}$	0.4
Naphthalene.....................	1.15	1.618	~ 3500	$\sim 6 \times 10^{-8}$	0.25
Phenanthrene....................	1.03	....	~ 4100	$\sim 10^{-8}$	0.3
trans-Stilbene....................	1.16	1.622	4080	0.8×10^{-8}	0.5
Liquid solution:					
Terphenyl in xylene, benzene, or toluene (3–5 g to the liter)..........	0.87	1.500	~ 4000	Fast (about $<10^{-8}$)	0.28–0.46
Anthracene in phenylether..........	1.073	1.583	~ 4400		
Solid solution (plastic scintillator):					
Terphenyl in polystyrene............	1.062	1.595	~ 4000	0.8×10^{-8}	<0.6

* For details see Robert K. Swank, Characteristics of Scintillators, *Ann. Rev. Nuclear Sci.*, **4**:111 (1954), or G. F. J. Garlick, Luminescent Materials for Scintillation Counters, *Progr. Nuclear Phys.*, **2**:51 (1952).

† Noble gases (He, Ne, A, Kr, and Xe) can also be used as scintillators with the aid of a wavelength shifter (quaterphenyl or tetraphenyl butadiene).   For details see J. A. Northrup and R. Nobles, *Nucleonics*, **14**(4):36 (1956), and C. Eggler and C. M. Huddleston, *Nucleonics*, **14**(4):34 (1956).

Actually one does not observe the pulses due to each light quantum, but rather the exponential building up of the voltage pulse of the photomultiplier with the time constant τ (Fig. 4-13).

The decay time of the fluorescence should be short to permit a high counting rate and good coincidence resolution. Most of the organic

TABLE 4-2. VARIOUS TYPES OF PHOTOMULTIPLIERS

	Tube				
	RCA 5819*	Du Mont 6292†	RCA 1P21‡	British EMI 4588	British EMI 5311
Number of dynodes......	10	10	9	9	11
Cathode shape..........	Circular (tube end), 1½ in. (diam), slightly curved	Circular (tube end) 1⅝ in. (diam) flat	Rectangular (internal) $1\frac{5}{16} \times \frac{5}{16}$ in.	Internal 3 in.²	Tube end 1 in. (diam)
Wavelength of maximum response, A..........	4800 ± 500	4800 ± 500	4000 ± 500	4000 ± 500	4800 ± 500
Current amplification....	1.5×10^5 (75 volts per dynode), 6×10^5 (90 volts per dynode)	2×10^5 (105 volts per dynode), 2×10^6 (145 volts per dynode)	3×10^5 (75 volts per dynode), 2×10^6 (100 volts per dynode)	10^6 (150 volts per dynode)	10^7 (160 volts per dynode)
Recommended operating voltage per dynode....	75	125–150	75	150	160
Anode dark current, μa...	0.05 (90 volts per dynode)	0.05 (105 volts per dynode)	0.1 (100 volts per dynode)	0.03	0.1
Dimensions:					
Length, in............	5⅞	5⅝	$3\frac{11}{16}$	10	838
Diameter, in..........	2¼	2	$1\frac{5}{16}$	2	2

* The new RCA 6342 is similar to RCA 5819 except that the tube end is flat and current amplification is higher. The RCA 6810 has 14 stages of dynodes and is capable of delivering pulse currents up to 0.5 amp.

† Du Mont 6291 has same characteristics as Du Mont 6292 except that the diameter of the cathode surface is 1⅛ in.

‡ RCA 931 A is similar to 1P21 except that the current amplification is lower ($\sim 1 \times 10^6$ for 100 volts per dynode). 1P22 and 1P28 are similar to 1P21 except that the wavelengths of maximum response are 4200 and 3400 A, respectively.

phosphors have decay times of the order of 10^{-8} sec or less. The inorganic phosphors all have decay times longer than 10^{-7} sec and some over 10^{-4} sec (Table 4-1).

Among the inorganic crystals, NaI (activated by Tl) and LiI (activated by Eu) have been widely used in detecting γ radiation and slow neutrons respectively. Among the organic crystals, stilbene is preferred because of its short decay time. When a fairly large piece of fluorescent material is required, as in cosmic-ray work, liquid or plastic scintillators are often used.

The fraction of the energy lost in a phosphor by an ionizing particle that is emitted as light quanta has been named by Kallman the "physical efficiency." Not all the light quanta excited in a phosphor can emerge, because of self-absorption. For instance, ZnS is almost opaque at a thickness >80 mg/cm². A ZnS screen is therefore useful for α particle detection, but not for β or γ radiation, which need thicker screens for adequate absorption. Most of the organic phosphors and some inorganic phosphors such as NaI are practically transparent to their own radiation, which enables one to use thick slabs.

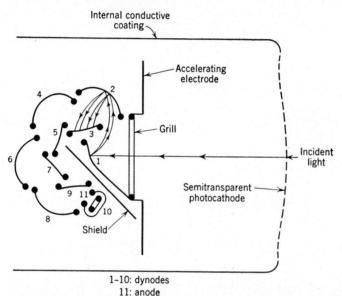

FIG. 4-14. A schematic diagram of a photomultiplier (1P21 or 5819).

Since the photocathode of a photomultiplier is sensitive to only a narrow band of light frequency, it is important that the frequency of the light quanta from a phosphor fall within this sensitive band.

4-13. Photomultiplier Tubes [6]. The photomultiplier tube (Table 4-2) consists of a photocathode followed by several stages of dynodes made of materials of high secondary emission (Fig. 4-14). The photocathode converts a fraction of the photons falling on it into photoelectrons, which are multiplied in successive collisions with the dynodes. For a multiplication factor of 4 for secondary electron emission per stage,* 10 stages, and almost 100 per cent focusing, the gain amounts to about $(4)^{10} = 10^6$.

An electron of 50 kev stopped in a phosphor will produce some 2000

---

* The multiplication factor increases rapidly with increasing dynode voltage. For accurate counts, therefore, a very constant voltage supply is required, with fluctuation and drift under 0.1 per cent at least, and preferably 0.01 per cent.

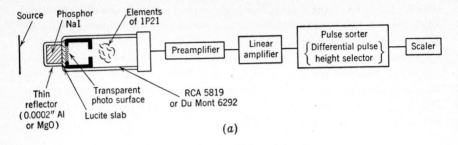

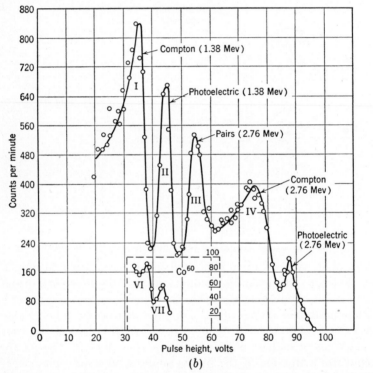

Fig. 4-15. (a) General arrangement of a scintillation-counter spectrometer. (b) Na²⁴ γ-ray spectra as obtained through a scintillation-counter spectrometer.

photons, of which about 1000 may be collected on the photocathode, producing about 100 photoelectrons. "Noise" is being continually produced by thermal emission at the cathode, but if the signal exceeds 20 photoelectrons at the cathode, it is possible to discriminate against the noise pulses by proper biasing of the electronic circuits. Cooling the multiplier by dry ice has been observed to decrease the noise background by a factor of 100. Liquid nitrogen is also used. Another ingenious method of reducing the noise background is to place the phosphor between the cathode surfaces of two multipliers and to register only the coincidences.

The rise time of a multiplier pulse is of the order of 10^{-9} sec for the focusing-type multipliers (931-A, 5819, etc.). The British type EM1-5311, with a "venetian blind" structure, has a rise time of some 10^{-8} sec.

Because the energy of the secondary electrons is usually low, a weak magnetic field (even the earth's field) may destroy the focusing properties of the tube. A shield of an iron alloy of high permeability such as Mumetal should be used around the tube.

The scintillation counter has an advantage over the Geiger-Müller counter in that the pulse heights are proportional to the energy of the particles. With suitable means of biasing or pulse-height selection, one can investigate the energy distribution of the β or γ radiations (Fig. 4-15a, b). Most of the β spectra investigated with scintillation phosphors showed more slow electrons than was predicted. This may be due to the fact that some of the electrons are scattered out of the crystal before they have lost all their energy, thus leaving a preponderance of lower pulse heights. The recent use of split crystals or one hollow crystal has removed most of these deviations.

In conclusion, a scintillation counter is superior to a Geiger-Müller counter in its high detecting efficiency for γ radiations, the energy-dependence of its pulse height, and above all, its very short resolving time. With fast amplifiers, coincidence circuits, and scalers, a high counting rate of 10^9 counts per second and a resolving time of 10^{-9} sec can be used.

4-14. Cerenkov Counter. For detecting fast charged particles, the Cerenkov radiation (see Sec. 2-12) which they emit on passing through a transparent material is extremely useful. It is emitted in directions making acute angles with the motion of the charged particles and is in effect an electromagnetic shock wave. It includes the blue visible haze due to β particles in the MTR, due to all the charged particles in HRE,* etc. It is detected by photomultiplier tubes or photographic plates so placed as to record the intensity only, or also the direction.

4-15. Photographic-emulsion Technique [7]. Following Becquerel's discovery of the activation of silver halide emulsion by radiations from uranium ore, the photographic technique was used extensively in comparative studies of the relative intensities of radioactive sources and in mapping the distribution of radioactive elements in rocks. However, because it yielded no quantitative information, it soon gave way to

---

* See App. N for reactor designations.

electroscopes, electrometers, and ionization-chamber methods of detection. But its potentiality of registering individual particles and furnishing information on their energy, mass, charge, and direction of propagation was recognized and has been continually improved. By now, thick "nuclear" photographic emulsion has established itself as yielding quantitative and versatile results of high precision and reproducibility in nuclear research.

In nuclear emulsion each particle should produce a well-defined track which stands out above the fog background. Thin or "optical" film is used when total radiation is to be measured but individual tracks need not be studied. In general, the silver halide content of the nuclear emulsion is three or four times as high as that of optical emulsion. The thickness of optical emulsion is about 25 μ; that of nuclear emulsion ranges from 25 to 2000 μ so that energetic particles may terminate their tracks inside the emulsion.

The probability that a grain struck by the particle will be rendered developable is roughly proportional to the cube of the grain diameter. Thus, emulsions with the smallest grain size are sensitive only to the heavily ionizing particles such as fission fragments or slow α particles, and coarse-grain minimum ionization plate has been developed which will detect as low a rate of energy loss as 0.0022 Mev/cm. The grain size of nuclear emulsion is 0.1 to 0.6 μ, and the grains are well separated. In optical emulsion the size is as large as 3.5 μ and the grains almost interlock.

Film is particularly valuable in cosmic-ray investigation because of its cumulative properties; even the rarest phenomena may be recorded by this method. A small film badge of dental-film size ($1\frac{1}{4}$ by $1\frac{3}{4}$ inch) carried by people exposed to radioactivity has been generally used as a standard method of determining the doses of radiations received. If the emulsion is specially impregnated with boron and lithium, then scanning the plate for α-particle tracks from $B^{10}(n,\alpha)$ or $Li^{6}(n,\alpha)$ reactions offers a sensitive means of monitoring slow-neutron dosages.

Photographic emulsions are also used to great advantage in studying the distribution of radioactive materials in mineral, tissue, fuel elements, neutron-irradiated specimens, etc. The surface or cross section to be studied is placed in direct contact with the film and kept dark for a few hours. The resolution obtained in such an "autoradiograph" is far superior to that with a counter. Figure 4-16 shows autoradiographs of the distribution of radioactive zinc in the fruit of a tomato plant.

4-16. Electronic Instruments [8, 9]. The ionization produced by the passage of charged particles through a gaseous medium is so small in magnitude that a highly sensitive device such as an electroscope, electrometer, or vacuum-tube electrometer is usually employed to measure the total ionization current.   When detecting single charged particles, the voltage pulse produced at the collecting electrodes of an ionization chamber or a proportional counter is generally only a small fraction of a volt.   An electronic *linear amplifier* is required to amplify the pulse to a size sufficient to operate a discriminator, a coincidence circuit, or a scaler followed by an electromechanical register.

The electromechanical register can be made to respond to evenly distributed pulses of frequencies less than about 100 per second.   For

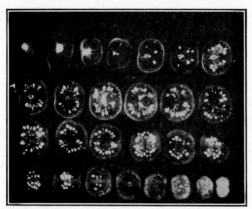

FIG. 4-16. Autoradiograph of the distribution of radioactive zinc in the fruit of a tomato plant which had been fed radioactive zinc chloride.   (*By Dr. Perry Stout of the University of California.*)

random pulses the rate of occurrence should be limited to only a few counts per second to avoid counting losses.   Since most pulse detectors can respond to much faster counting rates, it is desirable to use a fast-response *scaling circuit* which selects every mth pulse and passes it on to the register.   Thus the scaling circuit reduces the output counting frequency by a factor m and also regularizes the random rate so as to enable the register to follow.

In health surveying, background monitoring, etc., a direct indication on a *counting rate meter* is preferred.

It is often desired also to analyze the pulse height (energy) distribution by a *discriminator* or a *differential discriminator*.

Furthermore, in the investigation of a nuclear-decay scheme or level diagram, it is desirable to know whether two or more radiations are emitted simultaneously or not.   *Coincidence* or *anticoincidence circuits* are timing circuits specifically designed for this purpose.

4-17. Linear Amplifier.   When charge Q produced by the ionization by a nuclear particle in an ionization chamber or a proportional counter is collected and applied to the grid of the input tube of an amplifier, it causes a sudden change of the input grid potential by Q/C, where C is the input capacitance, including that of the ionization detector, the associated wiring, and the input grid of the tube. The input signal can therefore be approximated by a step function, as shown in Fig. 4-17. The output pulse of an amplifier depends on its frequency response and is usually shaped as shown in Fig. 4-17.

The delay time T_D is the time interval between the application of a step signal and the time at which the output response has reached the half-value point. The rise time T_R is the reciprocal of the slope at this half-value point. Thus, if a tangent is drawn at the half-value point and it intersects the time axis at a and the horizontal tangent through

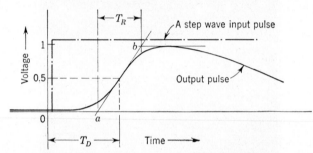

FIG. 4-17. A typical amplifier response to a voltage step function. The dot-dash curve represents the step wave input pulse. T_D is the delay time, and T_R is the rise time.

the maximum at b, then $T_R = t_b - t_a$. The clipping time is a measure of the rate at which the output potential returns to its normal state.

4-18. The Rise Time. To amplify the input step function without distortion, the rise time and the delay time would have to be zero and the clipping time infinite. In reality, the frequency response of an amplifier at both low and high frequency ends is limited. To attain a very short rise time, the high-frequency response of an amplifier must be properly boosted. In a resistance-coupled amplifier, a rise time as short as 0.03 μsec has been achieved. Nevertheless, this cannot be accomplished without sacrifice of gain per stage. It can be seen that $T_R = \sqrt{2\pi} RC$ and $T_D = RC$, the plate-circuit time constant. On the other hand, the gain is $g_m R$ where g_m is the tube transconductance and R the plate load resistor. Therefore one cannot improve the rise time without impairing the gain of the amplifier.

The ratio of gain to rise time for the stage equals

$$\frac{g_m R}{\sqrt{2\pi RC}} = \frac{g_m}{\sqrt{2\pi C}}$$

and measures the excellence of the tube for amplifying fast transients.

The minimum rise time can be improved by inductance shunt compensation of the plate load, as customarily used in video amplifiers. The introduction of the inductance in series with the plate-load resistance will make the rise time slope much steeper. However, to amplify pulses of extremely short rise time, *distributed amplifiers* should be used; a rise time of the order of 10^{-9} sec at a gain of 8 has been obtained. An amplifier of extremely short rise time should be used only for input pulses of very fast rise time and where good time resolution is required, as for organic scintillation counters connected in coincidence in a telescope arrangement. The most favorable condition for high signal-to-noise ratio is obtained with the rise times of the input pulse and of the amplifier equal to clipping time.

However, stability and linearity of an amplifier are even more important than the time constant in most cases, and generally some feedback is added to increase them in spite of the consequent increase in rise time.

4-19. Clipping. The long tail of a pulse should be clipped so that no overlapping or piling up will result at high counting rates. Therefore the low-frequency response is deliberately limited by introducing a single interstage coupling of very short time constant in the amplifier train. The clipping is preferably introduced at a late stage so that it also reduces the 60-cycle hum due to the a-c heaters of previous stages.

A superior method of clipping employs a delay line which gives an almost square pulse with very little tailing. It is even possible to shorten a pulse so that its duration is equal to its rise time and the highest possible resolution attained.

4-20. Typical Amplifier. There are quite a few linear amplifiers commercially available. They can be used for the amplification of pulses from ionization chambers, proportional counters, or scintillation counters. In general, a preamplifier containing two to three stages of amplification with a cathode follower output is built on a small chassis, close to a detector to minimize the lead capacitance. The gain is usually around 20 to 30. The main amplifier (Fig. 4-18) contains about four stages with some inverse feedback to gain stability and linearity. The gain per stage is between 20 and 100, and the maximum gain for the shortest rise time, 0.2 μsec, is around 3000. The rise time and clipping time can be varied by a control over a range from some tenths of a microsecond to a few microseconds. Gain control is effected by a coarse and a fine attenuator. Two outputs are provided. The low-level output will drive a coaxial cable to a maximum of about 5 volts at 150 ohms, and from the high level a maximum signal of about 100 volts can be obtained without appreciable amplitude distortion. A precision discriminator which gives uniform square pulses is also provided at the high-level output.

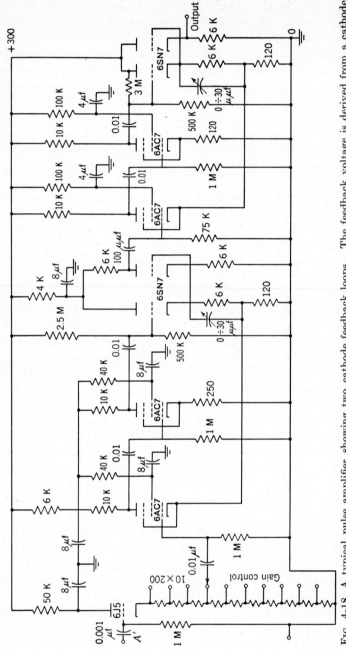

FIG. 4-18. A typical pulse amplifier showing two cathode feedback loops. The feedback voltage is derived from a cathode follower and is directly coupled to the input stage.

122

4-21. Scaling Circuit. The fundamental component of a scaling circuit is a flip-flop circuit, with two stable positions. An incoming pulse causes the circuit to flip or flop between these two stable positions. However, only one of these transitions is passed on to the subsequent scaling unit or register. Therefore the scaling ratio is commonly a power of 2. Scaling circuits based on powers of 10 have also been developed and are called "decade scalers." Figure 4-19 shows the Higinbotham scaling unit of 2, which is widely used.

Suppose that T_1 is conducting and T_2 is nonconducting. A negative pulse of sufficient size applied to the cathode of the twin diode 6H6 will

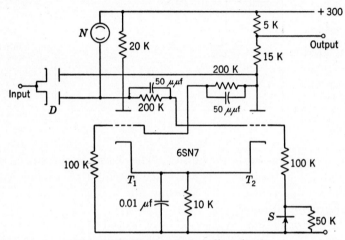

FIG. 4-19. Higinbotham scaling circuit.

make the upper half of 6H6 conducting but not the lower one, because its anode is at a lower potential than the cathode. Since the anode of T_2 is connected to the grid of T_1 by the big cross-coupling capacitor, the negative signal will cause T_1 to be momentarily nonconducting. Meantime, through the cross-coupling capacitor between the plate of T_1 and grid of T_2 the positive pulse will make T_2 conducting. T_1 will then stay nonconducting. Because of the perfect symmetry of the circuit, the next input pulse will flop the stable condition back to T_1 conducting and T_2 nonconducting. The output of a stage can be directly fed to the input of the next stage. Since the twin diode 6H6 in this circuit is completely insensitive to positive pulses, the next stage responds only when T_2 is flipping to conducting. Small neon lamps connected across the plate load resistors of T_1 indicate the state of the system. These are called "interpolation indicators."

4-22. Counting-rate Meter. A counting-rate meter (Fig. 4-20) is used to determine the rate of occurrence of pulses, irrespective of their sizes.

It is often used in portable instruments for surveying purposes or in obtaining continuous monitoring of radiation intensity in the vicinity of a nuclear reactor. Although it is inherently less accurate than a counting method, it is more convenient in use. The basic principle of its operation is as follows: A univibrator transforms pulses of all sizes to square pulses of equal magnitude and duration. These square pulses are then fed into a resistor-condenser arrangement so that each pulse puts a definite amount of charge on a capacitor in a time short compared

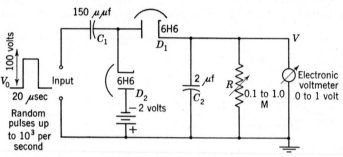

FIG. 4-20. Counting-rate meter.

with the average spacing of the pulses. The charge on the capacitor leaks off through a high resistor. The average potential across the capacitor is therefore proportional to the counting rate and can be measured with a vacuum-tube voltmeter.

4-23. Discriminators. A discriminator is used to prevent the counting of any pulse below a certain height. The simplest way to do this is to use the sharp-cutoff characteristics of a gas-filled vacuum-tube triode. The gas tube gives uniform output pulses but has too long a resolving time, which in turn limits the counting rate. The resolving time of a vacuum-tube triode is much smaller than that of the gas-filled triode, but the output pulses vary in height from zero to a certain value. To attain both fast resolving time and uniform pulses, the Schmitt trigger circuit (Fig. 4-21) is used. Since the grid potential of T_1 is lower than that of T_2, normally only T_2 is conducting. When an input positive pulse equal to or greater than the difference of the two grid potentials occurs at the grid of T_1, then the condition reverses. T_1 becomes conducting and T_2 nonconducting. This condition lasts until the potential of T_1 is lowered sufficiently to make T_2 conducting again. The resolving time of this circuit is of the order of 1 μsec.

One could analyze a pulse distribution by taking measurements with the discriminator set at different bias potentials and then taking the differences between consecutive readings. However, it is far better statistically to use a differential discriminator, which lets through only those pulses whose heights are within the desired range. This is accomplished electronically by operating two integral discriminators in anticoincidence with the bias, one set at the lower and the other at the upper level of the channel. Only pulses which trigger the lower level but fail to trigger the upper one will be recorded. Pulse-height analyzers with as many as 100 channels selecting successive pulse-height bands have been designed and built. They are called "multichannel pulse-height analyzers" or, in Great Britain, by the more descriptive name "kicksorters." Their use is no longer limited to research laboratories, but includes routine chemical analysis and many other industrial applications. Their development has paralleled that of analogue-to-digital converters. For reviews of current developments, see Ref. 8.

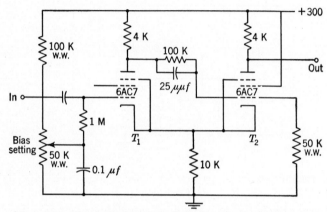

Fig. 4-21. Discriminator of the Schmitt trigger-circuit type.

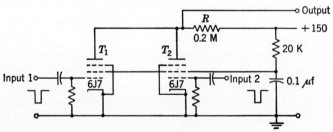

Fig. 4-22. Rossi coincidence circuit.

4-24. Coincidence and Anticoincidence. A coincidence circuit is used when it is desirable to determine whether two or more pulses occurring at different detectors occur simultaneously. The basic principle can be illustrated by Rossi's coincidence circuit (Fig. 4-22). The operation of this circuit is such that unless the two pulses occur at the same time, no pulse appears at the output. T_1 and T_2 have a common plate resistor R

which is large compared with the tube resistance. Normally both these tubes are conducting, and therefore the voltage at the output is nearly at ground potential. If now two large negative pulses are impressed on both tubes simultaneously, both tubes will be cut off at the same time and the voltage at the output point will jump up to the full plate supply voltage. However, if a pulse occurs at one of the two tubes and makes it nonconducting, since the other tube is still conducting there will be very little change in the output potential of the coincidence circuit. To have a reasonably large ratio between the height of a coincidence pulse and that of a single pulse, the plate resistor R must be large, which therefore limits the resolution time to a few tenths of a microsecond. One can replace the common plate resistor by a common cathode resistor and

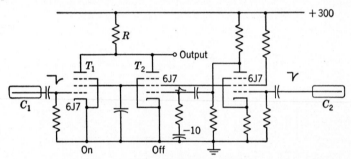

FIG. 4-23. Anticoincidence circuit.

obtain a resolution as short as 10^{-8} sec. Many circuits have been designed using diodes instead of triodes and giving a resolution as high as 10^{-9} sec. Recently, a very fast coincidence circuit employing a 6BN6 (gated-beam) tube has been successfully designed to give a resolution of the order of 10^{-10} sec.

Anticoincidence is widely used in the detecting of neutral particles and in reducing background counts. The basic circuit is shown in Fig. 4-23, where the negative pulse of detector C_1 is fed to T_1. The negative pulse of detector C_2 is inverted to positive pulse first, then fed to T_2 which is negatively biased at cutoff at normal condition. When a negative pulse from detector C alone appears, a large output pulse will occur, since both tubes will be cut off simultaneously. However, if pulses from both detectors occur simultaneously, the inverted positive pulse from detector C_2 will render the tube T_2 conducting, therefore no pulse will result at the output.

4-25. Nuclear Instruments for a Nuclear Reactor [1, 11–14]. To operate and maintain a nuclear reactor one must rely on sensitive and dependable instruments to measure and control the various factors. The most important factors are the neutron flux, the period of neutron-flux change, the fission rate, and gamma radiation. There are various methods of measuring the neutron flux at different intensity levels: The ones already discussed at length are the boron-lined or BF_3-filled ionization chambers or proportional counters, fission chambers, and LiI (Eu or Sn) or organic scintillation counters. With properly designed associated circuits, these instruments give extremely rapid response time, which is essential in the reactor control mechanism.

The BF_3 *Proportional Counter.* Well-designed metal proportional counters filled with enriched BF_3 gas can be obtained commercially (see Fig. 4-24). The pressure of the BF_3 gas can range from a few centimeters to as much as 120 cm Hg, depending on its usage. The detection efficiency of the enriched BF_3 counter for slow neutrons is very satisfactory. An ordinary counter 25 cm long and filled with 120 cm of BF_3 gas should give nearly 100 per cent efficiency for thermal neutrons. These counters are best suited for measuring a low-level neutron flux [up to 10^4 neutrons/$(cm^2)(sec)$] in the presence of γ radiation as great as 10^2 r/hr. However, in order to operate a high-pressure BF_3 proportional counter, a well-stabilized high-voltage supply of 1000 to 5000 volts must be used.

The Fission Ionization Chamber. It is sometimes more convenient to use a fission ionization chamber, where a collecting voltage of a few hundred volts is quite sufficient. The fission material is generally electroplated on the inner surface of the chamber or the electrode. Since the useful thickness of the fission layer is thinner than the range of the fission fragments, this limits the amount of fission material and therefore the efficiency of neutron detection.

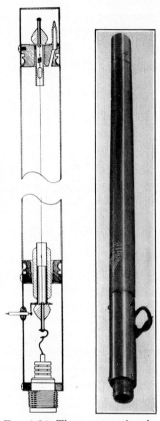

Fig. 4-24. The construction details and the photograph of an all-aluminum BF_3 neutron counter. The end sections of this counter are aluminum-brazed. The guard-ring construction prevents spurious counts. (*Radiation Counter Laboratories, Inc.*)

To increase the detecting efficiency, one could increase the surface area by increasing the number of electrodes. The multiple-plate fission chamber shown in Fig. 4-25 is often used. This instrument is used for the measurement of a neutron flux from 10 to about 10^5 neutrons/$(cm^2)(sec)$ in the presence of a γ intensity as high as 10^5 r/hr.

The Logarithmic Amplifier [14, 15]. When a wide range of the neutron flux, as much as six to eight decades, is to be measured, a logarithmic amplifier must be used. The principle of its operation is based on the logarithmic voltage-current characteristic of a thermionic diode. In this instrument a diode is used in place of the fixed, input resistor of the ordinary amplifier. The input signal is either the current of an ionization chamber or the output of a counting rate meter of some neutron detector. The neutron flux Φ or nv is thus converted to a logarithmic

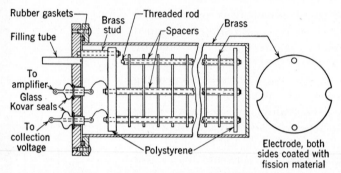

FIG. 4-25. A multiple-plate fission ionization chamber.

scale log Φ through the diode. The output of the amplifier thus obtained will be proportional to log Φ and can be displayed on current or voltage indicating meters. This is also known as a "log n" meter.

The Period Meter [14, 15]. The determination of the reactivity of a reactor is of great importance in predicting the changes in the power level. The best-known procedure for measuring the reactivity is to determine the time rate of change of $\ln \Phi$, which is the reciprocal of the stable reactor period. That is,

$$\frac{d}{dt} (\ln \Phi) = \frac{1}{T}$$

An instrument which is designed for the measurement of the period is called a "period meter." It differentiates the output signal of a logarithmic amplifier ($\ln \Phi$) by an RC circuit to give $(d/dt) (\ln \Phi)$, which is proportional to the reciprocal of the period and therefore to the reactivity. The output is generally in the microampere region and can be read on a meter or made to drive a recorder or to set off an alarm.

If the time element is irrelevant, then there are the methods of slow-neutron detection by neutron thermopiles and by foils.

The Neutron Thermopile [16]. This simple and compact unit for measuring the neutron flux is based on the well-known thermoelectric phenomenon. When two dissimilar metallic wires are connected together at their two ends and a temperature difference exists between these two joints, then an electric current will flow in the circuit. The neutron thermopile generally consists of 10 or more elements of thermopile connected in series electrically. The alternate hot junctions are coated with a neutron absorber (boron or uranium), and the whole assembly is housed in a compact container the size of a fountain pen. The sensitivity is around 1 mv at 10^{11} neutrons/(cm²)(sec) flux and the time constant is about 4 sec.

Slow-neutron Detection by Foils [10]. This method of measuring neutron distribution is based on the fact that many elements become radioactive when exposed to a neutron flux. This activation is sensitive only for a narrow region of neutron energy. Thereby one can simply measure the induced activities of various elements to determine the distribution of neutron flux. The frequently used elements which have strong resonance absorptions and yield induced radioactivity of suitable periods and radiations are In, Rh, Ag, Au, I, and Dy (see App. Q).

To measure a flux, (1) a thin foil of appropriate material is first irradiated at the desired location for a time interval t; (2) the foil is removed from the neutron beam and the induced activity measured on a counter for a counting period t_c after a certain waiting period t_1. From the number of disintegrations that occur during the counting period t_c, the resonance neutron flux can be calculated (see Probs. 4-6 and 4-7).

If two activities are induced at the same time, one can either adjust the waiting period to eliminate the shorter period or use suitable absorbers to screen out the softer radiation.

To remove the thermal neutrons from the beam, one generally uses Cd foil of a thickness of 0.02 in. or more wrapped around the detecting foil. In this case, one must correct for the absorption of resonance neutrons by the Cd foil. Besides, the depression of the thermal-neutron flux by the detecting foil must also be corrected for.

Compensated Ionization Chamber. Since the β and γ radiation in the vicinity of a reactor is usually extremely high, one is faced with the problem of separating the neutron signal from the intense β or γ background. For low-level neutron intensities, the separation can be accomplished by counting only the large pulses due to neutrons and discriminating against the low pulses from the β or γ radiations as described above. For high-level neutron intensities [10^4 to 10^{10} neutrons/(cm²)(sec)] in the presence of $\sim 10^5$ r/hr γ radiation, one could use the current-compensation method. Two ionization chambers, one sensitive to both neutron and γ radiation and the other sensitive to γ radiation only, are connected together electronically so that only the difference of the ionization currents is measured. If the volumes of the two chambers are properly designed, the net current should be approximately independent of γ-radiation background. These compensated ion current chambers are commercially available upon special order.

There is a multitude of problems on instrumentation applicable to reactors waiting to be solved. Because of the rugged industrial plant conditions, these instruments must be not only sensitive, wide-ranged, and dependable but also rugged, remote-controlled, and foolproof. This

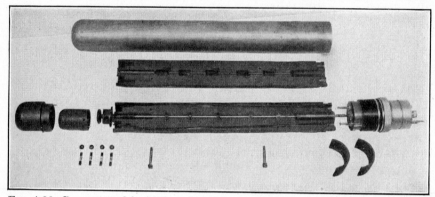

FIG. 4-26. Compensated ionization chamber (Oak Ridge design Q-1045). The construction is such that nitrogen gas can be passed through the active volume of the chamber. The majority of the metal components are made of magnesium. The electrodes are graphite, and the insulators are quartz. (*Radiation Counter Laboratories, Inc.*)

work will require a joint effort by both plant engineers and research physicists.

Prob. 4-1. In a radiation field of 200 mr/hr, what is the saturated ionization current of an ionization chamber of a volume of 1 liter filled with air at atmospheric pressure?

Prob. 4-2. The capacitance of a Lauritsen electroscope is 5 $\mu\mu$f. How long will it take for the electroscope to discharge from 15 volts if a beam of α particles of 5 Mev enters the electroscope at a rate of 100 per second?

Prob. 4-3. Show that, if a cosmic-ray particle penetrates a Geiger-Müller counter filled with argon at 7.6 cm Hg of pressure very close to the walls so that the path length is only $\frac{1}{3}$ cm, the efficiency = 63 per cent.

Prob. 4-4. Dead time of a counter can be measured by making separate measurements of the net counting rates due to two (or more) radioactive sources and of the combined sources. Assume the following procedure:

1. Measure the background rate B (assumed low) with both samples removed.

2. Move sample 1 near the counter to give a counting rate $n_1 + B$ of about 100 to 1000 counts per second.

3. Without disturbing sample 1, move sample 2 to a position where the counting rate $n_{12} + B$ approximately doubles.

4. Remove sample 1 without disturbing sample 2, obtaining counting rate $n_2 + B$. Show that the dead time τ can be calculated from the expression

$$\tau = \frac{n_1 + n_2 - n_{12}}{2n_1 n_2}$$

Prob. 4-5. Find the dead time of a Geiger-Müller counter from the following data:

Source A alone gives 109.3 counts per second.
Sources A and B give 214.0 counts per second.
Source B alone gives 112.5 counts per second.

If each of these counting rates was obtained through a 30-min counting interval, what is the standard deviation of the dead time measured?

Prob. 4-6. In the activation method of neutron detection, the foil is exposed to the neutron flux for a certain time interval t. After the termination of the irradiation, the neutron-induced activity is measured on a counter. If the total number of counts measured between t_1 and t_2 taking the end of irradiation as zero time is N, show that the initial counting rate at $t = 0$ under saturation condition $C_{0,sat}$ can be calculated by the following formula:

$$C_{0,sat} = \frac{N}{\tau(e^{-t_1/\tau} - e^{-t_2/\tau})(1 - e^{-t/\tau})}$$

where τ is the mean lifetime of the activity.

Prob. 4-7. From the initial saturation counting rate $C_{0,sat}$, show that the resonance neutron flux $n_v v$ is

$$n_v v = \frac{C_{0,sat}}{\epsilon \sigma N_0}$$

where N_0 = number of nuclei in foil being activated
σ = capture cross section
ϵ = counting efficiency

Prob. 4-8. A proportional counter 30 cm long is placed lengthwise in a beam of thermal neutrons. The pressure of the BF_3 gas inside the counter is $\frac{1}{2}$ atm. What is its efficiency of detection for thermal neutrons if the BF_3 gas used is a natural isotopic mixture? What is the factor of increase of its efficiency if a 96 per cent enriched B_{10} isotope is used?

REFERENCES

1. Rossi, B., and H. Staub: "Ionization Chambers and Counters," NNES, div. V, vol. 2, McGraw-Hill Book Company, Inc., New York, 1949.
2. Wilkinson, D. H.: "Ionization Chambers and Counters," Cambridge University Press, New York, 1950.
3. Wilson, R. R., D. R. Corson, and C. P. Baker: "Particle and Quantum Detectors," Preliminary Report 7, Nuclear Science Series, National Research Council, Washington, 1950.
4. Curran, S. C., and J. D. Craggs: "Counting Tubes," Academic Press, Inc., New York, 1949.
5. Staub, H. H.: "Experimental Nuclear Physics, Part I," John Wiley & Sons, Inc., New York, 1953.
6. Scintillation counters:
   Kaltman, H.: *Natur u. Tech.*, July, 1947.
   Deutsch, M.: *Nucleonics*, **2**(3):58–59 (March, 1948).
   Jordon, W. H., and P. R. Bell: *Nucleonics*, **5**(10):30–41 (October, 1949).
   Hofstadter, R., and J. A. McIntire: *Phys. Rev.*, **80**:631 (1950).
   Hofstadter, R.: *Nucleonics*, **6**(6):70 (June, 1950).
   Curran, S. C.: "Luminescence and the Scintillation Counter," Butterworth & Co. (Publishers), Ltd., London, 1953.

Zworykin, V. K., and E. G. Ramberg: "Photo-electricity," John Wiley & Sons, Inc., New York, 1949.

7. Yagoda, H.: "Radioactive Measurements with Nuclear Emulsions," John Wiley & Sons, Inc., New York, 1949.

8. Elmore, W. C., and M. Sands: "Electronics: Experimental Techniques," NNES, div. V, vol. 1, McGraw-Hill Book Company, Inc., New York, 1949.

Van Rennes, A. B.: *Nucleonics*, **10**(7):20 (July, 1952); **10**(8):22 (August, 1952); **10**(9):32 (September, 1952); **10**(10):50 (October, 1952).

Kelley, G. G.: "Methods of Pulse Analysis," Paper P/66, Geneva Conference, 1955.

Wilkinson, D. H.: *Proc. Cambridge Phil. Soc.*, **46**, 508 (1955).

Hutchinson, G. W.: *Nucleonics*, **11**(2):24 (February, 1953).

Hutchinson, G. W., and G. G. Scarrott: *Phil. Mag.*, **42**:792 (1951).

Byington, P. W., and C. W. Johnstone: *IRE Convention Record*, **10**:204 (1955).

Higinbotham, W. A.: "Time-of-Flight Instrumentation for Neutron Spectrometers," Paper P/806, Geneva Conference, 1955.

Higinbotham, W. A.: *Nucleonics*, **14**(4):61 (April, 1956).

9. Elmore, W. C., H. Kallman, and C. E. Mandeville: Practical Aspects of Radioactivity Instruments. 1. Construction, *Nucleonics*, **8**(6):S3–12 (June, 1951).

Ladnick, W. J.: Practical Aspects of Radioactivity Instruments. 2. Testing and Servicing, *Nucleonics*, **8**(6):S13–19 (June, 1951).

Marinelli, L. D.: Practical Aspects of Radioactivity Instruments. 3. Calibration, *Nucleonics*, **8**(6):S20–32 (June, 1951).

10. Tittle, C. W.: *Nucleonics*, **8**(6):5–9 (June, 1951); **9**(1):60–67 (July, 1951).

11. Cochran, O., and C. A. Hansen, Jr.: *Nucleonics*, **5**(8):4–11 (August, 1949).

12. Parsegian, V. L.: *Nucleonics*, **5**(10):76–78 (October, 1949).

13. Trimmer, J. D., and W. H. Jordan: *Nucleonics*, **9**(4):60–68 (October, 1951).

14. Harrer, J. M.: *Nucleonics*, **11**(6):35–40 (June, 1953).

15. Weill, J.: *Nucleonics*, **11**(3):36–39 (March, 1953).

16. Lapsley, A. C.: *Nucleonics*, **11**(8):62–64 (May, 1953).

17. Buyer's Guide, *Nucleonics*, **11**(11):D2–73 (November, 1953).

CHAPTER 5

BASIC CONCEPTS OF RADIATION PROTECTION

By Gioacchino Failla

INTRODUCTION

5-1. The main purpose of this chapter is to provide a certain background of information in order that the student entering the field of nuclear engineering may acquire an intelligent appreciation of the problems involved in the protection of personnel from the injurious effects of ionizing radiation. Rules and regulations are promulgated by the National Committee on Radiation Protection in the form of Bureau of Standards Handbooks, which may be purchased from the Government Printing Office. The nuclear engineer is concerned with the application of these rules not as a "health physicist" but as a designer and operator of plants in which radiation hazard is an important factor. Therefore, he must be acquainted with the fundamental aspects of the problem.

In general the physical, chemical, biological, and medical considerations involved in radiation protection are unfamiliar to the engineer. This requires, in the first place, a certain mental readjustment on his part, especially as regards the biological and medical aspects. Because knowledge of the biological effects of ionizing radiation is insufficient at present, he will read from time to time conflicting statements made by specialists of presumably equal competence. Such statements, in general, involve opinions rather than observations, and the engineer should be able to distinguish between the two. It is admittedly true that permissible limits of exposure recommended by the National Committee on Radiation Protection involve some assumptions and extrapolations. However, the basic concepts are derived from the experience gained during the last 50 years in the long-term occupational exposure of radiologists and technicians to X rays. The recommendations embody the combined judgment of specialists in different fields who have made a study of the subject. They should be considered, therefore, as the most reliable estimates that present knowledge permits, but as subject to change at some future time.

133

CHARACTERISTICS OF THE BIOLOGICAL EFFECTS OF IONIZING RADIATION

5-2. Historical. The discovery of X rays by Roentgen in 1895 was immediately followed by a rapidly increasing use of the new rays for medical diagnostic purposes. The early workers soon became aware of severe "burns" on their hands caused by exposure to the radiation. Recognition of the tissue-destroying property of X rays soon led to their use for the treatment of cancer. Thus the study of the biological effects of X rays was given greater impetus, and at the same time the number of people working with X rays increased rapidly. Partly through ignorance and partly through enthusiasm, most of the early workers were grossly overexposed to radiation. What happened to them subsequently provides us with the best direct information applicable to the protection problem. The number of such individuals is very large, because the full consequences of overexposure were not realized for many years. Mild X-ray burns were not considered to be serious. The harm of continued overexposure of internal organs did not become apparent for many years, because (1) it develops slowly, and (2) it is not marked when the radiation is of low penetrating power, as was the case in the early days. In fact the danger of whole-body overexposure was not fully realized until about 25 years ago.

5-3. Latent Period. The late manifestation of an injury following exposure to a sufficient dose of radiation is perhaps the most striking characteristic of the biological effects of ionizing radiation. To a much smaller extent sunlight exhibits the same property, i.e., a sunburn occurs —or becomes most marked—some time after the exposure (a few hours later). Ionizing radiation in sufficient amount produces something similar to a sunburn,* but the skin redness (erythema) appears 2 to 4 weeks after the exposure. If the amount received by the skin is very large, the redness may appear within a few days, but the reaction attains its peak in 2 to 4 weeks. When a sufficient dose has been administered in one treatment, most of the changes that may be produced become apparent within the first few weeks. If the dose has not been too large, complete or almost complete recovery takes place within a few months. Sometimes complications occur much later in a tissue that apparently had recovered almost completely. In general, therefore, there are short-term and long-term effects of radiation to be considered. Following moderate doses of radiation administered at one time, the short-term effects appear, reach a maximum, and disappear more or less completely

---

* In general, ionizing radiation is much more penetrating than the ultraviolet light responsible for the sunburn, and therefore the tissues underneath the skin are also damaged.

within a period of a few months.   Under the same conditions (moderate doses) long-term effects occur only rarely, but the delay may be very long—even 25 or more years.   In the interval between the healing of the first reaction and the onset of the long-term effect, the individual may be free of symptoms attributable to the exposure to radiation.

In connection with these statements it should be borne in mind that they apply to cases receiving a *moderate* dose in a single treatment.   A dose of radiation is moderate when the short-term injury is definite but can heal (almost) completely in the course of a few months.   As the dose becomes larger the interval between short-term and long-term effects becomes shorter, and the two may merge when the dose is extremely high [1].   When exposure occurs daily or at frequent intervals for many years, the situation is quite different.   For a time no injury is discernible.   This symptom-free period may be a few months or many years, depending on the average weekly dose received.   The injury will gradually get worse if exposure to radiation continues.   It may get better if exposure is stopped, but depending on the total dose received and the length of time during which the exposure occurred, long-term effects may occur years later, too.   Obviously, the problem here is much more complicated than in the case of exposure occurring at one time or within a short interval, such as a few days.

5-4. Radiosensitivity.   All living organisms may be killed by radiation if the dose is large enough.   However, some are much more *radiosensitive* than others in the sense that they may be killed or injured by very much smaller doses.   If all living organisms are included, the range of radiosensitivity encountered in nature is greater than 100,000, virus particles being by far the most radioresistant.   The same thing applies to different cell types in the body of a multicellular organism except that the range is not so great.   Starting with a fertilized ovum, a multicellular organism is built up by repeated cell divisions.   However, much more than this takes place.   By the process of "differentiation" certain cells acquire different properties and form the organs of the body.   Thus in the adult organism cells differ enormously in structure, physiological function, etc.   Some, like germ cells, continue to divide; others, such as brain cells, never divide once they are formed.   Those of the former type are referred to as "embryonal" or "undifferentiated," while those at the other extreme are said to be "highly differentiated," "adult" cells.   In between these two there are all gradations.   The term "young" or "adult" in this connection does not refer to the actual age of a particular cell but to the degree of differentiation.   In other words, it refers to the distance in the evolutionary scale that a cell has traveled from the initial germ-cell stage.   In this sense "old" cells are in general more radioresistant than "young" ones.   Thus the blood-forming organs are radiosensitive because they contain "young" cells which divide and constantly replenish the supply of the cellular elements of the blood as they "wear" out. For the same reason the cells in the germinal layer of the skin epidermis are also radiosensitive. Adult brain is considered to be radioresistant, but irradiation of the embryo (in utero) at the time that the brain is being formed may cause severe brain abnormalities with doses that are essentially harmless to adult brain.

When the entire body is exposed to very penetrating radiation, the distribution of dose throughout the body can be nearly uniform, but some organs are affected more than others because of differences in radiosensitivity. The over-all effect on the individual is the resultant of all these organ effects and may be death when the dose is large enough. It has been estimated that the lethal dose for the average man is 400 to 500 r administered in a short time, when the whole body is exposed to high-energy X rays [2]. The latent period in this case would be approximately one month, i.e., the person would die within one month. To cause death within a few hours the dose would have to be of the order of 100,000 r, and the mechanism of the lethal effect would be entirely different.

5-5. Biological Variability. In the preceding section the estimated lethal dose was given for the *average man*. This is because all individuals are not affected equally by the same dose of radiation. Differences in response occur among individuals and even among apparently identical cells of a given type. This variation is not a unique characteristic of radiation effects, since it occurs in all cases in which a deleterious agent of any kind (physical, chemical, or biological) acts on a population of living organisms.

The extent of "biological variability" may be estimated from the usual survival curves showing the relation between the number of survivors in a large group of animals and the radiation dose administered to the group. These curves have a characteristic shape showing that a few animals are killed by relatively small doses whereas some are able to survive much larger doses. In general, the dose required to kill 10 per cent of the individuals in the large group is roughly one-half the dose required to kill 50 per cent, and the dose required to kill 90 per cent is about twice this dose. In other words, there is a fourfold spread in the lethal dose for an individual in the group even when the most sensitive 10 per cent and the most resistant 10 per cent of the individuals are excluded. If these extremes are included, the spread in lethal dose (and, therefore, in individual radiosensitivity) may well be 10-fold. The spread is larger the less homogeneous is the group, but it is always present. It is to be expected that the larger the group the more likely it is to include individuals of very high or very low radiosensitivity.

Most of the information on biological variability has been obtained from experiments in which marked injury or death was the criterion of effect. However, it is found also in the case of much less severe effects. Hence it is generally assumed that it applies to people exposed to very small doses of radiation, in which case hardly perceptible effects may be expected. In other words, in the protection of personnel from radiation, account must be taken of differences in individual radiosensitivity. It is important to note, however, that there is no true idiosyncracy to ionizing radiation; one need not fear that a very small dose, harmless to others, will cause serious injury to him.

5-6. Recovery. A living cell can withstand injury of one sort or another and can recover from its effects, provided the injury is within certain limits. Injury produced by radiation is no exception to this rule. (Genetic changes in which there is no recovery will be discussed in a separate section.) A tissue or organism may recover not only because its component cells may recover individually, but also because damaged cells may be replaced by new ones. A tissue, therefore, has greater power of recovery than the component cells, many of which may be destroyed but replaced later. Recovery of a tissue or repair of an injury by the process of cellular multiplication depends largely on the type of tissue. In the case of the brain, for instance, one cannot expect this kind of recovery, because the cells do not multiply. On the other hand, in skin, which is constantly growing to renew the worn-out surface layers of cells, marked recovery does take place.

If the dose is not too large, a tissue or an organism will recover from the effects of radiation. Remembering that the latent period for some of the effects may be very long, it is difficult at any given time to say whether complete recovery has occurred. Recovery from the effects of fairly large doses that appear within the first few weeks, may take place within a few months and appear complete. Whether complications will develop much later depends on many factors and is generally impossible to predict. It may be taken for granted, however, that some permanent changes in some tissues have occurred, if for no other reason than the irreversible effect of radiation on chromosomes and genes. Thus, some of the wartlike growths that appear in overirradiated skin are remarkably permanent and may well be the result of somatic mutation of one or more of the other cells of the skin in that region.

Recovery in the sense used above refers to the healing of an injury after it has been produced. There is another type of recovery which tends to counteract the action of ionizing radiation and decreases the degree of injury. This may be studied by determining the dose necessary and sufficient to produce a certain degree of effect when administered in one lump in a short time; then determining the total dose required to produce the same effect in kind and degree when two treatments separated by an interval of time are given. In general the longer the time interval the larger is the sum of the two doses. Thus, in the case of the skin, a single dose of 600 r of ordinary X rays produces a mild erythema, but if the treatment is given in two exposures separated by 24 hr, the total dose must be $2 \times 450 = 900$ r to produce the same degree of erythema. Therefore, the skin must have recovered partially from the effect of the first dose by the time that the second dose was administered. It is important to note that in either case the skin redness appears 2 or 3 weeks later, and therefore "recovery" in this case applies to an initial effect that eventually causes the erythema. If the interval is greater than 24 hr, the total dose increases further but at a slower rate. A similar increase in total dose is required when equal daily treatments are given. In this case the total dose with daily treatments for 1 week is twice that required when given all at once. With equal daily treatments for 1 month the total dose is nearly three times that of the single treatment.

It should not be assumed that other tissues recover at the same rate as the skin. In general, recovery is most marked in tissues in which rapid turnover of cells occurs. Very little is known about the recovery process in other tissues. When a complex organism is irradiated different tissues are affected differently because they differ in radiosensitivity. Then they recover at different rates, and the over-all effect is the resultant of numerous interactions and variables. Therefore, while the final result (e.g., death) may be the same, the mechanism is not necessarily the same, when the conditions of exposure (e.g., the length of time) are very different.

5-7. Time Factor. Because of the recovery process just described the time during which a dose of radiation is administered has, in general, a considerable influence on the biological effects produced. This is usually referred to as the "time factor." The importance of this factor in any particular case depends, among other things, on the biological properties of the cells and tissue under consideration and on the kind and degree of effect studied. One effect—gene mutations—is known not to be influenced by the time factor. That is, the same dose produces the same number of gene mutations whether it is given in a short or a long time. The effect is then said to be "dose-rate-independent" or "completely cumulative."

The time factor obviously must play an important part in radiation protection. If the time of exposure is very long, the total dose received by a person may be quite large, without appreciable harm to the individual. Or, if the daily rate at which the radiation is received is sufficiently small the exposure may be continued indefinitely. It is presumed in this case that the recovery rate keeps pace with the rate at which damage may be produced.   It is not certain that such complete compensation takes place when the daily rate of exposure is at the presently accepted permissible levels.   As already pointed out, in the case of gene mutations the effects are completely cumulative and proportional to the dose, irrespective of the time pattern of administration.   Under present conditions, however, gene mutations do not constitute the limiting factor in setting up permissible limits of exposure (see later under Genetic Effects).   When recovery is present some compensation does take place. If now account is taken of the latent period—which for long-term effects is longer the smaller the dose—one may readily envisage a situation in which the level of chronic exposure is such that the latent period for the manifestation of an injury is longer than the remaining life span of the individual.   Hence, even though there be no complete balance between damage and recovery, no injury becomes evident in the lifetime of the individual.

CRITICAL TISSUES AND EFFECTS

5-8. General.   At present the danger of overexposure to ionizing radiation is generally recognized.   Therefore, a worker is not apt to be exposed to single large doses except through accident or reckless disregard of safety procedures.   Accordingly the problem of most practical importance is the protection of personnel exposed to small daily or weekly doses of radiation for many years.   It will be seen from the foregoing discussion that the effects to be guarded against are peculiar to this mode of exposure because of the interplay of the relative radiosensitivities of the organs, rates of recovery, and latent periods for different effects.   Because X rays have been used extensively for the diagnosis and treatment of disease in man for over 50 years, many radiologists and technicians have been occupationally exposed for many years and a great deal of information on the long-term effects of such exposure is now available.   It is possible, therefore, to pick out certain organs and effects as the critical ones.

5-9. Skin and Cancer.   Continued exposure of the hands to radiation in daily doses much above present permissible limits may not produce readily visible skin changes for one or more years.   Later certain abnormalities may become apparent, especially near the nails.   The skin may be slightly redder and shiny in this region.   Later it may become less pliable and less resistant to mechanical injury.   Such changes, however, can be detected only by comparison with the appearance and behavior of the same skin before exposure started, because many individuals not exposed to radiation present similar symptoms.   Readily visible changes in the skin ridges of the fingers and striations of the nails (which are well-known effects of radiation) do not occur early unless the accumulated dose is quite large. Later wartlike protuberances may appear here and there. As a rule these remain localized and grow only in a direction perpendicular to the skin surface. The skin in other areas may become thicker and leathery. It may then develop more or less persistent cracks. In other areas the skin may remain thin but devitalized, so that slight abrasion causes it to break repeatedly. These changes as a rule cause little discomfort, but they persist even after exposure to radiation has stopped. The real danger is that eventually cancer may develop in one of these abnormal skin areas. The latent period may be as long as twenty-five years or even longer. Accordingly, in so far as the skin is concerned, the aim is to limit exposure to a level that will not produce appreciable visible changes in the lifetime of the individual, in order to prevent the eventual development of cancer.

5-10. Blood-forming Organs and Leukemia. It is known that the incidence of leukemia in radiologists is significantly higher than in other physicians. Since the blood-forming organs are known to be very radiosensitive and leukemia has been produced in experimental animals by whole-body exposure to X rays, it may be taken for granted that the higher incidence of leukemia in radiologists resulted from continued exposure to radiation many years ago, before this hazard was recognized. As in the case of cancer of the skin, the latent period is very long, and it may be inferred that some permanent alteration of the blood-forming organs must have occurred long before the leukemic process became evident. Therefore, continued exposure to radiation should be limited to a level that causes no appreciable permanent deleterious changes in the blood-forming organs which may later produce leukemia.

Alterations of the blood-forming organs cause changes in blood counts, if they are sufficiently marked. However, since wide fluctuations in blood counts occur normally among different individuals and in the same person from time to time, without exposure to radiation, no great reliance can be placed on the detection of slight damage to the blood-forming organs by the usual blood-count determinations. In fact, if the level of exposure is low enough to prevent appreciable damage to the blood-forming organs, no significant changes in blood count should be expected. By the same token, routine blood counts cannot be used to control exposure to radiation. In a properly operated laboratory or plant, periodic blood counts are largely of psychological and legal value.

5-11. Gonads and Sterility. The gonads in men and women are quite radiosensitive, and overexposure may lead to temporary or permanent sterility.* Again the effect may be long delayed. Since it is estimated

* Genetic effects will be discussed later.

that the sensitivity of the gonads with respect to sterility or impaired fertility is not so high as that of the blood-forming organs with respect to the eventual development of leukemia, the level of exposure that is satisfactory for the blood-forming organs is also satisfactory for the gonads.

5-12. Lens of the Eye and Cataracts.   Radiological experience in the treatment of patients or in occupational exposure does not indicate that the lens of the eye is particularly sensitive to X rays.   However, some physicists exposed to neutrons have developed eye cataracts (cloudiness of the lens) without showing appreciable skin changes or permanent loss of hair.   Also, the incidence of cataracts among the survivors of Hiroshima and Nagasaki has been high.   Hence the lens of the eye must be considered to be a critical tissue in the case of exposure to ionizing radiation in general.   For occupational exposure it is assumed that the radiosensitivity of the lens of the eye is no greater than that of the blood-forming organs.

5-13. Whole Body.   The above-mentioned organs are particularly sensitive to radiation, and the long-term effects produced therein are serious.   For these reasons they constitute the critical organs and set the limits of exposure.   It should not be assumed, however, that other organs and tissues are not affected.   When the whole body is irradiated, all tissues are affected to some extent.   With small doses no tissue may be observably damaged but there may be some consequent over-all effect. It is suspected that one such effect may be a shortening of the life span.

Experiments with laboratory animals exposed daily to radiation at different dose levels show a shortening of the average life span.   However, this is statistically significant only when the daily dose is much in excess of permissible limits.   Of course, assuming that the same relation holds for the lower levels of exposure, one can always estimate by extrapolation the shortening of the life span at a low level of exposure. This procedure seems hardly justified at the present time, in view of the almost complete lack of knowledge of the *mechanism* of the body reaction to long-continued exposure at very low levels.   It has been thought that at some point the rate of recovery and repair essentially balances the rate of injury and no appreciable shortening of the life span occurs.   It is generally believed that this is the case in the range of presently accepted permissible limits of exposure.

Recently, however, the deaths of 82,441 physicians reported in the *Journal of the American Medical Association* from Jan. 1, 1930, until Dec. 31, 1954, were reviewed by Shields Warren.   In physicians grouped according to possible radiation exposure the average age of death was as follows:

Years

No known contact with radiation... 65.7
Some exposure (dermatologists, gastroenterologists, tuberculosis specialists, urologists)... 63.3
Radiologists.. 60.5
U.S. population over 25 years of age.................................... 65.6

Of course, it is not known what total dose of radiation these radiologists received in the course of their work, but it may be assumed to be several hundred roentgens on the average. Therefore, life shortening must now be taken into account in setting up permissible limits of exposure to radiation.

GENETIC EFFECTS

5-14. It has been experimentally demonstrated in a variety of living organisms, including mammals, that genes and chromosomes can be damaged by ionizing radiation. When damage occurs in the germ cells, the result of the injury manifests itself in the descendants of the individual in some future generation. Radiation-induced mutations are essentially the same as those that occur spontaneously in nature, but their frequency is increased. Accordingly, the problem is to limit exposure up to and during the reproductive period to a low enough level to prevent an unwarranted increase in the mutation rate.

As far as a given individual is concerned the total dose to the gonads accumulated up to the time of conception of a child may be quite high without apparent damage to the child or his offspring. However, he will carry a somewhat increased load of undesirable genes, which he transmits to all his descendants. The result is that the number of undesirable genes (both spontaneous and radiation-induced ones) in the general population is increased. In the course of many generations the proportion of people with slight or marked abnormalities will be greater than at present, other conditions being equal. Since we are dealing here with the effect on a large population, it is immaterial whether the undesirable genes are introduced into it by a small number of individuals whose gonads have received large doses or by a larger number who have received correspondingly smaller doses.* Up to now the number of persons occupationally exposed to radiation is small in comparison to the entire population, and therefore genetic damage to the race has not been the controlling factor in setting up permissible limits of exposure to ionizing radiation. However, as the nuclear-energy field expands, a reappraisal of the situation becomes more necessary and the permissible

---

* This is because the number of damaged genes increases linearly with the dose. It should be remembered in this connection that the effect is dose-rate-independent.

limits of exposure have to be lowered.   Thus, as a result of discussion
by committees of experts in the United States and in England, limits
have recently been set on the average per capita dose to the gonads
accumulated up to age thirty (see Sec. 5-35).   It behooves the engineer
to bear this in mind in the design of nuclear reactors and similar devices
producing ionizing radiation.

TERMINOLOGY AND UNITS

5-15. An understanding of the technical terms used in radiation protection is obviously essential.   Since the field is relatively new, some of
the terms are unfamiliar; some have been defined differently by different
authors and even by the same author at different times.   It is well, therefore, to review the basic concepts involved in the commonly used terms
and units.

5-16. Intensity of Radiation.   The intensity of radiation at a given
point in a beam of radiation is the amount of energy passing through unit
area per unit time, the area being perpendicular to the line of propagation
of the radiation at the point in question.   It is usually expressed in ergs
per square centimeter per second or in watts per square centimeter.   If
the number of photons passing per square centimeter per second is I
and the energy of each photon is E, the intensity of radiation is IE.

5-17. Quantity of Radiation.   Quantity of radiation is the time
integral of intensity of radiation.   It is the total energy that has passed
through unit area perpendicular to the beam at the point in question.
It is expressed in ergs per square centimeter or watt-seconds per square
centimeter.

It is important to note that both intensity and quantity of radiation
refer to *unit area*.   Therefore, the cross section of the beam is not
involved.   That is, a beam with a cross-sectional area of 1 mm^2 and one
of infinite cross section having the same intensity deliver the same quantity
of radiation at an irradiated point.

5-18. Dose of Radiation.   The term "dose" is loosely used to represent
an "amount" of radiation.   The "amount" is variously defined, and
often it is impossible from the context to tell which is meant, but it always
represents an amount per unit area, per unit volume, or per unit mass.
Thus, the area through which the radiation passes or the volume of the
irradiated tissue is not involved in a simple statement of the dose.   For
example, the dose may be numerically the same whether one finger or the
entire body of a person is irradiated.   The concepts that lead to different
definitions of "amount," and hence of dose, will be discussed below.

5-19. The Roentgen r.   To produce any change—physical, chemical,
or biological—in a medium, energy must be available.   Therefore, the

biological changes produced in a tissue must be attributed ultimately to the energy absorbed by the tissue from the radiation traversing it. Since X rays vary greatly in penetrating power, the energy absorbed by tissue from different beams having the same *intensity* (as defined above) but different wavelengths may be entirely different in magnitude. Hence, if dose is defined in terms of *quantity of radiation*—i.e., energy passing through the tissue at any given point—the absorbed energy may be very different and the biological effect produced will be strongly dependent on the wavelength of the radiation. On the other hand, if the dose is expressed in terms of energy absorbed by the tissue, the correlation between dose and effect may be expected to be less dependent on wavelength.

The determination of the energy absorbed by tissue is a complicated procedure that was not practical some 25 years ago. Therefore, it was decided to use ionization in air as the basis for a unit of dose of X rays. Since ionization is related to the energy absorbed by air and the absorbing properties of air and soft tissue are not very different in the wavelength range that was of radiological interest at the time, it was thought that this would be a good practical compromise.

The roentgen was first adopted internationally in 1928. Its definition was modified in 1937 and is as follows: The roentgen is the quantity of X or γ radiation* such that the associated corpuscular emission per 0.001293 g of air produces, in air, ions carrying 1 esu of quantity of electricity of either sign. The terminology is rather involved, but the concept is simple.

It may be pointed out first that 0.001293 g is the mass of 1 cm³ of air at STP. "Ions carrying 1 esu of quantity of electricity of either sign" means that the total charge carried by the positive (or negative) ions is 1 esu. Whence the number of ion pairs is the reciprocal of the electronic charge, $1/(4.80 \times 10^{-10}) = 2.08 \times 10^9$. "Associated corpuscular emission" refers to the secondary electrons liberated per 0.001293 g (1 cm³) of air, which then produce a total of 2.08×10^9 ion pairs in their passage through air (anywhere, not in the same cubic centimeter of air). The fundamental concept then is that a quantity of X radiation interacting with air liberates secondary electrons at such a rate per cubic centimeter and of such total energy that they eventually produce 2.08×10^9 ion pairs.

The interaction process is obviously that of true absorption whereby the photon delivers all or part of its energy to an electron. Therefore, a definite amount of energy is associated with the roentgen. This is per-

---

* The properties of X rays and γ rays are the same (see p. 38). Only the term "X rays" is used in this chapter.

haps more evident from the fact that a definite amount of energy must be spent in producing 2.08×10^9 ion pairs. The ionization potentials of the constituents of air are in the neighborhood of 15 volts (see page 28). So the energy *required* to produce an ion pair in air is about 15 ev. On the other hand, experiment shows that the average energy *lost* by a fast electron per ion pair it produces in air is approximately 34.0 ev.* Consequently somewhat more than one-half the energy lost by the electron goes into excitation of atoms and molecules, or processes other than ionization. Since probably this energy, or at least the major portion thereof, is also biologically active, it is customary to include it in assessing the energy equivalent of the roentgen. *Assuming* that the average energy lost by an electron in producing an ion pair in air is 34.0 ev and that it is independent of the energy of the electron,

$$1 \text{ r} = 34.0 \times 2.08 \times 10^9$$
$$= 7.07 \times 10^{10} \text{ ev per cubic centimeter of air at STP}$$
$$= 5.47 \times 10^{13} \text{ ev per gram of air}$$
$$\doteq 0.113 \text{ erg per cubic centimeter of air at STP}$$
$$= 87.6 \text{ ergs per gram of air}$$

In the photoelectric process the secondary electron loses some energy in coming out of the atom, and therefore less than that imparted to it by the photon is available for ionization. However, for light elements and X rays of ordinary wavelength the difference is small and is made up in part by the energy of the Auger electrons set free by internal conversion. Therefore, 87.6 ergs may be taken to represent the energy imparted by 1 r of X radiation to the secondary electrons per gram of air in which the interaction occurs. This is the energy *abstracted* from 1 r of X rays per gram of air at a given point in the radiation beam. When the radiation is in equilibrium with its secondary electrons, this is also the energy *locally absorbed* per gram of air at the point in question.

The foregoing applies to X radiation of such photonic energy that the interaction with air is largely through photoelectric and Compton processes. When the photon energy is so high that the pair-formation process becomes important, a considerable fraction of the energy of the interacting photons is utilized to create electron-positron pairs and hence is not available to produce ions (and register as roentgens). Since the energy required to create the electron-positron pair is known (1.022 Mev), it is possible to calculate the kinetic energy of the electron and positron that is available for ionization, excitation, etc. However, the practical difficulties of making measurements in terms of the roentgen increase as the photon energy gets above 2 or 3 Mev, and results are uncertain.

* Until recently the accepted value was 32.5 ev, which made 1 r = 83.7 ergs per gram of air and 1 rep = 93 ergs per gram of tissue.

It will be seen from this discussion that if Q is the *quantity of radiation* at a given point in a beam of X rays, in ergs per square centimeter, and if μ_r represents the fraction of the energy abstracted from the photons per centimeter of STP air they traverse *and* appearing as kinetic energy of the secondary electrons (and positrons),* then $Q\mu_r$ is the energy of the associated corpuscular emission per 0.001293 g (1 cm³) of air available to produce ions. Since when this energy is 0.113 erg we have 1 r of radiation, the amount of radiation in roentgens at the point in question is $Q\mu_r/0.113$. Therefore, the relation between *dose in roentgens* and *quantity of radiation* is $D_r = Q\mu_r/0.113$ when μ_r is in terms of reciprocal centimeters and Q is in ergs per square centimeter.

This relation is useful to the nuclear engineer in the design of protective barriers, since he can calculate the dose in roentgens per unit time when he knows the intensity of radiation and its spectral distribution, i.e., the number and energy of the photons passing through unit area per unit time. The coefficient μ_r is not strongly dependent on photon energy in the range of 0.08 to 3 Mev. Numerical values are available in the literature [3]. Conversely he can determine the photon flux if he knows the photon energy and measures the γ radiation in roentgens.

5-20. Absorbed Dose. Although the definition of the roentgen is not in terms of energy absorbed by air, one may think of a dose expressed in roentgens as corresponding to a certain amount of kinetic energy that the radiation is capable of delivering per cubic centimeter (or per gram) of air if present at the point of interest. In this sense then the roentgen is a unit of "absorbed dose." In the ordinary range of X-ray wavelengths, air and soft tissue absorb about the same amount of energy on a mass basis. The respective values per roentgen generally given in the literature are 84 and 93 ergs/g, based on 32.5 ev per ion pair in air. Because air and tissue do not have the same atomic composition, the absorption of tissue differs considerably from 93 ergs/(g)(r) in certain wavelength regions. Since in radiobiology the energy absorbed in tissue is obviously more significant than that absorbed in air, it is desirable to use energy absorbed in tissue as the basis of dose. To avoid ambiguity this is called "absorbed dose."

5-21. The Rad. All ionizing radiations produce the same kinds of biological effect, although there may be differences in *degree*. The comparison is generally made in terms of absorbed dose in the tissue and at the locus of interest. It is desirable, therefore, to have a unit of

---

* It will be seen that μ_r is a special kind of linear absorption coefficient appropriate to the definition of the roentgen and varies with the wavelength of the radiation. μ_r is (practically) equal to $\tau + \sigma_a + [(E - 1.022)/E]\pi$, in which τ is the photoelectric absorption coefficient, σ_a is the true absorption coefficient in the Compton process, and π is the pair formation absorption coefficient.

absorbed dose.   The International Commission on Radiological Units in 1953 adopted the rad as such a unit, in the following words: "Absorbed dose of any ionizing radiation is the amount of energy imparted to matter by ionizing particles per unit mass of irradiated material at the place of interest.   It shall be expressed in 'rads.'   The *rad* is the unit of absorbed dose and is 100 ergs per gram."

5-22. The Rep.   In the absence of an internationally accepted unit of dose of ionizing radiation other than X and γ rays (to which the roentgen was limited by definition), the "rep" (roentgen-equivalent-physical) was used extensively, especially in this country.   Several definitions of the rep have appeared in the literature, but in essence it may be considered to be a unit of absorbed dose with a magnitude of 93 ergs per gram of tissue.

For purposes of radiation protection, the difference in magnitude between the rep and the rad (93 and 100 ergs/g, respectively) is insignificant.   Therefore, the numerical values of permissible doses expressed in rads are the same as those expressed in reps heretofore.

5-23. Dosage Rate.   Dosage rate, or dose rate, is the time rate at which a dose of radiation is administered.   It is expressed in roentgens or rads per unit time, depending on the kind of dose involved.   In the case of X rays, the difference between intensity of radiation and dosage rate is the same as that between quantity of radiation and dose in roentgens. That is,

$$\text{Dosage rate} = \frac{J\mu_r}{0.113} \qquad \text{r/hr}$$

when J = intensity of radiation, ergs/(cm^2)(hr), and μ_r is the same as before.

Intensity of radiation and dosage rate are often used interchangeably. This is obviously incorrect, since they are dimensionally different. Intensity of radiation is ergs per *square* centimeter per second; dosage rate is (essentially) ergs per *cubic* centimeter per second.

5-24. Integral Dose and Total Dose.   As already explained, the fundamental concept of dose is that of volume or mass *concentration* of energy, i.e., ergs per cubic centimeter or per gram.   This is appropriate because the biological effect produced locally is more dependent on concentration than on the volume of tissue irradiated.   Obviously, however, from the point of view of injury to the individual the body region and total volume of irradiated tissue are of capital importance.   Therefore, in addition to the dose (as energy concentration) other pertinent factors must be specified in order to portray the true significance of an exposure to radiation.

For some purposes it is desirable to consider the energy absorbed in an organ or in the entire irradiated volume. This is called the "integral dose" and is expressed in gram-rads:

$$1 \text{ rad} = 100 \text{ ergs/g} \qquad 1 \text{ g-rad} = 1 \text{ rad} \times 1 \text{ g} = 100 \text{ ergs}$$

This expression was coined to avoid confusion with "total dose," which usually represents summation of dose (i.e., energy concentration) with respect to time. Sometimes "total dose" is used as the summation of doses delivered to the same tissue by different kinds of radiation, simultaneously or in succession. In this case the expression "combined total dose" might be preferable.

5-25. Specific Ionization. In the preceding section "energy concentration" was used in a macroscopic sense, i.e., as an average concentration in a substantial volume or mass of tissue. Actually the energy is released along the paths of the ionizing particles and is initially highly localized in a microscopic sense. It has been found experimentally that this essentially linear concentration of energy influences the biological effects produced by a given dose of radiation. In general the higher the concentration along the paths of the ionizing particles the greater is the biological effect. Since this linear concentration is demonstrated best in a cloud chamber by the condensation of water droplets on ions, it is generally called "specific ionization" and is expressed in ion pairs per centimeter of air. In radiobiology specific ionization is expressed in ion pairs per micron of water. This is really fictitious, since at present there is no way of counting ions in water. What is done is to determine the energy lost by the ionizing particle per micron of water traversed and then to calculate the number of ion pairs that the appropriate particle with this amount of energy would produce in air. This constitutes the number of ion pairs per micron of water. The difficulty may be avoided by expressing this attribute of an ionizing particle in terms of the energy loss per unit length of its path in the material of interest, e.g., thousands of electron volts per micron of water. This notation is gaining favor, and the term "linear energy transfer" (LET) has been coined.

In practice it is quite difficult to determine the specific ionization (or LET) associated with a tissue dose of radiation. Since ionizing particles of different energies are always present, one has to deal with a spectrum of specific ionization. The usual procedure is to calculate an average or mean value [4]. This is satisfactory so long as the range of specific ionizations involved is not too wide or is in the region in which biological effects are not strongly dependent on specific ionization.

The theoretical minimum specific ionization for a singly charged particle (electron) is 5.7 ion pairs per micron of water. For the secondary electrons of X rays produced at voltages of 30 to 200 kv, the average specific ionization does not vary much and in round figures may be taken to be 100 ion pairs per micron of water.*   In this region also biological effects are not strongly dependent on specific ionization.   For the recoil protons produced in tissue by moderately fast neutrons the average specific ionization may be between 500 and 1500 ion pairs per micron.   For naturally occurring α particles the value may be 3000 to 7500 depending on the part of the range considered.

5-26. Relative Biological Effectiveness (RBE).   Because X rays have been available for a long time and have been used extensively in medicine, most of the information on the biological effects of ionizing radiation has been obtained by the use of this type of radiation.   This is particularly true of information directly applicable to the protection problem. Therefore, it is convenient to use the biological effects of ordinary X rays as a point of reference in dealing with the effects produced by other types of ionizing radiation.   It is customary to make the comparison in terms of the absorbed dose required to produce the same degree of a given effect, when all other conditions are the same and only the specific ionization is different.   Since specific ionization is a property of the radiation, this procedure determines the relative biological effectiveness (RBE) of different types of radiation—or the same type with different specific ionizations—with respect to the X rays used as reference point.   For protection purposes it is convenient to take X rays with a specific ionization of 100 ion pairs per micron of water as the standard.

If, under otherwise comparable conditions, the lethal doses for mice are 600 rads of the standard X rays and 150 rads of fast neutrons of a particular energy, the RBE of these neutrons is 4.   Or, if the X-ray dose and the appropriate RBE are known, one may calculate the equivalent neutron dose in rads by the same relation:

$$\text{RBE} = \frac{D_x}{D_n} \qquad \text{or} \qquad D_n = \frac{D_x}{\text{RBE}}$$

It is very important to bear in mind that the RBE of radiation *with the same specific ionization* varies with the kind and degree of effect; the type of cell, tissue, organism; their physiological states; etc.; and with extrinsic factors such as the time distribution of the dose ("time factor"). Therefore, it is not possible to assign a unique value of RBE to a given type of radiation applicable to all effects and all conditions of irradiation, even when the specific ionization is the same.   This is best illustrated by an experiment [5] in which large groups of mice were exposed either to X rays or to fast neutrons.

* This is according to values generally quoted in the literature heretofore.   The subject is undergoing revision at the present time (see Ref. 4).

When the doses were adjusted so that the mortality in a period of several months was the same, most of the animals exposed to neutrons developed lens cataracts before they died, whereas very few of those exposed to X rays showed this effect. Accordingly, under these conditions the RBE of fast neutrons for cataract formation is higher than that for lethal action. Furthermore, when similar mice received daily treatments over a long period of time, the difference between the RBE's for lethality and for cataract formation was even greater. This is an example of the influence of the time factor on the RBE. In man the lens of the eye is, also, particularly sensitive to radiation of high specific ionization (fast neutrons) and for this reason it constitutes a critical organ, as has already been mentioned.

5-27. The Rem. In practice a person may be exposed to radiation of widely different specific ionizations (e.g., γ rays and fast neutrons) simultaneously or successively. The combined total dose in rads in this case is not a true index of the biological damage that may be produced, because the portion contributed by radiation of high specific ionization is biologically more effective. A simple expedient is to reduce all contributions to the dose to a "common denominator" by taking as a basis X rays with a specific ionization of 100 ion pairs per micron of water and using the appropriate RBE's. Thus, if the combined total dose is D rads and the component parts of different specific ionizations are A, B, and C rads, the biologically effective combined total dose is

$$D_{eff} = A \times (\text{RBE})_a + B \times (\text{RBE})_b + C \times (\text{RBE})_c$$

where D_{eff} represents the X-ray dose biologically equivalent to D and $(\text{RBE})_a$ represents the proper value of the RBE for the radiation contributing portion A of the dose, etc.

It has been found convenient to express doses of radiation of different specific ionizations in terms of a unit that embodies both the magnitude of the dose and its biological effectiveness. The unit has been called the "rem" and in the past has been defined in terms of the rep. The adoption of the rad to replace the rep necessitates a slight change in the magnitude of the rem, which may be defined as follows: The rem is the quantity of any ionizing radiation such that the energy imparted to a biological system (cell, tissue, organ, or organism) per gram of living matter by the ionizing particles present in the locus of interest has the same biological effectiveness as an absorbed dose of 1 rad of X radiation with average specific ionization of 100 ion pairs per micron of water, in the same region. Accordingly,

$$\text{Dose in rem} = (\text{dose in rads}) \times \text{RBE}$$

The rem is, therefore, a unit of effective absorbed dose of varying magnitude in terms of rads, depending on the RBE.   A unit of "varying magnitude" sounds peculiar, but it is an acceptable unit nevertheless. In this sense it is analogous to the curie, defined as follows: The curie is a unit of effective quantity of radioactive material ("radioactivity") of varying magnitude in terms of grams, depending on the disintegration constant and the atomic weight of the material.

$$\text{Amount in curies} = (\text{amount in grams}) \times K\frac{\lambda}{A}$$

where $K = \dfrac{6.02 \times 10^{23}}{3.7 \times 10^{10}} = 1.627 \times 10^{13}$

λ = disintegration constant, $\sec^{-1}$

A = atomic weight

The internationally accepted definition of the curie is as follows: "The curie is a unit of radioactivity defined as the quantity of any radioactive nuclide in which the number of disintegrations per second is 3.700×10^{10}." The practical advantage of this definition is that one need not know the disintegration constant, the atomic weight, and the mass of the material to determine its amount in curies; but in essence the two definitions are equivalent.

PERMISSIBLE LIMITS OF EXPOSURE TO RADIATION FROM EXTERNAL SOURCES

5-28. Permissible Dose.   The concept of a "tolerance dose" involves the assumption that if the dose is lower than a certain value—the threshold value—no injury results.   Since it seems well established that there is no threshold dose for radiation-induced gene mutations, it is evident that there is no such thing as a tolerance dose of radiation, when all possible effects on the individual and future generations are included. Accordingly, it is preferable to speak of a "permissible dose" (rather than a tolerance dose), and this is now the generally accepted terminology.

Permissible dose is defined as a dose of ionizing radiation that, in the light of present knowledge, is not expected to cause *appreciable bodily injury* to a person at any time during his lifetime.   As used here "appreciable bodily injury" means any bodily injury or effect that the average person would regard as being objectionable and/or competent medical authorities would regard as being deleterious to the health and well-being of the individual.   It should be noted that the *possibility* of some minor injury is not excluded, but the *probability* of such injury occurring during the lifetime of the individual is intended to be extremely low.   The maximum value of permissible dose that just meets the requirements of the definition is called a "maximum permissible dose."

In view of the dependence of biological effects of radiation on numerous factors besides the dose, it is not practical to stipulate maximum permissible doses for the innumerable conditions under which exposure may occur.   In practice the most important case is that of whole-body intermittent or continuous exposure at essentially constant rate over an indefinite period of years.   In this case the limit must be set on the rate at which the person receives radiation.   For practical reasons 1 week has been chosen as the unit of time.   When the exposure continues for many years, variations of fractional doses and dosage rates occurring within 1 week may be assumed to be unimportant.

A *permissible weekly dose* is a dose of ionizing radiation accumulated in 1 week of such magnitude that, in the light of present knowledge, exposure at this weekly rate for an indefinite period of time, is not expected to cause *appreciable bodily injury* to a person at any time during his lifetime.   By analogy with the previous terminology there is in this case a "maximum permissible weekly dose."

5-29. Basic Permissible Limits.   The concept of critical organs and effects requires that limits for the doses received by these organs be stipulated.   Some organs, such as the skin and the blood-forming organs, are widely distributed in the body, and it is difficult to decide what constitutes the organ dose when the organ is not uniformly irradiated. Obviously, an averaging process is involved whereby at every point of the organ account is taken of the local dose and its potentiality for harm to the organ and consequently to the individual.   This is impractical if not impossible.

Cancer of the skin arises locally in areas that have been damaged by radiation.   Leukemia possibly originates locally also and becomes widespread later.*   Therefore, a limit must be set to the above-mentioned averaging process to prevent the local accumulation of harmful doses. This is done by assigning a "significant volume" to each critical organ and restricting the averaging process to this volume.   Of course, the significant volume of interest is in the functional region of the organ subjected to the highest dosage rate during the exposure.

When the entire body is exposed from all directions to very penetrating radiation, all the organs receive throughout approximately the same dose and the risk is greatest.   Accordingly, in this case the permissible critical

---

* Experimentally it has been found that leukemia develops in mice when the whole body is exposed to X rays.   If part of the body (even a leg) is shielded from the radiation, the incidence of leukemia is less.   It would seem, then, that in this case the local dose is not so important; or at least, something other than the local dose is involved.   Assuming that this is also true for man, it provides justification for allowing some blood-forming tissue in the hands, feet, etc., to receive weekly doses that are much in excess of the basic permissible weekly dose for the blood-forming organs in general (see Table 5-4 for local doses to the hands, etc.).

organ doses should be at their minimum values.   They are referred to as "basic permissible weekly doses" for the critical organs.   Then the corresponding *maximum* permissible weekly doses are either equal to or greater than the basic ones, depending on the stipulated conditions of exposure.   When they are higher than the basic ones, some restrictions as to the conditions of exposure are always involved (e.g., exposure limited to the hands, exposure to X rays only, or exposure to radiation of very low penetration).

Basic permissible weekly doses for the critical organs have been set essentially on the basis of radiological experience acquired over a period of more than 50 years.   However, this experience is limited to exposure to X rays.   Therefore, for other types of radiation, the following procedure has been followed.   The logical objective is to make the potential risk of exposure to radiation of any type and energy equal to that involved in exposure to ordinary X rays under comparable conditions.   Having decided what the basic permissible weekly doses in *roentgens* should be for exposure to ordinary X rays, the corresponding permissible weekly doses in *rads* are obtained by assuming numerical equality of the two. In other words, if the permissible weekly dose for the gonads is 300 mr, it is assumed that the absorbed dose is 300 mrad.   The assumption is entirely justified, since the actual difference is of no significance for purposes of protection.   Now, in order to make the potential risk of exposure to radiation of any type and energy *equal to* that involved in exposure to ordinary X rays, it is only necessary to say that the permissible weekly doses *in rems* shall be numerically equal to the corresponding doses in rads.   This follows directly from the definition of the rem, and no approximations are involved.   However, in any practical application it is necessary to know the dose in rads that corresponds to the specified dose in rems, and this requires a knowledge of the appropriate RBE. For true equality of risk the RBE's for the critical organs should probably be different.   However, if one takes the highest RBE and applies it to all critical organs, then the over-all risk is *no greater* than that involved in exposure to ordinary X rays under comparable conditions.

In Table 5-1 are given the basic permissible weekly doses for the critical organs in millirems, and other pertinent data.   The average depth of the critical organ below the surface of the skin is useful for purposes of calculation, when the body is exposed to radiation from external sources. Thus, if the weekly skin dose is 600 mrem and the geometric conditions of exposure are such that the absorbed dose at a depth of 5 cm is one-half the skin dose, it is assumed that the blood-forming organs receive on the average a weekly dose of 300 mrem.   Since some parts of the blood-forming organs (e.g., some lymph nodes) are close to the body surface, it is evident that in this case they may receive nearly twice the basic

TABLE 5-1. BASIC PERMISSIBLE WEEKLY DOSES FOR THE CRITICAL ORGANS,
SIGNIFICANT VOLUMES, AND ASSUMED AVERAGE DEPTHS

Organ	Basic permissible weekly dose, mrem	Significant volume (or area) in the region of highest dosage rate	Assumed average depth (for purposes of calculation)
Skin....................	600	1 cm²	7 mg/cm²
Blood-forming organs....	300	1 cm³	5 cm
Gonads:			
Ovaries..............	300	10% of total volume	7 cm
Testes...............	300	10% of total volume	Variable depending on conditions of exposure; minimum, 1 cm
Lenses of the eyes.......	300	Volume of either lens	3 mm

permissible weekly dose in a volume that is much larger than the significant volume. It should be noted, however, that in such cases no part of the blood-forming organs can receive very large doses because the permissible dose for the skin sets a limit to it. Hence, the concept of an "average depth" is satisfactory and serves a useful purpose in practice.

Table 5-2 gives the RBE values for different specific ionizations (or LET's). Note that these values are intended to apply to all critical organs. Note, also, that the RBE is assumed to be equal to 1 for X rays, electrons, and positrons of any specific ionization. These values should

TABLE 5-2. RECOMMENDED VALUES OF THE RELATIVE BIOLOGICAL EFFECTIVENESS
(RBE) FOR DIFFERENT SPECIFIC IONIZATIONS APPLICABLE TO EXPOSURE TO
RADIATION FROM EXTERNAL SOURCES

Average specific ionization, ion pairs per micron of water	RBE	Average linear energy transfer to water, kev per micron of water
X Rays, Electrons, and Positrons		
Any specific ionization.........	1	(Any)
Heavy Ionizing Particles		
100 or less	1	3.5 or less
100–200	1–2	3.5–7.0
200–650	2–5	7.0–23
650–1500	5–10	23–53
1500–5000	10–20	53–175

be considered to be rough estimates (hopefully on the safe side) based on available data.   The present state of the art does not warrant fine distinctions.   In view of this it is perhaps inappropriate to give RBE values for narrow ranges of specific ionization.   This was done intentionally by the Subcommittee on Permissible Dose from External Sources of Ionizing Radiation of the National Committee on Radiation Protection* in order to avoid the heretofore common practice of assigning RBE values to "fast neutrons," "alpha particles," etc.   Modern particle accelerators can produce heavy ionizing particles (particularly protons and deuterons) of such high energy that the specific ionization in most of their range is comparable to that of electrons.

As previously indicated,

$$\text{Permissible dose in rads} = \frac{\text{permissible dose in rem}}{\text{RBE}}$$

In shielding design it may be assumed that the basic permissible weekly doses for the critical organs are not exceeded when the individual is exposed for 40 hr to the fluxes listed in Table 5-3 (but see page 168).

TABLE 5-3. PERMISSIBLE NEUTRON FLUXES IN 40 HR/WEEK EXPOSURE

Neutron energy, Mev	Neutron flux, neutrons/(cm²)(sec)
Thermal	2000
0.0001	1550
0.01	1400
0.1	250
0.5	90
1	60
2	52
3–10	50

5-30. Modifying Factors.   Since the risk of potential radiation damage depends on the conditions of exposure, maximum permissible weekly doses substantially higher than the basic ones may be set up for special cases, without appreciably increasing the risk.   This is desirable for practical reasons.   A good example is the handling of relatively small amounts of radioactive material.   In this case the dosage rate near a "point" source may be quite high, but since it decreases rapidly with distance, the whole-body dose would be much smaller than the local dose.   If the basic permissible weekly dose of 600 mrem in a significant area of the skin (1 cm²) in the region of highest dosage rate were to govern the exposure, only a small portion of the hand would be permitted to receive this dose.   Obviously, the risk would be much less than in the case in which all the skin of the body receives 600 mrem.   Therefore, the

---

* Hereafter this will be referred to as Subcommittee I of NCRP.

maximum permissible weekly dose has been set at 1500 mrem in the skin, for cases in which only the hands and forearms are exposed. It should be noted that this applies to the significant area of the skin in the region of highest dosage rate and is not an average for the whole skin of the hands and forearms.

Another special case of practical importance is that of exposure to radiation of very low penetrating power (e.g., β rays) for which the half-value layer (HVL) is less than 1 mm of soft tissue. Obviously, the blood-forming organs receive practically no radiation, and the skin is essentially the only critical organ involved. In this case the maximum permissible weekly dose in the skin of the whole body or any portion thereof has been set at 1500 mrem. Other special cases are considered in Handbook 59, to which the reader is referred.

A modifying factor of controversial nature is *age*. From the point of view of genetics it is obvious that after the normal reproductive period (say age forty-five) the exposure could be at a much higher rate than before. Also, if exposure starts late in life, it is reasonable to suppose that a higher permissible weekly dose would be satisfactory. The trouble is that it is impractical to apply a higher rate only to those who have not been exposed before age forty-five. If no such distinction is made and exposure continues at a substantially higher rate after age forty-five, the question arises as to whether the increased risk of radiation damage manifestable in the lifetime of the individual is still within the range envisaged in the definition of permissible weekly dose. In the absence of factual data, opinions differ. In the ultimate analysis the matter hinges on the magnitude of the "factor of safety" included in the existing permissible limits of exposure, for adults of any age. It is the opinion of the author that the factor of safety is not large but nevertheless is sufficient to permit doubling the permissible weekly dose after age forty-five. For one thing, while genetic considerations did not play a dominant part in setting up the present limits of exposure, they did play some part, and therefore the limits are lower than they would otherwise be. Then, from a different angle, the long latent period for the manifestation of injury tends to discount exposure occurring late in life.

This may become a matter of considerable importance in the future and deserves further exploration. As the uses of nuclear energy increase and more and more people are exposed to ionizing radiation, genetic factors become more important and the permissible limits of occupational exposure for *young persons* will have to be lowered. Since the cost of radiation protection in nuclear energy installations and in the use of devices emitting ionizing radiations is considerable, additional restrictions will be of substantial economic significance. The solution may very well be the assignment of older persons to tasks involving exposure to radiation at a higher rate than will be permitted for young people.   As a start in this direction Subcommittee I of NCRP has recommended doubling the permissible weekly doses at age forty-five and over (with some minor restrictions).

5-31. Weekly-dose Fluctuations.   For many years in this country the permissible limit of continued exposure (the so-called "tolerance dose") was 0.1 r/day.   Strictly speaking a dose in excess of 0.1 r in 1 day constituted "overexposure," with the implied connotation (at least in the mind of the layman) that it was harmful.   Obviously even marked fluctuations of such a small dose could not be dangerous if the average over a reasonable period of time did not exceed 0.1 r/day.   Soon after the war it was decided by the NCRP to lower the exposure level and at the same time increase the time unit to 1 week.   This permits a person to receive the entire weekly dose (e.g., 300 mr) in a short time, provided no further exposure occurs during the rest of the week, which is defined as seven consecutive days rather than a calendar week.   This has obvious practical advantages especially from the point of view of the employer.   Naturally this change led to the question, Why not make the reckoning time 1 month or 1 year?   The answer is that, in intermittent exposure for an indefinite period of years, chief reliance is placed on an approximate balance between the rate at which damage is produced and the rate at which recovery takes place.   When the dose is received in large "lumps," the balance is certainly affected adversely.   Therefore, it is prudent not to make them too large.

The period of dose reckoning may be prolonged considerably by taking compensatory measures.   One such scheme has been recommended by Subcommittee I of NCRP in the following words: "In exceptional cases in which it is necessary for a person to receive in one week more than the permissible dose, the unit of time may be extended to 13 weeks, provided that the dose accumulated during a period of any seven consecutive days does not exceed the appropriate permissible weekly dose by more than a factor of three and provided further that the total dose accumulated during a period of any 13 consecutive weeks does not exceed 10 times the permissible weekly dose."   In an extreme case this means that the person may be exposed at three times the weekly rate for $3\frac{1}{3}$ consecutive weeks if in the remaining $9\frac{2}{3}$ weeks of the period he is not exposed at all.   In compensation the total accumulated dose during the period of 13 weeks has been reduced in the ratio of 10:13.

This procedure provides a sound basis for dealing with situations in which occasionally the permissible weekly dose may be exceeded, namely, with cases of "technical overexposure."   If it is found that a person in a week has received considerably more than the appropriate maximum permissible base, steps can be taken to reduce the exposure in the next few weeks to make up for the excess. Since the compensation scheme outlined above allows a temporary threefold increase in the permissible weekly dose, it provides all the latitude justifiedly needed in practice. Compliance with the spirit and the letter of protection rules is thus facilitated without additional hazard to the individual, since in this case the total dose he receives in a period of 13 weeks is considerably less. If dose compensation in the manner outlined above cannot be accomplished in a period of 13 weeks, the protective measures employed must be glaringly deficient.

5-32. Summary of Maximum Permissible Weekly Doses. The values listed in Table 5.4 have been culled from Handbook 59 and include some

TABLE 5-4. MAXIMUM PERMISSIBLE TOTAL WEEKLY DOSES IN
CRITICAL ORGANS UNDER VARIOUS CONDITIONS OF
EXTERNAL EXPOSURE

Conditions of exposure			Weekly dose in significant volume in region of highest dose rate, mrem				Air dose,* mr
Part of body	Radiation	Adult age	Skin	Blood-forming organs	Gonads	Lens of the eye	
Whole body	Any radiation	Under 45	600	300	300	300	
		Over 45	1200	600	600	600†	
Whole body	X rays only (up to 3 Mev)	Under 45	450 mrad‡	400 mrad‡	300 mrad	450 mrad‡	300
		Over 45	900 mrad‡	800 mrad‡	800 mrad‡	600 mrad	600
Whole body	Any radiation with HVL < 1 mm of soft tissue	Under 45	1500	300§	300§	300	
		Over 45	1500	600§	600§	600†§	
Hands and forearms	Any radiation	Under 45	1500	300¶	300	300	
		Over 45	1500	600¶	600	600†	
Feet and ankles	Any radiation	Under 45	1500	300¶	300	300	
		Over 45	1500	600¶	600	600†	
Head and neck	Any radiation	Under 45	1500	300¶	300	300	
		Over 45	1500	600¶	600	600†	

* "Air dose" means that the dose is measured by an appropriate instrument in air in the region of highest dosage rate to be occupied by a person, without the presence of the human body or other absorbing and scattering material.

† Including not more than 300 mrem of radiation of high specific ionization.

‡ Rough estimates of maximum values resulting from specified air doses, under practical conditions of exposure.

§ Limits for concurrent exposure to penetrating radiation.

¶ In main portion of body, not in region of highest dosage rate (hand, foot, or head).

values not specifically stated therein but implied in the text. Some of these deserve comment.

In the case of exposure to radiation of very low penetrating power (HVL < 1 mm of soft tissue) the values specifically stated in the handbook are 1500 mrem for the skin and 300 mrem for the lens of the eye, and they apply to adults of any age.   Since with this radiation the blood-forming organs and gonads would receive negligible doses, none are stipulated.   In practice, however, the same individual may be exposed also to penetrating radiation.   In this case the limits for these critical organs should be the same as usual.   Since they are different for the two age groups, they are given separately.   The lens dose for age forty-five and over has been included by analogous reasoning.   These additional values are marked with superscript § in the table.

In the case of local exposure of the extremities or head and neck, no age distinction is made in the handbook.   Again, since whole-body exposure may also occur, it is well to include the appropriate figures for all critical organs.   Blood-forming tissue (principally bone marrow) in the designated part of the body (hands, etc.) is in the "region of highest dosage rate," and the doses in the table should be for a significant volume in this region.   However, the high dose in this portion of the blood-forming organs is purposely disregarded in this case.   The figures given, marked with the superscript ¶, apply to the main portion of the body.

Under very special circumstances some tissues in the body may be subjected to high local concentrations of radiation, even when the weekly doses in the critical organs do not exceed the respective prescribed limits. Such situations are not likely to occur in practice at the present time, but for the sake of completeness provisions have been made for the protection of all body tissues.   This is done by adopting a "basic permissible dose distribution" according to the depth below the surface of the body at which the tissue is located.   Since the basic permissible weekly dose for the skin is 600 mrem and for the blood-forming organs it is 300 mrem at an average depth of 5 cm, the basic-permissible-dose distribution curve shows an almost linear drop of 50 per cent in the first 5 cm.   Beyond this depth there is no further drop.   This means that a tissue at a depth between 0 and 5 cm may receive a weekly dose of 100 to 50 per cent of the skin dose, depending on the depth, provided of course that the weekly dose at a depth of 5 cm does not exceed the permissible value for the blood-forming organs.   The basic permissible weekly dose for all organs and tissues below a depth of 5 cm is the same as for the blood-forming organs, namely, 300 mrem.

Provisions are made in Handbook 59 for unusual situations.   One such provision is for accidental whole-body exposure to X rays up to an air dose of 24 r in a short time.   *If this happens only once in a lifetime* it may be assumed it has no effect on the radiation tolerance status of the individual, and he may continue to be occupationally exposed to radiation at the normal permissible rate.   This should not be considered in the category of a permissible dose.   It is simply recognition and regularization of a vexing problem that comes up from time to time in spite of all precautions.   A whole-body dose of 25 r does not produce appreciable short-term effects.   In comparison to a permissible weekly dose, it is so large that any possible damage it may have caused with respect to long-term effects can hardly be expected to be wiped out by a reasonable period of nonexposure.   Therefore, little—if anything—is gained by enforcing such a period, and the person may just as well continue his professional activities, without interruption but *with much greater care.*

PERMISSIBLE LIMITS OF EXPOSURE TO RADIATION FROM INTERNAL SOURCES

5-33. Radioactive isotopes may enter the body in a variety of ways and forms.   The fate of the material introduced into the body depends on the physical and chemical properties of the isotope and on numerous physiological factors (e.g., metabolic processes).   If the half-life of the isotope is sufficiently long, appreciable concentrations in different tissues may be maintained for a sufficient time to cause radiation damage of structure or function of the tissue.   This may result from entrance of the material in a single episode or in a more or less continuous fashion. In the latter case the accumulation depends on the rate of introduction into the body, the rate of elimination, and the rate of natural decay of the isotope.   The radiation damage* to a tissue depends largely on the concentration therein, the radiosensitivity, the length of time during which the dosage rate is sufficiently high, and the recovery rate.

The concept of "critical organs" forms the basis for proetction in the case of sources internal to the body as well as for external sources. However, factors other than radiosensitivity must also be considered in this case.   For example, injury may result not because the tissue or organ is particularly sensitive to radiation but because the isotope selectively concentrates therein.   The local concentration may result from metabolic processes or from physical retention or passage of the material in nonabsorbable form (e.g., passage through the intestinal tract).   It so happens that cancer is often the critical effect in the case of "internal emitters" also.   This is particularly true of bone, in which many isotopes tend to concentrate.

The object of protection is again prevention of appreciable bodily injury manifestable in the lifetime in the individual.   In principle it is accomplished by the requirement that the weekly dose of radiation must not exceed the maximum permissible weekly dose for the critical organ and effect under consideration.   Since fine distinctions are not justified, in general 300 mrem per week is taken as the maximum value for all tissues.

* Damage due to chemical toxicity is outside of the present discussion.

For purposes of calculation it is usually assumed that the material is uniformly distributed in a significant volume of the tissue of interest. This volume must be chosen carefully because even in the same organ the concentration may vary widely, depending on the properties of the isotope in question.   Radium and plutonium, both of which concentrate in bone, provide a good example of this: Radium eventually becomes fairly uniformly distributed throughout bone in a macroscopic sense, whereas plutonium tends to remain more or less localized in layers.   This is because the chemical properties of radium are similar to those of calcium, a natural constituent of bone.   Nevertheless high local (microscopic) concentrations do occur even in the case of radium.   Presumably the worst damage takes place in these centers.

The choice of a significant volume constitutes the most difficult problem in the calculation of permissible internal doses.   Fortunately, the tragic sequelae of radium ingestion by painters of "luminous" dials during World War I provide invaluable data.   By felicitous coincidence, radium emits high-energy γ rays that can be measured outside of the body, and it changes into radon, a radioactive noble gas that diffuses to the blood, is carried to the lungs, and is exhaled.   Sensitive measuring devices make possible the accurate measurement of both the γ rays and the radon from a minute amount of radium in the body.   Thus it has been possible to correlate the *total amount* of radium in the body with the clinical and pathological findings in the radium-dial painters (and others).   Careful radiological and analytical studies of the bones of some of the victims have provided information on the *distribution* of the radium.

From exhaustive studies of this sort it has been established that the maximum permissible body burden, namely, the total amount of radium fixed in the body that is not expected to produce appreciable bodily injury in the lifetime of the individual, is 0.1 microcurie (or 0.1 μg, in this case).   Since most of the energy emitted by radium in equilibrium with its disintegration products* is in the form of α rays, the deleterious effects must be attributed to this type of radiation.   From these studies, animal experiments, and reasonable assumptions an estimate has been made of the RBE applicable to this situation.   The presently accepted value is 5.   This RBE may then be used to calculate maximum permissible local concentrations (and hence maximum permissible body burdens) for other bone-seeking isotopes when the pertinent conditions are comparable.   Permissible values for the other principal radioactive nuclear materials are listed in Table 5-5.

* It has been shown that more than 50 per cent of the radium fixed in the body is in equilibrium with its disintegration products.

All this is relatively simple. The trouble comes in making practical use of recommended maximum permissible body burdens or tissue concentrations. Obviously, one must be able to determine the amount of the radioactive material of interest in the appropriate critical organ, in order to prevent its accumulation beyond permissible limits. In general this cannot be done by measuring the radiation leaving the body, since in

TABLE 5-5. MAXIMUM PERMISSIBLE AMOUNT OF RADIOISOTOPE IN TOTAL BODY AND MAXIMUM PERMISSIBLE CONCENTRATION OF NUCLEAR MATERIALS IN AIR AND WATER FOR CONTINUOUS EXPOSURE*

Substance	Parti-cle	Effec-tive half-life in body, days	Organ	Weight of organ, g	Micro-curies in total body	Micro-curies per ml of water	Quantity per ml of air
Po^{210}, soluble..........	α	40	Spleen	150	0.02	3×10^{-5}	2×10^{-10} μc
Po^{210}, insoluble	α	31	Lungs	10^3	7×10^{-3}	....	7×10^{-11} μc
U, natural, soluble.....	α	30	Kidneys	300	0.2	7×10^{-5}	1.7×10^{-11} μc
U, natural, insoluble....	α	120	Lungs	10^3	0.009	....	1.7×10^{-11} μc
U^{233}, soluble..........	α	300	Bone	7×10^3	0.04	1.5×10^{-4}	1×10^{-10} μc
U^{233}, insoluble	α	120	Lungs	10^3	0.008	....	1.6×10^{-11} μc
Pu^{239}, soluble..........	α	43,000	Bone	7×10^3	0.04	1.5×10^{-6}	2×10^{-12} μc
Pu^{239}, insoluble	α	120	Lungs	10^3	0.008	....	2×10^{-12} μc
H^3....................	β^-	19	Total body	7×10^4	10^4	0.2	2×10^{-5} μc
Be^7....................	None	.....	Bone	7×10^3	670	1	4×10^{-6} μg
Be......................	....	.....	....	....	....	....	2×10^{-6} μg
Bi......................	....	.....	Kidneys	....	....	....	2×10^{-3} μg
Cd......................	....	.....	....	....	....	....	1×10^{-4} μg
Pb......................	....	.....	....	....	....	....	1.5×10^{-4} μg
Th......................	....	.....	....	....	....	....	5×10^{-4} μg
U, soluble..............	....	.....	Kidneys	....	....	....	5×10^{-5} μg
U, insoluble............	....	.....	Kidneys	....	....	....	2.5×10^{-4} μg

* Mainly from Handbook 52, National Bureau of Standards, and AEC Proposed Standards for Protection against Radiation, released July 11, 1955. The latter permit triple the above water and air concentrations in 40 hr/week occupational exposure, and one-tenth the above water and air concentrations for continuous nonoccupational exposure. This figure of one-tenth came from a misunderstanding of a statement in the published report of the International Commission on Radiological Protection [7]. For the whole population the permissible concentration *should be* less than one-tenth (see Sec. 5-35).

Permissible levels for long-lived fission products (AEC, Proposed Standards) are listed in App. O.

most cases the dosage rate is too low. For instance, the presence in bone of isotopes emitting only β rays could be revealed outside the body by the bremsstrahlung resulting from the interaction of β rays and bone. However, the procedure is not feasible for monitoring purposes because the relevant amounts of such material are too small. The only recourse is to make an estimate by indirect means.

One such procedure is to establish a relation between the amount present in the body or in an organ and the amount excreted daily—preferably via the urine—in a suitable period of time and under carefully controlled conditions. Animal experiments serve as a first approximation.   Not infrequently accidents provide human data that could not be obtained otherwise.   However, for monitoring purposes (in a different sense) the most practical scheme is to control the intake of the material. This is done by setting up maximum permissible concentrations of each isotope in air, water, and food, such that the rate of intake of the isotope is low enough to prevent accumulation in the critical organs beyond maximum permissible limits.   Many assumptions are involved in this procedure, but great care has been exercised by the Subcommittee on Permissible Internal Dose of the NCRP, and the values recommended in Bureau of Standards Handbook 52 must be regarded as the best available at the time of printing.   Since then, some slight revisions have been made [6].   The latest authoritative data will be found in the report of the International Commission on Radiological Protection, summarized in Sec. 5-35.

REPORTS OF NCRP

5-34. In addition to setting standards in the form of maximum permissible levels of exposure generally applicable to the conditions encountered in practice, the National Committee on Radiation Protection has issued several handbooks dealing in more detail with special situations. These provide valuable information needed for the design of protective barriers and for safety of operation.   A list of the pertinent handbooks is given in Table 5-6.   Tolerances of the main chemical poisons employed in reactors have also been included in Table 5-5.

TABLE 5-6. HANDBOOKS OF THE NATIONAL BUREAU OF STANDARDS*

No.	Title	Price
41	Medical X-ray Protection Up to Two Million Volts	$0.25
42	Safe Handling of Radioactive Isotopes	0.20
48	Control and Removal of Radioactive Contamination in Laboratories	0.15
49	Recommendations for Waste Disposal of Phosphorus-32 and Iodine-131 for Medical Users	0.15
50	X-ray Protection Design	0.20
51	Radiological Monitoring Methods and Instruments	0.15
52	Maximum Permissible Amounts of Radioisotopes in the Human Body and Maximum Permissible Concentrations in Air and Water	0.20
53	Recommendations for the Disposal of Carbon-14 Wastes	0.15
54	Protection Against Radiations from Radium, Cobalt-60, and Cesium-137	0.25
55	Protection Against Betatron-Synchrotron Radiations Up to 100 Million Electron Volts	0.25
58	Radioactive-Waste Disposal in the Ocean	0.20
59	Permissible Dose from External Sources of Ionizing Radiation	0.30

* Available from the Government Printing Office, at the prices indicated.

RECENT DEVELOPMENTS

5-35. The world-wide radioactive fallout following detonation of nuclear weapons and the acceleration of the peaceful applications of nuclear energy brought about by President Eisenhower's "Atoms for Peace" program have focused the attention of scientists all over the world to the problem of genetic damage, which assumes great importance when the whole population of a country (or a large portion thereof) is exposed to ionizing radiation. Several national and international bodies have undertaken studies of the problem, and some reports have been issued [8–10]. Of chief interest in connection with the subject of this chapter is the report [8] of the International Commission on Radiological Protection,* which met in Geneva, Switzerland, in April, 1956. The salient features of this report are quoted below:

Statistically Detectable Effects

The definitions of permissible dose and permissible weekly dose are based on possible bodily injury manifestable in the life time of an individual. Since any such injury to a particular person that might result from exposure at present permissible limits would be very slight, and in view of large biological variations that always exist, certain types of injury cannot be detected in a single individual. This is particularly true in the case of a possible shortening of the life span. Therefore, it becomes necessary to consider injuries that become significant only when large groups are examined statistically.

The recommended maximum permissible weekly doses and the modified values for special circumstances, permit a desirable degree of flexibility for their application. In practice it has been found that in order not to exceed these maximum limits and also to comply with the general recommendations of the ICRP "that exposure to radiation be kept at the lowest practicable level in all cases," a considerable factor of safety must be allowed in the design of protective devices and operating procedures. Therefore, under present conditions, it is expected that the average yearly occupational dose actually received by an occupationally exposed person, would be about 5 rems and the accumulated dose in the employment period up to 30 years of age would be about 50 rems. Accordingly, the ICRP recommends continuation of the present conservative practice as regards doses actually received by occupationally exposed personnel, to keep the accumulated dose as low as practicable, especially up to age 30.

For Population Groups

The recommendations of the ICRP formulated at its meeting in 1953 and published in 1955 dealt primarily with the protection of persons occupationally exposed to ionizing radiations and to radioactive material. The Commission recognized, also, its responsibility toward the protection of persons who might be exposed unknowingly in the neighbourhood of a plant or source of radiation and

* Hereinafter designated the ICRP.

recommended that "in the case of the prolonged exposure of a large population, the maximum permissible levels should be reduced by a factor of ten below those accepted for occupational exposure". It was envisaged at the time that in any practical case only a relatively small number of individuals in the immediate surroundings of an installation would be in the 10% zone and the bulk of the large population would receive much less radiation. It has come to the attention of the Commission that this provision may be misinterpreted and, therefore, a more explicit recommendation is now made in terms of "controlled" areas.

A *controlled area* is one in which the occupational exposure of personnel to radiation or radioactive material is under the supervision of a radiation safety officer.

For such personnel the maximum permissible levels of exposure are those specified for occupational exposure. In the case of prolonged exposure to radiation from external sources, the maximum permissible levels for occupational exposure are represented by weekly doses of 600 mrem in the skin and 300 mrem in the blood-forming organs, the gonads and the lenses of the eyes. In the case of prolonged exposure to radioactive material that may enter the body, the maximum permissible levels of exposure are represented by the concentrations in air and water given in Table C.VIII of the report of Sub-Committee II.

For any person in any place outside of controlled areas the maximum permissible levels of exposure are 10% of the occupational exposure levels. Under the conditions at present envisaged the regions in which the 10% level might be reached would contain a very small fraction of the population and, therefore, the average organ doses per individual of the whole population would be very much lower than 10% of the maximum permissible organ doses for occupational exposure. (See also statement on genetic effects.)

For the Entire Population

The use of nuclear reactors for power production involves waste disposal and dispersion of radioactive material that may affect large sections of the population. The rapidly expanding use of radioactive materials in science and industry subjects more and more people to exposure to radiation. Therefore, genetic damage assumes greater importance. Designers of nuclear power plants and others concerned with the peaceful application of atomic energy, cannot plan for the future in the present state of uncertainty as to what the genetic problem may mean in terms of a permissible level for the whole population. Realizing the importance and urgency of the matter, and cognizant of its responsibility to the public, the Commission has decided to accelerate its study of the problem in order to be able to recommend in the near future a maximum permissible "genetic dose" applicable to the whole population. In the meantime those concerned with the problem may be guided by the following statement:

When genetic aspects of the effects of radiation are considered, the dose received by the whole population is of importance. Scientific data derived from human as distinct from experimental animal populations are so scanty that no precise permissible dose for a population can, at present, be set. The available information is being assessed by the Commission and other groups, including geneticists.

Until general agreement is reached, it is prudent to limit the dose of radiation received by gametes from all sources additional to the natural background to an amount of the order of the natural background in presently inhabited regions of the earth.

Permissible Weekly Dose for Internal Radiation

It is proposed that the basic absorbed dose rate for occupational exposure continue to be 0.3 rem per week for all organs of the body with the exception of the gonads and the total body.   In cases in which the gonads are the critical body organs, the absorbed dose rate will be reduced to 0.1 rem per week in order to limit the absorbed dose to the gonads to not more than 50 rem to the age of 30 (assuming few persons are employed before the age of 20).   This reduction is intended to prevent damage to man during his most reproductive period.   In the cases where the total body is the critical body organ the absorbed dose rate will again be set at 0.1 rem per week.   This reduction in the absorbed dose rate will limit the accumulated absorbed dose in the total body up to the age of 60 to not more than 200 rem and is intended to reduce the probability of damage resulting from total body exposure, e.g., reduce the probability of leukemia and reduce the possibility of shortening the life span of an individual.   There will be no change in the permissible absorbed dose rate when other organs such as liver, spleen, thyroid, G.I. tract, and bone are the critical body organs.   In the case of Ra^{226} and other bone seeking radionuclides, e.g., Sr^{89}, Sr^{90}, Ca^{45}, Ba^{140}, the point of reference will continue to be the well established maximum permissible body burden for man of 0.1 μg Ra^{226}.

The recommendations of the Genetics Committee in the "Study of the Biological Effects of Atomic Radiations" undertaken by the National Academy of Sciences [9] are listed below:

A) That, in view of the fact that total accumulated dose is the genetically important figure, steps be taken to institute a national system of radiation exposure record-keeping, under which there would be maintained for every individual a complete history of his total record of exposure to X rays, and to all other gamma radiation.   This will impose minor burdens on all individuals of our society, but it will, as a compensation, be a real protection to them.   We are conscious of the fact that this recommendation will not be simple to put into effect.

B) That the medical authorities of this country initiate a vigorous movement to reduce the radiation exposure from X rays to the lowest limit consistent with medical necessity; and in particular that they take steps to assure that proper safeguards always be taken to minimize the radiation dose to the reproductive cells.

C) That for the present it be accepted as a uniform national standard that X-ray installations (medical and nonmedical), power installations, disposal of radioactive wastes, experimental installations, testing of weapons, and all other humanly controllable sources of radiations be so restricted that members of our general population shall not receive from such sources an average of more than

10 roentgens, in addition to background, of ionizing radiation as a total accumulated dose to the reproductive cells from conception to age 30.

D) The previous recommendation should be reconsidered periodically with the view to keeping the reproductive cell dose at the lowest practicable level. If it is feasible to reduce medical exposures, industrial exposures, or both, then the total should be reduced accordingly.

E) That individual persons not receive more than a total accumulated dose to the reproductive cells of 50 roentgens up to age 30 years (by which age, on the average, over half of the children will have been born), and not more than 50 roentgens additional up to age 40 (by which time about nine tenths of their children will have been born).

F) That every effort be made to assign to tasks involving higher radiation exposures individuals who, for age or other reasons, are unlikely thereafter to have additional offspring. Again it is recognized that such a procedure will introduce complications and difficulties, but this committee is convinced that society should begin to modify its procedures to meet inevitable new conditions.

Both reports introduce the concept of "accumulated dose," which in the past has not been considered explicitly. For occupational exposure the recommended limits are essentially one-third of the doses that would be accumulated in the same period of time at the presently accepted weekly doses. Since no change in maximum permissible weekly doses is suggested, this means that exposure at these weekly levels is permitted for only a limited period of time instead of "indefinitely."

One of the fundamental principles of radiation protection is the assumption that a person may be occupationally exposed for the rest of his working life. Therefore, it is necessary to make provisions to distribute a permissible accumulated dose over a suitable period of time. This problem (and in fact the whole new situation) is being studied by the National Committee on Radiation Protection, and formal recommendations will be made later. It appears likely that a limit of 5 rem will be set for the total doses in the gonads, the blood-forming organs, and the lenses of the eyes, accumulated in a period of 52 consecutive weeks, the corresponding accumulated dose for the skin being 10 rem. In the absence of a national radiation-exposure record-keeping system (recommended by the Genetics Committee), it is impractical to include, in the accumulated yearly dose, contributions made by nonoccupational exposures (e.g., medical X-ray exposures). Therefore, the limit of 5 rem/-year refers to the dose resulting from occupational exposure only. It is evident, however, that a local record-keeping system for occupational exposure is necessary in order to take advantage of the 300-mrem permissible weekly dose and at the same time to keep the accumulated yearly dose under 5 rem. This may be accomplished by different methods and means depending on the particular circumstances under which the exposure takes place.   The main requirement of a satisfactory dose-reckoning system is that it should furnish a reliable estimate of the accumulated dose to show (by measurements, records, etc.) that it does not exceed the maximum permissible weekly and yearly limits.   The simplest such system is, of course, one in which the weekly doses are kept below one-third of the maximum permissible values throughout any period of 52 consecutive weeks.   Since in the case of fixed sources (e.g., power reactors) a reduction of the radiation level by a factor of 3 adds relatively little to the shielding cost, it is suggested that this practice be followed whenever possible.   The obvious advantage is that it greatly simplifies the personnel monitoring problem and thus reduces operating costs.

From the genetics point of view any additional exposure of the gonads to ionizing radiation is undesirable.   The limit of 10 r for the "per capita genetic dose" recommended by the Genetics Committee represents a reasonable balance between possible harm and benefit accruing from the expansion of activities involving exposure to radiation.   This marks an important step toward the planning on a national basis for the "atomic age."   Engineers, industrialists, and government agencies could not make long-range plans on the indefinite basis that exposure of the population to radiation should be as "low as possible."   A great deal remains to be done to clarify and codify the implications of the new concept of limiting the accumulated doses received by workers, persons in noncontrolled areas, and the general population.

Prob. 5-1.   What is the difference between intensity of radiation and dose rate? In what units is each expressed?

Prob. 5-2.   When it is said that 1 r of X rays corresponds to an energy absorption of 87.6 ergs per gram of air, what assumptions are made?

Prob. 5-3.   Permissible limits of exposure to ionizing radiation are intended to prevent the occurrence of what injuries manifestable in the lifetime of the individual?

Prob. 5-4.   In 1 week a worker received on his skin 100 mr of X rays and 20 mrad of fast neutrons with a specific ionization of 650 ion pairs per micron of water.   What is the total dose in mrem?

Ans.   The RBE for a specific ionization of 650 is 5 (from Table 5-2).

$$20 \text{ mrads} = 5 \times 20 = 100 \text{ mrem}$$
$$100 \text{ mr} = 100 \text{ mrem}$$
$$\text{Total dose} = 200 \text{ mrem}$$

Prob. 5-5.   In 1 week a worker weighing 70 kg received on his skin 200 mr of X rays of such penetrating power and under such conditions that the distribution of radiation throughout the body may be considered to have been uniform.   What is the integral dose received by the worker?

Ans.   1 r of X rays corresponds to an energy absorption of 93 ergs per gram of soft tissue.   Since the radiation is very penetrating, the absorption in bone, on a mass basis, is nearly the same as for soft tissue.   Therefore the integral dose is

$$0.2 \times 93 \times 70{,}000 = 1.3 \times 10^6 \text{ ergs} = 1.3 \times 10^4 \text{ g-rads} = 0.031 \text{ cal}$$

Prob. 5-6.   In 1 week a young worker has received a skin dose of 50 mrad of fast neutrons for which the RBE may be assumed to be 10.   (a) What is the skin dose in millirems?   (b) If the whole body was exposed, did the worker receive a dose in excess of the maximum permissible weekly dose?

Ans.   (a) Skin dose = 50 × 10 = 500 mrem.   (b) Maximum permissible weekly dose *for the skin* = 600 mrem but *for the lens of the eye* = 300 mrem.   Since at the depth of the lens of the eye (3 mm) the fast-neutron dose would be essentially the same as the skin dose, the worker received a dose 67 per cent in excess of the maximum permissible weekly dose.

Prob. 5-7.   In one week a young worker received a whole-body dose of penetrating X rays of 750 mr measured in air.   Since this is in excess of the maximum permissible weekly dose (see last column of Table 5-4), what should be done?

Ans.   The reason for the overexposure should be determined to prevent its reoccurrence.   Greater care should be exercised to make sure that the worker does not receive a total dose in excess of 3000 mr in a period of 13 consecutive weeks—according to the special provision described in Sec. 5-32.

REFERENCES

1. Hempelmann, L. H., H. Lisco, and J. G. Hoffman: The Acute Radiation Syndrome: A Study of Nine Cases and a Review of the Problem, *Ann. Internal Med.*, **36**(2):279–510 (February, 1952).
2. Dowdy, A. H.: "Tabulation of Available Data Relative to Radiation Biology," UCLA-22, AECU-253, TIS.
3. Failla, G.: Protection against High Energy Roentgen Rays (Caldwell Lecture, 1945), *Am. J. Roentgenol. Radium Therapy*, **54**(6):553–582 (December, 1945).
4. Cormack, D. V., and H. E. Johns: Electron Energies and Ion Densities in Water Irradiated with 200 keV, 1 MeV; and 25 MeV Radiation, *Brit. J. Radiol.*, **25**(295): 369–381 (July, 1952).
5. Evans, T. C.: Effects of Small Daily Doses of Fast Neutrons on Mice, *Radiology*, **50**(6):811–834 (June, 1948).
6. AEC: Proposed Standards for Protection against Radiation, released July 11, 1955.
7. Recommendations of the International Commission on Radiological Protection, *Brit. J. Radiol.*, supplement 6, 1955.
8. International Commission on Radiological Protection: statement of the Commission to the Eighth International Congress of Radiology, Mexico City, July, 1956 (will probably be published in radiological journals).
9. "The Biological Effects of Atomic Radiations," summary reports from a study by the National Academy of Sciences, 1956.
10. Medical Research Council (British): "The Hazards to Man of Nuclear and Allied Radiations," Command 9780, Her Majesty's Stationery Office, London.

CHAPTER 6

ELEMENTARY REACTOR PHYSICS

By R. Wayne Houston

6-1. Significance. A majority of the typical engineering problems found in reactor work are intimately related to the task of achieving and controlling the neutron chain reaction. For this reason it is essential that the nuclear engineer be acquainted with the fundamentals of nuclear-reactor physics. In addition, it is quite possible that the maximum development and utilization of nuclear energy may be delayed until the engineer has acquired proficiency in this basic phase of reactor work.

NEUTRONS IN REACTORS

6-2. Introduction. A nuclear reactor may be regarded as a device in which free neutrons may be generated in quantities as large or small as desired. In reactor physics attention is focused largely on the neutrons and nuclei involved in reactors rather than on the energy released in fission. Since the number of neutrons produced in the fission reaction exceeds the number consumed, neutrons are the key to the self-sustaining character that the fission reaction can be made to assume.

It is neither possible nor desirable in a reactor to have all the neutrons produced in the fission process generate additional neutrons by the same reaction. Many reactions compete for the neutrons, depending upon the kinds and numbers of nuclei present, and as was seen in Chap. 3, very few of these reactions produce additional neutrons. In addition to their dissipation by absorption in the nuclei present, many neutrons physically escape before having an opportunity to react within the system. They also decay into protons, but their half-life is of the order of 12 min, so that relatively few are lost by this process.

6-3. Neutron Chain Reactions. The above comments can be mathematically summarized by defining a quantity which is fundamental to reactor physics, the multiplication factor k, the average number of neutrons produced per fission which subsequently initiate another fission. If N neutrons are regarded as a first generation, then the second generation will contain Nk neutrons, the third Nk^2, and the nth generation Nk^n. It will be seen that a reactor is regarded as subcritical if $k < 1$, since Nk^n then approaches zero and the reaction dies out. Similarly, if $k > 1$, the reactor is supercritical and Nk^n becomes very large when n is very great. Obviously then, the condition of criticality is defined by $k = 1$, since under this circumstance the reaction is just self-sustaining and neither diverges to infinity nor decays to zero.

The condition $k = 1$ is not the only one which permits a steady-state condition, however. If S neutrons per second are allowed to enter a reactor, e.g., from an external source such as a mixture of Ra-Be, then after $n + 1$ generations the rate of production of neutrons is

$$N = S(1 + k + k^2 + \cdots + k^n) \tag{6-1}$$

which is still divergent for $k > 1$ but also for $k = 1$. For $k < 1$ and $n = \infty$, Eq. (6-1) becomes

$$N = \frac{S}{1 - k} \tag{6-2}$$

which represents a steady-state condition and an amplification of the source production rate by a factor of $1/(1 - k)$. This condition of a steady state is used experimentally in so-called "exponential" reactor tests but does not offer a worthwhile opportunity for producing quantities of power or radioisotopes with a subcritical reactor. The amplification required practically would be over 10^6, which implies a value of k only 10^{-6} less than unity.

If a subcritical reactor is operating at a steady state with an external source according to Eq. (6-2) and the source is removed, and k increased to unity simultaneously, the neutron-production rate N will remain finite and constant at the existing level, i.e., the reactor will now be critical. That the condition of criticality does not, per se, imply any particular level of neutron production rate is mathematically evident from Eq. (6-2), since at $S = 0$ and $k = 1$, N is indeterminate. Changing the neutron-production rate is thus effected by changing k by a small amount until the desired level is reached, k then being returned to unity. Critical reactors can thus be operated in steady state at any desired level consistent with the ability to dissipate safely the heat generated.

6-4. Neutron Cycles. Measurements have shown that neutrons emitted in the fission process have kinetic energies ranging from about 0.1 to 20 Mev, with an average of about 2 Mev. Although many of these fast neutrons can and do produce fissions, in U^{238} for example, the relative rate at which this reaction occurs is not large ($\sigma_f = 0.29$ barn, average). If a fission reaction were to be self-sustaining in natural uranium it is clear that the total cross section for all other competing reactions would have to be small. Scattering collisions lower the energy of the neutrons, as was seen in Chap. 2, and thus lower the chances for fission to take place in U^{238}. Although elastic scattering in materials of such high mass number as U^{238} requires a great many collisions to change the neutron energy significantly, inelastic scattering will generally lower the energy below the fission threshold, about 0.9 Mev, in a single collision. The cross section for this process in natural uranium is about 2.5 barns, so that a U^{238} fission would have to release over $(2.5 + 0.29)/0.29$ or 9.6 neutrons to be self-sustaining on this basis alone, disregarding other possible capture processes and leakage. The release of this many neutrons has never been observed in fission at these or lower energies.

It is clear, therefore, that if a chain reaction is to occur, the ratio of the probability of fission to that of all other dissipative reactions, including leakage, must be enhanced to equal the reciprocal of the number of neutrons released in the fission process. Except for the leakage effect, then, the feasibility of a chain reaction depends only on inherent nuclear properties of the materials involved. At the present time the nuclides U^{233}, U^{235}, and Pu^{239} are the most stable isotopes known which possess sufficiently high fission cross sections to allow chain reactions. For each of these, in the pure form, the ratio of fission cross section to the total cross section for all capture processes is greater than the reciprocal of the average number of neutrons released in the respective fission reactions. Only the leakage factor noted above prevents a spontaneous chain reaction with small isolated quantities of these materials.

The nature of the variation of σ_f with neutron energy for the above nuclear fuels suggests that certain advantages accrue by slowing down the neutrons before they are allowed to cause fission. This is particularly true in the case of U^{235} when used in its naturally occurring mixture with U^{238}. As has been seen in Chap. 3, most nuclides do not have large cross sections at neutron energies around 1 Mev, but many exhibit high resonance cross sections at lower energies, and nearly all have appreciable cross sections for slow-neutron capture. In the case of natural uranium, the increased value of the fission cross section of U^{235} at low neutron energies makes it feasible to use this isotopic mixture for a controlled chain reaction. Even when using any of the fuels mentioned above in relatively pure form, there are advantages gained in moderating fission neutrons to lower energies. One of the most important of these bears on the relative ease with which the chain reaction may be controlled, since the time interval between successive neutron generations is substantially increased. In some cases, another advantage is the increase in reactor volume accompanying inclusion of a moderator. This may simplify many of the internal engineering design problems.

In Fig. 6-1 are arranged schematically the possible fates of neutrons in a reactor, along with some of the factors that will be used in the quantitative description. The cyclical character is evident. Although all these processes occur to a greater or lesser extent in all reactors, it is generally desirable or necessary to suppress some of them in favor of others. An important general classification of reactors focuses attention on the energy of the neutrons that cause the majority of fissions. A *fast* reactor employs neutrons at high energies and has little or no moderator to prevent the slowing down of neutrons. In an *intermediate* (sometimes called "resonance" or "epithermal") reactor, neutrons are partially

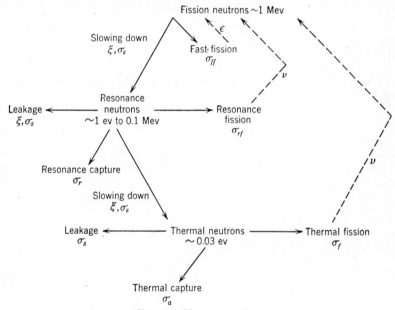

Fig. 6-1. Neutron cycles.

moderated and the fission reaction is initiated with neutrons of intermediate energies. A *thermal* reactor suppresses fission processes at intermediate and high energies and moderates neutrons over a maximum energy range. Although it is the most common type now in existence, it is not yet clear that it enjoys any undisputed advantage over the other types for power-producing purposes. This question is discussed further in Chap. 14.

6-5. Generalized Critical Equation. It is convenient to orient one's thinking of a critical reactor of finite size in terms of the effective multiplication factor k_{eff}. A quantitative description of the condition of criticality in terms of the processes noted in Fig. 6-1 may be made as follows.

Let E_0 represent the energy of fission neutrons, E_{th} the energy of thermal neutrons, E any energy in between, $p(E_0,E)$ the probability that a fission neutron will slow down to the energy E without undergoing resonance capture, and $P_r(E_0,E)$ the nonleakage probability during slowing down. Starting with N fission neutrons, there will be $Np(E_0,E)P_r(E_0,E)$ at the energy E and $N\left(p + \dfrac{dp}{dE}\,dE\right)\left(P_r + \dfrac{dP_r}{dE}\,dE\right)$ at the energy $E + dE$. The difference is the number of neutrons lost to the cycle in the energy range dE, namely, $NP_r\dfrac{dp}{dE}\,dE$ lost by resonance capture, and $Np\dfrac{dP_r}{dE}\,dE$ lost by leakage. Integrating over the range of energy from E_{th} to E_0 yields the total loss in slowing down to thermal energies. The number of the original N neutrons arriving at thermal energy is then

$$N\left(1 - \int_{E_{th}}^{E_0} P_r \frac{dp}{dE}\,dE - \int_{E_{th}}^{E_0} p \frac{dP_r}{dE}\,dE\right) = NP_r(E_0,E_{th})p(E_0,E_{th})$$

while the total loss due to resonance capture is $N\displaystyle\int_{E_{th}}^{E_0} P_r \frac{dp}{dE}\,dE$.

Denote by P_{th} the nonleakage probability of thermal neutrons, and assume that all thermal neutrons remaining in the reactor are eventually absorbed. Letting $k(E)$ represent the number of fission neutrons produced per neutron of energy E absorbed, the cycle may be completed and the number of new fission neutrons calculated. If the reactor is exactly critical, this must be equal to the number of neutrons at the start, namely, N. For criticality, then,

$$P_r(E_0,E_{th})p(E_0,E_{th})P_{th}k(E_{th}) + \int_{E_{th}}^{E_0} P_r \frac{dp}{dE}\,k(E)\,dE = 1 \qquad (6\text{-}3)$$

The first term on the left is k_{eff} for the thermal cycle, while the second term is k_{eff} for cycling at all higher energies. One of the essential purposes of reactor theory is to express each of the quantities appearing in Eq. (6-3) in terms of the nuclear properties noted on Fig. 6-1, and the size and shape of the reactor.

If a reactor is not critical, we may write

$$k_{ex} = k_{eff,th} + k_{eff,r} - 1 \qquad (6\text{-}4)$$

where k_{ex} is the excess effective multiplication factor. If k_{ex} is positive, the neutron population and power level rise with time, while if it is negative, they fall off with time.

In a fast or intermediate reactor, since most neutrons are not thermalized, $k(E_{th}) = 0$ and the lower limit on the integral defining $k_{eff,r}$ is somewhat higher than E_{th}. Generally, in a thermal reactor the quantity $k(E)$ is made negligible, but it will be noted that the condition $p = 1$ also can describe a thermal reactor, i.e., no resonance capture.

6-6. Theoretical Descriptions of Neutron Behavior. As it stands, Eq. (6-3) can hardly be regarded as useful. It merely suggests the existence of certain statistical factors for a chain-reacting system which presumably could be measured. It does not suggest how one might extrapolate such measurements to any other system. In order to accomplish this latter end, it is necessary to analyze the process in terms of more fundamental relationships, i.e., in terms of a theoretical model of neutron behavior. Such a model and the consequences which follow it are useful only in so far as they lead to confirming experimental evidence.

The behavior of free neutrons in materials is analogous in many respects to the behavior of molecules in a gas, so that many of the concepts of the kinetic theory of gases have been successfully applied to free neutrons. Such neutrons in thermal equilibrium with ideal nonabsorbing materials thus show a Maxwellian distribution of velocities. Just as gas molecules tend to move away from a region of higher concentration, neutrons also move away in zigzag paths from a region of higher neutron density. Such regions of high density are produced by sources of free neutrons such as the fission reaction. Diffusion of neutrons in this manner beyond the physical limits of a reactor is designated "neutron leakage" from the reactor.

Because of the manner in which neutron reaction rate constants (absorption cross sections) are defined, it is normally convenient to deal with neutron "flux" as a dependent variable rather than neutron density. Neutron flux, denoted by Φ, is the product of neutron density and speed, nv. Such a definition implies monoenergetic neutrons, and is conveniently thought of as the total distance traveled in 1 sec by all the neutrons in 1 cm³. The units are thus neutrons per square centimeter per second. In the mathematical development in Secs. 6-10ff., the neutron flux plays a role analogous to temperature in heat-conduction problems and to electromotive force in certain types of electrical problems. For polyenergetic neutron systems, the definition of flux may be generalized as an integrated value over the energy spectrum, namely, $\int n(E)v \, dE$, where $n(E) \, dE$ is the density of neutrons having energies between E and $E + dE$, and v is the velocity corresponding to E.

6-7. Transport Theory. The combined continuity and transport equation of kinetic theory is due to Boltzmann. When applied to neutron transport it is considerably simplified because neutron-neutron interactions may be neglected, neutron concentrations normally being many orders of magnitude less than ordinary concentrations of gas molecules. For example, the relatively high thermal neutron flux of 10^{14} neutrons/(cm²)(sec) corresponds to a neutron density of about 5×10^8 neutrons/cm³. It may be readily calculated that the density of gas molecules at standard conditions is about 3×10^{19} molecules/cm³. With this simplification and the added condition that free neutrons can be absorbed by nuclei, the transport equation relates the vector flux in a differential volume to scattering and absorption cross sections, the density of scattering and absorbing nuclei, the direction and speed of scattered neutrons, and also to the source of neutrons, if released within the volume element, as in fission. Details of transport theory in reactor physics are beyond the scope of this chapter but are treated elsewhere [1–4].

Reactor calculations which have physical significance necessarily involve three dimensions in space. A treatment on the basis of transport theory in such cases is often too involved to be of direct utility, but comparisons in hypothetical one-dimensional problems between transport theory and the less rigorous diffusion theory serve to define a region of validity for the latter. In addition, transport theory points out certain correction factors that can be applied to diffusion theory to make it more accurate. These will be noted in the appropriate sections of this chapter.

6-8. Diffusion Theory. If it is assumed that neutrons deflected in scattering collisions with nuclei may proceed in any direction with equal probability (spherically symmetric or isotropic scattering) and that all neutrons have a single constant velocity, then transport theory reduces to diffusion theory. If it is further assumed that the neutron density or flux, now the scalar quantity already defined, does not change rapidly with distance over a few mean free paths for scattering, usually a few centimeters, then diffusion theory reduces to elementary or first-order diffusion theory.

The latter assumption limits the elementary theory to systems in which the ratio of absorption to scattering cross sections is relatively small. These are usually called "weakly absorbing media." The assumption of isotropic scattering leads to correct estimates of the flux at distances of a few mean free paths from sources and from boundaries between materials having significantly different absorbing and scattering properties. Estimates of flux close to such sources or boundaries may be greatly in error if made on the basis of diffusion theory.

The relative simplicity with which elementary diffusion theory can be expressed mathematically lends itself ideally to the description of thermal neutron behavior in a reactor, and to groups of higher-energy neutrons as well. However, it tends to fall down in systems containing a high percentage of nuclear fuel, since these would be strongly absorbing and the physical boundaries for controllable systems may be only a few mean free paths apart.

6-9. Slowing-down Theory. In intermediate and thermal reactors, material is added to the fissionable fuel system to moderate or slow down the high-energy fission neutrons. It is the purpose of slowing-down theory to provide a rational method of selecting moderating material, to describe the transport characteristics of neutrons during moderation, and to describe the energy-dependence of neutron density during moderation. Since transport theory includes neutron velocity- (or energy-) dependence, it may be used to describe the slowing-down process. It may be simplified for many cases but should be employed when substantial moderation takes place by inelastic scattering collisions, i.e., in media of fairly high atomic weight, as in fast reactors.

For materials that moderate neutrons by elastic scattering collisions it is sometimes convenient to treat separately the energy-dependence and the transport characteristics. In this event a new dependent variable, the slowing-down density q, is introduced. This is defined as the number of neutrons in 1 cm³ which slow down past a given energy E in 1 sec. The manner in which q varies with neutron energy is one of the principal subjects of slowing-down theory. That this dependence is a complicated one may be appreciated by considering that elastic collisions involve discrete energy losses and may involve discrete resonance absorption losses, both of which tend to give q a discontinuous character. These two effects are generally separated by first calculating the slowing-down density without absorption, then multiplying the latter by the resonance escape probability (Sec. 6-5). Many of the details of such calculations have been published [3–5]. The transport characteristics of neutrons slowing down are then considered on the basis of diffusion theory.

An important special case of slowing-down theory arises when the average energy loss per collision is small enough that the slowing down may be regarded as a continuous process. This is the Fermi age theory that leads to a unique characterization of the transport properties, as seen in Sec. 6-23.

THE REACTOR DIFFUSION EQUATION

6-10. Derivation. As in the analysis of problems in mass, heat, or momentum transfer, an equation of continuity is sought. Neutrons may enter a small volume by diffusion, or may be "born" within it. Likewise they may leave by diffusion or be absorbed by nuclei in the volume element, thus losing their identity as free neutrons. The diffusion processes are conveniently described by elementary diffusion theory within the limitations noted in Sec. 6-8.

Consider an elemental volume $dx\, dy\, dz$ at any point (x,y,z) in a reactor. The rate of neutron transport into the element in the x direction across the surface $dy\,dz$ is assumed to obey the relation

$$J(x)\,dy\,dz = -D_v \frac{\partial n}{\partial x}\,dy\,dz \qquad (6\text{-}5)$$

J is the neutron *current* and has units of neutrons per square centimeter per second, i.e., the same as neutron flux but not to be confused with it. It is apparent that J is actually a vector quantity. The term D_v in Eq. (6-5) is a diffusion coefficient and may be regarded, for the moment, as being defined by this equation. Across the surface $dy\,dz$ at $x + dx$, the neutron transport out of the element in the x direction is

$$J(x + dx)\,dy\,dz = -D_v \left(\frac{\partial n}{\partial x} + \frac{\partial}{\partial x}\frac{\partial n}{\partial x}\,dx \right) dy\,dz \qquad (6\text{-}6)$$

The difference between Eqs. (6-5) and (6-6) is the net loss of neutrons in the x direction from the volume element by diffusion, namely, $-D_v(\partial^2 n/\partial x^2)$. Proceeding in like manner for the y and z directions, the total loss by diffusion is $-D_v(\partial^2 n/\partial x^2 + \partial^2 n/\partial y^2 + \partial^2 n/\partial z^2)$, or, more compactly, $-D_v\nabla^2 n$, where ∇^2 is the Laplacian operator. D_v is here assumed to be independent of position. The rate at which neutrons are lost from the volume element by absorption depends upon $\Sigma_a = N\sigma_a$, the probability of absorption per centimeter of path, and the flux nv, and is simply the product $nv\Sigma_a$.

The rate at which neutrons are generated in the volume element is less easily described but may be regarded as proportional to the rate at which fissions occur, if fission is the only source of neutrons. In the steady state the neutron density n is constant and a neutron-balance condition then gives

$$-D_v\nabla^2 n + nv\Sigma_a = \alpha nv\Sigma_f \qquad (6\text{-}7)$$

where α is a proportionality factor assumed independent of neutron density. Equation (6-7) actually applies only to neutrons of a single energy or velocity. It has already been mentioned that diffusion theory applies only to monoenergetic neutrons, and it will be seen in Sec. 6-18 that D_v is proportional to neutron velocity v. Since the combination nv appears together in the other two terms, it is convenient to include the velocity with the density in the first term, leaving a new diffusion coefficient D outside the operator. In terms of the flux of monoenergetic neutrons, Eq. (6-7) becomes

$$-D\nabla^2\Phi + \Sigma_a\Phi = \alpha\Sigma_f\Phi \qquad (6\text{-}8)$$

The details of the energy distribution of neutrons in a reactor are considered in Secs. 6-20ff. For the present it is sufficient to consider that most of the neutrons in a thermal reactor are in thermal equilibrium with the moderator atoms, and it is approximately correct to characterize all the neutrons as having a single energy in the thermal region. This energy is usually taken as about 0.026 ev, corresponding to the most probable value of neutron velocity for a Maxwellian distribution at 300°K, namely, 2200 m/sec. Equation (6-8) thus becomes a complete expression for the neutron balance in a thermal reactor.

The steady-state condition represented by Eq. (6-8) defines the critical state of a reactor if external neutron sources are absent. Rearranging and combining the coefficients into a single factor gives the steady state *reactor diffusion equation*

$$\nabla^2\Phi + B^2\Phi = 0 \tag{6-9}$$

where the factor B^2 is usually called the "buckling,"* since its value is a measure of the degree of curvature of a plot of flux vs. position in a reactor. It is an important constant for any critical thermal reactor which has an essentially uniform structure. Equation (6-9) may be regarded as the definition of the buckling, which is sometimes called the "geometric buckling." As will be seen in the next section, B^2 is related to the size and shape of a critical reactor.

6-11. Boundary Conditions. The reactor equation (6-9) is a homogeneous, linear, second-order partial differential equation. It is formally similar to the wave equation of quantum mechanics, and for this reason it is sometimes referred to as the "wave equation." In order to find a particular solution of Eq. (6-9) it is necessary to specify certain boundary conditions which the solution must satisfy. In this case the mathematical boundary conditions are concerned only with the behavior of neutrons at the physical boundaries of the reactor. In the absence of external neutron sources the possibilities are that none, all, or some of the neutrons that diffuse out across the boundaries diffuse back again into the reactor. It will be appreciated that an intimate knowledge of the nature of the material surrounding the reactor is necessary before any one of these conditions can be said to be applicable. If the surrounding medium is a perfect neutron absorber, e.g., a vacuum, the first condition is applicable. The second condition calls for a surrounding medium which is a perfect reflector, i.e., a material having $\sigma_a = 0$, $\sigma_s > 0$ (cf. Sec. 6-57). Only the third condition can be strictly realized. If the reactor is surrounded by a low-density medium, such as air, or by a strongly absorbing medium where $\sigma_a/\sigma_s \gg 1$, then the first condition is approximately true since the chances of a neutron being deflected back into the reactor by a scattering collision are then small. Such a reactor is generally called a "bare reactor."

---

* Some writers prefer to use the negative of this quantity and call it "the Laplacian of a reactor."

Unfortunately, elementary diffusion theory based on Eq. (6-5) does not furnish a means for stating the first condition mathematically. A convenient approach is to assume the flux vanishes at the boundaries of the reactor. It will be seen in Sec. 6-18 that this assumption is valid if the reactor dimensions are large compared to a scattering mean free path and, furthermore, the error can be corrected for in a relatively simple manner.

6-12. Rectangular Geometry. Equation (6-9), with the Laplacian operator expressed in rectangular coordinates, may be written

$$\frac{\partial^2 \Phi}{\partial x^2} + \frac{\partial^2 \Phi}{\partial y^2} + \frac{\partial^2 \Phi}{\partial z^2} + B^2 \Phi = 0 \qquad (6\text{-}10)$$

The variables x, y, and z are separable by assuming a solution of the form $\Phi = X(x)Y(y)Z(z)$. Substituting in Eq. (6-10) and dividing through by XYZ yields

$$\frac{X''}{X} + \frac{Y''}{Y} + \frac{Z''}{Z} + B^2 = 0 \qquad (6\text{-}11)$$

Each of the terms in Eq. (6-11) must be constant, since variations in x, y, or z cannot possibly affect the terms in which they do not appear. Denoting these constants by $-\alpha^2$, $-\beta^2$, and $-\gamma^2$, then

$$X'' + \alpha^2 X = Y'' + \beta^2 Y = Z'' + \gamma^2 Z = 0 \qquad (6\text{-}12)$$

and
$$\alpha^2 + \beta^2 + \gamma^2 = B^2 \qquad (6\text{-}13)$$

The solution of Eq. (6-10) is now dependent on solutions of the three second-order ordinary equations (6-12). By standard methods the solutions to these equations may be written as

$$\begin{aligned}
X &= C_1 \sin \alpha x + C_2 \cos \alpha x \\
Y &= C_3 \sin \beta y + C_4 \cos \beta y \\
Z &= C_5 \sin \gamma z + C_6 \cos \gamma z
\end{aligned} \qquad (6\text{-}14)$$

To introduce the boundary condition that the flux must vanish at the reactor surface, it is necessary to specify the origin of the coordinate system and the values of the coordinates at each point on the surface. For complex shapes these specifications are not simple, and recourse must be made to an approximate method of numerical integration. For a reactor in the form of a rectangular prism of sides a, b, and c in length, with the origin at the center, the boundary conditions are expressed as

$$\Phi = 0 \text{ at } x = \pm \frac{a}{2}, y = \pm \frac{b}{2}, z = \pm \frac{c}{2} \qquad (6\text{-}15)$$

Introducing these in Eq. (6-14) reveals that the constants C_1, C_3, and C_5 must be zero and that

$$C_2 \cos \frac{\pm \alpha a}{2} = C_4 \cos \frac{\pm \beta b}{2} = C_6 \cos \frac{\pm \gamma c}{2} = 0 \qquad (6\text{-}16)$$

which can be true, for nontrivial solutions, only if

$$\frac{\alpha a}{2} = \frac{\beta b}{2} = \frac{\gamma c}{2} = \frac{n\pi}{2} \qquad n = 1, 3, 5, \ldots \qquad (6\text{-}17)$$

It will be seen in Sec. 6-42 that the solutions for $n > 1$ vanish for a critical reactor in the steady state. The final solution may then be written as

$$\Phi = C \cos \frac{\pi x}{a} \cos \frac{\pi y}{b} \cos \frac{\pi z}{c} \qquad (6\text{-}18)$$

where the arbitrary constants of Eq. (6-16) have been lumped together in the C of Eq. (6-18). In agreement with the statement in Sec. 6-3, the condition of criticality does not fix the flux or power at any particular level.

From the relations (6-11) and (6-18) it is apparent that the buckling is related to the critical size by

$$B^2 = \frac{\pi^2}{a^2} + \frac{\pi^2}{b^2} + \frac{\pi^2}{c^2} \qquad (6\text{-}19)$$

For a cube of side a, the buckling is given by

$$B^2 = \frac{3\pi^2}{a^2} \qquad (6\text{-}20)$$

The units of the buckling are seen to be reciprocal length squared, so that small values will be associated with large reactors and vice versa.

Prob. 6-1. For a fixed value of the buckling, prove that the dimensions of a rectangular reactor having the least critical volume are those of a cube.

6-13. Cylindrical Geometry. For a cylindrical reactor of radius R and length or height H, it is convenient to express the Laplacian operator of Eq. (6-9) in cylindrical coordinates, since symmetry about the axis eliminates one variable. The reactor equation then becomes

$$\frac{\partial^2 \Phi}{\partial r^2} + \frac{1}{r} \frac{\partial \Phi}{\partial r} + \frac{\partial^2 \Phi}{\partial z^2} + B^2 \Phi = 0 \qquad (6\text{-}21)$$

The solution is obtained by methods similar to those used above. Taking the origin again at the geometric center, the flux vanishes at $r = R$ and at $z = \pm H/2$. The solution is

$$\Phi = C \cos \frac{\pi z}{H} J_0 \left(\frac{2.405r}{R} \right) \qquad (6\text{-}22)$$

where $J_0(x)$ represents the zero-order Bessel function of the first kind [6], which has its lowest positive root at $x = 2.405$. For this case the buckling is given by

$$B^2 = \frac{\pi^2}{H^2} + \frac{2.405^2}{R^2} \qquad (6\text{-}23)$$

One advantage of a cylindrical reactor is that, for the same buckling, the least critical volume is smaller than that of a cube. The relation between height and radius for a critical cylinder of minimum volume is

$$H = \frac{\pi R \sqrt{2}}{2.405} \qquad (6\text{-}24)$$

6-14. Spherical Geometry. The least critical volume for a given value of B^2 is, of course, obtained with a spherical shape. This case is conveniently treated by expressing the Laplacian in spherical coordinates, assuming symmetry about the origin. The reactor equation then becomes

$$\frac{d^2\Phi}{dr^2} + \frac{2}{r}\frac{d\Phi}{dr} + B^2\Phi = 0 \qquad (6\text{-}25)$$

and the solution satisfying the boundary condition $\Phi = 0$ at $r = R$ is

$$\Phi = \frac{C}{r} \sin \frac{\pi r}{R} \qquad (6\text{-}26)$$

where the buckling for this case is

$$B^2 = \frac{\pi^2}{R^2} \qquad (6\text{-}27)$$

The results for these three important geometries are summarized in Table 6-1.

Although there are other relatively simple geometries which can be treated analytically by the methods noted above [7], few seem of any practical significance. The treatment of homogeneous reactors with variously shaped holes, either concentric or eccentric, involves not only a modification of the geometry, but also generally a modification of the boundary conditions at the internal surfaces. Calculations of the effects of suitably shaped concentric control rods is superficially similar to the above, but with the important difference that it is generally desired that a reactor be subcritical with such a rod fully inserted.

TABLE 6-1. GEOMETRIC BUCKLING AND FLUX DISTRIBUTIONS FOR
BARE REACTORS

Geometry	Buckling	Flux distribution
Infinite slab of thickness a........	$B^2 = \dfrac{\pi^2}{a^2}$	$\Phi = C \cos \dfrac{\pi x}{a}$
Rectangular prism of sides a, b, c....	$B^2 = \dfrac{\pi^2}{a^2} + \dfrac{\pi^2}{b^2} + \dfrac{\pi^2}{c^2}$	$\Phi = C \cos \dfrac{\pi x}{a} \cos \dfrac{\pi y}{b} \cos \dfrac{\pi z}{c}$
Cylinder of radius R and height H	$B^2 = \dfrac{\pi^2}{H^2} + \dfrac{2.405^2}{R^2}$	$\Phi = C \cos \dfrac{\pi z}{H} J_0 \left(\dfrac{2.405 r}{R} \right)$
Sphere of radius R...............	$B^2 = \dfrac{\pi^2}{R^2}$	$\Phi = \dfrac{C}{r} \sin \dfrac{\pi r}{R}$

Many practical cases arise in which it is not possible to obtain complete analytical solutions to the boundary-value problem. Recourse must then be had to numerical and graphical solutions. Such cases generally arise from the desire to optimize a design in one or more directions and usually introduce more complex boundary conditions and sometimes a certain degree of asymmetry. The application of approximate solutions to the reactor equation is outside the scope of this book, but reference may be made to applicable procedures mentioned in Chap. 9 or any of the standard texts [8, 9], or, for example, the electric-circuit-model method of Kron [51].

Prob. 6-2. A proposed thermal reactor [37] for producing power and plutonium has a core in the form of an octagonal prism, 20 ft high, and 35 ft across flats. It would employ slightly enriched uranium fuel and graphite moderator. Neglecting any effect of a reflector, estimate the buckling for this reactor.

6-15. Application of Flux Expressions. Aside from relating the critical size to the buckling, the preceding equations are of importance in calculating specific neutron reaction rates in a reactor as a function of position and, by integration over the reactor volume, total reaction rates for the entire reactor. In the important case of the fission reaction this leads to expressions for the rate of heat generation as a function of position and the total heat generation rate for a reactor. This information is of obvious importance to the engineer interested in the heat-transfer problems involved. Similarly, since the fission reaction releases γ rays directly, as well as indirectly from the fission products, the problem of shielding cannot be solved rigorously without due regard to the source distribution of γ rays as well as the maximum operating power (flux) level. Finally, whether a reactor is used for generating heat, or for producing new materials by neutron reactions, it is important to know how the production rate varies throughout the reactor so as to estimate the best production schedule.

The rate at which fissions occur in a unit volume of reactor is given by $N_f \sigma_f \Phi = \Sigma_f \Phi$, so that H, the rate of heat generation per unit volume at a point where the neutron flux is Φ, becomes

$$H = \frac{\Sigma_f \Phi}{3.2 \times 10^{10}} \tag{6-28}$$

in watts per cubic centimeter, where Σ_f and Φ must be taken for neutrons of the same energy, and the constant 3.2×10^{10} is the number of fissions required to release an energy of 1 watt-sec, based on 195 Mev energy release per fission [19]. The total power P, in watts, is then given by

$$P = \int_0^{V_R} W \, dV = \frac{\Sigma_f}{3.2 \times 10^{10}} \int_0^{V_R} \Phi \, dV \tag{6-29}$$

where V_R is the critical volume in cubic centimeters and Σ_f is regarded as constant.

Prob. 6-3. Find an expression for the ratio Φ_{av}/Φ_{max} for a critical cubic reactor of side a. If this dimension is 300 cm, and the reactor contains 100 lb of U^{235} uniformly distributed, what is the total power output when the maximum thermal neutron flux is 10^{14} neutrons/(cm^2)(sec)?

Solution. From Eq. (6-18),

$$\frac{\Phi}{\Phi_{max}} = \cos \frac{\pi x}{a} \cos \frac{\pi y}{a} \cos \frac{\pi z}{z}$$

Then
$$\frac{\Phi_{av}}{\Phi_{max}} = \frac{1}{V_R} \iiint_{-a/2}^{+a/2} \cos \frac{\pi x}{a} \cos \frac{\pi y}{a} \cos \frac{\pi z}{a} \, dx \, dy \, dz$$

$$= \frac{8}{\pi^3}$$

$$H_{av} = \frac{8 N_f \sigma_f \Phi_{max}}{3.2 \times 10^{10} \pi^3} \quad \text{watts/cm}^3$$

Here
$$N_f = \frac{100 \times 454 \times 6.02 \times 10^{23}}{235 \times 300^3} = 4.3 \times 10^{18}$$

$$\sigma_f = 549 \times 10^{-24} \quad \text{(App. P)}$$

so that
$$H_{av} = \frac{8 \times 4.3 \times 10^{18} \times 549 \times 10^{-24} \times 10^{14}}{3.2 \times 10^{10} \pi^3}$$

$$= 1.96 \quad \text{watts/cm}^3$$

$$P = 1.96 \times 300^3 \times 10^{-6} = 53.1 \quad \text{Mw}$$

Prob. 6-4. Show that the ratio Φ_{av}/Φ_{max} for a spherical reactor is $3/\pi^2 = 0.304$.

6-16. Measurement of Buckling and Critical Size. It is quite possible and often expedient to construct a "pilot plant" in the form of a critical assembly before construction of a full-scale reactor. Since neutron flux can be maintained at an arbitrary level in a critical reactor, it is possible to assemble most of the proposed components of a reactor and perform many experiments at power levels so low that no provision need be made for removing the heat released other than natural convection cooling of the structure. Shielding is also considerably reduced although not eliminated entirely.

In principle, the measurement of the buckling is a simple task. The assembly is built up in the presence of an external source of neutrons, and neutron flux measurements are made by any of the methods noted in Chap. 4 at several locations within the structure. At each successive stage in the process of assembly, when the flux rises to a steady value, the system is known to be subcritical (Sec. 6-3). If the system just becomes critical, the flux does not level off but continues to rise until the external source is removed, remaining constant at the final level thereafter. Direct measurement of the dimensions would then allow a calculation of the buckling through one of the equations in Table 6-1, depending upon the geometry of the assembly.*

The operation of a critical assembly involves control problems which are more stringent than may be necessary to measure the buckling. Actually a very good estimate of B^2 can sometimes be made by means of so-called exponential experiments in which the assembly is built up in the presence of an external neutron source but not allowed to reach criticality. When the assembly is reasonably close to criticality, measurements of the flux at several points can be used to compute B^2 which will be applicable at criticality. This is discussed briefly in Sec. 6-54.

CRITICAL EQUATIONS

6-17. Introduction. In the preceding treatment the internal properties of a critical system were combined in a single parameter, the buckling, which is subject to direct experimental measurement. For preliminary design purposes it is desirable to be able to estimate the buckling in terms of cross sections and other available fundamental properties. Such a calculation is made by means of a critical equation, such as Eq. (6-3), when the terms are expressed as known functions of known properties. The form of a critical equation will be seen to depend somewhat upon the simplifying assumptions involved in its derivation. Although it is essential to understand these assumptions before using the equations, it is not always clear in which direction a calculated result is affected. Where corrections can be applied, these will be indicated in the following text.

6-18. Neutron Diffusion and Transport Theory Corrections. An analysis of the assumptions inherent in elementary diffusion theory leads

---

* Actually a small correction is necessary (Sec. 6-18).

to the conclusion that it is more likely to be accurate in describing the behavior of thermal neutrons than that of higher-energy neutrons, so that the following may be regarded as applying primarily to thermal reactors.

The simplification of transport theory leads to a ready interpretation of the diffusion coefficient in terms of the mean free path for scattering [1, 4]. For a hypothetical medium which does not absorb neutrons

$$\frac{D_v}{v} = D = \frac{\lambda_s}{3} = \frac{1}{3\Sigma_s} \tag{6-30}$$

where λ_s is the scattering mean free path. For the diffusion of neutrons in real media in which forward scattering of neutrons is preferred and in which absorption of neutrons occurs, D should be calculated from

$$D = \frac{1}{3\Sigma(1 - \bar{\mu})\left(1 - \frac{4}{5}\frac{\Sigma_a}{\Sigma} + \frac{\Sigma_a}{\Sigma}\frac{\bar{\mu}}{1 - \bar{\mu}} + \cdots\right)} \tag{6-31}$$

where Σ is the total macroscopic cross section, and $\bar{\mu}$ is the average cosine of the scattering angle per collision. It was seen in Chap. 2 that $\bar{\mu}$ is dependent on the mass of the scattering nucleus as $\bar{\mu} = 2/3A$, where A is the mass number of the nucleus in an unbound atom.

For weakly absorbing systems ($\Sigma_a \ll \Sigma \sim \Sigma_s$), Eq. (6-31) becomes

$$D = \frac{1}{3\Sigma_s(1 - \bar{\mu})} = \frac{\lambda_s}{3(1 - \bar{\mu})} = \frac{\lambda_t}{3} \tag{6-32}$$

which is a form commonly employed. The quotient $\lambda_s/(1 - \bar{\mu})$ is an important property of a given medium and is referred to as the "transport mean free path" λ_t. It represents the mean free path between scattering collisions, corrected for preferred angle scattering. The correction for light nuclei is seen to be particularly important. For homogeneous mixtures of materials having concentrations of $N_1, N_2, \ldots$ nuclei/cm³, the appropriate average value of D is calculated as

$$D_{av} = \frac{1}{3[\Sigma_{s1}(1 - \bar{\mu}_1) + \Sigma_{s2}(1 - \bar{\mu}_2) + \cdots]} \tag{6-33}$$

Although it does not have a bearing on the derivation of critical equations, with the above definition of a transport mean free path it is now possible to define the correction that must be applied at a reactor surface to admit the use of the boundary condition that $\Phi = 0$ at the surface. This correction was mentioned in Sec. 6-11 and is called the "extrapolation distance" d, where

$$d = 0.7104\lambda_t \tag{6-34}$$

It may be regarded as a distance which must be added to a true linear dimension in order that the flux is properly represented within the medium in which neutrons are diffusing (Fig. 6-2). For example, Eq. (6-18) should be written as

$$\Phi = C \cos \frac{\pi x}{a + 2d} \cos \frac{\pi y}{b + 2d} \cos \frac{\pi z}{c + 2d} \qquad (6\text{-}35)$$

where a, b, and c are the actual, or physical, dimensions. The extrapolation distance concept still assumes that the medium into which neutrons diffuse is a perfect absorber. As was pointed out in Sec. 6-8, diffusion theory does not strictly apply near boundaries between dissimilar media. The extrapolation distance is actually a correction such that the flux distribution predicted by diffusion theory is the same as that predicted by the rigorous transport theory except within distances of a few mean

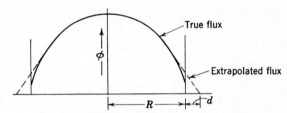

FIG. 6-2. Extrapolation distance concept.

free paths from the boundary. Equation (6-34) applies strictly to a plane boundary, the value of the coefficient increasing to a maximum of $\frac{4}{3}$ for zero radius of curvature. In many thermal reactors d is quite small relative to the dimension to which it is added, so that it is unnecessary to evaluate it with a high order of accuracy.

6-19. One-group Theory. The simplest approximation in reactor theory is that all neutrons are generated, diffuse, and are absorbed at a single energy level. Since $\Sigma_f \Phi$ represents the rate at which fissions occur, in this case the factor α becomes simply the average number of neutrons released in a single fission reaction, usually denoted by ν. Making this substitution in Eq. (6-8) and dividing through by Σ_a,

$$\frac{D}{\Sigma_a} \nabla^2 \Phi + \left(\frac{\nu \Sigma_f}{\Sigma_a} - 1 \right) \Phi = 0 \qquad (6\text{-}36)$$

The units of the coefficient of the term on the left are seen to be length squared. This coefficient has a special significance in neutron diffusion in absorbing materials. Its square root is called the "diffusion length,"

$$L = \sqrt{\frac{D}{\Sigma_a}} \qquad (6\text{-}37)$$

It may be shown, by letting $\nu\Sigma_f = 0$ in Eq. (6-36) and solving for the case of a point source of neutrons in an infinite medium, that the flux decreases exponentially with relaxation distance L. In addition, this solution can be used to show that L^2 is just one-sixth of the mean square distance that monoenergetic neutrons travel from a point source in an infinite medium to a point where they are absorbed. Since a reactor is a superposition of point sources, L^2 is important in characterizing critical size.

In the second term of Eq. (6-36) the quantity $\nu\Sigma_f/\Sigma_a$ is the number of neutrons released per neutron absorbed, or k, the multiplication factor for a reactor of infinite size in which no leakage occurs.* No neutron leakage implies $\nabla^2\Phi = 0$, so that Eq. (6-36) yields $\nu\Sigma_f/\Sigma_a = k = 1$, in agreement with Sec. 6-3.

Introducing k and L^2 into Eq. (6-36) and rearranging

$$\nabla^2\Phi + \frac{k - 1}{L^2}\,\Phi = 0 \qquad (6\text{-}38)$$

By comparison with Eq. (6-9), it is apparent that $(k - 1)/L^2$ may be identified with the buckling. In terms of k_{eff}, this relationship becomes

$$k_{eff} = \frac{k}{1 + L^2B^2} = 1 \qquad (6\text{-}39)$$

Equation (6-39) is the critical equation in one-group theory. The factor $1/(1 + L^2B^2)$ may be regarded as the nonleakage probability of thermal neutrons, P_{th} of Sec. 6-5.

The validity of the assumptions entering into one-group theory depends upon the particular system under study. On the basis of the neutron energy distribution, it may be argued that it should apply to fast reactors, where the majority of fission neutrons are absorbed or escape before they have an opportunity to lose much energy. If a fast reactor contained many light nuclei, the dimensions could not be more than a few mean free paths for elastic scattering. In a system of predominately heavy nuclei, inelastic scattering would be likely to occur with substantial frequency. In either case, as was pointed out in Sec. 6-8, elementary diffusion theory is a poor assumption. In intermediate reactors, the energy spread of neutrons would be relatively wide, so that a "one-velocity" theory would hardly seem applicable.

Thermal reactors generally have neutron-energy distributions extending over a range of 10^8 or greater. However, a simple calculation will show that it is easy to devise systems in which neutrons are slowed down to thermal energies very quickly compared to their mean lifetime as thermal neutrons (cf. Sec. 6-44). To the extent that this means that

---

* k as used here is often written as k_∞.

most neutrons released in fission neither escape nor are absorbed until they reach thermal energies, one-group theory is well adapted to a description of thermal reactors. The appropriate properties need be averaged only over the thermal-energy spectrum, and when the critical dimensions are relatively large in terms of scattering mean free paths, the conditions for the validity of diffusion theory are met.

6-20. Neutron-energy Spectrum. If it were possible to measure the neutron density or flux at a given point in a reactor as a function of neutron energy, one could plot a curve of the neutron flux per unit energy interval $\Phi(E)$ vs. energy E, extending up to the maximum energy of fission neutrons produced. For a thermal reactor such a curve might have the appearance shown in Fig. 6-3, although the relative magnitudes depend very strongly on the kinds of nuclei present. The distribution

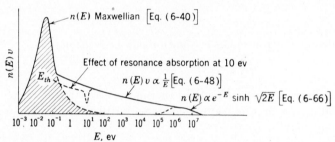

FIG. 6-3. Neutron-flux energy spectrum in a thermal reactor (hypothetical).

at the lowest energies represents thermal equilibrium between neutrons and the atoms or molecules in the system. If the system may be classed as weakly absorbing, this energy distribution of neutrons is very nearly that given by the Maxwell-Boltzmann equation of kinetic theory, namely,

$$n(E) = \frac{2\pi n}{(\pi k T)^{3/2}} E^{1/2} e^{-E/kT} \tag{6-40}$$

where n = total neutron density
  k = Boltzmann constant
  T = absolute temperature

Equation (6-40) is normalized so that $\int_0^\infty \frac{n(E)}{n} dE = 1$. Discussion of the distribution of higher-energy neutrons is reserved until Secs. 6-22 and 6-26.

6-21. Average Properties in Thermal Spectrum. It is possible to give somewhat more precise significance to the quantities entering one-group theory of thermal reactors by considering thermal neutrons as a group characterized by a distribution such as Eq. (6-40) and shown as the crosshatched area in Fig. 6-3.   In a bare homogeneous reactor at uniform temperature, the energy spectrum of neutrons is independent of position, so that the flux in any small energy interval dE may be written as $n(\mathbf{r})vm(E)\,dE$.   If this expression for the flux is substituted in Eq. (6-36), and the resulting equation integrated over all energies, it is readily seen that an average diffusion coefficient must be defined by

$$\bar{D} = \frac{\int_0^\infty D(E)m(E)v\,dE}{\int_0^\infty m(E)v\,dE} \tag{6-41}$$

and an average absorption cross section by

$$\bar{\sigma}_a = \frac{\int_0^\infty \sigma_a(E)m(E)v\,dE}{\int_0^\infty m(E)v\,dE} \tag{6-42}$$

The resulting neutron-balance equation can be written as

$$\bar{D}\nabla^2 n(\mathbf{r}) + (\nu\bar{\Sigma}_f - \bar{\Sigma}_a)n(\mathbf{r}) = 0 \tag{6-43a}*$$

It is an arbitrary custom to base thermal-neutron fluxes on a velocity v_0 of 2200 m/sec, as was mentioned in Sec. 6-10.   Thus, if $\Phi_{th} = n(\mathbf{r})v_0$, Eq. (6-43a) may be written

$$\bar{D}\nabla^2\Phi_{th} + (\nu\bar{\Sigma}_f - \bar{\Sigma}_a)\Phi_{th} = 0 \tag{6-43b}$$

Over the major portion of the thermal spectrum most absorption cross sections vary inversely as the velocity, i.e., $\sigma_a = b/v$.   For this case Eq. (6-42) reduces to

$$\bar{\sigma}_a = \frac{b}{\bar{v}} \tag{6-44}$$

the cross section at the average neutron velocity $\bar{v}$.   If $m(E)$ is Maxwellian [Eq. (6-40)], it can be shown that

$$\bar{\sigma}_a = \sqrt{\frac{\pi}{4}}\,\sigma_a(v_0) = \frac{\sigma_a(v_0)}{1.128} \tag{6-45}$$

Since thermal-neutron cross sections are usually stated for the point value of velocity $v_0 = 2200$ m/sec, Eq. (6-45) is convenient for calculating the correct value to use in thermal-neutron diffusion calculations.

For absorption cross sections that are not inversely proportional to v, a factor f may be computed such that the true integrated reaction rate $\bar{\Sigma}_a\Phi_{th}$ is given by $fN\sigma_a(v_0)\Phi_{th}$, where $\sigma_a(v_0)$ is the point value at velocity

---

* The number of neutrons produced per fission ν is assumed independent of energy here.

v_0 [10].   Such factors are given in Chap. 3, Ref. 2, along with the values of $\sigma_a(v_0)$, for those reactions marked "not $1/v$."   It should be stressed that the f factors given assume the applicability of Eq. (6-40) and are thus slightly temperature-dependent.

Scattering cross sections do not generally vary significantly over the range of thermal energies, so that the values of $\bar{\sigma}_s$ also given in Chap. 3, Ref. 2, may be used to calculate D directly from Eq. (6-32), although strictly it is the average of $1/\sigma_s$ that should be employed.   In many cases, it is more appropriate to calculate $\bar{D}$ from measured diffusion lengths and absorption cross sections employing Eq. (6-37).

6-22. Absorption in Resonance Region.   In one-group theory the slowing-down process is neglected.   If any loss of neutrons occurs during moderation, it is apparent that one-group critical-size calculations will yield low results.   The present section considers means of correcting for losses by resonance absorption.

Although it is possible to write down rigorous balance equations for neutrons undergoing moderation, they are not easily solved.   This situation arises because in the epithermal region the neutron density per unit energy interval is dependent on the difference between the slowing-down density (cf. Sec. 6-9) into and out of the energy interval,* and because the epithermal neutron-energy spectrum is not accurately known.   For calculations it is convenient to assume that this spectrum is independent of position, although this would strictly be true only for an infinite medium.   This assumption allows a computation of the resonance escape probability (Sec. 6-5) without reference to size or shape of a particular system.   The principal problem remaining is to find a suitable expression for the neutron-energy spectrum.

It has been seen in Chap. 2 that neutrons are moderated in discrete energy jumps, which may be characterized by a quantity ξ, the average change in the natural logarithm of the energy for a single collision.   If $q(E_0)$ neutrons are produced at high (fission) energy per cubic centimeter per second in an infinite medium and $q(E)$ is the slowing-down density at some lower energy, then

$$q(E) = q(E_0)p(E) \tag{6-46}$$

where $p(E)$ is the resonance-escape probability.   In the energy interval between E and $E + dE$, the change in slowing-down density is entirely accounted for by absorption

$$\frac{dq}{dE}\, dE = n(E)v\Sigma_a\, dE \tag{6-47}$$

---

* Note that for neutrons in thermal equilibrium with their surroundings this difference vanishes.

Furthermore, the number of scattering collisions in the interval is just $n(E)v\Sigma_s\,dE$, which also is equal to the slowing-down density out of the interval $q(E)$, times the average number of scattering collisions per neutron while in the interval, $dE/\xi E$. Thus

$$\frac{q(E)dE}{\xi E} = n(E)v\Sigma_s\,dE \tag{6-48}$$

which demonstrates that <u>the neutron flux per unit energy interval is approximately inversely proportional to the energy in the slowing-down region.</u> Dividing Eq. (6-47) by (6-48), integrating between limits of E and E_0, and substituting into Eq. (6-46) yield

$$p(E) = \exp\left(-\int_E^{E_0} \frac{\Sigma_a}{\xi\Sigma_s}\frac{dE}{E}\right) \tag{6-49}$$

For mixtures, the sum of terms $\xi\Sigma_s$ for each type of nucleus present is employed.

This simple derivation masks some of the assumptions actually inherent in it. It is best applied only in cases where the resonance capture is weak. A much more important case is that in which resonances are strong but widely separated in energy (cf. Fig. 6-3). In this event it may be shown [4] that the resonance-escape probability is more correctly calculated from

$$p(E) = \exp\left[-\frac{N_0}{\xi\Sigma_s}\int_E^{E_0} \frac{\sigma_a}{1+\dfrac{N_0}{\Sigma_s}\sigma_a}\frac{dE}{E}\right] \tag{6-49a}$$

where N_0 is the number of atoms of resonance absorber per unit volume, and $\xi\Sigma_s$ is assumed independent of energy. The integral in Eq. (6-49a) is known as the "effective resonance integral," and it takes into account the fact that the presence of resonance absorber lowers the neutron flux in the energy region in which absorption occurs. This integral is subject to direct experimental measurement [33]. For natural uranium in several moderators, the effective resonance integral has been correlated with the ratio Σ_s/N_0, the scattering cross section per uranium atom, as [19]

$$\int (\sigma_a)_{eff}\frac{dE}{E} = 3.85\left(\frac{\Sigma_s}{N_0}\right)^{0.415} \tag{6-50}$$

which is valid up to values of Σ_s/N_0 of 1000 barns. The limiting value of the integral at infinite dilution is 240 barns. No effect of the mass of the scattering nucleus was found, in agreement with the development of Eq. (6-49a). Limiting values $(N_0/\Sigma_s \to 0)$ of the integral for other materials have also been measured [33]. Note that N_0/Σ_s is also outside the integral in Eq. (6-49a), so that $p(E) = 1$ at infinite dilution. These equations are, of course, restricted essentially to homogeneous reactors. Their application to nonhomogeneous systems is discussed in Sec. 6-35.

The effect of resonance absorption may be incorporated in one-group theory by calculating k as

$$k = p(E_{th}) \frac{\Sigma_f}{\Sigma_a} \nu = p(E_{th})k_{th} \tag{6-51}$$

where E_{th} is taken as the lower limit of integration in Eq. (6-49a).

Prob. 6-5. Prepare plots of k vs. fuel-moderator ratio for natural-uranium fuel and for H_2O, D_2O, and graphite moderators. Show that only D_2O can be used with natural uranium as fuel in a homogeneous thermal reactor.

6-23. Fermi Age Theory. Turning now to the alternate problem of leakage during slowing down, let q_0 be the slowing-down density in a finite but nonabsorbing medium. In this case, the neutron-balance equation for any energy interval dE becomes

$$\frac{\partial q_0}{\partial E} dE = -D(E)\nabla^2 n(E)v \, dE \tag{6-52}$$

where the change in q_0 in a volume element is due to diffusion into or out of the element. Assuming the energy-dependence of $n(E)$ is that given by Eq. (6-48), then substitution of the latter into Eq. (6-52) yields

$$\frac{\partial q_0}{\partial E} = \frac{-D(E)}{\xi \Sigma_s E} \nabla^2 q_0 \tag{6-53}$$

This may be simplified by substituting a new variable $\tau(E)$ such that

$$\tau(E) = \int_E^{E_0} \frac{D \, dE}{\xi \Sigma_s E} \tag{6-54}$$

Equation (6-53) then becomes

$$\nabla^2 q_0 = \frac{\partial q_0}{\partial \tau} \tag{6-55}$$

the Fermi age equation, where the quantity τ is referred to as the "age" by analogy to the unsteady-state heat-conduction equation where the same position is occupied by the time variable. Consequently, solutions to the age equation are formally identical, with analogous boundary conditions, to solutions of the unsteady-state heat-conduction equation. The age equation expresses the manner in which neutrons undergoing moderation are distributed in space and energy. It may be shown that absorption during the slowing-down process has no effect on the form of this distribution if the absorption is relatively weak. The slowing-down density with absorption $q(E)$ is then given by $q_0(E)p(E)$. The assumptions which underlie age theory have been carefully listed and discussed by Marshak et al. [3]. One of the most important of these is that the number of collisions experienced by a neutron after leaving the source must be relatively large. Equation (6-48) is seen to imply the possibility of fractional collisions, since $dE/\xi E$ does not have an integral value. The slowing-down process is thus regarded as a continuous one rather than as one occurring by discrete energy jumps. For this reason it is frequently referred to as "continuous slowing-down theory." A clearer insight into the meaning of the age equation and the quantity τ is obtainable by examining its solution for a point source of neutrons of some relatively high energy E_0. A reactor is made up of a superposition of such sources of varying strengths.

Prob. 6-6. Solve the age equation for a source emitting Q_0 neutrons/sec at the center of a very large sphere of moderating material, and find the mean square distance traveled by neutrons from the source for any energy.

Solution. In spherical coordinates and assuming symmetry about the source, the age equation becomes

$$\frac{\partial^2 q_0}{\partial r^2} + \frac{2}{r}\frac{\partial q_0}{\partial r} = \frac{\partial q_0}{\partial \tau}$$

The solution to this equation with its appropriate boundary condition may be obtained by operational methods or from the equivalent problem in heat conduction [11, 12]. The solution is

$$q_0 = \frac{Q_0 e^{-r^2/4\tau}}{(4\pi\tau)^{3/2}}$$

For each value of τ corresponding to energy E, the spatial distribution of neutrons about the point source is seen to be of the form of the normal or Gauss error curve

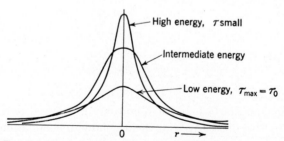

FIG. 6-4. Gaussian distribution of slowing-down density.

(Fig. 6-4). For a given age τ, the mean square slowing-down distance traveled by neutrons is given by

$$\overline{r_s^2} = \frac{\int_0^\infty r^2 q_0\, dV}{\int_0^\infty q_0\, dV} = \frac{\int_0^\infty r^2 e^{-r^2/4\tau} \times 4\pi r^2\, dr}{\int_0^\infty e^{-r^2/4\tau} \times 4\pi r^2\, dr} = 6\tau$$

Thus τ plays the same role in the slowing-down process as L^2 in the diffusion process (Sec. 6-19), namely, one-sixth of the mean square distance from the point of generation as fast neutrons to the point where the energy corresponds to τ. Because it has units of length squared, τ is sometimes called the "slowing-down area," and $\sqrt{\tau}$, the "slowing-down length." For slowing down to thermal energy the age is usually designated as τ_0 corresponding to a lower limit in the integral of Eq. (6-54) of E_{th}. It is this value of τ_0 which is important in determining the leakage effect during slowing down in thermal reactors.

For homogeneous mixtures of nuclei the value of the diffusion coefficient to use in Eq. (6-54) is that given by Eq. (6-33), and the terms $\xi\Sigma_s$ for each constituent nucleus are additive.

At very large distances from a source of fast neutrons it can be shown that age theory does not hold well. An approximate correction to the definition of the age is often used [5]:

$$\tau(E) = \frac{1}{3[\Sigma_s(E_0)]^2} + \int_E^{E_0} \frac{D\,dE}{\xi\Sigma_s E} + \frac{1}{3[\Sigma_s(E)]^2} \qquad (6\text{-}56)$$

where $\Sigma_s(E_0)$ is taken at energy E_0 and $\Sigma_s(E)$ at the final energy E. The two additional terms account for the distances traveled before the first and after the last slowing-down collisions.

Prob. 6-7. Plot the slowing-down density $q_0(E)$ and the flux $n(E)v$ as functions of energy down to 0.025 ev for a point source of 1-Mev neutrons in an infinite medium of Be metal. Assume ξ, Σ_s, and $\bar{\mu}$ are energy-independent.

6-24. Reactor Equation with Fermi Age Theory. From the results of the two preceding sections, it is possible to formulate a more realistic theory for thermal reactors. The correct source term in the continuity equation for thermal neutrons becomes $q_0(E_{th})p(E_{th})$, i.e., the number of neutrons slowing down to thermal energy per cubic centimeter per second. The reactor equation may then be written as

$$-D\nabla^2\Phi + \Sigma_a\Phi = q_0(E_{th})p(E_{th}) \qquad (6\text{-}57)$$

The additional space-dependent variable q_0 is determined by the age equation (6-55). The boundary conditions on these simultaneous equations include the source term for the age equation in terms of the thermal flux. Since the fission reaction is the source of fast neutrons, then $q_0(E_0) = \nu\Sigma_f\Phi$, which shows that the spatial dependence of q_0 will be the same as that of the thermal flux if the fuel concentration is uniform.

Assuming a solution of the age equation in the form

$$q_0(E) = \nu\Sigma_f\Phi F(\tau) \qquad (6\text{-}58)$$

where $F(0) = 1$, then substitution in Eq. (6-55) yields

$$\frac{\nabla^2\Phi}{\Phi} = \frac{F'(\tau)}{F(\tau)}$$

Since each term is independent of the other, then both are constant, or

$$\nabla^2\Phi + B^2\Phi = 0 \tag{6-59}$$

and

$$\frac{dF}{d\tau} = -B^2F \tag{6-60}$$

Equation (6-59) is the familiar reactor diffusion equation (6-9). The solution of Eq. (6-60) is readily obtained as

$$F = e^{-B^2\tau} \tag{6-61}$$

Substituting Eqs. (6-61) and (6-58) into (6-57) and rearranging,

$$\nabla^2\Phi + \frac{(ke^{-B^2\tau_0} - 1)}{L^2}\Phi = 0 \tag{6-62}$$

where the quantity $\nu\Sigma_f p(E_{th})/\Sigma_a$ has been set equal to k, as in Eq. (6-51). On comparing Eq. (6-62) with (6-59) it is seen that the constant B^2, which is the buckling, is related to the properties of the system through the critical equation

$$k_{eff} = \frac{ke^{-B^2\tau_0}}{1 + L^2B^2} = 1 \tag{6-63}$$

The factor $e^{-B^2\tau_0}$ is the nonleakage probability during slowing down, referred to in Sec. 6-5 as $P_r(E_0,E)$. If the values of k, τ_0, and L^2 are known for a given system of fissionable and moderating materials, the buckling may be calculated from Eq. (6-63) and related to the critical size through Eq. (6-59). Solution for B^2 in Eq. (6-63) must be made by a trial-and-error or iterative procedure.

Prob. 6-8. A uniform mixture of 1 part by weight of U^{235} to 15,000 parts of graphite is to be used to construct a bare spherical reactor. Estimate (a) the critical size, (b) the mass of U^{235} required, (c) the total leakage rate of neutrons from the reactor, and (d) the rate of fuel burnup for continuous operation at an average power output of 1000 kw.

6-25. Large Thermal Reactors. It has been noted previously that diffusion theory gives a valid description of neutron behavior only for systems whose dimensions are large compared to a scattering mean free path. Under these circumstances the buckling B^2 is relatively small. If τ_0 is not too large, the exponential term in Eq. (6-63) may be placed in the denominator and expanded in series, yielding

$$e^{B^2\tau_0} = 1 + B^2\tau_0 + \cdots$$

so that Eq. (6-63) becomes

$$k_{eff} = \frac{k}{1 + B^2(L^2 + \tau_0)} = 1 \tag{6-64}$$

neglecting terms of order B^4 and higher. The quantity $L^2 + \tau_0$ is usually referred to as the "total migration area" M^2. For large thermal reactors, then, the critical equation is given by Eq. (6-64) which allows a direct solution for B^2 and may be used to obtain the first trial value of B^2 when it is necessary to employ the more rigorous Eq. (6-63). It is of interest to note that the size of large thermal reactors is directly dependent on the additive mean square distances traveled by neutrons from the point of generation as fast neutrons to the point of absorption as thermal neutrons.

6-26. Allowance for Fission-neutron-energy Spectrum [39]. The foregoing treatment of thermal reactors utilizing continuous slowing-down theory may be readily expanded to include the knowledge that fission neutrons are created over a spectrum of energies (cf. Fig. 6-3). If $F(E)$ represents the normalized distribution of fission neutrons in energy, then the critical equation (6-63) becomes

$$k_{eff} = \frac{k}{1 + L^2B^2} \int_{E_{th}}^{E_0} \exp\left[-B^2\tau(E,E_{th})\right] F(E)\, dE = 1 \tag{6-65}$$

The arguments on τ represent the limits on the defining integral, Eq. (6-54). For U^{235} and Pu^{239} fission, it has been shown [13,14] that the energies of fission neutrons are distributed according to the normalized equation

$$F(E) = 0.484e^{-E} \sinh \sqrt{2E} \tag{6-66}$$

representing the range from 0.075 to 17 Mev.

6-27. Generalized Reactor Equation with Age Theory. Weinberg [15] has given a further generalization of age theory to include resonance absorption which leads to fission. The most general critical equation which defines the buckling B^2 is shown to be

$$\int_{E_{th}}^{E_0} \int_{E'}^{E_0} \frac{k(E')\Sigma_a(E')}{\xi\Sigma_s(E')} p(E,E') \exp\left[-B^2\tau(E,E')\right] F(E)\, dE\, dE'$$
$$+ \frac{k_{th}}{1 + L^2B^2} \int_{E_{th}}^{E_0} p(E,E_{th}) \exp\left[-B^2\tau(E,E_{th})\right] F(E)\, dE = 1 \tag{6-67}$$

where the two terms represent k_{eff} for the resonance and thermal-energy regions. In the first term E is the energy at which neutrons are released in fission, variable up to a maximum energy of E_0, while E' represents the energy at which neutrons are absorbed. The second term includes all neutrons absorbed at E_{th}.

6-28. Intermediate Reactors. This treatment is also directly applicable to intermediate reactors, for which $k_{th} = 0$. Since all the known fissionable fuels have appreciable values of σ_f at thermal energy, the only way $k_{eff,th} = 0$ can be physically attained is to prevent neutrons from slowing down all the way to thermal energies within the reactor. Such a condition is realizable largely by keeping $\xi\Sigma_s(E)$, the so-called "slowing-down power," small.

The use of age and diffusion theory for resonance reactors is subject to the limitations previously noted. For most calculations there is little loss in accuracy by assuming all neutrons are produced at a single energy E_0. In this case the critical equation reduces to

$$k_{eff} = \int_{E_{th}}^{E_0} \frac{k(E')\Sigma_a(E')}{\xi\Sigma_s(E')}\ p(E_0,E')\ \exp\ [-B^2\tau(E_0,E')]\ dE' = 1 \qquad (6\text{-}68)$$

In the resonance region the neutron flux is highly energy-dependent. A fast or resonance neutron flux may be defined by integrating the flux per unit energy interval over the entire energy range.

$$\Phi_f = \int_0^{E_0} \Phi(E)\ dE \qquad (6\text{-}69)$$

The limits need include only those energies for which $\Phi(E)$ is significantly greater than zero. Similarly neutron reaction rates in the resonance region must be calculated by an equation such as the following for the rate of heat generation per unit volume in a resonance reactor:

$$H = c \int_0^{E_0} \Sigma_f(E)\Phi(E)\ dE \qquad (6\text{-}70)$$

In turn, the flux per unit energy interval is separable into space- and energy-dependent factors as $\Phi(\mathbf{r},E) = n(\mathbf{r})m(E)$ where the space-dependent portion follows the usual reactor diffusion equation (6-9) and the energy-dependent portion is given by

$$m(E) = p(E_0,E)\ \exp\ [-B^2\tau(E_0,E)] \qquad (6\text{-}71)$$

For $E = E_{th}$, this is the result previously found for thermal reactors. Combining the above relations, the fast flux is given by

$$\Phi_f = n(\mathbf{r}) \int_0^{E_0} p(E_0,E)\ \exp\ [-B^2\tau(E_0E)]\ dE \qquad (6\text{-}72)$$

and the arbitrary flux level is incorporated in the solution of Eq. (6-9), as previously. From the foregoing it is clear that a knowledge of a fast flux in a reactor is of limited use as far as reaction-rate calculations are concerned, unless the relevant cross sections are essentially constant over the energy range, or appropriate average cross sections are known.

6-29. Hydrogen-moderated Reactors. The assumption of a continuous slowing-down process is least justified in the case of reactors moderated with water, hydrocarbons, or other predominantly hydrogenous materials. As shown in Chap. 2, neutrons can be thermalized in a single collision with protons, although on the average about 18 collisions are required. Although it is possible to make very precise calculations for hydrogen-moderated reactors by an extension of the group method given in the next section, it may be shown [15] that the following critical equation represents a better approximation than age theory for hydrogen-moderated reactors.

$$k_{eff} = \frac{k\Sigma_0 \tan^{-1}(B/\Sigma_0)}{B(1 + L^2 B^2)} = 1 \qquad (6\text{-}73)$$

The factor $(\Sigma_0/B) \tan^{-1}(B/\Sigma_0)$ represents the nonleakage probability for nonthermal neutrons. The cross section Σ_0 is assumed to characterize a process whereby fission neutrons are thermalized in single scattering

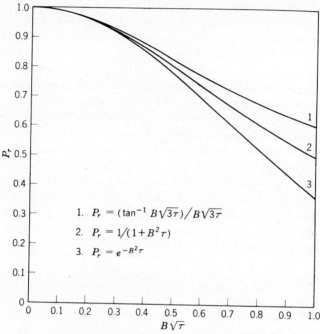

FIG. 6-5. Nonleakage probabilities during slowing down.

collisions. The physical interpretation of Σ_0 is facilitated by expanding $(\Sigma_0/B) \tan^{-1}(B/\Sigma_0)$ in infinite series. It may then be shown that $1/3\Sigma_0^2$ should be identified with one-sixth of the mean square slowing-down distance from fission to thermal energies

$$\frac{1}{3\Sigma_0^2} = \frac{\overline{r_s^2}}{6} = \tau_0 \qquad (6\text{-}74)$$

Since $\overline{r_s^2}$ is a physically measurable quantity, Eq. (6-74) is the definition of Σ_0 to be used in the critical equation (6-73).

For comparison, the nonleakage probabilities in Eqs. (6-73) and (6-63) are plotted against $B\sqrt{\tau}$ in Fig. 6-5. For a hydrogen-moderated reactor of a given size it is seen that Eq. (6-73) predicts less leakage during slowing down, particularly for small reactors, i.e., large B^2. This is in accord with the expectation that neutrons moderated in hydrogen do not possess high energies long enough to move very far as high-energy neutrons. Although a one-collision theory presumes no intermediate-energy neutrons exist in the reactor, the resonance-escape probability can be separately computed and included in the multiplication factor k.

6-30. Multigroup Theory [40]. In addition to the inapplicability of age theory in very light moderators, for composite media it is difficult to fit appropriate solutions of the age and diffusion equations to the boundary conditions. This fact arises because the age equation is a partial differential equation. In both these situations it is advantageous to extend the one-group theory of Sec. 6-19 but retain only space-dependent differential equations. For simplicity, the treatment here is restricted to thermal reactors in which nonthermal absorption is assumed negligible and fission neutrons are all produced at the single energy E_0.

The basic assumption of multigroup theory is that, in any energy interval, neutrons diffuse at constant energy until they have undergone the number of scattering collisions which, on the average, are required to lower the energy to the next group. As an example, consider a three-group theory in which all "thermal" neutrons form one group, all neutrons having energies from E_{th} to E_1 form a second group, and all neutrons having energies from E_1 to E_0 form the third group. On the average, the number of collisions required to change a neutron's energy from E_0 to E_1 is $\dfrac{\ln (E_0/E_1)}{\xi}$, and similarly the change from E_1 to E_{th} requires $\dfrac{\ln (E_1/E_{th})}{\xi}$ collisions. A neutron-balance equation for the highest-energy group may be written as

$$\int_{E_1}^{E_0} D(E)\nabla^2\Phi(E)\, dE - q(E_1) + q(E_0) = 0 \qquad (6\text{-}75)$$

with a similar equation using $q(E_{th})$ and $q(E_1)$ for the second group.

For the thermal group

$$\int D(E)\nabla^2\Phi(E)\, dE - \int \Sigma_a(E)\Phi(E)\, dE + q(E_{th}) = 0 \qquad (6\text{-}76a)$$

The integration in Eq. 6-76a is understood to be over the thermal spectrum, so that it becomes (cf. Sec. 6-21)

$$\bar{D}\nabla^2\Phi_{th} - \bar{\Sigma}_a\Phi_{th} + q(E_{th}) = 0 \qquad (6\text{-}76b)$$

where $\bar{D}$ and $\bar{\Sigma}_a$ are given by Eqs. (6-41) and (6-42).

The source of neutrons for the fast group is the fission process, so that $q(E_0) = \nu\Sigma_f\Phi_{th}$. The rate at which neutrons leave the fast group is assumed to be characterized by a fictitious cross section Σ_1 such that

$q(E_1) = \Sigma_1\Phi_1 = \Sigma_1 \int_{E_1}^{E_0} \Phi(E)\ dE$ is the rate at which neutrons leave unit volume by passing into the next lower energy group. By the original hypothesis this would be given by the rate at which scattering collisions occur within the energy group, divided by the number of collisions required to reduce neutron energy to E_1. Thus

$$\Sigma_1\Phi_1 = \frac{\int_{E_1}^{E_0} \Sigma_s(E)\Phi(E)\ dE}{[\ln\ (E_0/E_1)]/\xi} = \frac{\xi\bar{\Sigma}_s\Phi_1}{\ln\ (E_0/E_1)} \qquad (6\text{-}77)$$

with a similar definition for Σ_2, and $q(E_{th}) = \Sigma_2\Phi_2$.

The appropriate average diffusion coefficient is defined by

$$\bar{D}_1 = \frac{\displaystyle\int_{E_1}^{E_0} \frac{D(E)\ dE}{\xi\Sigma_s E}}{\displaystyle\int_{E_1}^{E_0} \frac{dE}{\xi\Sigma_s E}} \qquad (6\text{-}78)$$

and a similar form for $\bar{D}_2$, so that

$$\bar{D}_1\nabla^2\Phi_1 - \Sigma_1\Phi_1 + \nu\Sigma_f\Phi_{th} = 0 \qquad (6\text{-}79)$$
and $\qquad \bar{D}_2\nabla^2\Phi_2 - \Sigma_2\Phi_2 + \Sigma_1\Phi_1 = 0 \qquad (6\text{-}80)$

while Eq. (6-76b) becomes

$$\bar{D}\nabla^2\Phi_{th} - \bar{\Sigma}_a\Phi_{th} + \Sigma_2\Phi_2 = 0 \qquad (6\text{-}81)$$

If solutions in the usual form of Eq. (6-9) are assumed and substituted in Eqs. (6-79), (6-80), and (6-81), there results the following set of simultaneous algebraic equations:

$$\begin{aligned}
(\bar{D}_1B^2 + \Sigma_1)\Phi_1 & & -\nu\Sigma_f\Phi_{th} &= 0 \\
-\Sigma_1\Phi_1 + (\bar{D}_2B^2 + \Sigma_2)\Phi_2 & & &= 0 \qquad (6\text{-}82) \\
-\Sigma_2\Phi_2 + (\bar{D}B^2 + \bar{\Sigma}_a)\Phi_{th} &= 0
\end{aligned}$$

In accord with previous reasoning that the flux or power level be arbitrary, it is necessary that the determinant of the coefficients in Eq. (6-82) vanish. This requirement readily leads to a critical equation which can be written in the form

$$k_{eff} = \frac{k_{th}}{(1 + L_1{}^2B^2)(1 + L_2{}^2B^2)(1 + L^2B^2)} = 1 \qquad (6\text{-}83)$$

Here L is the thermal diffusion length previously defined, while

$$L_1{}^2 = \frac{\bar{D}_1}{\Sigma_1}$$

and $L_2{}^2 = \bar{D}_2/\Sigma_2$. From Eqs. (6-77) and (6-78), these are seen to be partial Fermi ages if $\xi\Sigma_s$ is constant in each group [cf. Eq. (6-54)]. The effect of resonance absorption can be separately computed and k_{th} replaced by $k = k_{th}p$. For large reactors, Eq. (6-83) goes over into the form previously found [Eq. (6-64)], where the migration area

$$M^2 = L_1{}^2 + L_2{}^2 + L^2$$

For a very large number of groups, it can be shown that the critical equation for multigroup theory becomes identical with that for age theory.

The application of these relations to critical reactor calculations is complicated by the fact that the critical equation is a cubic in B^2, or in general, of the nth degree for an n-group theory. All real roots of the critical equation must be accepted and the solutions for the various group fluxes written as linear combinations of functions satisfying Eq. (6-9) for each root. Thus, for the three-group illustration, the solutions may be written

$$\begin{aligned} \Phi_1 &= A_1X + A_2Y + A_3Z \\ \Phi_2 &= A_1'X + A_2'Y + A_3'Z \\ \Phi_{th} &= A_1''X + A_2''Y + A_3''Z \end{aligned} \qquad (6\text{-}84)$$

where each of X, Y, and Z satisfy Eq. (6-9) replacing B^2 successively by each of the real roots. If the cubic has only one real root, the terms in Y and Z will not have physical significance. The constants, A, A', and A'' are interrelated through Eqs. (6-82).

The significance of the group method may be more clearly understood by computing the manner in which the slowing-down density falls off from a point source of fast neutrons. The result $e^{-r/L_1}/4\pi rL_1{}^2$ for a single fast group should be compared with the Gaussian distribution of Prob. 6-6, Sec. 6-23. On further calculating the mean square slowing-down distance from a point source, it may be shown that $\overline{r_s{}^2} = 6L_1{}^2$ for a single fast group, or $\overline{r_s{}^2} = 6L_1{}^2 + 6L_2{}^2 + \cdots$ for any number of groups. Consequently, both age and group theories yield the same value of the mean square slowing-down distance, but the latter yields a more rapidly diminishing slowing-down density in the vicinity of the source. This is precisely the situation obtaining for very light moderators, hence its usefulness for criticality calculations for light- and heavy-water-moderated reactors. By the same token, measured slowing-down densities in light moderators cannot be fitted well to an equation containing a single constant, although the exponential form gives a better fit than the Gaussian. It has been found experimentally that at least three constants L_1, L_2, and L_3, are required to fit measured slowing-down distributions in light water [4].

Prob. 6-9. Estimate the critical radius of a bare homogeneous spherical reactor consisting of an aqueous solution of uranyl (U^{235}) sulfate. Compare the predictions based on (a) age theory, (b) the critical equation for large reactors, (c) two-group theory, and (d) the method of Sec. 6-29, if the atomic ratio of hydrogen to U^{235} is 1000:1.

Solution. The expression for k may be rearranged as follows:

$$k = \frac{\nu\Sigma_f}{\Sigma_a} = \frac{\Sigma_{aU}}{\Sigma_{aU} + \Sigma_{aH}} \frac{\nu\sigma_f}{\sigma_{aU}} = f\eta$$

where f is the first fractional term, known as the thermal utilization (cf. Sec. 6-32). The second fractional term is designated η. In terms of the moderator-fuel ratio $\alpha = N_H/N_U$, the expression for f becomes $f = \sigma_{aU}/(\sigma_{aU} + \alpha\sigma_{aH})$. Here, then, $f = 650/650 + 1000 \times 0.32 = 0.670$, where the thermal cross sections are taken from pages 86 and 98 and are assumed to be proportional to $1/v$. Also

$$\eta = 2.5 \times 549/650 = 2.1 \quad \text{and} \quad k = 0.670(2.1) = 1.41$$

The thermal diffusion length in the mixture is given by Eq. (6-37). Neglecting the effect of U on the diffusion coefficient, it may be rearranged to give $L^2 = L_1^2(1 - f)$ where L_1 is the diffusion length in pure water. From Table 6-5, $L_1 = 2.85$ cm, so that $L^2 = 2.85^2(1 - 0.670) = 2.68$ cm². Assuming slowing down to be due entirely to water, then $\tau_0 = 33$ cm² from Table 6-5.

a. Substitution of the above in the critical equation with age theory yields

$$1.41 \exp(-33B^2) = 1 + 2.68B^2$$

Trial-and-error solution for the buckling yields

$$B^2 = 0.0096 \text{ cm}^{-2}$$

The critical radius (extrapolated) is then

$$R = \frac{\pi}{\sqrt{0.0096}} = 32.1 \text{ cm}$$

The extrapolation distance is, from Table 6-4,

$$d = 0.71 \times 0.48 = 0.34 \text{ cm}$$

The actual critical radius of the bare sphere is thus predicted to be

$$32.1 - 0.3 = 31.8 \text{ cm}$$

b. Substitution in the critical equation for large reactors yields

$$1.41 = 1 + (33 + 2.68)B^2$$
or
$$B^2 = 0.0115 \text{ cm}^{-2}$$

from which $R = 29.0$ cm.

c. Substitution in the critical equation for two-group theory gives

$$1.41 = (1 + 33B^2)(1 + 2.68B^2)$$
or
$$B^2 = 0.0112 \text{ cm}^2$$

from which $R = 29.4$ cm.

d. The fictitious cross section Σ_0 is given by

$$\Sigma_0{}^2 = \frac{1}{3 \times 33} = 0.0101 \quad \text{or} \quad \Sigma_0 = 0.101 \text{ cm}^{-1}$$

Substitution in the critical equation (6-73) gives

$$1.41 \frac{0.101}{B} \tan^{-1} \frac{B}{0.101} = 1 + 2.68B^2$$

from which, by trial, $B^2 = 0.0142$ cm^{-2}; then, $R = 26.1$ cm.

HETEROGENEOUS THERMAL REACTORS

6-31. Modifications in Theory. The theory outlined in the preceding sections is strictly applicable only to reactors in which all materials are uniformly dispersed throughout the structure. If a reactor is to operate at sufficiently high power levels that some form of cooling is required, it is necessary to introduce some degree of heterogeneity unless a fuel-moderator solution is recirculated and cooled externally. Some advantages may also accrue when the fuel and moderator are physically separated. The usual technique has been to employ solid fuel elements dispersed in a regular matrix of moderating material. The governing factor that first led to this arrangement was that it made possible the use of *natural* uranium fuel with graphite moderator. Relatively simple calculations will show that homogeneous natural-uranium–moderator mixtures cannot have a multiplication factor in excess of unity unless the moderator is heavy water.

The increase in k that is brought about by the "lumping" of uranium is due largely to a decrease in the resonance absorption of neutrons by U^{238}. Since neutrons are slowed down only in the moderator, those having energies corresponding to strong U^{238} absorption peaks can enter the fuel only from the outside and are rapidly absorbed in the surface layer of fuel. Uranium 238 atoms in the inner part of the fuel elements thus have little opportunity to react with neutrons in the resonance region. This phenomenon is often called "self-protection." A factor of lesser importance is the production of 2 to 3 per cent of the neutrons by fast fissions in U^{238} due to fission neutrons produced within the fuel elements. These two advantages are partially offset by a decrease in the relative proportion of thermal neutrons absorbed by the fuel.

The disposition of fuel into lattice cells complicates the theoretical problems of estimating critical size. It is beyond the scope of this chapter to develop fully the methods for making such calculations, but some of the more important features will be summarized. In heterogeneous reactors with a regularly recurring lattice cell, the multiplication factor k, for the infinite reactor, is dependent on the dimensions, geometry, and composition of a single cell.   If the cells are small relative to the over-all size for a finite reactor, the over-all neutron flux distribution may be regarded as basically that obtaining in a homogeneous reactor as predicted by the methods of Secs. 6-10ff.

Superimposed upon this will be comparatively minor variations within cells (Fig. 6-6).   The critical equations for homogeneous reactors may thus be applied to heterogeneous reactors, with modifications employed in calculating the individual terms entering the critical equation [20, 41–43].

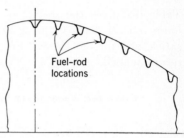

Fuel-rod locations

Fig. 6-6. Flux distribution in a heterogeneous reactor with uniform matrix.

6-32. The Multiplication Factor. It has proved convenient to express the multiplication factor k as a four-factor product

$$k = \epsilon p f \eta \qquad (6\text{-}85)$$

where ϵ = fast-fission factor ($\epsilon - 1$ = number of fission neutrons produced by fast fission of U^{238} per fission neutron produced by thermal fission of U^{235})

p = resonance-escape probability

f = thermal utilization (relative number of thermal neutrons absorbed in fuel)

η = number of fission neutrons produced per thermal neutron absorbed in fuel

The theoretical calculation of each of these factors constitutes what is frequently referred to as "microscopic pile theory," since they are independent of over-all reactor size.

6-33. Fast-fission Factor. Consideration of the various reactions which fission neutrons may undergo with U^{238} leads to an expression for ϵ in terms of the cross sections for the reactions and a quantity P which is the probability that a fission neutron, formed anywhere inside the fuel element, will make a single collision with a uranium nucleus before escaping into the moderator [4, 16, 44].   This expression takes the form

$$\epsilon - 1 = \frac{(\nu - 1 - \sigma_c/\sigma_f)(\sigma_f/\sigma)P}{1 - (1/\sigma)(\nu\sigma_f + \sigma_e)P} \qquad (6\text{-}86a)$$

where subscripts c, f, and e refer to nonfission capture, fission, and elastic scattering respectively, while σ is the total cross section, including that for inelastic scattering.   For natural uranium this expression becomes

$$\epsilon - 1 = \frac{0.0952P}{1 - 0.521P} \qquad (6\text{-}86b)$$

Table 6-2 gives the results of calculations for P for solid natural-uranium rods of density 18.7 g/cm³, for hollow cylinders of various radius ratios, and for flat slabs.   For rough calculations a value of $\epsilon = 1.03$ is recommended.

TABLE 6-2. PROBABILITY FUNCTION P IN CALCULATION OF FAST-FISSION EFFECT*

r_e, cm	r_i, cm	P	
		Solid or hollow cylinders	Slab of half-thickness r_e
1.0	0.0	0.2096	0.398
	0.2	0.2013	
	0.4	0.1743	
	0.6	0.1418	
	0.8	0.0893	
2.0	0.0	0.3543	0.560
	0.4	0.3428	
	0.8	0.3074	
	1.2	0.2546	
	1.6	0.1665	
3.0	0.0	0.4566	0.657
	0.6	0.4424	
	1.2	0.4044	
	1.8	0.3389	
	2.4	0.2253	
4.0	0.0	0.5368	0.722
	0.8	0.5239	
	1.6	0.4836	
	2.4	0.4123	
	3.2	0.2844	

* H. Castle, H. Ibser, G. Sacher, and A. M. Weinberg, CP-644, 1943; R. L. Murray and A. C. Menius, Jr., *Nucleonics*, **11**(4): 21–23 (April, 1953).

6-34. Thermal Utilization.   Computation of the thermal utilization f for an infinite heterogeneous reactor depends not only on the nature and geometry of the fuel elements but also upon all other constituents of the lattice cell such as moderator, coolant, and structural materials. Since each of the constituents of a cell is not exposed to the same thermal neutron flux, the definition of the thermal utilization becomes

$$f = \frac{\Sigma_{a0} V_0 \bar{\Phi}_0}{\Sigma_{a0} V_0 \bar{\Phi}_0 + \Sigma_{a1} V_1 \bar{\Phi}_1 + \Sigma_{ac} \bar{\Phi}_c V_c + \Sigma_{as} \bar{\Phi}_s V_s} \tag{6-87}$$

where subscripts 0, 1, c, and s, refer to fuel, moderator, coolant, and structural materials, respectively, and the terms $V\bar{\Phi}$ are the products of volume and mean thermal flux pertaining to each particular portion of a lattice cell.

For calculation purposes, however, it is more convenient to write

$$\frac{1}{f} = \frac{\int_0^{V_{\text{cell}}} q_{th} \, dV}{\int_0^{V_0} \Sigma_{a0} \Phi_0 \, dV_0} \tag{6-88}$$

where q_{th} is the rate of production of thermal neutrons per unit volume in the lattice cell. Equation (6-88) is valid only for the case of an infinite reactor where there is no net transport of thermal neutrons between adjacent cells. Solutions of Eq. (6-88) in terms of cell geometry and nuclear properties may be obtained by making appropriate simplifying assumptions, namely, (1) idealized cell geometry, (2) applicability of elementary diffusion theory, and (3) q_{th} constant in the moderator and zero elsewhere.* Details of the method for cells containing only moderator and fuel have been given by Weinberg [4, 18]. A more rigorous approach [34] has shown that this method gives good results for lattices near the optimum, i.e., maximum k.

The absorption effects of coolant and structural materials can be included in this type of calculation, but they are generally small and are readily handled more easily by perturbation methods (cf. Sec. 6-51), the use of which is recommended in conjunction with calculations based on Table 6-3. If coolant and structural materials occupy an appreciable volume of a lattice cell, there is a purely geometrical effect on the calculated values of f. A slight extension of the results of Weinberg is given in Table 6-3 for four different geometries which may be of importance in large-scale heterogeneous thermal reactors. These results are obtained by assuming that the coolant is a nonabsorbing medium and has no appreciable moderating effect. They should, therefore, be most accurate for gaseous or weakly absorbing liquid-metal coolants.

In Table 6-3, κ_0 and κ_1 are the inverse diffusion lengths in fuel and moderator, respectively; and R_0, R_1, R_c, and R_a are radial or half-thickness distances shown on the respective figures. In the cylindrical geom-

* It is possible to make allowance for thermal-neutron production in a coolant such as H_2O, which might also act as a moderator, but it complicates the treatment considerably.

TABLE 6-3. THERMAL UTILIZATION IN HETEROGENEOUS LATTICES

Geometry	$1/f$

Cylindrical:

1. Externally cooled fuel element (Fig. 6-7a)

$$\frac{\Sigma_{a1}V_1}{\Sigma_{a0}V_0} G(\kappa_0 R_0, 0) + F(\kappa_1 R_1, \kappa_1 R_c)$$

$$+ \frac{\Sigma_{a1}}{2D_c} (R_1{}^2 - R_c{}^2) \ln \frac{R_c}{R_0}$$

2. Internally cooled fuel element (Fig. 6-7b)

$$\frac{\Sigma_{a1}V_1}{\Sigma_{a0}(V_0 + V_c)} G(\kappa_0 R_0, \kappa_0 R_c) + F(\kappa_1 R_1, \kappa_1 R_0)$$

3. Internally and externally cooled fuel element (Fig. 6-7c)

$$\frac{\Sigma_{a1}V_1}{\Sigma_{a0}(V_0 + V_c)} G(\kappa_0 R_0, \kappa_0 R_c) + F(\kappa_1 R_1, \kappa_1 R_a)$$

$$+ \frac{\Sigma_{a1}}{2D_a} (R_1{}^2 - R_a{}^2) \ln \frac{R_a}{R_0}$$

Plane:

4. Externally cooled plate fuel element (Fig. 6-7d)

$$\frac{\Sigma_{a1}V_1}{\Sigma_{a0}V_0} \kappa_0 R_0 \coth \kappa_0 R_0 + \kappa_1 (R_1 - R_c) \coth \kappa_1 (R_1 - R_c)$$

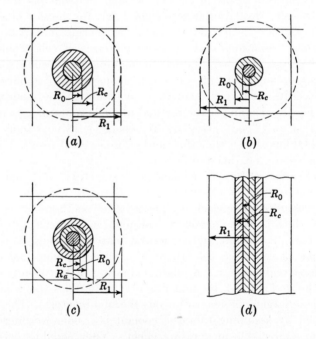

(a)　　　　　　　　　　　　(b)

(c)　　　　　　　　　　　　(d)

Key: ▨ Fuel　　▨ Coolant　　⬭ Moderator

FIG. 6-7. Geometries of lattice cells.

etry, R_1 is defined such that $\pi R_1{}^2$ is the total volume of the cell per unit length (cf. Fig. 6-7). D_c and D_a are diffusion coefficients in core coolant and annular coolant, respectively. For gaseous coolants, unless under very high pressure, the third terms in (1) and (3) will be negligibly small. The functions G and F are defined in terms of Bessel functions as follows:

$$G(x_0, x_c) = \frac{x_0}{2} \frac{I_0(x_0)K_1(x_c) + K_0(x_0)I_1(x_c)}{I_1(x_0)K_1(x_c) - K_1(x_0)I_1(x_c)} \tag{6-89}$$

and
$$G(x_0, 0) = \frac{x_0}{2} \frac{I_0(x_0)}{I_1(x_0)}$$

$$F(y_1, y_2) = \frac{y_1{}^2 - y_2{}^2}{2y_2} \frac{I_1(y_1)K_0(y_2) + K_1(y_1)I_0(y_2)}{I_1(y_1)K_1(y_2) - K_1(y_1)I_1(y_2)} \tag{6-90}$$

The derivations of these expressions lead also to the equations for the variation of the flux through the fuel element, namely,

$$\Phi = A I_0(\kappa_0 r) \tag{6-91a}$$

for the externally cooled fuel element (1),

$$\Phi = A'[K_1(\kappa_0 R_c)I_0(\kappa_0 r) + I_1(\kappa_0 R_c)K_0(\kappa_0 r)] \tag{6-91b}$$

for internally cooled elements (2), and for elements cooled on both sides (3), and

$$\Phi = A'' \cosh \kappa_0 r \tag{6-91c}$$

for the plate-type fuel elements (4). Since the power generation is proportional to the flux in a fuel element, these equations are of considerable significance in estimating temperature gradients through fuel elements, as will be shown in Chap. 9.

6-35. Resonance-escape Probability. The computation of the resonance-escape probability rests upon a combination of theory and empiricism. The resonance-escape formula for heterogeneous lattice assemblies takes the form [4]

$$p(E) = \exp\left[-\frac{N_0 V_0 \bar{\Phi}_{0r}}{V_1 \xi_1 \Sigma_{s1} \bar{\Phi}_{1r}} \int (\sigma_{a0})_{eff} \frac{dE'}{E'}\right] \tag{6-92}$$

where subscripts 0 and 1 refer to U^{238} and moderator, respectively. The mean fluxes are for neutrons having resonance energies, not thermal energies, and the flux ratio is assumed independent of energy. The effective resonance absorption integral in a finite mass of uranium is that given by Eq. (6-50) plus a contribution proportional to the surface exposed to the source of resonance neutrons. For natural uranium metal [19]

$$\int (\sigma_{a0})_{eff} \frac{dE'}{E'} = 9.25 + 24.7 \frac{S}{M} \qquad \text{barns} \tag{6-93}$$

where S/M = surface area per unit mass of fuel element, cm^2/g.

Evaluation of the ratio of mean resonance fluxes in uranium and moderator takes exactly the same form as in the calculation of the thermal utilization. In fact, if f_r is defined as the resonance utilization, it may be shown that Eq. (6-92) reduces to

$$p(E) = \exp\left(-\frac{f_r}{1 - f_r}\right) \tag{6-94}$$

where f_r is given by the same equations as f, namely, Eqs. (6-87) and (6-88). In calculating f_r, however, group constants appropriate to the resonance energy region must be employed. For uranium metal [19]

$$\kappa_0 = 0.022\rho \tag{6-95}$$

where $\rho =$ density of metal, g/cm^3. The macroscopic absorption cross section is calculated from

$$\frac{\bar{\Sigma}_{a0}}{N_0} = \frac{\int_{E_2}^{E_1} (\sigma_{a0})_{eff} \frac{dE'}{E'}}{\ln (E_1/E_2)} \tag{6-96}$$

where $\ln (E_1/E_2)$ has a value of 5.6 for uranium metal. Group constants appropriate for various moderators with uranium metal are shown in Table 6-4.

TABLE 6-4. RESONANCE GROUP CONSTANTS FOR MODERATORS WITH NATURAL-URANIUM-METAL FUEL*

Moderator	Σ_1, cm^{-1}	κ_1, cm^{-1}
Water....................	0.241	0.583
Heavy water..............	0.0313	0.155
Beryllium................	0.0276	0.237
Beryllium oxide...........	0.0150	0.138
Graphite.................	0.0108	0.1075

* S. Glasstone and G. Edlund, "The Elements of Nuclear Reactor Theory," D. Van Nostrand Company, Inc., New York, 1952.

6-36. Neutrons per Thermal Neutron Absorbed. This quantity is a characteristic property of the fuel employed and is independent of its manner of dispersion.

$$\eta = \frac{\Sigma_{f0}\nu}{\Sigma_{a0}} = \frac{\sigma_{f0}\nu}{\sigma_{a0}} \tag{6-97}$$

6-37. Diffusion in Heterogeneous Lattices. The diffusion coefficient D entering the neutron-balance equations should be a weighted mean value. Since D is the same order of magnitude for most moderators, coolants, and fuels and since the relative volume of moderator is normally very large in a thermal reactor, little error is involved in taking D as D_1 for the moderator alone. For gas-cooled reactors where the coolant ducts are relatively large, the effective value of D parallel to the ducts may be appreciably greater than perpendicular. This anisotropy can be accounted for by using separate values of D in the basic diffusion equations [1].

TABLE 6-5. DIFFUSION AND SLOWING-DOWN PROPERTIES AT 20°C*

Material	Density, g/cm³	λ_t, cm	L, cm	$\bar{\lambda}_s$, cm*	$\bar{r}_s^2/6$, cm²†
H₂O.....................	1.00	0.48	2.85	1.1	33
D₂O.....................	1.10	2.40	171	2.6	120
Be......................	1.85	1.43	20.8	1.6	98
C (graphite).............	1.60	2.74	50.8	2.6	350

* S. Glasstone and G. Edlund, "The Elements of Nuclear Reactor Theory," D. Van Nostrand Company, Inc., New York, 1952; D. J. Hughes, "Pile Neutron Research," Addison-Wesley Publishing Company, Cambridge, Mass., 1953.
† Fission neutrons moderated to thermal energies.

The mean effective absorption cross section is defined by

$$\bar{\Sigma}_a = \frac{V_0 \Sigma_{a0} \bar{\Phi}_0 + V_1 \Sigma_{a1} \bar{\Phi}_1 + V_c \Sigma_{ac} \bar{\Phi}_c}{V_0 \bar{\Phi}_0 + V_1 \bar{\Phi}_1 + V_c \bar{\Phi}_c} \tag{6-98}$$

Rearrangement of this equation and combination with the expression (6-87) for f yields

$$\bar{\Sigma}_a = \frac{V_1 \Sigma_{a1} + V_c \Sigma_{ac}}{V_1 + V_c} \frac{1}{1 - f} \tag{6-99}$$

where it has been assumed that $\bar{\Phi}_c/\bar{\Phi}_1 \approx 1$ and $V_0 \bar{\Phi}_0/\bar{\Phi}_1 \ll V_1 + V_c$. The resulting expression for the diffusion length is

$$L^2 = \frac{L_1^2}{1 + V_c \Sigma_{ac}/V_1 \Sigma_{a1}} \frac{V_1 + V_c}{V_1} (1 - f) \tag{6-100}$$

where L_1 is the diffusion length for the pure moderator. If the coolant is nonabsorbing, the correction factor for it is merely a volumetric one. In the absence of coolant, or if coolant and moderator are identical, Eq. (6-100) becomes (cf. Prob. 6-9 for homogeneous reactors)

$$L^2 = L_1^2(1 - f) \tag{6-101}$$

The diffusional properties of some common moderators are given in Table 6-5.

6-38. Fermi Age in Heterogeneous Lattices. The presence of heavy atoms, such as uranium and perhaps the coolant, in a moderator, tends to increase the number of collisions required to thermalize a neutron, hence to increase the age. Counterbalancing this effect, however, is the inelastic scattering occurring in the fuel elements, which aids the slowing-down process. The net result is that the age for many lattices is not very different from that for the pure moderator. Values of the age for some common moderators are given in Table 6-5.

For air-cooled graphite–natural-uranium lattices, the age may be calculated from [20]

$$\tau = \left(\frac{V_t}{V_1 + \frac{1}{2}V_0}\right)^2 (387 - 90P) \qquad (6\text{-}102)$$

where V_t is the total volume of a lattice cell and P is the probability term alluded to under the fast-fission effect and given in Table 6-2. Its coefficient involves the inelastic scattering cross section in uranium.

Prob. 6-10. An air-cooled, natural-uranium graphite-moderated reactor is to be constructed in the form of a right circular cylinder with its axis horizontal. A tentative design proposes the use of 1.0-in.-diameter solid uranium rods centered in a regular hexagonal lattice 8.0 in. across flats. Air flows in through a 0.20-in. annulus surrounding the fuel rods. Neglecting absorption in air or any structural materials, estimate the quantities of uranium and graphite required in the core for criticality at 20°C. What would be the value of k for a homogeneous mixture of the same proportions?

6-39. Effect of Structural Materials. The effect of any other materials in a reactor such as tubes or channels for coolant, jacketing or cladding of fuel elements, or supporting members is most readily calculated by applying perturbation theory as shown in Sec. 6-51. The calculation often takes the form of estimating the amount of fuel that must be added to compensate for neutron absorption in the additional materials.

NONCRITICAL REACTORS AND REACTOR KINETICS

6-40. Introduction. The actual operation of a reactor brings up many types of problems involving transient states of neutron flux and density. In general, any factor which tends to change the leakage, absorption, or production rate of neutrons during operation will produce a transient condition. The present section deals with the time-dependent reactor diffusion equation and means for calculating various types of transient effects that are important in the operation of reactors. The methods here are restricted largely to thermal reactors.

6-41. Unsteady-state Reactor Equation. In the treatment of the thermal reactor utilizing age theory to describe the slowing-down process, it was seen that the steady-state thermal-neutron-balance condition could be expressed as

$$-D\nabla^2\Phi + \Sigma_a\Phi = \nu\Sigma_f e^{-B^2\tau_0}p\Phi \qquad (6\text{-}103)$$

It is immediately apparent that if the flux is changing with time the neutron-generation rate at time t depends upon the flux at a time earlier than t by an amount l_f, equal to the average time required for neutrons to slow down from fission to thermal energy.   If the instantaneous rate of change of neutron density is denoted by $\partial n/\partial t = (1/v)(\partial\Phi/\partial t)$, then Eq. (6-103) becomes, in unsteady-state form at time t,

$$\nu\Sigma_f e^{-B^2\tau_0}p\Phi' - \Sigma_a\Phi + D\nabla^2\Phi = \frac{1}{v}\frac{\partial\Phi}{\partial t} \qquad (6\text{-}104)$$

where Φ' is the flux at time $t - l_f$.   The first term gives the rate at which fission neutrons appear in the thermal range, and the tacit assumption has been made that the fission neutrons are released instantaneously on fission.   Fortunately, this is not quite true, but this assumption simplifies the analytical treatment to a certain extent.

Of the ν neutrons released in fission, a small fraction are released not instantaneously but at times up to many seconds after fission.   Such delayed neutrons are emitted instantaneously from unstable fission-product nuclei, which in turn are the result of at least one radioactive disintegration of certain primary fission fragments.   The rate at which the delayed neutrons appear is thus dependent on the radioactive disintegration rate preceding the neutron emission.   Experimentally determined decay constants governing delayed-neutron generation are given in Table 6-6.

TABLE 6-6. DELAYED-NEUTRON DATA*

Half-life, sec	λ, sec^{-1}	Mean energy, kev	Yield in U^{235} fission, per cent	Yield in Pu^{239} fission, per cent	Yield in U^{233} fission, per cent
55.6	0.0124	250	0.025	∼0.014	0.018
22.0	0.0315	560	0.166	∼0.105	0.058
4.51	0.151	430	0.213	∼0.126	0.086
1.52	0.456	620	0.241 ⎫		0.062
0.43	1.61	420	0.085 ⎬	∼0.119	
0.05	14	...	0.025 ⎭		0.018
			0.755	∼0.364	0.242

* D. J. Hughes, J. Dobbs, A. Cohn, and D. Hall, *Phys. Rev.*, **73**:111–124 (1948); F. deHoffman and B. T. Feld, *ibid.*, **72**:567–569 (1947); W. C. Redman and D. Saxon, *ibid.*, **72**:570–575 (1947).

Taking the delayed neutrons into account, the unsteady-state form of the reactor equation becomes

$$\left[(1 - \beta)\nu\Sigma_f\Phi' + \sum_{i=1}^{m}\lambda_i C_i'\right]pe^{-B^2\tau_0} - \Sigma_a\Phi + D\nabla^2\Phi = \frac{1}{v}\frac{\partial\Phi}{\partial t} \qquad (6\text{-}105)*$$

* Differences in energy of delayed and prompt neutrons are neglected.

The first term in brackets represents the rate of appearance of prompt neutrons, and the second term gives the rate of appearance of delayed neutrons. C_i' is the concentration of delayed-neutron precursor at $t - l_f$, β_i represents the fraction of neutrons delayed according to the decay constant λ_i, and β is the sum of the β_i's. In turn, the concentration of fission products yielding delayed neutrons is related to the flux Φ' by the equation

$$\frac{dC_i'}{dt} = \beta_i \nu \Sigma_f \Phi' - \lambda_i C_i' \tag{6-106}$$

there being one equation for each of the delayed-neutron emitters.

Before investigating the solution to Eqs. (6-105) and (6-106) it is instructive to simplify the problem by assuming (1) that there are no delayed neutrons, $\beta = 0$; and (2) that the slowing-down time l_f is negligibly small. With these assumptions, Eq. (6-105) may be rearranged to give

$$ke^{-B^2\tau_0}\Phi - (\Phi - L^2\nabla^2\Phi) = l_0 \frac{\partial\Phi}{\partial t} \tag{6-107}$$

where $l_0 = 1/v\Sigma_a$ is the thermal-neutron mean lifetime, neglecting leakage. The space and time variables in Eq. (6-107) are separable. Substitution of a solution in the form $\Phi = \Psi(\mathbf{r})\chi(t)$ yields the two equations

$$\nabla^2\Psi + B^2\Psi = 0 \tag{6-108}$$

and

$$\frac{d\chi}{dt} = \frac{k_{ex}}{l}\chi \tag{6-109}$$

where $l = l_0/(1 + L^2B^2)$ is the thermal-neutron mean lifetime with leakage, and $k_{ex} = k_{eff} - 1$ as defined in Sec. 6-5.

A solution to Eq. (6-109) is

$$\chi = \chi_0 e^{(k_{ex}/l)t} \tag{6-110}$$

If a quantity T be defined as the reactor period, i.e., the time for which the flux changes by a factor of e, then $T = l/k_{ex}$. Since the thermal-neutron lifetime in many thermal reactors is of the order of 10^{-3} sec, then a 1 per cent increase in k_{eff} gives $k_{ex} = 10^{-2}$ and $T = 0.1$ sec. Thus in 1 sec the reactor power level would rise by a factor of e^{10}, or about 22,000. Such a situation would lead to extremely difficult control problems. The presence of the delayed neutrons will be seen to alleviate this situation considerably.

6-42. Neutron Modes. Returning now to the space-dependent portion of the solution, Eq. (6-108), consider the application to a rectangular reactor having sides of length a, b, and c, including the extrapolation distance. It was noted in Sec. 6-12 that the boundary condition requiring the flux to be zero at the reactor boundaries could be met by writing

$$\Psi = \sum_{n,n',n''=1}^{\infty} C_{n,n',n''} \cos \frac{n\pi x}{a} \cos \frac{n'\pi y}{b} \cos \frac{n''\pi z}{c} \tag{6-111}$$

where n, n', and n'' may take on values of 1, 3, 5, ...   The geometric buckling is then

$$B^2_{n,n',n''} = \frac{n^2\pi^2}{a^2} + \frac{n'^2\pi^2}{b^2} + \frac{n''^2\pi^2}{c^2} \tag{6-112}$$

The complete solution should then be written

$$\Phi = \Psi_{n,n',n''} e^{t/T_{n,n',n''}} \tag{6-113}$$

Thus for each mode, i.e., each combination of n, n', and n'', there is a definite reactor period.   This may be written expressly in terms of the buckling as

$$T_{n,n',n''} = \frac{l_0}{ke^{-B^2_{n,n',n''}\tau_0} - (1 + L^2 B^2_{n,n',n''})} \tag{6-114}$$

For a reactor of fixed composition and dimensions, it is apparent from Eq. (6-112) that the fundamental mode, $n = n' = n'' = 1$, gives the smallest value of B^2.   If this fundamental mode yields a value of $k_{ex,1,1,1} = 0$, then the period $T_{1,1,1}$ is infinite. Since all the harmonics of B^2 are larger, it is apparent that all the harmonics of k_{ex} are negative, so that the periods of the harmonics are also negative.   These contributions to the flux die out with increasing time, leaving the form Eq. (6-18) as the expression for the steady-state flux.   If the fundamental mode yields a negative period, then all modes die out and the reactor is completely subcritical.   If the fundamental mode yields a positive period, the reactor is supercritical.   Although some of the harmonics could then yield positive periods, such a condition is generally to be avoided in controllable reactors.

6-43. Delayed Neutrons.   Returning now to a consideration of Eqs. (6-105) and (6-106) to include delayed neutrons, if the slowing-down time l_f is still regarded as negligibly small it may be shown that delayed neutrons produce additional reactor periods.   If solutions of the form

$$\nabla^2 \Phi_0 e^{t/T} + B^2 \Phi_0 e^{t/T} = 0 \tag{6-115}$$

and
$$C_i = C_{i0} e^{t/T} \tag{6-116}$$

are substituted into Eqs. (6-105) and (6-106), it will be seen that there are $m + 1$ periods corresponding to the roots of the equation

$$\frac{k_{ex}}{k_{eff}} = \frac{l}{T+l} + \frac{T}{T+l} \sum_{i=1}^{m} \frac{\beta_i}{1 + \lambda_i T} \tag{6-117}$$

The quantity k_{ex}/k_{eff} has considerable significance in reactor physics.

It is called the "reactivity" and is denoted by ρ. From the definitions of k_{ex} and k_{eff},

$$\rho = 1 - \frac{1 + L^2 B^2}{k e^{-B^2 \tau_0}} \qquad (6\text{-}118)$$

In many practical problems k_{eff} is not greatly different from unity, so that $\rho \approx k_{ex}$.

In Fig. 6-8 is shown, qualitatively, the character of Eq. (6-117). It will be observed that for all positive values of ρ there is one and only one positive period, while for negative values, all periods are negative. (Note that only $\rho < 1$ has physical significance.) The curve on the extreme right, representing the algebraically largest periods, is the curve of greatest importance. The largest period is frequently referred to as the "stable reactor period." For positive reactivities it is the only one contributing to the rise of power level, while for negative reactivities, it is the period which limits the rate of decrease of power level.

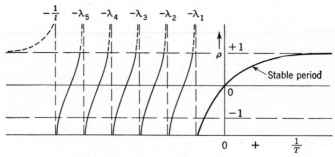

FIG. 6-8. Inverse reactor periods vs. reactivity. Prompt and delayed neutrons.

The significance of delayed neutrons is most easily clarified by assuming a single group of delayed neutrons representing the fraction β of the total produced and having an average decay constant defined by

$$\frac{\beta}{\lambda} = \sum_{i=1}^{m} \frac{\beta_i}{\lambda_i}$$

For U^{235}, the data in Table 6-6 yield $\lambda = 0.08 \text{ sec}^{-1}$. With this simplification the reactivity equation becomes a quadratic in T

$$\rho \lambda T^2 + [\rho - \beta + l\lambda(\rho - 1)]T + l(\rho - 1) = 0 \qquad (6\text{-}119)$$

For very small reactivities, $|\rho| \ll 1$, and the two roots of Eq. (6-119) are given by

$$T = \frac{\beta - \rho + l\lambda}{2\rho\lambda} \left\{ 1 \pm \left[1 + \frac{4\rho\lambda l}{(\beta - \rho + l\lambda)^2} \right]^{1/2} \right\} \qquad (6\text{-}120)$$

Employing a binomial expansion of the square-root term under the conditions that $|4\rho\lambda l| \ll (\beta - \rho + l\lambda)^2$ yields

$$T_0 \approx \frac{\beta - \rho + l\lambda}{\rho\lambda} \tag{6-121}$$

and

$$T_1 \approx \frac{-l}{\beta - \rho + l\lambda} \tag{6-122}$$

Equation (6-121) gives the positive or stable reactor period for positive reactivities, while Eq. (6-122) yields a transient period. For negative reactivities, both periods are negative, with T_0 being the larger numerically. The following example illustrates the effect of delayed neutrons on the rate of change of neutron flux.

Prob. 6-11. A natural-uranium–graphite reactor is operating at a constant average flux of 10^{13} neutrons/(cm²)(sec). An instantaneous increase in reactivity of 0.10 per cent causes the flux to increase for 30 sec, at which time a control rod is rapidly inserted, reducing the reactivity by 0.20 per cent. Assuming the thermal-neutron lifetime for this reactor is 10^{-3} sec, calculate and plot the flux as a function of time for 60 sec after the initial increase in reactivity. Assume all delayed neutrons form a single group.

Solution. For the initial rising period Φ and C are given by

$$\Phi = A_0 e^{t/T_0} + A_1 e^{t/T_1} \tag{1}$$
$$C = C_0 e^{t/T_0} + C_1 e^{t/T_1} \tag{2}$$

At $t = 0$,
$$\Phi_0 = 10^{13} = A_0 + A_1 \tag{3}$$

Noting that up to zero time $dC/dt = 0$, from Eq. (6-106), at $t = 0$,

$$C = \frac{\beta\nu\Sigma_f\Phi}{\lambda} = C_0 + C_1 \tag{4}$$

Furthermore, substitution of Eqs. (1) and (2) in Eq. (6-106) yields

$$\frac{C_0}{T_0} e^{t/T_0} + \frac{C_1}{T_1} e^{t/T_1} = \beta\nu\Sigma_f(A_0 e^{t/T_0} + A_1 e^{t/T_1}) - \lambda(C_0 e^{t/T_0} + C_1 e^{t/T_1})$$

Equating coefficients of like exponentials gives the two equations

$$C_0 = \frac{\beta\nu\Sigma_f A_0}{1/T_0 + \lambda} \tag{5}$$

$$C_1 = \frac{\beta\nu\Sigma_f A_1}{1/T_1 + \lambda} \tag{6}$$

The four constants in Eqs. (1) and (2) are thus determined by the remaining four equations. Solving for A_0 and A_1, and substituting in Eq. (1),

$$\Phi = \Phi_0 \frac{1/\lambda + T_0}{T_0 - T_1} e^{t/T_0} - \frac{1/\lambda + T_1}{T_0 - T_1} e^{t/T_1} \tag{7}$$

For U^{235}, $\lambda = 0.08 \text{ sec}^{-1}$ and $\beta = 0.0075$, so that

$$T_0 = \frac{0.0075 - 0.001 + 0.001 \times .08}{0.0010 \times .08} = 82.3 \text{ sec}$$

$$T_1 = -0.152 \text{ sec}$$

$$A_0 = \frac{1/0.08 + 82.3}{82.3 + 0.152} \Phi_0 = 1.15\Phi_0$$

Then $$\Phi = 10^{13}(1.15e^{t/82.3} - 0.15e^{-t/0.152}) \qquad (8)$$

which describes the flux as a function of time for the first 30 sec.

A similar procedure for the falling period gives

$$\Phi = 10^{13}(1.12e^{-t'/59.8} + 0.537e^{-t'/0.209}) \qquad (9)$$

where $t' = t - 30$.

The results are shown in Fig. 6-9. Included for comparison are the results that would be obtained if the delayed neutrons were disregarded. Equation (6-110) would predict a rising period of 1.0 sec and a falling period of 0.5 sec for this example. More complex reactivity changes are considered elsewhere [38]. See also Chap. 12.

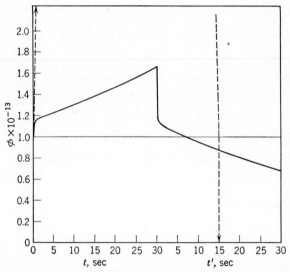

FIG. 6-9. Time-dependent flux in Prob. 6-11.

Prob. 6-12. A small homogeneous reactor has an active core volume of 10 ft³. The reactor is so designed that the fuel-moderator solution is continuously recirculated at a rate Q ft³/sec. Outside the reactor the solution passes through a considerable length of piping, a pump, and a heat exchanger. In such a system some of the delayed neutrons will inevitably be emitted outside the active core and cannot then aid in the control of the reactor.

Assuming a single group of delayed neutrons having a decay constant $\lambda = 0.08 \text{ sec}^{-1}$, and occurring in 0.75 per cent of the fissions, determine the maximum circulation rate permissible if not over 20 per cent of the delayed neutrons are to be emitted outside the active core. Assume that, on the average, 10 sec is required for the solution to pass through the entire external system and that the concentration of the delayed-neutron precursor is constant and uniform in the active core volume.

6-44. Effect of Slowing-down Time.   An expression relating reactivity to reactor period accounting for a finite, but small, slowing-down time l_f, may be obtained from Eqs. (6-105) and (6-106) by expanding Φ', C'_i, and $\partial C'_i / \partial t$ in a Taylor's series, retaining only the first two terms.   For $l_f \ll 1$ and $l_f(1 - \beta) \approx l_f$, the resulting expression is

$$\rho = \frac{l + l_f}{T + l} + \frac{T}{T + l} \sum_{i=1}^{m} \frac{\beta_i}{1 + \lambda_i T} \tag{6-123}$$

Prob. 6-13.   Estimate the slowing-down time to thermal energy for 2-Mev fission neutrons in light water and in graphite.   Compare these figures with their mean lifetimes as thermal neutrons.   Neglect leakage effects.

Solution.   In Eq. (6-48), the time interval dt between (fractional) collisions is

$$dt = \frac{dE}{\xi \Sigma_s v E} = \frac{2\,dv}{\xi \Sigma_s v^2}$$

Integrating,

$$l_f = \frac{2}{\xi} \int_{v_{th}}^{v_0} \frac{dv}{\Sigma_s v^2} = \frac{2\bar{\lambda}_s}{\xi} \left(\frac{1}{v_{th}} - \frac{1}{v_o} \right)$$
$$\approx \frac{2\bar{\lambda}_s}{\xi v_{th}}$$

For light water, $\bar{\lambda}_s = 1.1$ cm, $\xi = 1$, so that

$$l_f = \frac{2 \times 1.1}{2.2 \times 10^5} = 1 \times 10^{-5} \text{ sec}$$

For graphite, $\bar{\lambda}_s = 2.6$ cm, $\xi = 0.158$, and

$$l_f = \frac{2 \times 2.6}{0.158 \times 2.2 \times 10^5} = 1.5 \times 10^{-4} \text{ sec}$$

The mean lifetime of thermal neutrons is $l_0 = 1/v_{th}\Sigma_a$.   For water,

$$\Sigma_a = 1.0 \times \tfrac{2}{18} \times 6.023 \times 10^{23} \times 0.32 \times 10^{-24}$$
$$= 0.0214 \text{ cm}^{-1}$$

neglecting absorption in oxygen and taking 1.0 as the density of water.   Then,

$$l_0 = \frac{1}{2.2 \times 10^5 \times 0.0214} = 2.1 \times 10^{-4} \text{ sec}$$

For graphite,

$$\Sigma_a = 1.60 \times \tfrac{1}{12} \times 6.023 \times 10^{23} \times 0.0045 \times 10^{-24}$$
$$= 3.6 \times 10^{-4} \text{ cm}^{-1}$$

where 1.60 is the bulk density of graphite.   Then,

$$l_0 = \frac{1}{2.2 \times 10^5 \times 3.6 \times 10^{-4}} = 1.2 \times 10^{-2} \text{ sec}$$

6-45. Reactivity Expressions.   It has become common to adopt either of two additional methods for specifying reactivity.   Inspection of Fig.

6-8 shows that the stable period decreases with increasing positive reactivity, gradually becoming insensitive to the presence of delayed neutrons. As a mnemonic aid in fixing the magnitude of ρ for which delayed neutrons cease to play a predominant part in determining the reactor period, a condition known as "prompt criticality" may be defined.   From Eq. (6-105), for a steady-state flux due to prompt neutrons only, it follows that

$$\rho = \beta \qquad (6\text{-}124)$$

In a U^{235} reactor then, if the reactivity is 0.00755, or greater, it is critical, or supercritical, respectively, on the prompt neutrons alone, and the reactor periods become more and more dependent upon the thermal-neutron lifetime alone.   In fact for $\rho \gg \beta$, it is readily shown that $T \approx l/k_{ex}$, in agreement with Eq. (6-110).   For $\rho = \beta$ and for reasonable values of l, the stable reactor period at prompt criticality is not likely to be greater than about 1 sec.

A reactivity unit based upon the relationship expressed in Eq. (6-124) is known as the "dollar."   It may be defined by

$$\delta = \frac{\rho}{\beta} \qquad (6\text{-}125)$$

so that a reactivity of 1 dollar represents the condition of prompt criticality.   A unit of 1 cent then represents one-hundredth the reactivity of a dollar.

It is frequently more convenient to use a reactivity expression more closely related to the reactor period, particularly for small reactivities. For this purpose, the inverse hour, or inhour, expression is commonly employed.   By definition, a reactivity of 1 inhour is that value of ρ for a given reactor which yields a stable period of 1 hr.   From Eq. (6-117),

$$1 \text{ inhour} = \rho = \frac{l}{3600 + l} + \frac{3600}{3600 + l} \sum_{i=1}^{m} \frac{\beta_i}{1 + 3600\lambda_i} \qquad (6\text{-}126)$$

For the delayed neutrons in U^{235} fission, the summation, using the data from Table 6-6, is 2.59×10^{-5}.   Since $l \ll 3600$, Eq. (6-126) becomes

$$1 \text{ inhour} = \frac{l}{3600} + 2.59 \times 10^{-5} \approx 2.6 \times 10^{-5} \text{ reactivity unit} \qquad (6\text{-}127)$$

For any other period, the reactivity in inhours is

$$\rho = \frac{\dfrac{l}{T + l} + \dfrac{T}{T + l} \displaystyle\sum_{1}^{m} \dfrac{\beta_i}{1 + \lambda_i T}}{\dfrac{l}{3600 + l} + \dfrac{3600}{3600 + l} \displaystyle\sum_{1}^{m} \dfrac{\beta_i}{1 + 3600\lambda_i}} \qquad \text{inhours} \qquad (6\text{-}128)$$

For periods $T \gg l$, and for U^{235}, this becomes the inhour equation

$$\rho = \frac{38{,}600l}{T} + \frac{20.4}{T + 0.62} + \frac{204}{T + 2.19} + \frac{545}{T + 6.62} + \frac{2030}{T + 31.7}$$
$$+ \frac{778}{T + 80.6} \quad \text{inhours} \quad (6\text{-}129)$$

Equation (6-129) is plotted for four values of l in Fig. 6-10. It is appar-

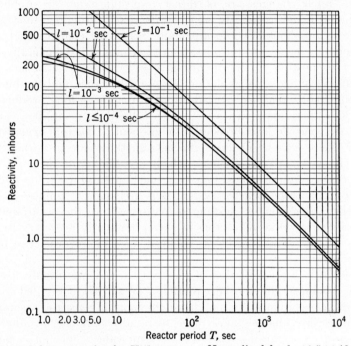

FIG. 6-10. Inhour equation for U^{235} reactors. Normalized for $l = 1.5 \times 10^{-3}$ sec.

ent that, for very large periods, the reactivity in inhours becomes inversely proportional to the period. The asymptotic form is

$$\rho = \frac{38{,}600l + 3577}{T} \approx \frac{3600}{T} \quad \text{inhours} \quad (6\text{-}130)$$

Prob. 6-14. An air-cooled graphite-uranium reactor with a thermal-neutron mean lifetime of 10^{-3} sec is operating at a steady power level. A small cylinder containing slightly enriched uranium is inserted in the pile and the recorder on the control panel indicates a rising flux with a period of 8.33 min. Calculate the pile reactivity in normal units, in dollars and cents, and in inhours.

Solution. The period in seconds is $8.33 \times 60 = 500$ sec. Assuming $\lambda = 0.08$ sec^{-1} adequately represents the grouped delayed neutrons, solving Eq. (6-121) for ρ yields

$$\rho = \frac{\beta + l\lambda}{\lambda T_0 + 1} = \frac{0.00755 + 10^{-3} \times 0.08}{0.08 \times 500 + 1} = 0.00019, \text{ or } 0.019 \text{ per cent}$$

From Eq. (6-125),

$$\delta = \frac{0.00019}{0.00755} = 0.025 \text{ dollar, or 2.5 cents}$$

From Fig. (6-10) the reactivity in inhours is read directly as 6.8 inhours.

6-46. Transient Reactivities. The reactivity of an operating reactor may change with time as a result of any condition that upsets the neutron balance. Such changes are of direct importance in the problem of reactor control and frequently have considerable bearing on other reactor engineering problems. There are many reactivity changes that are a direct consequence of reactor operation, their importance generally being in direct proportion to the operating level. These are due largely to temperature changes, the depletion of the nuclear fuel, and the build-up of fission-product poisons. In addition, in reactors which produce new nuclear fuel there may exist a tendency to build up reactivity, particularly if the reactor is a breeder. In gas-cooled reactors, particularly with air, there may be pressure effects, e.g., those due to normal barometric changes. All such effects which are the result of physical or nuclear changes in the system may be termed "natural reactivity effects." In order to control a reactor system it is necessary to impose physical or nuclear changes on the system to counterbalance the natural effects. In experimental work on reactors it is also frequently necessary to force changes on the system. These changes may be regarded as forced transient effects.

The quantitative calculation of reactivity transients usually leads to insoluble nonlinear differential equations [23] so that one is led to consider limiting behavior. On one hand this may involve the assumption that a reactivity change takes place so rapidly that the flux does not change appreciably until ρ reaches a constant value. The effect on flux and power level is then calculated on the basis of an instantaneous or step change in reactivity. On the other hand, many changes come about sufficiently slowly that compensating control measures can maintain criticality and constant flux. In either case, then, the problem resolves itself into means for calculating reactivity changes at constant flux. In addition, such calculations must take into account whether the effect is caused by local changes in the system, which will then depend upon the actual position, or whether it is a homogeneous effect over the entire system. The details of the calculation of homogeneous reactivity changes due to temperature and pressure changes, as well as to changes in the concentrations of reactive nuclei, follow directly from the definition of reactivity. In the following paragraphs are summarized methods for making some of these calculations.

6-47. Temperature Effects. It is often convenient to design a reactor for criticality at ambient temperature, T_0. As it is brought up to operating power level, however, its temperature rises and this causes a change in reactivity which must be counterbalanced by control reactivity to maintain criticality. The total amount of control reactivity required is given by

$$\rho_c = - \int_{T_0}^{T} \beta_t \, dT \qquad (6\text{-}131)$$

where β_t is the temperature coefficient of reactivity of the reactor and T is the reactor operating temperature, assumed uniform here. If β_t is negative, the control reactivity must be positive and vice versa. A reactor with a negative temperature coefficient is inherently stable, since any sudden increase in power and temperature causes it to become subcritical, thereby lowering the power until criticality returns. A reactor with a positive temperature coefficient requires more stringent control to avoid costly, and perhaps hazardous, runaways.

In terms of the theory of Sec. 6-25, for large thermal reactors, the reactivity is given by the expression

$$\rho = \frac{1 + (L^2 + \tau_0)B^2}{k} \qquad (6\text{-}132)$$

The quantities k, L^2, and τ_0 may be affected by temperature in two ways. A nuclear effect results because a change in temperature shifts the thermal-neutron-energy spectrum, upon which the various microscopic cross sections are dependent. The macroscopic cross sections are further affected by changes in density. If the use of Eq. (6-132) is restricted to near critical reactors, B^2 is very nearly the geometric buckling, which is dependent only on reactor geometry and size, which may change with temperature. It is convenient therefore to separate the contributions to β_t from each of these three causes. Regarding ρ as a function of the density γ, the cross sections σ_a and σ_s, and the geometric buckling B^2, then

$$\beta_t = \left(\frac{\partial \rho}{\partial T}\right)_{\gamma, B^2} + \left(\frac{\partial \rho}{\partial T}\right)_{\sigma_a, \sigma_s, B^2} + \left(\frac{\partial \rho}{\partial T}\right)_{\sigma_a, \sigma_s, \gamma} \qquad (6\text{-}133)$$

The first term is usually called a "nuclear temperature coefficient." Its value may be estimated from Eq. (6-132) once k, L^2, and τ_0 are written as functions of temperature.

If all thermal-neutron-absorption cross sections exhibit $1/v$ behavior and if scattering cross sections are assumed independent of energy, then it may be shown that [4] k_{th} is temperature-independent, and

$$L^2 = L_0{}^2 \left(\frac{T}{T_0}\right)^{\frac{1}{2}} \tag{6-134a}$$

$$\tau = \tau_0 - \frac{D}{\xi\Sigma_s} \ln \frac{T}{T_0} \tag{6-134b}$$

$$\ln p = \frac{-N_{ar}}{\xi\Sigma_s} [I_{r0} + b(T - T_0)] \tag{6-134c}$$

The subscript 0 refers to values taken at the ambient temperature T_0. In Eq. (6-134c) the temperature coefficient b on the resonance absorption integral I_r does not follow from slowing-down theory but must be determined experimentally. It is usually positive. For the resonance integral in U metal, Eq. (6-93), $b \approx 9 \times 10^{-4}$ °C^{-1} [19]. The most predominant nuclear temperature effect is that due to the change in diffusion length which leads to increasing values of L^2 with temperature, and thus greater leakage and lowered reactivity.

The second term in Eq. (6-133) is the density-temperature coefficient. The multiplication factor k involves only ratios of densities and is therefore independent of density changes. The definitions of L^2 and τ_0 show that both are inversely proportional to the square of the density, so that it is readily shown from Eq. (6-132) that

$$\left(\frac{\partial \rho}{\partial T}\right)_{\sigma_a, \sigma_s, B^2} = \frac{2B^2(L_0{}^2 + \tau_0)}{k} \frac{\gamma_0{}^2}{\gamma^3} \frac{d\gamma}{dT} \tag{6-135}$$

Since density normally decreases with increasing temperature, the density-temperature coefficient of reactivity is normally negative also.

The third term in Eq. (6-133) is a gross volume coefficient of reactivity. By direct differentiation of Eq. (6-132)

$$\left(\frac{\partial \rho}{\partial T}\right)_{\sigma_a, \sigma_s, \gamma} = \frac{-(L_0{}^2 + \tau_0)}{k} \frac{dB^2}{dT} \tag{6-136}$$

Increasing temperature normally tends to increase the volume, thus decreasing the buckling. The gross volume coefficient will therefore usually be positive, although it is not normally a large number, particularly if the reactor structure is constrained.

Prob. 6-15. Assume the "water boiler" of Prob. 6-9, Sec. 6-30, is operating at essentially zero power level, when the solution is at 20°C, and a small perturbation suddenly increases the reactivity by 15 cents. If it is equipped with an overflow device to maintain the contents at constant volume, what is the maximum adiabatic temperature rise it would experience? Use the modified one-group theory approximation.

6-48. Fission-product Poisoning. Many of the fission-product nuclei formed directly or indirectly in the fission process have substantial thermal-neutron-absorption cross sections. During reactor operation they tend to build up sufficient concentrations to cause an appreciable increase in the total macroscopic absorption cross section Σ_a, thereby acting to "poison" the chain reaction. The nucleus of dominant importance in this respect is Xe^{135}, occurring in the following fission-product decay chain:

$$Te^{135} \xrightarrow{2\ min} I^{135} \xrightarrow{6.7\ hr} Xe^{135} \xrightarrow{9.2\ hr} Cs^{135} \xrightarrow{20,000\ years} Ba^{135}\ (stable)$$

The importance attached to this isotope is due to its high thermal-neutron cross section of 3.5×10^6 barns, the relatively large fission yield of 5.6 per cent for this chain, and the relatively short time, of the order of a day, for this isotope to come to an equilibrium concentration. The second most important fission-product poison is the stable Sm^{149} formed in the following chain:

$$Nd^{149} \xrightarrow{1.7\ hr} Pm^{149} \xrightarrow{47\ hr} Sm^{149}\ (stable)$$

This chain has a yield of 1.4 per cent, and Sm^{149} has an absorption cross section for thermal neutrons of about 50,000 barns. Equilibrium with respect to Sm poisoning may occur in about 2 weeks of steady operation.

In both cases the concentrations of the poisons are built up during reactor operation by the decay of the primary fragments. The neutron absorption processes tend to diminish these concentrations, and in the case of Xe, its radioactive decay also tends to reduce its concentration. The result of these processes is that each of the poisons approaches an equilibrium concentration when the neutron flux is steady. Because of the large number of fission products formed, it is not practicable to consider each one separately. The total concentration of fission products, of course, increases linearly with fuel burnup, so that, on the whole, there is a gradual increase in the poisoning effect of the fission products other than Xe^{135} and Sm^{149}.*

Since poisoning lowers the reactivity of a reactor, for steady operation it is necessary to provide compensating control measures. For Xe^{135} and Sm^{149} the excess control reactivity required can be estimated by computing the steady-state concentrations of these nuclei, at a given flux, then incorporating the additional absorption cross section in the reactivity expressions. Results of some approximate calculations are shown in Table 6-7. Because Sm^{149} is stable, its rate of production and loss are both proportional to the flux, so that the equilibrium concentra-

---

* For remote areas it may be desired to operate either solid- or liquid-fuel reactors by adding U^{235} as required, without removing the fission products (except probably Xe in the latter case). Excluding Xe and Sm, the excess reactivity required for steady operation at a flux of 10^{14} for up to 10 burnups F of the original U^{235} charge is given approximately [52] by $0.09(1 - e^{-0.367F}) + 1.832(1 - e^{-0.0385F})$.

tion is independent of flux. At fluxes below 10^{15} neutrons/(cm²)(sec) the rate of loss of Xe^{135} by radioactive decay becomes an important factor in limiting its equilibrium concentration.

The excess, or control, reactivity required to compensate for the other poisons is dependent on the degree of fuel burnup desired. Although cross-section data are not available for each of the fission products, an examination of the cross sections of the naturally occurring elements falling in the fission-product range indicates that the average fission-product nucleus might have an absorption cross section of 20 to 30 barns for thermal neutrons. Since two fission fragments are formed per fission, it is possible to make a rough estimate of the total absorption cross section of the fission-product poisons as a function of fuel burnup, and consequently of the excess reactivity required.*

A transient effect of extreme practical importance arises in connection with a build-up of poison concentrations following a reactor shutdown. This situation arises because the poisons of importance are products of radioactive decay processes which continue after shutdown and the poisons are no longer removed by neutron reaction. The importance of this effect, which is called an "override," is highly depend-

TABLE 6-7. LIMITING STEADY-STATE POISONING EFFECTS ON REACTIVITY*

Thermal flux Φ	Reactivity due to presence of	
	Xe^{135}	Sm^{149}
10^{10}	-0.000081	
10^{11}	-0.00080	
10^{12}	-0.0069	-0.012, independent of flux
10^{13}	-0.030	
10^{14}	-0.046	
10^{15}	-0.048	

* Rates of build-up can be computed by rate equations as in Chap. 3, plus the fission yields, decay chains, flux, and cross section. For Sm, for instance, the time to almost reach its limiting effect of 1.2 per cent is approximately $5/\Phi\sigma_a$, where σ_a is 5.3×10^{-20} cm². At $\Phi = 10^{14}$ this is 11 days.

ent on the flux level obtaining prior to shutdown. The concentration of Xe^{135} may rise to many times the equilibrium value held during steady operation. It may be shown [4, 36] that for a thermal flux of about 2×10^{14} neutrons/(cm²)(sec) prior to a complete shutdown,† the maximum build-up of Xe^{135} concentration occurs in about 10 hr after shutdown

* See previous footnote.
† This is the average thermal flux in the Materials Testing Reactor at 30,000-kw operation.

and results in a reactivity loss of the order of 30 per cent. This would mean that considerable excess reactivity would have to be available to "override" the xenon if it were desired to start up again at any moderate time interval after a shutdown. This may pose a serious problem for mobile power plants (see Fig. 9-39).

6-49. Fuel Burnup. It is self-evident that some provision must be made to replace continuously the fuel that is burned up in an operating reactor in order to maintain criticality. This may be done, in principle, by the physical addition of new fuel at the same rate at which it is being used up. If different fuel nuclei are added, due allowance must be made for differences in σ_f, σ_a, and ν. In many cases, practical considerations make it advisable to add excess fuel at the start to take care of that which will be burned up during steady operation over a period of time. The excess reactivity must then be reduced to avoid supercriticality by suitable control procedures. The control requirements for this case are easily calculated if the power level and operating time between reactor shutdowns are specified.

6-50. Fuel Production. Reactors containing appreciable quantities of U^{238} or Th^{232} will unavoidably produce additional nuclear fuel according to the following reaction schemes:

$$U^{238} + n \rightarrow U^{239} \xrightarrow{23\ m} Np^{239} \xrightarrow{2.3\ d} Pu^{239}$$

$$Th^{232} + n \rightarrow Th^{233} \xrightarrow{23\ m} Pa^{233} \xrightarrow{27.4\ d} U^{233}$$

In so-called regenerative and breeder reactors, one of the main purposes is to carry out reactions such as these. These fuels are produced by neutron reactions in the intermediate neutron-energy region as well as the thermal region, so that an estimate of the rate of fuel production and consequently its effect on reactivity is somewhat more difficult to calculate than for effects of purely thermal reactions. Problems in the estimation of fuel production are considered in Sec. 6-62. The effect of new fuel production is, of course, to offset fuel burnup. In breeder-type reactors it may actually result in a long-term reactivity gain.

6-51. Localized Changes; Perturbation Theory. As has been seen above, the reactivity of an operating reactor may tend to change because of various types of neutron reactions as well as physical changes in the concentrations of nuclei. The importance of all such effects depends to a large extent on the neutron flux. Inasmuch as the flux actually varies with position in a reactor, these reactivity effects vary with position. In addition, any changes in a localized region or regions affect the neutron balance of the reactor as a whole.

Consider the effect of changing the buckling B^2 in any region of a thermal reactor by changing k or Σ_a (but not D). For an initially critical reactor,

$$\nabla^2\Phi + B^2\Phi = 0 \tag{6-9}$$

where $B^2 = (k - 1)/L^2$ and one-group theory is assumed. After making a small change in B^2 to, say B_0^2, the neutron-balance condition becomes

$$\nabla^2\Phi' + B_0^2\Phi' = \frac{1}{Dv}\frac{\partial\Phi'}{\partial t} \tag{6-137}$$

where Φ' is the perturbed flux and delayed neutrons are neglected. Multiplying Eq. (6-9) by Φ' and Eq. (6-137) by Φ and subtracting yields

$$(\Phi\nabla^2\Phi' - \Phi'\nabla^2\Phi) + (B_0^2 - B^2)\Phi\Phi' = \frac{\Phi\Phi'}{DvT} \tag{6-138}$$

where the term on the right introducing the reactor period T is a result of assuming the time-dependent solution in the form $e^{t/T}$ (cf. Sec. 6-41). On multiplying through by dV and integrating over the entire reactor volume, the first term vanishes. (This is true as long as Φ and Φ' form complete orthogonal sets, as in one-group theory. Φ and Φ' are then said to be "self-adjoint." In multigroup theory it is necessary to define fictitious fluxes, known as "adjoint fluxes," to satisfy this condition.) The result may be written

$$\frac{1}{T} = \frac{Dv\int_R (B_0^2 - B^2)\Phi\Phi'\, dV}{\int_{V_R} \Phi\Phi'\, dV} \tag{6-139}$$

where the integration in the numerator need extend only over that region R of the reactor in which B_0^2 is different from B^2. As a first-order approximation, if the perturbation is small, then $\Phi \approx \Phi'$ and Eq. (6-139) becomes

$$\frac{1}{T} = \frac{Dv\int_R (B_0^2 - B^2)\Phi^2\, dV}{\int_{V_R} \Phi^2\, dV} \tag{6-140}$$

The effect of any change in B^2 due to change in k or Σ_a is thus seen to be weighted as the square of the flux, a conclusion of considerable practical importance.

Prob. 6-16. The introduction of a small disk of cadmium into a bare cubic reactor at a point halfway between the center of one face and the center of the reactor is found experimentally to produce a falling period of 0.60 hr. Calculate the period if the disk were placed at the center of the reactor.

Solution. The flux is given by

$$\Phi = A \cos\frac{\pi x}{a} \cos\frac{\pi y}{a} \cos\frac{\pi z}{a}$$

so that at a point $(x,y,z) = (0,0,a/4)$, $\Phi = A \cos\pi/4$, while at the center $\Phi = A$.

Assuming the disk sufficiently small that the flux in the region R is the point value, then from Eq. (6-140)

$$\int_V \Phi^2 \, dV = 0.6 Dv (B_0^2 - B^2) V_D A^2 \cos^2 \frac{\pi}{4}$$

where V_D is the volume of the disk, so that the period for the disk at the center, by substitution in Eq. (6-140) and canceling terms, is

$$T = 0.6 \cos^2 \frac{\pi}{4} = 0.3 \text{ hr}$$

The result expressed in Eq. (6-140) can be extended [4] to include possible changes in D. The expression for the inverse reactor period becomes

$$\frac{1}{T} = \frac{\int_R \{ \delta[(k - 1)\Sigma_a v]\Phi^2 - \delta(Dv) \, |\text{grad } \Phi|^2 \} \, dV}{\int_V \Phi^2 \, dV} \tag{6-141}$$

where the symbol δ represents a *small* change in the indicated quantity. Note that changes in the diffusion coefficient are weighted as the square of the flux gradient.

Since delayed neutrons have not been considered in Eq. (6-141), the calculated period cannot be highly accurate. In most instances the information desired is the amount of reactivity required to counteract any change. If Eq. (6-141) is set equal to zero, compensatory increments can be calculated which leave the reactor just critical.

$$\int_R \{ \delta[(k - 1)\Sigma_a v]\Phi^2 - \delta(Dv) \, |\text{grad } \Phi|^2 \} \, dV = 0 \tag{6-142}$$

Results obtained from Eq. (6-142) are not subject to error introduced by neglecting delayed neutron periods.

Prob. 6-17. Find expressions for the two periods in one-group perturbation theory when a single delayed-neutron period is assumed.

6-52. Reactor Control Methods. One of the main purposes of the study of reactor kinetics is to elucidate some of the requirements that must be met for successful control of a reactor. Although the subject is sufficiently broad to encompass a separate chapter, it is fitting to note here some of the methods that can be employed to effect changes in reactivity, aside from the automatic control effect with a negative reactivity temperature coefficient. In a reactor operating at relatively high power levels, at least three types of control will normally be necessary: (1) emergency shutdown (scramming); (2) coarse, or shim, control; and (3) fine control. These categories are roughly indicative of the order of magnitude of the reactivity changes which must be effected. Rapid decreases in reactivity of a few hundredths are normally sufficient for emergency shutdown to avoid hazards to personnel or to the reactor structure itself. Coarse control may involve reactivity changes over a wide range and is necessary to compensate for temperature, poisoning, fuel depletion, and other similar effects. Reactor start-up, power-level changes, and partial or complete shutdowns may require relatively large but not rapid reactivity changes, and it is the function of the coarse-control element to effect the necessary changes. The category of fine control is necessary because the operation of a reactor at criticality is only nominally a steady state. Minor variations in reactivity will always accompany reactor operation and must be subject to continuous control to minimize fluctuations of power level. Fine control involves very small reactivity changes.

It is apparent that these reactivity changes can be brought about in many ways. Removing fuel, inserting neutron-absorbing materials, or increasing the effective-surface–volume ratio of a reactor (e.g., by removing reflector) all tend to decrease reactivity, while the reverse procedures increase reactivity. These three methods have been called "fuel control," "absorber control," and "configuration or reflector control." All may be used in an almost endless variety of ways depending only on the ingenuity of the designer and the engineering specifications that must be met.

There are two fundamental principles that should serve as a guide in the selection of control methods. First, reactivity changes should be made as homogeneously as possible, for two reasons. Highly skewed and variable flux patterns may cause severe thermal stresses or require a degree of flexibility in the heat-transfer system which is difficult or impossible to provide. Also, as was seen in the previous section, localized changes have an effect on reactivity dependent upon the square of the relative flux in the region of change. If the flux pattern varies, a given control change will not always result in the same effect. Operationally, this could be a confusing feature. The second principle may be regarded as a matter of neutron economy. In so far as compatibility with control specifications allows, consideration should always be given to control methods which do not waste neutrons. Fuel and absorber control may thus be favored over configuration control. In absorber control it may often be possible to utilize nuclear reactions that yield isotopes having considerable value. In any event, it is clear that breeder reactors would not be successful if insufficient attention were paid to this point.

The ultimate choice of methods for control of a particular reactor depends upon many features characteristic of that reactor. Such features would include reactor type, i.e., thermal, intermediate, or fast;

purpose of reactor; and physical state of fuel and moderator. Absorber control appears to be more successful in thermal reactors because of the high neutron-reaction rates attainable with certain types of materials, notably boron, cadmium, and hafnium.

In many existing thermal reactors, absorber control is utilized in the form of rods containing strongly absorbing nuclei. The effects of introducing or removing absorber in this manner are not easily calculated with accuracy. Perturbation methods outlined above suffice for small reactivity changes but fail to predict gross reactivity effects. Recourse may then be had to modification of the reactor diffusion equations [4], although this treatment becomes somewhat complex when the geometry involved loses its symmetry. For the simple case of a concentric cylindrical control rod in a cylindrical bare reactor, the change in k from that existent in the absence of the rod may be estimated by

$$\Delta k = \frac{7.5M^2}{R^2} \left[0.116 \left(1 + \frac{\tau_0}{L^2} \right) + \frac{\tau_0}{L^2} \ln \frac{L \sqrt{\tau_0}}{MR'} + \ln \frac{R}{2.4R'} \right]^{-1} \quad (6\text{-}143)$$

where $M^2 = L^2 + \tau_0$

R = radius (extrapolated) of reactor

R' = radius (extrapolated toward axis) of control rod

The equation is based on a two-group reactor theory and assumes the rod absorbs all thermal neutrons entering it but does not absorb epithermal neutrons. Estimates of the change in multiplication based on Eq. (6-143) will normally be too large, because the removal of a control rod usually leaves an open duct which results in a loss of neutrons for the critical reactor not accounted for in the derivation. The effect of a partially inserted control rod on the flux actually observed in the Argonne pile CP-2 is illustrated in Fig. 6-11.

Control rods usually contain a sufficiently large number of absorbing nuclei that the fractional rate of consumption is extremely small. Such rods can be calibrated, and the calibration remains constant with time over several years provided they work in the same flux pattern. In certain types of power reactors it is desirable to operate for long periods of time without having to replace fuel elements. Since this means high percentage burnup of fuel and consequent build-up of fission-product poisons, the large excess of fuel initially present must be compensated by excess absorption. Although this could be done by using a sufficient number of control rods which are gradually removed, another procedure might be to use a control absorber that is burned up at a rate which just compensates for the build-up of the fission product poisons other than Xe^{135} and Sm^{149}. Such absorbers, which have been called "burnable poisons," might even be fabricated homogeneously with the fuel elements.   High percentage fuel burnup does not, of course, lead to high neutron economy, so that this procedure would be useful principally for mobile power reactors, or reactors in locations remote from chemical processing facilities.

For fast reactors, absorber control is less effective because of the low cross sections at high neutron energies. Configuration control has been used in the EBR, where a linear motion of the reflector (see Secs.

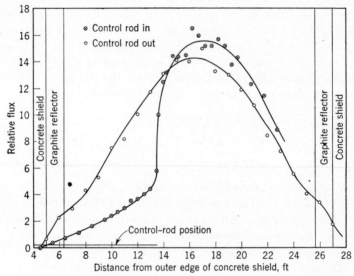

FIG. 6-11. Measured flux distributions in Argonne pile CP-2 [25].

6-57ff.) effects the necessary changes in reactivity.   A proposed fast reactor [37] employs a unique combination of configuration and absorber control in the form of rotating drums.   Each drum contains reflector and fertile absorber (e.g., U^{238}) in a concentration pattern that varies with position about the axis.

6-53. Photoneutron Sources [45].   Although the fission process is by far the most prolific source of neutrons in a reactor, occasionally materials may be used which react with some of the γ rays and produce additional neutrons.   Most notable are $H^2(\gamma,n)H^1$ and $Be^9(\gamma,n)Be^8$, since deuterium and beryllium are frequently excellent choices for moderator or reflector materials.   These reactions are threshold reactions, however, and require γ rays having minimum energies of about 2.2 and 1.6 Mev respectively, in order for them to proceed.

If either deuterium or beryllium is used as a moderating material in a homogeneous bare reactor, strictly speaking an additional neutron-source term should be added to the neutron-balance equation.   In an operating reactor the γ rays contributing to this term will consist of contributions from the fission process directly, possibly from n,γ reactions, and from a few of the fission-product nuclei.   The latter yield delayed photoneutrons, dependent upon the decay periods of the radioactive nuclei.   Lanthanum 140 has been identified as one of the fission products emitting γ rays leading to photoneutron reactions. It occurs as the decay product of Ba^{140}, which has a half-life of 12.8 days. Thus over a month of steady operation is required to build up a steady contribution of photoneutrons from this source.

Measurements show that the number of photoneutrons produced per fission neutron is of the order of 10^{-6}, so that the contribution is essentially negligible during steady-state operation. Delayed photoneutrons, however, constitute a source after shutdown that can extend the shutdown period to the longest decay period of the responsible fission product. The shutdown flux in a heavy-water-moderated reactor due to La^{140} photoneutrons, after steady operation for a few months at flux Φ, followed by t days of shutdown, may be estimated by the equation [26]

$$\Phi_{\text{shutdown}} = \frac{5 \times 10^{-7}\Phi}{1 - k_{eff}} e^{-0.054t} \tag{6-144}$$

6-54. External Neutron Sources.

There are two important circumstances in reactor work where it is necessary to use external sources of neutrons. In starting up a reactor a source is sometimes employed to provide a statistically significant neutron flux for measurement of the approach to criticality. At criticality, of course, the source is removed. The other circumstance is in reactor-design work where a subcritical reactor mock-up may be operated to measure lattice properties and buckling. Such a mock-up is usually called an "exponential assembly" or "exponential pile" because the thermal flux falls off approximately exponentially with distance from the neutron source. A treatment of these situations is outside the scope of this text [4, 27, 46].

OPTIMUM REACTOR SYSTEMS

6-55. Introduction.

The primary purposes for which a reactor is built often dictate the desirability of optimizing the nuclear design in one or more directions. For many reactors prime considerations will be to minimize the mass of fuel required and the total volume of the system. Since nuclear fuel tends to be limited and expensive, particularly if it is enriched uranium, the economic desirability of maintaining a low fuel inventory is obvious. The size of the system is directly reflected in the cost of shielding and housing for the structure, so that again the desire for minimum size appears obvious. Another feature that may be of importance, particularly in power reactors, is the distribution of power generation. For more uniform burnup of fuel and for more efficient heat transfer, it may be desirable to achieve a more uniform flux distribution, or one that is distorted somewhat from the usual symmetric distribution. In isotope-producing reactors another form of optimization may be the maximizing of the production rate of some particular nuclide. In the remaining sections of this chapter are summarized some of the methods whereby these ends may be achieved [47].

6-56. Minimum Critical Mass and Volume. It has already been shown that, for a given value of the buckling and for a given shape, a minimum critical volume exists. These results are summarized in Table 6-8. If the effect of composition, i.e., fuel-moderator ratio, is investi-

TABLE 6-8. DIMENSIONS FOR MINIMUM CRITICAL VOLUMES

Geometry	Dimensional relation	β	γ
Spherical............	R = radius	π	$\dfrac{4\pi}{3}$
Rectangular	R = side of cube	$\pi \sqrt{3}$	1
Cylindrical	$H = \dfrac{\pi \sqrt{2}\, R}{2.405}$	$2.405 \sqrt{\dfrac{3}{2}}$	$\dfrac{\sqrt{2}\, \pi^2}{2.405}$

gated, it will be found that there may exist one particular composition for which the critical mass of fuel is a minimum. This important fact is the result of the relative rates at which k and the fast-neutron non-leakage probability change with varying fuel-moderator ratio. It is most easily seen by rearranging the critical equation in terms of the critical mass, as follows:

Substituting $k = pf\eta$ and the definition of f for a homogeneous reactor, the critical equation employing age theory, Eq. (6-63), may be rearranged to give

$$1 + \frac{\alpha\sigma_{a1}}{\sigma_{a0}} = \frac{L_1^2 B^2 + p\eta e^{-B^2 \tau_0}}{1 + L_1^2 B^2} \tag{6-145}$$

where $\alpha = N_1/N_0$, the ratio of moderator to fuel, and L^2 has been replaced by $L_1^2(1 - f)$ [cf. Eq. (6-101)]. The buckling may be replaced by a quantity β^2/R^2 where β is a geometric parameter as shown in Table 6-8. In addition, the critical mass of fuel M_c may be written in terms of α and the reactor dimension R as

$$M_c = \frac{N_1 \gamma R^3 A_0}{\alpha N_{Avo}} \tag{6-146}$$

where A_0 = atomic weight of fuel

N_{Avo} = Avogadro number

γR^3 = total volume of (bare) reactor

It is not possible to obtain an explicit form for M_c in terms of α by eliminating R. However, if α is eliminated from the critical equation (6-145) by means of Eq. (6-146), there results an explicit equation for M_c in terms of R.

$$M_c = \frac{aR^3 + bR}{p\eta \exp \dfrac{-\beta^2 \tau_0}{R^2} - 1} \tag{6-147}$$

where $a = \gamma A_0 \Sigma_{a1}/N_{Avo}\sigma_{a0}$ and $b = \gamma A_0 D_1 \beta^2/N_{Avo}\sigma_{a0}$. If the concentration of moderator is held constant, a plot of M_c vs. R will show a minimum, since at small R the numerator increases with R less rapidly than the denominator, while the situation is reversed at large R. The unique value of α for minimum M_c may be found by substituting $M_{c,\min}$ and the corresponding R in Eq. (6-146).

It is apparent from Eq. (6-147) that a minimum value of R is given by

$$p\eta \exp \frac{-\beta^2 \tau_0}{R_{\min}^2} = 1 \tag{6-148}$$

In order for M_c to be positive, any actual R must be greater than this minimum. It is further apparent from Eq. (6-145) that as R becomes very large, α approaches a limiting value given by

$$\alpha_{\max} = \frac{\sigma_{a0}}{\sigma_{a1}} (p\eta - 1) \tag{6-149}$$

The foregoing relationships are strictly applicable only to homogeneous thermal reactors. In heterogeneous reactors, for a given fuel-moderator ratio it is still possible to vary lattice and fuel-element dimensions. In the absence of resonance absorption it is clear that any heterogeneous assembly will be likely to require more fuel because of the effect of self-protection in fuel elements, which lowers the thermal utilization. When resonance absorption does occur, as in U^{238}, it is quite possible that any lowering of the thermal utilization is more than compensated by an increase in the resonance-escape probability. Thus, at any given fuel-moderator ratio, there should exist optimum lattice dimensions for which k is a maximum resulting in minimized volume and critical mass.

Prob. 6-18. Calculate the minimum critical mass of a U^{235} Li^7-moderated liquid-metal homogeneous reactor at 750°F. Assume the absorption cross sections are proportional to $1/v$, the scattering cross sections are constant, and the solution density is that of lithium at the same temperature.

Prob. 6-19. In the final stages of chemical or isotope separations processes for obtaining pure fissionable material, it is essential to avoid accumulation of a mixture which might become critical. For cylindrical containers there is presumably a maximum safe diameter such that, regardless of the height of the container and of the concentration of fuel, the mixture could never become critical. Estimate this diameter for aqueous solutions of a soluble plutonium salt. Assume the vessels are stored in the open air and at large distances from one another. How would this solution be modified if the vessel were completely surrounded by water; if several vessels were stored close to one another [35]?

6-57. Reflectors. A more powerful method for minimizing critical mass is available with the use of a neutron reflector. In a bare reactor the flux gradient at the boundaries has a very large value, so that the neutron diffusion current out of the reactor is proportionately large. This loss of neutrons can be avoided by surrounding the active core with a material that minimizes the flux gradient at its external boundary. If the material is strongly absorbing, little is gained. But if the material is a weak absorber and has good scattering properties, it tends to reflect neutrons back into the active core. This reduction in loss of neutrons by leakage means that less fuel is required for the generation of sufficient neutrons to sustain a chain reaction.

6-58. The Albedo. A convenient criterion for characterizing reflectors is called the "albedo" or reflectivity. This property, frequently designated by β, represents the fraction of neutrons entering a reflector which are returned to the core. The value of β depends upon the properties and geometry of the reflector and upon the energy of the neutrons. Good reflectors generally have an albedo for thermal neutrons in excess of 0.8. It should also be noted that materials making good moderators also make good reflectors.

In another sense the albedo concept may be regarded as a generalization of the linear extrapolation distance d, defined by Eq. (6-34), which now becomes

$$d = 0.7104\lambda_t \frac{1 + \beta}{1 - \beta} \tag{6-150}$$

In certain simple cases (one-dimensional geometry) values of β may be predicted in terms of diffusion theory. They are also subject to direct experimental measurement. The usefulness of the albedo in reflector calculations arises through its use as a boundary condition. It is of particular utility when it is necessary to resort to numerical methods of solving the reactor diffusion equation. Placzek [18] has given a thorough discussion of the albedo concept in elementary diffusion theory, in which it is noted that it is equally applicable to a case where a reflector may contain fissionable fuel, e.g., in a breeder blanket.

6-59. Critical Equations with Reflectors. One-group Theory. The development of critical equations for reflected reactors follows from the methods employed in Secs. 6-17ff. for bare reactors. However, convenient analytical solutions may be obtained only for one-dimensional geometries. In the one-group approximation (thermal neutrons only), the neutron-balance equations in the core c, and reflector r are

$$-D_c\nabla^2\Phi_c + \Sigma_{ac}\Phi_c = k\Sigma_{ac}\Phi_c \tag{6-151}$$
$$\text{and} \qquad -D_r\nabla^2\Phi_r + \Sigma_{ar}\Phi_r = 0 \tag{6-152}$$

if there is no fuel in the reflector. The boundary conditions require continuity of flux and neutron current at the core-reflector interface, and the vanishing of Φ_r at the extrapolated boundary of the reflector.

Solutions for symmetrically reflected infinite-slab, infinite-cylinder, and spherical geometries are summarized in Tables 6-9 and 6-10.

TABLE 6-9. CRITICAL EQUATIONS FOR SYMMETRICALLY REFLECTED REACTORS—ONE-GROUP THEORY

Geometry	Critical equation*
Infinite slab	$D_c B_c \tan B_c R = D_r \kappa_r \coth \kappa_r T$
Sphere	$\cot B_c R = \dfrac{1}{B_c R}\left(1 - \dfrac{D_r}{D_c}\right) - \dfrac{D_r \kappa_r}{D_c B_c} \coth \kappa_r T$
Infinite cylinder	$\dfrac{D_c B_c}{D_r \kappa_r} \dfrac{J_1(B_c R)}{J_0(B_c R)} = \dfrac{K_1(\kappa_r R) I_0[\kappa_r(R + T)] + I_1(\kappa_r R) K_0[\kappa_r(R + T)]}{K_0(\kappa_r R) I_0[\kappa_r(R + T)] - I_0(\kappa_r R) K_0[\kappa_r(R + T)]}$

* R = core radius or half-thickness; T = extrapolated reflector thickness.

TABLE 6-10. FLUX DISTRIBUTIONS IN SYMMETRICALLY REFLECTED REACTORS—ONE-GROUP THEORY

Geometry	Flux distribution*	
	Core	Reflector
Infinite slab	$A \cos B_c x$	$A' \sinh \kappa_r \left(\dfrac{H}{2} + T - x\right)$
Sphere	$\dfrac{A \sin B_c r}{r}$	$\dfrac{A' \sinh \kappa_r (R + T - r)}{\kappa_r r}$
Infinite cylinder	$A J_0(B_c r)$	$A'[K_0(\kappa_r r) I_0[\kappa_r(R + T)] - I_0(\kappa_r r) K_0[\kappa_r(R + T)]]$

* A is the (arbitrary) maximum flux, and A' is related to A by flux continuity at core-reflector interface.

The critical equations in Table 6-9 relate the radius R of the active core of a critical reactor to the reflector thickness T and the properties D_c, B_c, of the core and D_r, κ_r of the reflector. Except for the slab geometry it is not possible to solve explicitly for the core size, and recourse must be made to trial-and-error solution. Examination of the equations lead, in all cases, to the following important qualitative results: (1) the critical core size is always less than that of a bare critical reactor having the same material buckling; (2) as the reflector thickness increases beyond about two diffusion lengths ($2L_r$) the core radius becomes independent of T. The following example illustrates the extent of fuel savings made possible by utilizing reflectors.

Prob. 6-20. Estimate the effect on critical dimensions of a uranium-graphite reactor in the form of a sphere by surrounding it with an 18-in. graphite reflector. In the active core the material buckling is 2×10^{-4} cm^{-2}.

Solution. For the unreflected reactor, the extrapolated radius is

$$R = \frac{\pi}{\sqrt{2 \times 10^{-4}}} = 222 \text{ cm}$$

For the reflected reactor, the critical equation of Table 6-9 gives

$$\cot 0.01414R = \frac{-1}{0.01414 \times 50.2} \coth \frac{18 \times 2.54}{50.2} = -1.95$$

from which $R = 189$ cm.   This represents a fuel savings of approximately

$$\frac{222^3 - 189^3}{222^3} = 38 \text{ per cent}$$

In the above solution the extrapolation distance is neglected and the diffusion coefficient in the core is assumed essentially unaffected by the presence of uranium.

Prob. 6-21.   For the graphite-moderated U^{235} reactor of Prob. 6-8, calculate (a) the critical size, (b) the mass of U^{235} required, and (c) the total leakage rate of neutrons from the reflector, if it is provided with 25 cm of graphite surrounding the core. Repeat the calculation for a case in which the graphite in both core and reflector contains 5 ppm of boron, and for the case in which only the graphite in the reflector contains the boron.

Before passing to more refined methods for criticality calculations with reflectors, it should be pointed out that one-group theory can be readily applied to two- or three-dimensional problems by using numerical methods.   In particular, the technique of relaxation may be quite easily applied to two-dimensional problems.   In using this technique, a reasonable distribution of flux of arbitrary magnitude is chosen for a specified reactor size.   If the reactor is too large, the relaxation process quickly reveals a steadily rising flux, while if it is too small, the flux falls everywhere to zero.   Other numerical techniques have been employed to improve upon the one-group approximation [28].   The application to heterogeneous cores has also been discussed [48].

6-60. Two-group Theory of Reflectors.   Although the previous results with one-group theory are qualitatively correct, the failure to account for fast neutrons introduces considerable error.   Inasmuch as many fast neutrons diffuse into and are moderated in the reflector, the gross current of thermal neutrons returning to the core is actually greater than would be predicted by one-group theory.   The effect may be seen in Fig. 6-11, showing an experimentally determined thermal flux distribution in a graphite-reflected reactor.   It is, in fact, possible for the maximum thermal flux in the reflector to be higher than that in the core.   Extreme cases are illustrated by the KAPL Thermal Test Reactor (TTR) [29] and the MTR [36].   The active lattice in the TTR forms a cylindrical shell with graphite acting as reflector both inside and outside the lattice. The measured thermal flux distribution is shown in Fig. 6-12.   It is clear that one-group theory applied to this reactor would result in a completely erroneous picture of the flux distribution.

The application of multi-group theory to reflected reactors predicts even greater fuel savings than are calculated with the one-group method.

This is the combined result of the additional scattering of fast neutrons back into the core plus the fact that neutrons having resonance energies in the reflector are physically removed from any strong resonance absorbers that may be in the core. The theory outlined in Sec. 6-30 is applicable here but must be expanded to include neutron-balance equa-

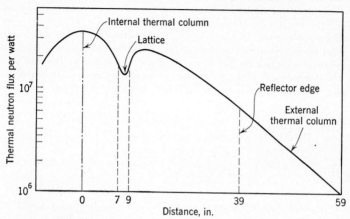

FIG. 6-12. Flux distribution in thermal test reactor. (*From Nucleonics, May*, 1953.)

tions in the reflector. Because of their considerable practical importance, the results of applying two-group theory to symmetrically reflected reactors are summarized here.

1. General critical equation:

$$\frac{k}{(1 + \tau_c B^2)(1 + L_c^2 B^2)} = 1 \qquad (6\text{-}153)$$

2. Neutron flux in core (cf. Eq. 6-84):

Fast flux: $\Phi_{1c} = A_1 X + A_2 Y$ (6-154)

Thermal flux: $\Phi_{2c} = S_1 A_1 X + S_2 A_2 Y$ (6-155)

where A_1 may be regarded as arbitrary and A_2 determined by boundary conditions. S_1 and S_2 are "coupling" coefficients defined by

$$S_1 = \frac{D_{1c}}{\tau_c D_{2c}} \frac{1}{\kappa_{2c}^2 + B_1^2} \qquad (6\text{-}156)$$

$$S_2 = \frac{D_{1c}}{\tau_c D_{2c}} \frac{1}{\kappa_{2c}^2 - B_2^2} \qquad (6\text{-}157)$$

B_1^2 and $-B_2^2$ are the two real roots of the critical equation (6-153), and X and Y satisfy Eq. (6-9). The forms of X and Y depend on the geometry. For X, the expressions for the one-group-theory core-flux distributions apply with $B_c = B_1$, Table 6-10. For Y, these become $\cosh B_2 X$, $(\sinh B_2 r)/r$, and $I_0(B_2 r)$ for the slab, sphere, and cylindrical geometries, respectively.

3. Neutron flux in reflector:

$$\Phi_{1r} = C_1 Z_1 \qquad (6\text{-}158)$$

$$\Phi_{2r} = S_3 C_1 Z_1 + C_2 Z_2 \qquad (6\text{-}159)$$

where C_1 and C_2 are related to A_1 and A_2 through the necessary boundary conditions. S_3 is a third "coupling" coefficient defined by

$$S_3 = \frac{D_{1r}}{\tau_r D_{2r}} \frac{1}{\kappa_{2r}{}^2 - \kappa_{1r}{}^2} \tag{6-160}$$

Z_1 and Z_2 are functions satisfying the expression $\nabla^2 Z_i - \kappa_{ir}{}^2 Z_i = 0$ and have the forms given in Table 6-10 for flux distributions in the reflector using the corresponding values of κ_{ir}.

4. Boundary conditions: The requirements of continuity of flux and current for both fast and thermal neutron groups at the core-reflector interface leads to a set of four algebraic equations in A_1, A_2, C_1, and C_2. In order for the flux to remain arbitrary, the determinant of the coefficients of these constants must vanish, yielding the following equation expressing the specific criticality condition for a reflected reactor:

$$\left(D_{1c} \frac{X'}{X} - D_{1r} \frac{Z_1'}{Z_1} \right) \left[S_2 D_{2c} \frac{Y'}{Y} - S_3 D_{2r} Z_1' - (S_2 - D_{2r}) \frac{Z_2'}{Z_2} \right]$$
$$- \left(D_{1c} \frac{Y'}{Y} - D_{1r} \frac{Z_1'}{Z_1} \right) \left[S_1 D_{2c} \frac{X'}{X} - S_3 D_{2r} \frac{Z_1'}{Z_1} - (S_1 - S_3) D_{2r} \frac{Z_2'}{Z_2} \right] = 0 \tag{6-161}$$

where X, Y, Z_1, and Z_2 are evaluated at the core-reflector boundary and X', Y', Z_1', and Z_2' are the values of their first derivatives at the same boundary. If the properties of core and reflector are known, R (or H) and T must be found by trial to fit this equation. It is usually convenient to adjust the results found using one-group theory to make the first approximation. Expressions for the various fluxes may then be plotted by evaluating the ratios of the constants, say A_2/A_1, C_1/A_1, and C_2/A_1 from any three of the boundary-condition equations and assigning an arbitrary value to A_1. Computations in two-group theory are sufficiently lengthy that special forms have been devised to facilitate the work [30].

6-61. Optimizing Power Density Distribution. The desirability of changing the macroscopic neutron-flux distribution from that which normally would prevail in a reactor of essentially uniform make-up depends very strongly on the purpose of the reactor and on the nature and extent of other limitations that may be imposed upon it. It is not within the scope of this section to discuss a great many variations in details but some of the more important considerations that have a bearing on flux distribution will be pointed out.

In high-power reactors a limitation may be imposed on the total power produced by the existence of a maximum power density at some point in the reactor. This limitation may be due, for example, to a maximum internal fuel-element temperature to prevent melting or phase change, or to a maximum fuel-element surface temperature to prevent excessive corrosion. To increase total power production, the flux distribution can be altered such that many more fuel elements operate nearer to maximum internal (or surface) temperatures. A significant step in this direction would be to provide a more nearly uniform distribution of flux over the entire reactor.   Provision of a reflector is a material aid in this direction, as is seen by the following example.

Prob. 6-22.   Compare the ratio of average to maximum flux for the reflected reactor of Prob. 6-20 to that of the bare reactor.

Solution.   For the bare reactor, the average flux is

$$\Phi_{av} = \frac{\int \Phi 4\pi r^2 dr}{\int 4\pi r^2 dr} = \frac{3A}{4\pi R^3} \int_0^R \sin \frac{\pi r}{R} (4\pi r)\, dr = \frac{3A}{\pi R}$$

The maximum flux at the center is

$$\Phi_{max} = \lim_{r \to 0} \frac{A}{r} \sin \frac{\pi r}{R} = \frac{A\pi}{R}$$

so that
$$\frac{\Phi_{av}}{\Phi_{max}} = \frac{3}{\pi^2} = 0.304$$

For the reflected reactor, the average flux is

$$\Phi_{av} = \frac{3A}{R^3 B_c{}^2} (\sin B_c R - B_c R \cos B_c R)$$

The maximum flux at the center is $A B_c$, so that

$$\frac{\Phi_{av}}{\Phi_{max}} = \frac{3}{R^3 B_c{}^3} (\sin B_c R - B_c R \cos B_c R)$$

Substituting $R = 189$ cm and $B_c = 0.01414$ from Prob. 6-20 yields

$$\frac{\Phi_{av}}{\Phi_{max}} = 0.447$$

Another important step that can be taken is to provide a nonuniform fuel loading [49].   A flat flux can be achieved by a symmetrical loading of fuel, with the minimum concentration at the center of the reactor. Since power density is proportional to the product of flux and fuel concentration, however, a flat flux obtained in this manner does not yield a uniform power density.   To obtain the latter by variation in fuel loading requires further adjustment of fuel toward the outer faces of the reactor.

Prob. 6-23.   Determine specifications for the spatial distribution of fuel in a bare cylindrical thermal reactor that leads to uniform power distribution in the axial direction.   Assume that the length and diameter of the cylinder are the same and that one-group theory is applicable.

If heat-transfer considerations are now taken into account, it is clear that locations of higher-power densities should be associated with lower coolant temperatures.   This could be achieved by skewing the fuel concentration so that it is higher at the coolant inlet end.   Another way that the flux distribution can be altered is by the insertion of neutron-absorbing materials.   As seen in Fig. 6-11, the insertion of a control rod pushes the region of maximum flux away from it.   This method represents poor neutron economics but may be justifiable on the basis of expediency or if the absorbed neutrons produce a worthwhile by-product.

Alteration of the flux or power density may also affect reactor operation in other ways.   The maximum permissible percentage of fuel burnup will normally be limited by any one of the following: (1) reactivity loss due to loss of fuel and build-up of fission product poisons, (2) weakening of fuel element structurally, or (3) approach to a maximum or optimum concentration of some nuclide.   Since the rate of fuel burnup is proportional to the power density, any alteration in the latter will affect the fuel-reloading schedule.   This is particularly true if each fuel element is to be used to the maximum burnup.   In turn, maximum burnup of each element is desirable because it minimizes the number of times the unused fuel must go through chemical separations and refabrication into fuel elements.   The relationships involved here are obviously complex and require careful analysis for each situation before any conclusions may be drawn as to optimizing neutron-flux distribution.

6-62. Production of Fissionable Materials [50].   The relative importance of conversion, regeneration, and breeding of nuclear fuels merits some consideration here also.   Any nuclear reaction process may be substantially affected by the occurrence of side reactions.   These may tend to decrease the conversion to a desired product, a situation entirely analogous to a vast number of organic chemical reactions.   Reaction schemes for the production of U^{233} and Pu^{239} are shown in Fig. 6-13 employing the grid scheme of Van Wye and Beckerley [31].   Known unclassified and declassified data are indicated thereon [10].   The production rate of some particular nuclide may sometimes be optimized by recycling with chemical removal of one or more constituents, which may be particularly convenient if the fuel is in solution or suspension. Since fission-product poisons, for example, represent unproductive neutron losses, recycling with chemical removal of poisons, which would lower the steady-state concentration of poison in the reactor, leaves more neutrons for productive reactions.   For this case there would exist an optimum recycling rate.

The additional factor of neutron energy can be quite important in nuclear-fuel production.   In a natural-uranium heterogeneous reactor, for example, the fraction of neutrons absorbed in U^{238} might be increased somewhat by allowing the resonance-escape probability to assume a value less than that leading to a maximum k.   In other words, a design may be compromised in the direction of a larger reactor by shifting some of the thermal absorption into the resonance region where the relative probability of capture in U^{238} may be much greater.   In a graphite-moderated reactor there may not be sufficient leeway in k to do this, but with heavy-water moderator it would be possible.

Irrespective of such modifications it is of value to compute the limits placed on fuel production by the fuel itself. All fuels possess appreciable

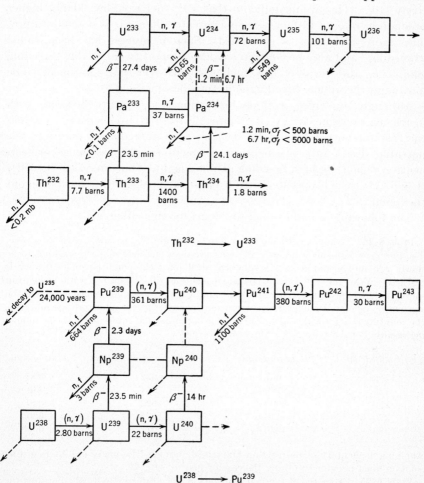

FIG. 6-13. Reaction schemes in nuclear-fuel production.

nonfission-capture cross sections. In general, such neutron captures neither contribute neutrons nor lead to appreciable quantities of any other fuel. From this point of view, then, the maximum number of neutrons per fission available for producing new fuel is given by

$$\frac{\sigma_f(E)}{\sigma_f(E) + \sigma_c(E)} \, \nu(E) - 1 = \eta(E) - 1 \qquad (6\text{-}162)$$

where the 1 represents the neutron required to propagate the chain reaction. The dependence of the cross sections and ν, the number of neutrons released per fission, on energy is emphasized. For thermal neutrons η has values of about 2.1 and 1.94 for U^{235} and Pu^{239}, respectively [10]. This would indicate that U^{235} fuel would yield the higher production rates. In fact, since η is greater than 2 it is possible to produce more fuel than is consumed. Such a process has been called "breeding" and may be characterized by a quantity called the "breeding gain," i.e., the *excess* of new fuel produced over that consumed. The maximum theoretical breeding gain thus becomes $\eta(E) - 2$.

Although the cross sections for fission and capture decrease with increasing neutron energy, it is possible for $\eta(E)$ to increase. Such is the case for both U^{235} and Pu^{239} [32]. This immediately points to the potential desirability of operating a reactor in the resonance or fast-neutron region if a high breeding gain is sought. The success of the EBR (Chap. 14), which operates with a fast-neutron spectrum, has borne out this point.

The following are additional problems for this chapter:

Prob. 6-24. A homogeneous reactor is to be brought to criticality for the first time. The fuel is an aqueous solution containing uranium enriched in U^{235}, and it is to be slowly run into a stainless-steel spherical shell until the condition of criticality is reached. As an aid to determine how much solution will be required, a Ra-Be neutron source is placed directly under the shell and a fission chamber is placed above the shell in the reflector to determine the neutron density. As the filling process proceeds, the counting rate from the fission chamber is determined at intervals, with the following results:

Fuel added, ml	Counting rate	Fuel added, ml	Counting rate
0	667	622	1,960
125	769	764	3,330
250	909	822	5,000
370	1110	875	8,620
485	1330	917	20,000

Applying elementary chain-reaction kinetics, determine how many milliliters of fuel solution must be added to bring the reactor just to criticality.

Prob. 6-25. A source of thermal neutrons is placed in a cylindrical hole along the axis of a graphite cylinder whose length may be considered very long with respect to its outer diameter. (*a*) If the radius of the axial hole is r_0, the extrapolated radius of the graphite cylinder is R_0, and the source emits Q_0 neutrons/(sec) per centimeter of axial length, determine the relation between the flux and the radial distance from the axis. (*b*) How is this solution modified if the source becomes a line source along the axis?

Prob. 6-26. A cubical reactor is to be built of natural uranium and graphite with a graphite reflector 10 in. thick on all sides. To provide for exposing special samples to the high neutron flux in the center of the pile, it is decided to provide a central column, 20 by 20 in. in cross section, and extending horizontally through the reactor, from which uranium will be omitted. The material buckling of the active portion is 2×10^{-4} cm^{-2}, and the inverse diffusion length in the graphite may be taken as 0.02 cm^{-1}. (*a*) Sketch a vertical cross section through the reactor. (*b*) Determine the critical dimensions of the reactor. (*c*) Plot the thermal flux from the center to the outer edge of the pile along a direction perpendicular to the axis of the central column. (*d*) How does the presence of the central column affect the neutron leakage from the sides of the pile as compared to that from a cubical pile without such a column?

Prob. 6-27. The fuel-element assembly of the Argonne CP-5 research reactor is described [53] as follows: "Each fuel element is an assembly 3×3 inches square and about 30 in. long. The active section is made from 10 aluminum 'sandwiches' 3 in. $\times$ 24 in. $\times$ 0.060 in. thick. Each sandwich consists of a sheet of aluminum-uranium alloy fabricated between two 2S-Aluminum sheets into a solid plate. . . . Each fuel assembly contains about 75 gm. of U^{235}." This reactor is cooled and moderated by heavy water, which is also used as a reflector approximately 60 cm in thickness. Making reasonable assumptions, estimate the critical mass of U^{235} of this reactor in the cold, clean condition. Plot the thermal neutron flux as a function of distance from the center of the core to the outer edge of the reflector, using two-group theory, and taking the maximum flux at the center as 3×10^{13} neutrons/(cm^2) (sec). Reference 53 gives 850 g of U^{235} as estimated critical mass and a flux plot for comparison.

Prob. 6-28. Employing perturbation theory, estimate the per cent enrichment of U^{235} in the fuel required for operation of the reactor of Prob. 6-10 if water replaces air as the coolant, no other changes being made.

Prob. 6-29. Estimate the direction of, and per cent change in, reactivity if a water-cooled reactor is suddenly thrust into a film-boiling range of heat transfer. Assume 1.00-in.-diameter natural-uranium fuel elements in a graphite lattice having an effective diameter of 8 in., a 0.250-in. coolant annulus, and a film of steam 0.002 in. thick.

Prob. 6-30. Investigate the stability of a natural-uranium graphite-moderated light-water-cooled reactor toward the loss of coolant, and show that proper design can result in self-regulation of such a reactor. In other words, show that the reactivity gain in losing neutron-absorbing coolant may be offset by increases in resonance absorption and neutron leakage. Discuss the significance of this with respect to boiling-water reactors.

NOMENCLATURE

B^2 = buckling, cm^{-2}
d = extrapolation distance, cm
D = diffusion coefficient, cm
D_v = diffusion coefficient, cm^2/sec
f = thermal utilization
f_r = resonance utilization
$k, k(E)$ = multiplication factor for infinite reactor
k_{eff} = effective multiplication factor for finite reactor
k_{ex} = excess multiplication factor for finite reactor
l = mean lifetime of thermal neutron in finite reactor, sec
l_0 = mean lifetime of thermal neutron in infinite reactor, sec
L = diffusion length, cm
M = migration length, cm
$n, n(E)$ = neutron density, neutrons/cm^3

N = concentration, nuclei/cm^3

p, $p(E)$ = resonance-escape probability

q, $q(E)$ = slowing-down density, neutrons/(cm^3)(sec)

$\mathbf{r}$ = position vector

v = neutron velocity, cm/sec

V_R = volume of reactor, cm^3

V_0 = volume of fuel, cm^3

V_1 = volume of moderator, cm^3

V_c = volume of coolant, cm^3

β = fraction of delayed neutrons, or albedo

ϵ = fast-fission factor

η = fission neutrons produced per neutron absorbed in fuel

κ = inverse diffusion length, cm^{-1}

κ_0 = inverse diffusion length in fuel, cm^{-1}

κ_1 = inverse diffusion length in moderator, cm^{-1}

λ = mean free path, cm

λ_s = mean free path for scattering, cm

λ_t = mean free path for transport, cm

ν = neutrons produced per fission

ξ = mean logarithmic energy decrement in a scattering collision

ρ = reactivity

σ = microscopic cross section, cm^2/nucleus or barns

σ_s = microscopic cross section for scattering, cm^2/nucleus or barns

σ_a = microscopic cross section for absorption, cm^2/nucleus or barns

Σ = macroscopic cross section ($N\sigma$), cm^{-1}

Σ_s = macroscopic cross section for scattering, cm^{-1}

Σ_a = macroscopic cross section for absorption, cm^{-1}

τ = Fermi age, cm^2

Φ = neutron flux, for thermal neutrons unless otherwise specified, neutrons/-(cm^2)(sec)

REFERENCES

1. Weinberg, A. M.: "Elementary Theory of Neutron Diffusion," AECD-3405, 1952.
2. Weinberg, A. M., and L. C. Norderer: "Second Order Diffusion Theory," AECD-3410, 1952.
3. Marshak, R. E., H. Brooks, and H. Hurwitz, Jr.: *Nucleonics*, 4(5):10–22 (May, 1949); 4(6):43–49 (June, 1949); 5(7):53–60 (July, 1949); 5(8):59–68 (August, 1949).
4. Glasstone, S., and G. Edlund: "The Elements of Nuclear Reactor Theory," D. Van Nostrand Company, Inc., New York, 1952.
5. Weinberg, A. M., and L. C. Norderer: "Slowing Down of Neutrons," AECD-3411, 1952.
6. Gray, A., G. B. Mathews, and T. M. MacRobert: "A Treatise on Bessel Functions," 2d ed., Macmillan & Co., Ltd., London, 1922.
7. Merrill, L. C.: "Solutions of Diffusion Equations for Generalized Reactor Shapes," NEPA-1075, 1949.
8. Southwell, R. V.: "Relaxation Methods in Theoretical Physics," Oxford University Press, New York, 1946.
9. Milne, W. E.: "Numerical Solutions of Differential Equations," John Wiley & Sons, Inc., New York, 1953.

10. "Neutron Cross Sections," AECU-2040, May 15, 1952, and supplements 1 and 2.
11. Churchill, R. V.: "Modern Operational Mathematics in Engineering," McGraw-Hill Book Company, Inc., New York, 1944.
12. Carslaw, H. S.: "Introduction to the Mathematical Theory of the Conduction of Heat in Solids," Dover Publications, New York, 1945.
13. Nereson, N.: *Phys. Rev.*, **88**:823–824 (1952).
14. Watt, B. E.: *Phys. Rev.*, **87**:1037–1041 (1952).
15. Weinberg, A. M.: *Am. J. Phys.*, **20**:401–412 (1952).
16. Castle, H., H. Ibser, G. Sacher, and A. M. Weinberg: CP-644, 1943.
17. Murray, R. L., and A. C. Menius, Jr.: *Nucleonics*, **11**(4):21–23 (April, 1953).
18. Goodman, C. (ed.): "Science and Engineering of Nuclear Power," vol. II, Addison-Wesley Publishing Company, Cambridge, Mass., 1949.
19. AEC: technical release, April, 1952.
20. Guggenheim, E. A., and M. H. L. Pryce: *Nucleonics*, **11**(2):50–60 (February, 1953).
21. Hughes, D. J., J. Dobbs, A. Cohn, and D. Hall: *Phys. Rev.*, **73**:111–124 (1948).
22. DeHoffman, F., and B. T. Feld: *Phys. Rev.*, **72**:567–569 (1947).
    Redman, W. C., and D. Saxon: *Phys. Rev.*, **72**:570–575 (1947).
23. Lansing, N. F. (comp.): "The Role of Engineering in Nuclear Energy Development," TID-5031, OTS, 1951.
24. Goodman, C. (ed.): "Science and Engineering of Nuclear Power," rev. ed., vol. I, Addison-Wesley Publishing Company, Cambridge, Mass., 1952.
25. Seren, L., W. Sturm, and W. Mayer: N-656, declassified 1952.
26. Lundby, A.: *Nucleonics*, **11**(4):16–17 (April, 1953).
27. Masket, A. V.: *Am. J. Phys.*, **21**:151–159 (1953).
28. Thompson, A. S.: *J. Appl. Phys.*, **22**:1223–1235 (1951).
29. Stewart, H. B., et al.: *Nucleonics*, **11**(5):38–41 (May, 1953).
30. Spinrad, B. I., and D. Durath: ANL-4352, 1952.
31. Van Wye, R. F., and J. G. Beckerley: *Nucleonics*, **9**(4):17–21 (October, 1951).
32. Weinberg, A. M.: *Nucleonics*, **11**(5):18–20 (May, 1953).
33. Hughes, D. J.: "Pile Neutron Research," Addison-Wesley Publishing Company, Cambridge, Mass., 1953.
34. Plass, G. N.: AECD-3214 and AECD-3217, 1943.
35. Macklin, R. L.: AECD-3170.
36. Huffman, J. R.: AECD-3587.
37. "Reports to the U.S. Atomic Energy Commission on Nuclear Power Reactor Technology," Government Printing Office, May, 1953.
38. Hurwitz, H., Jr.: *Nucleonics*, **5**(7):61–67 (July, 1949).
39. Murray, R. L., M. R. Keller, and D. E. Hostetter: *Nucleonics*, **12**(9):64–65 (September, 1954).
40. Ehrlich, R., and H. Hurwitz, Jr.: *Nucleonics*, **12**(2):23–30 (February, 1954).
41. Atkinson, I. C., and R. L. Murray: *Nucleonics*, **12**(4):50–53 (April, 1954).
42. Persson, R.: *Nucleonics*, **12**(10):26–29 (October, 1954).
43. Fornaguera, R. O., A. Corbo, and T. Iglesias: *Chem. Eng. Progr., Symposium Ser.*, no. 12, pp. 88–95, 1954.
44. Hostetter, D. E., A. C. Menius, Jr., and R. L. Murray: *Nucleonics*, **12**(7):76–77 (July, 1954).
45. Lundby, A.: *Nucleonics*, **12**(8):25–27 (August, 1954).
46. Cohen, E. R.: *Chem. Eng. Progr., Symposium Ser.*, no. 12, pp. 72–81, 1954.
47. Miles, F. T., and I. Kaplan: *Chem. Eng. Progr., Symposium Ser.*, no. 11, pp. 159–176, 1954.

48. Medina, A., and F. E. Prieto: *Chem. Eng. Progr., Symposium Ser.*, no. 13, pp. 147–148, 1954.
49. Goertzel, G., and W. A. Loeb: *Nucleonics*, **12**(9):42–45 (September, 1954).
50. Bogaardt, M., and M. Bustraan: *Nucleonics*, **12**(12):32–35 (December, 1954).
51. Kron, E.: *Trans. AIEE, Communications and Electronics Section*, July, 1954.
52. Robb, W. L., J. B. Sampson, J. R. Stehn, and J. K. Davidson: KAPL, 1955.
53. Untermyer, S.: *Nucleonics*, **12**(1):14 (January, 1954).

CHAPTER 7

SHIELDING OF POWER REACTORS

By David C. Peaslee

SHIELDING—THE GENERAL PROBLEM

7-1. Shields and Shielding Materials. Shielding of a nuclear reactor is vital for the protection of the operating personnel. This function is not directly related to the mechanical and nuclear operation of the pile. Given any reactor, it is generally possible to tailor a shield to fit without substantially altering the original reactor design. Complete power plants including shield should be compared, however, when space or weight limitations are important, as in the selection of aircraft or small naval propulsion units. In fact, with mobile reactors, the shield design and associated radiation- and radioactivity-protection problems should be considered in the initial solution of a reactor type and integrated with the other aspects of the reactor design as they unfold. Research reactors are frequently shielded to cut the emergent radiation to one-tenth or less of the biologically permissible intensity, to decrease instrument background.

The inherently harmful radiations of a pile are α particles, β particles, thermal (slow) neutrons, fast neutrons, and γ rays. Only the last two are of serious concern for pile shielding. A fraction of an inch of a solid substance will stop the electrically charged α and β particles. The β rays upon coming to rest emit a weak γ radiation known as "bremsstrahlung" (see Sec. 2-11). However, this radiation is so weak in comparison with the nuclear γ radiation that it may be neglected as far as shielding is concerned. Thermal neutrons are rapidly absorbed by matter, because cross sections for nuclear capture of neutrons vary as $1/v$ (see page 88), which is largest for thermal neutrons. Any shield thick enough to provide protection against γ rays and fast neutrons will automatically remove the thermal neutrons coming from the pile. If special shielding against thermal neutrons is required, a boron* compound will reduce

---

* "Boral" sheets can be rolled of aluminum cast with 50 per cent B_4C (40 per cent B) by volume [14–16]. Density is 2.53 and tensile strength over 5000 psi. Sol-

the thermal-neutron flux by a factor of about 10^{10} per 0.1 in. of boron equivalent.

To a first approximation the effectiveness of a nuclear shield depends only upon its mass. The chief virtue of high-density materials is to save space where it is costly, as in mobile installations. For stationary power reactors where space is not at a premium, the best shielding materials from this point of view are those with the lowest cost per ton, e.g., ordinary concrete [1].

Closer examination of shielding materials reveals some intrinsic differences not taken into account by mass alone. For γ-ray shielding heavy elements such as lead are somewhat more effective per unit weight than light elements such as carbon. For fast-neutron shielding, light elements, particularly hydrogen, are much more effective per unit weight than heavy elements. Some compromise between these extremes must generally be made in practice; and if a single material is to be used for the shield, concrete appears as a favorable compromise (see Prob. 7-21).

There are three main factors to consider in the design of a reactor shield: the total amount of radiation produced by the pile, the amount of radiation that can be tolerated outside the shield, and the shielding properties (attenuation) of the material. These factors are next considered.

7-2. Reactor Radiations: Fast Neutrons. To specify the strength of the radiation at any point in a quantitative manner, the flux I is defined as the number of particles per second crossing an area of 1 cm². This definition is used for neutrons. It can also be applied to γ-ray photons, but the energy flux J is more convenient. For a beam of photons of a single energy E per photon the energy flux is

$$J = EI \tag{7-1}$$

and is commonly expressed in millions of electron volts per square centimeter per second; J is known as the "intensity."

The production of fast neutrons and γ rays in a pile at constant power is proportional to the rate of fission and hence to the power. Approximately 3×10^{13} fissions per second yield 1 kw of thermal power. Ura-

---

uble borax ($Na_2B_4O_7 \cdot 10H_2O$) or insoluble calcium borate (Colemanite, 10 per cent B) may be added to concrete shields.

Cadmium is another substance that strongly absorbs slow neutrons and is frequently considered for reactor use. For bulk shielding it has two disadvantages in comparison with boron. Upon neutron capture it releases energetic γ rays (boron usually does not), against which further shielding must be provided. Furthermore, cadmium is roughly three times as heavy and ten times as expensive as boron for equivalent amounts of slow-neutron shielding. See page 86 for the neutron-absorption cross sections.

nium 235 releases about 2.5 neutrons per fission, of which 1 is needed for a subsequent fission and about 0.5 is absorbed by the materials present. Thus, the number of neutrons S emitted per second by the pile is about

$$S = 3 \times 10^{13} K \qquad (7\text{-}2)$$

where K = the reactor power level, kw. For any specific reactor a more precise figure can be obtained by the methods in Chap. 6.

To a first approximation these neutrons proceed uniformly in all directions from the pile, and at a distance R from the pile center they pass through a spherical surface of area $4\pi R^2$. The flux I_0 in neutrons/-(cm²)(sec) at distance R' in feet with no intermediate absorption is

$$I_0 = \frac{2.6 \times 10^9}{(R')^2} K \qquad (7\text{-}3)$$

The energy distribution of the neutrons released in fission is well represented [2] by the formula (cf. Sec. 6-26)

$$f(E)\, dE = \sqrt{\frac{2}{\pi e}}\, e^{-E} \sinh \sqrt{2E}\, dE \qquad (7\text{-}4)$$

where E is in millions of electron volts. This distribution is normalized to have $\int_0^\infty f(E)\, dE = 1$. The average energy of the distribution is $\int_0^\infty Ef(E)\, dE = 2$ Mev. Of course, when the neutrons collide with nuclei in the reactor, their energies will decrease. On the other hand, the fast neutrons to be attenuated by the shield are mainly those which escape the reactor with few if any collisions. Equation (7-4) accordingly represents a reasonable approximation to the spectrum of epithermal neutrons incident on the shield.

7-3. Reactor Radiation: γ Rays. There are three principal sources of γ radiation in a pile (see Table 9-1): (1) about 5 Mev of "prompt" γ rays accompany the fission; (2) in equilibrium operation the γ decay of fission fragments produces another 5 Mev or so per fission; (3) in a critical reactor employing U^{235}, 1.5 neutrons of the 2.5 neutrons produced per fission are eventually captured in nonfissionable material, releasing about 7 Mev of γ radiation per capture. This makes a total of about 20 Mev per fission. The unshielded intensity J_0 of γ radiation, in millions of electron volts per square centimeter per second, at R' feet from the center of the pile follows from multiplying Eq. (7-3) by the factor 20:

$$J_0 = \frac{5.2 \times 10^{10} K}{(R')^2} \qquad \text{Mev/(cm²)(sec)} \qquad (7\text{-}5)$$

Although each γ ray has a specific energy, there are so many different

γ-ray energies that together they comprise a practically continuous distribution. Unfortunately there appears to be no direct experimental measurement of a distribution formula for pile γ radiation similar to Eq. (7-4) for neutrons. A useful approximation is to represent the γ-ray spectrum by a small number of discrete lines. For the present discussion these lines are assumed to be at 1, 3, and 7 Mev. It is now necessary to determine the distribution of γ-ray energy among these lines.

The highest-energy γ rays come mainly from neutron capture. Studies of these energies and their intensities for certain substances have been made [3]. Table 7-1 reviews some of the findings. The rows list the

TABLE 7-1. NEUTRON-CAPTURE γ-RAY ENERGIES*

Mev	Element										
	Be	C	O	Na	Mg	Al	Fe	Ca	Zn	Pb	Bi
0–5	0	2.5	4.1	4.7	6.6	3.7	3.7	4.5	6.3	0	4.2
5–10	6.8	2.5	0	1.7	2.4	4.0	4.1	3.4	1.7	7.3	0
Total..........	6.8	5.0	4.1	6.4	9.0	7.7	7.8	7.9	8.0	7.3	4.2

* B. B. Kinsey, G. A. Bartholomew, and W. H. Walker: *Phys. Rev.*, **89**:375, 386 (1953); **83**:519 (1951); **82**:380 (1951); *Can. J. Phys.*, **29**:1 (1951).

amount of energy emitted per neutron capture in the form of γ radiation with energy in the ranges 0 to 5 Mev and 5 to 10 Mev. The sum of these two is the total γ-ray energy emitted per neutron capture, which is the binding energy of the captured neutron in the final nucleus. The values given in Table 7-1 are necessarily approximate but serve to show the order of magnitude involved. The carbon γ-ray spectrum consists predominantly of a single 5-Mev line, which has been equally divided between the two groups. The capture γ rays in oxygen have not been measured, but the total binding energy is known to be only 4.1 Mev (see Prob. 7-20). An average of the elements in Table 7-1 shows that about 60 per cent of the energy is in the 0- to 5-Mev region, which will be represented by a single line at 3 Mev; the remaining 40 per cent of the capture energy is in the 5- to 10-Mev region, which will be approximated by a single line at 7 Mev. For any specific reactor the distribution could in principle be estimated more accurately from the materials and amounts actually present; however, the above averages are convenient for general purposes.

The γ radiation from fission-product decay is of relatively low energy and will be represented by a single line at 1 Mev. The prompt γ rays in fission will be equally divided between the 1-Mev and 3-Mev lines. The total distribution of γ energies is then 7.5, 8.5, and 4 Mev in the 1-, 3-, and 7-Mev lines, respectively. To one significant figure, this fractional energy distribution f_γ is

$$
\begin{array}{llll}
E_\gamma = 1 \text{ Mev} & 3 \text{ Mev} & 7 \text{ Mev} \\
f_\gamma = 0.4 & 0.4 & 0.2
\end{array}
\tag{7-6}
$$

This energy distribution evidently has an average energy $\Sigma E_\gamma f_\gamma = 3$ Mev. Gamma rays of 3 Mev energy are about the most difficult to shield against with heavy metals, and 7-Mev rays with lighter materials (see Fig. 7-1), so that use of the distribution (7-6) should lead to conservative shield design. It is to be hoped that experimental data will presently allow replacement of (7-6) with a more accurate formula.

7-4. The Attenuation Factor. The intensity of a beam of γ rays or fast neutrons will gradually diminish on passage through matter (see Sec. 2-13). This attenuation is generally described by Beers' law:

$$
dI = -\mu I \, dx
\tag{7-7}
$$

Here dI is the change in flux I of a beam in passing through an infinitesimal thickness dx of material, where μ is the "linear attenuation coefficient" of the material.* The minus sign in Eq. (7-7) indicates that intensity is lost in passing through the thickness dx. Equation (7-7) also applies with dI and I replaced by the energy fluxes dJ and J.

If μ is a constant, the integral of Eq. (7-7) for a finite thickness of material is

$$
\frac{I}{I_0} = e^{-\mu x} = e^{-x/\lambda}
\tag{7-8}
$$

Equation (7-8) gives the ratio of emergent (I) and incident (I_0) beam intensities on passage through a finite thickness x of the material. This ratio is called the "attenuation factor." The quantity $\lambda = 1/\mu$ is the "mean free path"; if the beam consists of particles, it is the average distance a particle travels between successive collisions with atoms of the material, or between generation and collision.

The coefficient μ or λ is related to the cross section σ per atom for attenuation of the particles. If n_0 is the number of atoms per unit volume of the material, N_0 is Avogadro's number 6.02×10^{23}, ρ the density, and A the atomic weight of the material,

$$
\mu = \frac{1}{\lambda} = n_0 \sigma = N_0 \rho \sum \frac{\sigma}{A} = N_0 \rho \frac{\Sigma a \sigma}{\Sigma a A}
\tag{7-9}
$$

The expressions with summations apply if the shield is a compound or mixture of several elements, where a is the atom fraction of each.

Equation (7-9) shows that the linear attenuation coefficient is proportional to the density of the material. In some applications, particu-

* As in Chap. 5, not the mass attenuation coefficient of Sec. 2-13.

larly theoretical, it is convenient to have an attenuation parameter that depends only on the composition of the material and not on the denseness with which it is packed. For this purpose the "mass attenuation coefficient" is employed:

$$\mu' = \frac{\mu}{\rho} = N_0 \frac{\Sigma a \sigma}{\Sigma a A} \qquad (7\text{-}10)$$

If the units of μ are reciprocal centimeters, those of μ' are square centimeters per gram. In practice the symbols μ' and μ are not very carefully distinguished, so that it is important to note the units employed in every case. The linear attenuation coefficient μ will be used throughout this chapter.

Equations (7-9) and (7-10) show the meaning of the first and second approximations mentioned in Sec. 7-1. The attenuation per unit mass of the material depends only on $\Sigma a \sigma / \Sigma a A$, according to Eq. (7-10). To a first approximation this quantity is roughly constant for all materials, both for γ rays and for fast neutrons. More precisely, however, σ/A for γ rays increases slightly with increasing A, so that materials like lead provide the greatest γ attenuation per unit weight.

7-5. Attenuation of γ Rays. The linear attenuation coefficient μ for γ radiation can be computed from Eq. (7-9) by inserting the known cross sections (cf. Sec. 2-13). This theoretical μ requires correction for the fact that the Compton effect does not entirely absorb the γ radiation but also scatters it as γ radiation of lower energy. Although the first such Compton scattering will deflect the γ radiation from the original beam direction, after several Compton encounters some of the scattered radiation will return to the original beam. This scattered radiation is degraded by the Compton collisions, so that it consists of photons of lower energy than those of the incident beam. For this reason it is most convenient not to count the individual photons and their energies emerging from the shield but instead to measure intensity in terms of the total energy flux J.

The secondary Compton-scattered radiation is included in the attenuation formula (7-8) by writing

$$\frac{J}{J_0} = (1 + B)e^{-\mu x} \qquad (7\text{-}11)$$

Here the "build-up factor" B is a function of the distance x through the material, being zero at $x = 0$ and increasing continuously with x. Extensive calculations of the build-up factor are available for a number of materials [4].

If the build-up factor is written in the form

$$(1 + B) = e^{\Delta \mu x} \tag{7-12}$$

Eq. (7-11) becomes

$$\frac{J}{J_0} = e^{-(\mu - \Delta \mu x)} = e^{-\mu_B x} \tag{7-13}$$

Strictly speaking, $\Delta\mu$ and hence μ_B are functions of x, but over moderate ranges of x it is sufficient to approximate μ_B by a constant. Approximate values of μ_B are plotted in Fig. 7-1 for lead, iron, water, and con-

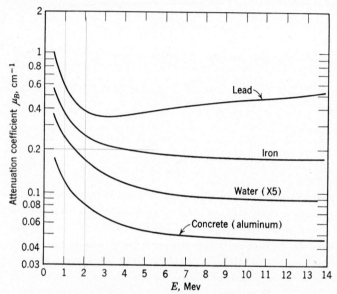

FIG. 7-1. Effective γ-ray attenuation coefficient.

crete (equivalent to aluminum for γ-ray shielding). They are obtained from a combined theoretical and experimental approach [5]. Comparison with the explicit calculations of B [4] indicates that the approximate μ_B values of Fig. 7-1 are slightly conservative.

The curves of Fig. 7-1 show a marked dependence of μ_B on the energy of the incident γ radiation. Formula (7-13) must be applied separately to each of the three components (7-6) assumed for the pile radiation, choosing the appropriate μ_B for each energy. Figure 7-2 plots the mass attenuation coefficient $\mu_B/\rho = \mu'_B$ against atomic number Z for the energies of distribution (7-6). For any compound or mixture of total density ρ the attenuation coefficient is $\mu_B = \rho\mu'_B = \rho\Sigma_j a(j)\mu'_B(j)$. Here the sum is over all the elements in the mixture, and $a(j)$ is the weight fraction of each. It should be emphasized that the correction for build-up included in the curves of Fig. 7-2 is very approximate, since they are obtained from interpolation of μ_B with the assumption that $\Delta\mu$ in Eq. (7-12) is a constant. The build-up effect is more pronounced for light elements, further accentuating the relative effectiveness of heavy elements for γ attenuation.

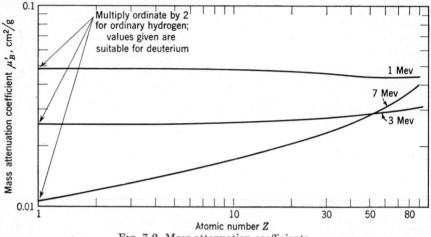

FIG. 7-2. Mass attenuation coefficients.

7-6. Attenuation of Fast Neutrons. Very little direct capture of fast neutrons occurs; the attenuation of fast neutrons depends mainly on slowing them down to thermal energies, where they are captured. For the attenuation of fast neutrons it is practically essential that the shielding material contain hydrogen. In the first place, the proton has about as large a neutron-collision cross section as any nucleus for average fission neutrons. Furthermore, the loss of energy per collision is greatest with hydrogen, since the fractional energy loss of a neutron on collision with a nucleus of atomic weight A is proportional to $A/(A+1)^2$, a maximum for $A = 1$. Thus hydrogen is by far the most effective element per unit mass for slowing down fast neutrons. The two neutron-shielding materials considered here are water and concrete; the latter has enough water of crystallization to qualify also as a hydrogen-containing material.

After a fast neutron makes its first collision in the shield, its motion can be described as a diffusion in which the net forward progress through the shield is relatively slight. The diffusion usually ends with thermalization and capture of the neutron in the vicinity of its first collision. In a simple approximation which neglects the diffusion, the attenuation of fast neutrons is obtained by considering each neutron to be effectively removed from the beam by its first collision with a nucleus in the shield.

The attenuation coefficient for removal of fast neutrons in this approximation is obtained from Eq. (7-9), inserting for hydrogen the cross section

$$\sigma_H = \frac{11}{E + 1.65} \tag{7-14}$$

where σ_H is in barns (10^{-24} cm^2) and E is the neutron energy in millions of electron volts. For heavier materials the so-called "geometrical" cross section is used. The nucleus is considered to be a sphere of constant density, having a radius $R = 1.4 \times 10^{-13} A^{1/3}$ cm (see page 66). The geometrical cross section is πR^2 for elastic scattering and πR^2 for other processes, including inelastic scattering and absorption. The total removal cross section, in barns, is then

$$\sigma = 2\pi R^2 = 0.12 A^{2/3} \tag{7-15}$$

From Eqs. (7-14) and (7-15) the neutron attenuation constant for water, in reciprocal centimeters, is

$$\mu(E) = 0.03 + \frac{0.7}{E + 1.65} \tag{7-16}$$

Assuming normal concrete, except for the water it contains, to have the properties of aluminum leads to an attenuation constant of

$$\mu(E) = 0.05 + \frac{0.1}{E + 1.65} \tag{7-17}$$

These attenuation coefficients are functions of the neutron energy E. To determine the attenuation produced by a slab of thickness x requires an integral over the neutron-energy distribution (7-4):

$$\frac{I}{I_0} = \int_0^\infty f(E) e^{-\mu(E)x} \, dE \tag{7-18}$$

This integral has been performed numerically, and the result may be approximated by the sum of two exponential terms:

Water: $$\frac{I}{I_0} = 0.98 e^{-0.19x} + 0.02 e^{-0.10x} \tag{7-19}$$

Concrete: $$\frac{I}{I_0} = 0.96 e^{-0.08x} + 0.04 e^{-0.06x} \tag{7-20}$$

For each formula, x is in centimeters.

These results are only a rather crude first approximation. A build-up factor similar to that for γ radiation should be included to account for neglected diffusion effects and for the fact that the scattering on the first collision tends to be in the forward direction. For attenuation factors

I/I_0 down to about 10^{-5}, this build-up factor does not seem to be large; for $I/I_0 \ll 10^{-5}$, no experimental information is available. The design of a shield is illustrated in Prob. 7-1.

7-7. Tolerance Limits. To calculate the thickness of a shield, it is necessary to specify some standard intensity that can be tolerated in the working space outside the shield. The present conventional tolerance limit, based on a 40-hour work week, is 7.5 mrem/hr, as discussed in Chap. 5. The mrem = millirem = 10^{-3} roentgen-equivalent-man is a radiation unit defined in terms of biological damage. The roentgen (r) is that amount of X or γ radiation which releases 93 ergs of energy on passage through 1 g of biological tissue.* The roentgen-equivalent-man (rem) is defined for other radiations to be the amount of radiation that does as much biological damage as 1 r of γ radiation. Fast neutrons are considered to be ten times as damaging as γ radiation, so that 1 rem of fast neutrons releases about 9 ergs per gram of tissue (Table 5-2).

According to Eq. (7-13) the rate at which a γ-ray beam of energy E releases energy in tissue is

$$H = \frac{-dJ}{dx} = \mu(E)J \tag{7-21}$$

For tissue the flatness of the curve of $\mu(E)$ over a wide range of E makes it suitable to assume that $\mu/\rho \approx 0.03$ cm^2/g, roughly independent of E for $E > 0.1$ Mev. If the tolerance limit* for H is substituted in Eq. (7-21), the corresponding limit for J is

$$J = \frac{H\rho}{\mu} = \frac{(93 \text{ erg/g} \times \text{roentgen})(7.5 \times 10^{-3} \text{ r/hr})}{(0.03 \text{ cm}^2/\text{g})(3600 \text{ sec/hr})(1.6 \times 10^{-6} \text{ erg/Mev})}$$
$$= 4.0 \times 10^3 \text{ Mev/(cm}^2)(\text{sec}) \tag{7-22}$$

Fast neutrons in tissue lose energy predominantly through collisions with the nuclei of hydrogen atoms in the tissue. The hydrogen cross section is given by Eq. (7-14). The composition of tissue is roughly that of water, in which there are $n_0 = 2N_0/18$ hydrogen atoms per cubic centimeter (or per gram, since the density of water is $\rho = 1$ g/cm^3). To a good approximation a neutron loses $1/e$ of its energy in an average collision with hydrogen and essentially nothing in other collisions. The rate H of energy release per unit volume in the tissue by a neutron flux I is then

$$H = n_0 I \int_0^\infty \frac{E}{e} f(E)\sigma_H \, dE = 5.23 \frac{n_0 I}{e} \times 10^{-24} \quad \text{Mev/cm}^3 \tag{7-23}$$

Equation (7-23) has been evaluated numerically, using Eq. (7-4) for $f(E)$.

* See also Sec. 5-19.

With this value the tolerance dosage rate of 9 ergs/g becomes equivalent to a fast-neutron intensity* of

$$I = \frac{e \times 9 \times 7.5 \times 10^{-3}}{\tfrac{1}{9} \times 6.02 \times 10^{23} \times 5.23 \times 10^{-24} \times 3600 \times 1.6 \times 10^{-6}}$$
$$= 91 \approx 10^2 \text{ neutrons/(cm}^2\text{)(sec)} \tag{7-24}$$

SPECIAL ASPECTS OF SHIELDING

7-8. Geometrical Corrections and Self-absorption. The discussion above has assumed the idealized situation of a point source of radiation at the center of a shield in the form of a thin spherical shell. Reactor-shielding problems in practice do not usually approach this ideal very closely, since the reactor is a large source of radiation rather than a point, and the shield is not usually spherical. It is possible to write down a simple general formula for computation of shielding in an arbitrary geometry. However, this general formula usually must be integrated numerically for each geometry, and no general analytic formula can be given.

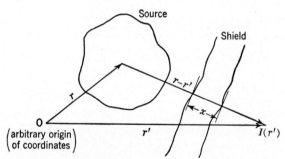

FIG. 7-3. Shielding in an arbitrary geometry.

Consider an irregular source of radiation surrounded by an arbitrary shield, absorption in which follows Beers' law [Eq. (7-8)]. The total radiation at any point is obtained by regarding the source as a collection of infinitesimal point sources, applying the point-source formula, and integrating over the entire source volume. Thus, using boldface symbols for vectors,

$$I(\mathbf{r}') = \int_{\text{source}} S(\mathbf{r}) \frac{e^{-\mu x}}{|\mathbf{r} - \mathbf{r}'|^2} \, d^3r \tag{7-25}$$

This formula is illustrated in Fig. 7-3. The quantity $I(\mathbf{r}')$ is the intensity coming straight from the source to the point of observation at $\mathbf{r}'$. A source function representing the radiation per second produced per

* For greater detail, see Table 5-3, Sec. 5-29.

volume element d^3r in the source is $S(\mathbf{r})\, d^3r$; the total rate of radiation from the entire source is $\int S(\mathbf{r})\, d^3r = S$. The distance between the source element and point of observation is $|\mathbf{r} - \mathbf{r}'|$. Along this straight path a segment of length x lies within the shield, resulting in an attenuation factor $e^{-\mu x}$. This case evidently requires numerical integration or summation.

An important specific case is the self-absorption of γ radiation by a spherical pile. In this simple geometry the integral (7-25) is not prohibitively difficult. The distribution of fuel and control rods would be generally fixed so that the neutron flux in the reactor is approximately uniform; the source distribution of γ radiation is thus approximately uniform. The γ radiation is assumed to be absorbed in the pile material with a single (average) attenuation coefficient μ. The numerical integration of (7-25) for this case has been carried out and can be summarized by the approximate formula

$$I = \frac{S_0}{4\pi R^2} \frac{1}{1 + 0.72\mu R_1} \tag{7-26}$$

Here I is the unshielded flux at a point outside the sphere a distance R from its center, S_0 is the total uniformly distributed source strength of the pile, and R_1 is the pile radius. The factor $(1 + 0.72\mu R_1)^{-1}$ corrects for self-absorption in the pile.

For piles of nonspherical shape, a simple approximation is to use R_1 in Eq. (7-26) for an "equivalent" sphere having the same volume-surface ratio as the actual pile. For a flat slab this is a maximum of 10 per cent in error. Calculations for other geometries are available [6].

Equation (7-26) should be applied only to γ radiation. For fast neutrons, the self-absorption is already included in the estimate of the fast-neutron leakage from the pile; it is just this fast leakage against which shielding must be provided.

Another geometrical question in shield design is the introduction of ducts and passages for pipes, electrical lines, and mechanical access to the pile. These ducts should be arranged to prohibit not only direct paths from the pile out through the shield but also paths in which radiation from the pile can travel down the duct with only one or two reflections on the walls of the duct. The latter arrangement involves oblique or stepped ducts. Access holes that are not open during the operation of the pile can run straight through the shield and be filled with removable plugs during reactor operation.

7-9. Gamma Production in the Shield. Even if the reactor could be represented as a point source, the γ radiation from it could not. About half the γ radiation produced by a pile is associated with the capture of neutrons in nonfissionable material. A certain fraction of this neutron capture must occur in the shield, which thus becomes a source of γ radiation. Approximately half this radiation escapes outward through the shield; the other half returns to the interior region and enters the shield on the opposite side of the reactor. The outward-escaping half of the radiation does not have to penetrate the full thickness of the shield, so that additional shielding will be required over that calculated for a γ source entirely interior to the shield.

To calculate this secondary effect, Beers' law [Eq. (7-8)] will be assumed to hold with constant attenuation coefficients μ_n and μ_γ for fast neutrons and γ rays. The γ flux that escapes a shield of thickness x_0 from a production point at a depth x in the shield is

$$dJ = \tfrac{1}{2}E\, dI e^{-\mu_\gamma(x_0-x)} \tag{7-27}$$

where $\tfrac{1}{2}$ = factor for outgoing radiation
E = average γ energy emitted per fast neutron captured
Also $-dI = \mu_n I_0 e^{-\mu_n x}$ = rate of fast-neutron absorption at depth x. The total γ flux emergent at the thickness x_0 is

$$J = \int_0^{x_0} dJ = \frac{E}{2}\,\mu_n e^{-\mu_\gamma x_0} I_0 \int_0^{x_0} e^{(\mu_\gamma-\mu_n)x}\, dx$$

$$= \frac{1}{2}\,E I_0\,\frac{\mu_n}{\mu_\gamma-\mu_n}\,(e^{-\mu_n x_0} - e^{-\mu_\gamma x_0}) \tag{7-28}$$

When $\mu_n \approx \mu_\gamma \approx \mu$, as is the case in an efficiently designed shield, this becomes

$$J \approx I_0 E\,\frac{\mu x_0}{2}\,e^{-\mu x_0} \tag{7-29}$$

The total γ flux produced by neutron capture in the shield is $I_0 E$; and if all this γ radiation had to pass through the entire shield, the emergent flux would be

$$J_0 = I_0 E e^{-\mu x_0} \tag{7-30}$$

The increase in the shield-generated γ flux due to neutron penetration of the shield before capture is therefore

$$\frac{J}{J_0} \approx \frac{1}{2}\,\mu x_0 \tag{7-31}$$

which is generally a factor of 10 or less but is not negligible.

If α is the total fraction of the reactor γ flux that is generated by fast neutrons in the shield, the γ intensity outside the shield computed by the method of Sec. 7-5 should be multiplied by the factor

$$1 - \alpha + \alpha(\tfrac{1}{2}\mu x_0) \qquad (7\text{-}32)$$

In case this factor raises the γ intensity above the tolerance level, the shield must be made thicker to bring the resultant γ flux to the tolerance level. The value of α will depend on the details of the reactor design— in particular, the fast-neutron leakage—but must be limited by $\alpha \lesssim \tfrac{1}{2}$.

The γ radiation in the shield comes mainly from capture of the neutrons after they have been thermalized. Boron is unique with regard to neutron capture, for it does not emit γ radiation but splits into an α particle and a lithium nucleus. It also has a very large capture cross section for thermal neutrons. Addition of 1 per cent by weight of boron to concrete, which has proved feasible in practice, would reduce γ production in the shield by a factor on the order of 10. This reduction would apply to the last term in Eq. (7-32), reducing the entire expression to a factor about equal to 1, so that no extra shielding would be needed to compensate for γ production in the shield.

Prob. 7-1. The Commonwealth Edison–Public Service gas-cooled reactor proposal [1] generates 350 Mw of heat in a horizontal cylinder of graphite 35 ft in diameter and 20 ft long. It is surrounded by a SA 204 low-alloy steel spherical vessel 44 ft in diameter and 1.67 in. thick. The whole is enclosed in a Quonset hut of concrete 8 ft thick, the closest outer approachable surface of which is the ends, 40 ft from the center of the reactor. Compute the γ and fast-neutron intensity there and compare with the accepted working levels. Assume boron concrete is used, so Eq. (7-32) is unnecessary.

Solution. a. Fast neutrons:

$$I_0 \text{ by Eq. (7-3)} = 2.6 \times 10^9 \times \frac{350{,}000}{40^2} = 5.7 \times 10^{11} \text{ neutrons/(cm}^2)\text{(sec)}$$

σ for the core (assume all graphite) by Eq. (7-15) $= 0.12(12)^{2/3} = 0.629$ barns

$$\mu \text{ by Eq. (7-9)} = 6.03 \times 10^{23} \times 2 \times 0.629 \times \frac{10^{-24}}{12} = 0.0632 \text{ cm}^{-1}$$

$$\text{Self-attenuation by Eq. (7-26)} = \frac{1}{1 + 0.72 \times 0.0632 \times 10 \times 30.48} = 0.0675$$

σ for the steel (assume all iron) by Eq. (7-15) $= 1.75$ barns

μ by Eq. (7-9) $= 0.146$ cm^{-1}

$$\text{Attenuation by the steel by Eq. (7-8)} = \frac{I}{I_0} = e^{-0.146 \times 1.67/2.54} = 0.9$$

$$\text{Attenuation for the concrete by Eq. (7-20)} = 0.96 e^{-0.08 \times 8 \times 30.48} + 0.04 e^{-0.06 \times 8 \times 30.48}$$
$$= 2.08 \times 10^{-8}$$

σ for the air (assume all nitrogen) by Eq. (7-15) $= 0.693$ barn

μ by Eq. (7-9) $= 3 \times 10^{-5}$

Attenuation for the air by Eq. (7-8) $= 0.98$ \qquad (negligible)

$$I = 5.7 \times 10^{11} \times 0.0675 \times 0.9 \times 0.98 \times 2.08 \times 10^{-8} = 710 \text{ neutrons/(cm}^2)\text{(sec)}$$

This is larger than Eq. (7-24) or Table 5-3; another foot of concrete is needed.

b. Gamma rays:

$$J_0 \text{ by Eq. (7-5)} = 1.14 \times 10^{13} \text{ Mev/(cm}^2\text{) (sec)}$$

	1 Mev	3 Mev	7 Mev
J_0 [Eq. (7-6)]...................	4.56×10^{12}	4.56×10^{12}	2.28×10^{12}
Graphite μ_B' (Fig. 7-2).............	0.049	0.026	0.015
$\mu_B = 2\mu_B'$......................	0.098	0.052	0.030
$(1 + 0.072\mu r)^{-1} = J/J_0$..........	0.318	0.467	0.603
Iron μ_B (Fig. 7-1).................	0.38	0.23	0.20
J/J_0 [Eq. (7-8)].................	0.200	0.377	0.427
Air (nitrogen) μ_B' (Fig. 7-2)........	0.049	0.026	0.015
$\mu_B = 0.001\mu_B'$..................	4.9×10^{-5}	2.6×10^{-5}	1.5×10^{-5}
J/J_0 [Eq. (7-8)].................	0.998	0.999	0.999
Concrete μ_B (Fig. 7-1)..............	0.12	0.068	0.049
J/J_0 [Eq. (7-8)].................	1.84×10^{-13}	6.12×10^{-8}	6.4×10^{-6}
External J......................	0.0533	49,000	3.53×10^6
Total J......................	....	....	3.58×10^6

By comparison with Eq. (7-22) the shielding is inadequate; another 5 ft of concrete is needed.

7-10. Coolant Shielding. The primary coolant material circulates outside the main reactor shield and will require additional shielding if it becomes appreciably radioactive by neutron capture. The β radioactivity creates no problem, but it is frequently accompanied by γ radiation, for which shielding may be necessary.

The strength of this γ radiation may be estimated as follows: Assume that the coolant and the fuel are distributed in the same way throughout the pile, and that fission and coolant activation are both produced mainly by neutrons of thermal energy. Suppose that certain nuclei of atomic weight A_n in the coolant produce radioactivity by capture of thermal neutrons with a cross section σ_n. Let M_n be the total mass of such coolant atoms contained within the reactor at any one time (this is only a fraction of the corresponding mass in the total volume of the coolant), and let A_f, σ_f, M_f be the corresponding quantities for the fissionable material in the pile. The rate at which these nuclei throughout the pile capture thermal neutrons and become radioactive is

$$Q_n = \frac{A_f}{M_f \sigma_f} \frac{M_n \sigma_n}{A_n} \times 3 \times 10^{13} K \qquad (7\text{-}33)$$

where K = reactor power, kw.

The radioactive nuclei in the coolant decay with a mean life $\tau_n = 1.44 t_n$, where t_n is the usually quoted half-life of the decay (see Sec. 9-4 also).

While the coolant is in the reactor, the number of radioactive nuclei satisfies the equation

$$\frac{dN_n}{dt} = Q_n - \frac{N_n}{\tau_n} \tag{7-34}$$

The solution of Eq. (7-34) with the initial condition that $N_n = 0$ at $t = 0$ is

$$N_n(t) = Q_n \tau_n (1 - e^{-t/\tau_n}) \tag{7-35}$$

Let the rate of coolant flow through the reactor be M_i/T_i, where M_i is the total coolant mass inside the reactor, and T_i is the time required for any element of the coolant to pass through the reactor. The activity per gram of coolant, just as it emerges from one pass through the reactor, is

$$Y_0 = \frac{N_n(T_i)}{\tau_n M_i} = \frac{Q_n}{M_i}(1 - e^{-T_i/\tau_n}) \tag{7-36}$$

If an element of coolant reaches a position x in the system a time T_x later, the residual activity from Y_0 is $Y_0 e^{-T_x/\tau_n}$.

If the coolant has completed an entire cycle in time T and is emerging from the reactor a second time, the activity per gram is

$$Y_1 = Y_0 + Y_0 e^{-T/\tau_n} \tag{7-36a}$$

Here the activity Y_0 is generated in the second pass through the reactor, and $Y_0 e^{-T/\tau_n}$ is the residual activity from the first pass. Likewise after three passes

$$Y_2 = Y_0 + Y_0 e^{-T/\tau_n} + Y_0 e^{-2T/\tau_n} \tag{7-36b}$$

and in general, for very many passes

$$Y = Y_0 \sum_{j=0}^{\infty} e^{-jT/\tau_n} = \frac{Y_0}{1 - e^{-T/\tau_n}} \tag{7-36c}$$

The total source of radiation in the heat exchanger, for example, is

$$S_n = M_{exch} Y e^{-T_x/\tau_n} = S_n^0 \frac{1 - e^{-T_i/\tau_n}}{1 - e^{-T/\tau_n}} e^{-T_x/\tau_n} \tag{7-37}$$

where M_{exch} is the mass of coolant in the heat exchanger, T_x is the time required for the coolant to reach the exchanger from the pile, and

$$S_n^0 = \frac{M_{exch}}{M_i} Q_n$$

Equation (7-37) simplifies in the extreme cases of long or short half-lives:

$$S_n \approx S_n^0 \frac{T_i}{T} \qquad \tau_n \gg T$$

$$\approx S_n^0 e^{-T_x/\tau_n} \qquad \tau_n \ll T_i \qquad (7\text{-}38)$$

Equation (7-37) is based on the assumption of equilibrium conditions; namely, that the sum over a finite number of passes is well approximated by the infinite sum in Eq. (7-36). This is not true in case $\tau_n \gg T_0$, the total operating time of the reactor. Then $S_n \approx (T_i/T)(T_0/\tau_n) S_n^0 \ll S_n^0$. Therefore in considering coolant activities one may drop those with $\tau_n \gtrsim 10$ years as negligible.

The relevant information for sodium, potassium, and the most prominent sources of activity in ordinary water are given in Table 7-2. For any element, only certain isotopes produce any γ activity on neutron capture. It is therefore important to know the relative abundance r_n of the activity-producing isotope among all naturally occurring isotopes of the element. Only the important activity-producing isotopes are listed in Table 7-2 (natural elements are listed on page 86). The total γ energy released per average decay is listed as $E_{n\gamma}$; for shielding purposes the attenuation coefficient corresponding to an energy per γ ray of about 2 Mev can be used.

TABLE 7-2. COOLANT ACTIVATION PARAMETERS

Capturing isotope	Relative abundance r_n, in element	σ_n, barns	τ_n	$E_{n\gamma}$, Mev
O^{18}...........	2.0×10^{-3}	2×10^{-4}	0.7 min	1.1
Na^{23}..........	1.00	0.5	0.9 day	4.1
Mg^{26}..........	0.11	0.05	0.2 hr	1.0
Cl^{37}..........	0.25	0.56	1 hr	1.5
K^{41}	0.069	1	0.7 day	0.4
Ca^{48}..........	1.8×10^{-3}	1.1	0.2 hr	2.7

For a mixture of activities the unshielded γ intensity at a distance R from the heat exchanger is given by

$$J_0 = \frac{1}{4\pi R^2} \Sigma_n S_n E_{n\gamma} \qquad (7\text{-}39)$$

Here S_n is obtained from Eq. (7-33) by using $M_n = r_n M$, where M is the mass of the element and r_n is the relative abundance of the particular isotope involved. From Eqs. (7-33) and (7-39) a suitable figure of danger for any isotope is $r_n \sigma_n E_n$. This coefficient indicates that, from the point of view of shielding, K is a preferable coolant to Na.

Fast neutrons can give rise to certain additional coolant activities that are not produced by slow neutrons. Although these activities will be relatively weak, they may be of rather high energy and inherently difficult to shield. For example, radioactive N^{16} and N^{17} are produced by (n,p) reactions on O^{16} and O^{17}, which will be present in an oxygen-containing coolant (water). The (n,p) reactions have a threshold of about 10 Mev, and not very precisely known cross sections on the order of $\sigma = 10^{-2}$ barns above this threshold. The relative abundance of O^{16} is approximately 100 per cent, of O^{17} is about 4×10^{-4}. The mean lives are $\tau_{16} \approx 10$ sec, $\tau_{17} \approx 6$ sec. The radiations are $E_\gamma = 5.3$ Mev for N^{16}, and N^{17} emits a delayed neutron of mean energy $E_n \approx 1$ Mev. The intensity of these activities may be computed as above if one multiplies Eq. (7-33) by a factor F, which is the ratio of fast flux ($E_n \gtrsim 10$ Mev) to thermal flux in the pile.

For a water coolant sufficiently free of impurities, especially sodium, these nitrogen activities will constitute the chief coolant-shielding problem.*

7-11. Fission-product Shielding. A semiempirical formula has been obtained to represent the radiation produced by fission products as a function of time [7]. Under the assumption that the neutrinos carry half and the β particles and γ rays each carry one-quarter of the total energy, the formula becomes

$$SE_\gamma = 6P \frac{M_f}{A_f} (d^{-1.2} + 3d^{-1.4}) \times 10^{17} \qquad \text{Mev/sec} \qquad (7\text{-}40)$$

where $d > 1$ is the average time in days since the fission products were produced. To a first approximation $d = \frac{1}{2}b + c$, where b is the burning time of the fuel in the reactor and c is the subsequent cooling time. In Eq. (7-40) M_f is the mass of fuel in the slug in grams, A_f its atomic weight, and P is the fraction of the fuel burned up before extraction from the reactor. See Sec. 9-4 for further details.

The total unshielded intensity a distance R from a fuel slug is

$$J_0 = \frac{SE_\gamma}{4\pi R^2} \qquad (7\text{-}41)$$

To compute shielding, an attenuation coefficient corresponding to an average γ-ray energy of 1.0 to 1.5 Mev may be used.

---

* It is predicted (Chap. 9, Ref. 90) that an attenuation to 0.001 will prove ample for water-cooled reactors and that little if any shielding around steam lines from boiling reactors will be needed.

7-12. Heat Production in Shield. The absorption of neutrons and γ rays releases energy in the shield, which is ultimately converted to thermal energy. This heat release is mostly concentrated on the inner side of the shield, where special cooling may become necessary. Heat production in the shield comes from three primary sources: the γ radiation incident from the pile, the fast neutrons and the γ radiation they emit upon capture, and the energy released by thermal neutrons upon capture. Although the thermal neutrons are not important in shielding considerations, they may contribute appreciably to the energy release in the first foot or so of the shield.

The rate of energy release per unit volume in millions of electron volts per cubic centimeter per second, at a depth x in the shield, is

$$H'' = \frac{I_0 E}{\lambda} e^{-x/\lambda_n} + \frac{I_0' E'}{\lambda'} e^{-x/\lambda_{n'}} + \frac{J_0}{\lambda_\gamma} e^{-x/\lambda_\gamma} \qquad (7\text{-}42)$$

Here I_0, I_0' are the intensities of fast and thermal neutrons incident on the shield, and λ_n, λ_n' the mean free paths for removal of these neutrons. $E \approx E' \approx 8$ Mev are the energies released on their capture; λ, λ' are the total mean free paths for absorption of this energy in the shield. J_0 is the γ-energy flux incident from the pile, and λ_γ its absorption mean free path. Multiplying Eq. (7-42) by 1.56×10^{-8} converts H'' to H' in Btu per cubic foot per hour.*

For concrete, $\lambda_\gamma \approx 20$ cm from Fig. 7-1. The slow-neutron mean free path is $\lambda_n' \approx 5$ cm, and the fast-neutron mean free path is $\lambda_n \approx 13$ cm from Eq. (7-20). The fast neutrons will generally be thermalized before capture, so that their effective mean free path for penetration into the shield will be somewhat larger than λ_n. If neutron capture gives rise to γ radiation, the associated mean free paths for energy absorption are $\lambda' = \sqrt{(\lambda_n')^2 + \lambda_\gamma^2} \approx 20$ cm, $\lambda = \sqrt{\lambda_n^2 + (\lambda_n')^2 + \lambda_\gamma^2} = 25$ cm. For the fast neutron the path length λ_n' is included as a minimum estimate for the thermalizing process. If boron is added to the shield in sufficient quantities (about 1 per cent by weight) to capture most of the thermalized neutrons, there will be essentially no γ radiation ($\lambda_\gamma = 0$), so that λ and λ' will be considerably reduced. The intense heat production would be limited to a narrower region of the shield, simplifying the problem of heat removal. This is another advantage of the boron-containing shield.

---

* The inner portion of shield that most needs cooling is designated the "thermal shield" and is generally constructed of metal, such as steel plate. The remainder, usually of concrete, is known as the "biological shield." It may also require some cooling, since concrete heated to over 200°F tends to lose H_2O, and therefore strength and neutron shielding effectiveness. The cooling of shields is covered in Chap. 9, particularly Sec. 9-7.

7-13. Permanent Fission-product Disposal. A serious problem akin to shielding arises when one envisions a large-scale nuclear power industry, as opposed to pilot installations. This is the permanent disposal of fission products. Activities with half-lives exceeding 10 years are present, and an expanding nuclear power industry would produce radioactive wastes faster than they decay. If the total average nuclear power production is K kw, the waste products produced will average about $500K$ curies per year [10]. The electric power production in the United States (1953) is on the order of 10^8 kw (see Chap. 13). If even 2 per cent of this were produced by nuclear reactors, a billion curies of radioactivity per year would require disposal.

There are two approaches to the problem of disposal: (1) the fission products can be localized in specific, isolated dumps; (2) they might be dispersed so uniformly over a large area that the increase in activity at any point is negligible. The only obvious way to effect (2) is to release the (liquid) fission products into the ocean. The natural radioactivity of all the water in the oceans (roughly 10^8 cu miles) of the earth is something approaching a trillion (10^{12}) curies, so that the addition of a billion curies per year would not seem serious. The results of such a program are essentially unknown, however: there is no assurance that the wastes would disperse uniformly throughout the ocean rather than concentrating in dangerous amounts, perhaps in biological matter. Furthermore, if the nuclear power industry should increase by a factor of 100 over the modest 2 million kw allowed here, the annual addition of radioactivity in the form of wastes would approach the total activity of the oceans. Ocean disposal thus appears as a temporary expedient in which any real difficulties are just postponed to form a problem for later generations. Since the dispersal is essentially irreversible, any presently unforseen difficulties might amount to a catastrophe. To judge from the lack of foresight shown in the past on such matters as industrial pollution of inland waterways, the establishment of a policy of ocean dispersal seems prohibitively dangerous. The earth's atmosphere, assumed 10 miles high, has a volume of 2×10^9 cu miles. Appendix O shows that atmospheric disposal is entirely out of the question, even if fallout could not occur.

Under alternative 1, the chief problem is to assure that the fission products remain localized over long periods of time. If they are liquid, the storage tanks must be subject to constant inspection and guard against corrosion. If they are solid, they must not be carried through the earth by seepage. Although further research is needed, a favorable process for at least some of the long-lived activities is to contain them in certain natural clays, such as montmorillonite. The clay will readily exchange cations, adsorbing metallic fission products to the extent of about 1 milliequivalent/g. When fired to 1000°C it produces a very durable ceramic, which is inactive to further ion exchange. The ceramic would be resistant to heat, solution by water, and in pebble form would not be disturbed by major shocks, such as those produced by earthquakes. Burial in desert areas would provide isolation and relative freedom from water.

The optimum locations of a nuclear power plant, a reactor-fuel reprocessing plant, and an ultimate waste-disposal site will in general not coincide. As radioactivity decays in storage, the cost of shipping decreases because less shielding must be shipped, but storage charges continue. The solution of a processing and disposal problem will thus generally include an optimized cooling period before each transportation step. For instance, SR waste concentrated to a reasonable specific activity costs 30 cents per gal to store. It is then shipped some 500 to 1000 miles. The optimum cooling period is about six years. Similarly, spent MTR fuel elements are allowed to decay for some three months before being shipped as γ-ray sources. The radioactive-shipment business alone will be considerable when nuclear power is well established.

A large dump of waste products would generate considerable heat (see Prob. 9-10) which might be utilized as a source of secondary power. Further study is needed of the problems involved in the collecting and handling of radioactive wastes, and particularly the disposal of long-lived activities [11]. Possible magnitudes and principal isotopes involved are listed in App. O.

Prob. 7-2. Wolman and Gorman [11] quote the following costs for radioactive-waste-disposal processes:

1. Permanent underground steel or reinforced concrete tanks for highly active wastes, with cooling system and external drains leading to a monitoring pit for leak detection—first cost = $0.35 to $1.75 per gal.

2. Disposal on the ocean bottom of low and intermediate-level wastes, by mixing in concrete and setting in steel drums—cost = $0.30 per lb, including shipping to seaport, for 20 tons per year.

3. Complete incineration of combustible radioactive animal and vegetable matter, with dust removal from the gaseous products by glass-fiber filters—up to $5 per ft^3 when oxygen is used.

4. Present costs of handling all normal radioactive production, processing, and operating wastes exceed 0.03 cent per kwhr.
Check these costs with original designs or methods of your own.

The following are general problems for this chapter:

Prob. 7-3. Show that the distribution by Eq. (7-4) is normalized to unit integral, and that the average energy is $\bar{E} = 2$ Mev. What is the root-mean-square energy $\sqrt{\overline{E^2}}$, and the average "spread" of the energies about the mean, $\sigma_E = \sqrt{\overline{E^2} - (\bar{E})^2}$?

Prob. 7-4. Find the unshielded γ-radiation intensity at distances of 10, 30, and 100 ft from a 60,000-kw pile.

Prob. 7-5. Find the distance from a 60,000-kw pile at which the unshielded γ intensity is at the 40 hr/week tolerance level, including the effect of attenuation in air. This is the closest distance of working approach to the unshielded reactor.

Prob. 7-6. A radioactive source of 1 curie is defined as undergoing 3.7×10^{10} decays per second. Radioactive Co^{60} emits two γ rays, of energies 1.33 and 1.17 Mev, in each decay. Suppose a 100-curie point source of Co^{60} to be inserted in the center of a solid spherical shield. What must be the radius of the shield for attenuation to the tolerance level at the outside surface of the shield, if it is made of (a) lead, (b) iron, (c) concrete? Compare the weights of the shields in the three cases. Neglecting the expense of fabrication, compare the costs if lead, iron, and concrete cost respectively $150, $50, and $2 per cu ft.

Prob. 7-7. Suppose a 100-curie source of γ radiation with the spectral distribution (7-6) to be placed in the spherical shield of Prob. 7-6. What must now be the shield thickness for lead, iron, and concrete?

Prob. 7-8. (a) Using the two-component expressions, Eqs. (7-19) and (7-20), find the average penetration depth of fast neutrons into water and concrete. (b) Using the three-component distribution, Eq. (7-6), find the average penetration depth of pile γ rays into lead, iron, and concrete.

Prob. 7-9. A 60,000-kw reactor is to have a concrete shield in the shape of a Quonset hut. The outer surface of the shield is 45 ft from the center of the reactor. How thick should the shield be, considering both γ rays and fast neutrons, but neglecting secondary γ production in the shield? Compare with Prob. 7-1.

Prob. 7-10. The reactor in Prob. 7-9 is a sphere of 18-ft radius. Assuming its γ-ray absorption to be roughly that of aluminum, calculate the reduction in the γ flux into the shield, as compared with that from a reactor without self-absorption. How much does this reduce the thickness of the shielding needed for γ radiation? What is the final thickness of the shield, also taking into account secondary γ production in the shield?

Prob. 7-11. A radioactive source of 1-Mev γ rays is in the form of a horizontal disk of radius r cm and uniform density σ curies/cm². Calculate the unshielded intensity at a point a distance R directly above the center of the disk. Compare this intensity with that from a point source of the same total intensity concentrated at the center of the disk.

Prob. 7-12. Analysis of a sample of untreated fresh water shows the following impurities in parts per million by weight: Ca, 90; Mg, 30; Na + K, 10; Cl, 10. This water is used to cool a 60,000-kw reactor, where the water in the reactor has one-fifth as much mass as the U^{235} ($\sigma_f = 550$ barns [8]), and the heat exchanger holds one-half the water in the coolant system. The heat exchanger is not to be shielded but is to be located in a pit of sufficient size that no injury will come to personnel outside the pit. How much bigger than the heat-exchanger system should the pit be?

Prob. 7-13. Chapter 9, Ref. 111, describes a hypothetical accident to a 1000-Mw reactor after operation. It is assumed that 1 per cent of the fission products are released to the air during the pessimistic condition of atmospheric temperature inversion, so that the fission product cloud remains at ground level and the rate of expansion is limited. (This assumption is reasonable for a reactor using either solid fuel of which only a fraction vaporized or liquid fuel continually scrubbed to low fission-product content.) Fallout and intake are neglected. The total γ-ray dose received by a person in the path of the cloud center is estimated at 1000 r at a distance of 0.1 mile, 178 r at 1 mile, 16 r at 10 miles, and 0.9 r at 100 miles. Estimate the safe radius and a reasonable exclusion area on this basis, and compare with the norms of 1 electrical or 5 thermal kw/acre, which have been suggested for certain reactors and conditions.

Prob. 7-14. A horizontal section through the MTR reactor [9] core shows a rectangle of 40 by 70 cm (60 cm high) in which the U^{235} fuel elements are placed and the main cooling-water stream flows. Surrounding this is Be reflector to an OD of 54.4 in. and a 60-in. OD Al tank of 1-in. wall. This tank is surrounded to a 7-ft 4-in. square with 1-in. graphite reflector spheres. There follow in turn solid graphite reflector to a distance of 6 ft 3 in., two 4-in. steel-plate thermal shields, and 9 ft of concrete employing $BaSO_4$ as the gravel in the mix. The thermal power is 30 Mw, but the shields were designed to reduce all radiations at 60 Mw to less than 10 per cent of the accepted 8-hr tolerance, to provide low instrument background around the reactor. Check this result for both γ and neutron flux.

Prob. 7-15. The only shielding above the MTR reactor (see Prob. 7-14) during unloading is the 20-ft depth of cooling water to the center of the core. This keeps γ radiation below 0.1 r per 8 hr. Verify this value theoretically for a reasonable burnup.

Prob. 7-16. During operation of the MTR an additional 11-in. layer of Pb shot in a steel top plug is employed over the reactor (see Prob. 7-15). Verify this design.

Prob. 7-17. Holes go through the MTR shields to the surface of the core (see Prob. 7-14). When not in use they are filled with composite plugs having the same thicknesses of Be, Fe, graphite, and concrete as the shield. When these are removed during shutdown to insert an experiment they are received into a shielding "coffin," primarily to protect against γ rays from impurities in the Be. 10 in. of Pb is employed. Check the need of this thickness after 1 month of operation at 30 Mw. Assume that (1) the impurities are 3 per cent and consist primarily of Na; (2) the Be tends to oxidize.

Prob. 7-18. The General Electric TTR employs a 5-ft graphite cube as the moderator. Centrally located in it is a cylindrical annulus 18 in. long, 18 in. OD, and 12 in. ID. Enriched U disks are mounted on 16 equidistant rods in the annulus, parallel to its axis. Around the graphite is a $\frac{1}{16}$-in. sheet of Cd and a $\frac{1}{4}$-in. sheet of steel. It is stated that operation at 10 kw for 40 per cent of the normal working day requires an outer shield of 3 ft of magnetite concrete (assume 50 per cent Fe_3O_4 by volume). Check this shield thickness, and compute what it should be for 30-kw operation over the full working day.

Prob. 7-19. The following approximate shielding has been suggested (Chap. 9, Ref. 103) for a 250-Mw reactor:

1. A thermal shield attenuation of gammas of 10^3, obtained by 1 to 2 ft of iron slabs
2. A neutron attenuation of 10^3 obtained by 2 ft of water, serving also as the thermal shield coolant
3. Further attenuation of gammas of 10^8, obtained by a 9-ft concrete biological shield

The outside diameter of such a reactor and shield is evidently some 36 ft. (a) Check the above design by the methods of this chapter. (b) Estimate the first cost of the shielding, and the incremental fixed cost due to the shielding per kilowatthour of electrical output at 24 per cent over-all efficiency, 90 per cent load factor, and 10 per cent total annual fixed charges. Compare with estimates in Chap. 9, Ref. 103:

1. Concrete at $200 per yd³ installed—2000 yd³ for the core and 500 yd³ for outer piping and equipment
2. Iron at $33\frac{1}{3}$ cents per lb installed—150 tons
3. Building at $2 per ft³

Prob. 7-20. A Bulk Shielding Reactor (BSR), if operated indefinitely at 1000 kw with 50 ppm of Na_2CrO_3 as corrosion inhibitor in the water is said to exceed the γ-tolerance level at the surface of the pool. Check this calculation if the tank is a 20-ft cube with the core near the bottom.

Prob. 7-21. The following materials have been used as coarse aggregate in concrete shields [12]:

Material	Specific gravity	Per cent H_2O	Per cent Fe	Cost, $/ton	Concrete specific gravity
Magnetite............	4.6	0	55+	20	3.5
Limonite-goethite......	3.4	11	50+	40	
Ferrophosphorus.......	6.5	0	75	~100	4.8
Calcite..............	2.7	0	0	~10	2.4
Steel punchings........	7.8	0	100	~100	

Estimating that the cement and sand alone, set to a specific gravity of 2.3, costs $50 per yd³ and equals in volume the coarse aggregate, show which coarse aggregate would give the least total-shielding-material first cost for a reactor.

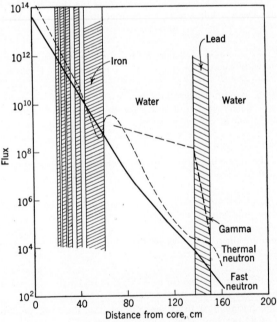

FIG. 7-4. STR shield and approximate flux densities.

Prob. 7-22. Figure 7-4 shows the attenuation in the shield of STR (Chap. 9, Ref. 103). Check this curve by the methods of this chapter.

Prob. 7-23. It has been estimated, within a factor of 10, that 100,000 lethal doses of fission-product radioactivity exist within an operating power reactor for each kilowatt [13]. Verify this estimate.

Prob. 7-24. A 250,000-kw reactor would consume about 2 lb of fuel per day and produce almost an equal weight of radioactive fission products. What exposure to a day's fission products after 2 hr of "cooling" would be lethal at a distance of 15 ft?

Ans. 10^{-3} sec.

REFERENCES

1. "Reports to the U.S. Atomic Energy Commission on Nuclear Power Reactor Technology," Government Printing Office, May, 1953.
2. Bonner, T. W., R. A. Ferrell, and M. C. Rinehart: *Phys. Rev.*, **87**:1032 (1952). Hill, D. L.: *Phys. Rev.*, **87**:1034 (1952). Watt, B. E.: *Phys. Rev.*, **87**:1037 (1952).
3. Kinsey, B. B., G. A. Bartholomew, and W. H. Walker: *Phys. Rev.*, **89**:375, 386 (1953); **83**:519 (1951); **82**:380 (1951); *Can. J. Phys.*, **29**:1 (1951).
4. Spencer, L. V.: NBS Report 1726, 1952. Goldstein, H., J. E. Wilkins, and L. V. Spencer: *Phys. Rev.*, **89**:1150 (1953). Goldstein, H., et al.: NYO-3075 (1954), NYO-3079 (1953). Peebles, G. H.: Gamma-ray Transmission through Finite Slabs, The RAND Corporation, Santa Monica, Calif., 1952.
5. Ruddy, J. M.: AECU-1211.
6. For example, J. L. Balderston, J. J. Taylor, and G. J. Brucker: "Nomograms for the Calculation of Gamma Shielding," KLX-24, Kellex Corporation, 1948; also J. Dwork et al., KAPL-1262, 1955.
7. Way, K., and E. P. Wigner: *Phys. Rev.*, **73**:1318 (1948).
8. Hughes, D. J.: "Pile Neutron Research," Addison-Wesley Publishing Company, Cambridge, Mass., 1953.
9. Huffman, J. R.: "The Materials Testing Reactor," Phillips Petroleum Co., IDO-16121 and 16122, 1953.
10. Hatch, L. P.: *Am. Scientist*, **41**:410 (1953).
11. Wolman, A., and A. E. Gorman: *Mech. Eng.*, **77**:321–324 (1955).
12. Davis, H. S.: *Gen. Elec. Rev.*, September, 1954, pp. 43–45.
13. Benedict, M.: *Chem. Eng. Progr.*, **51**:53–66 (February, 1955).
14. McKinney, V. L., and T. Rockwell III: ORNL-242, 1949.
15. Rockwell, T.: "Reactor Shielding Design Manual," McGraw-Hill Book Company, Inc., New York, 1956.
16. "Boron Carbide and Elemental Boron," Norton Company, Worcester, Mass., 1955.

CHAPTER 8

THE FLOW OF FLUIDS

By Charles F. Bonilla

The heat generated in the majority of conceivable nuclear reactors is removed by a fluid flowing through the reactor. It is desirable to be able to design to the optimum pressure drop and to select the exact size of pump required. It is absolutely indispensable that the coolant distribute itself properly among the many parallel channels usually employed. It is necessary to investigate in detail the effect on coolant flow of all possible changes in regime, so as to ensure economy, reliability, and long life in routine operation and minimum hazard in any accidents. It is evident that many phases of fluid flow are involved.

PHYSICAL PROPERTIES

8-1. Viscosity. Steady horizontal flow of a fluid along a pipe (Fig. 8-1a) requires a corresponding pressure gradient, $-dP/dN$. The fluid in a cylinder of radius r and length dN thus has a net axial force of $\pi r^2\, dP$ acting on it in the direction of flow. To keep the flow from accelerating, an equal and opposite axial force must be acting on the cylinder. The only possibility is a shear stress of magnitude

$$\tau = \frac{\pi r^2\, dP}{2\pi r\, dN} = \frac{r}{2}\frac{dP}{dN}$$

at the cylindrical surface.

A fluid is a substance in which the molecules are free to move past each other and which thus has no shear strength. However, fluid which is stationary or in streamline flow undergoes continual random thermal agitation of its individual molecules. In turbulent fluids large groups of molecules, called "eddies," move at random over longer distances. Whenever the separate molecules or the eddies move normal to the flow from a slow streamline to a faster streamline, they convey net backward momentum to the latter, and vice versa. Molecular or eddy crossflow through a surface parallel to the flow is thus equivalent to a shear stress at the surface. Let us designate the average crossflow velocity over the whole surface AA in Fig. 8-1b as v, in both directions at once. The molecules are assumed to travel a distance Δy perpendicular to AA, on the average, and then lose their difference in axial velocity to their new neighbors. If du/dy is substantially constant within $\pm\Delta y$ of AA, the molecules starting anywhere within Δy from AA will transfer

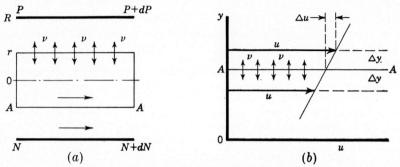

FIG. 8-1. Crossflow and shear stress in fluid flow in a pipe.

on crossing AA the same amount of relative axial momentum, on the average. The shear stress at AA is then

$$\tau = \frac{v\rho\,\Delta u}{g_c} = \frac{v\rho\,\Delta y}{g_c}\frac{du}{dy} \qquad (8\text{-}1)^*$$

When turbulence is absent the product $v\rho\,\Delta y$ is designated the viscosity μ of the fluid, and

$$\tau = \frac{\mu}{g_c}\frac{du}{dy} \qquad (8\text{-}2)$$

is its defining equation. Any fluid for which μ does not vary with τ or du/dy is said to be a "Newtonian fluid."

By integrating Eq. (8-2) and applying the boundary conditions of any desired flow apparatus, a method for the determination of viscosity from the observed force, velocity, and dimensions is obtained. Contrariwise, knowing the dimensions of a conduit and the viscosity and flow rate of a fluid in streamline flow, the necessary pressure drop may be computed. The viscosity is also involved in predicting pressure drops in turbulent flow.

To obtain the true viscosity, independent of container dimensions, every dimension of the fluid body and curvature of the velocity gradient should be many times larger than Δy, say 1000 or more times the mean free path. The cgs unit of viscosity is the "poise," or gram per centimeter per second. Viscosity in poises must be multiplied by 0.0672 to convert

* g_c is the dimensionless ratio F/ma of the unit of force to the unit of ma employed. $g_c = 1$ for any "absolute" system of units.

it to pounds per foot-second, by 242 for pounds per foot-hour, and by 0.00209 for slugs per foot-second.

8-2. Viscosity of Gases. For any one gas, the product $\rho \, \Delta y$ in Eq. (8-1) is constant at moderate densities and v varies only with temperature. Thus, the viscosity of gases is independent of pressure at normal pressures, such as up to 10 per cent of the critical pressure while above the critical temperature, and is also independent of shear stress. Experimental data on the effect of P and T on μ have been summarized [1] for the common gases.

Correction of the low-pressure viscosity to a different temperature may be carried out by several methods. For rapid approximate interpolation, a log-log plot of viscosity vs. temperature may be assumed to be straight. The slopes observed for the principal gases [2] range from 0.65 to 1.0 compared with the simple kinetic-theory slope of 0.5. For extrapolations to high temperatures or for long interpolation well above the critical temperature, Sutherland's empirical equation is generally reliable [3]:

$$\mu = \frac{BT^{3/2}}{T + C} \tag{8-3}$$

Values of B and C may be computed for any gas or gas mixture from the viscosity at two or more temperatures. Values for the common gases of possible nuclear-power significance are listed in Table 8-1.

TABLE 8-1. EMPIRICAL CONSTANTS FOR SUTHERLAND'S EQUATION [Eq. (8-3)]
FOR LOW-PRESSURE VISCOSITY OF GASES*

Gas	Range, °C	B, poises × °K$^{-1/2}$	C, °K	Reference
Nitrogen............	227–1420+	0.1405 × 10⁻⁴	109.08	†
Air................	16–1560+	0.1460 × 10⁻⁴	109.58	†
Argon..............	296–1595+	0.19071 × 10⁻⁴	135.38	†
Carbon dioxide.......	280–1413+	0.15237 × 10⁻⁴	219.44	†
Oxygen............	0–827+	0.1700 × 10⁻⁴	126.8	†
Hydrogen‡.........	−21–302	0.06415 × 10⁻⁴	71.7	§
Helium¶............	15–185	0.1464 × 10⁻⁴	80.3	§

* For the effect of pressure on these gases, see J. Hilsenrath and Y. S. Touloukian, *Trans. ASME*, **76**:967–985 (1954); also "NBS-NACA Thermal Tables," National Bureau of Standards. The low-pressure viscosity of steam does not follow Sutherland's equation. To over 1200°C it is approximately $(88.4 + 0.361t)$ micropoises, where t is in degrees Centigrade. (C. F. Bonilla, R. D. Brooks, and P. L. Walker, "General Discussion of Heat Transfer," London, 1951; ASME, New York, 1953.)

† Bonilla, Brooks, and Walker, *op. cit.*

‡ For higher temperatures, use Tables 8-2 and 8-3 (Hilsenrath and Touloukian, *op. cit.*).

§ J. H. Perry, "Chemical Engineers' Handbook," 3d ed., pp. 370–374, McGraw-Hill Book Company, Inc., New York, 1950.

¶ See Prob. 8-1.

An alternate procedure employs Hirschfelder's collision integrals for nonpolar spherical molecules and has been shown [4] to agree within several per cent over the full temperature range even for long molecules such as n-C_9H_{20} and polar molecules such as CH_3CH_2I. The calculated results are expressable by any two of the three dimensionless ratios in Table 8-2.

TABLE 8-2. DIMENSIONLESS VISCOSITY-TEMPERATURE CORRELATION FOR
LOW-PRESSURE VISCOSITY OF GASES, FROM HIRSCHFELDER'S
COLLISION INTEGRALS (BASED ON LENNARD-JONES POTENTIAL)*

kT/ϵ	$\mu r_0^2/(M\epsilon)^{1/2} \times 10^{12}$	$\mu r_0^2/(MkT)^{1/2} \times 10^{12}$
0.3	0.04473	0.08171
0.6	0.08522	0.1100
1	0.1454	0.1454
2	0.2737	0.1935
4	0.4707	0.2354
7	0.6936	0.2622
10	0.8783	0.2755
30	1.790	0.3268
60	2.800	0.3615
100	3.894	0.3894
200	6.089	0.4306
400	9.520	0.4760

* k = Boltzmann constant, 1.38048×10^{-16} erg/°K
   ϵ = energy difference between molecules when separated and when at maximum energy of attraction
   r_0 = low-velocity collision diameter

Nomographs and interpolation tables for much closer values of kT/ϵ are available [4, 5].

If no values of r_0, ϵ, or μ are available for a desired gas, any two of the following approximations may be employed:

$$\epsilon = 0.75 T_c k = 1.39 T_b k \tag{8-4}$$
$$r_0 = 0.833 \times 10^{-8} V_c^{1/3} \tag{8-5}$$
$$\frac{\epsilon^{1/2}}{r_0^2} = 1.58 \times 10^{15} \frac{k^{1/2} P_c^{2/3}}{T_c^{1/6}} \tag{8-6}$$

where T_c = critical temperature
   T_b = boiling point
   P_c = critical pressure
   V_c = molar volume at critical point

Equations (8-4) and (8-5) are dimensionally consistent; Eq. (8-6) requires energy in ergs, r_0 in cm, P_c in atmospheres, and T_c in °K. An average deviation of some 5 to 8 per cent may be expected by these methods [5].

If the viscosity is known at one temperature, Eq. (8-4) can be employed to give ϵ, and thus kT/ϵ, at that temperature. The corresponding value of $\mu r_0^2/(MkT)^{1/2}$ is then read by interpolation in Table 8-2 and r_0 computed, or else the relative viscosity at any other temperature obtained with the two values of $\mu r_0^2/(MkT)^{1/2}$. If the viscosity is known at two or more temperatures, by a trial-and-error procedure the values of ϵ and r_0 can be found which best satisfy the table. These values of ϵ and r_0 have been determined from available viscosity data for 45 common gases [4, 5]. Table 8-3 gives the constants for the principal gases.

TABLE 8-3. EMPIRICAL CONSTANTS FOR TABLE 8-2

Constant	Gas							
	N$_2$	Air	A	CO$_2$	O$_2$	H$_2$	He	Hg
ϵ/k, °K..............	91.46	97.0	124.0	190	113.2	33.3	6.03	851
$(\sqrt{M}/r_0^2) \times 10^{-17}$, cm^{-2}.............	3.907	4.116	5.409	4.154	4.800	1.612	2.751	0.134

Though low pressures have little effect on the viscosity of a gas, it rises rapidly above the critical pressure, particularly near the critical temperature. Knowing the critical temperature and pressure, the ratio of viscosity under pressure to that at the same temperature and atmospheric pressure may be obtained within several per cent from Fig. 8-2 [6].

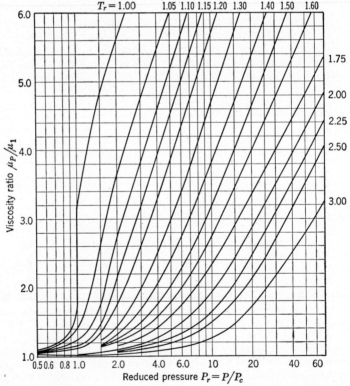

FIG. 8-2. Correlation of viscosity ratio vs. reduced pressure.

If T_c and P_c are not known, they may be estimated by several methods. Correction to high pressure may also be made with good accuracy by Enskog's theoretical method, requiring the density and $(\partial P/\partial T)_V$ at the high temperature and pressure [2].

The viscosity of a mixture of gases of similar molecular weight, such as oxygen and nitrogen, may be accurately obtained by interpolation by mole fraction between the viscosities of the pure components at the same temperature and pressure as the mixture. With other gases the viscosity of the mixture may range up to 25 per cent or more higher than the interpolated value, and may roughly be estimated graphically [3], or calculated for nonpolar gases by Hirschfelder's method [4]. For most practical cases Eq. (8-7), which is simpler and good to about 2 per cent, is adequate [5].

$$\mu_{\text{mixture}} = \sqrt{8} \sum_{i=1}^{i=n} \frac{\mu_i}{\displaystyle\sum_{j=1}^{j=n} \frac{x_j[1 + (\mu_i/\mu_j)^{1/2}(M_j/M_i)^{1/4}]^2}{x_i(1 + M_i/M_j)^{1/2}}} \tag{8-7}$$

Each summation is over the n components, x being the mole fraction and M the molecular weight. In addition, Fig. 8-2 applies within several per cent to mixtures by employing their pseudocritical properties (average P_c and T_c weighted by mole fraction).

Prob. 8-1. Predict the atmospheric pressure viscosity of He at 450°F and at 742°F by Table 8-1 and by Tables 8-2 and 8-3. Compare with the empirical expression [1]: $\mu = 4.2295 T^{3/2}/(T^{0.826} - 0.409)$ micropoises for $T = {}°K$. Predict μ at these temperatures and 150 psia by Fig. 8-2 (see Prob. 8-26).

8-3. Viscosity of Liquids. When the molecules of a fluid are close together, as in a liquid or in a gas under high pressure, the mean free path is short and momentum transfer as per Eq. (8-1) is not the only factor contributing to the viscosity. At these high densities, according to one view [7], the slipping of a molecule past its neighbors requires that a "hole" be present for it to move into and that the molecule possess a certain energy of activation E. From the Maxwell-Boltzmann distribution law, the number of molecules having an energy E or greater is proportional to $e^{-E/RT}$. Thus, at a given density, the higher the temperature the larger the fraction of the molecules that has the necessary energy to move into the available holes, and the lower the viscosity. At a given pressure, as the temperature is raised, the density will decrease, the number of holes will increase, and the viscosity will decrease. Finally, however, the molecules will be far enough apart that the momentum-transfer process will predominate, and the viscosity will start to increase again as the temperature is raised still further.

Another view is that liquid viscosity is the sum of the momentum-transport term and a term due to direct intermolecular forces [8]. In Appendixes A to E are listed viscosities of a number of liquids of possible interest in nuclear power.

Many methods are available for the interpolation or short extrapolation of liquid viscosities against temperature from one or more known values:

1. A plot of μ vs. t yields a smooth curve with a maximum deviation from a straight line of the order of 1 per cent per 20°C interval over useful ranges. Thus linear interpolation over short intervals is permissible.

2. A plot of fluidity $1/\mu$ vs. t is closely linear over fairly long temperature intervals, and thus suitable for reasonably long interpolations or extrapolations.

3. Duhring's rule applies quite well to liquid viscosities, particularly for related liquids [2]; a plot against each other of the temperatures at which two liquids have the same viscosity yields substantially a straight line over reasonably long temperature intervals.

4. The Arrhenius equation

$$\mu = Ae^{E/RT} \tag{8-8}$$

can be interpreted theoretically [7], E being the energy of activation of the hole formation. A depends on molecular weight and molar volume and is 0.0005 poise for many unassociated liquids. E is about one-third of the molar latent heat of vaporization for unassociated liquids and one-fourth for unsymmetrical molecules. For liquid metals E is only 4 to 12 per cent of the heat of vaporization, indicating that the ion, which requires less space than the atom, is the unit that offers the main resistance to flow. This is the cause of the unusual constancy of viscosity of liquid metals. A plot of log μ vs. $1/T$ is straight over temperature ranges having a constant E, and is useful for interpolation or extrapolation with two or more known values of μ and t. If viscosity is known at only one temperature, it can be extrapolated by a line parallel to that for a similar compound.

5. From the above observations concerning Duhring's rule, Eq. 8-8, and the near constancy of the molar latent heat of vaporization of most liquids, an approximate general correlation for nonmetallic liquids is indicated. This is available as a plot of μ against $t + a$, where a is the quantity that must be added to the temperature so that all nonmetallic liquids agree as closely as possible. The value of a may be found from the generalized plot [2] and one known value of the viscosity.

6. A procedure which is analogous to method 5 may be based on Bingham's equation for the fluidity $1/\mu$ of water [2]. Correcting to $\mu = 0.010019$ at 20°C [9] for μ in poises,

$$\frac{1}{\mu} = 2.155[t + a + \sqrt{8078.4 + (t + a)^2}] - 120.4 \tag{8-9}$$

For water, $a = -8.435°C$. For any other liquid a can be computed having μ at one temperature.

7. Viscosity may be related to the critical properties. Batschinski's equation [7], which holds within an average of 1 per cent for unassociated liquids, may be written

$$v - w_1 = \frac{c}{\mu} \tag{8-10}$$

where v is the specific volume and w_1 and c are constants. This method is evidently related to method 2, and w_1 and c may be evaluated if μ and v are known at two conditions. w_1 averages 0.307 v_c at the critical point, so that if v_c is known, μ and v at only one condition are necessary to establish the constants. $v - w_1$ is the free space or "holes" in the liquid.

8. Walther's equation

$$\frac{\mu}{\rho} + a' = Ae^{B/T^C} \tag{8-11}$$

has been found to apply up to the high viscosities reached by petroleum oils at low temperatures. μ/ρ is the kinematic viscosity. The cgs unit, square centimeters per second, is designated the "stoke." Almost straight lines are obtained by using μ/ρ in centistokes and setting $a' = 0.6$ and $A = 1$, and plotting log log $(\mu/\rho + 0.6)$ vs. log T, as would be indicated by twice taking logarithms of Eq. (8-11). The slight deviations from straight lines are at low temperatures and viscosities and have been eliminated in the plotting paper employed [10] by altering the log-log scale slightly. This paper covers the range of 3 to 2×10^7 centistokes and -30 to $450°F$ and permits long interpolations and extrapolations within these ranges with reasonable accuracy.

High pressure increases the viscosity of liquids, though relatively not as much as that of gases. For most organic liquids the increase is exponential, and a plot of log μ against pressure is substantially straight. For petroleum oils about 3000 psi is required to double their viscosity at room temperature and 6000 psi at $100°C$, but for smaller organic molecules 15,000 psi is necessary. However, for ammonia μ is linear in pressure [11], and for water the change is very small [3]. A short time lag in the response of the viscosity to pressure has also been reported [12]. In all cases the changes are very small below the critical pressure. Above the critical pressure and temperature the methods for gases in Sec. 8-2 may be employed.

Since the mean free path of the molecules in a liquid [similar to Δy in Eq. (8-1)] is extremely short, no effect of nonuniform velocity gradient on the viscosity is encountered. For long molecules high velocity gradients may tend to orient the molecules, or possibly alter their average extension. Evidence both pro [13, 14] and con [15] a decrease in viscosity of long hydrocarbons when at high shear stress has been presented, doubt being due to the high heat generation. The drop has initiated for a number of liquids at a shear rate of about 2500 sec^{-1} [13] or at a shear stress of about 15 to 20 psi [14, 16], and the viscosity has dropped 75 per cent by 10 times this threshold shear.

The viscosity of a mixture of liquids is intermediate between the viscosities of the ingredients. The two principal methods of estimating it follow:

1. For similar compounds the values of E in Eq. (8-8) are additive, and log μ for a mixture equals the summation of x log μ for the constituents, where x is the mole fraction. If the components are dissimilar the actual viscosity is lower, and if they form addition complexes it is higher [7].

2. For petroleum oils it is recommended [17] to interpolate by volume between the ordinates for the ingredients on the ASTM viscosity chart [10]. Thus log log $(\mu/\rho + a')$ is taken as equal to the summation of C log log $(\mu/\rho + a')$ for the constituents, where C is the volume fraction and a' is 0.6 centistoke.

Prob. 8-2. Predict the viscosity of D_2O at 750 psia and 388.8 and 440°F by methods of Sec. 8-3 and data in Appendixes C and E (see Prob. 8-10).

8-4. Flow Properties of Suspensions. The properties of suspensions are of interest in nuclear engineering in view of the possibility of "slurry piles" with the fuel and/or fertile element suspended in a liquid. Dilute suspensions of solid particles tend to be Newtonian, following the Einstein equation

$$\frac{\mu_s}{\mu_m} = 1 + 2.5C \tag{8-12}$$

up to a volume fraction C of solid of the order of 0.05. μ_s and μ_m are the viscosities of the suspension and medium, respectively. The coefficient 2.5 was derived for rigid spheres, but it holds for any reasonably equiaxed shape and is constant or increases only slightly in going to the smallest sizes. For particles having a ratio f of longest to shortest axis, Huggins adds the term $(f/4)^2$ to the coefficient 2.5.

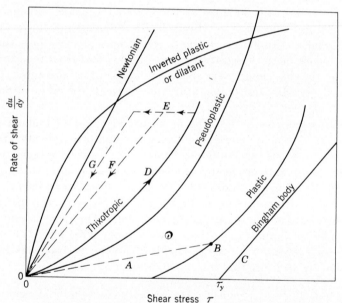

FIG. 8-3. Types of shear diagrams in streamline flow.

A straight line through the origin on a plot of du/dy vs. τ indicates Newtonian behavior [Eq. (8-2)], the slope of the line being the fluidity $1/\mu$. Suspensions with equiaxed particles which interfere more with each other the higher the shear rate show a downward concavity in Fig. 8-3, and are "dilatant," or inverted plastics. Long particles which line up more closely parallel with each other the higher the shear show the opposite curvature, and are "pseudoplastic"; the shear diagram usually becomes straight at high shear. If the lining up requires an appreciable amount of time or work, the substance is "thixotropic." In Fig. 8-3 the line D represents a rapid increase in shear and G the infinitely slow or equilibrium increase or decrease. If the rate of shear is held constant after a rapid increase, the shear stress will gradually decrease along a line such as E, reaching F and finally G.

Solids and some concentrated suspensions require a definite yield shear stress, τ_y, before any motion occurs. These materials are not therefore fluids and are designated "plastic substances." Many plastic suspensions approach an idealized straight-line behavior, that of the "Bingham body."

The ratio of the shear stress to the velocity gradient at any point on the shear curve of a substance is the "effective viscosity" μ' at that point. For instance, the slope of line A is $1/\mu'$ for point B. The reciprocal of the local slope at any point, such as B, is designated the rigidity η at that point.

"Slip" occurs if the containing walls do not possess a specific degree of roughness such that there is no discontinuity in the velocity gradient near the walls. The effect of slip particularly shows up [18] when the particle size is a perceptible fraction of the container dimensions. The apparent viscosity of a suspension in a smooth-wall capillary may be only a fraction of the true value.

With non-Newtonian materials an added difficulty occurs. In test instruments the shear stress is not uniform. Thus, the average viscosity obtained is an integrated value over a range of shear stress. Correct methods are outlined in Sec. 8-39.

Several equations are available for predicting the effective viscosity of suspensions at practical shear stresses. For volume concentrations up to the order of 25 per cent, reasonable agreement for equiaxed particles is obtained by retaining more terms in Einstein's derivation, as in Vand's equation [19]

$$\frac{\mu_s'}{\mu_m} = 1 + 2.5C + 7.17C^2 + 16.2C^3 \tag{8-13}$$

Orr [20] has found substantially Newtonian behavior in suspensions of different types and sizes of particles with 10 to 20 per cent solids by volume. In streamline flow they roughly follow Eq. (8-14) up to the fraction of solids by volume, C_∞, to which the particles settle in infinite time.

$$\frac{\mu_s}{\mu_m} = \left(\frac{1 - C}{C_\infty}\right)^{-1.8} \tag{8-14}$$

Many other equations have been proposed for high concentrations [19–21], but it is not safe in general to rely on such relations without experimental confirmation at the identical conditions.

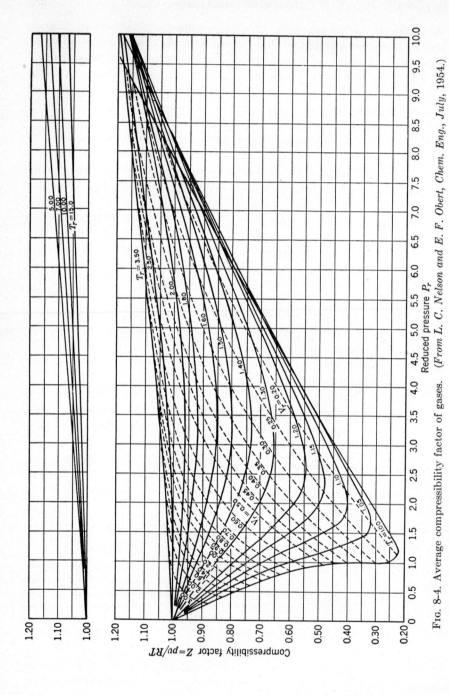

Reduced pressure P_r

Compressibility factor $Z = pv/RT$

FIG. 8-4. Average compressibility factor of gases. (*From L. C. Nelson and E. F. Obert, Chem. Eng., July, 1954.*)

8-5. Density of Fluids. The density of fluids in nuclear reactors is of importance in determining flow velocity, friction in turbulent flow, absorption of radiation, etc.   Sufficiently accurate values of the density of a gas may usually be obtained from the ideal gas law, namely, that the "compressibility factor" PV/RT equals 1.   Density changes most rapidly with temperature and pressure near the critical point.   In this region, lacking actual data for a given gas or vapor, the density can be obtained approximately from a generalized chart of PV/RT against P_R and T_R, as given in Fig. 8-4.   Kay has shown [22] that the chart is also useful for a mixture, by using for P_c and T_c the "pseudocritical" pressure and temperature of the mixture, the summations of the values of P_c and T_c, respectively, for the ingredients multiplied by their mole fractions. R is 0.082054 liter-atm/(°K)(g mole) or 0.7302 ft³-atm/-(°R)(lb mole) or 1545 ft-lb/(°R)(lb mole).

An alternate source for the density of a gas or vapor is a more exact equation of state than the ideal gas law, for which the coefficients have been determined for the particular gas.   No such equations so far are accurate over the whole range, and particularly near the critical point.   A suitable one is the Beattie-Bridgeman [23] equation

$$P = \frac{RT}{V^2}\left(1 - \frac{C}{VT^3}\right)\left[V + B_0\left(1 - \frac{b}{V}\right)\right] - \frac{A_0}{V^2}\left(1 - \frac{a}{V}\right) \tag{8-15}$$

Average values of the coefficients for P in atmospheres, V in liters per gram mole, and $T = °C + 273.13$ are in Table 8-4.   A generalized form of the Beattie-Bridgeman

TABLE 8-4. BEATTIE-BRIDGEMAN EQUATION COEFFICIENTS

Gas	A_0	a	B_0	b	C
He...........	0.0216	0.05984	0.01400	0	40
A............	1.2907	0.02328	0.03931	0	9,900
H_2..........	0.1975	−0.00506	0.02096	−0.04359	504
N_2..........	1.3445	0.02617	0.05046	−0.00691	42,000
O_2..........	1.4911	0.02562	0.04624	0.004208	48,000
Air...........	1.3012	0.1931	0.04611	0.01101	43,400
CO_2.........	5.0065	0.07132	0.10476	0.07235	660,000

equation in terms of the pressure and temperature and volume at the critical point is available [24] which is accurate to within 1 per cent nearly up to the critical density.

With liquids well below the critical temperature, the density can usually be interpolated or extrapolated within reason from any available values.   Increasing pressure causes a linear increase in the logarithm of the density.   Increasing temperature has a more complicated effect, which can be correlated against reduced temperature and pressure by Watson's "expansion factor" ω [25], given in Fig. 8-5.   The ratio ρ/ω at a given pressure is independent of temperature and need only be known for the desired pressure.

Thermal expansion of fluids is of special interest, in that it is the driving force of natural convection. The coefficient $dV/V\,dt = -d\rho/\rho\,dt$, and by the ideal gas law equals $1/T$. This is usually adequately accurate for real gases. It is also possible to differentiate the appropriate equation of state, or isobaric points may be read from Fig. 8-4, plotted against temperature, and the slope measured. For liquids a similar graphical procedure is possible with Fig. 8-5, since $\partial\omega/\omega\,\partial T = \partial\rho/\rho\,\partial T$ at the same pressure. A simpler method, for reduced temperatures below 0.65 and pressures up to 10 atm, uses Gamson and Watson's relation [25] that $\omega = 0.1745 - 0.0838 T_R$. By differentiation:

$$-\left(\frac{\partial\rho}{\rho\,\partial T}\right)_P = (2.082 T_c - T)^{-1} \tag{8-16}$$

Prob. 8-3. Predict the density of helium at the conditions of Prob. 8-1 by Eq. (8-15), and compare with the ideal gas law.

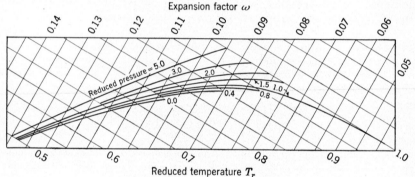

Fig. 8-5. Expansion factor for liquids.

8-6. Vapor Pressure and Surface Tension.

It is normally undesirable to permit boiling in a nuclear reactor or a pump unless the equipment is specially designed for it. Thus, vapor pressures are of interest.

Knowing the vapor pressure P of a liquid at two or more temperatures, reliable interpolation or short extrapolation is possible by plotting $\log P$ vs. $1/T$. A straight line may be expected over a considerable range by plotting against each other the temperatures at which another similar liquid and the desired one exert the same vapor tension, or the vapor tensions of the two liquids at the same temperature on log-log paper [26].

The interfacial tension between two liquids measures the contracting tendency common to all interfaces and has the units of surface energy per unit area or force per unit length. It controls liquid level rise or fall in capillaries and cracks, and droplet size and shape in drop or spray columns. It decreases with rising temperature, but is little affected by pressure. It can decrease greatly due to impurities which, having this property, concentrate at the interface, but impurities which increase interfacial tension and surface energy therefore have little effect [27]. When one phase is a gas it is called "surface tension." Knowing the surface tension σ and liquid density ρ at two temperatures t, σ can be interpolated or reasonably extrapolated by the Eötvös equation:

$$\sigma\left(\frac{M}{\rho}\right)^{\frac{2}{3}} = A_2 - B_2 t \approx B_2(t_c - t - 6)$$

by plotting $\sigma/\rho^{\frac{2}{3}}$ against t. If the critical temperature t_c is known or estimated, only one point is needed, since $\sigma/\rho^{\frac{2}{3}}$ extrapolates to zero approximately at $t = t_c - 6°C$. If no values of σ are available B_2 may be set $= 2.1$ in cgs units for nonpolar liquids or 1.2 for polar liquids. Interfacial tension σ_i between two liquids 1 and 2 can be predicted by Antonoff's rule [27] that $\sigma_i = \sigma_{1\text{-}2} - \sigma_{2\text{-}2}$, provided that $\sigma_{1\text{-}1} > \sigma_{1\text{-}2}$, where $\sigma_{1\text{-}1}$ is the higher σ of a liquid under its own vapor, $\sigma_{2\text{-}2}$ is the lower, and $\sigma_{1\text{-}2}$ is that of liquid 1 under vapor 2.

MASS, FORCE, AND ENERGY BALANCES

8-7. Material or Mass Balance. Applying the law of conservation of mass to a flow process yields a material balance, or "continuity equation." The interchange between mass and energy is so small, even in nuclear reactors, that for engineering purposes the change in flow may be neglected. Thus, the steady-state flow w in mass per unit time is the same through any surface S completely cutting the stream. Standard methods of computation are

$$w = Q\bar{\rho} = \overline{u\rho}S = \bar{G}S = \sum_0^S (\overline{u\rho}\ \Delta S) = \int_0^S u\rho\ dS = \int_0^S G\ dS \quad (8\text{-}17)$$

The bars designate averages over S. The summation is commonly employed to evaluate w in a test when velocity and temperature vary irregularly over a large duct. The integrals are useful for radially symmetrical flow in a pipe; the area under the curve of ρu vs. r^2 is w/π and under $\rho u r$ vs. r is $w/2\pi$.

8-8. Static Pressure and Force Balances. Any static pressure in a fluid exerts a force normal to the containing walls. Such forces may be very large in high-pressure or large reactors and must be considered during design. If z is the elevation above a datum surface at which the pressure is constant, the difference in pressure between elevations z_2 and z_1 in a static fluid (no friction nor acceleration) is

$$P_2 - P_1 = \int_{z_2}^{z_1} \frac{\rho g\ dz}{g_c} = \frac{\overline{\rho g}(z_1 - z_2)}{g_c} \quad (8\text{-}18)$$

This expression is useful in computing free convection driving pressure in loops, nonisothermal manometer lead corrections, etc. Integrating completely around a closed circuit gives

$$\Delta P_B = \frac{g}{g_c} \left(\int_{z_2}^{z_1} \rho_a \, dz + \int_{z_1}^{z_2} \rho_b \, dz \right)$$

$$= \frac{g}{g_c} \int_{z_2}^{z_1} (\rho_a - \rho_b) \, dz = \frac{g \rho_0 \bar{\beta}_0}{g_c} \int_{z_2}^{z_1} (t_b - t_a) \, dz \quad (8\text{-}19)$$

Prob. 8-4. Compute the force against the reactor in Prob. 9-19 due to the air pressure, with unidirectional and with split flow.

Important forces may be exerted on the fuel elements by the flowing coolant. For flow around isolated objects the usual correlations (Sec. 8-45) yield the total (shear plus impulse) force directly. For short flat surfaces the wall shear stress distribution is predicted (Sec. 8-27) and integrated along the surface to yield the shear force. For longer channels (Secs. 8-13 to 8-18 and 8-23 to 8-26), banks of cylinders (Sec. 8-46), and packed beds (Sec. 8-47) the correlations yield the fluid pressure drop ΔP_F. A force balance then gives the total shear force on the walls as equal to $\Delta P_F S$, where S is the total cross section of the stream. For known deflections of the flow the impulse force is obtainable by Sec. 8-11.

8-9. Total Energy Balance. The law of conservation of energy is particularly useful in cases in which the velocity, expansion, and heat flow are all high, as in many optimized nuclear-reactor designs. Following unit mass of a stream in steady flow along a differential length of its path,

$$di + d\left(\frac{P}{\rho}\right) + \frac{g \, dz}{g_c} + \frac{\alpha \, d\bar{u}^2}{2g_c} = dh + \frac{g \, dz}{g_c} + \frac{\alpha \, d\bar{u}^2}{2g_c} = dQ + dW \quad (8\text{-}20)$$

All terms must be in the same units of energy per unit of mass. dQ is heat gained from the outside, and dW is work done on the fluid from the outside in addition to the "flow work" $d(P/\rho)$. α is a factor to obtain the kinetic energy from the average velocity $\bar{u}$ (see Secs. 8-13 and 8-28 and Table 8-5). Tables of i and h are available for many fluids, and generalized plots of h vs. reduced temperature and pressure for gases. C_V and C_P are reasonably constant over a limited ΔT except near the critical point, and $\Delta i = C_V \Delta T$ and $\Delta h = C_P \Delta T$. Equation (8-20) may be integrated between any two positions by merely replacing each differential by the finite increment.

Prob. 8-5. A tube in a nuclear reactor is to dissipate 70.93 Btu/sec. It is to be cooled by a stream of 0.08155 lb/sec of water entering at 70°F and the necessary P_1 and leaving at 14.7 psia. Compute the enthalpy h_1 and the quality x_2 of the stream at the outlet. Take enthalpies of water and steam from Steam Tables.

Solution. Assume $h_1 = 38.05$ Btu/lb (at saturation pressure, from Steam Tables). (a) If the pipe is large enough so that $\bar{u}^2/2g_cJ$ is negligible compared with h, $h_2 = h_1 + \Delta Q/w = 38.05 + 70.93/0.08155 = 907.82$ Btu/lb. But $h_2 = 180.07 + 970.3x_2$. Thus $x_2 = 75$ per cent. (b) If the pipe is of 0.5-in. ID, $h_2 = h_1 + \Delta Q/w + (\bar{u}_1{}^2 - \bar{u}_2{}^2)/2g_cJ = 907.82 + [(V_1')^2 - (V_2' + V_2''x_2)^2]G^2/2g_cJ$, where V' is the specific volume of the liquid and V'' is the increase in specific volume on vaporization. x_2 is obtained directly as 72.2 per cent.

In either case, starting with x_2, P_1 can be calculated by Sec. 8-41, and a more accurate h_1 obtained. The correction is negligible with liquid feed, as in this case.

8-10. Mechanical Energy Balance.

Applying Newton's law $F = ma$ to a steady stream yields the mechanical energy balance, or Bernoulli's principle [28]. This relation is more convenient than the total energy balance when thermal effects are small, and it is necessary in evaluating the friction in a test, predicting coolant pressure drop in a reactor, etc. However, the relation between ρ and P must be known before it can be integrated. It may be written

$$\frac{g\,dz}{g_c} + \frac{dP}{\rho} + \frac{\alpha\,d\bar{u}^2}{2g_c} = dW - dF \tag{8-21}$$

dF here is the energy lost as friction.

These energy balances simplify in many applications. For instance, in stationary fluid the last three terms are zero, and in horizontal flow $dz = 0$. If ρ is constant, the integral of the second term is $\Delta P/\rho$, and if in addition the stream cross section is constant $d\bar{u}^2 = 0$. If ρ varies only moderately it is usually sufficiently accurate to replace it by its average $\bar{\rho}$. $W = 0$ unless there is a pump or fluid motor in the system between the limits of integration.

For constant or moderately changing ρ or for isothermal gas flow Eq. (8-21) can be integrated analytically (see Sec. 8-26). For flow through a nozzle or orifice dF may be neglected, and Eq. (8-21) integrated by obtaining the relation between P and ρ from Eq. (8-20) or the adiabatic expansion relation [29]. For the general case in a reactor of heat generation and a considerably expanding fluid, it is convenient to proceed in succession over short sections of the channel such that ρ does not vary much. Knowing the conditions at one end of a section, Eqs. (8-20) and (8-21) are solved simultaneously for P and ρ at the other end. This is continued over the whole channel. If stated conditions at the second end of the channel are not matched, the process is repeated with a different flow or different assumed first-end conditions. Some 10 to 20 increments usually yield an accuracy comparable to that of the friction calculation. The method is applied to a gas in Prob. 9-19, and to a boiling liquid in Prob. 8-21. It is evidently also applicable to adiabatic flow on setting $\Delta Q = 0$, and to isentropic flow on setting $\Delta F = 0$.

Prob. 8-6. It is planned in a Liquid Metal Fuel Reactor (LMFR) for molten bismuth at approximately 400°C to flow through the tubes of a heat exchanger at 30 fps. It is proposed to flare the discharge end of each tube from its ID of 0.25 in. to 0.375 in. to save pressure. (*a*) What is the maximum possible saving in pressure? (*b*) Would the flaring be profitable if it costs $5 per tube, total annual fixed charges are 25 per cent, and the pumping energy saving is 2 cents per kwhr delivered to the fluid? Operation is continuous.

8-11. Momentum Balance.

Applying Newton's law over an interval of time $\Delta\theta$ yields the equality between impulse and momentum change, $Fg_c\,\Delta\theta = m\,\Delta u$, where F is the net or resultant force and Δu is the vectorial change in velocity. Dividing by $\Delta\theta$ for a steady flow process,

$$Fg_c = \Delta u\,\frac{m}{\Delta\theta} = \Delta u\,\frac{dm}{d\theta} = w\,\Delta u \qquad (8\text{-}22)$$

In certain problems requiring the calculation of a force or pressure due to a change in direction or magnitude of the velocity of a stream, the momentum balance is more convenient than an energy balance.

8-12. Comparison and Use of the Mechanical-energy and Momentum Balances in Steady Flow.

Equations (8-21) and (8-22) both hold for any steady flow process. Some cases exhibit an apparent discrepancy between the two methods, but this is due to an incorrect application. This is illustrated in Prob. 8-7.

Prob. 8-7. Given horizontal uncompressed flow through a gradual expansion in cross section so that the stream fills it at every point. Compute P_2 as a function of P_1, S_1, S_2, w, and ρ. Neglect friction and nonuniformity of velocities.

Solution 1. Mechanical-energy Balance—Correct. By Eq. (8-21),

$$P_2 = P_1 + \frac{\rho}{2g_c}\,(u_1{}^2 - u_2{}^2) = P_1 + \frac{w^2}{2g_c\rho}\left(\frac{1}{S_1{}^2} - \frac{1}{S_2{}^2}\right)$$

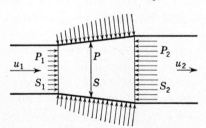

Solution 2. Momentum Balance—Incorrect. Applying Eq. (8-22) at first glance only to the stream cross sections,

$$F = P_2S_2 - P_1S_1 = \frac{w}{g_c}\,(u_1 - u_2)$$

$$P_2 = P_1\frac{S_1}{S_2} + \frac{w^2}{g_c\rho}\left(\frac{1}{S_1S_2} - \frac{1}{S_2{}^2}\right)$$

which evidently does not agree with solution 1.

Solution 3. Momentum Balance—Correct. See Fig. 8-6. The horizontal com-

Fig. 8-6. Stream with a gradual expansion in cross section (Prob. 8-7).

ponent of the force by the walls on the fluid in the expansion was neglected in solution 2. Computing the local pressure P by Eq. (8-21),

$$P = P_1 + \frac{w^2}{2g_c\rho}\left(\frac{1}{S_1{}^2} - \frac{1}{S^2}\right)$$

Since P acts equally in all directions, the contribution of P to F can be evaluated as

$$\int_{S_1}^{S_2} P\, dS = P_1(S_2 - S_1) + \frac{w^2}{2g_c\rho}\left(\frac{S_2 - S_1}{S_1{}^2} - \frac{1}{S_1} + \frac{1}{S_2}\right)$$

This is to the right. Subtracting it from $P_2 S_2 - P_1 S_1$ above will be seen to yield the correct result.

STREAMLINE FLOW

8-13. Isothermal Newtonian Streamline Flow in Ducts of Regular and Constant Cross Section. Streamline flow in symmetrical channels such that the relation between the shear stress and the distance y from the wall is known may be analyzed directly by Eq. (8-2). These include the cylindrical tube and annulus and thin wide channels. The relation for local velocity u is obtained by expressing the shear stress τ in terms of y and friction pressure gradient dP/dN, separating the variables u and y, and integrating between the limits of $0 \to y$ and $0 \to u$, since $u = 0$ at $y = 0$. To illustrate, for a horizontal tube of radius R, $-dr = dy$, and

$$\int_0^u \mu\, du = g_c \int_R^r \tau(-dr) = g_c \int_r^R \frac{\pi r^2}{2\pi r} \frac{dP}{dN}\, dr$$
$$u = g_c \frac{dP}{dN} \frac{R^2 - r^2}{4\mu} \tag{8-23}$$

The total volumetric flow rate Q and the average velocity $\bar{u}$ are obtained by integrating u over the cross section S

$$Q = \frac{w}{\rho} = \int_0^S u\, dS = \frac{\pi g_c}{4\mu} \frac{dP}{dN} \int_0^{R^2} (R^2 - r^2)\, dr^2 = \frac{\pi g_c R^4}{8\mu} \frac{dP}{dN} \tag{8-24}$$

With gases or liquids undergoing significant changes of density or temperature as a known function of length or pressure, total pressure drop across a pipe may be obtained by integrating Eq. (8-24) (see Sec. 8-26).

The total momentum and kinetic energy of the stream per unit time are useful in steady-state flow calculations, and equal $(\pi\rho/g_c)\int_0^R u^2\, dr^2$ and $(\pi\rho/2g_c)\int_0^R u^3\, dr^2$, respectively. Integrated results are given in Table 8-5.

8-14. Streamline Flow in Irregular Ducts. The complete force balance on an element of a viscous fluid is given by the Navier-Stokes equations [30]. The x-component balance is

$$F_x\rho g_c - \frac{\partial P}{\partial x} g_c + \frac{\mu}{3}\frac{\partial}{\partial x}\left(\frac{\partial u}{\partial x} + \frac{\partial v}{\partial y} + \frac{\partial w}{\partial z}\right) + \mu\left(\frac{\partial^2 u}{\partial x^2} + \frac{\partial^2 u}{\partial y^2} + \frac{\partial^2 u}{\partial z^2}\right) = \rho\frac{du}{d\theta} \tag{8-25}$$

Similar expressions hold for the y and z directions. F_x is the total external force in the x direction per unit of mass, such as that due to gravity, and u, v, and w are the velocity components in directions x, y, and z respectively. For steady uncompressed flow along a horizontal conduit of constant cross section, setting $dx = dN$,

$$\frac{dP}{dN} g_c = \mu \left(\frac{\partial^2 u}{\partial y^2} + \frac{\partial^2 u}{\partial z^2} \right) = \mu \left(\frac{\partial^2 u}{\partial r^2} + \frac{1}{r} \frac{\partial u}{\partial r} \right) \qquad (8\text{-}26)$$

TABLE 8-5. UNCOMPRESSED FLOW IN LONG HORIZONTAL CHANNELS

	f_F	Circular pipe, radius $= R$	Concentric annulus, radii $= R_1, R_2$				Wide flat channel, spacing $= b$	Square channel, side $= b$
Streamline flow:								
$\dfrac{\bar{u}\mu N}{\Delta P g_c}$ (for average velocity)	......	$\dfrac{R^2}{8}$	$\dfrac{1}{8}\left[R_2{}^2 + R_1{}^2 - \dfrac{R_2{}^2 - R_1{}^2}{\ln (R_2/R_1)} \right]$				$\dfrac{b^2}{12}$	$\dfrac{b^2}{28.6}$
$\dfrac{u\mu N}{\Delta P g_c}$ (for local velocity)	......	$\dfrac{R^2 - r^2}{4}$	$\dfrac{1}{4}\left[R_1{}^2 - r^2 + \dfrac{R_2{}^2 - R_1{}^2}{\ln (R_2/R_1)} \ln \dfrac{r}{R_1} \right]$				$\dfrac{yb - y^2}{2}$	Sec. 8-14
			R_2/R_1					
		∞	3	2	1.2	1		
$\dfrac{u_{\max}}{\bar{u}}$	......	2	1.5180	1.5077	1.5019	1.500		2.07
γ	......	1.333	1.207	1.204	$\sim$1.2	1.200		1.36
α	......	2	1.568	1.552	$\sim$1.543	1.543		2.22
Turbulent flow:*								
$\dfrac{u_{\max}}{\bar{u}}$	0.01	1.286	......	......	....	1.18		
	0.005	1.202	......	......	....	1.14		
	0.0025	1.143	......	......	....	1.13		
γ	0.01	1.055	......	......	$\sim$1.035	1.035		1.078
	0.005	1.030	......	......	$\sim$1.018	1.018		1.038
	0.0025	1.017	......	......	$\sim$1.010	1.010		1.02
α	0.01	1.135	......	......	....	1.229		
	0.005	1.073	......	......	....	1.055		
	0.0025	1.040	......	......	....	1.022		

$$\gamma = \frac{\text{momentum per unit time}}{\text{momentum per unit time at } \bar{u}} = \frac{\text{kinetic energy per unit length}}{\text{kinetic energy per unit length at } \bar{u}}$$

$$\frac{\text{Momentum per unit length}}{\text{Momentum per unit length at } \bar{u}} = 1 \qquad \alpha = \frac{\text{kinetic energy per unit time}}{\text{kinetic energy per unit time at } \bar{u}}$$

* W. M. Kays, *Trans. ASME*, **72**:1067–1074 (1950).

For a channel of any given cross section it is possible to write an arbitrary equation for u such that it fits Eq. (8-26) to any desired degree. For instance, for a square cross section [31] with its center at the (y,z) origin and sides of length 2 we may assume

$$u = (1 - y^2)(1 - z^2) + A(1 - y^4)(1 - z^4) + B(1 - y^6)(1 - z^6)$$
$$+ C(1 - y^8)(1 - z^8) \quad (8\text{-}27)$$

This expression is of a reasonable form, since u equals zero at all four walls, is a maximum at the center, and shows central symmetry. Taking the partial second derivatives,

$$\frac{\partial^2 u}{\partial y^2} + \frac{\partial^2 u}{\partial z^2} = -2(1 - z^2 + 1 - y^2) - 12A[(1 - z^4)y^2 + (1 - y^4)z^2]$$
$$- 30B[(1 - z^6)y^4 + (1 - y^6)z^4] - 56C[(1 - z^8)y^6 + (1 - y^8)z^6] \quad (8\text{-}28)$$

Substituting $y = z = 0$ at the center of the duct, evidently Eq. (8-28) equals -4. It is now possible to take any three points, such as $y = 0.2$, 0.5, and 0.8 at $z = 0$, substitute them into Eq. (8-28), and obtain three simultaneous equations from which $A, B,$ and C can be determined. Evidently the other nine symmetrical points are automatically included. The degree of constancy of Eq. (8-26) can then be checked by substituting other points.

A somewhat better procedure is to equate to -4 the *average* value of Eq. (8-28) for z between -1 and $+1$, at three given values of y. Integrating Eq. (8-28) with respect to z between -1 and $+1$, dividing by 2, substituting the three above values of y, and solving these equations simultaneously, $A, B,$ and C are obtained as -0.029, $+0.423$, and -0.210, respectively [31]. Substituting them into Eq. (8-27) and performing the double integration, $\iint u\, dy\, dz$ from -1 to $+1$ yields the total flow. This result is within 1 per cent of the exact value listed in Table 8-5. Evidently a similar method can be applied to any cross section. A number of unusual cross sections have already been studied [32] by various methods.*

8-15. Streamline Flow in Short Channels. When a fluid stream is accelerated by a sudden drop in pressure, such as at a nozzle, the pressure drop, thus the increase in u^2, is uniform over the stream. Thus, if the increase is high compared to the initial velocity, a substantially uniform final velocity results. This method is commonly employed to obtain a region at uniform velocity, as in the throat of a wind tunnel.

A trumpet-shaped inlet nozzle attached to a long pipe will accordingly show a drop in pressure of $\rho \bar{u}^2/2g_c$, neglecting the velocity of approach and friction in the nozzle. At the end of the pipe, the kinetic energy is seen by Table 8-5 to be twice $\bar{u}^2/2g_c$; thus an additional drop in pressure of $\rho \bar{u}^2/2g_c$ occurs along the pipe—in addition to friction—as the velocity distribution changes from flat to parabolic. Furthermore, the friction is

---

* For instance, covering the cross section with a grid of fine squares Δy on a side, the average velocity $\bar{u}_0$ for one square is related to $\bar{u}_n$ in the four surrounding squares, from Eq. (8-26), by

$$\bar{u}_0 = \frac{\bar{u}_1 + \bar{u}_2 + \bar{u}_3 + \bar{u}_4}{4} + \frac{(\Delta y)^2}{4\mu}\left|\frac{\Delta P}{\Delta N}\right|$$

By repeated "iteration" or "relaxation" (see Sec. 9-39), approximately correct values of $\bar{u}$ over the whole cross section can be computed and the total flow and the velocity distribution determined.

higher than normal near the inlet, because of the steeper velocity gradient at the wall [33]. These two effects have been recalculated for pipes and flat ducts to average velocity pressures for Table 8-6.

TABLE 8-6. EXCESS STREAMLINE PRESSURE DROP* IN SHORT TUBES AND WIDE SLOTS DUE TO UNIFORM INLET VELOCITY

Pipe diameters or slot widths downstream of nozzle outlet, divided by Re	Excess velocity pressure, $\dfrac{\Delta P}{\rho \bar{u}^2 / 2 g_c}$	
	Pipe	Slot
0	0	0
0.000125	0.14	0.104
0.00025	0.20	0.148
0.0005	0.288	0.208
0.0010	0.396	0.290
0.0015	0.464	0.339
0.0020	0.522	0.370
0.0025	0.570	0.39
0.0050	0.74	0.448
0.010	0.99	0.522
0.015	1.14	0.560
0.030	1.32	0.601
0.040	1.37	0.601
0.060+	1.41	0.601
1.000†	64.00†	48.00†

* To be added to pressure drop by Table 8-5 (first line) to obtain total pressure drop from start of tube or slot. Does not include pressure drop across the nozzle.

† Pressure drop in long channel with parabolic density distribution (from first row in Table 8-5), due to friction alone.

8-16. Annular Channels and Nonparallel Walls in Streamline Flow.

For a concentric annulus and constant viscosity the exact streamline flow equation can be derived as for a pipe. The equation for $\bar{u}$ in Table 8-5 is seen to reduce to that of a pipe when $R_1 = 0$, as it should. For a narrow annulus with an eccentricity, or distance between the two axes, of ϵ, an approximate flow equation is obtained by taking the annular width as $R_2 - R_1 + \epsilon \cos \theta$, substituting into the relation for flow in a parallel wall channel, and integrating around the semicircle. If $R_2/R_1 < 1.5$, the formula obtained $\bar{u} = (g_c/12\mu)(dP/dN)[(R_2 - R_1)^2 + 0.5\epsilon^2]$ is correct to 1 per cent or better.

For a flat wide channel of nonparallel walls and constant cross section, integrating across the channel from b_1 to b_2 gives

$$\bar{u} = \frac{g_c}{24\mu} \frac{dP}{dN} (b_1{}^2 + b_2{}^2)$$

Prob. 8-8. By what ratios do (a) the total coolant flow and (b) the maximum velocity in streamline flow at a given friction pressure drop vary in the annulus if an inner cylindrical fuel element moves from the concentric to the tangential position with respect to the outer tube?

Solution.   *a.* In the concentric position $\epsilon = 0$, and in the tangential $\epsilon = R_2 - R_1$. From the previous equation for a narrow eccentric annulus $w_{\text{tang}}/w_{\text{conc}} = 1.5$. The flow evidently varies appreciably with eccentricity.

b. From Table 8-5, for a channel with parallel walls $\bar{u}$ is proportional to b^2, other quantities being constant. Thus $(\bar{u}_{\text{tang}})_{\text{max}}/\bar{u}_{\text{conc}} = 2^2/1 = 4$, approximately. Along the line of contact, however, the velocity will be zero, and dangerous overheating may occur.

8-17. Effect of Wall Roughness in Streamline Flow.

When flow normal to a flat disk reaches a Reynolds number of about 30 or higher, a vortex wake forms downstream and the character of the flow evidently changes considerably. From this it is possible to estimate [34] that roughness of height e on a pipe wall will disturb the flow if e exceeds $2D/\sqrt{\text{Re}}$ or $\sqrt{4D\mu/\bar{u}\rho}$. For a channel with parallel walls a distance b apart, the critical e is $3b/\sqrt{\text{Re}}$ or $\sqrt{4.5b\mu/\bar{u}\rho}$. Both these cases correspond to a critical e of $\sqrt{30\mu/\rho\tau_w g_c}$, where τ_w is the friction shear stress at the wall. This analysis agrees satisfactorily with experiment.

8-18. Nonisothermal Streamline Pressure Gradient.

If the viscosity at the wall of a pipe μ_w is higher than that at the average temperature of the stream μ_b, as in cooling a liquid or heating a gas, the pressure gradient $-dP/dN$ will be higher than that computed for isothermal flow at the average stream temperature t_b and vice versa. Theoretical calculation of the correction is difficult, but empirically it is found satisfactory [35] to divide the isothermal pressure drop by $(\mu_b/\mu_w)^{0.23}$ when cooling the stream and by $(\mu_b/\mu_w)^{0.32}$ when heating—or by $1.1(\mu_b/\mu_w)^{0.25}$ for either case, though with less accuracy.

Another method is to employ the isothermal formula with the viscosity at a temperature t' intermediate between t_b and the wall temperature t_w. For oils $t' = 0.25t_w + 0.75t_b$ is used. For gases $t' = 0.58t_w + 0.42t_b$ has been recommended [36], and for liquid metals $t' = 0.54t_w + 0.46t_b$.

8-19. Potential Theory.

The Navier-Stokes equations (Sec. 8-14) apply to all cases of streamline fluid flow, but certain simpler methods are frequently applicable.

In this and the next two sections, methods will be described which apply to "viscous" streamline flow, in which inertia forces are negligible. They can yield the streamlines and isobars, from which the viscous pressure drop may be calculated. They also hold for "potential flow," in which viscous forces are negligible, provided the main stream follows the walls (i.e., flow in a converging nozzle or between tangential vanes) or its contour is otherwise known (as in some cases with flow separation from the walls). The inertia pressure change is then obtainable from the widths of a channel by Eq. (8-21) or (8-22).

For two-dimensional streamline flow between parallel walls or through a porous body, conformal mapping by complex variable [37] is a useful technique.

Another method of getting approximate results in certain cases is the use of real and fictitious sources and sinks and their potentials. The first step is to lay out point sources and sinks so as to fit the given boundary conditions as well as possible, namely, the real source and sink isobars and any constant-pressure and zero-flow lines. For instance, assume that fluid at a constant pressure enters a thin annular channel of thickness b between two concentric cylinders of average perimeter $2c$, travels up between the cylinders, and then leaves through two circular holes of radius e 180° apart near the closed end of the annular channel. The distance from the inlet to the center of the outlet is m, and from there to the end is n. The rectangle $ABCD$ in Fig. 8-7 represents half the channel, AB being the inlet, and BC and DA being lines through which no flow passes because of symmetry, and CD because it is closed off. As a first approximation, the point sink corresponding to the outlet is taken at the center of the

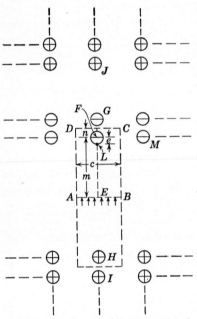

real sink F (or slightly closer to AB). Opposite F is placed an equal imaginary source H, equidistant from AB. This will cause AB to be a constant-pressure line, with flow normal. On the other side of CD and equidistant therefrom is placed an equal imaginary sink G, which will cancel any flow from F through CD. However, for H to have flowlines symmetrical with those at F, it will need another source at I. Similarly G will need another source at J. Evidently, to obtain proper symmetry, the row of two sources and two sinks alternating must extend to infinity in both directions. In order to achieve zero flow through DA and CB, respectively, additional rows of sources and sinks identical to that through F will have to be drawn to left and right, extending to infinity in all directions.

The basic principle of the method is that pressure change for any number of forced sources and sinks is additive. In addition the pressure can be arbitrarily fixed at any one point. Thus at any location the pressure is the sum of the pressures due to each of the sources and sinks computed separately, and the velocity can be computed from the net pressure gradient. By integrating Eq. (8-2) for viscous flow between parallel

Fig. 8-7. Portion of infinite array of sources and sinks to give flow from source AB to sink F with nonflow boundary $ADCB$ (Prob. 8-9).

planes the pressure P at any distance r from a single point source of volumetric flow rate Q is $P_1 - 6\mu Q/\pi g_c b^3 \ln (r/r_1)$, if P_1 is the pressure at distance r_1. Similarly, the pressure at a distance r from a sink is $P_1 + 6\mu Q/\pi g_c b^3 \ln (r/r_1)$. These pressures may be abbreviated $P_1 \mp K' \log (r/r_1)$. Calling P_1 the pressure at point E and setting it at 0, the pressure may be computed due to each source and sink at any other point, say L on the periphery of the outlet. Thus, for sink F, $P_L = 0 + K' \log (e/m)$; for G, $P_L = K' \log [(2n + e)/(2n + m)]$; for H, $P_L = K' \log [m/(2m - e)]$; for M, $P_L = K' \log \sqrt{(c^2 + e^2)/(c^2 + m^2)}$; etc. The net value of P_L/K' due to the horizontal row of sinks containing F would be $\log (e/m) + \sum_{s=1}^{s=\infty} \log \{[(sc)^2 + e^2]/[(sc)^2 + m^2]\}$.

The contribution of every row as far away as it is appreciable could similarly be evaluated. Instead, adjacent rows of sources and sinks should be combined into one series, which will converge more rapidly. The final value of P_L found will be the approximate pressure to maintain the flow Q through each of the two orifices in the cylindrical shell. If it is desired to find the flow distribution, it is necessary to find the pressure gradient. This can be accomplished by setting up by the same method the expression for the pressure at a point (x,y) with the origin at E and differentiating. For the flow distribution along AB one would differentiate with respect to y, then substitute $y = 0$.

If the same procedure is carried out for other points on the perimeter of the outlet, P_L will in general not be obtained, for two reasons. The first is that the best location for the point sink is not in the center of the circular outlet, but somewhat farther toward point L. The second is that, in general, the isobars are not perfect circles, so that an exact solution for a circular outlet could not be obtained in any case. For such situations special methods are available [38]. In the present problem a more exact solution could be obtained by averaging P_L with the value similarly computed for the opposite side of the outlet. The pressure 60 to 90° away from P_L might also be somewhat more correct than P_L for the outlet pressure. It should be pointed out

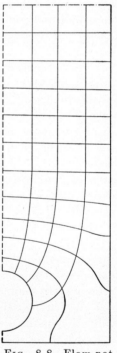

that the problem is much simplified if m, c, and/or n in Fig. 8-7 are infinite, or substantially so, with respect to the other dimensions. For instance, if m is several times larger than c, substantially the correct solution is obtained by leaving out all the sources and sinks except for the horizontal rows containing F and G. On the other hand, if c is several times larger than m, only the vertical row containing F and H need be retained. Finally, if n and m are both much smaller than c, only F and G need be retained. This would also be the case if one were the inlet and the other the outlet, and n were small compared to the size of the flow area. In this last case it also happens that round isobars are obtained, so that a more exact solution is possible [39]. Inertia and thermal buoyancy effects must be neglected.

Potential theory is similarly applied in the case of two- or three-dimensional flow in homogeneously porous solids. In these cases the pressure follows, respectively, the relations $P = P_1 \mp Q/2\pi bK''' \ln (r/r_1)$ or $P_1 \mp Q/4\pi K'''(1/r_1 - 1/r)$, where K''' is the ratio of "superficial" or average velocity through the porous body to pressure gradient. The same method applies to isothermal compressible flow between parallel planes or in porous bodies on replacing P by P^2 and Q by $2R'Tw$, where R' is the gas constant per unit mass.

8-20. The Flow Net. An alternate approximate method of solving two-dimensional problems is to draw by hand the "flow net." First, curved streamlines are drawn from the source to the sink, estimated to enclose the same flow between each pair of adjacent lines. Then curved iso-

FIG. 8-8. Flow-net solution for Prob. 8-9.

bars are drawn, intersecting the streamlines and closed walls normally and forming a row of curvilinear squares between each pair of adjacent streamlines. If the isobars can be drawn so as to cover the whole field and simultaneously fit all boundaries, the paths have been prop-

erly chosen; if not, they are redrawn until close enough agreement is
obtained (see Fig. 8-8). The viscous pressure drop per curvilinear square
is $12\mu Q'/g_c b^3$ (for porous bodies $Q'/\pi b K'''$), as for a true square, where Q'
is the flow per channel and b the spacing of the parallel planes. The viscous
pressure drop is then $12\mu Q n/g_c b^3 m$ (or $Q n/\pi b K''' m$), where Q is the total
flow, n is the number of squares per channel, and m is the number of
channels.

8-21. Electrical Analogues. A convenient experimental method
exploits the similarity between electrical conduction and streamline flow.
In two-dimensional problems a sheet conductor such as metal-coated
paper is cut out, or a bath of uniform depth of an electrolyte such as
acidified copper sulfate solution is constructed, to scale with the given
closed flow boundaries. Low-resistance electrodes are provided at the
conducting boundaries, and cut or nonconducting edges at the nonflow
boundaries. The electrical resistance between the electrodes is then
measured and compared with that of a square sheet or bath of the conductor. The ratio of $\Delta P/Q$ for the complete problem, or $\Delta P^2/w$ for
compressible flow, to that for any sized square equals the ratio of the
observed electrical resistance for the complete profile and for a square
of the same sheeting or depth. With the electrolytic bath three-dimensional problems can also be solved. Copper electrodes and alternating
current should be employed to avoid electrode polarization. Any solid
sheet conductor or semiconductor should be tested for uniformity of
resistance in all directions, and for contact resistance, since most sheet
materials are deficient in one or both respects.* If several sources or
sinks of different strengths are involved they can be represented by
currents in the proper ratio, fed in or out at the proper locations. Voltages may be read with a potentiometer for accurate results.

Alternatively a "lumped network" may be employed (Fig. 8-9), in
which the flow field is represented by electrical resistors in a two- or
three-dimensional grid. Each resistor is set proportional to the ratio of
distance to transverse area between the points or "nodes" it joins. The
ratio of the resistance of a square (or cube) to that between two isopotential bus bars equals the ratio of the pressure drop across a square (or
cube) flow path to the actual pressure drop at the same flow rate.†

Resistors for a lumped network can be obtained more accurate than

---

* Metallized Teledeltos paper from Western Union Telegraph Co. is suitable. It is
uniform within 8 per cent in all directions, and Du Pont Electrochemicals Department
conductive lacquer No. 4922 is employed for contacts and isopotential edges (see
Sec. 9-41).

† Nonlinear resistors (Fluistors) suitable as lumped network elements for turbulent
flow are also available [109]. These generate a voltage drop proportional to (current)$^{1.85}$, thus an element with the proper ratio $\Delta E/I^{1.85}$ can closely simulate Eq. (8-33),
also Eq. (8-35) on using the correct density.

sheet resistors, but there is appreciable geometrical error unless a fine grid is used [40]. Unsteady flow can also be analyzed [108]. Electrical analogues are described further in Secs. 9-41 and 9-42. Equivalent computational methods can also be used, such as those employed for heat conduction [i.e., Eq. (9-108)], on replacing temperature by pressure.

Prob. 8-9. A reactor employs several concentric steel cylinders as structural elements and γ-ray shields. A closed annular space between two adjacent cylinders is 28 in. long and ⅜ in. thick. Coolant flows slowly through a circumferential row of 2.5-in.-diameter holes 9.3 in. between centers and out through an opposite identical row at the other end of the annulus. The center of each hole is 1.75 in. from the nearer end of the annulus. Find the ratio $b^3 g_c \, \Delta P / Q \mu$ from inlet to outlet holes, neglecting thermal buoyancy and inertia effects.

Solution 1. This problem fits the layout of sources and sinks in Fig. 8-7, with ΔP double that computed for the problem described in Sec. 8-19. Summing the various terms which give the potential at point L until they have become negligible, the total value of $b^3 g_c \, \Delta P / Q \mu$ from H to F is obtained as 31.5. (This result is probably slightly low, as only the nearest point L was used.)

Solution 2. Drawing a fairly coarse flow net, Fig. 8-8 is obtained. Inasmuch as $b^3 g_c \, \Delta P / Q \mu$ for each square equals 12, that from source to sink equals $12 \times 2\frac{2}{8}$ = 33. Note that each bottom corner is contained in one side of a "square."

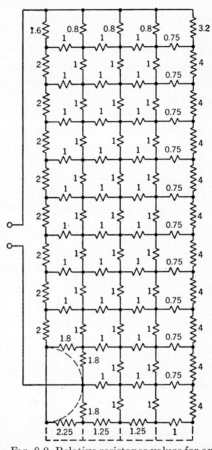

FIG. 8-9. Relative resistance values for an electrical network solution for Prob. 8-9.

Solution 3. Relative resistances for an electrical lumped network for this problem are given in Fig. 8-9. The distance between nodes was set equal to the radius of the hole. Each full resistor is labeled 1, others equaling the ratio of length to average width for the element of area they represent. Actual ohms might conveniently be 100 to 1000 times the numbers shown. $b^3 g_c \, \Delta P / Q \mu$ would equal 12 times the ratio of total ohms between terminals to ohms of each full resistor.

TURBULENT FLOW

8-22. Dimensional Analysis—Buckingham's Pi Theorem. Dimensional analysis rests on the axiom that all true and complete physical laws can be expressed in terms of appropriate numbers which are dimensionless and therefore not dependent on the particular units used to measure the variables involved [41]. If this is so, these units should cancel completely in all correctly stated physical laws. If a self-consistent set of units is employed it is sufficient to represent them by the "dimensions" of length (L), mass (M), time (θ), temperature (T), etc., which should also cancel completely. It must be recognized, however, that no specific set of dimensions is mandatory. For instance, F, for force, may readily be employed in place of either M or L or θ, or even in addition in some cases. It is also convenient at times to employ both T and H, heat, as well as the mechanical dimensions [42]. It has also been recommended to employ separate dimensions of L_R and L_T for radial and tangential lengths, as in torque (Fl) and work (Fs), respectively [43].

The principal value of dimensional analysis is in suggesting compact methods of correlating experimental results over a broad range of a phenomenon which is too difficult to analyze theoretically. Another similar application is in fixing test conditions for a model to simulate a prototype at specific operating conditions. The dimensionless numbers in the dimensionless physical laws can of course be determined or checked in some cases by theoretical analysis of the phenomenon. In other cases, theory can yield limiting asymptotes. In all cases, the conclusions of dimensional analysis should be checked against experimental results if possible, to determine whether the correct variables and proper dimensions have been employed.

Buckingham's "pi theorem" says that the dimensionless statement of any physical law requires a number of dimensionless ratios (which he designated π's) equal, at the most, to the number of variables minus the number of dimensions involved. In addition, each variable must appear in at least one dimensionless ratio, and none of the ratios can be derivable exclusively from others that are employed in the statement of the law. In general, it is preferable for the convenient solution of later problems to employ the simplest dimensionless ratios that meet the above requirements, rather than more complicated combinations thereof. However, an example to the contrary is the plotting of orifice calibration results— the discharge coefficient *with* velocity-of-approach correction is more complicated but more constant and therefore usually more convenient than *without* it.

In applying the pi theorem, the first step is to decide on the variables believed to be significant. The next step is to select known dimensionless ratios and/or devise new ones involving these variables until all the variables have been included and the predicted number of ratios obtained. Finally, experimental points covering as wide ranges of the variables as possible are correlated by computing the dimensionless ratios for each point and plotting them against each other. If fairly good agreement on one line (for two ratios) one family of lines (for three ratios), etc., is obtained, the dimensional analysis and its results are probably correct. The principal simple dimensionless ratios are given in Table 8-7.

TABLE 8-7. COMMON DIMENSIONLESS RATIOS*

Name	Symbol	Definitions
Reynolds No.	Re	$D_e\bar{u}\rho/\mu$, $(4w/\pi D\mu)_{\text{pipe}}$
Fanning friction factor	f_F	$\Delta P_F\, g_c D_e/2N\rho\bar{u}^2$, $F g_c D_e/2N\bar{u}^2$
Von Karman No.	Ka	$\Delta P_F\, g_c D_e^3/N\mu^2 = \text{Re}^2\, 2f_F$
Size factor	SF	$\Delta P_F\, g_c w^3\rho/N\mu^5 = (\pi\,\text{Re}/4)^5\, 2f_F$
Diameters	....	N/D
Froude No.	Fr	u^2/Ng
Mach No.	M	u/c
Drag coefficient	C_d	$2\Delta P\, g_c/\rho u^2$
Rotational Reynolds No.	....	$D^2 n\rho/\mu$
Nusselt No.	Nu	hD/k, hN/k
Prandtl No.	Pr	$C\mu/k$
Stanton No.	St_H; St_M	$h/CG = \text{Nu}/(\text{Re Pr})$; $F'/u = \text{Sh}/(\text{Re Sc})$
Graetz No.	Gz	wC/kN
Grashoff No.	Gr	$(D^3\rho^2 g/\mu^2)\beta\,\Delta t$
Peclet No.	Pe	$DGC/k = \text{Re Pr}$
Fourier No.	Fo	$k\theta/C\rho N^2$, $\alpha\theta/N^2$
Schmidt No.	Sc	$\mu/\rho D_v$
Sherwood No.	Sh	$F'D/D_v$
j factor	j_H; j_M	$St_H\,\text{Pr}^{2/3}$; $St_M\,\text{Sc}^{2/3}$

* Additional nomenclature:

n = rotational velocity k = thermal conductivity
c = velocity of sound D_v = molecular diffusivity
h = heat-transfer coefficient F' = mass-transfer coefficient in velocity units

8-23. Turbulent Friction in Ducts.

Turbulent friction in pipes is profitably studied by dimensional analysis. For an uncompressed Newtonian fluid, the pressure gradient in a long straight pipe seems reasonably to depend on the variables listed in Table 8-8.

According to the pi theorem, 6 variables − 3 dimensions = 3 dimensionless ratios are required. Evidently the "roughness factor" e/D is a simple ratio and may be selected at once. For the other two ratios, considerable latitude exists. It is seen that the first four ratios of Table 8-7 are all applicable, and any two will include all the variables except e and with e/D will thus be adequate. Common practice is to employ the Reynolds number, which is well known, and the friction factor, which is the most constant of these ratios. It is customary to define f as either the Fanning friction factor,

$$f_F = \frac{\Delta P_F g_c D}{2 N \rho \bar{u}^2} \quad \text{or} \quad f_w = \frac{2 \Delta P_F g_c D}{N \rho \bar{u}^2}$$

the Weisbach friction factor, instead of the simpler definition without any 2. In employing a graph or equation for f it is therefore necessary

TABLE 8-8. VARIABLES IN TURBULENT FLOW IN PIPES

Variable	Symbol	Dimensions
Friction pressure gradient.........	$\Delta P_F/N$	$(F/L^3) = M/L^2 T^2$
Fluid viscosity....................	μ	M/LT
Fluid density.....................	ρ	M/L^3
Average velocity.................	$\bar{u}$	L/T
Pipe diameter....................	D	L
Pipe wall roughness..............	e	L

to verify which f was employed, or an error of 4 is possible. f_F will be employed throughout this volume, thus

$$\frac{dP_F}{\rho} = \frac{g \, dF}{g_c} = \frac{2 f_F \bar{u}^2}{g_c D_e} dN = \frac{2 f_F \bar{G}^2}{g_c D_e \rho^2} dN = \frac{2 f_F w^2}{g_c D_e \rho^2 S^2} dN \quad (8\text{-}29)$$

Many turbulent-pressure-drop results have been accumulated for widely different fluids and conditions, including wetting and nonwetting liquid metals. For any one pipe they are successfully correlated as f vs. Re, which indicates that the dimensional analysis outlined is correct. The values of e adopted are the diameters of sand particles glued to cover uniformly the inner wall of the pipe. Commercial pipe must be tested for pressure drop to evaluate the equivalent e. Recommended values [44] of equivalent e are: drawn tubing, 0.000,005 ft; commercial steel pipe, 0.00015 ft; galvanized iron, 0.0005 ft; cast iron, 0.00085 ft; and riveted pipe, 0.003 to 0.03 ft.

The following relations, which are plotted in Fig. 8-10, are generally accepted [44]:

Smooth pipe:

$$(f_F)^{-\frac{1}{2}} = 4 \log [\text{Re } (f_F)^{\frac{1}{2}}] - 0.4 = 4 \log [0.797 \text{ Re } (f_F)^{\frac{1}{2}}] \quad (8\text{-}30)$$

Rough pipe:

$$(f_F)^{-\frac{1}{2}} = -4 \log \left[\frac{e}{3.7D} + \frac{1.255}{\text{Re } (f_F)^{\frac{1}{2}}} \right] \quad (8\text{-}31)$$

High-flow limit of transition zone:

$$(f_F)^{-1/2} = 2.28 - 4 \log \frac{e}{D} \qquad (8\text{-}32)$$

Equations explicit in f are more convenient for design purposes. The smooth-pipe line is closely approximated [35] by

$$f_F = 0.00140 + 0.125 \, \text{Re}^{-0.32}$$

and for $5000 < \text{Re} < 200{,}000$ by

$$f_F = 0.046 \, \text{Re}^{-0.20} \qquad (8\text{-}33)$$

Average clean commercial pipe above 1 in. in size may exceed Eq. (8-33) by some 10 per cent [41], following the curve so labeled in Fig. 8-10.

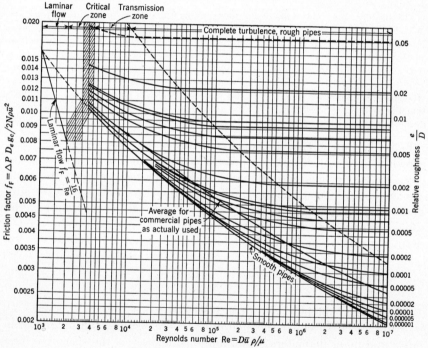

Fig. 8-10. Fluid friction in pipes.

The usual problem in this field is the calculation of the pressure drop for a given fluid flow, pipe size, and roughness. Evidently, e/D and Re can be immediately calculated, f read from the curve, and ΔP_F computed. Direct solution for unknown viscosity can also be obtained with Fig. 8-10, since μ appears only in Re. This case is in general impractical, since because of the small or zero slope of the curve in the turbulent region, the value of Re and therefore μ will be unreliable or indeterminate, respectively. However, where the use of the viscosity is for the reverse process, of computing turbulent pressure drop under different conditions, not much accuracy is needed.   Thus the equivalent viscosity of turbulent suspensions is on occasion determined by turbulent pipe-flow tests.

An alternate problem is determining the velocity or flow of a given fluid for a given pressure gradient and pipe.   It is evident that a direct solution is not immediately possible, since $\bar{u}$ is in both Re and f.   Several methods are available [45]: (1) trial-and-error assumption of values of $\bar{u}$ in the calculation of Re and f until a point on the curve of correct e/D is obtained; (2) assumption of one trial value of $\bar{u}$, plotting the point as Re vs. f on the log-log plot (Fig. 8-10) and computing the correct $\bar{u}$ from the intersection of a straight line of slope -2 with the curve of correct e/D; (3) replotting Fig. 8-10 in terms of e/D, the Von Karman number (the ratio which does not contain velocity), and any one of the other three dimensionless ratios.   As defined in Table 8-7, Ka for any point on a curve can be calculated as $2f_F(\mathrm{Re})^2$ for that point.   Direct solution for $\bar{u}$ becomes possible with this plot.   In particular, the plot of $f^{-\frac{1}{2}}$ vs. log Ka for commercial pipe, similarly to Eq. (8-30) for smooth pipe, is convenient in that it has been found to be straight for turbulent flow, and permits convenient extrapolation.

The problem may also come up of determining the correct or minimum pipe size to yield a certain flow, or less frequently a certain velocity, with a given fluid, pressure gradient, and roughness.   This problem is more involved, in that D is present in all three dimensionless ratios of Fig. 8-10.   Trial and error can be employed.   Or a trial assumption of D_1 can be made, f_1 plotted vs. Re_1 on Fig. 8-10, and a line of slope -5 (for w or Q fixed) drawn.   The solution is its intersection with $e/D = (e/D_1)$ $(\mathrm{Re}_2/\mathrm{Re}_1)$, obtained by trial.   It is also possible to obtain direct solution by replotting Fig. 8-10 so as to eliminate D from all but one of the three dimensionless ratios. Suitable ratios without D would be SF, also Re, f, and Ka with D replaced by e.

8-24. Turbulent Flow in Noncircular Cross Sections.   Experience has shown that in substantially equiaxed cross sections and in rectangular cross sections or concentric annuli in which the turbulence in the main stream would extend almost equally over the cross section, the pressure gradient for a given fluid and average velocity will be substantially the same as for a circle having the same ratio of cross section to wetted perimeter, known as the hydraulic radius m.   Since a circle has

$$m = \frac{\pi D^2}{4\pi D} = \frac{D}{4}$$

the equivalent diameter D_e for any section to use Fig. 8-10 is $4m$. D_e is readily seen to equal the side of a square section, the difference in outer and inner diameters for an annulus, and twice the spacing b for a wide duct with parallel walls.*   For a concentric annulus, however, if the inner diameter is less than some two-thirds of the outer diameter, the pressure drop starts to exceed that given by the usual hydraulic radius [46].   If the hydraulic radius is computed for the cross section and wetted perimeter outside of the radius of maximum streamline velocity, which agrees with the radius of maximum turbulent velocity

* However, better agreement is obtained using $4b$ for D_e in Re [113].

$[(R_2{}^2 - R_1{}^2)/2 \ln (R_2/R_1)]^{1/2}$, agreement with Fig. 8-11 is obtained.* This effect may be due to the higher shear stress at the inner wall. Equation (8-33) using D_e holds also for flow parallel to tube or rod bundles at axial spacing/diameter ratios below 1.3 [112]. At a higher ratio of 1.46 the friction may be up to 65 per cent higher (Chap. 9, Ref. 171).

Flow through a thin eccentric annulus, or a thin duct of wide nonparallel sides, could be obtained by finding the unknown local average velocity $\bar{u}$ as a function of duct thickness b at different values of the duct width or perimeter L, and computing Q as $\int_0^L (\bar{u}b)\, dL$. If Eq. (8-33) holds over the complete channel it can be shown that $\bar{u}$ is proportional to $b^{2/3}$, and the uniform $\bar{b}$ which would give the same total flow is obtained by averaging $b^{5/3}$ along L and taking the five-thirds root. If b is linear in L, $(\bar{b})^{5/3} = 3(b_2{}^{8/3} - b_1{}^{8/3})/8(b_2 - b_1)$, for instance.

8-25. Nonisothermal Turbulent Flow. Isothermal flow and low temperature gradients are minimized in nuclear power plants, as they represent unproductive weight, space, cost, and pumping power. Thus highly nonisothermal flow is important for gases in pipes. Figure 8-10 holds [52] if μ and ρ are used at $t' = (t_b + t_w)/2$, or [60] μ at $t' = 0.4t_b + 0.6t_w$ and ρ at t_b. For liquids [35] μ and ρ may be taken at t_b and the computed pressure drop divided by $1.02(\mu_b/\mu_w)^{0.13}$. In high-speed flow [47] the pipe walls will heat up by an amount $\gamma'\bar{u}^2/2g_cC_pJ$ due to frictional heat. J is the mechanical equivalent of heat and γ', the "recovery factor," approximately equals $(\mathrm{Pr})^{1/3}$.

8-26. Turbulent Flow of Compressible Gases. It has been observed experimentally that Fig. 8-10 holds at any point in a pipe for fluids under all isothermal conditions, including velocities up to several times that of sound [47], providing that $D \gg$ the mean free path.

In isothermal or substantially isothermal flow, Eqs. (8-21) and (8-29) can be combined to eliminate dF and integrated analytically. Setting dW and $dz = 0$, for gases,

$$\frac{P\, dP}{R'} = \frac{\alpha G^2 T\, d\rho}{g_c\rho} - \frac{2f_F G^2 T\, dN}{g_c D_e}$$

$$P_1{}^2 - P_2{}^2 = \frac{2R'G^2}{g_c}\left(\alpha\bar{T} \ln \frac{\rho_1}{\rho_2} + \frac{2\overline{f_F T}N}{D_e}\right) \quad (8\text{-}34)$$

Either unknown pressure is usually obtainable on the second trial, by substituting for ρ_1/ρ_2 the first trial value of P_1/P_2.

Equation (8-34) is also applicable in spite of large temperature variation, providing the friction term considerably exceeds the velocity term. In this case the average of $f_F T$ over N is employed. Normally μ and therefore Re do not vary with P, and with only about the three-fourths power of T. By Eq. (8-33), f will vary only as the one-seventh power

---

* Thus $D_e = 2R_2 - (R_2{}^2 - R_1{}^2)/[R_2 \ln (R_2/R_1)]$.

of T, approximately, and thus may be taken, for simplicity, as constant at an average value $\bar{f}$. $\bar{T}$ then is averaged over N; thus it is the midplane temperature for a nuclear reactor with symmetrical heat generation and a coolant of constant specific heat.   Problem 9-19 in Sec. 9-26, part f, applies this method (though the velocity term is appreciable and the method is not entirely appropriate).

Multiplying Eq. (8-34) through by V before carrying out the integration yields, for any fluid,

$$P_1 - P_2 = \frac{G^2}{g_c}\left[(V_2 - V_1)\alpha + \frac{f_F N (V_2 + V_1)}{D_e} \right] \qquad (8\text{-}35)$$

The quantity $2g(z_2 - z_1)/(V_2 + V_1)g_c$ should be added outside the brackets if appreciable.   This relation is usually sufficiently accurate if $2 > (V_2/V_1) > 0.5$, and particularly useful in an integration by finite increments of channel length (see Prob. 9-19, part g).   Special methods are available for adiabatic flow [29].   Equations (8-34) and (8-35) apply to streamline flow as well as turbulent if f does not vary too much to permit the use of $\bar{f}$.

Cooling a reactor with through-flow of a large volume of gas may involve high pumping energy.   This can be greatly cut (to one-eighth, if f and V remain constant) by splitting the reactor, as in the Brookhaven reactor or the Commonwealth Edison–Public Service 1953 Gas-cooled Reactor proposal, and admitting the stream at the middle to flow toward both ends.

Prob. 8-10. The Commonwealth Edison–Public Service 1953 Liquid-cooled Reactor proposal [72] uses 18 steam generators in which D_2O at about 750 psia flows inside of 656 return-bend $\frac{5}{8}$-in. OD No. 20 BWG tubes in parallel in each boiler. The tubes average 39.2 ft long including the bend and a double tube sheet requiring 1.1 ft at each end.   The coldest one-seventh, approximately, of each tube is specially baffled in the shell to preheat the H_2O, which supposedly boils from the rest of each tube.   The D_2O flow through each steam generator is 3,580,000 lb/hr, averaging 15.4 fps.   It enters at 440°F, reaching the preheating section at 401.3°F, and leaving the steam generator at 388.8°F.   The H_2O enters at condensation temperature, 109°F, and boils at 373°F, each unit handling 180,000 lb/hr.   (a) Compute the D_2O pressure drop, and compare with the design value of 35 psi.   (b) It is suggested to insert a 16.5-ft-long 0.25-in. OD stainless rod concentrically into each straight portion of each return-bend tube to decrease D_2O holdup.   Compute ΔP for this case. (c) If D_2O costs $28 per lb, the rods $1.50 per lb installed, annual fixed charges on both are 15 per cent, and pumping power delivered to the D_2O is 1.3 cents per kwhr net, show whether the rods are justified by economy alone.   *Ans.* (a) 20 psi. (b) 52 psi. (c) No. $220,000 worth of D_2O is saved, but at a net cost of $175,000 per yr.*

* This calculation neglects the economic advantage of earlier start-up if D_2O is in short supply.

8-27. The Formation of Boundary Layers. At the inlet of a pipe or the leading edge of a plate parallel to a flowing fluid, a boundary layer of slowed-down fluid is initiated. It begins with zero thickness and high velocity gradient and shear stress, and it becomes thicker as it travels downstream. Since the boundary layer is thin at the start, flow in it is streamline. Velocity, friction, and heat-transfer measurements all indicate that at a certain distance downstream the nature of the boundary layer suddenly changes. A hot-wire anemometer would show that only an inner sublayer remains streamline thereafter, the balance of the sloweddown fluid becoming turbulent. The Reynolds number at the transition,

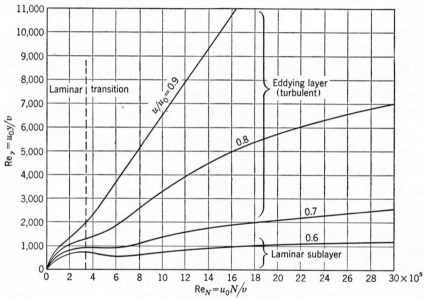

FIG. 8-11. Velocity distribution in the boundary layer.

$\mathrm{Re}_N = N u_o \rho / \mu$, where N is distance downstream from the leading edge and u_o is the undiminished velocity of the stream, may range from about 10^5 to 10^6, depending on the initial turbulence of the stream. It is, however, frequently about 400,000.

Since the slowing down extends asymptotically away from the plate, there is no sharp outer boundary; it has become customary to define it as the position at which u is 90 or 99 per cent of u_o. A generalized profile of the layers is shown in Fig. 8-11 [48]. It is well established theoretically and experimentally that the thickness δ of the streamline boundary layer to 99 per cent of u_o is given by $\delta / N = 5(\mathrm{Re}_N)^{-\frac{1}{2}}$ and the local wall shear stress $\tau_w = 0.332 \rho u_o^2 (\mathrm{Re}_N)^{-\frac{1}{2}}$. The average τ_w from the leading edge to N is shown by integration to be double the τ_w at N. Similarly, for the turbulent boundary layer $\delta/N = 0.377(\mathrm{Re}_N)^{-1/5}$, the local

$$\tau_w = 0.0295\rho u_o{}^2(\mathrm{Re}_N)^{-1/5}$$

and the average τ_w is 25 per cent higher. At high speeds or with simultaneous heat transfer the physical properties may be evaluated [49] at $t' = 0.75t_w + 0.25t_b$.

If the spacing of fuel elements and other solid walls is definitely less than 2δ by the appropriate expression above, the boundary-layer relations rather than long-channel relations will apply in the calculation of wall shear stresses and forces.

8-28. Steady Velocity Distribution in Isothermal Turbulent Flow in Channels. As a fluid flows downstream in a duct, the boundary layer becomes thicker, finally meeting from all sides in the center. Thereafter a substantially constant velocity distribution is obtained.

The steady local velocity gradient du/dy may be expected in any streamline region very near the wall to depend on τ_w and μ. In any turbulent region near the wall it will also depend on ρ. The local velocity $\int_0^y (du/dy)\,dy$, will depend as well on y. Thus there are five variables and $5 - 3 = 2$ dimensionless ratios, by the pi theorem, needed to represent the velocity distribution near the wall. The most convenient ratios for plotting the velocity distribution are evidently the ones with u but not y, and vice versa. The "velocity parameter" is found by canceling dimensions among the variables to be $u\sqrt{\rho/\tau_w g_c}$, and is commonly abbreviated u^+. The "distance parameter" y^+ is found to be $y\sqrt{\tau_w \rho g_c}/\mu$. For a pipe $\tau_w = \rho\bar{u}^2 f_F/2g_c = (R/2)(dP/dN)$, and for a flat channel R is replaced by the spacing b.

Many investigators have measured the isothermal velocity distribution of gases and liquids in pipes and wide flat channels [50], including for liquid metal [51]. On a plot of u^+ vs. y^+ the data for Re above about 30,000, except near the center of each duct, tend to plot on a single curve regardless of fluid, velocity, or roughness. The portion nearest the wall would be expected to exhibit streamline flow, as per Fig. 8-11. Making this assumption, replacing du/dy in Eq. (8-2) by u/y, and eliminating u and y by the above definitions of u^+ and y^+, the "streamline layer" relation is obtained:

$$u^+ = y^+ \tag{8-36}$$

A few of the experimental points have been as close to the wall as $y^+ = 1.5$ to 2.0 [52, 53] and approximately verify Eq. (8-36). The turbulent points beyond $y^+ = 25$ to 30 and up to about half the radius, when plotted as u^+ vs. $\log y^+$ tend to fall near the straight line

$$u^+ = 5.5 + 2.5 \ln y^+ = 5.5 + 5.75 \log y^+ \tag{8-37}$$

The intersection of Eqs. (8-36) and (8-37) is at $u^+ = y^+ = 11.6$, and originally Eq. (8-36) was used up to 11.6 and Eq. (8-37) beyond.

Noticing that most experimental points near the intersection fall below both curves, Von Karman suggested [54] adding an intermediate line. This he drew straight from $y^+ = 30$ on Eq. (8-37) to tangency with Eq. (8-36), which occurred at $y^+ = 5$. Its equation is

$$u^+ = -3.05 + 5.0 \ln y^+ = -3.05 + 11.5 \log y^+ \qquad (8\text{-}38)$$

This line is commonly interpreted as describing a "buffer zone," though actually no clearcut boundaries or even zones probably exist. Many

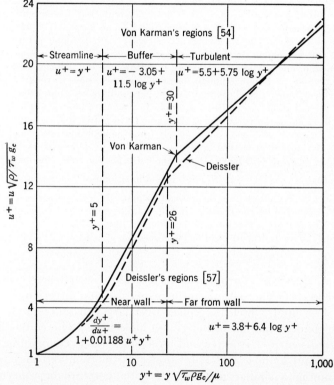

FIG. 8-12. Isothermal velocity-distribution curves, according to Von Karman and Deissler.

applications have been made of the velocity profile represented by Eqs. (8-36), (8-37), and (8-38) and shown in Fig. 8-12, partly because of their simple form and ease of mathematical handling. Other empirical or semiempirical equations have been proposed [55–57, 113; Chap. 9, Refs. 99, 100] which fit the data better but are mathematically more inconvenient. Deissler's is shown in Fig. 8-12.

Under certain conditions Eqs. (8-36), (8-37), and (8-38) are less reliable. At low Reynolds numbers the assumption made in deriving Eq. (8-36) of a thin streamline layer and therefore a constant shear stress across the layer is not true [58]. Integrating Eq. (8-2) from 0 to y and 0 to u with $\tau = \tau_w(R - y)/R$ yields $u = (\tau_w g_c/\mu)(y - y^2/2R)$, the same as for streamline flow throughout a pipe or wide flat duct. Eliminating u and y with the definitions of u^+ and y^+ yields for the velocity distribution over a thick streamline boundary layer

$$u^+ = y^+ - \frac{(y^+)^2}{2R^+} = y^+ - \frac{(y^+)^2}{(\text{Re } \sqrt{f_F/2})} \tag{8-39}$$

At the center of a duct, u and therefore u^+ reach a maximum, so that the slope of u^+ vs. y^+ should $= 0$ at $y = R$, or $y^+ = R^+$. Equation (8-36) does not hold to the center for streamline flow, but Eq. (8-39) does.

For the turbulent zone Eq. (8-37) does not hold completely to the center, as it also yields a definite slope at $y = R$. A modification by Von Karman [52] including variation of τ with y is a better fit, giving, at the center, half the slope of Eq. (8-37). It is evident that agreement on the true isothermal relation between u^+ and y^+ has not been reached, or on equations to represent it. However, the sequence of Eqs. (8-36), (8-37), and (8-38) is reliable enough for most purposes, such as roughly predicting friction or heat-transfer coefficients under special conditions [52, 60; Chap. 9, Ref. 99, etc.]. These equations are difficult to apply to irregular channel sections, since the variation of τ_w along the perimeter of the section will not be known, even though its average value can be obtained approximately from the hydraulic radius and Fig. 8-10.

Prob. 8-11. (a) Show that for a pipe $y^+ = (y/R)(\text{Re}) \sqrt{f_F/8}$, $u^+ = (u/\bar{u}) \sqrt{2/f_F}$ and $du/dy = (\tau_w g_c/\mu)(du^+/dy^+)$. Derive the corresponding relations for a flat duct. (b) Show that $\int_0^{R^+} u^+(1 - y^+/R^+) \, dy^+$ is a dimensionless integral proportional to the flow in a pipe, and that it theoretically equals $\text{Re } \pi/2 = Q\rho/R\mu$. Check how closely this is obeyed by the relations in Fig. 8-12 at any desired value of Re.

Sample Solution (b). The integral, and the calculated flow, are 0.8, 6, and 18 per cent too high at $\text{Re} = 10^6$, 10^4, and 5,000, respectively, using Eqs. (8-36), (8-37), and (8-38).

8-29. Analysis of Turbulence in Fluids.

Turbulence consists of random eddies, or fluctuations in the magnitude and direction of the local velocity in a fluid, which take place over larger distances and volumes than the random motion of individual molecules. The eddies have an average cross-current velocity and an average range comparable respectively to v and Δy in Eq. (8-1) for molecular motion, and they transmit momentum in a similar manner. If a temperature or concentration gradient exists in the fluid, it is evident that heat or matter, respectively, will be simultaneously transferred by the same eddies. Measurements of heat- and mass-transfer coefficients are frequently difficult or time-consuming. Thus an important application of turbulence theory is the prediction of such coefficients from flow data. This is usually based on the assumption that the effective size and velocity of the eddies are identical for the transfer of momentum, heat, and matter. This "geometrical" effectiveness of the eddies is designated the eddy diffusivity ϵ

and is measured by the product $v \, \Delta y$ in Eq. (8-1). ϵ is evidently proportional to the volumetric capacity to transport matter. The capacity of the eddies to transport momentum or shear stress is evidently proportional to $\epsilon\rho$, and their capacity to transmit sensible heat is proportional to $\epsilon\rho C_P$. In the second case ϵ is usually written ϵ_M and designated the eddy viscosity, and in the last as ϵ_H the eddy conductivity, to allow for possible differences among the values of ϵ actually observed.

Von Karman assumed [54] that the streamline and turbulent shear stresses are independent and therefore additive. Thus, adding Eq. (8-1) in terms of ϵ_M to Eq. (8-2) to give the total shear stress

$$\tau g_c = \tau_w g_c \left(1 - \frac{y}{R}\right) = (\mu + \rho\epsilon_M) \frac{du}{dy} \tag{8-40}$$

The comparable relations for heat and mass transfer are

$$\frac{q}{A} = (k + \rho C_P \epsilon_H) \frac{dt}{dy} \tag{8-41}$$

$$N_M = (D_v + \epsilon) \frac{dc}{dy} \tag{8-42}$$

Equation (8-40) can be employed to compute ϵ_M from experimental values of the pressure and velocity gradients, or to predict ϵ_M from the correlations of Figs. 8-10 and 8-12. For instance, from Eqs. (8-37) and (8-40) for y^+ from 30 up to a considerable fraction of R^+

$$\frac{\epsilon_M \rho}{\mu} = \frac{\tau_w g_c}{\mu} \frac{1 - y/R}{du/dy} - 1 = \frac{1 - y/R}{du^+/dy^+} - 1 = 0.4 y^+ \left(1 - \frac{y}{R}\right)$$
$$- 1 \approx 0.4 y^+ \tag{8-43}$$

Usually y/R is neglected near, and even distant from the wall—its effect is also neglected in Eq. (8-37)—and the final 1 because viscous forces are small compared to turbulent forces. Similarly, for $30 > y^+ > 5$, $\epsilon_M \rho/\mu = 0.2 y^+ (1 - y/R) - 1$, and below $y^+ = 5$, $\epsilon_M = 0$, as postulated, from Eqs. (8-38) and (8-39), respectively.

8-30. The Structure of Turbulence. Von Karman has also developed an alternative approach to eddy viscosity [59], derivable by dimensional analysis. It is reasonable that at a considerable distance from the wall the eddies are independent of y and are due to curling of the flow, caused by the velocity gradient with possible additional influence of the second and/or higher derivatives of u with respect to y. The turbulent shear stress is considered a result of the eddies and the velocity gradient, rather than a cause. The absolute velocity u should not create local effects. Experience also indicates that μ [which cancels out of Eq. (8-43), for instance] does not affect ϵ_M. Accordingly, ϵ_M and the quantities that may determine it are the following:

Variable....................	ϵ_M	ρ	du/dy	d^2u/dy^2	d^3u/dy^3
Dimension................	L^2/T	M/L^3	$1/T$	$1/LT$	$1/L^2T$

Evidently ρ must be dropped, since no other variable contains M. The most direct way that the dimensions of ϵ_M can be canceled is seen to be $(L^2/T)(1/LT)^2T^3$, without involving the third derivative.

Accordingly, the simplest dimensionless result is

$$\epsilon_M \frac{(d^2u/dy^2)^2}{(du/dy)^3} = \epsilon_M \frac{\rho}{\mu} \frac{(d^2u^+/dy^{+2})^2}{(du^+/dy^+)^3} = C' \qquad (8\text{-}44)$$

where C' is a dimensionless constant. Taking $\epsilon_M = 0.4y^+\mu/\rho$ from Eq. (8-43), and using Eq. (8-37) to compute the derivatives yields $C' = 0.16$. Thus the result of the dimensional analysis is verified by experimental data, and higher derivatives are not required. Evidently if the flow follows Eq. (8-37) and the shear stress is known, Eqs. (8-43) and (8-44) with $C' = 0.16$ will yield the same value of ϵ_M. In effect, Eq. (8-44) with $C' = 0.16$ is a method of using Eq. (8-43), or Eqs. (8-37) and (8-40), in terms of d^2u/dy^2 rather than τ_w.

Further understanding of turbulence would be facilitated if it were possible to distinguish the intensity v and range Δy in ϵ_M. For this purpose Prandtl defined a "mixing length" L as the value of Δy such that $\Delta u = v$. (This definition is based on the assumption that fluid moving normally to the wall a distance Δy would create a temporary or turbulent $\Delta u = \Delta y \, du/dy$, which should equal v if the turbulence is isotropic, or equal in all directions.) Rewriting Eq. (8-1) for the turbulence alone, $\tau_t g_c = \Delta u \, \rho L(\Delta u/\Delta y) = \rho L^2(\Delta u/\Delta y)^2 = \rho L^2(du/dy)^2$. Adding the streamline shear stress to obtain the total τ and solving for L give

$$L = \frac{\tau g_c/\rho - (\mu/\rho)(du/dy)^{1/2}}{du/dy} \qquad (8\text{-}45)$$

The value of v corresponding to this definition of L is

$$v = L\frac{du}{dy} = \left(\frac{\tau g_c}{\rho} - \frac{\mu}{\rho}\frac{du}{dy}\right)^{1/2} = \frac{\epsilon_M}{L} \qquad (8\text{-}46)$$

In the turbulent core, the streamline shear stress and thus the terms containing μ/ρ in Eqs. (8-45) and (8-46) are usually neglected. Obtaining du/dy from Eq. (8-37) and substituting it into Eqs. (8-45) and (8-46), it is seen that near the wall $L = 0.4y$ and $v = \sqrt{\tau_w g_c/\rho}$. At the center of the pipe Eqs. (8-43) and (8-45) are indeterminate, but L extrapolates to $0.14R$.

8-31. Velocity Distribution in Nonisothermal Turbulent Flow. During heat transfer or high-speed flow, μ and ρ are not constant throughout a cross section of the flow. Since μ and ρ are absent from Eq. (8-44), Deissler [60] assumes that it holds in nonisothermal as well as isothermal flow (above $y^+ = 26$ and with $C' = 0.1296$). In both cases, his suggestion for $0 < y^+ < 26$ is $\epsilon_M = 0.01188uy = 0.01188(\mu/\rho)_w u^+ y^+$, except at high Prandtl numbers. Numerically integrating these relations simultaneously with the corresponding heat-transfer relations, he showed good agreement at high Re between the calculated and experimental u^+ vs. y^+ curve, pressure drop, and heat transfer for air at various heat velocities.

For nonisothermal liquids with wide variation in physical properties, Goldmann suggests [61] that Fig. 8-12 be still applied, with u^+ defined as $\int_0^u \sqrt{\rho/\tau_w g_c}\, du$ and y^+ as $\int_0^y (\sqrt{\tau_w g_c \rho}/\mu)\, dy$. Moderate agreement was obtained between the theory based on these assumptions and the heating of supercritical water under extreme conditions.

8-32. The Statistical Theory of Turbulence. Direct measurements of the intensity and of the range or "scale" of turbulence can be made by means of sensitive hot-wire anemometers [62]. Wires under 0.01 mm in diameter and 1 mm in length are employed, and circuits which can follow frequencies from 2 to 70,000 cps are available [63]. At any one location, turbulence shows up as random instantaneous velocity fluctuations, u', v', w', where u' is in the direction of the flow velocity $\bar{u}$, and v' and w' are at right angles. The time average value of u' (or of v' or w') is zero; the intensity of the turbulence in the x direction is defined as the root-mean-square value of u', or $[\overline{(u')^2}]^{1/2} = \left[\dfrac{1}{\theta}\int_0^\theta (u')^2\, d\theta\right]^{1/2}$ and is reported as a percentage of $\bar{u}$. $\sqrt{\overline{(v')^2}}$ is in each direction half of the time and thus is twice v in Eq. (8-46).

The turbulence in a turbulent stream can conveniently be decreased to below 0.5 per cent by forcing the stream through a smooth nozzle or fine screens, or increased to above 20 per cent by passing it through a coarse screen [64].* Shortly after such an obstruction the turbulence has again become isotropic, i.e.,

$$\overline{(u')^2} = \overline{(v')^2} = \overline{(w')^2}$$

Isotropic turbulence also exists at the center of a pipe [115], where

$$\sqrt{\overline{(v')^2}} = 0.75\bar{u}\,\sqrt{f_F/2}$$

The axial turbulence thus ranges from about 3 to 4 per cent for practical turbulent flow. The turbulence increases and becomes anisotropic as the walls are approached.

The "scale" of the turbulence can be measured by the simultaneous

* For $DG/\mu > 100$, the per cent turbulence between 2.5 mesh lengths and 1000 wire diameters x/D down stream of a screen equals $112\ (D/x)^{5/7}$.

use of two hot wires.   Several definitions of the scale of turbulence have
been proposed; the commonest is the Eulerian scale L':

$$L' = \int_0^\infty R_y \, dy = \int_0^\infty \frac{\overline{u'_0 u'_y}}{(\overline{(u'_0)^2} \; \overline{(u'_y)^2})^{1/2}} \, dy \qquad (8\text{-}47)$$

where R_y is the coefficient of correlation between u'_0 at the desired point
and u'_y at the same instant and a distance y away at right angles to u.
L' averages approximately 63 per cent of the Lagrangian scale L_θ [114],
which describes the distance traversed by the eddies in time and equals
the mixing length L.   Thus ϵ for Eqs. (8-41) and (8-42) is obtainable by

$$\epsilon = \tfrac{1}{2} L_\theta \; \sqrt{\overline{(v')^2}} = 0.8 L' \; \sqrt{\overline{(v')^2}}$$

as well as by Eq. (8-46).

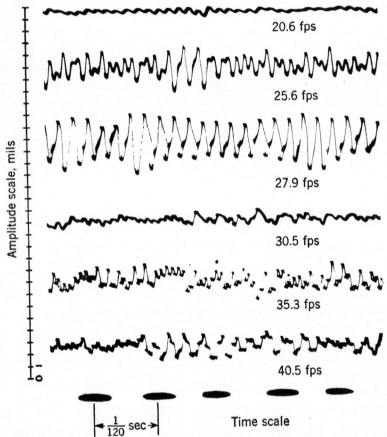

FIG. 8-13. Typical vibrographs of an experimental fuel assembly, measured on the
fuel plate near the inlet at various water velocities.   (*From S. McLain, ANL-5424,
1955.*)

The turbulence can also be measured by several indirect methods. If the wake from a heated wire is traversed with a thermocouple [65], ϵ_H can be computed from the local angle α' in degrees within which the fluid temperature has risen 50 per cent or more of the maximum rise, since $\epsilon_H = (\alpha'/190.8)^2 ux - k/C_P\rho$, where x is the distance downstream to the traverse. The intensity of the turbulence in per cent is obtainable as $142\sqrt{\epsilon_H/ux}$. The same theory can be applied to the spreading of a dye in a liquid, or a gas in another gas [64]. Another method [66], which gives the ratio of the intensity to $(L')^{1/5}$, involves finding the size of sphere for which there just occurs the characteristic drop in drag coefficient on incidence of turbulence in the boundary layer (Fig. 8-17). Besides this decrease in critical Re and drag due to turbulence, heat- and mass-transfer coefficients may increase. For flow normal to a cylinder, high turbulence shows no effect at a cylinder Re below 2000, but increases the coefficients up to 25 per cent at Re = 20,000 [67].

Components of random turbulence or fluctuations in coolant velocity can cause harmful vibration of objects such as fuel elements that are in the stream, particularly if they coincide with natural vibrational frequencies of the fuel element. In addition, regular "Karman" vortices form behind cylinders and strips during crossflow, with a frequency $0.22\bar{u}/D$, where D is the diameter or thickness. Natural frequencies of a simple structure can be predicted from formulas in texts on mechanical vibrations. Proposed power-reactor-core designs should be tested at full size for vibration (see Fig. 8-13) to assure a serviceable unit at expected conditions.

8-33. The Incidence of Turbulence and the Critical and Transition Regions.

From visual observation of threads of dye solution [58], turbulence in pipes first seems to occur at the axis, and at Re = 900, approximately, though with a smooth trumpet inlet much faster flow may remain streamline. An over-all phenomenon such as friction pressure drop (Fig. 8-10) generally first shows in long pipes observable departure from streamline flow [Eq. (8-24), or its equivalent, $f_F = 16/Re$] at about Re = 2100. In the "critical region," approximately 2100 < Re < 3500 if flow rate is given and 70,000 < Ka < 270,000 if pressure drop is given, flow is irregular, especially in short pipes. It is probably desirable to avoid this region in nuclear reactors, since pulsing may develop in which flow alternates between streamline and turbulent, the velocity of the stream controlling the period of the pulses. Such flow irregularities would cause undesirable temperature and thermal stress cycling. In other cross sections, such as annuli and square or flat ducts, turbulence initiates less abruptly and a dip is not generally obtained nor is pulsing encountered.

Departure from the smooth-pipe curve in Fig. 8-10 commences when the effective wall roughness e becomes appreciable compared to the boundary-layer thickness, and the region of constant f is reached when the boundary-layer thickness has become small compared to e.

Prob. 8-12. Roughness may be put on the generalized scale of Fig. 8-12, since $e^+ = 2R^+e/D = Re\sqrt{f_F/2}\,e/D$. (a) Defining appreciable departure from the smooth-pipe curve as occurring when $e/3.7D$ in Eq. (8-31) equals 5 per cent of the last term, show that e^+ at appreciable departure = 0.16. (b) Show by Eq. (8-34) that f_F becomes constant at $e^+ = 70$.

TABLE 8-9. FRICTION IN TYPICAL PIPE FITTINGS AND CHANGES*

Fitting or change	K	N_e/D	C_D	Convergence ratio $(D_2/D_1$ or $b_2/b_1)$	C_c†
Turbulent Flow					
One average-velocity head..............	1.0	45			
Elbow, 45° standard...................	0.35	15			
Elbow, 45° long radius................	0.2	10			
Elbow, 90° standard..................	0.82	35			
Elbow, 90° long radius................	0.53	25			
Elbow, 90° square....................	1.3	60			
Bend, 180° close.....................	2.0	90			
Bend, 180° medium...................	1.2	55			
Tee, 0° through......................	0.4	20			
Tee, 90° side outlet..................	1.3	60			
Tee, 90° side inlet...................	1.5	70			
Coupling, union, 0° through...........	0.04	2			
Valve, open, 0° gate..................	0.16	7			
Valve, open, 0° globe.................	6.	270			
Valve, open, 90° angle................	3.	135			
Valve, open, 0° swing check..........	2.	90			
Bend, 90°; $r/D = 0.5$.................	0.8	35			
Bend, 90°; $r/D = 1$.................	0.35	16			
Bend, 90°; $r/D = 4$.................	0.16	7			
Bend, 90°; $r/D = 10$................	0.16	7			
Coil, each turn (360°); $r/D = 1$........	2.2	100			
Coil, each turn (360°); $r/D = 4$........	0.55	25			
Coil, each turn (360°); $r/D = 10$.......	0.3	14			
Sudden contraction, $D_1/D_2 = 1.33$......	0.19†	9†			
Sudden contraction, $D_1/D_2 = 2$........	0.33†	15†			
Sudden contraction, $D_1/D_2 = 4$........	0.42†	19†			
Sudden expansion, $D_2/D_1 = 1.33$.......	0.19†	9†			
Sudden expansion, $D_2/D_1 = 2$..........	0.56†	25†			
Sudden expansion, $D_2/D_1 = 4$..........	0.92†	42†			
Small square-edged orifice..............	0.05	2	0.60		
Rounded nozzle.......................	0.05	2	0.98		
Excess loss in tube below nozzle........	0.02	1			
Short tube downstream of plate ($N < 3D$)	0.05	2	0.60		
Long tube downstream of plate ($N > 3D$)	0.5	23	0.82		
Short tube upstream of plate ($N < 3D$)..	0.06	3	0.53		
Long tube upstream of plate ($N > 3D$)..	0.8	35	0.72		
Gradual conical expansion in pipe†......	0.2–0.5				
Gradual conical convergence in pipe†...	0.06–0.15				
Gradual conical converging inlet†.......	0.13–0.4	....	0.35–0.94		
Streamline Flow					
Conical or straight slot convergence (135°)...............................	.....	....	....	0	0.746
	.....	....	....	0.4	0.749
	.....	....	....	0.8	0.789
	.....	....	....	1.0	1.0
Sharp-edged orifice or slot (90°).........	.....	....	....	0	0.611
	.....	....	....	0.4	0.631
	.....	....	....	0.8	0.722
	.....	....	....	1.0	1.0

* For additional data see particularly M. P. O'Brien and G. H. Hickox, "Applied Fluid Mechanics," p. 350, McGraw-Hill Book Company, Inc., New York, 1937.

† Based on the smaller diameter.

8-34. Pressure Losses in Fittings. Pressure and energy losses in fittings cannot be reliably predicted, since fittings are not highly reproducible, differ widely among different brands, and affect each other unless widely separated. However, approximate rules have been developed to estimate the losses for rough design purposes.

In turbulent flow, the irregularities in most fittings are analogous to large values of e in Fig. 8-10, and the friction loss is thus generally proportional to $\bar{u}^2$. A convenient measure is the number K of pipe average-velocity heads or pressures lost in the fitting. Another convenient unit is the length of straight pipe expressed in pipe internal diameters (N_e/D) having the same friction loss as the fitting. These quantities are related by $K = 4f_F N_e/D$. An average value of $f_F = 0.0055$ may be used to compute N_e/D approximately from K, instead of the actual f_F in the pipe. The friction computed from Table 8-9 is added to that for the same actual total distance if it is straight pipe. Screwed and flanged fittings give similar K values. The computed drop in pressure is in addition to any drop or rise by Bernoulli's principle.

Losses at an expansion may be estimated by

$$K_{\exp} = 1 - \frac{2\gamma_1 A_1}{A_2} + \left(\frac{A_1}{A_2}\right)^2 (2\gamma_2 - 1)$$

where γ is the momentum ratio from Table 8-5. At a contraction,

$$K_{\mathrm{con}} = \frac{1}{C_c} - 2 + C_c(2\gamma_2 - 1)$$

where C_c is the coefficient of stream area contraction from Table 8-9. In streamline flow these K values may average 19 per cent high, and in turbulent flow 11 per cent high [69].

When many fittings are near each other, as frequently occurs in reactor designs, the pressure drop may be up to several times that calculated by Table 8-9, which is for isolated fittings. Only actual experiment will then give accurate values.

PIPING DESIGN PROBLEMS

8-35. Design for Optimum Size of Pipes and Channels. To compute the optimum pipe or channel to conduct a given stream it is necessary to reduce all the factors involved to a common unit, preferably net cost per unit of total time, and determine the size at which the total cost is a minimum. Ideally, every cost item should include all advantages and disadvantages, such as availability, probability of unexpected troubles, etc. [70], as well as the money cost. Actually these may be difficult to assess and usually are merely kept in mind in making final decisions. The high cost concentration in nuclear-power equipment usually makes it preferable to compute precisely each reasonable alternative design, or at least closely bracket the minimum cost, before making the final selection.* However, formulas including approximate assumptions are valuable for preliminary orientation.

The total annual cost per unit length of pipe is $\pi\delta(D + \delta)\rho_w(1 + F')C_w$, where D = internal diameter, δ = wall thickness, ρ_w = density of pipe, F' = additional fractional cost for fittings and complete installation, and C_w = total annual cost of pipe per unit of mass. If C_w varies considerably with D, it should be evaluated at the optimum D as estimated from experience, or approximated [71] by the simplified turbulent flow equation $D''_{opt} = 2.2w_M^{0.45}/\rho_f^{0.31}$, in which D'' = inches, w_M = thousands of pounds per hour, and ρ_f = pounds per cubic foot density of the fluid. C_w includes interest, amortization, maintenance, taxes, profit, scarcity surcharge, net additional reactor cost due to neutron consumption, if inside the reactor, and any other charges against the metal, minus any credits such as moderating or shielding value.† If δ is small compared to D, and proportional to D, over a reasonable range of sizes the annual cost of the pipe may be approximated by XD^2, where X is constant. Any other relation between δ and D could of course be employed if desired.

The annual charge for the fluid in the pipes is usually negligible, but if the fluid is heavy water, an expensive heavy metal, or particularly a solution or suspension of a fissionable isotope, this charge may be significant or even predominant. The total annual cost per unit pipe length is $\pi D^2\rho_f C_f = YD^2$, where C_f is the total annual cost of the fluid per unit mass. C_f includes the same itemized charges as C_w, except for shielding value. If the fluid also is the fuel carrier, no charge for neutron consumption should be made against it.

The annual cost of the required pumping energy is ZD^{-n}, where Z is the total annual cost per unit length of the pipe computed at unit diameter. The exponent $n = 5 + (d \log f_F)/(d \log \mathrm{Re})$, from Fig. 8-10. In streamline flow $n = 4$. In nuclear-power applications, high values of Re are indicated, and $n = 5$ is likely. If unit diameter is well outside the optimum range, $Z = Z'(D')^n$, where Z' is computed for a reasonable diameter D'. Z includes the total cost of the energy consumed in pumping and the fixed charges on the pump, minus credit for recoverable heat dissipation. Alternatively, Z (or Z') could be taken as the total

---

* For instance, each pound per square inch saved in the main coolant cycle of a single medium-sized reactor (e.g., LSR) represents a total saving on the order of $50,000.

† Shielding value would strictly be proportional to δ, thus roughly to D, but for simplicity can be left as proportional to D^2, or neglected.

annual cost of operating the plant, multiplied by the ratio of the power used per unit length of pipe of unit (or D') diameter to the total power produced.

Thus the total annual cost per unit length of pipe is

$$C_t = (X + Y)D^2 + ZD^{-n} \tag{8-48}$$

Equating dC_t/dD to zero yields the optimum D:

$$D_{opt} = \left[\frac{nZ}{2(X + Y)}\right]^{1/(n+2)}$$
$$= 1.12 \left(\frac{Z}{X + Y}\right)^{1/6}_{\text{(for } n=4)} = 1.12 \left(\frac{Z}{X + Y}\right)^{1/7}_{\text{(for } n=5)} \tag{8-49}$$

After solving for D_{opt}, Re should be computed to check whether the flow is yet in the region that was employed when computing Z. Furthermore, new values of X, Y, and Z (from Z') can be computed at D_{opt} and a more accurate D_{opt} obtained by using Eq. (8-49) a second time. Finally, the complete cost equation can be written at D_{opt}, and the significance of varying D shown by plotting total annual cost against it, or differentiating partially with respect to it. This is important so that reasonable tolerances may be specified, or ranges for permissible subsequent modifications in design. Fortunately, at any minimum cost design, D can be changed appreciably with no significant change in total cost. Similar calculations can be carried out for ducts of other cross-sectional shapes.

Prob. 8-13. The Commonwealth Edison–Public Service 1953 Liquid-cooled Reactor proposal [72] employs D_2O as coolant and moderator, flowing at 64,500,000 lb/hr through pipes of type 347 stainless steel. P_{max} at pump discharge = 800 psia. At the reactor outlet the D_2O is at 440°F and flows through 18 pipes in parallel to 18 steam generators. It then flows from each pair of steam generators in one pipe to one of the nine pumps at 388°F. Assume in addition that fixed charges on the pipe total 20 per cent of installed cost per year and on the D_2O 15 per cent, the fabricated and installed cost of type 347 piping is $2.50 per lb, pipe-wall stress is to be 20,000 psi at P_{max}, operation is 8000 hr/yr, and pumping energy delivered to the D_2O costs 1.3 cents per kwhr in pump fixed charges and lost power for sale. Compute the optimum ID, wall thickness, and average velocity for the 18 and for the 9 parallel pipes for D_2O at $28 and $83 per lb. Compare with standard 12-in. and 18-in. pipe, respectively (schedule not listed), and 25 fps, specified in the proposal for $83 D_2O.

Prob. 8-14. Of the four 1953 Industry Team reactor proposals [72], three include Na-cooled reactors. Two employ a Na velocity of 30 fps and the other 20 fps. Assume that all Na piping has 3/8-in.-wall stainless steel pipe costing $2.50 per lb installed. Na costs 20 cents per lb delivered, all annual fixed charges are 20 per cent, power is 1 cent per kwhr, and d-c electromagnetic pumps of 50 per cent efficiency are used. Neglecting pump fixed charges, plot the total cost of conveying 100, 600, and 3000 cfm of Na per pipe for 8000 hr/yr at 800°F vs. $\bar{u}$ in the pipe. Note the optimum velocity and the per cent increase in cost for 20 per cent deviation from the optimum.

Prob. 8-15. In the Commonwealth Edison–Public Service 1953 Gas-cooled Reactor [72] each of 12 blowers handles 270,000 lb/hr of He at about 393°F and 150 psia. Compute the optimum pipe size and He velocity in the pipes to each blower if they are steel with ⅜-in. wall at $1.50 per lb installed, and other data are as in Prob. 8-28.

8-36. Flow in Complex Circuits. Complex circuits and networks in steady-state flow may be solved by principles analogous to Kirchhoff's laws for electrical circuits. Thus, the net flow to any location is zero, and the pressure change when integrated around any closed path is zero. If we write the flow balance at each node and the pressure balance around each small loop in terms of the flows in each branch, one equation less than the number of unknown flows is obtained. A relation with the specified total flow or over-all pressure drop completes the necessary number for simultaneous solution.

With streamline flow at known temperatures, the simultaneous equations are linear in mass flow rate w and may be solved directly. In turbulent flow the solution is more difficult, and trial and error is necessary for all but the simplest circuits.

Branched circuits without crossflow may be solved by computing the equivalent resistance of separate sections and finally that of the whole circuit. For laminar flow, Eq. (8-24) may be rewritten as a flow resistance $\Delta P/w = 128\bar{\nu}N/\pi g_c D^4$, where $\bar{\nu}$ is the average kinematic viscosity ν over the length N. The total equivalent $\bar{\nu}N/D^4$ of pipes in series is the sum of their separate values of $\bar{\nu}N/D^4$ and that of pipes in parallel is $1/\Sigma(D^4/\bar{\nu}N)$. Thus, the equivalent resistance of the complete network is readily computed.

A similar principle can be applied in turbulent flow. The exact solution requires f_F for each pipe, requiring trial and error unless Re is high enough for f_F to be constant. From the definition of f_F, for pipes in parallel,

$$w_{\text{total}} = \sum w_p = \left(\frac{\pi^2 g_c \Delta P}{32}\right)^{1/2} \sum \left(\frac{\bar{\rho}D^5}{f_F N}\right)^{1/2} \tag{8-50}$$

Thus, the value of $(f_F N/\bar{\rho}D^5)_p$ for a single pipe equivalent to pipes in parallel is $[\Sigma(\bar{\rho}D^5/f_F N)^{1/2}]^{-2}$. Similarly, for pipes in series,

$$\Delta P_{\text{total}} = \sum \Delta P_s = \frac{32w^2}{\pi^2 g_c} \sum \frac{f_F N}{\bar{\rho}D^5} \tag{8-51}$$

The value of $(f_F N/\bar{\rho}D^5)_s$ for a single pipe which is equivalent to pipes in series parallel is the sum of the separate $f_F N/\bar{\rho}D^5$ values and $(f_F N/\bar{\rho}D^5)_p$ values in series along the main path.

A more direct solution is possible with an empirical relation for f_F. Equation (8-33) yields

$$\sum w_p = \left(\frac{\Delta P g_c}{0.142}\right)^{1/1.8} \sum \left(\frac{\bar{\rho} D^{4.8}}{\bar{\mu}^{0.2} N}\right)^{1/1.8}$$

$$\sum \Delta P_s = \frac{0.142 w^{1.8}}{g_c} \sum \frac{\bar{\mu}^{0.2} N}{\bar{\rho} D^{4.8}} \tag{8-52}$$

Thus the pipe equivalent to several in parallel has its

$$\frac{\bar{\mu}^{-0.2} N}{\bar{\rho} D^{4.8}} = \left[\sum \left(\frac{\bar{\rho} D^{4.8}}{\bar{\mu}^{0.2} N}\right)^{1/1.8} \right]^{-1.8}$$

and for pipes, or parallel groups in series, their values or equivalent values, respectively, of $\bar{\mu}^{0.2} N/\bar{\rho} D^{4.8}$ are added. Evidently, if ρ, μ, or f_F is the same in all pipes it may be factored out of the summations. For a channel varying in cross section, Eqs. (8-51) and (8-52) may be graphically or analytically integrated over the length N. With minor modifications, ΔP due to fittings (from Table 8-9) or to velocity changes [by Eq. (8-21)] may be incorporated.

If heat generation, as in a nuclear reactor, or high pressure drop causes wide changes in ρ or μ, or there is significant thermal buoyancy change with flow, or precise results are required, one of the above methods should be used for orientation, and the accurate solution obtained by separate calculation of each pipe with subsequent flow adjustments, as needed, by trial and error or interpolation.

It seems likely that pulsing could develop in tubes in parallel operated in the critical region that is more aggravated than that for single tubes (Sec. 8-33). Square or other sections not having a dip in the f vs. Re curve should be stable.

For flat wide channels in streamline flow, $\Delta P/w = (12/g_c)\bar{\nu} N/b^3 W$, where W is the width and b the spacing. Actual and equivalent values of $\bar{\nu} N/b^3 W$ are handled in the same way as $\bar{\nu} N/D^4$ for pipes. In turbulent flow $(2b = D_{eq})$ Eq. (8-51) becomes $\Sigma \Delta P_s = (w^2/g_c)\Sigma(f_F N/\bar{\rho} b^3 W^2)$ and Eq. (8-52) becomes $\Sigma \Delta P_s = (0.0529 w^{1.8}/g_c)\Sigma(\bar{\mu}^{0.2} N/\bar{\rho} b^3 W^{1.8})$. The groups in the summations are handled in the same way as for pipes. It is possible to treat by these direct methods cross sections of different shapes, fittings, orifices, losses, etc., all in the same network, provided they all have the same power of w in the expression for ΔP. Values of N_e from Table 8-9 can also be employed. In these cases, the numerical constants do not factor out of the expressions for ΔP and for w.

Another type of complex circuit includes a pump, or other equipment of characteristics different from pipes and fittings. This may most conveniently be solved graphically by plotting on the w vs. ΔP characteristic of the pump that of the rest of the circuit, known as the "system characteristic." The operating point is the intersection (see Fig. 8-20).

8-37. Liquid Flow with Thermal Buoyancy.*

Applying Eqs. (8-18) and (8-29) to a channel with heat generation q and liquid in steady one-dimensional upward or downward flow,

$$P_1 - P_2 - \frac{\rho_1 g(z_2 - z_1)}{g_c} = \Delta P_F - \frac{\rho_1 g \bar{\beta}_1}{g_c} \int_{z_1}^{z_2} (t - t_1) \, dz$$

$$= \frac{2 f_F w^2}{\rho S^2 g_c} \frac{N}{D_e} - \frac{\rho_1 g \bar{\beta}_1 q(z_2 - z_1)}{2 w C g_c} \qquad (8\text{-}53)$$

Subscripts 1 and 2 refer to inlet and outlet, respectively, and z is elevation. Usually the vertical distribution of heat generation is symmetrical about the horizontal midplane, which permitted replacement of the integral by $(t_2 - t_1) \, \Delta z/2 = q \, \Delta z/2wC$, where q is the rate of heat removal for the channel.† Evidently Eq. (8-53) can be used to compute w if ΔP is given, or vice versa. For a vertical channel, the ratio of the buoyancy and the friction terms on the right side equal

$$\frac{\rho \beta g (t_2 - t_1) N_e}{2 g_c \Delta P_F} = \frac{\rho^2 \beta g (t_2 - t_1) D_e}{4 f_F G^2} = \frac{\overline{Gr}}{2 \bar{f}_F \overline{Re}^2} = \frac{\overline{Gr}}{\overline{Ka}}$$

where $\overline{Gr}$ is the Grashof number employing $(t_2 - t_1)/2$ for the usual Δt term and all the dimensionless ratios are averages over the channel. In streamline flow, $f_F = 16/Re$; thus $\overline{Gr}/\overline{Ka}$ simplifies to $\overline{Gr}/(32 \, \overline{Re})$.

Channels in parallel with upward flow will have the left-hand side of Eq. (8-53) identical, therefore also the right side. Two limiting cases can occur. If $\overline{Gr}/\overline{Ka} \ll 1$, say <0.05, buoyancy is minor and the methods of Sec. 8-36 apply. If $\overline{Gr}/\overline{Ka} \gg 1$, say >20, friction is minor. The flow will then distribute itself so that all channels have substantially the same t_2, which can be computed from total w and q by a heat balance. This method is useful in power reactors in maintaining uniform temperatures and low thermal stresses, particularly in thermal shields, but requires relatively large channels. In the intermediate case, if $P_1 - P_2$ is given the w in each channel may be directly computed by Eq. (8-53) and then each t_2 calculated. If instead the total flow is given, the most convenient method of solution is to assume several reasonable values of $P_1 - P_2$, compute w for each channel by Eq. (8-53), and plot Σw against $P_1 - P_2$. The solution is at $\Sigma w = $ the given total

---

* Internal thermal buoyancy due to a definite heat generation in closed channels of large N/D is treated in this section and in Sec. 9-29. External thermal buoyancy, in large volumes of fluid, is treated in Chap. 9, Ref. 27, etc.

† Strictly, β should be multiplied by $1 + \dfrac{\Delta t_{\text{flow mean}} - \Delta t_{\text{space mean}}}{t_2 - t_1}$, as per Fig. 9-7 and Table 9-6. This is generally not done because of the uncertainties in $\Delta t_{\text{space mean}}$ and in ΔP_F, and the additional inconvenience.

flow, and the separate values of w and t_2 are then obtained. Differentiating Eq. (8-53) with respect to w yields the metastable downflow condition as $\overline{Gr}/\overline{Ka} = 2N_e/(z_1 - z_2)$, which equals 2 for a vertical channel. Downward flow should be regarded with caution* unless $\overline{Gr}/\overline{Ka} \ll 2$.

This may be illustrated by a simplified sodium-cooled reactor design for downflow. Assume 4-ft vertical tubes with an ID of 0.336 in. The top header is at 700°F and the bottom header at 1030°F. Rated capacity is to be reached at $w = 1$ lb/sec through each central tube. At this flow, Re is calculated to be 250,000 and f_F by Fig. 8-10 is 0.0040. Friction pressure drop by Eq. (8-29) is 1500 lb/sq ft at $w = 1$. The variation of $P_{bottom} - P_{top} = \Delta P_t$ with w is calculated by Eq. (8-53). Figure 8-14 shows ΔP_t at 100 per cent and at 10 per cent of rated power for low flows. The curves have been terminated at w_b, at which the exit Na would just reach boiling. For still lower flows ΔP_t could be computed by Sec. 8-40.

With either upflow or downflow, stable operation in parallel of tubes at full power is obtained above w_b. If in upflow the applied ΔP_t increases, accelerating the coolant stream, a higher w is soon reached at which the required ΔP_t (as given in Fig. 8-14) equals the new applied ΔP_t. In downflow w would decrease, but again an equilibrium value would be reached. From the proximity of the 10 per cent power curve, it is also seen that the curves do not change much in position with a variation in q at high flow. Thus, operation above w_b is stable in spite of reasonable differences or fluctuations in ΔP_t, q, or tube dimensions. For rated q and w, the criterion $\overline{Gr}/\overline{Ka}$ is found to be 0.007, corroborating that values much smaller than unity indicate stability in downflow.

If rated flow is maintained at lower powers, $t_2 - t_1$ decreases. Thus Gr and $\overline{Gr}/\overline{Ka}$ decrease still further and stability is increased. However, to decrease pumping power, temperature changes, and thermal stress it might be desired to maintain $t_2 - t_1$ constant by decreasing w proportionally with the power demand. At 10 per cent of rated power and flow $\overline{Gr}/\overline{Ka}$ becomes 0.42 and instability in downflow is more likely. The upflow curve in Fig. 8-14 at 10 per cent power shows the same stability as the rated power curves. The downflow curve at 10 per cent power shows at B a maximum ΔP_t of 96 lb/sq ft for $w = 0.088$ lb/sec. This is just above the ΔP_t of 95 required for the desired operating flow of 0.1 lb/sec. At any ΔP_t slightly below 96 there are apparently three possible operating flows. Operation at A (or at D) is stable in that a small increase in applied ΔP_t accelerates the upflow (or decelerates the downflow), the change in flow being in the direction that requires an increase in ΔP_t. The flow thus stabilizes at the new ΔP_t. However, operation at B or smaller downflow is metastable (unstable). An infinitesimal decrease in applied ΔP_t at C is an increase in $P_{top} - P_{bottom}$ or $-\Delta P_t$ and thus accelerates the downflow. But the increased downflow requires a higher ΔP_t or lower $-\Delta P_t$. Thus an additional excess in $-\Delta P_t$ is generated, and the downflow continues to accelerate past B, stabilizing near A. On the other hand, an infinitesimal increase in applied ΔP_t at C decreases the downflow, which soon reverses and stabilizes near D providing burnout does not occur in the process. An analysis of the flow reversal could be attempted by Sec. 8-38

---

* Despite this objection, downflow is frequently used. One advantage, in case personnel, instrumentation, sodium freeze seals, etc., are to be located on top, is that the top is cooler (also less radioactive). Also no lifting forces develop due to the coolant flow, simplifying the holding of the fuel elements in place.

and Chap. 9.   The dashed curves show the trend only roughly.   High coolant specific
heat and thermal conductivity and short tubes would decrease burnout tendency
during reversal.

It is clearly unsafe to operate at 10 per cent power with an average downflow of

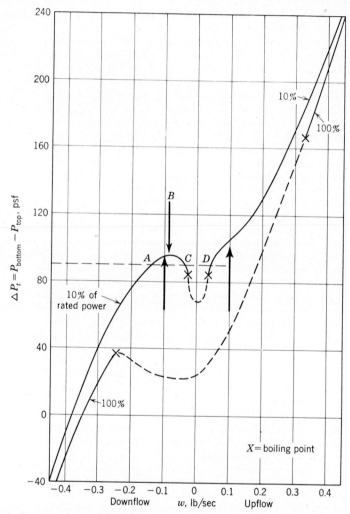

FIG. 8-14. Pressure drop with upflow and with downflow in a vertical reactor tube
with constant header temperatures and varying percentages of rated power.   Normal
flow is downward.

0.1 lb/sec, or at any lower power and rate.   Small fluctuations in operation of an
average tube might push it up to point B and subsequent burnout.   In addition,
other tubes would have a higher flow resistance or heat generation than average, and
thus a lower characteristic downflow curve, and reverse or burnout as 10 per cent
power and flow were being established.   If it is required to keep the header tempera-

tures constant in spite of power changes, it is possible to increase ΔP_F with nozzles or other flow resistances in series with the tubes. The stability criterion, $(\overline{\text{Gr}/\text{Ka}})$ $\Delta z/N_e$ in complete form, is evidently thereby reduced. Another procedure is not to operate the reactor below a safe power, possibly 20 to 25 per cent in this case, and discard the excess heat elsewhere. Alternatively, if at low power the header temperatures need not be constant, it is possible to use a safe minimum flow, possibly 0.2 lb/sec in this case, for all lower power rates.

The above analysis is only approximate at intermediate values of $\overline{\text{Gr}/\text{Ka}}$. One reason is that ΔP_F with appreciable superimposed buoyancy cannot be predicted exactly by the usual isothermal relations, or even by Sec. 8-18 or 8-25. This would be yet more so at high $\overline{\text{Gr}/\text{Ka}}$, but there ΔP_F is negligible. In addition, downflow in channels of small $(N/D)_e$, or nonuniform wall shear stress—as with small fuel elements in a large tube—will favor local circulation within a channel. This might cause burnout except at low values of $\overline{\text{Gr}/\text{Ka}}$. This situation could be analyzed theoretically by means of Eq. 8-18 and Sec. 8-31.

Prob. 8-16. (a) Show that w at the maximum of the downflow curve of Fig. 8-14 approximately equals $(\rho^2 S^2 \beta g D_e q \; \Delta z/8C_p N f_F)^{1/3}$. (b) Show that the lowest stable velocity in downflow when $t_2 - t_1$ is kept constant is approximately given by setting $d(P_1 - P_2)/dw$ from Eq. (8-53) equal to zero:

$$\frac{w_{crit}}{\rho S} = \bar{u}_{crit} = \left(\frac{\bar{\beta}_1 g (t_2 - t_1) D_e}{8 f_F} \frac{\Delta z}{N} \right)^{1/2}$$

(c) For a fixed flow downward through a nuclear reactor, show in the limiting case of no friction or velocity head in the tubes that the stable condition is downward flow in one tube and upward flow in all the others. Show also that the flow in each tube would be infinite.

If the heat source is horizontal instead of vertical, $t = t_2$ in Eq. (8-53), and the 2 in the denominator of the last term is eliminated.

Fluid in a large container with a lower horizontal heat source and an upper horizontal heat sink develops adjacent upward warmer and downward colder circulating currents. Theory shows that small "cellular" currents should start if $(\Delta z)^3 g \; \Delta \rho/k\mu$ exceeds 1700 for rigid source and sink, 1100 for one being nonrigid, and 660 for neither rigid. Actually, convection usually starts with larger "columnar" currents at a value of the ratio an order of magnitude lower [100]. Minimum dt/dz for convection in porous beds has also been studied [101]. Tests between parallel vertical insulating planes have shown currents of the order of $\frac{1}{3} \; \Delta z$ wide.

8-38. Unsteady-state Flow.
The calculation of unsteady-state fluid flow is of importance in evaluating temperatures and stresses in rapid start-up, load change, or scram of a nuclear reactor. Initial maintenance of the flow by the momentum of the stream and by any flywheel effects, and subsequently by thermal buoyancy, is also important or necessary in the event of sudden power or pump failure, to prevent fuel element burnout by the delayed heat generation (see Sec. 9-4). This characteristic is the "flow coast-down" curve.

The calculation can be carried out by equating the pressure drops around the flow loop which oppose the flow to those which maintain the flow. The few available data [104] on pressure drop during acceleration indicate that the correlations for steady-state flow through pipes and orifices apply in general with reasonable accuracy for the accelerations and decelerations likely to be encountered. Pipe friction ΔP_F by Eqs. (8-29) and (8-65) may be expressed by $R_F w^2$. N_e can be estimated by Table 8-9, though an actual flow test is more reliable for the usual short loops with close fittings. Similarly, the net pressure drop ΔP_o across orifices, valves, etc., can be expressed by $R_o w^2$, as per Eq. (8-65). Pressure recovery downstream should be subtracted. The pressure drop ΔP_d equivalent to the drag of a transverse vane, rod, pipe, axial pump rotor, etc., per Sec. 8-45 will equal $C_d \rho A \bar{u}^2/2g_c S = C_d A w^2/2\rho g_c S^3 = R_d w^2$. C_d can be multiplied by $(u^2)_{\mathrm{av}}/\bar{u}^2$ if this correction can be evaluated and is significant.

Inertia of the stream will tend to maintain the flow. For any conduit the axial momentum per unit length $= \int_0^S (u\rho/g_c)\, dS = w/g_c$. Accordingly, the total pressure rise ΔP_i due to deceleration $= -(N/Sg_c)dw/d\theta$ for each length N of constant cross section S.

The thermal buoyancy ΔP_B, due to the effective center of the reactor being Z below the level of the effective center of the heat sink, can be evaluated in several ways

$$\Delta P_B = \int_0^Z \frac{(t - t_0)\beta_0\rho_0 g}{g_c}\, dz = \frac{\beta_0\rho_0 g Z(t - t_0)_{\mathrm{av}}}{g_c} = \frac{\beta_0\rho_0 g Z\bar{q}}{g_c C\bar{w}} \quad (8\text{-}54)$$

The cold-leg temperature t_0 could be substantially constant after a scram if the secondary coolant flow to the heat sink (normally a boiler) is maintained at a constant temperature, since the heat generation falls rapidly to a low per cent of rated power (Fig. 9-1). The last expression applies over a time interval $\Delta\theta$ and a height Z short enough that q and w do not change greatly. The pressure in the direction of flow generated by the pump may be designated ΔP_P. The pressure balance may be written $\Delta P_P + \Delta P_B - \Sigma(N/Sg_c)\, dw/d\theta = (\Sigma R_F + \Sigma R_0 + \Sigma R_d)w^2$, and can be separated and integrated to

$$\theta_2 - \theta_1 = \Delta\theta = \left(\sum \frac{N}{Sg_c}\right) \int_{w_2}^{w_1} \frac{dw}{w^2 \Sigma R - \Delta P_P - \Delta P_B} \quad (8\text{-}55)$$

If q is substantially constant, as in a constant-power start-up or a pump failure without scram, the denominator can be calculated for various values of w and Eq. (8-55) graphically integrated between any desired w_1 and w_2. If most of the flow resistance is localized in one size of pipe (thus at the same Re), Eq. (8-55) can be generalized by replacing w by $(\mu S/D_e)$Re:

$$\frac{N_e\mu}{ND_e^2\rho}\,\Delta\theta = \frac{1}{2}\int_{\mathrm{Re}_2}^{\mathrm{Re}_1} \frac{d\,\mathrm{Re}}{f_F \mathrm{Re}^2 - E/\mathrm{Re}} \quad (8\text{-}56a)$$

where $E = \rho^2 \beta g D_e^4 Zq / 2CS\mu^3 N_e$. Equation (8-56a) has been evaluated graphically for smooth pipes and is given in Fig. 8-15 from $\mathrm{Re}_1 = 10^7$ to lower Re values for several values of E. Limiting values are in Table 8-10. If the pipe is rough, or f is affected by acceleration, the correct $\Delta\theta$ in Eq. (8-56a) is obtained by multiplying E and the integral from Table 8-10 both by the ratio $(\bar{f}_{\mathrm{smooth}}/f_{\mathrm{actual}})_{\mathrm{av}}$.

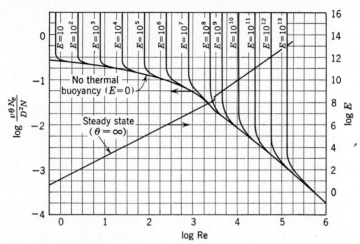

FIG. 8-15. Generalized solution for steady- and unsteady-state natural convection flow in a pipe loop with thermal buoyancy.

For a horizontal or isothermal loop $E = 0$, and Eq. (8-56a) can be directly integrated for either turbulent or streamline flow, yielding, respectively,

$$\Delta\theta = \frac{ND_e\rho S}{2N_e\bar{f}_F}\left(\frac{1}{w_2} - \frac{1}{w_1}\right) \quad \text{or} \quad \frac{D^2\rho N}{32\mu N_e}\ln\frac{w_1}{w_2} \qquad (8\text{-}56b)$$

TABLE 8-10. LIMITING CASES FOR "THERMAL-SIPHON" FLOW IN A LOOP

$E = 0$:	Re_θ	10^7	10^6	10^5	10^4	3000	1000	100	10	1	0.1
	$\dfrac{\bar{\nu}\theta N_e}{D^2N}$	0	0.000176	0.00134	0.00842	0.0209*	0.0552*	0.127*	0.199*	0.271*	0.343*

$\theta = \infty$:	E	10^{13}	10^{12}	10^{11}	10^{10}	10^9	10^8†	10^7†	10^6†	10^5†	10^4†
	Re_∞	133,000	58,200	25,400	11,000	4720	2500	790	250	79	25

$$* \frac{\bar{\nu}\theta N_e}{D^2N} = 0.0209 + 0.072\log\frac{3000}{\mathrm{Re}}$$
$$\dagger\ E = 16\ \mathrm{Re}_\infty{}^2$$

The second (streamline) expression holds only for a pipe.

In a normal operational power transient, q would vary more widely. In that case Eq. (8-55) or (8-56a) might be employed over several successive intervals, estimating in advance for each interval the average value of ΣR, ΔP_P, $w \Delta P_B$, and E and later correcting the computed $\Delta \theta$ for the amount by which the estimate was off, if desired, before going on to the next interval.

In a scram or power surge, q varies very rapidly and widely. Here numerical integration using many small increments of θ and w is employed, either manually or with a high-speed digital computer. The increments employed could be small enough that average conditions over the increment differed only slightly from initial conditions. In this case Eq. (8-55) becomes

$$w_2 = w_1 - \frac{\Delta \theta}{\Sigma(N/Sg_c)} [w_1{}^2 (\Sigma R)_1 - (\Delta P_P)_1 - (\Delta P_B)_1] \quad (8\text{-}55a)$$

From w_1 and the coefficients in Eq. (8-55a) at w_1 and θ_1, w_2 is calculated, the coefficients are modified as needed for w_2 and θ_2, the calculation repeated to yield w_3, and thereafter as many times in succession as required. If the calculated w oscillates considerably, a smaller $\Delta \theta$ should be used.

The emergency cooling characteristics of a system which is scrammed due to power or pump failure can be shown by a plot of q/wC, the instantaneous temperature rise across the reactor, vs. θ. For alternative piping or equipment dimensions or layouts, that showing the lowest instantaneous, or maintained, maximum q/wC would be the best for emergency cooling. Other things being equal, the design with the highest $ZD_e{}^4/SN_e$ will run the coolest after ΔP_i has become negligible. Similarly, that with the highest $ND_e{}^2/N_e$ or $\Sigma(N/Sg_c)/\Sigma R$ will take the longest for the flow to slow down. If one alternative design should excel in both respects it is, without further calculations, evidently the best for emergency cooling.

As is always the case with finite increment calculations, the average per cent error due to using initial conditions instead of mean conditions over each interval is approximately half the average per cent change per interval. This correction increases the time required to drop from w_1 to w_n by roughly $[200 \ln (w_1/w_n)]/(n - 1)$ per cent.

Prob. 8-17. Bismuth at 500°C flows at $\bar{u} = 40$ ft/sec through an idealized 12-in. ID closed path consisting of two straight tubes 20 ft long and two 180° return bends of 4-ft radius, with a linear electromagnetic pump that offers no obstruction. If pump power suddenly fails, how long will it take for the velocity to fall to 4 and to 0.4 fps? The loop is horizontal.

8-39. Flow of Slurries (Bingham Bodies). Slurries may be useful in nuclear power on account of poor solubility of a fuel or fertile element in an otherwise desirable fluid medium, provided settling out does not occur or can be controlled. They may also offer processing advantages over a solution or a stationary solid.

Direct experimental determination of the flow characteristic curve (Fig. 8-3) of a non-Newtonian fluid can in general be carried out only in a pipe, and with streamline flow. This is the only shape (besides the infinitely wide flat duct) for which the wall shear stress is uniform, and which thus yields a single point on the curve. A complete curve is obtainable from a series of tests at different flows. For each flow test, the volumetric flow Q, pressure gradient dP/dN, and radius R are known. The wall shear stress τ_w is $(R/2)dP/dN$. The wall velocity gradient may be computed by the Rabinowitsch-Mooney equation [73] from the slope of the curve of Q vs. τ_w.

$$\left(\frac{du}{dy}\right)_w = \frac{3Q + \tau_w(dQ/d\tau_w)}{\pi R^3} \tag{8-57}$$

Each pair of values of τ_w and $(du/dy)_w$ is a point on Fig. 8-3.

Prob. 8-18. (a) Show that the Rabinowitsch-Mooney equation can be derived as follows: $Q = \displaystyle\int_0^{u_{max}} \pi r^2 \, du = \int_R^0 \pi r^2(du/dr) \, dr$, changing variable from r to τ by means of $r = \tau R/\tau_w$ (but leaving the term du/dr unchanged), and differentiating with respect to τ_w. (b) Show that the equation for flow in a duct of wide parallel walls and thickness b (and thus approximately for thin concentric annuli) comparable to the Rabinowitsch-Mooney equation is $(du/dy)_w = 2 \, [2Q' + \tau_w(dQ'/d\tau_w)]/b^2$, where Q' is the volumetric flow per unit width.

If the graphical (or analytical) relation between du/dy and τ is known for any fluid, it is possible to obtain $\bar{u}$ for a specific value of dP/dN for any channel for which the distribution of τ is known, such as a pipe or a wide flat duct. The double graphical (or analytical) integration is

$$\bar{u} = \frac{2}{R^2} \int_0^R \left(\int_R^r \frac{du}{dy} \, dr \right) r \, dr = \frac{2}{b} \int_0^{b/2} \int_0^y \frac{du}{dy} \, dy \, dy \tag{8-58}$$

for the pipe, and the flat duct of thickness b, respectively.

Conversely, if any relation between du/dy and τ is assumed for a given fluid, it is possible to integrate over any viscometer or channel for which the distribution of τ is known, and check the assumed relation by the extent of agreement between the integrated result and actual tests.

For instance, the relation for a Bingham body in streamline flow is $\tau = \tau_y + \eta(du/dy)/g_c$. Including slip as due to a film of viscosity μ and small thickness δ, the principal cases are:

1. Wide thin flat duct of spacing b:

$$\bar{u} = \tau_w g_c \left\{ \frac{b}{6\eta} \left[1 - \frac{3}{2} \frac{\tau_y}{\tau_w} + \frac{1}{2} \left(\frac{\tau_y}{\tau_w} \right)^3 \right] + \frac{\delta}{\mu} \right\} \qquad \text{with } \tau_w = \frac{b}{2} \frac{dP}{dN} \tag{8-59}$$

Plotting $\bar{u}/b$ vs. τ_w, curves at different b will coincide if δ is negligible, their asymptote at high τ_w having a slope of $g_c/6\eta$ and an intercept at $\bar{u} = 0$ of $3\tau_y/2$.

2. Pipe:

$$\bar{u} = \tau_w g_c \left\{ \frac{D}{8\eta} \left[1 - \frac{4}{3} \frac{\tau_y}{\tau_w} + \frac{1}{3} \left(\frac{\tau_y}{\tau_w} \right)^4 \right] + \frac{\delta}{\mu} \right\} \qquad \text{with } \tau_w = \frac{D}{4} \frac{dP}{dN} \tag{8-60}$$

Plotting $\bar{u}/D$ vs. τ_w, curves at different D will agree if δ is negligible, their asymptote at high τ_w having a slope of $g_c/8\eta$ and an intercept at $\bar{u} = 0$ of $4\tau_y/3$.

3. Concentric cylinder rotational viscometer [74] with $\tau_0 > \tau_y$:

$$\frac{\omega}{g_c} = \frac{T'\delta}{2\pi L\mu}\left(\frac{1}{R_i{}^3} + \frac{1}{R_0{}^3}\right) \pm \left[\frac{T'}{4\pi L\eta}\left(\frac{1}{R_i{}^2} - \frac{1}{R_0{}^2}\right) - \frac{\tau_y}{\eta}\ln\frac{R_0}{R_i}\right] \qquad (8\text{-}61)$$

where L = length of cylinders
R_0 and R_i = radii of cylinders
  ω = angular speed of moving cylinder, radians
  T' = required torque, usually measured at stationary cylinder

The $+$ sign applies when the inner cylinder is stationary and the $-$ sign when the outer is stationary. T' is usually obtained from the twist of an elastic torsion wire, calibrated by the above equation with a Newtonian fluid of known μ. Neglecting δ, if ϕ is the twist in any units, and subscript c refers to the average of calibration runs: $\eta\omega = \pm(\mu\omega/\phi)_c\phi \mp g_c\tau_y \ln(R_0/R_i)$. Plotting ω vs. ϕ yields a straight line of slope $(\mu\omega/\phi)_c/\eta$; τ_y is obtainable from the intercept. Turbulence can occur at high speed and must be avoided for reliable results.

The three equations above yield the Newtonian fluid forms if δ and τ_y are equated to zero. Formulas can be derived which correct b, D, or R_i and R_0 for δ, in case it is appreciable in magnitude compared to them. Vand [19] reports that $\delta = 0.54D_p$ for spheres of diameter D_p, which may hold approximately for all shapes. Thus tests in a capillary may be unreliable because of the much greater relative effect of wall slip than in larger tubes. Accordingly, accurate pressure-drop data in small channels require flow tests with the same channel size and wall roughness.

It is evident that these equations can be used to determine whether a slurry is a Bingham body, and if so to find η and τ_w. Subsequently they can be used in reverse, to compute flow or pressure drop for such a fluid in streamline flow.

In all these cases turbulence develops at high speeds. In channels the lower limit of the critical Reynolds number range is usually taken as 2100. With a Bingham body the rigidity η has been used for the μ in Re [75], and with a pseudoplastic the limiting μ at $\tau = 0$ [76]. The upper limit of the critical range is near Re = 3000.

In the turbulent flow of suspensions Fig. 8-10 has been shown to hold, with the effective viscosities at infinite shear usually being employed for the μ in Re, namely, η for Bingham bodies and μ_∞ for pseudoplastics. However, extreme precision is not important in viscosity, on account of its small effect on f and ΔP. For Newtonian suspensions the density of the slurry and the viscosity of the pure medium seem best for computing Re, and thence ΔP by Fig. 8-10 [20, 75].

8-40. Friction Pressure Drop in Mixed Flow of Two Fluid Phases. In general, high gas or vapor flow rates tend to carry a coexistent liquid phase along as slugs in a small pipe, or as a film on the wall of a large pipe, whereas high liquid rates tend to carry the gas or vapor as bubbles. When both rates are low, fluids in a horizontal pipe stratify [77].

Surprisingly, the friction pressure gradient in such varied flows near atmospheric pressure correlates with an average deviation of some 25 per cent [78] against the friction pressure gradients of the two phases flowing separately, each at its own flow rate.   To find the two-phase

TABLE 8-11. FRICTION PRESSURE GRADIENT CORRELATION FOR FLOW
OF LIQUID-GAS MIXTURES*†

log $(\Delta P_L/\Delta P_G)$	log $(\Delta P_{2P}/\Delta P_L)$							Volume fraction liquid, R_L		
	Liquid and gas turbulent		P_c	P_a	Liquid viscous, gas turbulent	Liquid turbulent, gas viscous	Liquid and gas viscous	Up	P_a	P_c
	Up	Down								
−4.0	.....	.....	4.000	4.214	4.158	4.098	4.042	.....	....	0.005
−3.5	.....	.....	3.505	3.762	3.705	3.624	3.556	.....	....	0.010
−3.0	.....	.....	3.012	3.340	3.249	3.167	3.084	0.010	....	0.019
−2.5	.....	.....	2.526	2.930	2.800	2.732	2.630	0.019	0.03	0.036
−2.0	.....	.....	2.052	2.534	2.364	2.323	2.187	0.034	0.05	0.067
−1.5	.....	.....	1.598	2.172	1.978	1.954	1.774	0.06	0.08	0.12
−1.0	.....	2.042	1.180	1.832	1.635	1.616	1.404	0.11	0.12	0.21
−0.5	1.790	1.727	0.817	1.519	1.328	1.321	1.086	0.17	0.17	0.34
0.0	1.389	1.473	0.527	1.246	1.083	1.083	0.833	0.30	0.23	0.50
0.5	1.050	1.267	0.317	1.027	0.879	0.879	0.663	0.45	0.30	0.66
1.0	0.771	1.088	0.181	0.831	0.696	0.723	0.537	0.60	0.37	0.79
1.5	0.567	0.925	0.099	0.651	0.534	0.580	0.440	0.81	0.45	0.88
2.0	.....	.....	0.0529	0.486	0.403	0.440	0.352	0.91	0.53	0.93
2.5	.....	.....	0.0279	0.366	0.311	0.338	0.281	.....	0.63	0.96
3.0	.....	.....	0.0146	0.262	0.227	0.236	0.219	.....	0.72	0.98
3.5	.....	.....	0.0076	0.169	0.159	0.159	0.159	.....	0.81	0.99
4.0	.....	.....	0.0040	0.091	0.091	0.091	0.091	.....	0.90	0.995

* R. W. Lockhart and R. C. Martinelli, *Chem. Eng. Progr.*, **45**:39 (1949); R. C. Martinelli and D. B. Nelson, *Trans. ASME*, **70**:695 (1948).

† Values are for pressures near atmospheric (P_a) and horizontal flow unless marked otherwise.

friction pressure drop ΔP_{2P}, Re is first calculated for each fluid separately. Under 1000 is assumed streamline and over 2000 turbulent. ΔP is then computed from Fig. 8-10 for each fluid alone and Table 8-11 entered with the ratio $\Delta P_L/\Delta P_G$ at the proper column.   The corresponding $\Delta P_{2P}/\Delta P_L$ is read off or interpolated, and multiplied by ΔP_L to give ΔP_{2P}.   The fraction by volume of liquid and of gas also correlate reasonably well.   That for the liquid R_L is listed in the table; that for the gas is $1 - R_L$.   The columns headed P_c are for liquid and vapor at the critical pressure [98]. For intermediate pressures log $(\Delta P_{2P}/\Delta P_L)$ may be interpolated linearly with P. The column headed "Down" is for downward flow of boiling water in an $\frac{1}{8}$-in. annulus [105], and the columns headed "Up" for upward flow in a 1-in. tube (Chap. 9, Ref. 26).

An alternative method of predicting the friction pressure drop employs Fig. 8-10 and considers the two-phase mixture as a single fluid. Specific volume and fluidity are assumed additive on a mass-flow-rate basis. A straight-line plot of $1/\mu$ vs. fraction vapor x does not change rapidly with pressure above the boiling point. The former method is more accurate, but the latter is more convenient for repeated calculations.

Slow vertical two-phase flow is of importance in computing the buoyancy pressure ΔP_B in natural convection boilers and boiling reactors and in estimating the volumes of the phases for designing liquid expansion tanks, computing moderation or shielding, etc. Above a reduced pressure P_r of about 0.7 the vapor and liquid phase densities are close enough that their natural convection velocities are substantially equal [106]. As low as $P_r = 0.5$ the relative velocity is still small compared to a circulation velocity of several feet per second, so that the circulation rate and liquid volume can be found by a simple trial-and-error calculation. A liquid circulation rate is assumed, the fractional volume of vapor and ΔP_B are computed from the heat to be dissipated, and then the opposing friction pressure drop in the loop, calculated by the latter method just above, and acceleration pressure drop (Sec. 8-41) added to see if they equal ΔP_B (see also Sec. 9-17).

In wall or "local boiling" of subcooled liquid coolant (Sec. 9-18) the pressure gradient exceeds that for nonboiling at the same heat and coolant flows. However, the expansion and therefore pressure gradient are less than for "bulk boiling" at the saturation temperature. In a $\frac{3}{8}$-in. ID pipe the local to nonboiling pressure gradient ratio for water at 30 to 85 psig [102] is

$$\cosh\left[\left(4.6 \times 10^{-6}\frac{q}{A} + 1.2\right)\frac{N}{N_t}\right] = \cosh\frac{aN}{N_t}$$

where q/A is in Btu per hour per square foot, N is the distance downstream from the incidence of local boiling, and N_t is the total length that would be required in local boiling before the liquid reached the boiling point. Integrating over the local boiling length, the total pressure drop ΔP_{LB} at constant q/A is obtained as the nonboiling pressure gradient multiplied by $(N_t/a)(\sinh aN/N_t)$. The point of incidence of local boiling is obtained as in Fig. 9-12. For other channels for which the local boiling h is known (Sec. 9-18) ΔP_{LB} can be approximated as the nonboiling ΔP_{NB} multiplied by $(h_{LB}/h_{NB})^{2.3}$, the same power dependance as in nonboiling flow in pipes [107]. The vapor volume in subcooled boiling at high circulations is usually negligible; the average density of local-boiling water has been determined for evaluation as a moderator [103], and decreases very little.   Local-boiling results cannot as yet be extrapolated safely beyond test conditions.

8-41. Pressure Drop in a Boiling Tube.   Bulk boiling of the coolant in a nuclear reactor has several possible disadvantages, including fluctuation in quantity of moderator or fuel and in wall temperature and thermal stress, poor flow stability, liquid hammer, etc.   However, under certain conditions boiling is feasible, and in addition it is necessary to consider it as a possible emergency condition in a nonboiling reactor due to accidental decrease in flow or increase in heat generation.   In boiling, x, P, t, V, and Q vary along the channel, and a stepwise calculation is generally desirable.   Going downstream* over an increment of length from m to n with $dw = 0$, Eq. (8-20) may be written

$$E_n = E_m + \Delta Q_{m/n} = \frac{z_n g}{g_c} + h_n + \frac{G^2 \alpha}{2 g_c} V_n{}^2 = \frac{z_n g}{g_c} + h_n' + x_n h_n''$$
$$+ \frac{G^2 \alpha}{2 g_c} (V_n' + x_n V_n'')^2 \quad (8\text{-}62)$$

where h' and V' are the enthalpy and specific volume of the saturated liquid and h'' and V'' are the increases on complete evaporation.   It is desirable that $G^2 \alpha \, \Delta V^2 / 2 g_c$ be small compared to E.   Thus smaller length increments are preferable near the outlet.   One increment may be adequate for the nonboiling portion.

A chart of h vs. V for the two-phase region (Fig. 8-16) is drawn by joining the points for each saturated phase at a given P with straight lines.   These isobars are then subdivided into equal fractions of their lengths, and curves of constant x drawn.   The following procedure is followed:

1. E_n is evaluated as $E_m + \Delta Q_{m/n}$.
2. V_n is estimated (somewhat $> V_m$).
3. h_n is calculated by Eq. (8-62) and x_n read from the chart.
4. $\overline{Re}_{m/n}$ is computed as $D_e G[(1/\mu)' + \bar{x}_{m/n}(1/\mu)'']$, where prime and double prime are defined as above and $\bar{x}_{m/n}$ is the average of x_n and x_m.
5. f_F is read from Fig. 8-10.
6. P_n is calculated by Eq. (8-35).
7. The chart is consulted to see if the estimated V_n and computed h_n and P_n agree.   If not, a new V_n is estimated and steps 3 through 7 repeated.
8. The next increment is started.

* If the "sonic velocity" is reached at the outlet or at an earlier restriction, or outlet rather than inlet conditions are given, one works upstream from the outlet by the same sequence of steps.   See Sec. 8-42 and Prob. 8-21.

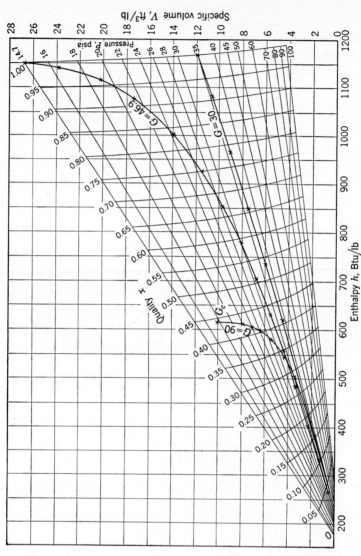

FIG. 8-16. Thermodynamic properties of wet steam. The curves of constant G (pounds per second per square foot) are for Prob. 8-21.

334

It is evident that f will not change rapidly, so that Re and f_F will not ordinarily need to be determined for every increment.

Use of Table 8-11 replaces steps 4, 5, and 6. If many intervals with a significant kinetic-energy term are to be computed it may be advantageous to replace step 3 by a graphical solution. For this purpose the h-V plot is best replotted as h vs. V^2. Equation (8-62) may be summed to give $E_n - z_n g/g_c = E_i + Q_{i/n} - z_n g/g_c = h_n + (G^2 \alpha/2g_c) V_n^2$. Thus each location in the channel must fall on a specific straight line on the h vs. V^2 plot, which gives the trial value of h_n directly at its intersection with the trial value of V_n. For any G the lines will have the slope $-2g_c/G^2\alpha$ and will be spaced $h_n - h_m = z_m - z_n + \Delta Q_{n/m}$ at constant V. The lines for each G can be drawn on a transparent graph-paper overlay, if preferred.

Prob. 8-19. Show how the V-h plot may be avoided, and time probably saved, if only one G is desired.

Solution. x_n may be computed from Eq. (8-62) by

$$x_n = \frac{[B_n^2 - h_n' - (AV_n')^2 + E_n - zg/g_c]^{1/2} - B_n}{AV_n''} \tag{8-63}$$

where $A = G^2\alpha/2g_c$ and $B_n = h_n''/2AV_n'' + V_n'$. For a given G, the quantities AV'', B, and $B^2 - h' - (AV')^2$ can be plotted against P over the expected range of P in two-phase flow. The following procedure is employed:

1. Evaluate E_n as before, assume P_n, and compute x_n by Eq. (8-63).
2. Compute V_n and $V_{m/n}$, substitute into Eq. (8-35) and calculate P_n.
3. Check this calculated P_n against the assumed P_n.

Equation (8-62) assumes "fog" flow, namely, that the liquid and vapor phases are at the same temperature and velocity at any one location. However, the R_L values of Table 8-11 generally indicate unequal velocities (the vapor higher) or "slip" flow. Slip flow could also be employed in the calculation by increments. However, the fog-flow calculation is simpler and agrees well with experimental data.

A quick approximate solution for over-all pressure drop when it is small compared to the absolute pressure has been developed for q linear in tube length [98]. ΔP_F is obtained by integration of the data in Table 8-11. The acceleration ΔP for fog or slip flow is computed separately by Eq. (8-22) and added to ΔP_F. The predicted ΔP for fog and for slip flow differ appreciably; actual runs usually show an intermediate ΔP closer to fog flow [79]. The over-all equations are much simpler but less accurate than the incremental method with fog flow, and do not yield the temperature and pressure distribution along the channel, which are usually desired.

The pressure-drop characteristics of a boiling liquid in turbulent flow in a heated channel are conveniently shown in a plot of P vs. w^2, as in Fig. 8-17a. Assume that P_o in the outlet header is fixed, as when many pipes feed into a common header. At high flow with constant heat generation a decrease in w raises the average fluid temperature and decreases the average viscosity. Thus Re and particularly f_F do not change much, and a straight line toward P_o may be plotted through one computed point of inlet pressure P_i vs. w^2. If preferred for maximum accuracy, however, the curve may be drawn through several computed points. Decreasing w further, the exit liquid reaches the boiling point at w_B in Fig. 8-17a. Stepwise calculation backward from the outlet at lower values of w shows that P_i decreases to a minimum slightly below w_B, then rises. In actual practice, either because of local boiling or because boiling does not occur simultaneously all across the channel, a flatter bend is obtained, as shown by the dots. A maximum P_i is reached

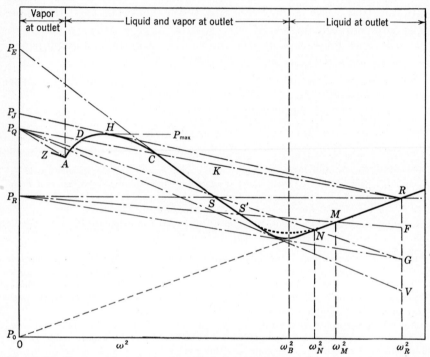

FIG. 8-17a. Pressure drop in a tube with heat generation, and its protection with an inlet series nozzle. Outlet header pressure constant at P_o.

at about 25 to 50 per cent of the liquid vaporized. Further decrease in w again decreases P_i. A sharp bend in the curve is obtained at A, when 100 per cent vaporization is reached at the outlet. As w is decreased further, the vapor becomes more and more superheated. Finally the heated tube or the fuel element burns out, i.e., at Z. Burnout of steel tubes is usually at above 75 per cent vaporization at high power levels and not until considerable superheat at low power levels [79], though lower-melting metals and fuel elements might burn out sooner. In either case, burnout flow rate depends on pressure, heat velocity g/A, coolant flow, and other factors (see Sec. 9-19).   It may even occur before w_B is reached, i.e., in "local boiling" (see Sec. 9-18), if the heat velocity is high enough.

8-42. Maximum Flow Rate.   A compressible fluid in a pipe can nowhere exceed the mass velocity G_c at which all the energy obtained from a differential pressure drop is used up in the corresponding increase in kinetic energy of the stream.*   At a point at which G_c is reached, the expansion is isentropic, as it will take place so rapidly that heat transfer and friction may be neglected.   On further expansion the velocity would rise even more, which is not possible in a pipe of constant cross section; thus G_c is reached, if at all, at the outlet of the pipe [80].   If in a downstream stepwise calculation at a given P_i, G, and h_i with unknown P_o, as per Sec. 8-40, G_c for the local values of P and x is reached before the pipe outlet, too low a P_i was employed.   The calculation should be repeated starting from the downstream end at the necessary P_o to yield G_c = the desired G.   If there is a constriction in the pipe upstream of the outlet, the local maximum velocity may be reached there as well as, or instead of, in the outlet.

If G is a maximum at the "critical" conditions, its derivative with respect to any other variable = 0.   Thus, at G_c,

$$\frac{dG}{d\bar{u}} = 0 = \frac{d(\bar{u}\rho)}{d\bar{u}} = \rho + \bar{u}\frac{d\rho}{d\bar{u}} \quad \text{and} \quad \frac{d\bar{u}}{\bar{u}} = -\frac{d\rho}{\rho}$$

The isentropic energy balance is $\bar{u}d\bar{u}/g_c = -dP/\rho$.   Eliminating $d\bar{u}$, at G_c,

$$\bar{u}_c = \left(\frac{g_c dP}{d\rho}\right)_s^{\frac{1}{2}}$$

$$G_c = \rho\left(\frac{g_c dP}{d\rho}\right)_s^{\frac{1}{2}} = \left(\frac{-g_c dP}{dV}\right)_s^{\frac{1}{2}}$$

(8-64)

These relations are familiar as yielding the velocity of sound for a homogeneous fluid.

Prob. 8-20.   Show theoretically that "sonic" velocity will not be encountered at the outlet of a liquid-cooled tube at any lower flow rate if it is not present when the stream is just 100 per cent vaporized.   (Similarly, sonic velocity cannot appear at the outlet of a gas-cooled tube on decreasing the flow.)

To apply the above considerations to two-phase mixtures it is simplest to assume fog flow (see Sec. 8-40).   This approach has received some experimental support for steam-water mixtures [81].   A rough table of G_c vs. P and x can be quickly prepared from a Mollier chart, or accurate values can be computed from thermodynamic tables for the substance. At a given P and x the corresponding entropy is read from the chart or

---

* By Eq. (8-21), neglecting dz, dW, and dF.

computed from the table. For a small (say 1 psi) increase and/or decrease in P, the corresponding values of x at constant s and of the corresponding specific volume V are read or computed.* The value of $(\Delta P/\Delta V)_s$ thus obtained is substituted for the derivative in Eq. (8-64). Values for water are given in Table 8-12. Each point also has listed the corresponding total† energy $E = h + \bar{u}_c^2/2g_c$ above standard temperature (32°F) and zero velocity. Interpolation for other pressures is facilitated by computing G_c/P, which is almost constant at high x.

TABLE 8-12. CALCULATED MAXIMUM MASS VELOCITY G_c IN POUNDS PER SECOND PER SQUARE FOOT FOR STEAM-WATER MIXTURES IN A TUBE, AND ENERGY E IN BTU PER POUND

Quality x, vapor fraction	Pressure, psia					
	14.7		50		200	
	G_c	E	G_c	E	G_c	E
0.995	53.8	1186.6	177.0	1214.4	680.9	1242.1
0.75	61.4	938.2	201.2	976.1	761.3	1021.8
0.50	73.7	684.7	240.7	733.2	918.2	799.1
0.25	98.9	431.4	317.5	490.3	1151.0	575.3
0.10	136.9	279.8	430.7	345.3	1522.6	442.6
0.005	221.0	184.9	609.1	254.7	2153.3	359.7

Complete equilibrium between the two phases is usually not attained, particularly when x is small and the pipe size large. For instance, experience indicates that 221 lb/(ft²)(sec) = 3.7 fps of boiling water at atmospheric pressure with a trace of steam can be readily exceeded in a pipe. Preliminary results show that in ¼-in. to 1-in. pipes [99] the ratio of observed to theoretical G_c for fog flow is roughly $2/(x + 1)$, though rising much higher for x under 2 per cent. For a ⅛-in. annulus it is similar at low x, but falls rapidly to about 1.1 at $x = 0.1$. Nevertheless, reasonable agreement between calculated and observed values of P_1 is obtained, so that the general outline of the curves in Fig. 8-17a is still obtainable by the above theoretical methods. For all practical purposes, data for H_2O (Fig. 8-16 and Table 8-12, etc.) are valid also for D_2O for pressure-drop calculations, except that the density is some 10 per cent higher.

* $V = V' + V''(s - s')/s''$, where s = entropy of the desired mixture and V' and s' are for the liquid phase and V'' and s'' are the changes on vaporization at the higher or lower pressure.

† Neglecting potential energy.

Prob. 8-21*a*.   A 0.5-in. ID tube 15 ft long is to dissipate 70.93 Btu/sec to a stream of water which enters at 70°F and leaves into a manifold at 14.7 psia.   Compute the required inlet pressure for the exit stream ranging from subcooled water to super-heated steam.   Neglect local boiling.

Method.   From Eq. (8-35) the pressure drop in psi across each increment of length N ft is $\Delta P'_F = 0.0002372G^2\,\Delta V + 0.005176G^2N(V_m + V_n)f_F$, where G is in pounds per second per square foot and V is in cubic feet per pound.   In the boiling and super-heat regions h_i is taken at 38 Btu/lb (see Prob. 8-5) and h computed at each foot of N by Eq. (8-62).   Then, working backstream from the outlet by the method of Sec. 8-40 gives the inlet pressure.   The additive fluidity method is used for Re and f_F.   Whenever the desired G is higher than G_c by Table 8-12 for 14.7 psia, the exit P is found graphically.   An auxiliary plot is made of P vs. x for lines of constant G for the data

TABLE 8-13. INLET AND OUTLET CONDITIONS FOR WATERCOOLED TUBE WITH CONSTANT HEAT GENERATION AT SEVERAL FLOW RATES (PROB. 8-21*a*)

G, lb/(sec)(ft²)	w^2, (lb/sec)²	P_i, psi	t_o, °F	P_o, psi	x_o, fraction vapor
30	0.00166	39.2	1518	14.7	Superheated
46.9	0.00406	36.9	212	14.7	1.000
60	0.00665	40.5	212	14.7*	0.750
90	0.01495	45.7	275.1	17.5†	0.446
120	0.0266	47.4	277.4	19.7†	0.288
150	0.0415	47.3	277.1	21.2†	0.196
180	0.0599	44.8	273.5	22.1†	0.136
240	0.1068	38.4	263.5	23.1†	0.050
300	0.1661	29.1	246.2	23.5†	0.010
367.5	0.250	16.5	212	14.7*	0
372.7	0.257	16.5	210	14.7	Subcooled
401.4	0.298	16.8	200	14.7	Subcooled
434.9	0.350	17.2	190	14.7	Subcooled
474.4	0.416	17.6	180	14.7	Subcooled
521.9	0.504	18.2	170	14.7	Subcooled
572.5	0.622	18.8	160	14.7	Subcooled
652.3	0.786	19.9	150	14.7	Subcooled

* Threshold of sonic velocity at outlet.

† Sonic velocity at outlet; $P_o = P_c$ and $x_o = x_c$.

of the problem and of P_c vs. x_c for lines of constant G_c by Sec. 8-42.   The intersections at which $G = G_c$ give the necessary P_c and x_c.   Typical stepwise results at 1-ft intervals are shown on Fig. 8-16 for a sonic, a subsonic, and the two-phase region of a superheated outlet stream run.   The results are in Table 8-13.   The range of x_o from 0 to 0.75 is found to be sonic at the outlet.   The values of P_i vs. w^2 are plotted in Fig. 8-17*b*.

Prob. 8-21*b*.   A horizontal 0.375-in. ID tube 5 ft long is to dissipate heat at a uniform rate $q/A = 250,000$ Btu/(hr)(sq ft).   The inlet temperature is to be 70°F and the outlet pressure 80 psig.   Compute the necessary pressure drop when local boiling occurs in the tube.

Solution.   Using Eq. (9-44*a*) to give the nonboiling Δt_f, Eq. (9-51*d*) to indicate start of local boiling, Fig. 8-10 for the nonboiling ΔP, and Sec. 8-40 for the local boiling

ΔP yields the following:

w, lb/hr	N_{LB}, ft	ΔP_{LB}, psi	ΔP_{total}, psi	Comments
484.0	2.44	0.597	0.713	Incipient bulk boiling at outlet
580.8	1.35	0.262	0.502	
629.2	0.79	0.106	0.429	
677.6	0.33	0.012	0.420	
726.0	0	0	0.494	
968.0	0	0	0.836	

Evidently from the rise in ΔP with decreasing w, local boiling in parallel tubes can be unstable in the same way as is bulk boiling (see Sec. 8-43). Additional flow resistance, such as pipe connections or a nozzle, will cause the instability to decrease or disappear. Since the vapor volume in local boiling is negligible, the same instability would be present in upflow or downflow.

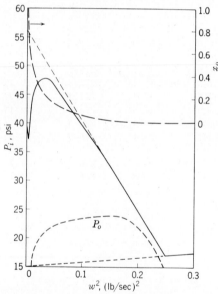

FIG. 8-17b. Stability and outlet conditions in a boiling tube (Prob. 8-21a).

8-43. Protection with Orifices against Burnout.

A heated tube with a fixed outlet pressure and liquid coolant in its boiling range has been shown to have a P_i vs. w^2 characteristic as in Fig. 8-17a. A somewhat flatter curve would be obtained with a gaseous coolant. The full inlet header pressure, say P_R, may be available at the tube inlet and a flow of w_R would result. However, if there is an initial fixed orifice or nozzle, a flow-regulating valve, a length of pipe, a temporary accidental obstruction, or merely pressure loss at the inlet of the pipe, P_i will be less than P_R. In this case, P_i to the heated tube when computed from the inlet header will be given by a substantially straight line of negative slope, such as P_RF, which yields a flow of w_M.

Operation at R or M will be stable, since if w should drop momentarily for external or statistical reasons, a net accelerating ΔP becomes available, and if it should rise a net decelerating ΔP is produced, restricting the change in flow, and returning the flow to the original value when the cause of the flow change disappears. However, S and S' are metastable (unstable) operating points. If w rises slightly above that at S or S', it will accelerate slowly, then rapidly, then slowly again to R or M, respectively. If w starts or falls slightly below S or S' it will drop to the next intersection of the orifice line and the pipe curve, unless burnout is reached sooner. Burnout of reactor tubes in this way is sometimes designated "boiling disease." With line K stable operation is obtained at D as well as at R. It is seen that a positive value of $d(\Sigma\Delta P)/dw$ at the intersection indicates stability, and a negative value instability.

The inlet resistances mentioned above (except for a regulating valve or an accidental obstruction, which have a variable cross section) will have a constant or almost constant ratio $\Delta P/w^2$, which may be called their "resistance" R:

$$\frac{\Delta P}{w^2} = R = \overset{\substack{\text{Orifices}\\\text{nozzles}}}{\frac{1}{2g_c\rho_i C_D{}^2 S_0{}^2}} = \overset{\substack{\text{Pipes,}\\\text{etc.}}}{\frac{2f_F N_e}{g_c\rho_i D_e S^2}} = \overset{\text{Table 8-9}}{\frac{K}{2g_c\rho_i S^2}} \tag{8-65}$$

where S is the smaller cross section involved. The slope of any one of the inlet pressure lines, such as P_RF, is $-\Sigma R$, summed up for all the inlet resistances that are in series.

It may be desirable to protect a nuclear-reactor tube by a fixed nozzle* against some particular value of accidental momentary or persistent additional resistance or restriction. For instance, assume that the desired or rated flow is w_R. If P_R is directly applied with no appreciable fixed inlet resistances, an accidental additional restriction up to

$$R = \frac{RG}{w_R{}^2}$$

is possible without boiling and immediate burnout. However, if the inlet header pressure is increased to P_Q, which requires fixed inlet resistances with a total R of $(P_Q - P_R)/w_R{}^2$, a more severe accidental additional restriction having $R = RV/w_R{}^2$ is possible without boiling starting, or the additional resistance $RG/w_R{}^2$ will only decrease the flow to w_N. With P_Q there is the further important advantage that burnout at Z will not occur without a more severe restriction yet. Operation can

---

* A nozzle is preferable to an orifice because its C_D is more independent of flow rate and wear.

jump to and continue on the arc DA and almost to Z without burnout, presumably, if instruments make the trouble apparent and the reactor is shut down rapidly. Evidently the protective orifice can be an adjunct to an automatic scram installation responsive to a brief but steady change in pressure and/or temperature difference across each tube. Less orifice protection is called for the more sensitive the automatic controls.

P_J was obtained by drawing a tangent in Fig. 8-17a from R to the boiling curve. An important advantage of P_J or a higher inlet header pressure is that w can return automatically from the HDA arc to w_R if and when the accidental restriction is flushed out or otherwise removed. If a lower pressure, say P_Q, were employed, it would be necessary to drop the power to move from D to C and lower.

P_E is the highest intercept reached by any tangent to the curve. A higher inlet manifold pressure than P_E is necessary in order to be able to regulate to any desired flow rate by an automatic or manual control valve. This feature might be particularly useful in an experimental unit.

In designing protective nozzles, the accidental restriction to be protected against should first be decided on, from the estimated probability, extent, and duration of the possible accidental restrictions, the fixed and pumping charges of the protective nozzles, required reliability and other characteristics of operation, hazards and expense of a burnout, etc. The lowest flow in the tube, or point on its curve, to be reached must also be selected. Assuming, for instance, that this is A, it is then necessary to find the point P_Q such that the slope of its line (K) to R, plus the slope due to the accidental restriction decided on, equals the slope from P_Q to A. From the slope of K, the R values due to any permanent inlet resistances are subtracted, leaving the R for the protective nozzle, from which its size is computed by Eq. (8-65).

Prob. 8-22. (a) Show that P_Q may be computed by

$$P_Q \frac{1}{w_A{}^2} - \frac{1}{w_R{}^2} = R_i + \frac{P_A}{w_A{}^2} - \frac{P_R}{w_R{}^2}$$

where A is the desired lowest operating point and R_i is the maximum accidental inlet resistance to be protected against. (b) Show that, under the above conditions, the force F tending to push out the accidental obstruction R_i is $(w_A{}^2/C_D)(R_i/2g_c\rho_i)^{1/2}$. (c) Why would a Venturi tube be less desirable than a nozzle or orifice for burnout protection?

Instead of carrying out a detailed analysis, however, the inlet header pressure is commonly obtained from central coolant channel pressures by a simple arbitrary reasonable relation. For instance, 100 per cent of $P_R - P_o$ above P_R or 30 per cent of $P_{max} - P_o$ above P_{max} have been used, the inlet nozzle and piping dissipating the excess ΔP above P_R.

If a protective nozzle and/or other resistances exist between the heated tube and the outlet header at P_o, the flow curve of the tube is considerably altered. The advantage is obtained that boiling will not start until a lower flow, since the heated-tube outlet pressure and therefore boiling point will be raised by the ΔP across these outlet resistances. However, if boiling starts, the downstream pressure drop will rise more rapidly and burnout is more likely. Detailed calculation of several such boiling flow curves and comparison with the desired protection will indicate whether downstream restrictions are desirable.

Besides accidental obstructions, it is desirable to consider any remotely possible accidental local or general heat surges, such as produced by sudden control rod movements or by control equipment failure. The characteristic curves of the tube at normal and peak heat should be plotted, and the same methods shown in Fig. 8-17a employed.

In addition the effect on flow of all possible variations in coolant-channel dimensions should be checked, such as by plotting the characteristic curve for the smallest diameter and greatest length channel. The fractional flow stability* may be expressed mathematically as $(\partial \log w/\partial \log q)_{\Delta P}$, $(\partial \log w/\partial \log D)_{\Delta P}$, etc. The closer these derivatives are to zero, the more stable is the flow. Decreasing ΔP per pass, such as by multipass flow with intermediate mixing headers or by use of high recycling of coolant, is the most effective means of increasing stability, but it is usually inconvenient or uneconomical.

In Sec. 8-37 the greater stability of nonboiling upflow over downflow was shown. This is also the case with boiling. Figure 8-17c qualitatively shows the pressures that might be obtained in a low-velocity natural-convection boiling reactor. A reasonably wide range of stable boiling operation at low exit quality is obtainable with upflow, but not with downflow. The power can jump to a higher power curve as shown, and stable operation will be obtained at B' because there is excess $P_i - P_o$ to accelerate the flow as required. To obtain stable boiling in downflow with low x_o would evidently require heavy orificing.†

Besides the over-all instability of downflow and even horizontal flow compared to upflow, there is at low velocities a local instability in the possible separation and fall-through of liquid, which can only be evaluated by test under the exact conditions. At higher velocities Carter reports (Chap. 9, Ref. 84) downflow may be slightly more stable for local reasons. However, his tests show that in parallel downflow channels q cannot safely exceed about half that for a single channel.

* The corresponding deviations from thermal stability, $\partial t_o/\partial q$, $\partial t_o/\partial D$, etc., are related and are usually more important.

† Figure 8-17c also shows that an inlet nozzle can be detrimental in boiling upflow; for the orifice line $C'A'$ and the power jump shown, operation at C' instead of B' would result, which might be harmful.

The effectiveness of a protective nozzle is diminished if it serves two or more tubes in parallel, or if the tube is subdivided by long axial heat-transfer or centering vanes, or if the channel is an annulus or slot of high perimeter–hydraulic-radius ratio. In view of all factors, a protective nozzle is not by any means always justifiable, and if economical power is the first objective, it should in fact probably be replaced under most conditions by design adequate to prevent obstructions.

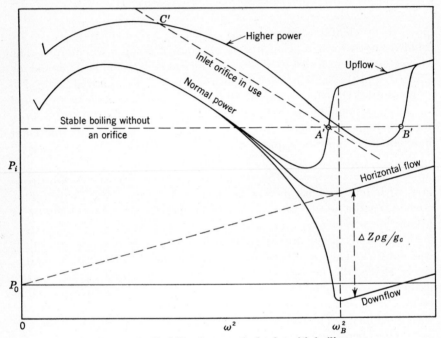

FIG. 8-17c. Stability in a vertical tube with boiling.

Prob. 8-23. Derive the relation for inlet header pressure P_N required to protect N tubes in parallel against boiling due to an accidental inlet obstruction in one of them, when the protective resistance is in their common inlet.

Nomenclature. The nonboiling pressure drop across each free channel is assumed to equal $a + bw^2$; $R_N = \Delta P / w^2$ for the common inlet protective nozzle; ΔP_x = pressure drop across the accidental inlet obstruction at w_B; w_B and ΔP_B = mass flow rate and channel pressure drop in the obstructed channel; w_A = mass flow rate in each of the $N - 1$ unobstructed channels; w_R and ΔP_R = rated mass flow and pressure drop for any one channel.

Relations. 1. Before any accidental obstruction:

$$P_N - P_o = \Delta P_R + R_N (N w_R)^2$$

2. Through the obstructed channel:

$$P_N - P_o = a + b w_B^2 + \Delta P_x + R_N [w_B + (N - 1) w_A]^2$$

3. Through an unobstructed channel:

$$P_N - P_o = a + bw_A^2 + R_N[w_B + (N - 1)w_A]^2$$

Solution. Show that the necessary P_N is

$$P_N = P_o + \Delta P_R + \cfrac{\Delta P_B + \Delta P_x - \Delta P_R}{1 - \left(\dfrac{w_B}{w_R}\right)^2 \left\{1 + (N - 1)\left[\dfrac{(w_R^2 - w_B^2)\,\Delta P_x}{w_B^2(\Delta P_R - \Delta P_B)}\right]^{\frac{1}{2}}\right\}^2} \qquad (8\text{-}66)$$

R_N may subsequently be computed from the first relation above and w_A from the others.

Problem 8-24. A reactor is to be cooled by water flowing in a bundle of parallel tubes 0.25-in. ID and 9 ft long. The velocity head adds 1 ft of equivalent length (Table 8-9). Rated flow per central tube is 1 lb/sec, entering at 40°C and leaving at 85°C and atmospheric pressure. It is desired to prevent bulk boiling of the coolant, which evidently could initiate at 100°C, or 0.75 lb/sec. It is considered to increase the inlet header pressure to protect against an accidental inlet obstruction having the same R as the pipe at rated flow. Except for the required inlet flow-controlling nozzle, neglect any inlet and outlet flow resistances. Compute the necessary header pressure if each central tube has its own flow-controlling nozzle or nozzles, and if each two or four tubes have a common nozzle.

Solution. The following quantities are first determined:

Lb/sec	$\bar{t}$, °C	$\bar{\mu}$, cp	$\overline{Re}$	f_F (Fig. 8-10)	$1/\bar{\rho}$	ΔP_F, psi [Eq. (8-29)]
$w_R = 1$	62.5	0.446	199,000	0.0051	0.01632	$148 = \Delta P_R$
$w_B = 0.75$	70	0.407	164,000	0.00525	0.01638	$86 = \Delta P_B$

Also, $R_x = 148/1^2 = 148$; $\Delta P_x = 148 \times (\frac{3}{4})^2 = 83.25$ psi. By Eq. (8-66),

N	1	2	4
P_N, psig	197	273	1373

The use of a header pressure of 148 psig with no inlet nozzle permits an inlet obstruction of $R = (148 - 86)(\frac{4}{3})^2 = 110.3$, or 75 per cent of the tube resistance at w_R, to develop before the flow falls to w_B. An additional $197 - 148 = 49$ psi is required for the permissible inlet obstruction to equal the tube in resistance. While the use of any inlet orifice is debatable, and a common one for each two tubes is uneconomical, one per four tubes is out of the question. Also, instability of the last case is shown by the fact that if f were constant and $\Delta P_B = 83.25$ psi instead of 86, P_N would only need to be 680 psi. If the rated outlet temperature were closer to the boiling point, nozzles would be relatively more effective.

Prob. 8-25. The Commonwealth Edison–Public Service 1953 Liquid-cooled Reactor proposal [72] employs 25 fps D_2O velocity in its central channels, average design conditions being 388°F and 800 psia at inlet and 440°F and 775 psia at outlet for 1064 Mw of heat generation. The following P-T saturation data are given: at 478, 496, and 515°F, P is 540, 650, and 775 psia, respectively. (a) What fraction of the channel cross section would need to plug up (assume $C_D = 0.8$) at the inlet for the coolant in the central channels to reach the boiling point at the above t_i and P_o? (b) If it is assumed that heat generation in the peripheral channels is 40 per cent of that in the central channels, t_0 being maintained at 440°F in all channels by appropriate inlet flow-control nozzles (see Sec. 8-44), predict the fraction of the coolant channel cross section that would have to plug at the inlet to reach boiling at the outlet. (c) Same as (a) and (b) for plugging at the outlet ends. (d) Estimate G_c for saturated liquid D_2O at 775 psia with a small amount of vapor.

8-44. Use of Nozzles for Flow Control.

Optimum design in a reactor for power generation will normally require the coolant to issue at the same temperature from all tubes, or at otherwise related temperatures throughout the reactor. Thus the flow through the outer, lower-power tubes must be decreased by using smaller channels, separate pumps and inlet headers, or series nozzles. Whether or not protective nozzles are used in the central tubes, nozzles are usually the most practical flow control for the outer tubes.

Prob. 8-26. The Commonwealth Edison–Public Service 1953 Gas-cooled Reactor proposal [72] employs 3,270,000 lb/hr of He to remove the 350 Mw of heat generated. The He enters at a midplane normal to the cylindrical axis, splitting and flowing out through 664 identical reactor channels in each half. Inlet conditions are 153 psia and 450°F, and outlet conditions 150 psia and an average of 742°F. (a) Assuming the outlet temperature is uniform, what is the He flow to axial and peripheral channels based on an average heat generation per channel of 73.5 per cent of the axial channels, and heat generation per peripheral channel of 25 per cent of the axial ones (as per Prob. 9-19)? (b) Assuming the axial channels have no nozzles, compute the necessary inlet nozzle area (take $C_D = 0.98$) for each peripheral channel. Assume f does not change appreciably between the axial and peripheral flow rates.

FLOW AROUND OBJECTS (EXTERNAL FLOW)

8-45. Flow past Isolated Objects.

The force F exerted on a given object by fluid flowing past it at relative velocity u may be expected to depend on fluid density and viscosity, the relative velocity, and the size of the object, expressible by a characteristic dimension L. On this basis the pi theorem (Sec. 8-22) predicts that $5 - 3 = 2$ dimensionless ratios may be required. The possible ratios are $Lu\rho/\mu$, $Fg_c/\rho L^2 u^2$, $Fg_c/Lu\mu$, and $Fg_c\rho/\mu^2$. The first, a Reynolds number, is commonly employed with D for L if there is a diameter involved. The other ratio commonly used is the "drag coefficient," $C_d = 2Fg_c/\rho Au^2$, a variation of $Fg_c/\rho L^2 u^2$. A is the area projected on a plane perpendicular to the flow, so C_d is the ratio of the actual average drag pressure, F/A, to the dynamic pressure of the stream.

Figure 8-18 shows C_d vs. Re for several shapes. Direct solution is possible for unknown F or μ. For unknown L, u, or ρ, trial and error can be employed. It is also possible to calculate Re and C_d from an assumed L, u, or ρ, plot this trial solution on log-log paper, and draw a line of slope $-\frac{1}{2}$, $-\frac{1}{2}$, or -1, respectively, to the true solution at the intersection with the Re vs. C_d curve. One can also replot the C_d vs. Re curves in terms of Re $\sqrt{C_d}$, equivalent to $Fg_c\rho/\mu^2$ above, which permits direct solution for L and u, and of Re C_d, equivalent to $Fg_c/Lu\mu$, which does the same for ρ.

Yet another method of direct solution is the use of theoretical or empirical relations between C_d and Re. Stokes' law for spheres at Re below

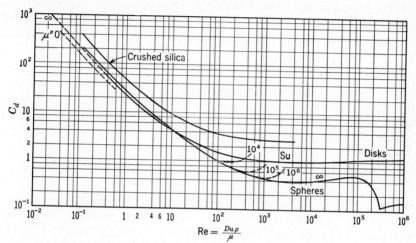

FIG. 8-18. Drag coefficients for steady motion in a fluid. Solid lines, solid particles; broken lines, bubbles and drops. The parameters are as follows:

$$\mu'' = \mu_p - \mu_0 \qquad \mathrm{Su} = \frac{g_c\sigma_i D\rho_0}{\mu_0^2}$$

0.1 yields $C_d = 24$ Re; for $1000 < \mathrm{Re} < 200{,}000$, $C_d = 0.44\pm$, and in between $C_d = 18.5\,\mathrm{Re}^{-0.6}$. Irregular particles tend to agree with Stokes' law to Re $= 50\pm$. Above Re $= 1000$, C_d averages 1.2, as for cylinders and disks.

Frequently F is a gravitational or centrifugal force, as in a suspension of particles, rather than a directly applied one. F should then be replaced by $V'g\Delta\rho/g_c$ where g is the acceleration, V' the volume of the object, proportional to L^3, and $\Delta\rho$ is the difference in the densities. In this case a slope of $+1$ solves graphically and a plot of Re/C_d solves directly for unknown L; the other methods are not altered.

Corrections are necessary for molecular motion for particles under about 0.003 mm in diameter, for the walls or other particles if they are not far away [82], and for acceleration [83]. In addition, suspended particles may flocculate by collision and adhesion, and then settle out more rapidly.

To carry out heat and/or mass transfer, one liquid might be dropped through another. Several limiting cases are shown in Fig. 8-18. The relative motion is complicated by internal circulation and shape changes in the drops. Initial time of fall may be greatly increased by the build-up of these changes [83].

Prob. 8-27. How fine would a particle of metallic uranium (assumed spherical to be conservative) need to be to settle at 1 cm/min in lead at 500°C?

8-46. Flow through Tube Banks. Flow parallel to the tubes in heat-exchanger shells can be handled by Secs. 8-14, 8-18, 8-24, and 8-25. For flow perpendicular to tube banks the average velocity of the stream $\bar{u}_{max}$ in passing between the tubes at their closest approach is employed. Each of five or more banks of tubes [68] has a loss K of approximately 0.72 velocity head or pressure for staggered tubes and 0.32 for tubes in line (rectangular lattice). For greater accuracy, with 10 or more banks of staggered tubes spaced up to $D/2$ between tubes, the pressure drop per bank is, within 25 per cent,

$$g_c \, \Delta P = \frac{1.5\rho^{0.8}\bar{u}_{max}^{1.8}\mu^{0.2}}{D_s^{0.2}} \tag{8-67}$$

where D_s is the spacing between tubes [84]. The nonisothermal corrections of Sec. 8-25 may be applied. Nonwetting mercury has shown 40 per cent lower ΔP than Eq. (8-67) (Chap. 9, Ref. 36, etc.).

If baffles are employed in the shell, two additional velocity heads may be lost at each. Actual pressure drops with baffles may be much lower at a given flow due to leakage around or through them, but for the same heat transfer, baffle leakage will require a larger flow, and may actually thus increase the pressure drop [85]. Special methods are available for other spacings and for greater accuracy [84, 86, 111].

8-47. Flow through Fixed Particulate Beds. Fixed packed beds of the normal 35 to 45 per cent voids follow within ± 25 per cent, on the average, the following expressions for streamline and turbulent flow ($D_p\bar{u}_s\rho/\mu = \mathrm{Re}_p$ less and greater than 40, respectively):

$$\frac{\bar{u}_s\mu L}{g_c \, \Delta P} = K'' = \frac{D_p^2}{1700A_f} \quad \text{or} \quad \frac{D_p^2}{76A_f \, \mathrm{Re}_p^{0.85}} \tag{8-68}$$

where L = bed depth
D_p = particle diameter
$\bar{u}_s$ = average velocity disregarding packing
A_f = wall leakage factor
The wall leakage factor varies linearly from 0.81 in streamline and 0.7 in turbulent flow at $D_p/D_{bed} = 0.1$ to 1 at $D_p/D_{bed} = 0$.

Five- to sixty-mesh-per-inch wire screens, as well as beds of spheres, fall (Chap. 9, Ref. 49) within some 20 per cent of the following correlation:

$\bar{u}_s \rho / \mu \alpha$	0.5	2.5	10	25	125
$\Delta P \, g_c \alpha^3 / \rho N \beta \bar{u}_s^2$	10	2.5	0.85	0.5	0.25

where α = fractional volume of voids in the full screen (usually twice the wire diameter) or bed thickness N, and β = total surface area per unit screen or bed volume.

Methods are available for abnormal voids or roughness [84]. If the mean free path becomes appreciable compared to the pore size as a result of low pressure, high temperature, and/or small pores, the permeability K'' increases linearly with $1/P$ because of "slip" or molecular flow [87].*

8-48. Fluidization of Particulate Solids. If a particulate bed is not confined and the flow rate is such that the pressure drop across it (Sec. 8-47) roughly equals its net weight per unit area, but below the transport velocity for the separate particles (Sec. 8-45), the bed will become "fluidized." Good fluidization exhibits fast and thorough mixing of the bed and temperature uniformity throughout the bed and fluid, and it offers possibilities for nuclear fuel processing and reactor operation. Best fluidization is obtained with uniform particles fluidized by a liquid of low viscosity, and in the fluidized bed the particles occupy some 5 per cent of the total volume. Accurate predictions are difficult [88], and tests are required on the specific system for quantitative design purposes.

PUMPING

8-49. Mechanical Pumps and Blowers. Centrifugal and axial-flow pumps and blowers are suitable for most fluid moving in nuclear-power installations. They are available in a wide range of sizes, can stand reasonably high temperatures and pressures, and can be sealed well in case leakage is expensive or dangerous. Erosive fluids would damage them but would be ruled out of other equipment also. These pumps are available with an inert-gas seal on the shaft, or without emergent shaft thanks to a built-in motor or to a thin sealed nonmagnetic diaphragm through which magnetic drive occurs [89]. Their flow characteristics are widely described [90–94] and will only be reviewed here.

Variables in substantially constant-density operation of a pump or blower of a given geometrical design include rotational speed n in revolutions, a dimension such as the rotor diameter D, fluid pressure rise ΔP,

* See also Sec. 9-6.

density ρ, viscosity μ, and volumetric flow rate Q. The six variables indicate by the pi theorem that $6 - 3 = 3$ dimensionless ratios are required to represent the operation. The rotational Reynolds number $D^2 n\rho/\mu$ from Table 8-7 is evidently called for. Other obtainable ratios are the "capacity coefficient" $C_Q = Q/nD^3$, the "head coefficient" $C_H = \Delta P\, g_c/\rho n^2 D^2$, the "specific speed" $n_s = nQ^{1/2}(\rho/\Delta P\, g_c)^{3/4}$, and $\Delta P\, g_c D^4/\rho Q^2$, a Von Karman number.

In common practice, the characteristics of a pump are plotted as constant-speed curves of ΔP [or of pressure head $(\Delta P\, g_c/\rho g)$], shaft HP, and efficiency E vs. Q for the desired fluid. By instead employing the

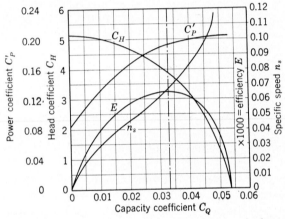

FIG. 8-19. Dimensionless characteristics of a general-service single-stage centrifugal pump rated at 60 gpm and 70 ft at 3450 rpm.

above dimensionless ratios, operation for any fluid and conditions can be predicted, or the size of a geometrically similar pump selected for a different Q, etc. In so handling pump data it is found that unless $D^2 n\rho/\mu$ is unusually small, it has no significant effect on the other ratios. Viscous friction losses are evidently relatively small. Thus, over its practical range, the characteristics of a centrifugal pump or blower can be expressed as any one of the four above ratios plotted against any other one of them, as in Fig. 8-19.

Cavitation occurs with a given liquid when the absolute pressure at any location in the pump falls to the vapor pressure at the same location. Noise, harmful hammering, and fall-off in efficiency ensue. If the absolute static pressure at the inlet is P_1 and the absolute vapor pressure of the inlet fluid P_v, the "cavitation ratio" $(P_1 - P_v)/\Delta P$ below which cavitation develops has been found to be roughly constant for a given pump shape. For centrifugal pumps handling water it is recommended that $P_1 - P_v$ be at least some 7 or 8 psi (15 to 20 ft) [92]. Centrifugal pump and blower tip speeds are usually kept under 200 fps, and inlet opening velocities commonly below 10 fps for water and 100 fps for air.

The shaft power $\mathbf{P}$ can be expressed dimensionlessly as the "power coefficient" $C_P' = \mathbf{P}g_c/\rho n^3 D^5$, which also plots as a function of any one of the four groups above, provided viscous and cavitation effects are absent.

Efficiency $\Delta P \, g_c Q/\mathbf{P}$ is also a dimensionless ratio dependent on any one of the others. The maximum efficiency of a pump is obtained at some one point on the generalized efficiency curve, and thus at one value of each of the four other ratios above. The specific speed is particularly convenient in that D is absent and the other quantities in it are usually specified in a given pumping job (except for any possible alternate values of n). Thus, for any given pumping operation the specific speed can be computed and a pump or blower of the desired type having the desired specific speed at its maximum efficiency selected from manufacturers' data or scaled up or down from data on a given shape, such as Fig. 8-19. For commercial single-stage centrifugal pumps and fans of various designs the dimensionless coefficients usually fall in the ranges indicated in Table 8-14.

TABLE 8-14. DIMENSIONLESS PERFORMANCE COEFFICIENTS FOR COMMERCIAL
SINGLE-STAGE CENTRIFUGAL PUMPS AND FANS

Coefficient	Pumps		Fans	
	At $E_{\max}$	At $Q = 0$	At $E_{\max}$	At $Q = 0$
C_H	3.5–5.8	4.4–6.5	3.5–11	3.6–12
C_Q	0.01–0.4	0	0.2–1.3	0
n_s	0.03–0.3	0	0.1–0.3	0
E	0.4 –0.9	0	0.4–0.9	0
C_P'	....	0.06–0.8	....	0.5–10

For single-stage axial-flow pumps, n_s at $E_{\max}$ usually ranges from 0.3 to 0.8. Actually, the optimum operating point is usually not exactly at maximum efficiency. Minimum pump size, investment, and total pumping cost are obtained by operation slightly beyond maximum efficiency (higher Q); high emergency-reserve pumping capacity is obtained by normal operation below maximum efficiency (lower Q).

The combined characteristic ΔP vs. Q curve of dissimilar pumps in series can be obtained by plotting the sum of the separate ΔP values at the same Q vs. the value of Q. Similarly, for pumps in parallel ΣQ at each ΔP is plotted vs. ΔP.

Prob. 8-28. The Commonwealth Edison–Public Service 1953 Gas-cooled Reactor proposal [72] employs He at 10 atm abs as the coolant. The reactor is in a 44-ft-diameter low-alloy steel sphere with 1.67-in. walls. The total weight of He, including 250 per cent stored reserve, is 23,000 lb, worth $150,000. ΔP across the blowers is 6 psi. Blower inlet and outlet temperatures are 384 and 403°F. Blower and motor were selected to obtain an over-all efficiency of 74 per cent under design conditions. Assume in addition that energy at the bus bars is worth 1 cent per kwhr, the steel shell costs $2 per lb installed, regardless of thickness, operation is for 8000 hr/yr, and total annual fixed charges are 20 per cent of installed cost on the steel sphere and 15 per cent on the He.

a. Show that for thin-walled spherical tanks (uniform s) the weight per unit weight of contained gas is $1.5\rho'R'T(P - P_a)Z'/sP$, where ρ' = metal density, s = wall stress, P_a = atmospheric pressure, P and T = storage absolute pressure and temperature, and Z' is the compressibility factor. (For long cylinders 1.5 is replaced by 2.)

b Assume that, if design P is changed, the weight and cost of the sphere vary as per above expression, but the total He volume required when measured at P, and the He temperatures and mass velocities, all are kept constant. Find the optimum P if only the fixed charges on the sphere and He and the pumping charges are taken into account.

Prob. 8-29. More dimensionless ratios are mentioned in Sec. 8-49 than the necessary minimum; thus some can be computed from others for any given case. Derive the interrelationships among all of them.

Examples. $C_Q = (n_s)^2(C_H)^{3/2}$; $E = C_Q C_H / C'_P$.

Prob. 8-30. The pump characteristics of Fig. 8-19 were obtained with cold natural water. (*a*) The rating is stated to be at the maximum efficiency. Compute n_s and check the curve in the figure. Also compute D from C_H. (*b*) If the same pump is to deliver cold D_2O at the same ΔP and n, what will be the per cent change in E, in Q, and in **P**?

8-50. Electromagnetic Equipment for Liquid Metals. Liquid metals usually present special handling problems on account of their reactivity

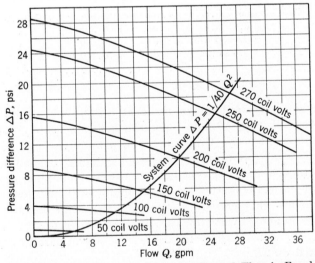

FIG. 8-20a. Typical set of operating curves for General Electric Faraday-type a-c electromagnetic pump.

with most gases, packing materials, and containers, poor lubricating value, possibilities of plugging lines, high operating temperatures, expansion on melting or freezing, etc. [89, 95, 96]. In most cases all pipe joints must be soundly welded, valves should be all-metal bellows type, and flow meters should be electromagnetic, or else rotameters. At low temperatures conventional liquid pumps may be useful, particularly if run well

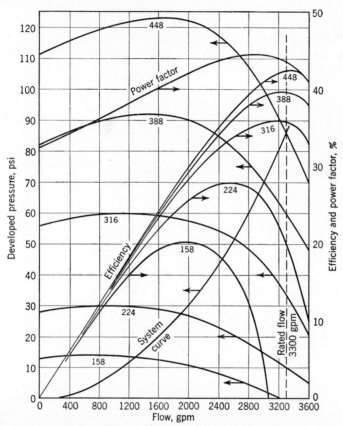

FIG. 8-20b. Characteristics of the three-phase 60-cycle linear-induction EM pumps originally on the submarine *Sea Wolf*, with sodium at 580°F. Parameter = volts. Over-all pump length is 8 ft. (*Courtesy R. G. Rhudy, KAPL.*)

under normal speed and with inert seal gas. However, electromagnetic (EM) pumps have proved very valuable, particularly for the alkali metals, which wet well and have high electrical conductivity. The EM pumps are the only absolutely leakproof ones so far developed. A second EM pump connected in opposition is at times useful as an EM "throttle valve."

EM pumps depend on the normal force exerted on a current by a magnetic field at right angles to it. They are of two principal types: (1)

"conduction," in which the current is passed through the tube walls and the stream by external electrodes normally to the magnetic field, as in a d-c or a-c series motor; and (2) "induction," in which it is induced by the magnetic field, which moves in the flow direction.   Conduction EM pumps require many thousands of amperes at 1 to 2 volts.   D-c pumps thus require a high-current transformer and rectifier.   They may give over 60 per cent efficiency with Na.   A-c conduction pumps contain

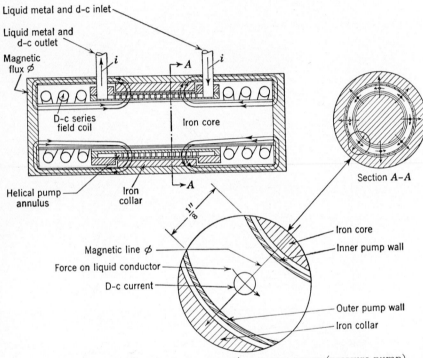

Fig. 8-21. General Electric electromagnetic pressure gauge (pressure pump).

their own transformer, but so far only give about 20 per cent efficiency. The characteristics of a commercial a-c conduction pump are given in Fig. 8-20a.   The system curve is for a circuit requiring 10 psi at 20 gpm; the intersections give the flow that would be obtained at the corresponding voltage.

Linear induction EM pumps employ a flat straight channel with a three-phase a-c winding developed along it.   Efficiencies of over 40 per cent are attained on Na (see Fig. 8-20b).   Helical induction EM pumps employ a three-phase field rotating about a stator, as in an induction motor.   Pressures of 100 psi are readily available by using only one or two parallel paths of several turns each, and lower pressures but higher flows with more parallel paths of larger pitch. An efficiency of 20 per cent with Na has been attained so far.

EM flow meters employ a permanent magnet clamped on the outside of a nonmagnetic metal pipe. Two leads from the pipe wall, at the ends of the diameter normal to the magnetic flux, go to a potentiometer. Good wetting in the pipe and absence of net thermoelectric emf's in the leads are necessary for accuracy.

An EM pressure gauge [97] is outlined in Figs. 8-21 and 8-22. The current automatically is varied to maintain the liquid metal at a reference level in the off-stream leg from the helical channel. A reliable calibration curve is obtainable of current or shunt IR drop vs. pressure.

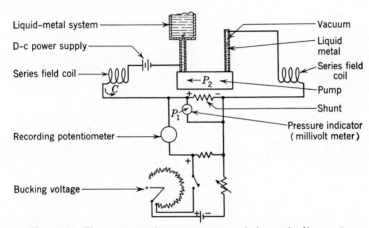

FIG. 8-22. Electromagnetic pressure gauge (schematic diagram).

The use of bimetallic constantan-chromel sodium tubes in a water-cooled heat exchanger has been proposed [110]. The thermoelectric current produced serves to generate the required magnetic flux. It is also the current for a self-excited EM pump which pumps the sodium through the attached reactor. The pump would be self-regulating—at higher temperature differences, higher currents (thus higher EM pressures and flows) would be produced. Low thermal efficiency of the thermoelectric generator and EM pump would not matter, since the energy utilized is merely heat.

NOMENCLATURE

C = specific heat
D = diameter
G = mass velocity
g = acceleration of gravity
g_c = ratio of the unit of F to that of ma ($g_c = 1$ for all absolute systems)

h = enthalpy
M = molecular weight
N = distance along pipe axis
P = absolute pressure
Q = volumetric fluid flow rate
R = radius to wall
r = local radius
S = stream cross section
T = absolute temperature
t = temperature, °F or °C
u = local velocity
$\bar{u}$ = flow average velocity
V = specific volume
w = mass flow rate
x = weight fraction
y = local distance from wall
z = elevation against gravity
β = coefficient of volumetric thermal expansion
θ = time
μ = absolute viscosity
ν = kinematic viscosity
ρ = density

Symbols for dimensionless ratios are in Table 8-7. Other symbols are defined when used.

REFERENCES

1. Hilsenrath, J., and Y. S. Touloukian: *Trans. ASME*, **76**:967–985 (1954). "NBS-NACA Thermal Tables," National Bureau of Standards.
2. Perry, J. H.: "Chemical Engineers' Handbook," 3d ed., pp. 370–374, McGraw-Hill Book Company, Inc., New York, 1950.
3. Bonilla, C. F., R. D. Brooks, and P. L. Walker: "General Discussion of Heat Transfer," London, 1951; ASME, New York, 1953.
4. Hirschfelder, J. O., R. B. Bird, and E. L. Spotz: *Chem. Revs.*, **44**:205–231 (1949). Hirschfelder, J. O., C. F. Curtis, and R. B. Bird: "Molecular Theory of Gases and Liquids," John Wiley & Sons, Inc., New York, 1954.
5. Bromley, L. A., and C. R. Wilke: *Ind. Eng. Chem.*, **43**:1641 (1951).
6. Meissner, H. P.: *Chem. Eng. Progr.*, **45**:149 (1949).
7. Glasstone, S.: "Physical Chemistry," p. 501, D. Van Nostrand Company, Inc., Princeton, N.J., 1946.
8. Sage, B. H., and W. N. Lacey: *Ind. Eng. Chem.*, **30**:831 (1931).
9. Swindells, J. F., R. Coe, Jr., and T. B. Godfrey: *J. Research NBS*, **48**:1 (1952).
10. Standard D341–39, charts C and D, American Society for Testing Materials, Philadelphia.
11. Carmichael, L. T., and B. H. Sage: *Ind. Eng. Chem.*, **44**:2731 (1952).
12. Charron, F.: *Publ. sci. et tech. ministère air*, no. 268 (Paris), 1952.
13. Morris, W. J., and R. Schnurmann: *Nature*, **167**:317 (1951).
14. Weltmann, R. J.: *Ind. Eng. Chem.*, **40**:279 (1948).
15. Needs, S. J.: Special Technical Publication 111, American Society for Testing Materials, Philadelphia, 1951.

16. Lapple, C. E., et al.: "Fluid and Particle Mechanics," p. 120, University of Delaware Press, Newark, Del., 1951.
17. *Lubrication*, **31**:40 (1945).
18. Bonilla, C. F., A. Cervi, T. J. Colven, and S. J. Wang: *Chem. Eng. Progr.*, *Symposium Ser.*, no. 5, 1953.
19. Vand, V.: *J. Phys. & Colloid Chem.*, **52**:300 (1948).
20. Orr, C.: Ph.D. dissertation, Georgia Institute of Technology, 1952.
21. Brailey, R. H.: "The Viscosity of Concentrated Suspensions," preprints, Division of Paint, Varnish and Plastics Chemistry, American Chemical Society, September, 1951.
22. Ref. 2, p. 354.
23. Partington, J. R.: "An Advanced Treatise on Physical Chemistry," vol. I, p. 724, Longmans, Green & Co., Inc., New York, 1949.
24. Joffe, J.: *J. Am. Chem. Soc.*, **69**:540 (1947).
25. Ref. 2, pp. 535, 536.
26. Ref. 2, p. 294.
27. Harkins, W. D.: "The Physical Chemistry of Surface Films," Reinhold Publishing Corporation, New York, 1952.
28. Ref. 16, p. 14.
29. Ref. 2, pp. 379–381.
30. Streeter, V. L.: "Fluid Dynamics," p. 222, McGraw-Hill Book Company, Inc., New York, 1948.
31. Purday, H. F. P.: "Streamline Flow," p. 13, Constable & Co., Ltd., London, 1949.
32. Ref. 2, p. 385.
33. Goldstein, S.: "Modern Developments in Fluid Dynamics," vol. I, pp. 299–309, Oxford University Press, New York, 1938.
34. Ref. 33, p. 311.
35. Ref. 2, pp. 382–383.
36. Deissler, R. G.: *NACA TN* 2410, 1951.
37. Ref. 30, pp. 93ff.
38. Ref. 30, chap. VIII, etc.
39. Ref. 30, p. 110.
40. Klein, E. O. P., et al.: Paper 52-A-65, ASME, New York, 1952.
41. Bridgman, P. W.: "Dimensional Analysis," Yale University Press, New Haven, Conn., 1931.
42. McAdams, W. H.: "Heat Transmission," 2d ed., chap. IV, McGraw-Hill Book Company, Inc., New York, 1942.
43. Moon, P., and D. E. Spencer: *J. Franklin Inst.*, **248**:495 (1949).
44. Moody, L. F.: *Trans. ASME*, **66**:671 (1944).
45. Bonilla, C. F.: *Ind. Eng. Chem.*, **31**:618 (1939).
46. Rothfus, R. R., et al.: *Ind. Eng. Chem.*, **42**:2511 (1950).
47. Keenan, J. H., and Neumann, E. P.: *J. Appl. Mechanics*, **68**:A91–100 (June, 1946).
48. Dryden, H. L.: *NACA Rept.* 562, 1936.
49. Rubesin, M. W., and H. A. Johnson: *Trans. ASME*, **71**:383 (1949).
50. Corcoran, W. H., et al.: *Ind. Eng. Chem.*, **44**:417 (1952).
51. Isakoff, S. F., and Drew, T. B.: "General Discussion on Heat Transfer," Institution of Mechanical Engineers, London, 1951; ASME, New York, 1953.
52. Deissler, R. G.: *NACA TN* 2138, 1950.
53. Sherwood, T. K.: *Ind. Eng. Chem.*, **42**:2078 (1950).

54. Von Karman, T.: *Trans. ASME*, **61**:705 (1939).
55. Miller, B.: *Trans. AIChE*, **33**:486 (1937).
56. Kennison, R. G.: AECU-2010.
57. Deissler, R. G., and C. S. Eian: *NACA TN* 2629, 1952.
58. Rothfus, R. R., and R. S. Prengle: *Ind. Eng. Chem.*, **44**:1686 (1952).
59. Von Karman, T.: *J. Aeronaut. Sci.*, **1**:1–20 (1934).
60. Deissler, R. G.: *Trans. ASME*, **76**:73–85 (1954).
61. Goldmann, K.: *Chem. Eng. Progr., Symposium Ser.*, no. 11, pp. 105–113, 1954.
62. Dryden, H. L.: *Quart. Appl. Math.*, **1**:7–42 (1943).
63. Kovasznay, L. S. G.: *NACA TN* 2839, 1953.
64. Baines, W. D., and E. G. Peterson: *Trans. ASME*, **73**:467–480 (1951).
65. Schubauer, G. B.: *NACA Rept.* 524, 1935.
66. Dryden, H. L., et al.: *NACA Rept.* 581, 1937.
67. Comings, E. W., et al.: *Ind. Eng. Chem.*, **40**:1076–1082 (1948).
68. Fluid Friction Standards, Hydraulic Institute, New York.
    Technical Paper 409, Crane Co., Chicago.
    K. H. Beij, R.P. 1110, National Bureau of Standards.
    M. P. O'Brien and G. H. Hickox, "Applied Fluid Mechanics," p. 350, McGraw-Hill Book Company, Inc., New York, 1937.
    Ref. 16, p. 40.
    R. J. S. Pigott, *Trans. ASME*, **72**:679 (1950).
69. Kays, W. M.: *Trans. ASME*, **72**:1067–1074 (1950).
70. Ref. 16, p. 170.
71. Ref. 2, p. 384.
72. "Reports to the U.S. Atomic Energy Commission on Nuclear Power Reactor Technology," Government Printing Office, May, 1953.
73. Alves, G. E., D. F. Boucher, and R. L. Pigford: *Chem. Eng. Progr.*, **48**:385 (1952).
74. Green, H.: "Industrial Rheology," John Wiley & Sons, Inc., New York, 1949.
75. Caldwell, D. H., and H. F. Babbitt: *Ind. Eng. Chem.*, **33**:249 (1941).
76. Winding, C. C., W. I. Kranich, and G. P. Baumann: *Chem. Eng. Progr.*, **43**:621 (1947).
77. Johnson, H. A., and A. H. Abou-Sabe: *Trans. ASME*, **74**:979 (1952).
78. Lockhart, R. W., and R. C. Martinelli: *Chem. Eng. Progr.*, **45**:39 (1949).
79. Weiss, D. H.: ANL-4916.
80. Hunsacker, J. C., and B. G. Rightmire: "Engineering Applications of Fluid Mechanics," pp. 176–178, McGraw-Hill Book Company, Inc., New York, 1947.
81. Bottomley, W. T.: *Trans. North East Coast Inst. Engrs. & Shipbuilders*, **53**:65 (1936–1937).
82. Ref. 16, p. 285.
83. Hughes, R. R., and E. R. Gilliland: *Chem. Eng. Progr.*, **48**:497 (1952).
84. Ref. 2, p. 391–394.
85. Tinker, T.: "Shell Side Heat Transfer Characteristics," Ross Heater & Mfg. Co., Buffalo, N.Y., 1947.
86. Bergelin, O. P., A. P. Colburn, and H. L. Hull: "Heat Transfer and Pressure Drop during Viscous Flow across Unbaffled Tube Banks," University of Delaware Press, Newark, Del., 1950.
87. Biles, M. B., and Putnam, J. A.: *NACA TN* 2783, 1952.
88. Weintraub, M., and M. Leva: *Ind. Eng. Chem.*, **45**:76–77 (1953).
89. "Liquid Metals Handbook," 2d ed. (R. N. Lyon, ed.), Government Printing Office, 1952.
90. Taylor, I.: *Chem. Eng. Progr.*, **46**:637–642 (1950).

91. Dolman, R. E.: *Chem. Eng.*, **59**:155–169 (March, 1952).
92. "Standards of the Hydraulic Institute," 10th ed., New York, 1953.
93. Marks, L. S.: "Mechanical Engineer's Handbook," 5th ed., McGraw-Hill Book Company, Inc., New York, 1951.
94. Baumeister, T.: "Fans," McGraw-Hill Book Company, Inc., New York, 1935.
95. Cage, J. F., Jr.: AECU-2282.
96. Vanderberg, L. B., and G. P. Hendricks: KAPL Memo LBV-5, 1952.
97. Carr, N. L.: Research Bulletin 23, Institute of Gas Technology, Chicago.
98. Martinelli, R. C., and D. B. Nelson: *Trans. ASME*, **70**:695 (1948).
99. Isbin, H. S.: personal communication, 1953.
100. Pellew, A., and R. V. Southwell: *Proc. Roy. Soc. (London)*, **176A**:312–343 (1940).
101. Rogers, F. T., et al.: *J. Appl. Phys.*, **22**:1476–1478 (1951).
102. Reynolds, J.: ANL-5178, 1954.
103. Jens, W. H., and P. A. Lottes: ANL-4627, 1951.
104. Daily, J. W., and W. L. Hankey: Report 10, Massachusetts Institute of Technology, Hydrodynamic Laboratory, 1953; *Trans. ASME*, **76**:87–95, January, 1954.
105. Stein, R. P., et al.: *Chem Eng. Progr., Symposium Ser.*, no. 11, pp. 115–126, 1954.
106. Styrikovich, M. A., and G. E. Kholodovskii: *Otdel. Tekh. Nauk*, no. 4, pp. 506–528, 1951.
107. Jelinek, R. V.: private communication, 1954.
108. Isakoff, S. E.: *Ind. Eng. Chem.*, **47**:413–421 (1955).
109. McIlroy, M. S.: *J. New Engl. Water Works Assoc.*, **65**:299–318 (1951); American Gas Association Pamphlet DMC-52-10 (Standard Electric Time Co., Springfield, Mass.).
110. Luebke, E. A., and L. B. Vanderberg: Proceedings of the 1955 Conference on Nuclear Engineering, University of California at Los Angeles, p. A-106.
111. Boucher, D. F., and C. E. Lapple: *Chem. Eng. Progr.*, **44**:117–134 (1948).
112. Dingee, D. A., et al.: BMI-1026, 1955.
113. Rothfus, R. R., and C. C. Monrad: *Ind. Eng. Chem.*, **47**:1144–1149 (1955).
114. Mickelsen, W. R.: *NACA TN* 3570, 1955.
115. Laufer, J.: *NACA TN* 2954, 1953.

CHAPTER 9

HEAT REMOVAL

By Charles F. Bonilla

INTRODUCTION

9-1. Significance of Heat Removal from Nuclear Reactors. Before a nuclear reactor can reach criticality the geometrical and material purity requirements outlined in Chap. 7 must be met. Any such reactor can then be operated at very low power or for very brief intervals at high power with no provision for cooling. In steady operation of moderately low-power research reactors, cooling must be provided, but the most convenient or economical coolant and fuel-element shape for the case are usually adequate. For instance, the 3000-kw Oak Ridge X-10 reactor [1] employs air as a coolant and has relatively large (see Prob. 9-2) solid cylindrical fuel elements lying in the bottom corner of a square coolant channel with no provision for extended cooling surface, centering in the channel, etc.

Powerful research or production reactors need more effective heat-removal methods, including one or more of the following: liquid coolant, high coolant velocity, extended surface, thin fuel elements, etc. They are in general designed and operated as hot as is reasonably safe in their thermally limiting location. This is normally the axis of the fuel element, which should not melt or deform, or the surface of the fuel-element sheath, which should not vapor-bind or be corroded by the coolant.

Reactors for the production of economical power (see Chap. 13) will normally employ high coolant temperatures, as well as enriched fuel and high specific heat generation. This will be particularly true in a propulsion reactor, in which small size is necessary. In such reactors the stress-bearing materials of construction as well as the fuel elements and moderator may be operating near their highest safe temperatures, and proper heat transfer design and coolant flow are even more necessary.

It is evident that the development of materials which can operate at higher temperatures will increase thermodynamic efficiency and/or simplify design. However, currently available materials are already ade-

360

quate for many reasonable operating conditions and designs. Keeping in mind also that the neutron-flux density is limited only by the rate at which the generated heat can be removed safely, nuclear-power-reactor design has repeatedly been classed as primarily a problem of heat removal.*

HEAT GENERATION IN NUCLEAR REACTORS

9-2. Distribution of Steady-state Heat Generation in Nuclear Reactors. The heat generated in nuclear reactors is directly or indirectly derived from the energy released in the nuclear fissions. The energy of an average U^{235} fission is distributed roughly as in Table 9-1. The distribution is similar for U^{233} and Pu^{239}. The range of the radiation in different fuels, moderators, blankets, coolants, structural materials, and shields could in principle be estimated from the probability of each nuclear reaction, the energies of their particles or radiations, and the appropriate

TABLE 9-1. APPROXIMATE AVERAGE HEAT DISTRIBUTION IN THERMAL REACTORS, MEV PER ATOM OF U^{235}

	Range			Total
	Short	Medium	Long	
Prompt energy of fission:				
1. Kinetic energy of fission fragments..........	168*	...	...	168
2. Kinetic energy of fast neutrons.............	....	5	...	5
3. Energy of prompt γ.......................	....	...	5	5
Radioactive decay of fission products:				
4. β............................	7	...	...	7
5. γ..	....	...	6	6
6. Neutrinos (unabsorbable)...................	....	...	...	11
Nonfission reactions with neutrons:				
7. $\beta + \gamma$...................................	....	...	7	7
				209
Total absorbed in reactor and shield.........	175	5	18	198
Total recoverable by primary coolant.............	175	5	12 ±	192 ±
Per cent of total useful heat....................	91	3	6	

* The average range in metals is under 0.001 in.

cross sections or attenuation factors. Actually this is impractical. For heating calculations, α, β, and fission-product particles are assumed to

---

* The economic significance is indicated by the fact that for a number of completed reactors the cost of the heat removal equipment has been about 18 per cent of the total cost. For power reactors some 25 per cent may be a more economic allocation of funds [14].

have zero range. For γ and neutron radiation, detailed calculations involve the energy as well as the material (Chap. 7).

In general, therefore, the fuel elements in a heterogeneous reactor dissipate all of energy items 1 and 4, as well as a fraction (possibly one-third) of 3, 5 and 7 that depends on the mass of fuel elements. The distribution among and locally within the fuel elements of items 1 and 4 is proportional to the local fissioning neutron-flux density Φ^* and the local fuel concentration. The distribution of γ heating would be perceptibly flatter, but for simplicity, in normal operation it may be assumed in the core to follow the same distribution, because of its relatively minor importance. Thus the total heating density H_f in the fuel may be computed from the given distribution of Φ by

$$H_f = Q'\Phi \sum (\sigma_f N_f) = 6.02 \times 10^{23} Q'\Phi\rho \sum \frac{\sigma_f x_f}{M_f} \qquad (9\text{-}1)$$

where σ_f = microscopic cross sections

$\quad\;\; N_f$ = number of atoms per unit volume

$\quad\;\; x_f$ = fraction by weight

$\quad\;\; M_f$ = atomic weight of the fissionable nuclide or nuclides in the fuel element†

From Table 9-1, an average value for Q' is $175 + 1\%\!_3 = 181$ Mev, 0.000290 erg, 2.90×10^{-11} watt-sec, 6.93×10^{-12} cal, or 2.75×10^{-14} Btu per fission.

The moderator and reflector in general dissipate item 2, and a fraction of 3, 5, and 7 that depends on their relative mass. The heating might average $5 + 1\%\!_2 = 14$ Mev, or 7 per cent of the total recovered heat for a large reactor, or less than this figure for a small reactor. Heating of the moderator is frequently sufficiently uniform to be so considered. The neutron reaction heating 7 includes the excess reactivity (see Sec. 9-47 and Chap. 12), dissipated by the control rods. It can be computed by equations like Eq. (9-1) with the appropriate Q and σ values. The total heat Q'' removed by the primary coolant and available to the power plant thus approximates 192 Mev per fission, depending somewhat on the design of the reactor and cooling system.

The thermal and biological shields dissipate most of the remaining energy, about 1 to 3 per cent of the total heat. A small amount of heat is also generated in the coolant, in particular an appreciable part of 2 if it is a good moderator and of the γ energy if it is a heavy metal (Chap. 6). According to Table 9-1, the total γ energy generated is about 18 Mev for some 192 Mev of recoverable energy throughout the reactor. For

---

* Also frequently designated nv.

† With the given value of Avogadro's number, ρ must employ grams.

simplicity with adequate accuracy it may usually be assumed for heat-generation purposes that all the γ radiation in the core originates in the fuel elements. Thus if any specified portion of fuel-element volume is generating total heat at rate Δq_t, the fraction that is γ radiation Δq_γ is approximately $18/192$, or 9 per cent. At a distance r, large compared to the dimensions of the volume element, the γ-radiation intensity J_0 with no intervening absorption is

$$J_0 = \frac{\Delta q_\gamma}{A} = 0.09 \frac{\Delta q_t}{4\pi r^2} = 0.0074 \frac{\Delta q_t}{r^2} \tag{9-2}$$

If there are a number of γ sources or frequencies, each with one or more intervening absorbing layers with a total attenuation of the corresponding $e^{-\Sigma \mu_B \Delta r}$ to a given location, the total resultant intensity for each at the given location is

$$J = J_0 e^{-\Sigma \mu_B \Delta r} \tag{9-3}$$

The total local heating density from γ radiation is then

$$H_\gamma = \Sigma \mu_B J \tag{9-4}$$

μ_B being for each radiation at the given location. For the total heating density in a shield, see Sec. 7-12. H in a shield is usually only predicted within a factor of 2 or 3 because of the approximations involved. Shields are generally subdivided into an outer or "biological" shield of concrete (see Chap. 7), made as thick as possible without requiring forced cooling, and one or more steel "thermal" shields with forced- or free-convection cooling, made as thick as possible while yet avoiding serious stresses. If the attenuation is under about 3, it is usually adequate for temperature calculations to assume H uniform at its average value (see Table 9-5), and if higher to assume H varies as a simple exponential.

Approximate heating distributions for several specific reactors are as follows (see App. N for descriptions):

Reactor	Size, ft	Heating distribution, %			
		Fuel	Moderator	Reflector	Shield
EBR-1 (fast)........	1	72	....	28 (blankets)	Unrecovered
SIR...............	4	95 (including Be)		4	1
MTR..............	2	94.3 (including H_2O)	3.9 (Be)	1.6	0.2
BNL..............	20	92	6	>1	<1
OMRE............	3	91	5.5		3.5
SRE..............	6	91.5	7		1.5
SGR..............	7	92.3	7		0.7

Prob. 9-1. Show that Eqs. (9-2) and (7-6) agree approximately.

Prob. 9-2. Determine the fraction of the q_γ from fission products that appears in the graphite and in the uranium in the Oak Ridge X-10 reactor. The channels have a square cross section, 1.75 in. on a side, and are on a square lattice with 8-in. spacing of centers. The uranium slugs have a diameter of 1.1 in. (a) Compute as for a homogeneous reactor with Eq. (9-4). (b) Compute approximately for the square around one channel, with the actual disposition of materials.

9-3. Calculation of Steady-state Heat Generation with Uniform Loading.

If the nature of the fissioning neutron-flux distribution is known, H throughout the fuel elements is usually computed by proportion from the total desired heat generation and the relative distribution of the neutron flux, without employing the actual values of Φ. As in Eq. (9-1), the difference in the distribution of the relatively minor γ contribution is overlooked.

The simplest case would be a uniform flux distribution. This is fairly closely approached by small water-reflected reactors such as the swimming-pool reactors, in which the reflector serves as a source of thermal neutrons by moderating fast neutrons. In these reactors the fuel elements are thin, and H everywhere is substantially equal to the total thermal power divided by the volume of the active fuel elements.

Other ideal or simple cases, as listed in Table 6-9, are as follows:

1. The large homogeneous reactor (or the heterogeneous reactor with many equal and uniformly spaced fuel elements). In this case there is net production of slow neutrons at a local rate (or average local rate) proportional to their local flux, and $\nabla^2\Phi = -K_1\Phi$, also thus

$$\nabla^2 H = -K_1 H \qquad (9-5)$$

2. The homogeneous fuel element (small compared to the average neutron diffusion length).* In this case there is consumption of thermal neutrons at a rate proportional to their local concentration, and $\nabla^2\Phi = +K_2\Phi$, or

$$\nabla^2 H = +K_2 H \qquad (9-6)$$

The same relation yields the distribution of the (n,γ) radiation generation (line 7, Table 9-1) in homogeneous moderators, reflectors, and breeding blankets.

Usually the moderator, if it is not also the coolant, is as near the coolant as is the fuel element. Also H in the moderator is much smaller, so normally there is no important thermal problem. H_γ is strongest nearest the fuel element (and coolant) and H_n also. Therefore in a heterogeneous reactor a simple conservative calculation of the maximum moderator temperature that is usually sufficiently accurate is obtained by assuming H uniform at $\bar{H}_m$, its average throughout the cross-sectional element of

* Fast power reactors will have substantially constant H.

moderator surrounding the given channel at the desired position along it. $\bar{H}_m$ is thus most readily obtained from $\bar{H}_f$ calculated for the fuel at the same location, and the ratio $\bar{H}_m V_m / \bar{H}_f V_f$ of the total heat generations, which is frequently about 0.05. V stands for total volume. For an accurate value of the ratio or for the variation of H_m over the cross-sectional element, methods of Chap. 6 are employed.

In computing H_f in the fissionable element at any given location, its average over-all value $\bar{\bar{H}}_f$ is first obtained by dividing the desired fissionable element thermal power q_f by its volume V_f. $\bar{\bar{H}}_f$ is then multiplied by the ratio $H/\bar{H}$ for each dimension over which H varies (four dimensions for a heterogeneous rectangular reactor, down to one for a homogeneous spherical, infinite slab or infinite cylindrical reactor), equal to $\dfrac{H/H_i}{\bar{H}/H_i}$. These ratios, as obtained by integration of Eqs. (9-5) and (9-6), are given in Table 9-2 for the usual simple shapes.

Prob. 9-3. The X-10 reactor may be considered a bare rectangular reactor, and employs 35 tons of natural uranium in slugs 1.1 in. in diameter, lying in process tubes 24 ft long. For a thermal power of 3800 kw, compute (a) the average uranium "specific power" in watts per gram; (b) the number of process tubes; (c) the largest kilowatts per tube; (d) the maximum heating density H_f in Btu per cubic foot; (e) the maximum local heat velocity q/A at a slug surface in Btu per hour per square foot. Assume that the edges operate at 25 per cent of the maximum thermal flux.

Prob. 9-4. The X-10 reactor contains 620 tons of graphite. Estimate the over-all average $\bar{\bar{H}}_m$ for the moderator and the maximum average $\bar{H}_m$ at the center of the pile.

Prob. 9-5. The MTR employs 4 kg of 90 per cent U^{235} + 10 per cent U^{238}. Compute the specific power at its peak of 30 Mw.

Prob. 9-6. The maximum thermal-neutron flux in the MTR is given as 5×10^{14} neutrons/(sec)(cm^2). Compute the heating density H_n for U^{235} and for natural U at that flux. Compute the specific power at 30 Mw and compare with Prob. 9-5.

Prob. 9-7. The maximum γ heating in the MTR at 30 Mw is approximately 5 watts/g. Compute the corresponding γ flux, and check it from the power level.

9-4. Distribution of Unsteady-state Heat Generation.

When the thermal power of a reactor is altered, the neutron flux and thus the heating density H throughout the reactor change almost exactly in direct proportion to their original definitions. The main reasons for a falloff in proportionality in steady-state operation are temperature coefficients and fission-product poisoning (Secs. 6-46 to 6-48). Additional transient causes are delayed neutrons (Table 6-5) and the radioactive decay of fission products (Table 9-1).

These transient effects occur on start-up or change in power level, but they are minor compared to the energy of fission and thus unimportant with full coolant flow. However, they become dangerous if the coolant flow fails or decreases greatly, even if the reactor is immediately scrammed, also on spent-fuel-element removal, handling, and storage

TABLE 9-2. LOCAL AND AVERAGE RATE OF FISSION HEAT GENERATION IN SOLID SHAPES OF HOMOGENEOUS MEDIA

Geometrical shape	Homogeneous reactor, $H_i = H_0 > H$	Homogeneous fuel, $H_i = H_0 < H$	External bare reflector, $H_i = H_2 > H$
1. Infinite slab:			
H/H_i	$\cos(\pi x/2X_e) = \cos x'$	$\cosh(x/L) = \cosh x''$	$\dfrac{\sinh(x_3'' - x'')}{\sinh(x_3'' - x_2'')}$
$\bar{H}/H_i$	$\dfrac{\sin(\pi x_1/2X_e)}{\pi x_1/2X_e} = \dfrac{\sin x_1'}{x_1'}$ (Bare $= 2/\pi = 0.6366$)	$\dfrac{\sinh(x_1/L)}{x_1/L} = \dfrac{\sinh X''}{X''}$	$\dfrac{\cosh(x_3'' - x_2'') - 1}{(x_3'' - x_2'')\sinh(x_3'' - x_2'')}$
2. Parallelepiped:			
H/H_i	$\cos x' \cos y' \cos z'$	....	As above for any one direction
$\bar{H}/H_i$	$\dfrac{\sin x_1'}{x_1'}\dfrac{\sin y_1'}{y_1'}\dfrac{\sin z_1'}{z_1'}$ (Bare $= 8/\pi^3 = 0.2580$)	....	As above for any one direction
3. Infinite cylinder:			
H/H_i	$J_0\left(\dfrac{2.4048r}{R_e}\right) = J_0(r')$	$I_0\left(\dfrac{r}{L}\right) = I_0(r'')$	$\dfrac{K_0\left(\dfrac{r}{L}\right) I_0\left(\dfrac{r_3}{L}\right) - I_0\left(\dfrac{r}{L}\right) K_0\left(\dfrac{r_3}{L}\right)}{K_0\left(\dfrac{r_2}{L}\right) I_0\left(\dfrac{r_3}{L}\right) - I_0\left(\dfrac{r_2}{L}\right) K_0\left(\dfrac{r_3}{L}\right)}$
$\bar{H}/H_i$	$1 - \dfrac{(r_1'/2)^2}{1!2!} + \dfrac{(r_1'/2)^4}{2!3!} - \cdots$ (Bare $= 0.4313$)	$1 + \dfrac{(r_1''/2)^2}{1!2!} + \dfrac{(r_1''/2)^4}{2!3!} + \cdots$	$\dfrac{r_2(r_3 - L)}{r_3^2 - r_2^2}$ within 5% for $0.5 < r_2/L < r_3/L < 3.5$

As above for each direction

4. Finite cylinder:

H/H_i .... : $(\cos z')J_0(r')$

$\bar{H}/H_i$ .... : $\dfrac{\sin z_1'}{z_1'}\left(1 - \dfrac{(r_1'/2)^2}{12!} + \cdots\right)$

(Bare = 0.2746)

5. Sphere:

H/H_i .... : $\dfrac{\sin(\pi r/R_e)}{\pi r/R_e} = \dfrac{\sin r'}{r'}$; $\quad \dfrac{L}{r}\sinh\dfrac{r}{L}$; $\quad \dfrac{L}{r}\sinh\dfrac{r_3 - r}{L}$

$\bar{H}/H_i$ .... : $\dfrac{3(\sin r_1' - r_1'\cos r_1')}{(r_1')^3}$; $\quad 3\left(\dfrac{1}{1!3} + \dfrac{(r_1/L)^2}{3!5} + \dfrac{(r_1/L)^4}{5!7} + \cdots\right)$; $\quad 3\,\dfrac{\sinh\left(\dfrac{r_3}{L} - \dfrac{r_2}{L}\right) + \dfrac{r_2}{L}\cosh\left(\dfrac{r_3}{L} - \dfrac{r_2}{L}\right) - \dfrac{r_3}{L}}{(r_3/L)^3 - (r_2/L)^3}$

(Bare = $3/\pi^3$ = 0.0968)

0 = center, axis, or midplane of solid shape
1 = outer surface of solid shape
2 = inner surface of hollow shape
3 = outer surface of hollow shape
i = innermost
H = local heating density
$\bar{H}$ = volume average H over the cross section
r, x, y, z = local radius or distance from midplane
R_e, X_e, Y_e, Z_e = radius or half-thickness of a bare reactor with same H
$r', x', y', z', r'', x'', y'', z''$ = as defined in table above
L = average diffusion length

367

unless ample cooling is provided. Accordingly, only the case following shutdown will be discussed. Time lags in operation of the control rods and in falloff of the neutron flux will not be considered here.

Way and Wigner [2] estimated the average β and γ energy of the fission products as 11 Mev per fission and computed the instantaneous rate of its release $F(\theta)$ as a function of time θ after the fission, as plotted in Fig. 9-1.* The fractional rate is $F(\theta)/Q''$, and the total rate at time θ after

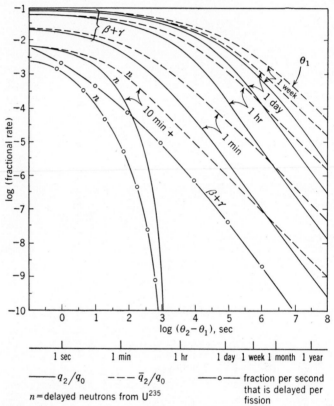

FIG. 9-1. Delayed heat generation by fission products after reactor scram. $\theta_1 =$ time of operation at power q_0.

a fission heat generation of $dQ = q\,d\theta$ is $q\,F(\theta)\,d\theta/Q''$. If a given fuel element, or the whole reactor, operates at variable power level $q(\theta)$ from time 0 to θ_1 and then is scrammed, the instantaneous power at time θ_2 ($\theta_2 > \theta_1$) due to the fissions that took place over the interval 0 to θ_1 is

* This curve may range up to 50 per cent high over certain intervals but is advisable for safe design. Equations for under 1 sec and over 1 day [Eq. (7-40)] and an approximate equation for 10 sec to 100 days are also available [2]. Additional heat generated in natural uranium by decay of U^{239} and Np^{239} adds 10 per cent more to Fig. 9-1 [32].

$$q_2 = \int_0^{\theta_1} \frac{F(\theta_2 - \theta)\, q(\theta)}{Q''}\, d\theta \tag{9-7}$$

A second integration gives the total fission-product heat Q_2 or the average power $\bar{q}_2$ from θ_1 to θ_2:

$$\bar{q}_2 = \frac{Q_2}{\theta_2 - \theta_1} = \frac{\int_{\theta_1}^{\theta_2} \int_0^{\theta_1} F(\theta_2 - \theta)\, q(\theta)\, d\theta\, d\theta}{Q''\,(\theta_2 - \theta_1)} \tag{9-8}$$

These equations are illustrated in Fig. 9-2. For $q(\theta)$ constant at q_0 and $Q'' = 185$ Mev, the ratios q_2/q_0 and $\bar{q}_2/q_0$ obtained by graphical integra-

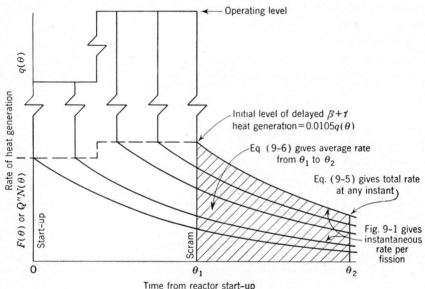

FIG. 9-2. Analysis of delayed heat generation after reactor scram.

tion are given in Fig. 9-1 for typical values of θ_1.*   Of this heat, most of the β and part of the γ, or some 60 to 80 per cent of the total for low to high ratios of fuel to total weight, would appear in the fuel element and the balance in moderator, coolant, and shield, as can be calculated by Sec. 9-2 and Chap. 7.†

* To convert to the more recent value of 13 Mev of fission-product energy and to another Q'' figure than 185 (Table 9-1), values from Fig. 9-1 should be multiplied by $13 \times 185/11Q'' = 219/Q''$.

† The local or the average rate of heat generation in a circulating-fluid-fuel loop outside the reactor core can be computed from Fig. 9-1 by summing the effect of several previous passes of the fuel through the core until each term has become equal to the heat generation which would be obtained after continuous operation of the fuel at actual power multiplied by the ratio of the fuel volume in the core to the total fuel volume.   The remainder of H is then obtained in one step (see also Sec. 7-10).

The delayed neutrons produce additional fissions, though their importance is small. By Table 9-1, as with all fissions, about 95 per cent of the heat generated by them appears in the fuel element. From Table 6-5 it is possible to compute $N(\theta)$, the instantaneous rate of release of delayed neutrons per neutron produced in fission. $N(\theta)$ can replace $F(\theta)/Q''$ in Eqs. (9-7) and (9-8) to yield the corresponding delayed neutron heat. This is also given in Fig. 9-1 for U^{235}. For Pu^{239} the rates are about half as high. In an actual scram condition the reactor k (see Chap. 6) would be slightly under 1.0 and delayed-neutron heating from prescram fissions proportionally less. Removing the fuel element from the reactor would make the neutron heating negligible and would decrease the fission-product heating somewhat since γ radiation from other fuel elements would disappear.*

Prob. 9-8. Show that $N(\theta)$ in per cent of total fission neutrons per second is obtainable from Table 6-6 as $\Sigma \lambda$ (per cent yield) $e^{-\lambda \theta}$; that

$$\frac{q_2}{q_0} = 0.01 \, \Sigma[(\% \text{ yield})(e^{\lambda \theta_1} - 1)e^{-\lambda \theta_2}]$$

and that

$$\frac{\bar{q}_2}{q_0} = \frac{0.01}{\theta_2 - \theta_1} \sum \left[(\% \text{ yield}) \frac{1 + e^{-\lambda \theta_2} - e^{-\lambda \theta_1} - e^{-\lambda(\theta_2 - \theta_1)}}{\lambda} \right]$$

for constant power.

Prob. 9-9. In the MTR after long operation at a total thermal power of 30 Mw, the rate of γ and neutron heat generation in the Be moderator is about 0.6 Mw. Immediately after scram the heat generation drops to 2 Mw in the fuel elements and 0.4 Mw in the Be moderator. Predict these heat generation values from the total power, Fig. 9-1 and Table 9-1, and explain any disagreement.

Prob. 9-10. Assume that a 300-Mw homogeneous reactor processes the fuel continuously, 8 hr being required to separate the fission products and pump them to storage tanks. What would be the rate of heat generation of the total combined stored fission products after continuous operation for a very long period?

Prob. 9-11. One gram of fission products has been stated to have a heat power level of about 1 watt for $\theta_2 - \theta_1 = 1$ year, somewhat irrespective of θ_1 [62]. Check this statement.

HEAT CONDUCTION IN NUCLEAR REACTORS

9-5. Heat-conduction Equations. Fourier's law gives the steady-state or instantaneous heat flow q through a plane area A in a stationary

---

* It should be noted that, on scramming a reactor, the rate of heat generation in the fuel elements (primarily due to fission) falls off percentagewise much more rapidly than in the thermal shield (where it is in considerable part due to fission-product γ rays). Thus the coolant outlet temperature generally cannot be kept constant, and coolant flow should not be decreased as rapidly as reactor power, to avoid overheating of the shield.

medium as

$$q = \frac{dQ}{d\theta} = -kA\frac{dt}{dx} \tag{9-9}$$

where x is normal to A. The negative sign is employed to make q positive when x is measured in the direction of heat flow, which is that of decreasing temperature. Equation (9-9) also serves as the defining equation for thermal conductivity k. If dt/dx is not uniform over A, the local heat velocity is

$$\frac{dq}{dA} = -k\frac{dt}{dx} \tag{9-10}$$

and the total q is obtained by integrating over A.

Usually k is a known function of t, and for many cases A and q are theoretically known or experimentally measured functions of x. Thus, to integrate Eq. (9-9) for steady heat conduction between isothermal surfaces at 0 and x_1 it is usually rewritten in the form

$$\int_x^{x_1} q\frac{dx}{A} = \int_{t_1}^t k\,dt \tag{9-11}$$

If q is constant (no heat generated between x and x_1), the integral $\int_x^{x_1} dx/A$ is designated the geometrical resistance, and $\dfrac{t - t_1}{q} = \dfrac{\int_x^{x_1} dx/A}{\bar{k}}$, the thermal resistance from x to x_1.

Since thermal conductivity is not easy to measure accurately, it is frequently taken as constant over a reasonable temperature interval, or at most to vary linearly with t. In this case it is readily shown that the right-hand integral in Eq. (9-11) equals $\bar{k}(t - t_1)$, where $\bar{k}$ is $(k + k_1)/2$ and also k at $(t + t_1)/2$.

If heat is being generated in the body, q is not constant and must be determined as a function of x by the integration

$$q = q_0 + \int_0^x HA\,dx \tag{9-12}$$

and substituted into Eq. (9-11) before it is in turn integrated. q_0 is any heat entering at $x = 0$. Results of this double integration are given in Sec. 9-7 for the usual geometries ("one-dimensional" cases, for which A is a known function of x) and the simple distributions of H.

If the heat velocity in the x direction changes over an interval dx by an amount $-d(k_x\,dt/dx)$, this represents an excess in heat out over heat in and must be due to heat generated in the volume under consideration, or net heat flow normal to x. In Cartesian coordinates the total heat balance in a stationary medium may be written

$$C_\rho \frac{\partial t}{\partial \theta} = \left[\frac{\partial}{\partial x}\left(k_x \frac{\partial t}{\partial x}\right) + \frac{\partial}{\partial y}\left(k_y \frac{\partial t}{\partial y}\right) + \frac{\partial}{\partial x}\left(k_z \frac{\partial t}{\partial z}\right) \right] + H \quad (9\text{-}13)$$

In extruded uranium, stacked graphite bars, etc., the average k varies with direction. However, the variation of k with direction and even with temperature seldom needs to be considered, and Eq. (9-13) is then rewritten in terms of the average physical properties.

$$\frac{\partial t}{\partial \theta} = \frac{k}{C_\rho}\left(\frac{\partial^2 t}{\partial x^2} + \frac{\partial^2 t}{\partial y^2} + \frac{\partial^2 t}{\partial z^2}\right) + \frac{H}{C_\rho} = \alpha(\nabla^2 t) + \frac{H}{C_\rho} \quad (9\text{-}14)$$

If the medium is in motion with respect to the coordinates the heat thereby transported must be included, yielding

$$\frac{\partial t}{\partial \theta} = \alpha(\nabla^2 t) + \frac{H}{C_\rho} - \left(u_x \frac{\partial t}{\partial x} + u_y \frac{\partial t}{\partial y} + u_z \frac{\partial t}{\partial z}\right) \quad (9\text{-}15)$$

Even in liquid metals at low velocities and over short distances heat is transported more rapidly by flow than by molecular conduction, and the conduction term in any direction of flow may usually be neglected. If turbulence is present the corresponding $\rho C \epsilon_H$ is added to each k, as per Eq. (8-41).

When cylindrical symmetry exists it is more convenient to employ cylindrical coordinates without the angular term. Equation (9-15) becomes

$$\frac{\partial t}{\partial \theta} = \alpha\left(\frac{\partial^2 t}{\partial r^2} + \frac{1}{r}\frac{\partial t}{\partial r} + \frac{\partial^2 t}{\partial z^2}\right) + \frac{H}{C_\rho} - \left(u_r \frac{\partial t}{\partial r} + u_z \frac{\partial t}{\partial z}\right) \quad (9\text{-}16)$$

Similarly, Eq. (9-15) in spherical coordinates with spherical symmetry is

$$\frac{\partial t}{\partial \theta} = \alpha\left(\frac{\partial^2 t}{\partial r^2} + \frac{2}{r}\frac{\partial t}{\partial r}\right) + \frac{H}{C_\rho} - u_r \frac{\partial t}{\partial r} \quad (9\text{-}17)$$

General solutions of some practical heat-conduction problems [4] can be obtained by analytically integrating the appropriate equations above and fitting the given boundary conditions. In many nuclear-reactor problems, however, the analytical solution is too involved or not possible. But it is always possible, given enough time, to apply the equation successively over small increments and solve any specific problem to any desired accuracy (see Sec. 9-39). The solution to a heat-transfer problem can of course also be obtained by performing the actual heat-transfer test, a scale model of it, or a more convenient analogous electrical or fluid-flow test (see Secs. 9-41 to 9-44).

Prob. 9-12. Show that the ratio of heat conducted to heat transported by flow normal to isothermal planes a distance Δx apart is $\alpha/(u\,\Delta x)$. Show that the heat conducted is only 1 per cent of the total for molten lead flowing at 1 fps when $\Delta x = \frac{1}{8}$ in.

9-6. Thermal Conductivity.

Data on the thermal conductivity, specific heat, enthalpy, and many other properties of gases over fairly wide temperature and pressure ranges have been correlated in the NBS-NACA Thermal Tables [5]. For extrapolation of thermal conductivities to higher temperatures or for interpolation, the Sutherland equation [Eq. (8-3)] is satisfactory; thus $\sqrt{T}/k$ plots straight vs. $1/T$. Keyes [8] has also employed a modified Sutherland equation

$$k = \frac{B'T^{3/2}}{T + 10^{D'/T}C'} \tag{9-18}$$

The constants for the common gases at atmospheric pressures are given in Table 9-3. The thermal conductivity can also be computed [6] for monatomic gases from the relation $k/\mu C_V = 2.4 + 0.016\sqrt{M}$, and for the common linear molecules except H_2 by $k/\mu C_V = 1.45 C_P/C_V - 0.13$. Theoretical prediction of k is more difficult than for μ (Sec. 8-2) because of the slower transfer of vibrational energy than of other forms. The thermal conductivity for a mixture of gases can be higher or lower than the molar interpolation from the pure components but can be predicted [7] within several per cent.

TABLE 9-3. CONSTANTS IN EQS. (9-18) AND (9-19) FOR THERMAL CONDUCTIVITY OF GASES

Gas	B', cal/(sec)(cm)(°C)	C', °K	D', °K	E', °K/atm	F', °K/atm
He.........	2.35×10^{-5}	43.5	10		
H_2........	3.76×10^{-5}	166.0	10	0.44	-1.0
O_2........	0.673×10^{-5}	266.0	10		
N_2........	0.451×10^{-5}	84.1	12	0.49	0.2
Air.........	0.632×10^{-5}	245.0	12		
CO_2.......	4.61×10^{-5}	6212.0	10	1.14	1.0
H_2O.......	1.546×10^{-5}	1737.3	12		
A..........	....	......	...	0.75	0.0

The thermal conductivity of gases increases slowly at first, then rapidly with pressure. For practical nuclear-reactor pressures the pressure correction can be made [9] by

$$\frac{k_P}{k_0} = 1 + E'\frac{P}{T}10^{F'P/T} \tag{9-19}$$

For high pressures above the critical temperature Comings's empirical correlation [10] vs. reduced temperature and pressure is useful.

Theories of the thermal conductivity of nonmetallic liquids require the sonic velocity and thus are not helpful for predictions. Observed values of k for the principal coolant liquids are given in Apps. A to E. H_2O and D_2O have the highest values, except for certain fused salts. In general, k of the saturated liquid increases with temperature to a maximum, then bends back toward that of the vapor at the critical point.

The thermal conductivity of solid pure metals varies more than 10:1, as shown in App. F, and experimental values should be used when available. It usually decreases slightly as temperature rises, and it drops on melting to roughly one-third of the solid value. Cold-worked metal usually has a somewhat lower k than annealed, and cast metal even lower. Impurities and solid-solution formation in solid alloys decrease k considerably. Thus the fission products in a fuel element, which in atom percentage amount to about twice the per cent burnup, may decrease the thermal conductivity and increase the axial temperature considerably. Alloys in general range in k from 50 to 100 per cent of the linearly interpolated value for the pure constituents.

Practically all the heat conducted in metals has been shown to be transported by the free electrons. The Lorenz relation between electrical and thermal conductivity for pure metals

$$kr_e = 23T \qquad \text{volts}^2/°\text{K} \qquad (9\text{-}20)$$

is useful for predicting k if the electrical resistivity r_e is known. The constant is 23 ± 1 for most solid or liquid pure metals, and ranges from two-thirds to twice this for alloys [11]. k is in watts per centimeter per degree Kelvin for r_e in ohm-centimeters.

The thermal conductivity k_m of two-phase mixtures in which the discontinuous-phase particles do not in general touch each other, such as suspensions of solids in liquids, or have the lower k, can be approximately predicted (Chap. 8, Ref. 20) by Maxwell's formula

$$k_m = k_c \frac{2k_c + k_d - 2x'_d(k_c - k_d)}{2k_c + k_d + x'_d(k_c - k_d)} \qquad (9\text{-}21)$$

from the thermal conductivities k_c and k_d of the continuous and discontinuous phases, respectively, and the volume fraction x'_d. For packed beds of solid particles permeated with a gas, k_m may be conservatively estimated [12] within some 20 per cent by Table 9-4, where k_s and k_g are for the solid and gas, respectively. k_m starts to fall off from the above values when the gas mean free path exceeds about 0.072 per cent of the linear average particle diameter, dropping gradually to some 1 to 2 per cent of k_s at zero pressure.

If the fluid is flowing through the bed, the effective conductivity normal to the flow is given [13] by $k_m = 3.7(\bar{\mu}\overline{C}_P/x'_f)(G\sqrt{a}/\bar{\mu})^{1/3}$, where

TABLE 9-4. THERMAL CONDUCTIVITY RATIO k_m/k_g FOR PACKED BEDS

k_s/k_g	Fractional gas volume							
	0	0.2	0.3	0.4	0.5	0.6	0.8	1
1	1	1.0	1.0	1.0	1.0	1.0	1.0	1
3	3	2.4	2.1	1.9	1.7	1.5	1.2	1
10	10	5.4	4.3	3.4	2.8	2.2	1.5	1
30	30	10.7	7.5	5.4	4.0	3.0	1.7	1
100	100	22	13	8.3	5.5	3.7	1.9	1
300	300	42	20	11	7.0	4.5	2.0	1
1000	1000	83	32	15	8.5	5.3	2.2	1

G is the over-all mass velocity and a the surface area per particle, and x_f' the fluid fractional volume.

Prob. 9-13. A fuel element is being considered consisting of UO_2 powder in a stainless-steel tube filled with He at 100 psia. Fraction of voids is 0.37 and the UO_2 particles average 0.004 in. in size. Assume k_s equals that for MgO. (a) Predict k_m for the core of the element and compare with the experimental result 0.90 Btu/(hr)(ft)(°F), constant from 100 to 1200°F [12].   (b) Repeat for argon and compare with $k_m = 0.25$ at 200°F to 0.45 at 1500°F.

9-7. The Infinite Slab—Steady State. In nuclear reactors any fuel-element or moderator surface in contact with the coolant is at the average temperature of the coolant stream at that location plus the temperature differential needed (Sec. 9-10) to transport the heat from the surface to the coolant.   Temperatures t within the solid are thus computed with respect to the cooled surface temperatures. Cases 2, 3, and 4 below are cooled to t_1 at x_1 and have no heat flow at $x = 0$.

1. *Constant q.* For a large flat homogeneous slab with $H = 0$ and constant q_0/A entering at $x = 0$, Eq. (9-11) yields

$$t - t_1 = \frac{x_1 - x}{\bar{k}} \frac{q_0}{A}$$

$$t_0 - t_1 = \frac{x_1}{\bar{k}} \frac{q_0}{A}$$

$$(9\text{-}22)$$

This applies to fuel-element cladding that does not generate appreciable heat itself. It causes the principal temperature drop of the fuel element when a thin layer of enriched fuel is employed with relatively thick cladding.

2. *Constant H.* The simplest case of internal heat generation is that of constant heating density H. This will occur with homogeneous fuel elements, slabs of moderator, cladding, radiation shields, etc., thin enough that there is very little attenuation of the heat-generating flux

(neutron and γ) density traversing them.   Measuring x from the plane of no heat flow, so that $q_0 = 0$, Eq. (9-12) yields $q = HAx$.   Substituting into Eq. (9-11),

$$t - t_1 = \frac{(x_1 - x)^2}{2\bar{k}} H$$

$$t_0 - t_1 = \frac{x_1{}^2}{2\bar{k}} H = 0.5 \frac{x_1}{\bar{k}} \frac{q_1}{A} \tag{9-23}$$

where q_1 is the heat generated between 0 and x_1.   If there is also through-flow of heat, Eqs. (9-22) and (9-23) are added to give the total $t - t_1$ or $t_0 - t_1$.

3. *Homogeneous Reactor.*   For a homogeneous slab reactor, from Table 9-2, H is directly proportional to $\cos (\pi x/2X_e)$, as shown in Fig. 9-3. As in Table 6-9, $\pi/2X_e$ may be designated B_c.   Frequently the ratio of surface to maximum heating density H_1/H_0 is provided or desired; evidently $B_c x_1 = \cos^{-1} (H_1/H_0)$.   Equation (9-12) yields

$$\frac{q}{q_1} = \frac{\sin B_c x}{\sin B_c x_1}$$

and Eq. (9-11)

$$t - t_1 = \frac{\bar{H} x_1 (\cos B_c x - \cos B_c x_1)}{B_c \bar{k} \sin B_c x_1}$$

$$t_0 - t_1 = \frac{\bar{H} x_1 (1 - \cos B_c x_1)}{B_c \bar{k} \sin B_c x_1} \tag{9-24}$$

Actually this temperature distribution is an impractical case, since it holds only for a homogeneous slab reactor thick enough to be critical and cooled only at the outer faces in steady state.   Such a reactor could not generate much power because of its thermal resistance.

4. *Homogeneous Fuel Element.*   As before,

$$\frac{q}{A} = \bar{H} x_1 \frac{\sinh (x/L)}{\sinh (x_1/L)} = \left(\frac{q}{A}\right)_1 \frac{\sinh (x/L)}{\sinh (x_1/L)} \tag{9-25}$$

$$t - t_1 = \frac{\bar{H} x_1 L \left(\cosh \dfrac{x_1}{L} - \cosh \dfrac{x}{L}\right)}{\bar{k} \sinh \dfrac{x_1}{L}}$$

$$t_0 - t_1 = \frac{\bar{H} x_1 L \left(\cosh \dfrac{x_1}{L} - 1\right)}{\bar{k} \sinh \dfrac{x_1}{L}} \tag{9-26}*$$

* Retaining only first terms in the expansions (App. L), a convenient first approximation is obtained:

$$t_0 - t_1 = \frac{\bar{H} x_1{}^2}{2\bar{k}} \left[1 - \frac{1}{12} \left(\frac{x_1}{L}\right)^2\right] = \left(\frac{q}{A}\right)_1 \frac{x_1}{2\bar{k}} \left[1 - \frac{1}{12} \left(\frac{x_1}{L}\right)^2\right] \tag{9-26a}$$

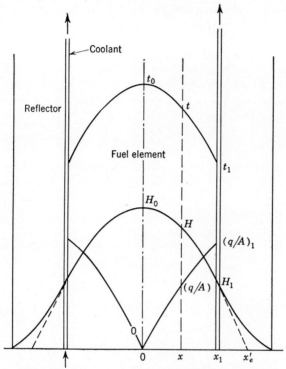

FIG. 9-3. Cosine distribution of heating density H, heat velocity q/A, and temperature t in homogeneous slab reactor with coolant and reflector (see Fig. 6-2).

5. *Bare Homogeneous Reflector (Thermal-neutron Heating)*. The heat generation by thermal-neutron absorption (including the γ as well as β radiation produced as having negligible range, but excluding fast-neutron-deceleration heat and the change in thermal-neutron flux due thereto) yields for the case of cooling at the inner surface, with x'' for x/L,

$$\frac{q}{A} = H_i L \frac{\cosh (x_3'' - x'') - 1}{\sinh (x_3'' - x_2'')} \tag{9-27}$$

$$t_3 - t_2 = \frac{H_i L^2}{\bar{k}} \left[1 - \frac{x_3'' - x_2''}{\sinh (x_3'' - x_2'')} \right] \tag{9-28}$$

If $\bar{H}$ is known instead of H_i, their ratio from Table 9-2 is employed to obtain H_i. Actually the flux Φ and H drop to zero not at x_3 but at an extrapolation distance of roughly twice the thermal-neutron diffusion coefficient. For conservative results this may merely be added to x_3.

Prob. 9-14. Show that, for any one-dimensional heat flow with constant k and any distribution of H, the sum of $t_3 - t_2$ for cooling at the inner surface alone and $t_2 - t_3$ for cooling at the outer surface alone equals $t_2 - t_3$ if all the heat generated passes from one to the other surface.

For cooling at the outer surface, by Eq. (9-28) and Prob. 9-14,

$$t_2 - t_3 = \frac{H_i L^2}{\bar{k}} \left[(x_3'' - x_2'') \coth (x_3'' - x_2'') - 1 \right] \qquad (9\text{-}29)$$

6. *Homogeneous Shield or Other Medium* (γ *Heating*). For enriched-fuel reactors, which are frequently small, an approximate solution for the shield assuming the γ-ray source localized at the center is usually adequate. For this case

$$H_\gamma = \frac{a}{r^2} e^{-\mu_B r} \approx b e^{-\mu_B x} \qquad (9\text{-}30)$$

The proportionality constant a can be computed from Eqs. (9-2), (9-3), and (9-4). The second form employing $a/\bar{r}^2 = b$ is adequate for the usual γ shields in which $\Delta r/\bar{r}$ per cooled slab or shell is small. For inner cooling, by Eqs. (9-30), (9-12), and (9-11),

$$\frac{q}{A} = \frac{H - H_3}{\mu_B}$$

$$t_3 - t_2 = \frac{H_2 - H_3}{\bar{k}\mu_B^2} - \frac{H_3(x_3 - x_2)}{\bar{k}\mu_B} \qquad (9\text{-}31)$$

$\bar{H} = (H_2 - H_3)/\mu_B(x_3 - x_2)$, and for outer cooling

$$\frac{q}{A} = \frac{H_2 - H}{\mu_B}$$

$$t_2 - t_3 = \frac{H_2(x_3 - x_2)}{\bar{k}\mu_B} - \frac{H_2 - H_3}{\bar{k}\mu_B^2} \qquad (9\text{-}32)$$

7. *Actual Reflectors, Shields, etc.* If heat generation occurs simultaneously by two or more of the previous methods, the necessary temperature differentials must be added to give the total temperature differential [i.e., Eq. (7-42)].

In general, shields with high H, such as steel "thermal shields," must be subdivided into a number of sheets with intermediate coolant streams to avoid extreme temperatures. When a slab is cooled through both sides an exact solution for the temperature distribution becomes more difficult. Assuming that the coolant temperature is known on both sides at the desired level, the distribution between the two sides of the total heat generated is assumed, the two wall temperatures computed from estimated heat transfer coefficients, and the intermediate plane located at which the temperature is the same when computed from either side with the known distribution of H. Then the assumed distribution on both sides of the heat dissipation must be checked.*

---

* The distribution of q can also be calculated as if the temperatures on both sides were equal (Table 9-5), then adding to the cold side and subtracting from the hot side the heat that would be conducted through the slab due to the actual temperatures, without heat generation.

Usually, however, all coolant streams should be proportioned so their temperature rise is about the same. Furthermore, heat-transfer coefficients generally are high compared to the solid conductances. In this case equal temperatures of the walls may be assumed, as shown in Fig. 9-4, and the problem is much simplified. For a simple distribution of H an analytical solution is then obtained by equating the temperature differentials to both surfaces from an intermediate position x_m. In Table 9-5 the location of the hottest plane, the distribution of the dissipated heat, and the temperature differential from hottest plane to wall are given for a large thin slab of material with straight-line and with exponential [Eq. (9-30)] distribution of H. In general, for any γ shields which are not too thick from a thermal-stress or convenience standpoint, any reasonable distribution of H for the same average heating density $\bar{H}$ will give adequate results, considering other uncertainties in the design. The formulas for temperature distribution are useful for thermal-stress or warpage estimation (see Chap. 11).

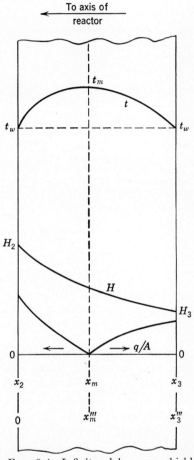

Fig. 9-4. Infinite slab γ-ray shield with exponential variation of heating density H. Cooled to same temperature on both sides.

Prob. 9-15. Boral sheet (Chap. 7, Ref. 14) contains 50 per cent B_4C by volume and has $k = 30$ Btu/(hr)(ft)(°F), roughly. (a) Compute the attenuation of thermal neutrons by a ¼-in. sheet (10^{10} is stated) if σ_a of B for thermal neutrons is 100 cm^{-1}. (b) If the n/α reaction yields 5 Mev per capture, and the incident flux on one side is 10^{11} neutrons/(cm²)(sec), compute the distribution of heating density H throughout the sheet. (c) Compute the maximum temperature in the Boral and its location if both walls are at 100°F.

9-8. The Infinite Cylinder—Steady State.

Cylindrical fuel elements are useful because of their ease of fabrication and replacement, and because of the desirability of tubes as fuel-element jackets and as coolant-channel walls.

TABLE 9-5. COOLING OF A THIN INFINITE SLAB THROUGH BOTH SIDES AT THE SAME TEMPERATURE WITH LINEAR AND EXPONENTIAL VARIATIONS OF HEATING DENSITY

(Symbols as per Fig. 9-4)

Ratio of extreme heating densities H_3/H_2*	1	0.6667	0.5	0.3	0.2	0.1
Location of hottest plane x_m'''/x_3''':						
Linear	0.5	0.4834	0.4725	0.4561	0.4463	0.4353
Exponential	0.5	0.4835	0.4714	0.4505	0.4344	0.4080
Distribution of heat $\dfrac{q_2}{q_2+q_3}=\dfrac{(q/A)_2}{\bar{H}X_3'''}$:						
Linear	0.5	0.5333	0.5555	0.5897	0.6111	0.6363
Exponential	0.5	0.5342	0.5576	0.5981	0.6288	0.6769
Maximum temperature $\dfrac{(t_m-t_w)k}{\bar{H}(X_3''')^2}$:						
Linear	0.125	0.1251	0.1254	0.1260	0.1265	0.1273
Exponential	0.125	0.1248	0.1241	0.1225	0.1207	0.1167

Temperature distribution:

Linear
$$t = t_w + \frac{H_2(x_3''')^2}{6\bar{k}}\left[\left(2+\frac{H_3}{H_2}\right)\frac{x'''}{x_3'''} - 3\left(\frac{x'''}{x_3'''}\right)^2 + \left(1-\frac{H_3}{H_2}\right)\left(\frac{x'''}{x_3'''}\right)^3\right]$$

Exponential
$$t = t_w + \frac{\bar{H}x_3'''}{\bar{k}\mu}\left(\frac{1-e^{-\mu_B x'''}}{1-e^{-\mu_B x_3'''}} - \frac{x'''}{x_3'''}\right)$$

* Equals $e^{-\mu_B(x_3-x_2)}$ for exponential.

Specific relations for cylindrical reflectors and shields can be worked out from Eqs. (9-11) and (9-12), also from Table 9-2, if applicable, and App. L. However, for a thickness $r_3 - r_2$ small compared with the radius, the relations derived for slabs hold approximately. Some of the relations for cylinders are as follows:

1. *Constant q in Hollow Cylinder.* For thin-walled fuel-element cladding and heat-exchanger tubes, Eq. (9-22), using the arithmetic mean cylindrical area $\bar{A} = \pi N(r_2 + r_3)$, is adequate [i.e., Eq. (9-22) is 0.3 per cent low for $r_3/r_2 = 1.2$]. For thicker walls Eq. (9-11) yields, for heat flow in either direction,

$$\Delta t = \frac{1}{2\pi\bar{k}}\frac{q}{N}\ln\frac{r_3}{r_2} = \frac{0.367}{\bar{k}}\log\frac{r_3}{r_2} \qquad (9\text{-}22a)$$

q/N is the local value of the power per unit length (more properly dq/dN).

2. *Constant H.* This case applies closely to reactors with solid cylinders of homogeneous fuel of radius r_1 under about half the neutron-diffusion length L (see Table 9-2), which would include most high-power

reactors with this shape of fuel element.   Previous methods yield

$$t - t_1 = \frac{H(r_1{}^2 - r^2)}{4\bar{k}}$$

$$t_0 - t_1 = \frac{Hr_1{}^2}{4\bar{k}} = \frac{0.0795}{\bar{k}} \frac{q}{N}$$

(9-33)

Evidently the permissible q/N is independent of diameter in the frequent case that t_0 and t_1 are specified.   The total required length of fuel elements can then be directly computed from the power.*

3. *Constant H in Hollow Cylinder.*   With external cooling at r_3,

$$t - t_3 = \frac{H}{4\bar{k}}\left(r_3{}^2 - r^2 - 2r_2{}^2 \ln \frac{r_3}{r}\right)$$

$$t_2 - t_3 = \frac{Hr_3{}^2}{4\bar{k}}\left[1 - \left(\frac{r_2}{r_3}\right)^2 \left(1 + 2\ln\frac{r_3}{r_2}\right)\right]$$

(9-34)

With internal cooling the same equations hold if r_3 is the inner radius and r_2 the outer radius.   A rough solution for maximum temperature of a solid moderator surrounding a coolant channel is obtainable by assuming that the "cell" of the moderator is a cylindrical shell of the desired r_2 and cross section (Fig. 6-7).

With cooling to the same temperature at both r_2 and r_3, the method of Sec. 9-7 yields

$$r_m{}^2 = \frac{r_3{}^2 - r_2{}^2}{2 \ln (r_3/r_2)}$$

(9-35)

The maximum temperature t_m is obtained as t_2 or t_3 by substituting r_m for r_2 or r_3 in Eq. (9-34).

4. *Homogeneous Reactor.*   Cooling only at the outer surface is impractical for power production, as in the case of the slab.   However, it may be applied for low power or in an emergency.   Employing the J_0 distribution,

$$t_0 - t_1 = \frac{H_i r_1{}^2}{4\bar{k}}\left[1 - \frac{(r_1')^2}{4^2} + \frac{(r_1')^4}{4^2 6^2} - \frac{(r_1')^6}{4^2 6^2 8^2} + \cdots\right]$$

(9-36)

for H_i and r' defined as in Table 9-2.   This agrees with Eq. (9-33) for $R_e = \infty$ or $r_1' = 0$.   For a bare reactor $t_0 - t_1 = 0.2348 H_i r_1{}^2/\bar{k}$.

---

* For a square bar $t_0 - \bar{t}_1 = (0.0833/\bar{k})(q/N)$ if t_1 is fairly uniform.

5. *Homogeneous Fuel Element.*   Employing the I_0 distribution (Table 9-2),

$$t_0 - t_1 = \frac{H_i r_1^2}{4\bar{k}} \left[1 + \frac{(r_1'')^2}{4^2} + \frac{(r_1'')^4}{4^2 6^2} + \frac{(r_1'')^6}{4^2 6^2 8^2} + \cdots \right] \quad (9\text{-}37)^*$$

This agrees with Eq. (9-33) for $r_1'' = 0$.

6. *Linear H in Hollow Cylinder.*   The actual variation of H from Table 9-2 can be employed, but many cases are close enough to the linear variation $H = a + br$.   For this case and external cooling,

$$t_2 - t_3 = \frac{ar_2^2}{4\bar{k}} \left[\left(\frac{r_3}{r_2} \right)^2 - 1 - 2 \ln \frac{r_3}{r_2} \right]$$
$$+ \frac{br_2^3}{9\bar{k}} \left[\left(\frac{r_3}{r_2} \right)^3 - 1 - 3 \ln \frac{r_3}{r_2} \right] \quad (9\text{-}38)$$

For internal cooling the same equation applies if r_2 and r_3 are interchanged. For a solid cylinder evidently $t_0 - t_1 = ar_1^2/4\bar{k} + br_1^3/9\bar{k}$.

9-9. The Sphere—Steady State.   Large spherical fuel elements have not been used since the CP-2 reactor, as they are hard to support and cool. However, the relations for spheres will generally apply also to rounded particles of fuel suspended or embedded in moderator or coolant or undergoing γ heating, etc.   For thin reflectors and shields the relations of Sec. 9-7 apply, and for thick ones special relations may be derived, as mentioned under Cylinders.

1. *Constant q in Spherical Shell.*   For thin-walled spherical cladding, Eq. (9-22), using $\bar{A} = \pi(r_2 + r_3)^2$, is adequate [i.e., Eq. (9-22) is 0.8 per cent low for $r_3/r_2 = 1.2$].   For thicker walls Eq. (9-11) yields

$$t_2 - t_3 = \frac{q}{2\pi\bar{k}} \left(\frac{1}{r_2} - \frac{1}{r_3} \right) \quad (9\text{-}22b)$$

2. *Constant H.*   With external cooling and uniform surface temperature,

$$t - t_1 = \frac{H}{6\bar{k}} (r_1^2 - r^2) \qquad t_0 - t_1 = \frac{H r_1^2}{6\bar{k}} \quad (9\text{-}39)$$

An externally cooled concentric spherical shell yields

$$t_2 - t_3 = \frac{H}{6\bar{k}} \left(2 \frac{r_2^3}{r_3} + r_3^2 - 3r_2^2 \right) \quad (9\text{-}40)$$

* Retaining first terms only and combining with $\bar{H}/H_i$ from Table 9-2, a convenient first approximation is obtained:

$$t_0 - t_1 = \frac{\bar{H} r_1^2}{4\bar{k}} \left[1 - \left(\frac{r_1}{4L} \right)^2 \right] = \frac{0.0795}{\bar{k}} \frac{q}{N} \left[1 - \left(\frac{r_1}{4L} \right)^2 \right]$$
$$= \frac{r_1}{2\bar{k}} \left(\frac{q}{A} \right)_i \left[1 - \left(\frac{r_1}{4L} \right)^2 \right] \quad (9\text{-}37a)$$

3. *Homogeneous Reactor.* The $(\sin r')/r'$ distribution gives

$$t_0 - t_1 = \frac{H_i r_1^2}{k} \left[\frac{1}{3!} - \frac{(r_1')^2}{5!} + \frac{(r_1')^4}{7!} - \cdots \right] \qquad (9\text{-}41)$$

which reduces to Eq. (9-39) for small r_1' or $\pi r_1/R_e$. As with the slab and cylinder, only a small steady power can be dissipated by surface cooling alone of a critical reactor.

4. *Homogeneous Fuel Element.* The $(\sinh r'')/r''$ distribution gives

$$t_0 - t_1 = \frac{H_i r_1^2}{k} \left[\frac{1}{3!} + \frac{(r_1'')^2}{5!} + \frac{(r_1'')^4}{7!} + \cdots \right] \qquad (9\text{-}42)$$

which reduces to Eq. (9-39) for small r_1''.

HEAT TRANSFER TO THE COOLANT

9-10. Heat-transfer Coefficients. The heat generated in nuclear reactors is removed by fluid flowing past solid surfaces. At any location on an interface through which heat flows, a difference Δt must exist between the temperature t_w of the solid surface and the "bulk" or mixing-cup temperature $\bar{t}$ of the stream. The ratio of the "heat velocity" q/A to Δt at any location is designated the "local film heat-transfer coefficient" h. Thus

$$\frac{q}{A} = \pm h(t_w - \bar{t}) = h \, \Delta t \qquad (9\text{-}43)$$

the $+$ sign applying if the solid is the heat source. [With unusual temperature gradients, however, q/A or h by Eq. (9-43) could be negative.]

A principal phase of any nuclear heat-transfer design is prediction of the local and average h values to be encountered. For streamline flow this can be done theoretically, in principle at least, by simultaneously solving applicable fluid- and heat-flow equations such as Eqs. (8-25) and (9-15). For turbulent flow, Eqs. (8-40) and (8-41) are also needed, and empirical data such as Eqs. (8-36), (8-37), and (8-38). A number of analytical and numerical solutions have been obtained, some taking into account the variation of temperature and physical properties. However, the bulk of the useful information consists of generalized dimensionless correlations of actual test values of h for each flow geometry against the most significant dimensions and properties. An adequate theoretical analysis naturally also takes all the significant dimensions and properties into account, but for simplicity not the unimportant ones. Thus both theoretical and empirical results are generally presented in terms of the same variables, and thus dimensionless ratios. Both types of information, when available, will be summarized here for each important geometry.

A simple dimensional analysis by the pi theorem (Sec. 8-22) yields the ratios employed. The dimensions are L, M, θ, and T. The size of a given shape or geometry is described by one length, frequently taken as a diameter D. The coefficient h would be expected to depend on the variables D, $\bar{u}$, ρ, and μ that define turbulence and boundary-layer thickness, on C_P which affects the heat-removing capacity of the eddies, and on k which affects the thermal resistance of the boundary layer. Thus there are $7 - 4 = 3$ ratios needed.

The most commonly employed are the Reynolds number Re, the Prandtl number Pr, and either the Nusselt number Nu or Stanton number St, this latter being more constant and therefore more convenient. Other ratios (Table 8-7) are used when additional variables are significant, as in entrance effects, natural convection, etc.

For usual values of Δt the physical properties of the coolant do not vary so much normal to the flow but that each can be represented adequately by an average value at each position along the channel in correlating test results or computing design values from correlations of h. These average properties are most commonly taken at the "film temperature" $t_f = (t_w + \bar{t})/2$. In many nuclear reactors and heat exchangers even the temperature rise of the coolant stream is not sufficient to affect the fluid properties greatly. In this case the lengthwise average $\bar{t}_f$ may be adequate for computing an over-all average h, or even local values of h. Although in design problems t_w is an unknown in Eq. (9-42) and also affects h (through t_f), a trial-and-error solution for t_w is seldom necessary, since h does not usually vary significantly from a first calculation made at any estimated t_f. All the physical properties employed in Secs. 9-11 through 9-14 are at t_f unless specifically designated by the subscript b or w to be at $\bar{t}$ or t_w, respectively.

9-11. Forced Convection in Channels of Uniform Cross Section. Turbulence generally occurs during heat transfer in substantially equiaxed or parallel wall cross sections when Re_f exceeds some 2000 to 10,000, depending on whether μ varies slightly or greatly. Re_f is computed from the equivalent diameter D_e (Sec. 8-24) and viscosity μ_f at t_f. Below $\mathrm{Re}_f = 2000$ flow is generally streamline. In both cases the velocity distribution departs from that in isothermal flow on account of nonuniformity of μ throughout the cross section. For nonequiaxed and nonparallel wall sections, such as wedges, some portions may be streamline and others turbulent simultaneously, and local values of h would vary widely. For such cross sections usually experimental heat transfer data are unavailable and theoretical analysis is almost impossible. Specific tests, or very approximate estimates of h based on roughly equiaxed portions of the cross section, are the alternatives. The critical region, and particularly $1500 < \text{Re} < 3000$, should be avoided in nuclear reactors because of poor flow stability and temperature fluctuations (see Sec. 8-33).

The local coefficient h decreases with distance downstream,* slightly in turbulent flow [Eq. (9-44b)] and more so in streamline flow [Eq. (9-45a)], since near the inlet the heat need not travel so far into the coolant as farther downstream. Also $\bar{t}$ increases with distance downstream. Thus with reasonably constant heat generation the highest fuel element temperatures are at the outlet of the channel, and with any symmetrical distribution of H they are downstream of the center. Accordingly the usual correlations yielding average or local downstream h for *long* tubes are conservative, and they are adequate unless an extremely precise temperature map of the fuel elements is desired. The local heat-transfer coefficient h_N at position N may be obtained from the expression for the average coefficient $\bar{h}_N$ from 0 to N by the relation $h_N/\bar{h}_N = n + 1$, where n is the power of N to which $\bar{h}$ and h are proportional.

If the beginning of the heated section is at or near a sudden contraction in the flow cross section, as from an inlet header, going downstream the velocity will vary gradually from uniform to the final distribution. Most theoretical and empirical predictions of h are for the final distribution (after an initial "calming length") and will thus again yield conservative metal temperatures (somewhat too high) near the inlet if the inlet velocity is uniform. The following equations all employ $D_e = 4$ (cross section/perimeter) in Nu, Re, and Pe, unless otherwise stated.

Turbulent Flow in Pipes, Equiaxed Ducts, Concentric Annuli, and Parallel-walled Channels. In turbulent flow h is not substantially affected by variation in q/A or t_w except as the physical properties may vary, nor by presence or absence of a previous unheated section. Figures 9-5 and 9-6 show typical results.

The Colburn equation for nonmetals in long channels [17], including fused salts:

$$\text{Nu} = 0.023 \ \text{Re}^{0.8} \ \text{Pr}^{1/3} = \text{Nu}^0 \qquad (9\text{-}44a)$$

See Fig. 9-5.

Local Nu at distance N downstream [18]:

$$\text{Nu}_N = \left[\frac{2.8}{\text{Re}^{0.06}} \left(\frac{N}{D} \right)^{-2.25 \ \text{Re}^{-0.3}} \right] \text{Nu}^0 \qquad (9\text{-}44b)$$

for $2000 < \text{Re} < 60{,}000+$. At higher Re, h_N is 50 per cent over the long-tube value at one diameter and down to 10 per cent over by five diameters.

---

* However, with liquid coolants h will generally rise again farther downstream according to the correlations given hereafter, owing to the changes in physical properties with $\bar{t}$, particularly the decrease in μ.

Nonmetals parallel to a plane: Fig. 9-5. $G = G_{max}$, beyond the boundary layer. $L =$ distance downstream from leading edge.

Nonmetals in coils [17]:

$$\mathrm{Nu} = \left(1 + 3.5\,\frac{D}{D_{coil}}\right)\mathrm{Nu}^0 \qquad (9\text{-}44c)$$

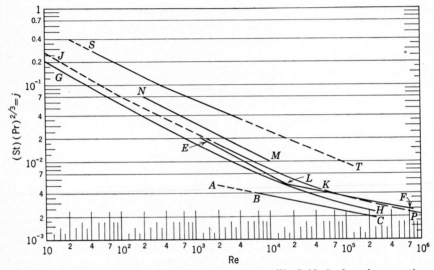

FIG. 9-5. Mean heat-transfer coefficients for nonmetallic fluids in forced convection.

Curve	Direction of flow and shape of solid	j	Re
NM........	Past short strip or pin fins	$(\bar{h}/CG)(C\mu_f/k)^{2/3}$	PG_{max}/μ_f
JKP.......	Normal to banks of staggered tubes*	$(\bar{h}/CG_{max})(C\mu/k)_f^{2/3}$	DG_{max}/μ_f
ELF.......	Parallel to a plane	$(\bar{h}/CG_\infty)(C\mu/k)_\infty^{2/3}$	LG_∞/μ_∞
GH	Normal to a simple cylinder	$(\bar{h}/CG)(C\mu/k)_f^{2/3}$	DG/μ_f
ABC.......	Turbulent flow inside tubes	$(\bar{h}/CG)(C\mu_f/k)^{2/3}$	DG/μ_f
ST........	Flow through beds of convex solids†	$(\bar{h}/CG)(C\mu/k)_f^{2/3}$	$D'G/\mu_f$

* O. E. Dwyer et al., Proceedings of the 1953 Conference on Nuclear Engineering, University of California (Berkeley), F-73, 1953; *Ind. Eng. Chem.*, **48**:1836 (1956).

† C. R. Wilke and O. A. Hougen, *Trans. AIChE*, **41**:445–451 (1945); Eqs. (9-47i, j).

Nonmetals in wide concentric annuli: Eq. (9-44a), using D_e for heating at the outer surface [23]. For heating at the inner surface [168]:

$$\mathrm{Nu} = 0.020\left(\frac{r_3}{r_2}\right)^{1/2}\mathrm{Re}^{0.8}\,\mathrm{Pr}^{1/3} \qquad (9\text{-}44d)$$

For nonmetals between parallel flat walls, Eq. (9-44a) is used for equal q/A at both walls. For q at one wall only, from Eq. (9-44d) for the inner surface, Nu is 13 per cent below Eq. (9-44a).   For through-flow of q, and other ratios, see Ref. 185.

Liquid nonmetal [16] plus p' per cent gas by volume:

$$\text{Nu} = (1 + 6.1G^{-0.53} \log p')\, \text{Nu}^0 \qquad (9\text{-}44e)$$

for upward flow with $25 < G < 100$ lb/(ft²)(sec).

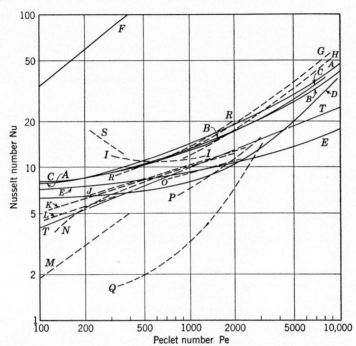

FIG. 9-6. Heating of liquid metals in turbulent pipe flow.   Solid lines A to F, theoretical relations; dashed lines G to T, experimental results.   U, unwetted; W, wetted.

A. Martinelli (Pr = 10^{-2})
B. Martinelli (Pr = 10^{-3})
C. Lyon [Eq. (9-44i)]
D. Deissler
E. Kennison
F. Colburn [Eq. (9-44a); Pr = 0.027]
G. Isakoff and Drew (x/d = 58) (U)
H. Isakoff and Drew (x/d = 138) (U)
I. Untermeyer (W)
J. Johnson et al. (1951)

K. Trefethen (W and U)
L. Doody and Younger (W)
M. Doody and Younger (U)
N. English and Barrett (U)
O. Johnson et al. (1950)
P. Lubarsky (U)
Q. Untermeyer (U)
R. Werner, King, and Tidball (W)
S. Keen
T. Lubarsky and Kaufman [147] [Eq. (9-44i)]

Air in long tubes at high Δt [15]:

$$\frac{hD}{k_{0.4}} = 0.236D \left[\bar{u} \left(\frac{\rho}{\mu} \right)_{0.4} \right]^{0.78} \qquad (9\text{-}44f)$$

where $t_{0.4} = 0.4t_w + 0.6\bar{t}$.

Gases at high $\bar{u}$ [17]: Eq. (9-44a) and

$$\Delta t = t_w - \bar{t} - \frac{\bar{u}^2}{2g_c J C_P} \text{Pr}^{1/2} \qquad (9\text{-}44g)$$

where the exponent $\frac{1}{2}$ decreases toward $\frac{1}{3}$ for $N\bar{u}\rho/\mu > 500{,}000$, roughly.

Solid particles in turbulent gas streams increase Nu^0 based on the gas properties by the ratio $(G_{\text{solids}}/G_{\text{gas}})^{0.45}$ when this exceeds 1 [184].

Nonmetals flowing parallel to tube or rod bundles: Eq. (9-44a) holds, with D_e in Nu and Re, for axial spacing–diameter ratios up to about 1.3 (Chap. 8, Ref. 112). However, Nu rises, with spacing, to 40 per cent above Eq. (9-44a) at a ratio of 1.46 [171]. Circumferentially h is quite uniform. For water, with D'_e in inches from 0.5 to 2 and with D from $\frac{3}{8}$ to $\frac{5}{8}$ in. [51] also:

$$\text{Nu}_D = 0.128(D'_e)^{0.6} \text{Re}_D{}^{0.6} \text{Pr}^{0.3} \qquad (9\text{-}44h)$$

Molten metals [29, 35]: Fig. 9-6 (for conservative design use one-half of these Nu values).

In a long tube with constant q/A:

$$\text{Nu} = 7 + 0.025 \text{Pe}^{0.8} \qquad (9\text{-}44i)$$

Also [147] for $50 < \text{Pe} < 20{,}000$:

$$\text{Nu} = 0.625 \text{Pe}^{0.4} \qquad (9\text{-}44i')$$

In a long tube with constant t_w:

$$\text{Nu} = 4.8 + 0.025 \text{Pe}^{0.8} \qquad (9\text{-}44j)$$

In a short tube with constant t_w [30]:

$$\text{Local Nu}_N = 1.31 \left[\text{Pe}\, (D/N) \right]^{0.455} \qquad (9\text{-}44k)$$

Heating through one wall of long flat channels at constant q/A:

$$\text{Nu} = 5.8 + 0.02 \text{Pe}^{0.8} \qquad (9\text{-}44l)$$

Heating through both walls of long flat channels at constant q/A:

$$\text{Nu} = 10.5 + 0.036 \text{Pe}^{0.8} \qquad (9\text{-}44m)$$

Heating at different constant q/A values through both walls: [29, 185]. Annulus with $r_3/r_2 < 1.4$: Obeys flat channel closely at either wall [29].

Annulus with $r_3/r_2 > 1.4$ [29]:

$$\frac{h_2(D_3 - D_2)}{k} = \left\{ 5.25 + 0.0188 \left[\frac{(D_3 - D_2)\bar{u}\rho C}{k} \right]^{0.8} \right\} \left(\frac{D_3}{D_2} \right)^{1/3} \qquad (9\text{-}44n)$$

Annulus [37] with constant q/A at r_2:

$$\text{Nu}_2 = (5.25 + 0.0175 \text{ Pe}) \left(\frac{r_3}{r_2}\right)^{0.53} \tag{9-44o}$$

Square and triangular sections, at constant q/A, with fair accuracy up to Pe = 300± : Table 9-6. Irregular sections can also be approximated by numerical calculation (see Sec. 9-39), neglecting the eddy conductivity [163].

Unbaffled shell side of heat exchanger [37, 129] (D = tube OD):

$$\text{Nu}_D = 11.6 \text{ Pe}^{0.6} \left(\frac{D_e}{N}\right)^{1.2} \tag{9-44p}$$

Streamline Flow. (Parabolic velocity distribution—as with constant μ —unless otherwise stated.*)

Nonmetals in horizontal or vertical tubes with constant t_w [17, 26]:

$$\text{Nu}_{\text{AM}} = 1.76 \text{ (Gz)}^{\frac{1}{3}} \left(\frac{\mu_b}{\mu_w}\right)^{0.14} \frac{2.25(1 + 0.01 \text{ Gr}^{\frac{1}{3}})}{\log \text{Re}} \tag{9-45a}$$

for Re > 10. Nu_{AM} employs $\bar{h}_{\text{AM}} = q/[A(\Delta t)_{\text{AM}}]$, where Δt_{AM} is the arithmetic mean of the inlet and local Δt.

For flat ducts heated on both sides: Nu_{AM} is about 15 per cent higher than in pipe at the same Gz (use D_e in Nu and Gz).

For slow upward flow in pipes: Eq. (9-46a) [64].

Local Nu in short channels (0 to 15 per cent low [25]):

Pipes, for Gz = $(CGD_e^2/kN) > 12$:

$$\text{Nu} = 1.078 \text{ (Gz}')^{\frac{1}{3}} \tag{9-45b}$$

Flat ducts, for Gz' > 70:

$$\text{Nu} = \frac{hD_e}{k} = 1.22 \text{ (Gz}')^{\frac{1}{3}} \tag{9-45c}$$

For constant q/A in pipes (neglecting axial conduction—see Prob. 9-12, also Table 9-6 footnote):

Gz	0–10±	34.4	73.7	124	191	378	632	955	2000	
Nu	4.36	5	6	7	8	10	12	14	19	(9-45d)

For local Nu in long channels (conservative for short channels): see Table 9-6. Because of viscosity change with temperature, a liquid being heated or a gas being cooled will fall between the parabolic and uniform cases.† The tabulated Nu yields $t_w - \bar{t}$, the usual or flow mean Δt. The

---

* With uniform velocity at the inlet of a tube [97], Nu for gases is about 15 per cent higher than with parabolic inlet velocity at $CGD^2/kN = 1$, and about 50 per cent higher at 10.

† For gases in long tubes at constant q/A, $\text{Nu}_{\text{AM}} = 4.36$, where k is at $1.27\bar{t} - 0.27t_w$. For liquid metals [Chap. 8, Ref. 36] $\text{Nu}_{\text{AM}} = 4.36(\mu_b/\mu_w)^{0.14}$.

next column yields the minimum fluid temperature, of interest in thermal-shock estimates, and the last yields the space mean temperature, for computing the mass or heat or buoyancy of fluid in the channel (see Fig. 9-7). Nonuniform wall shear stress (i.e., rectangular and triangular cross sections) yields higher Nu than expected, due to circulating currents, particularly for nonmetals in long and large channels.

TABLE 9-6. LOCAL NUSSELT NUMBER AND TEMPERATURE DIFFERENTIALS IN STREAMLINE FLOW IN LONG CHANNELS
(Axial conduction neglected)

Channel shape	Constant heating conditions	Velocity	Nu based on flow mean temperature*	$\dfrac{\Delta t_{max}}{\Delta t_{flow\ mean}}$	$\dfrac{\Delta t_{space\ mean}}{\Delta t_{flow\ mean}}$
Round pipe.......	q/A	Parabolic	4.36	1.636	0.727
	q/A	Uniform	8	2.013	1.000
	t_w	Parabolic	3.66		
	t_w	Uniform	5.80		1.000
Flat channel (infinite width)	q/A	Parabolic:			
		Heated 1 side	5.384	1.346	0.942
		Heated 2 sides	8.240	1.288	0.824
	q/A	Uniform:			
		Heated 1 side	6	1.500	1.000
		Heated 2 sides	12	1.500	1.000
	t_w	Parabolic:			
		Heated 2 sides	7.60		
	t_w	Uniform:			
		Heated 1 side	4.94†	1.571	1.000†
		Heated 2 sides	9.88†	1.571	1.000†
Square..........	q/A	Uniform‡	6§	2.00¶	
	q/A	Parabolic	3.68§		
Triangular:‡					
60° × 60° × 60°	q/A	Uniform	4§	3.00¶	
45° × 45° × 90°	q/A	Uniform	3§	4.10¶	
30° × 60° × 90°	q/A	Uniform	2§		
Rectangular:					
$a/b = 2$.......	q/A	Parabolic	4.16§		
$a/b = 3$.......	q/A	Parabolic	4.78§		
$a/b = 5$.......	q/A	Parabolic	5.55§		
$a/b = 10$......	q/A	Parabolic	6.77§		

NOTE: The length in Nu is D_e for all channels.

* R. H. Norris and D. D. Streid, *Trans. ASME*, **36**:525 (1940).

† To include axial conduction divide by $\{1 + [1 + (\pi\alpha/bu)^2]^{1/2}\}/2$.

‡ "Liquid Metals Handbook" (2d ed., R. N. Lyon, ed.; 3d ed., C. B. Jackson, ed.), Government Printing Office, Washington, D.C., 1955.

§ Based on mean wall temperature t_w.

¶ $[(t_w)_{max} - \bar{t}]/(\bar{t}_w - \bar{t})$.

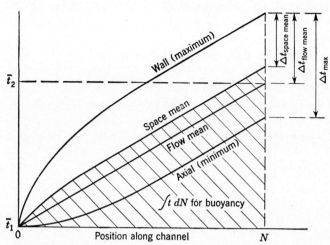

Fig. 9-7. Interrelationships among the temperatures in a stream of constant specific heat heated at constant q/N.

TABLE 9-7. LOCAL NUSSELT NUMBER FOR STREAMLINE FLOW IN LONG CONCENTRIC CYLINDRICAL ANNULI HEATED AT CONSTANT q/A ON ONE SIDE*

(Nu based on D_e; $D_e = D_3 - D_2$)

$\dfrac{D_{\text{insulated}}}{D_{\text{heated}}}$	Uniform velocity, Nu	"Parabolic" velocity (μ constant)	
		Nu	$\dfrac{\Delta t_{\max}}{\Delta t_{\text{flow mean}}}$
54.60	32.96	34.78	1.1
7.389	9.71	10.18	1.16
2.718	6.748	6.758	1.23
1.649	6.168	5.898	1.29
1.284	6.040	5.606	1.31
1†	6	5.384	1.35
0.7788	6.036	5.222	1.37
0.6065	6.132	5.104	1.40
0.3679	6.442	4.962	1.46
0.1353	7.144	4.854	1.55
0‡	8	4.364	2.18

* L. M. Trefethen, NP-1788, 1950.
† Flat duct.
‡ Round pipe.

For constant q/A in long concentric annuli, see Table 9-7.

Non-Newtonian fluids: For Bingham bodies and pseudoplastics h may be estimated from the fact that the velocity distribution is intermediate between constant u and constant μ. Thus $h_{\text{parabolic}} < h < h_{\text{uniform}}$ (see Table 9-6).

9-12. Free (Gravity) Convection. The distance in Nu and Gr is designated L. In turbulent flow the exponent of Gr Pr* is usually $\frac{1}{3}$, in which case L cancels out and local $h = \bar{h}$. Nu for liquid metals is lower than for nonmetals [see Eqs. (9-46u) and (9-46v)]. In streamline flow the

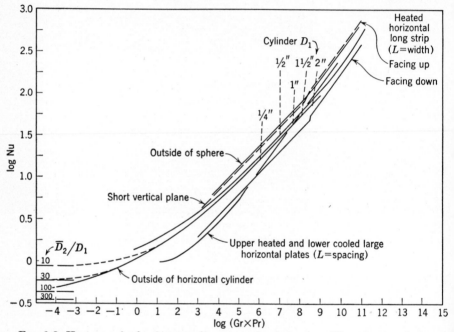

Fig. 9-8. Heat transfer by free gravity convection between solid surfaces and non-metallic liquids.

exponent of Gr Pr is usually $\frac{1}{4}$, and local h is three-fourths of $\bar{h}$ to that point. All physical properties are at t_f and all Nu values are averages over the surface, unless otherwise stated. Most theoretical equations are for constant t_w, and empirical correlations are intermediate between constant t_w and constant q/A. Actual applications are also intermediate, and any theoretical or empirical results available for the same shape may usually be applied with adequate accuracy. Some of the following cases are shown in Fig. 9-8:

Nonmetals at a vertical flat wall or large tube [11], or flowing slowly

* The product Gr Pr is sometimes abbreviated as the Rayleigh number Ra.

upward in a duct (L = height):

$$10^7 < (\text{Gr Pr}) < 10^{12}: \quad \text{Nu} = 0.13 \ (\text{Gr Pr})^{1/3} \qquad (9\text{-}46a)$$
$$3 \times 10^{10} < (\text{Gr Pr}) < 10^{12}: \quad \text{Nu} = 0.067 \ \text{Gr}^{1/3} \ \text{Pr}^{0.43} \qquad (9\text{-}46b)$$
$$10^8 < (\text{Gr Pr}) < 3 \times 10^{10}: \quad \text{Nu} = 0.73 \ (\text{Gr Pr})^{1/4} \qquad (9\text{-}46c)$$
$$10^3 < (\text{Gr Pr}) < 10^8: \quad \text{Nu} = 0.555 \ (\text{Gr Pr})^{1/4} \qquad (9\text{-}46d)$$
$$10^{-2} < (\text{Gr Pr}) < 10^3: \quad \text{Nu} = 1.12 \ (\text{Gr Pr})^{0.15} \qquad (9\text{-}46e)$$

Any fluid at a vertical wall [27]:

$$(\text{Gr Pr}) < 10^8 \pm: \qquad \text{Nu} = 0.677 \left(\frac{\text{Pr}^2 \ \text{Gr}}{\text{Pr} + 0.952}\right)^{1/4} \qquad (9\text{-}46f)$$

Total thickness δ of the laminar boundary layer = $2.66L/\text{Nu}$.

Nonmetals at a vertical wire [40] ($L = D$ = diameter. To D add twice mean free path, if appreciable):

$$10^{-11} < \left(\text{Gr Pr} \frac{D}{Z}\right) < 3 \times 10^{-5}:$$

$$\text{Nu}^{-1} = 1.15 \log \left[1 + 4.47 \left(\text{Gr Pr} \frac{D}{Z}\right)^{-0.26}\right] \qquad (9\text{-}46g)$$

Nonmetals with heat flow between enclosed parallel flat vertical walls [11]* (L = spacing, H = height; $3 < H/L < 42$):

$$10^5 < (\text{Gr Pr}) < 10^7: \quad \text{Nu} = 0.072 \ (\text{Gr Pr})^{1/3} \left(\frac{L}{H}\right)^{1/9} \qquad (9\text{-}46h)$$

$$10^4 < (\text{Gr Pr}) < 10^5: \quad \text{Nu} = 0.20 \ (\text{Gr Pr})^{1/4} \left(\frac{L}{H}\right)^{1/9} \qquad (9\text{-}46i)$$

$$\text{Gr Pr}) < 10^3: \quad \text{Nu} = 1 \qquad (9\text{-}46j)$$

Nonmetals at a horizontal strip facing upward (L = width):

$$(\text{Gr Pr}) > 10^9: \quad \text{Nu} = 0.162 \ (\text{Gr Pr})^{1/3} \qquad (9\text{-}46k)$$
$$10^3 < (\text{Gr Pr}) < 10^9: \quad \text{Nu} = 0.71 \ (\text{Gr Pr})^{1/4} \qquad (9\text{-}46l)$$

Nonmetals at a bounded horizontal disk facing upward, with appreciable $\Delta\rho/\rho$ [45] (L = diameter):

$$(\text{Gr Pr}) > 3 \times 10^{11}: \quad \text{Nu} = 1300 \qquad (9\text{-}46m)$$

$$10^9 < (\text{Gr Pr}) < 3 \times 10^{11}: \quad \text{Nu} = 0.174 \left[\left(\frac{L^3 \rho^2 g}{\mu^2}\right)_f \left(\ln \frac{\rho_b}{\rho_w}\right) \text{Pr}\right]^{1/3} \qquad (9\text{-}46n)$$

* For constant q/A and vertical temperature gradient dt/dz controlling [98],

$$\text{Nu} = A \ \frac{\sinh A \cosh A + \sin A \cos A}{\sin^2 A + \sinh^2 A} \qquad (9\text{-}46o)$$

where $A = (\text{Gr Pr}/64)^{1/4}$ with $L^4 \ dt/dz$ in place of $L^3 \ \Delta t$. Valid for $10^3 < A < 10^4+$.

Nonmetals at a horizontal strip facing downward (L = width):

$$50\% \text{ of Eqs. (9-46}m\text{) and (9-46}n\text{)} \qquad (9\text{-}46p)$$

Nonmetals between an upper cooled and a lower heated horizontal plate [39] (L = spacing):

$$(\text{Gr Pr}) > 10^5\text{:}\quad \text{Nu} = 0.126 \ (\text{Gr Pr})^{\frac{1}{3}} \qquad (9\text{-}46q)$$
$$10^3 < (\text{Gr Pr}) < 10^5\text{:}\quad \text{Nu} = 0.300 \ (\text{Gr Pr})^{\frac{1}{4}} \qquad (9\text{-}46r)$$
$$10^2 \pm (\text{Gr Pr}) < 10^3\text{:}\quad \text{Nu} \rightarrow 1.00 \text{ asymptotically} \qquad (9\text{-}46s)$$

Nonmetals at a sphere (L = diameter):

$$(\text{Gr Pr}) > 10^9\text{:}\qquad \text{Nu} = 0.15 \ (\text{Gr Pr})^{\frac{1}{3}} \qquad (9\text{-}46t)$$
$$10^4 < (\text{Gr Pr}) < 10^9\text{:}\qquad \text{Nu} = 0.75 \ (\text{Gr Pr})^{\frac{1}{4}} \qquad (9\text{-}46u)$$

Any fluid at a horizontal cylinder or bank of cylinders of any diameter [38] (L = diameter; add twice the mean free path, if appreciable):

$(\text{Gr Pr})^{\frac{1}{3}} < 5400D'$ ft:

$$\text{Nu}^{-1} = 1.15 \log \left[1 + 3.77 \left(\frac{\text{Pr} + 0.952}{\text{Pr}^2 \, \text{Gr}} \right)^{\frac{1}{4}} \right] \qquad (9\text{-}46v)$$

For $D > 0.01$ ft: $\text{Nu} = 0.53 \left(\dfrac{\text{Pr}^2 \, \text{Gr}}{\text{Pr} + 0.952} \right)^{\frac{1}{4}}$ $\qquad (9\text{-}46w)$

(Simplifications: For liquid metals the denominator = 1 and for nonmetals the parenthesis = Pr Gr.)

$(\text{Gr Pr})^{\frac{1}{3}} < 10^{-4}\text{:}$ $\text{Nu} = \dfrac{0.87}{\log \ (\bar{D}_2 / D_1)}$ $\qquad (9\text{-}46x)$

(Pure conduction to the walls of the container, of average diameter $\bar{D}_2$. For partial surrounding see Ref. 95.)

$(\text{Gr Pr})^{\frac{1}{3}} = 5{,}400D'$ ft:

$$\text{Nu}^{-1} \leqq 1.15 \log \left\{ 1 + 0.006 \left[\frac{1 + 0.952/\text{Pr}}{(D')^3} \right]^{\frac{1}{4}} \right\} \qquad (9\text{-}46y)$$

Nonmetals between concentric horizontal or vertical cylinders [11] ($L = D_2 - D_1$):

$$10^7 < (\text{Gr Pr}) < 10^9\text{:}\quad \text{Nu} = 0.26 \ (\text{Gr Pr})^{0.20} \qquad (9\text{-}46z)$$
$$5 \times 10^4 < (\text{Gr Pr}) < 10^7\text{:}\quad \text{Nu} = 0.060 \ (\text{Gr Pr})^{0.29} \qquad (9\text{-}46aa)$$
$$(\text{Gr Pr}) < 10^4\text{:}\quad \text{Nu} = 1 \qquad (9\text{-}46bb)$$

Fluids in porous beds between horizontal planes [41]:

$$(\text{Gr Pr}) < 10^2 \pm \text{:}\qquad \text{Nu} = 1 \qquad (9\text{-}46cc)$$

Prob. 9-16. The 1-ft-diameter spherical core of the homogeneous Super Power Water Boiler (SUPO) is cooled by water flowing in three 20-ft lengths in parallel of

¼-in. OD, $\frac{3}{16}$-in. ID stainless-steel tubing distributed uniformly throughout the sphere. The following thermal data were obtained at low to high power [149]:

Power, kw	Water flow, lb/hr	Temperature, °F		
		Fuel solution	Water in	Water out
3.85	355	144.7	64.4	100.2
11.6	573	141.4	59.0	124.2
25.8	1435	158.0	64.0	122.7

Fuel-solution properties average as follows: $k = 0.385$, $\rho = 60.6$, $\beta = 0.0002$, and $\mu = 0.84$, in pounds mass, feet, hours, and degrees Fahrenheit, respectively. Heat transfer to the coolant tube as γ rays was calculated as 0.1 per cent of the total, and heat generation in the free-convection boundary layer was estimated to increase h by less than 20 per cent. Compute $\bar{h}$ outside the tubing and compare with Eqs. (9-46).

Outline of Solution. Using Eq. (9-98) with h_i from Eq. (9-44a), $F = 1$, and $r_0 = r_i = 0$, h_0 comes out [149] about 25 per cent of Eq. (9-46v) or (9-46w) at the lowest power, and 200 per cent at the higher powers. This is attributed to radiolysis of H_2O to $H_2 + O_2$, which forms 0.44 l/min. of gas per kilowatt. At low power the main effect is blanketing of the tubes. At high power it increases circulation, as in slow boiling.

9-13. Forced Convection Normal to Cylinders, Tube Banks, Screens, Beds, etc.

Values of h and $\bar{h}$ are somewhat higher for constant q/A than for constant t_w (i.e., Table 9-6). Most experimental results and most nuclear applications are closer to constant t_w. Equations are for nonmetals unless otherwise stated. All Nu values are surface averages, yield ng $\bar{h}$.

Flow normal to single circular* or streamline cross-section cylinders (see Fig. 9-5):

Nonmetals:

$$0.1 < Re < 1000: \quad Nu = (0.35 + 0.47\ Re^{0.52})Pr^{0.3} \tag{9-47a}$$
$$1000 < Re < 50{,}000: \quad Nu = 0.26\ Re^{0.6}\ Pr^{0.3} \tag{9-47b}$$

Metals, turbulent [37]:

$$Nu = 0.80\ Pe^{0.5} \tag{9-47c}$$

Flow through banks of staggered tubes [26]: Fig. 9-5. Re based on $G_{max} = G$ through the smallest flow cross section. For the fifth and subsequent staggered banks, with nonmetals [26, 35, 187]:

$$500 < Re < 75{,}000: \quad Nu = 0.36\ (Re)_{max}^{0.6}\ Pr^{\frac{1}{3}} \tag{9-47d}$$
$$75{,}000 < Re\ [35]: \quad Nu = 0.38\ (Re)_{max}^{0.8}\ Pr^{\frac{1}{3}} \tag{9-47e}$$

* For gases at high Δt [156] and $Re'' = D\bar{u}\rho_f/\mu_f > 100$:

$$Nu = 0.46\ (Re'')^{\frac{1}{2}} + 0.00128\ Re''$$

The first, second, third, and fourth banks have Nu about 60, 75, 94, and 97 per cent, respectively, above Eq. (9-47d). Banks in line have about 60 per cent of Eq. (9-47d) for the first and second banks, 70 per cent for the third and fourth banks, and 80 per cent subsequently.

Irregular variation of local h, as $h/\bar{h}$ around a cylinder in a staggered bank at constant t_w [22]:

Row	Front	45°	90°	135°	Back
1	1.3	1.0	0.4	0.55	1.0
2	1.9	1.1	0.3	0.9	1.1
3	1.7	1.4	1.0	0.5	1.2
4+	1.5	1.2	0.8	0.7	0.8

Results are similar at constant q/A.

For mercury at $10^4 < \mathrm{Re}_{max} < 10^5$ flowing at 100°F past an equilateral triangular staggered bank of $\frac{1}{2}$-in. tubes $\frac{3}{16}$ in. apart, unwetted $\bar{h} = 11.6\ \mathrm{Re}^{0.52}$ and wetted $\bar{h} = 3.45\ \mathrm{Re}^{0.66}$ after the third row [36]. Alternatively [187] $\mathrm{Nu} = 4 + 0.023\ \mathrm{Pe}^{\frac{2}{3}}$. $\bar{h}$ dropped up to 20 per cent at high Re in the first and last several banks. h_{max} (at front) $= 1.44\bar{h}$ and h_{min} (at back) $= 0.72\bar{h}$, approximately. Local h can drop 30 per cent, and $\bar{h}$ can drop 10 per cent with less than 1 per cent gas by volume when unwetted, but much less when wetted [187].

For NaK under similar conditions [169] $\mathrm{Nu} = 1.17(\mathrm{Pe}_{max})^{0.8}\ \mathrm{Pr}^{\frac{1}{3}}$.

For NaK on the shell side of a heat exchanger with segmental baffles and $\frac{1}{8}$-in. tubes [50], also nonmetals with larger tubes [51]:

$$\mathrm{Nu}_D = 0.19(D'_e)^{0.6}\ \mathrm{Re}_D{}^{0.6}\ \mathrm{Pr}^{\frac{1}{3}}\left(\frac{\bar{\mu}}{\mu_w}\right)^{0.14} \qquad (9\text{-}47f)$$

D = tube OD, D' = inches of outer equivalent diameter, G = geometric mean of G parallel to the tubes and of average crossflow G between baffles.

Flow normal to square or hexagonal bar at $5000 < \mathrm{Re} < 10^5$ [27].

Flow normal to small strips (staggered or unstaggered, but parallel to the flow) and pins [117, 118], for $\mathrm{Re}_P = PG_{max}/\mu$ from 200 to 20,000 and P as the perimeter:

$$\mathrm{Nu}_P = \mathrm{Re}_P{}^{\frac{1}{2}}\ \mathrm{Pr}^{\frac{1}{3}} \qquad (9\text{-}47g)$$

Flow through fixed beds (p = particles; G is based on total cross section S):

Heat transfer between fluid and wall (D) around a bed of spheres (D_p) [21]:

$$\mathrm{Nu}_p = 0.813e^{-6D_p/D}\ \mathrm{Re}_p{}^{0.9} \qquad (9\text{-}47h)$$

Heat transfer between fluid and bed [34] (D' = sphere of same surface):

Re' > 350: $\qquad$ Nu' $= 1.06\ (\mathrm{Re}')^{0.59}\ \mathrm{Pr}^{\frac{1}{3}} \qquad (9\text{-}47i)*$

Re' < 350: $\qquad$ Nu' $= 1.96\ (\mathrm{Re}')^{0.49}\ \mathrm{Pr}^{\frac{1}{3}} \qquad (9\text{-}47j)*$

* Line ST of Fig. 9-5.

Heat transfer between fluid and randomly packed spheres ($\alpha = 0.35$ to 0.4 usually) and rough particles ($\alpha = 0.45$ to 0.6 usually) with $6 < G_s/\mu\beta < 13{,}000$ and Pr near unity [49]:

$$\frac{h\alpha}{G_s C} = B \ \mathrm{Pr}^{-\frac{2}{3}} \left(\frac{4G_s}{\mu\beta}\right)^{-m} \tag{9-47k}$$

where α = fraction voids
$\quad \beta$ = ratio of particle surface to total volume
$\quad G_s = w/\text{total } S$
m and B are the following functions of α:

α	0.35	0.45	0.55	0.65	0.75	0.85
m	0.28	0.35	0.40	0.44	0.47	0.51
B	0.21	0.29	0.40	0.65	1.10	1.80

Flow through fluidized bed (p = particles; G is based on total cross section):

Heat transfer between fluid and wall [19]:

$$\mathrm{Nu}_p = 0.5 \left(\frac{D}{N}\right)^{0.65} \left(\frac{D}{D_p}\right)^{0.17} \left(\frac{\rho_p C_p}{\rho C}\right) \mathrm{Re}_p^{0.8} \tag{9-47l}$$

Heat transfer between fluid and tubes [20]:

$$\mathrm{Nu}_p = 1.25 \left[\frac{G\nu}{(G\nu)_0}\right]^{0.35} \tag{9-47m}$$

Nu employs the mean volume/surface diameter and $(G\nu)_0 = G\mu/\rho$ at incipient fluidization.

Heat transfer between air and bed [28]:

$$\mathrm{Nu}_p = 0.0135 \ \mathrm{Re}_p^{1.3} \tag{9-47n}$$

Flow past a single sphere:

$$\mathrm{Nu} = 0.80 \ \mathrm{Re}^{\frac{1}{2}} \ \mathrm{Pr}^{0.31} \tag{9-47o}$$

Heat transfer between fluid and wire screens: The above method [Eq. (9-47k)] holds if α and β are computed for the screen alone (thickness usually = $2 \times$ wire diameter) and G_s is replaced by $G_{\max}$, at the minimum free cross section.

9-14. Combined Forced and Gravity Convection. Nonmetals in forced convection normal to horizontal cylinders [43]:

If $3 < (\mathrm{Nu}_{\text{forced}}/\mathrm{Nu}_{\text{natural}}) < 0.33$, neglect the smaller Nu.

If $2 > (\text{Nu}_{\text{forced}}/\text{Nu}_{\text{natural}}) > 0.5$, for forced and natural convection in

Counterflow: $\quad\quad\quad$ Nu $= 0.9 \times$ the larger Nu
Perpendicular flow: $\quad$ Nu $= 1.1 \times$ the larger Nu
Parallel flow: $\quad\quad$ Nu $= 1.2 \times$ the larger Nu

Nonmetals heated by turbulent forced upflow and natural convection in short vertical pipes [44] (L = vertical distance up from inlet):

$\text{Re}/(\text{Gr Pr})^{0.40} < 8.25$: $\quad$ Use Eqs. (9-46a) to (9-46d)
$\text{Re}/(\text{Gr Pr})^{0.40} > 15.0$: $\quad$ Use Eq. (9-44a)

Between these limits Nu dips to a minimum of 75 per cent of Eq. (9-44a).

Nonmetals heated by turbulent forced downflow and natural convection show a wider and flatter intermediate region. Nu exceeds $\text{Nu}_{\text{natural}}$ for $\text{Re}/(\text{Gr Pr})^{1/3} > 18$.

Nonmetals in laminar flow at constant q/A in vertical pipes: Over practical ranges ($500 < \text{Gz} < 2500$ and $10^4 < \text{Gr Pr } D/L < 10^7$) liquids heated in downflow yield double the h of liquids heated in upflow [63], contrary to simple theory [64].

9-15. Condensing Vapors. The theoretical formulas assume that the $(\Delta t)_f$ across the condensate film is uniform. At substantially all reasonable condensation rates [11] the outer condensate layer temperature may be assumed equal to the saturation temperature of the vapor, though "temperature jump" occurs at the vapor-condensate interface at high q/A and low pressure [47].

Streamline filmwise condensation on vertical surfaces:

$$\bar{h} = 0.943 \left(\frac{\lambda g \rho^2 k^3}{\mu N \, \Delta t}\right)^{1/4} = 0.925 \left(\frac{k^3 g \rho^2}{\mu \Gamma}\right)^{1/3} \tag{9-48a}$$

where λ is the latent heat of vaporization and Γ the condensate mass flow rate per unit width at distance N down. The local h at N is three-fourths of $\bar{h}$ down to N. For a plane ϕ off the vertical, multiply h by cos ϕ.

Turbulent filmwise condensation on vertical surfaces (actual for nonmetals):

$$\bar{h} = 0.0134 \left(\frac{k^3 \rho^2 g}{\mu^2}\right)^{1/3} \left(\frac{\Gamma}{\mu}\right)^{0.4} \tag{9-48b}$$

Streamline filmwise condensation on n horizontal tubes in a vertical bank:

$$\bar{h} = 0.724 \left(\frac{\lambda g \rho^2 k^3}{\mu \, nD \, \Delta t}\right)^{1/4} = 0.76 \left(\frac{k^3 g \rho^2}{\mu \Gamma}\right)^{1/3} \tag{9-48c}$$

$\bar{h}_n$ for the nth tube $= \bar{h}_1[n^{3/4} - (n - 1)^{3/4}]$.

Theory for metals [46]: Use Eq. (9-48a) for $0 < 4\Gamma/\mu < 10^4+$. Actual may be much lower at high Δt [47]; that is, q/A in Btu per hour per square foot for Hg vapor up to 15 psia $= 400,000 + 1000 \, \Delta t$ for Δt in degrees Fahrenheit and for Na vapor $= 22,000 \, \Delta t - 36,000$ whenever these Δt values $> \Delta t$ by Eq. (9-48a) or (9-48c).

Correction for nonuniform $(\Delta t)_f$ at same condensation rate:

$\bar{h}_{\text{actual}}/\bar{h}_{\text{const}\,t_w}$	0.96	1	1.05	1.09	1.15
$(\Delta t)_{f,\text{bottom}}/(\Delta t)_{f,\text{top}}$	0.5	1	2	3	5

Vertical streamline condensate film mean temperature:

For computing enthalpy of subcooling:

$$\bar{t} = t_{\text{vapor}} - 3 \, \Delta t/8$$

For correct $\bar{h}$ with subcooling [96] multiply λ in Eqs. (9-48a) and (9-48c) by $(1 + 3C_P \, \Delta t/5\lambda)$.

For computing h if $1/\mu$ is linear in t:

$$t_f = t_{\text{vapor}} - 3 \, \Delta t/4$$

High downward vapor velocity can increase h 10-fold [48,96].

Ripples start in vertical filmwise condensation when $(s^3\rho/\mu^4 g) < 0.3 \, \text{Re}^{0.8}$, increasing h appreciably (s = surface tension).

Dropwise condensation on vertical surfaces occurs on smooth surfaces when the vapor-phase wetting angle is appreciably less than 180°. It may be maintained by adding an unwetted nonvolatile compound that will adhere to the surface. With nonmetals h is usually increased 5 to 10 times over Eq. (9-48a). With metals h may be lower than in film condensation; i.e., for Hg vapor at 15 psia [47]

$$\frac{q}{A} = 500,000 \, \text{Btu/(hr)(ft}^2) \text{ at } \Delta t = 300°\text{F}$$

Presence of noncondensible gas even in traces [27] will greatly decrease h if it is not swept away effectively parallel to the surface, and somewhat even if it is. Small diameters decrease h on horizontal tubes below Eq. (9-48c), as the pendant drops cover appreciable fractions of the condensing surface. Yet smaller diameters increase h above Eq. (9-48c) because surface tension keeps the condensate film thin by squeezing it over to preferred spots, from which it drops off [59].

9-16. Boiling Liquids—General Characteristics. Boiling can occur at a heated surface in two ways according to the temperature of the liquid, three according to the temperature of the surface, and three according to the method of circulation of the liquid and vapor. If the liquid is at the boiling point, "saturated," "bulk," or "net" boiling is said to occur. If the liquid is below the boiling point, "subcooled."

"local," or "surface" boiling is obtained. If the temperature of the surface is not much above the boiling point, "nucleate" boiling occurs; if it is well above, "film" or "stable-film" boiling; and if it is intermediate, "partial-film," "mixed," "unstable-film," or "metastable-film" boiling. If the circulation is by natural convection in a large vessel, "pool" boiling is occurring; if it is in a restricted pipe or loop, "thermal-siphon" or "thermal-circulation" boiling. If a pump maintains the flow regardless of buoyancy, "forced-convection" boiling is present. The 18 possible combinations of these characteristics have almost all been studied, and almost all have actual or potential interest in nuclear power. All these, plus homogeneous boiling, are listed in Table 9-8.

TABLE 9-8. TYPES OF BOILING

Method of fluid circulation past heated surface	Liquid temperature	Method of vaporization	Type
Free or external natural convection (in a large volume or pool of liquid)	Subcooled	Nucleate	1
		Mixed	2
		Film	3
	Saturated*	Nucleate	4
		Mixed	5
		Film	6
Confined or internal natural convection (by "thermal siphon" in a closed loop or conduit)	Subcooled	Nucleate	7
		Mixed†	8
		Film	9
	Saturated*	Nucleate	10
		Mixed†	11
		Film	12
Forced convection (past open surfaces or inside of conduits)	Subcooled	Nucleate	13
		Mixed	14
		Film	15
	Saturated*	Nucleate	16
		Mixed†	17
		Film	18
No heated surface—homogeneous heat generation and boiling..........................	Saturated*	Nucleate	19

* Actually slightly superheated at the lower pressures. (M. Jakob, "Heat Transfer," John Wiley & Sons, Inc., New York, 1949.)

† These cases have not been studied experimentally so far.

The stages of saturated pool boiling, shown in Fig. 9-9, will be discussed qualitatively to illustrate certain general characteristics.

In most equipment low heat velocities can be dissipated without boiling, usually by natural convection of the liquid to a cooled wall, or to the free surface, vaporization occurring there because of the lower boiling point. When natural convection cannot maintain the heating surface below the boiling point (point A), bubbles begin to form at favored nuclei in the surface, which constitutes nucleate boiling.   Natural convection of the liquid is still the principal cause of circulation at low boiling rates (i.e., zone A-A').

If the liquid has not yet heated up to its boiling point or is kept below it by simultaneous cooling, type 1 boiling occurs, or even type 2 or 3 at high enough Δt.   If and when the liquid rises to its boiling point, type 4

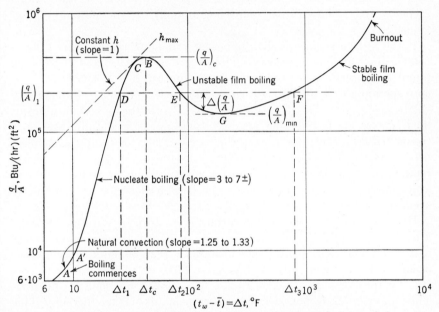

FIG. 9-9. Saturated pool boiling at a submerged horizontal plate or tube.

boiling commences.   Beyond A' the bubbles increase the natural convection circulation, so that beyond A' in strong nucleate boiling the whole mass of liquid is usually circulating rapidly at a very uniform temperature, except for a small amount of superheat near the heating surface. As t_w and thus Δt are increased, the number of nuclei for which Δt is sufficient superheat to initiate ebullition goes up rapidly, as well as the agitation by the bubbles formed.   The line $A'DC$ usually has a slope from 3.4 to 3.6 and is fairly straight.

Once the heating surface and immediately adjacent liquid film have reached the necessary superheats, vapor formation starts immediately, several milliseconds being adequate with a large power burst for expulsion of a small volume of liquid [90].

At some point B a maximum is reached in the rate at which the vapor bubbles can leave the boiling surface and the liquid can yet reach it by simultaneous counterflow. When liquid cannot reach the surface at the same rate at which it is boiling away, depletion of liquid at the surface occurs immediately, the surface temperature rises further (at least locally), and the increasing molecular velocity of the vapor finally forms a vapor "film" that holds the liquid away from the surface. This vapor binding or "film boiling," which causes the "spheroidal state" of water drops on a hot stove, irregularly but increasingly replaces nucleate boiling over the region BEG, of film boiling (type 5). Beyond G, film boiling (type 6), and radiation to an increasing amount, are the methods of heat transfer.

If the heated surface is maintained at a definite temperature t_w by a condensing vapor or a rapidly circulating fluid stream, and the boiling liquid temperature l is fixed by pressure control, Δt is constant and the corresponding q/A by Fig. 9-9 will be obtained. No over-all instability will be observed at any operating condition, except the local instantaneous fluctuations due to bubbling or to alternate nucleate and film boiling in region $BEGF$. Most design is in the region ADC, to obtain the smallest necessary Δt. Film boiling, however, is useful if it is desired to "waste temperature," as in cooling a high-temperature loop by boiling water at atmospheric pressure. It may also be employed at low q/A for more accurate control.

However, if q/A to the boiling surface is set at $(q/A)_1$ in Fig. 9-9, as in a nuclear-reactor fuel element, there are evidently three possible values of Δt. E is shown as follows to be unstable. If an instantaneous increase occurs in heat velocity to the boiling surface, its temperature will rise. But by the curve this means that q/A removed by boiling will drop. Thus an even larger net q/A remains to heat the surface further. This rate of heating of the surface is proportional to $\Delta(q/A)$ in Fig. 9-9 and accelerates to a maximum at G, then decelerates and again yields steady operation at F if the surface or the heat-generating element behind it have not melted or otherwise "burned out" in the process. By the same analysis, an instantaneous decrease in q/A carries Δt down to Δt_1 at D. A similar consideration shows that D and F are stable operating points. If operation at high film-boiling temperatures is to be avoided, the "critical" heat velocity $(q/A)_c$ must, of course, not be reached.

Available information on the 19 types of boiling is summarized in Secs. 9-17 to 9-20. Equations are in consistent units when dimensionless. Other quantities are in units of feet, pounds, hours, degrees Fahrenheit, and Btu, particularly if primed, unless otherwise stated.

9-17. Free-natural-convection Boiling. *Types 1, 2, and 3.* Subcooled free- or gravity-convection boiling could be significant in nuclear reactors cooled by internal natural convection or slow forced convection. However, very little information on it is available. The following can be used as a guide in design work, but tests under actual conditions are called for in most cases.

1. GENERAL.   This type of boiling has been studied over a wide range of Δt by quenching the desired shapes in cold water and following their temperature.   The results are probably also reasonably valid for steady-state operation.   Figure 9-10 shows such a curve for a horizontal 2-in. silver cylinder of $L/D = 1$ and a sphere, quenched at 0 psig in a large

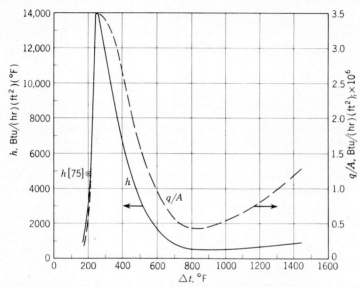

FIG. 9-10. Subcooled pool boiling (2-in. silver sphere or cylinder in water at 62°F and atmospheric pressure).

volume of water at 68°F.   The subcooled liquid condenses the vapor bubbles close to the heated surface, permitting a much higher $(q/A)_c$ and $(\Delta t)_c$ than in type 4, 5, and 6 boiling of saturated liquid.   With this high a subcooling (144°F) no true type 3 film boiling seems to occur, even up to the very high surface temperatures shown.   With less subcooling film boiling would appear.   A general correlation has been proposed [79] for the quenching of steel rods of radius R:

$$\text{Nu} = 407{,}000 \left(\frac{s}{\rho_L R^2} \right)^{1.275} - \text{Pr}^{-\frac{1}{3}} \left(\frac{\Delta t_{\text{sub}}}{\Delta t_{\text{sat}}} \right)^{2.466}$$

where s is the surface tension of the liquid.   However, other experience [61] shows very irregular effects of small changes in subcooling and gas content.   Considerable noise and vibration usually develop in subcooled boiling.   The average life of a bubble is of the order of 0.001 sec, and correlates with bubble size $(q/A)_c$ and the physical properties [160].

2. EFFECT OF DIAMETER ON $(q/A)_c$. Steady data for wires are also available. Platinum wire of 0.0039-in. diameter burns out [11] at a $(q/A)_c$ of 1,200,000 Btu/(hr)(sq. ft²), 55 per cent of that for the 2-in. cylinder,* and 0.0079-in. nickel wire [66] at about 1,320,000. Values of $(q/A)_c$ for other diameters in water at 14.7 psia and 144°F subcooling could be approximated by a log-log interpolation of $(q/A)_c$ vs. D. A tight horizontal coil with turns as little as $D/2$ apart still agrees with straight wire [66].

3. EFFECT OF LIQUID, SUBCOOLING, AND PRESSURE ON $(q/A)_c$. For other liquids, subcoolings, and/or pressures, $(q/A)_c$ may be obtained within some 20 per cent by the correlation [57]:

$$\left(\frac{q}{A}\right)_{c,sub} = \left(\frac{q}{A}\right)_{c,sat}\left[1 + \frac{C_P\,\Delta t_{sub}}{25\lambda}\left(\frac{\rho_L}{\rho_V}\right)^{0.923}\right] \qquad (9\text{-}49a)$$

Δt_{sub} is $t_{sat} - \bar{t}$; at t_{sat} pressure of the liquid plus any dissolved gas equals that on the system [66]. $(q/A)_{c,sat}$ is obtained from Fig. 9-13 and seems to vary little with diameter. $(q/A)_{c,sub}$ by Eq. (9-49a) is for small wires; thus it should be conservative for most fuel elements (according to 2, above). Values of $(q/A)_{c,sub}$ for type 1 boiling could also be approximated from reliable correlations for type 11 boiling by substituting zero velocity.

4. EFFECT OF SUBCOOLING ON Δt_c. Available data [11] show t_c for a wire constant as subcooling is changed all the way to zero. Thus Δt_c for any subcooling can be computed if it is known at one subcooling, or for saturated boiling (i.e., Fig. 9-12).

5. TYPE 1 BOILING ON WIRES. If $(q/A)_c$ and $(\Delta t)_c$ are known or can be estimated, Δt can be approximated for lower values of q/A because q/A falls off about as $(\Delta t_{sat})^{3.1}$. For 0.0079- and 0.01-in. wires in water at 0 psig [66] $q/A = 0.54(\Delta t_{sat})^{3.1}$ Btu/(hr)(ft²)(°F) for $\Delta t > 30$°F. No large effect of diameter on this curve would be expected, the same as in type 4 boiling, since the circulation is similar to that in liquid-phase natural convection, in which h is independent of L or almost so [Eqs. (9-46b), (9-46k), (9-46t), and (9-46w)].

6. EFFECT OF DISSOLVED GAS. Dissolved gas is of interest because of the dissociation of H_2O and D_2O, etc., in reactors. Increasing the dissolved-gas content decreases t_{sat} at a given pressure. Since Δt_{sat} is

---

* This lower $(q/A)_c$ must in part be due to the small heat capacity per unit surface for wires and thus high susceptibility at constant heat-generation rates to overheating from local instantaneous fluctuations in heat-removal rate caused by bubbles. It must also in part be due to nonuniformities in wires, which cause hot spots, and to the (usual) increase of electrical resistance with temperature, which causes an overheated spot to generate yet more heat. D-c heated wires thus yield conservative values of $(q/A)_c$ compared to nuclear heating and to larger shapes, and a-c even more so.

primarily a function of q/A for each liquid, t_{sat}, Δt and t_w decrease about equally [66]. Accordingly, a given q/A can be dissipated with a considerably lower t_w in the presence of gas. On the other hand, vapor phase would start to form at a lower q/A than with degassed water, which would be undesirable for smooth operation because of the change in moderation or neutron economy, though it might be desirable as a safety feature. Gas also decreases $(q/A)_c$ significantly [66], as is shown in Fig. 9-11.

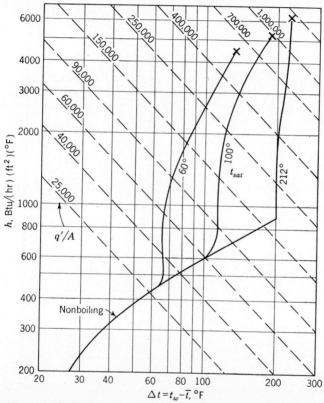

FIG. 9-11. Effect of dissolved gas on natural-convection subcooled boiling of water at atmospheric pressure from wires. X = burnout (nickel wire).

Prob. 9-17. Show that a 0.0039-in.-diameter Pt wire at q/A = 1,200,000 Btu/(hr) (ft²) is generating enough heat to raise itself from 267°F to the melting point 29 times per second, or to the 2-in. cylinder t_c of 374°F 780 times per second.

Type 4. Saturated nucleate pool-boiling data for useful reactor coolants are shown in Fig. 9-12. D₂O may be assumed to follow the H₂O curves.

1. EFFECT OF SHAPE. Cylinders and plates and even 10 vertical banks of tubes [26] yield the same coefficients at the same P and q/A. Addition of shallow fins decreases Δt for a given heat dissipation as would be predicted with the fin area and efficiency. Wires [54] give up to about one-fourth higher $(q/A)_c$ and h, because of the greater accessibility by surrounding liquid. At low q/A, h at wires also falls less, because of more effective natural convection and because of surface tension squeezing the vapor to fewer locations before release [59].

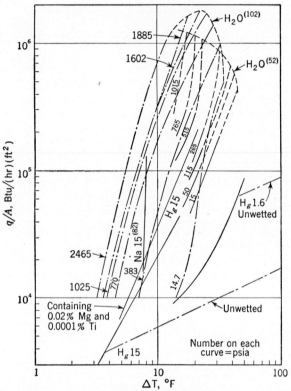

FIG. 9-12. Heat transfer to liquids in pool boiling at plates and tubes under pressure.

2. EFFECT OF PRESSURE. Increase of pressure and temperature always decreases Δt [52] or increases h in nucleate boiling. A correlation [54] for any one liquid is

$$\frac{C\,\Delta t}{\lambda} = K\left[\frac{q/A}{\mu\lambda}\left(\frac{g_c s}{g\,\Delta\rho}\right)^{1/2}\right]^{0.33} \mathrm{Pr}^{1.7} \tag{9-49b}$$

$K = 0.013$ for water on clean platinum and 0.0060 on clean brass. Other surfaces and liquids require other values of K. Uncertainty is about ± 30 per cent. $\lambda =$ latent heat of vaporization, and $s =$ surface tension in force per unit length.   Increase of pressure narrows down the range in which boiling can occur by bringing together the natural-convection limit below and the threshold of film boiling above [52].   Moderate increases in P at normal pressures [56] roughly increase h proportionally to $P^{0.25}$ to $P^{0.4}$.   The highest h values are near the critical pressure and are about $18P_c$ Btu/(hr)(ft^2)(°F) for P_c in pounds per square inch absolute.

3. EFFECT OF AGITATION.   At low q/A additional agitation by forced convection or by bubbles or particularly vapor jets may increase h several fold [53].   A liquid velocity $\bar{u}$ past a boiling surface multiplies the nonboiling h by $\left(1 + \dfrac{\bar{u}\lambda\rho_L}{q/A}\right)^{0.18}$ approximately [80].   At high $\dfrac{q}{A}$ there is no appreciable increase in h, in $(q/A)_c$, or in $(\Delta t)_c$.   Vapor flow past the top of a vertical plate, tube, or deep bundle may even decrease the local h.

4. EFFECT OF SURFACE ROUGHNESS.   Long operation can cause h to decrease as much as 70 per cent without apparent soiling.   Cleaning or aeration restores it.   Coarse scratches increase h more than fine ones, and separated deep scratches even more so; 0.25 in. apart is best for water at atmospheric pressure [55].

5. GENERAL CORRELATIONS.   A rough general correlation of type 4 boiling is Lukomskii's plot [26] of $(q/A)/(q/A)_c$ vs. $\Delta t/(\Delta t)_c$.   Averaging many data, he suggests a straight line on rectangular coordinates going upward from $q/q_c = 0.15$ and $\Delta t/\Delta t_c = 0.3$ to $(0.9, 0.9)$, then curving over to a slope of 0 at $(1,1)$.

6. LIQUID MIXTURES.   Both miscible and immiscible mixtures require a Δt up to several times that interpolated or estimated, respectively, from the pure components.

7. EFFECT OF METAL SURFACE.   There is a definite effect of the metal, due probably to wetting angle, centers for nucleation, and thermal diffusivity.   For instance, in boiling ethanol freshly polished Cu, Au, and Cr and aged Cr show successive 20 per cent increases in Δt at a given q/A [56].

8. LOW BOILING RATES.   If Δt is so small that $(\text{Gr Pr}) < 10^{10}$, liquid natural convection predominates and Eq. (9-46d) or equivalent applies for vertical surfaces and Eq. (9-46k) for horizontal surfaces [11].

9. BUBBLE VOLUME AND FREQUENCY [11, 56].   The theoretical bubble volume V on release as a function of surface tension for liquid-phase wetting angles ψ up to 140° in degrees is $0.0000048(\psi^2 s g_c/g\,\Delta\rho)^{3/2}$.   If ψ is unknown, 50° is a good average.   Above the boiling surface V in rapid boiling grows to an ultimate value some 50 per cent larger (several times larger in slow boiling) because of the liquid superheat.   Bubble frequency per second from a boiling nucleus at moderate pressures is

roughly $64V^{-0.215}$ for breakoff V in cubic millimeters. A bubble is stable in superheated liquid if its diameter $= 4s/\Delta p$ where Δp is the increase in vapor tension due to the superheat. Larger bubbles grow and smaller ones condense.

Vapor volume holdup is important in predicting the change in moderation it causes, and in selecting the volume of expansion tanks required. General experience is that up to 50 per cent of the total volume between closely spaced tubes or flat vertical plates can safely be, and will be, vapor at high heat-transfer rates when there is negligible restriction of the recirculating liquid.

The relative or "slip" velocity u_s of the vapor bubbles with respect to the liquid can be estimated by Fig. 8-18. For low q/A boiling, a conservatively high value of per cent vapor volume could be estimated by neglecting the liquid velocity. For higher q/A a liquid velocity $\bar{u}_L$ up from the heat source can be assumed, the vapor volume V_V at vapor velocity $u_s + \bar{u}_L$ computed, and then a check made as to whether the liquid velocity head $\bar{u}_L{}^2/2g_c$ equals the buoyant heat $V_V(\rho_L - \rho_V)/A_H\rho_L$, where A_H is the horizontally projected area of the heat source. This method is rough, partly because higher u_s is observed than given by Fig. 8-18 (i.e., Fig. 9-14), vapor momentum is neglected, and stream cross section in uncertain. For reliable design, tests should be carried out for the specific conditions of interest.

Although high t_w is required for rapid initiation of boiling, it is of importance in providing safety for boiling reactors that no additional time delay has been found to occur prior to steam formation [90].

10. MAXIMUM BOILING RATE $(q/A)_c$. The critical $(q/A)_c$ at which film boiling initiates is obtainable from Fig. 9-13 for liquids normally boiling above room temperature [52]. $(q/A)_c$ is increased slightly by stirring [161]. The following correlations also yield accuracy of about ± 20 per cent but require physical properties:

Addoms [in 26] for $P_r < 0.7 \pm$:

$$\left(\frac{q}{A}\right)_c = 2.2\lambda\rho_V \left(\frac{gk}{C_P\rho}\right)_L^{1/3} \left(\frac{\rho_L}{\rho_V} - 1\right)^{1/2} \tag{9-49c}$$

Rohsenow [162], using feet and hours:

$$\left(\frac{q}{A}\right)_c = 14.3\lambda\rho_V \left(\frac{\rho_L}{\rho_V} - 1\right)^{0.6} \tag{9-49c'}$$

Kutateladze [57]:

$$\left(\frac{q}{A}\right)_c = 0.16\lambda[sg_c g\rho_V{}^2(\rho_L - \rho_V)]^{1/4} \tag{9-49d}$$

Soiled surfaces average 25 per cent higher $(q/A)_c$. Though these relations include no effect of the metal, other data on wires show over 2-to-1 variation in $(q/A)_c$ under identical conditions with different metals [11].

11. Δt_c. The temperature difference Δt_c at or causing the termination of saturated pool nucleate boiling (see Fig. 9-13) is increased slightly by stirring [161]. It increases with viscosity [75], going for instance from 58° to 220°F on changing from pure water to pure glycerine boiling at atmospheric pressure from a platinum wire. It also varies with the

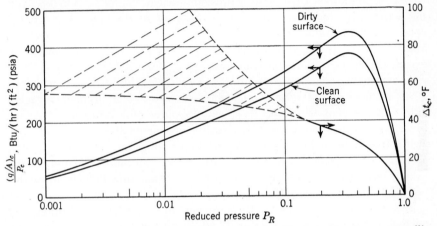

FIG. 9-13. Correlation of Δt_c and $(q/A)_c$ at peak of nucleate boiling for nonmetallic liquids under pressure at plates and tubes.

metal; for water at 1 atm Pt black gave 55°; shiny Pt, 58°; Fe, 59°; Ag, 72°; Ni, 83°; Cu, 87°; and Pb, 104°F (this series is the same as for hydrogen overvoltage). Of these metals, only Pt did not melt on going into film boiling at $(q/A)_c$.

12. SURFACE-ACTIVE ADDITIVES. h for boiling water at moderate pressures can be increased some 20 per cent by adding soluble wetting agents [26]. h for mercury is also improved by Na or Mg, but more important is the increase in $(q/A)_c$ (see Fig. 9-10). Effect of additives on $(q/A)_c$ for H_2O has not been reported. Soluble salts generally decrease h.

13. LIQUID METALS. As per Fig. 9-12, liquid metals give h values comparable to water when they wet the boiling surface. However, Hg and Cd tend not to wet stainless steel even with traces of Na and Mg, respectively, as dissolved wetting agents [82]. Pure Hg on stainless steel and on plain steel tends to shift on use and deaeration from type 6 to 4 boiling if $q/A > 10,000 \pm$ Btu/(hr)(ft²). On wetted steel or other surfaces it tends to superheat and bump, because of the high surface tension and bubble-formation superheat of liquid metals [181].

Type 5. On log scales the region *BE* (Fig. 9-9) is roughly a mirror image of *BD*. *BEG* may thus be approximated by drawing a mirror image of *DB* from $(q/A)_c$ and Δt_c, then curving it to intersect *FG* tangentially (see type 6 discussion). Alternatively, a Lukomskii plot (see paragraph 5 under type 4) can be extended on linear coordinates by a curve from $q/q_c = 1$ and $\Delta t/\Delta t_c = 1$ roughly to $(0.9, 1.1)$, then straight to $(0.2, 1.8)$. Strong stirring can raise the curve 100 per cent [161].

Type 6. Saturated natural-convection "stable-film" pool boiling at atmospheric pressure yields h values of only some 15 to 80 Btu/-(hr)(ft^2)(°F), about 1 per cent of those in nucleate boiling. The blanketing vapor film rises because of its buoyancy in the same way that a condensate film falls by gravity in condensation, and similar relations apply. There is no appreciable effect of surface metal or roughness except on the radiation contribution.

1. MINIMUM q/A. The minimum q/A (at G, Fig. 9-9) ranges from about 10 to 30 per cent of $(q/A)_c$, regardless of pressure or liquid, averaging about 20 per cent of $(q/A)_c$ [57]. At pressures at least up to $P_r = \frac{1}{4}$, $(q/A)_{\min}$ on tubes [86] is roughly directly proportional to P.

If a fuel element has gone over into film boiling (and not burned out), q/A must decrease to $(q/A)_{\min}$ to leave the film-boiling region. This takes place by a jump along the curve $GEBCD$ and down on the nucleate-boiling branch to the existing q/A.

2. h WITHOUT RADIATION. Bromley has shown [58], for the boiling film on horizontal cylinders 0.19 to 0.47 in. in OD at atmospheric pressure, that $\bar{h}_{con}$ for conduction is given within ± 15 per cent by

$$\bar{h}_{con} = 0.62 \left[\frac{k_V{}^3 \rho_V (\rho_L - \rho_V) g\lambda}{D\mu_V \, \Delta t} \right]^{\frac{1}{4}} = 0.53 k_V \left[\frac{\rho_V (\rho_L - \rho_V) g\lambda}{D\mu_V q/A} \right]^{\frac{1}{3}} \quad (9\text{-}49e)$$

where subscripts V and L represent vapor (at its average temperature) and liquid, respectively. The coefficient 0.62 is halfway between 0.724, theoretically derivable for no resistance to vapor flow by the liquid [see Eq. (9-48c)] and 0.512 if the liquid is stationary.

By the same method, for a vertical wall,

$$\bar{h}_{con} = 0.80 \left[\frac{k_V{}^3 \rho_V (\rho_L - \rho_V) g\lambda}{N\mu_V \, \Delta t} \right]^{\frac{1}{4}} = 0.74 k_V \left[\frac{\rho_V (\rho_L - \rho_V) g\lambda}{N\mu_V q/A} \right]^{\frac{1}{3}} \quad (9\text{-}49f)$$

where 0.80 is halfway between 0.943 [Eq. (9-48a)] and 0.666 for stationary liquid.

3. EFFECT OF SUPERHEAT. The sensible heat of the vapor stream is usually significant. Correction can be made [26] by multiplying λ by $(1 + 0.4\Delta t \, C_P/\lambda)^2$.

4. EFFECT OF DIAMETER. On small wires h is 30 to 100 per cent higher than predicted [26]. To fit wires as well as larger cylinders, Banchero et al. [60] replace $0.62D^{-1/4}$ in Eq. (9-49e) by $a_1/D + a_2$. From boiling O_2 and N_2, $a_1 = 0.00375$ ft$^{3/4}$ and $a_2 = 1.64$ ft$^{-1/4}$ for D from 0.025 to 0.75 in.

5. EFFECT OF RADIATION. Radiation transfers additional heat, but in so doing increases the film thickness, decreasing $\bar{h}_{con}$. The approximate effect is to add $\frac{3}{4}h_r$. If the other quantities are given, the wall temperature T_w can be most accurately computed [58] by trial and error from

$$\frac{q'}{A} = \left\{ 0.53k_V \left[\frac{\rho_V(\rho_L - \rho_V)g\lambda}{D\mu_V} \right]^{1/3} \right\}^{4/3} \frac{(T_w - T_L)^{4/3}}{(q/A)^{7/9}} + \frac{1.73 \times 10^{-9}(T_w^4 - T_L^4)}{1/\epsilon_w + 1/\epsilon_L - 1} \quad (9\text{-}49g)$$

ϵ_w is the wall emissivity, ϵ_L the effective liquid absorptivity (1 may be used for H_2O and D_2O and 0.1 for clean metals), and 1.73×10^{-9} is the Stefan-Boltzmann constant in Btu/(hr)(ft^2)($^\circ$R^4).

9-18. Confined Natural-convection Boiling. Thermal-siphon boiling has characteristics intermediate between the local-natural-convection and the forced-convection types. It differs from local natural convection in that the rest of the loop affects the circulating flow and h. It differs from forced convection in that the flow is directly dependent on the heat transferred. It occurs in water-tube boilers, evaporators, and free-boiling nuclear reactors with vertical tubes. Boiling between vertical parallel plates and in tube bundles in closed vessels also falls in this type, particularly if the circulation is reinforced by a vertical mixed-flow tube or chimney.

There are practically no data on subcooled or on film natural-convection boiling in tubes (types 7, 9, and 12). However, subcooled natural-convection boiling usually cannot be very effective by itself because of the small vapor volume, and film boiling probably also cannot promote strong circulation because of the film location of much of the vapor. Thus the relations for type 1, 3, and 6 boiling should apply, respectively [10].

Natural-convection nucleate boiling in tubes (type 10) has been studied frequently [26], but again no useful general correlations are available to give h or w directly from given dimensions and conditions. Accordingly, one of the following three methods must be used to obtain rough design information:

1. Rough interpolation between values estimated for free natural convection and for slow forced convection by the methods of this chapter.

2. Use of actual operating data for type 10 boiling in commercial

boilers, etc. Large 1200- to 1400-psi natural-circulation boilers [88] generally employ 2.5- to 3-in. ID tubes, and yield inlet water velocities of 1.5 to 2 fps. Outlet qualities are 3 to 10 per cent, rising to 15 to 20 per cent at 2400 psi. As with commercial type 16 boiling, their average q/A and other operating conditions are too low for optimum boiling nuclear-reactor design. The highest q/A is about 250,000 Btu/(hr)(ft²) of internal projected area but is seldom reached. The actual local maxi-

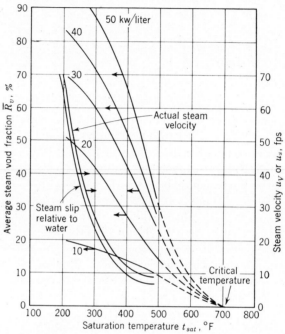

FIG. 9-14. Typical exit-steam velocity and exit-steam "slip" velocity, and over-all average steam volume, in natural-convection boiling of saturated water between vertical plates 5 ft high and ½ in. apart at 50 kw/liter.

mum is lower because of conduction around the tube wall. However, these conditions are useful as conservative lower limits. If the circulating flow is regulated to a desired value, the methods for type 16 boiling may be used to estimate h. Some direct data on h are also available [27] for unregulated flow.

Some experimental results are also available for possible boiling reactor geometries. Untermyer [90] reports a definite slip velocity u_s, decreasing with pressure but independent of liquid velocity. u_s and R_V for upflow between parallel plates ½ in. apart are given in Fig. 9-14. Bailey [91] also found u_s independent of u_L; for water at room temperature and pressure, u_s equaled $1/R_L$ ft/sec. Cook [93], for ¼-in. spacing at 600 psi, found instead that $\bar{u}_V/\bar{u}_L$ remains roughly constant at 2.5 ± 0.5. R_V was substantially proportional to boiling length at constant q/A, which facilitates prediction of buoyancy pressure ΔP_B from an assumed outlet R_V. However, with $\frac{1}{2}$-in. spacing Lottes [159] reports $\bar{u}_V/\bar{u}_L$ about 1.5 at 250 psia, 2 at 100 psia, and 8 at 25 psia.

3. Estimating the flow rate by a trial-and-error calculation. A flow rate w is first assumed. The over-all average "voids" or fractional volume $\bar{R}_V$ of vapor held up is next estimated. For the lower pressures $R_V = 1 - R_L$, from the upflow column of Table 8-11. For higher pressures R_V is obtainable by interpolation between the low pressure and the P_c columns of Table 8-11, or from the slip u_s, or from the relative velocity of vapor and liquid (see Sec. 8-40). The buoyant pressure

$$\Delta P_B = (\rho_L - \rho_V)\frac{g}{g_c}\bar{R}_V \Delta z = (\rho_L - \rho_V)\frac{g}{g_c}\int R_V \, dz \qquad (9\text{-}50)$$

is computed. Then the sum of the frictional and acceleration pressure drops ΔP_{FA} around the loop is evaluated as per Secs. 8-23, 8-41, etc. If L/D_e is low, friction is negligible, and ΔP_{FA} approximately equals 1 to 2 times the velocity pressure of the stream leaving the boiling channel. If $\Delta P_B < \Delta P_{FA}$, the trial w or G was too high, and vice versa.* After circulation has been determined, Eq. (9-52c) can be used to predict h, or the h by Eqs. (9-44a) and (9-49b) added [81].

Between closely spaced parallel cylinders or vertical plates, $\bar{R}_V$ increases approximately as the square root of q/S, the heat dissipation per unit horizontal cross section.† Vertical plates 5 ft high and $\frac{1}{4}$ in. apart in a tank, as an illustration, can safely reach over 90 per cent voids at the top and an average $\bar{R}_V$ of about 0.8 and can dissipate a q/S of about 20 million Btu/(hr)(ft²) [see also Eq. (9-52e)]. Figure 9-14 gives data for a $\frac{1}{2}$-in. spacing [90]. Burnout occurs when R_V reaches 1 at the top, but it is not readily predictable; tests at the desired conditions are necessary. For conservative design (EBWR, etc.) an exit quality of 20 per cent is not generally exceeded.

9-19. Forced-convection Boiling. The behavior of a liquid pumped through a pipe and heated at constant q/A is shown in Fig. 9-15. The

---

* Employing u_s from Fig. 9-14 and setting the buoyant head up to any level equal to two liquid-phase velocity heads at that level yields a close check on the observed $\bar{R}_V$ in Fig. 9-14.

† $G_L = u_L\rho_L(1 - R_V)$ and $G_V = (u_L + u_s)\rho_V R_V$ for upward flow. Eliminating u_L:

$$G_V = \left[u_s + \frac{G_L}{\rho_L(1 - R_V)}\right]\rho_V R_V$$

Since $G_V = q_V/\lambda S$ and $G_L = G_{\text{total}} - G_V$, R_V at any elevation z can be computed from the vaporizing heat q_V for trial values of G_{total}.

flow-mean temperature $\bar{t}$ of the subcooled liquid has a substantially constant axial gradient $d\bar{t}/dx = (q/A)\pi D/wC$. Wall temperature t_w rises at almost the same rate: $t_w - \bar{t} = q/(Ah)$. t_w rises until it is somewhat above the boiling point before type 11 boiling, commonly designated "local boiling," initiates. In this phenomenon small flattened bubbles form at nuclei on the wall, grow very rapidly to a small thickness, then are condensed by the subcooled liquid flowing past and collapse noisily. The total vapor volume is very small, so that no important loss in moderation (reactivity) would occur. The pressure gradient increases somewhat (Sec. 8-40). After the necessary length of channel in local boiling, the liquid has reached its boiling point and type 16 forced-convection "bulk" boiling starts. If q/A is high enough, type 18 film boiling occurs

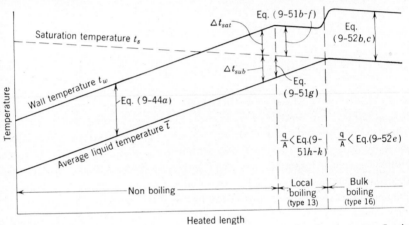

FIG. 9-15. Modes of heat transfer from a uniformly heated tube to water flowing within (nucleate boiling assumed).

and the tube generally burns out. Forced convection sweeps the bubbles away strongly, so the necessary q/A to initiate type 18 boiling exceeds that for type 6 for the same liquid and pressure.

Though not so far reported, short L/D may exert some effect since the radial distribution of subcooling (or of vapor content) will differ from low to high L/D (Fig. 9-7). Strong mechanical vibrations or "singing" of small tubes with internal cooling water flow operating at high pressure and subcooling near burnout have been reported [87]. For such reasons it is necessary to conduct tests under the exact desired operating conditions to obtain reliable design data. Dissolved gas has some effect; unless otherwise stated, t_{sat} refers to gas-free liquid.

Type 13. Forced-convection subcooled boiling has been studied under many conditions, but no fundamental correlation has as yet been proved. Results are usually given in dimensional equations in terms of liquid subcooling $\Delta t_{sub} = t_{BP} - \bar{t}$, and average linear velocity of the liquid before boiling $\bar{u}_L$. t' is in degrees Fahrenheit, q'/A in Btu per hour per square feet, and u'_L in feet per second. There are a number of relations available; the most applicable should be used in each case, with an appropriate safety factor.

1. q/A VERSUS Δt. McAdams et al. [67] give, for water in an annulus 0.09 to 0.26 in. thick heated at its ID of 0.25 in., clean 304 stainless steel 3.75 to 11.5 in. long, 20 to 150°F subcooling, 30 to 90 psia, 1 to 36 fps upward flow,

$$\Delta t'_{sat} = t'_w - t'_{sat} = a_1 \left(\frac{q'}{A}\right)^{0.26}_t \tag{9-51a}$$

where $a_1 = 1.97$ for degassed water and 1.54 when saturated with air. Δt_{sat} increased slightly with pressure. $(q'/A)_t$ is the total heat velocity.

Rohsenow and Clark [68] subtract the computed nonboiling $(q/A)_{nb}$ [by Eq. (9-44a) or (9-45a)–(9-45d)] from $(q/A)_t$ to obtain the boiling $(q/A)_b$. For water heated in a tube at 0 to 2000 psig, 0 to 30 fps upward flow, and 0 to 300°F subcooling,

$$\Delta t_{sat} = a_2 \frac{\lambda^{2/3}}{C_L} \left[\frac{(q/A)_b}{\mu_L} \sqrt{\frac{g_c s}{g(\rho_L - \rho_V)}}\right]^{0.33} \mathrm{Pr}_L^{1.7} \tag{9-51b}$$

within ± 50 per cent, where $a_2 = 0.006$ in a 0.18-in. ID nickel tube and 0.014 in a 0.54-in. ID 304 stainless tube. Given $(q/A)_t$, $\bar{t}$, and t_{sat} at a given location in a tube, t_w can be found by trial-and-error or other solution of the equation

$$\left(\frac{q}{A}\right)_t = \left(\frac{C_L}{a_2}\right)^3 \frac{\mu_L}{\lambda^2 \mathrm{Pr}_L^{5.1}} \left[\frac{g(\rho_L - \rho_V)}{g_c s}\right]^{1/2} (t_w - t_{sat})^3 + h_N(t_w - \bar{t}) \tag{9-51c}$$

Jens and Lottes [69] conclude, for water at $P' = 85$ to 2500 psia and $\bar{u}_L = 3$ to 40 fps,

$$\Delta t'_{sat} = 1.9 \left(\frac{q'}{A}\right)^{1/4} e^{-P'/900} \tag{9-51d}$$

Dissolved gas and nature of the surface have little effect.

Other approximate simple relations for water:

With 0.226-in. ID 347 stainless tube, 0 to 900 cm³ N_2 per liter, 5 to 40 fps, (q'/A) to 3,800,000, and $(\Delta t)_{sub}$ to 236°F [71]:

$$\Delta t'_{sat} = 123 - 35 \log P' \tag{9-51e}$$

for $1000 < P' < 2500$ psia, also with reasonable accuracy down to $P' = 15$ psia.

With 0.18-in. ID L-nickel tube, 10 to 30 fps, $P' = 1000$ to 2000 psia:

$$\Delta t'_{sat} = 28 - 0.012P' \qquad (9\text{-}51f)$$

2. INCIDENCE OF LOCAL BOILING. Forced-convection subcooled boiling commences at a subcooling $(\Delta t_{sub})_0$ at which the curve intersects that for nonboiling heat transfer (Fig. 9-15). From Eqs. (9-44a) and (9-51a), in units of degrees Fahrenheit, Btu, feet, and hours,

$$(\Delta t_{snb})_0 = \frac{43.5 D_e{}^{0.2}\mu^{0.8}}{kG^{0.8}\,\mathrm{Pr}^{\frac{1}{3}}}\frac{q}{A} - a_1\left(\frac{q}{A}\right)^{0.26} \qquad (9\text{-}51g)$$

For q/A low and G high, Eq. (9-51g) will be negative, indicating that nonboiling heating passes directly into saturated boiling without the appearance of local boiling. This can occur at moderate pressures but has not been observed at high pressures [68].

Dissolved gas decreases by up to 20 per cent the q/A at which local boiling initiates at 2000 psi if $\Delta t_{sat} < 36°F$, and at 500 psi if $\Delta t_{sat} = 100°F$ [71].

In any study of reactor cooling in which subcooled or saturated boiling is anticipated, complete plots of the nonboiling and boiling regimes should be made for all anticipated conditions. The feasible regime giving the lowest t_w for any particular value of $\bar{t}$ will be the one obtained.

Prob. 9-18. Reference 71 gives the following empirical conditions at initiation of local boiling of water in a 0.26-in. ID 347 stainless tube:

At 2000 psia:

$$\left(\frac{q'}{A}\right)_c = 0.0027 \frac{G^{0.8}}{D^{0.2}}(\bar{t})^{0.32}(646 - \bar{t}) \qquad (\pm 10\%)$$

At 500 psia:

$$\left(\frac{q'}{A}\right)_c = 0.00083 \frac{G^{0.8}}{D^{0.2}}(\bar{t})^{0.53}(487 - \bar{t}) \qquad (\pm 20\%)$$

in units of Btu, hours, feet, pounds, and degrees Fahrenheit. Compare these equations with Eq. (9-51g), etc.

3. MAXIMUM HEAT VELOCITY $(q/A)_c$.* McAdams [67] gives, for water in an annulus 0.26 in. thick, 0.25 in. ID, and 3.75 in. long $(L/D_e = 7.2)$, $\Delta t'_{sub} = 20$–$100°F$, $\bar{u}'_L = 1$ to 12 fps upward flow, and 30 to 90 psia:

$$\left(\frac{q'}{A}\right)_c = (400{,}000 + 4800\Delta t'_{sub})(\bar{u}'_L)^{\frac{1}{3}} \qquad (9\text{-}51h)$$

Gunther [70] gives, for water heated on one side of a ⅛-in.-square

---

* In general, and certainly at the higher velocities, no effect of direction of flow on $(q/A)_c$ is observed. $(q/A)_c$ is also the same for substantially steady state or for rapidly changing conditions. These equations are for degassed water; presumably dissolved gas decreases $(q/A)_c$ somewhat in subcooled boiling.

channel at $\bar{u}'_L$ = 5 to 40 fps, P = 14 to 160 psia, and $\Delta t'_{sub}$ 20 to 280°F:

$$\left(\frac{q'}{A}\right)_c = 7000(\bar{u}'_L)^{\frac{1}{2}}\,\Delta t'_{sub} \tag{9-51i}$$

Buchberg [71] gives, for water in a 0.226-in. ID tube at $\bar{u}'_L$ = 5 to 30 fps, L/D = 110, P = 250 to 3000 psia, and $\Delta t'_{sub}$ = 3 to 160°F, for G' in pounds per hour per square foot:

$$\left(\frac{q'}{A}\right)_c = 520(G')^{\frac{1}{2}}\,(\Delta t'_{sub})^{0.20} \tag{9-51j}$$

McGill and Sibbitt's results [69] under similar conditions for ID = 0.143 in. and L/D = 21 are represented by replacing 520 by 530 and 0.20 by 0.28.

Jens and Lottes [69] give, for Buchberg's data:

$$\left(\frac{q'}{A}\right)_c = 10^6 a_3\left(\frac{G'}{10^6}\right)^m\,(\Delta t'_{sub})^{0.22} \tag{9-51k}$$

where a_3 is 0.817, 0.626, 0.445, and 0.250 and m is 0.16, 0.28, 0.50, and 0.73, respectively, at 500, 1000, 2000, and 3000 psia.

For published data on water in round, square, and annular ducts from 14 to 3000 psia, $\bar{u}'_L$ from 1 to 54 fps, $\Delta t'_{sub}$ from 65 to 380°F, and $(q'/A)_c$ from 360,000 to 1,130,000 Btu/(hr)(ft²), Bernath obtained [92]

$$\left(\frac{q'}{A}\right)_c = \left[5710\left(\frac{D'_e}{D'_h}\right)^{0.6} + 48\,\frac{\bar{u}'_L}{(D'_e)^{0.6}}\right]\left[102.6\ln P' - 97.1\,\frac{P'}{P'+15}\right.$$
$$\left. - \frac{\bar{u}'_L}{2.22}\left(\frac{D'_h}{D'_e}\right)^{0.6} + 32 - \bar{t}'_b\right] \tag{9-51l}$$

The equivalent diameter D'_e of the stream and the heated diameter $D'_h(=$ (heated perimeter)$/\pi)$ are in feet and P' is in pounds per square inch absolute. The second bracket is the over-all Δt_c, and thus the first is h_c. This method correlates the data mentioned previously within ±15 per cent on the average, and ±30 per cent at worst.

These equations are compared in Fig. 9-16 for a possible power-reactor design. It is seen that even the relations stated to hold for the specific conditions do not agree particularly well among themselves. Thus a design with q/A at all locations and times less than some two-thirds of the lowest pertinent correlation, and no flow instability, seems necessary for safety. For any desired higher q/A, burnout and stability tests at the identical conditions and dimensions are called for.

For flow normal to tubes, $(q'/A)_c$ about double Eq. (9-51h) has been reached at comparable conditions. For upward flow of atmospheric-pressure water subcooled 50 to 125°F past $\frac{1}{16}$-in. wires, at 3 to 10 fps, $(q'/A)_c = 27,500\Delta t_{sub}(u)^{1/3}$ within an average of 10 per cent [165].

It is evident from the previous relations that a heated tube with uniform q/A and a given G, on increasing q/A will burn out at the outlet, where Δt_{sub} has its lowest value. Although the formulas do not correlate burnout at varying conditions very accurately, tests in a single tube show high stability; safe operation is usually obtainable indefinitely at constant q and G within several per cent of $(q/A)_c$. However, a small

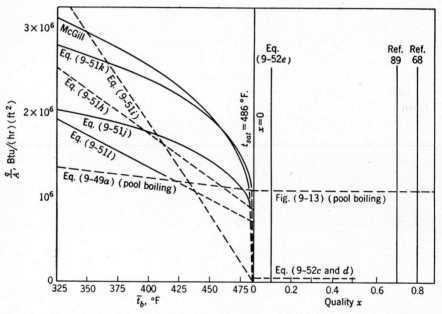

Fig. 9-16. Comparison of equations for critical heat velocity $(q/A)_c$. Water flowing at 10 fps and 600 psia in an annulus with heated ID = 1 in. and insulated OD = 1.25 in.

increase in q/A or decrease in G also decreases Δt_{sub} and will thus cause burnout to approach faster than otherwise. Furthermore, an increase in q/A will decrease G more or less (see Sec. 8-40), depending on orificing and other system characteristics.

Thus a careful calculation of coolant pressure and temperature throughout the reactor under all possible conditions is desirable in order to estimate the safety of the operation and the power at which boiling and/or burnout will occur. If bulk boiling occurs, calculation by finite increments is desirable (Sec. 8-41). A convenient procedure is to plot for each desired channel the q/A vs. position, then the P, $\bar{t}$, and t_{sat} from given inlet or outlet conditions. Next t_w is computed by the possible methods of heat transfer, the lowest t_w fixing the method and $(q/A)_c$ by that method.

Figure 9-17 qualitatively shows the two possible results when the calculation is carried out to determine the burnout conditions. If $(q/A)_2$ is comparable to $(q/A)_{av}$ or larger, as in fast or intermediate reactors with a neutron-thermalizing reflector, or thermal reactors having higher neutron losses in the core than in the reflector, on burnout the (q/A)

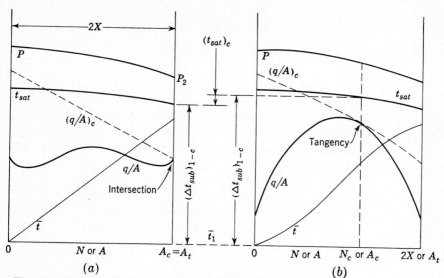

FIG. 9-17. Conditions at burnout by local boiling. (a) Burnout at outlet (heavily reflected reactor). (b) Burnout before outlet (bare or lightly reflected reactor).

and $(q/A)_c$ curves will intersect at the outlet (Fig. 9-17a). If $(q/A)_2$ falls to a small fraction of $(q/A)_{av}$, as in thermal reactors with low neutron losses in the core, the two curves will meet by tangency at some earlier position A_c (Fig. 9-17b).

In some cases it is possible to directly predict the burnout power of a reactor. In the case of Fig. 9-17a, $(q/A)_2$, which from the neutron-flux distribution is in known proportion to total power q_t, is equated to $(q/A)_c$ at the outlet. If $\bar{t}_2$ is given, any one of Eqs. (9-51h) to (9-51k) can be directly solved for w, P_2 [Eq. (9-51k) requires trial and error] or $(q/A)_2$ and q_t. If $\bar{t}_1$ is given or desired instead of $\bar{t}_2$,

$$(\Delta t_{sub})_2 = (\Delta t_{sub})_{1-c} - \frac{q_t}{w\bar{C}}$$

is substituted for Δt_{sub} and the unknown quantity obtained directly or by trial and error, as required. Equations (9-51h) and (9-51i) can also be solved directly for q_t, yielding, respectively,*

$$q'_{t,c} = \frac{400{,}000 + 4800(\Delta t_{sub})_{1-c}}{\dfrac{(q/A)_2}{q_t}\left(\dfrac{\rho_2 S}{w}\right)^{1/3} + \dfrac{1.33}{w\bar{C}}} \qquad (9\text{-}51m)$$

$$q'_{t,c} = \frac{7000(\Delta t_{sub})_{1-c}}{\dfrac{(q/A)_2}{q_t}\left(\dfrac{\rho_2 S}{w}\right)^{1/2} + \dfrac{1.95}{w\bar{C}}} \qquad (9\text{-}51n)$$

In the case of Fig. 9-17b the two simultaneous equations†

$$\frac{q}{A} = \left(\frac{q}{A}\right)_c \quad \text{and} \quad \frac{d(q/A)}{dA} = \frac{d(q/A)_c}{dA}$$

must be solved‡ to find the burnout conditions analytically. This is generally impractical, and in any case the full temperature profiles for a number of operating conditions are usually desirable for stress and other purposes. Thus a graphical solution is best.

4. DENSITY AND VAPOR VOLUME. The vapor quality x in the bubble boundary layer at $(q/A)_c$ has been estimated [68] to be near 0.8. However, this corresponds to a very small volume of vapor in any reasonable channel. The decrease in mass of water per unit of heated surface [69] is, within a ratio of about 2:1,

$$\frac{\Delta m}{A} = \frac{15{,}600\rho_L}{\bar{u}'_L(\Delta t'_{sub})^4}\left[\frac{q/A}{(q/A)_{nb}}\right]^{1.5} \qquad (9\text{-}51p)$$

in units of pounds, feet, seconds, and degrees Fahrenheit, $(q/A)_{nb}$ being at incipient local boiling.

Visual observation of subcooled boiling near atmospheric pressure [67] seems to show a significant and fluctuating volume of vapor. However, quantitative methods yield volumes low enough and fluctuations rapid

---

* ρ_2 is for the liquid phase only. $(\Delta t_{sub})_{1-c}$ is seen from Fig. 9-17 to be $t_{sat,c} - t_1$.

† For brevity, q/A has been used for the local heat velocity dq/dA, in which q is the total time rate of heat gain by the stream up to the given position. Similarly, $d(q/A)/dA$ is more properly d^2q/dA^2.

‡ For the simple case of a bare reactor with cosine distribution of q/A along the channel and substantially constant t_{sat}, Gunther's equation yields

$$\frac{\pi}{2}\frac{N_c - X}{X} = \tan^{-1}\frac{0.619 A_t}{\bar{C}(w\rho S)^{1/2}} = \tan^{-1} B$$
$$\frac{(\Delta t_{sub})_{1-c}}{q_t} = \frac{1 + \sin B}{7200 w\bar{C}_{1-c}} + \frac{\cos B}{4460\,A_t}\left(\frac{\rho_L S}{w}\right)^{1/2} \qquad (9\text{-}51o)$$

Units are the same as for Eq. (9-51m).

enough so as probably to cause no difficulty in reactors, if the subcooling is at least 20°F at high pressures [69] or somewhat more at low pressures for q/A of the usual magnitude desired in nuclear power reactors. For instance, water at 100 psia, 80°F subcooling, and 1 fps showed less than 0.0001 in. of vapor at $q'/A = 400,000$ [72]. Also [70] water at 23.5 psia subcooled 155°F, flowing at 10 fps with $(q'/A) = 2,330,000$ (60 per cent of $(q'/A)_c$) had only 4 per cent of the surface covered with bubbles, which had maximum radii of 0.01 in. and lifetimes of 0.0002 sec. Water at 2000 psia [71] in a ¼-in. ID tube showed less than 1 per cent $\Delta\rho$ with local boiling if $\Delta t_{sub} > 35$°F. A decrease of 3 per cent was observed at $\Delta t_{sub} = 4$°F. $\Delta\rho$ and its fluctuations increased the lower the pressure; at 100 psia the standard deviation of ρ was about 10 per cent.

Type 14. This case has been studied [160] for water at 16 to 60 psia, subcooled 50 to 100°F, flowing at 1 to 5 fps along the outside of a ¼-in. tube. $\Delta t'_{sat} = 600$°F, and agreed well with

$$\frac{q'}{A} = 880,000 \, (\Delta t'_{sat})^{-2.4} \tag{9-51q}$$

Type 15. Forced-convection subcooled film boiling in tubes has not yet been studied, except for the conditions required for its initiation, summarized earlier (type 13, paragraph 2). However, nonaqueous liquids subcooled 20 to 80°F have been flowed upward at 3 to 13 fps [73] past film-boiling horizontal ⅜- to ⅝-in. OD cylinders. The subcooling raised the values of h almost to those for nucleate boiling. The results are approximately correlated by the dimensionless relation (see type 6, paragraphs 2 and 5):

$h_{con} \left(\dfrac{D \, \Delta t_{sat}}{u \bar{k}_V \bar{\rho}_V \lambda} \right)^{½} - \dfrac{7.29}{h_{con}} \left(\dfrac{u \bar{k}_V \bar{\rho}_V \lambda}{D \, \Delta t_{sat}} \right)^{½} \dots\dots$	-2^*	0.5^*	3^*	5^*	7^*	9^*
$\Delta t_{sub} \, C_{L\rho L}{}^{0.95} \left[\dfrac{(u_0 D_0)^{0.90} \mu_L{}^{0.1}}{\Delta t_{sat} \, \bar{k}_V \rho_V \lambda} \right]^{½} \dots\dots\dots$	0	100	200	300	500	800

* Average deviation = ±1; maximum deviation = ±2.

where u is the velocity past the cylinder, u_0 and D_0 are for the duct creating the turbulence in the stream, and λ is the enthalpy decrease from $\bar{l}_V$ to saturated liquid. For ethanol, benzene, and hexane h_{con} was roughly $10u'$ Btu/(hr)(°F)(ft²) at $\Delta t_{sub} = 0$ and varied linearly with Δt_{sub} to $30u'$ at $\Delta t_{sub} = 80$°F. The total $h = h_{con} + ⅞ h_r$.

Type 16. Forced-convection saturated nucleate boiling in channels yields rather low values of Δt, which are fairly independent of dissolved gas content [71] and of velocity and channel dimensions. Because of the considerable pressure drop and its high negative slope vs. flow rate at constant q (Sec. 8-41), type 16 boiling in parallel channels in nuclear reactors may be dangerously unstable unless the channels are roughly equiaxed in cross section and heavily orificed individually at the inlet (Sec. 8-43), or have individual pumps, or flow is upward* and the channels are large so that the buoyant pressure of the vapor is significant compared to the friction (Sec. 8-37).

The negative pressure coefficient (along $S'SCH$ of Fig. 8-17a) is even important in burnout tests on individual tubes. In a system in which the flow drops appreciably if ΔP across the heated tube rises (as with a low-pressure centrifugal pump and no inlet orifice) the values of $(q/A)_c$ observed may be considerably lower than for constant flow and may be misleading. Nonuniformity of q/A or of velocity distribution, as in a square duct [84] also lowers $(q/A)_c$.

1. q/A vs. Δt. In general, h near the inlet of a tube containing boiling water is not very different from that in type 4 pool boiling if the velocity is low or from nonboiling flow if the velocity is high, and correlations for these cases may be accepted as conservative. It is generally observed at low pressures that h approximately doubles [26] as the quality (and thus velocity) increases downstream to about 50 per cent. Thereafter wetting is poorer and h starts to drop, particularly after 80 per cent vaporization. Turbulence promoters are not advantageous.

For light petroleum oils at 3 to 30 fps in an annulus of 0.225 in. heated ID and 0.012 in. width and $\Delta t'_{sat} = 50$ to $250°F$ [76]:

$$\left(\frac{q'}{A}\right) = 0.000097(\Delta t'_{sat})^2 \tag{9-52a}$$

For a number of liquids heated near 15 psia at constant q/A in slow vertical upflow, entering at 0 per cent quality and leaving at under 5 per cent, the average h in pipes is correlated within several per cent by [81]:

$$\text{Nu} = 0.0086\,(\text{Re}_m)^{0.8}\,\text{Pr}^{0.6}\left(\frac{s_{H_2O}}{s}\right)^{\frac13} \tag{9-52b}$$

where Re_m employs the log mean of the average inlet and exit velocities, neglecting vapor slip. s is surface tension and all physical properties are for the liquid.

For water entering 0.465-in. ID 347 stainless tubes at 1.5 to 6 fps and boiling at 45 to 200 psia and $q/A = 50,000$ to $250,000$ Btu/(hr)(ft²), the local Nu_L, in terms of the quality x and the total G, is [89]

$$\text{Nu}_L = \left[4.3 + 0.0005\left(\frac{V_V - V_L}{V_L}\right)^{1.64} x\right]\left(\frac{q}{AG\lambda}\right)^{0.464}\text{Re}_L^{0.808} \tag{9-52c}$$

* Carter [84] predicts that boiling downflow is more stable than upflow, but his tests show that in parallel downflow channels q cannot safely exceed about one-half that for a single channel.

Equation (9-52c) held within 10 per cent on the average and was independent of N/D, from $x = 0$ to 0.4. Above $x = 0.5$, h decreased toward the value for steam at $x =$ about 0.7. Above $x = 0.6$, temperatures fluctuated widely. Nu_L and Re_L employ k_L and μ_L, respectively.

2. MAXIMUM HEAT VELOCITY $(q/A)_c$. The highest values of $(q/A)_c$ are obtained [76] near or somewhat below a reduced pressure of one-third, as in saturated pool boiling (Fig. 9-13).

$(q/A)_c$ compared to nonboiling $(q/A)_{NB}$ by Eq. (9-44a) calculated for the same fluid and wall temperatures is given for JP-3 (gasoline) and JP-4 (80 per cent gasoline, 20 per cent kerosene) hydrocarbon jet fuels [76] of critical pressure P_c approximately $= 550$ psia, in tubes at $\bar{u}_L = 3$ to 80 fps, $P' = 30$ to 500 psia, $D_e = 0.15$ to 0.6 in. and G' in pounds per hour per square foot by

$$\frac{(q/A)_c}{(q/A)_{NB}} = 2000(P'G')^{-\frac{1}{3}} = \frac{h_c}{h_{NB}} \tag{9-52d}$$

In soiled tubes burnout may ensue [77] below $(q/A)_c$ for clean tubes. In square or other angular sections at constant q/A, $(q/A)_c$ may be only a fraction of that in tubes.

For $q'/A > 100,000$ burnout with water in general occurs when x reaches about 0.7 [89] to 0.8 [68]. Burnout x seems to increase slowly with pressure and with decrease in q/A.

Deissler has correlated all the available data for $(q/A)_c$ for water in tubes in saturated and slightly subcooled boiling, within ± 25 per cent on the average, by the ratios $G_V/G_L = (q/A)_c/\lambda G_L$ and $G_L D/\mu_V = Re_b$. If $Re_b < 20,000$, $G_V/G_L = 0.006$; and if $Re_b > 20,000$,

$$\frac{G_V}{G_L} = 1.25(Re_b)^{-0.54}$$

G_L is for liquid in the pipe and G_V for vapor generation at the wall (based on wall area).

A rough general correlation [78] of effect of velocity on the safe q/A for water (a lower limit of $(q/A)_V$ for saturated boiling in tubes), for quality x from 0.05 to 0.6, is

$$\bar{u}'_L x_2 = 0.8 \tag{9-52e}$$

This yields the following values of safe production rate of saturated vapor G'_V and heat of vaporization q'_V/S per second per square foot of boiling channel cross section S:

P', psia.....	14.7	100	500	1,000	1,500	2,000
G'_V..........	47.8	45.0	40.6	37.0	34.0	31.1
q'_V/S........	46,400	40,000	30,700	24,000	18,900	14,400

In general, these values of burnout quality do not hold below $N/D = 25$. At lower N/D the burnout x_2 can (conservatively) be considered to decrease proportionally to N/D.

3. COMMERCIAL BOILER OPERATION. Some commercial high-pressure boilers operate with forced convection [88]. The main purpose of the pumps is to maintain the desired distribution of the water among the tubes and thus decrease the danger of burnout by boiling dry or scaling, make up for low buoyancy at low heads or high pressures, permit smaller, lighter, and thinner-walled tubes, and obtain higher exit quality. It is also possible to force the inlet water along any desired path, such as first past the region of highest q/A, rather than being limited to substantially vertical upward flow. Each tube usually has a flow distribution orifice (Secs. 8-43 and 8-44) with ΔP_o approximately equal to the friction and acceleration ΔP_{FA} in the highest flow tubes, and greater in the others.

Recirculation is used commercially above about 1500 psi with 1 to $1\frac{1}{2}$-in. ID tubes, and 2.5 to 5 fps inlet liquid water velocities.* The maximum local q/A_i ranges up to about 200,000 Btu/(hr)(ft²). At the exit, quality is usually 10 to 30 per cent (95 or 100 per cent in the Sulzer and Benson boilers†), but q/A is much lower. Because of the presence of some scale-forming salts in the usual boiler water and the limited q/A values obtainable in fuel-fired furnaces, these operating conditions may be considered conservative lower limits for nuclear reactors.

4. LOCATION OF BUBBLES. Free bubbles tend to accumulate in the center of the stream because of shear stress distribution and Bernoulli forces. This decreases R_V and the effect of the bubbles on moderation. Whirling of the coolant as it passes through the tubes has the same effect, due to centrifugal acceleration, and should considerably increase boiling reactor stability and permissible heating density. Vapor volume or the total coolant mass in boiling coolant channels can be computed subject to any desired assumptions. For instance, for a slip velocity u_s and heat pickup distribution $q(x)$, the liquid velocity u_L can be computed as a function of x, the distance from start of boiling, by trial and error from

$$S = \frac{w}{\rho_L u_L} + \frac{q(x)}{\lambda}\left[\frac{1}{\rho_V(u_L + u_s)} - \frac{1}{\rho_L u_L}\right] \qquad (9\text{-}52f)$$

The total vapor volume in the channel is then the graphical integral

$$\int_0^X S_V \, dx = \frac{1}{\rho_V \lambda}\int_0^X \frac{q(x)\, dx}{u_L + u_s} \qquad (9\text{-}52f')$$

---

* It is found permissible to operate for an hour or more at about 75 per cent of these normal inlet velocities (half the pumps shut down).

† Superheater tubes are used at 2500 psi and 1050°F.

If $u_s = 0$ and q/A and pressure are substantially constant, the correct mean specific volume $\bar{V}$ over the boiling length is the log mean of the liquid and outlet specific volumes.   For a bare reactor with cosine distribution of heating and saturated inlet coolant $\bar{V}$ is the geometric mean. For this case and weighting the density by the flux distribution, as would be appropriate in computing coolant irradiation, the geometric mean is again correct, in conjunction with the mean neutron flux and the total residence time.

5. LIQUID METALS.   In a well-wetted tube, obtainable by Na or K additions or strong turbulence, h for mercury is reported [83] to be substantially the same function of dP/dx whether the mercury is below the boiling point, boiling, or superheated vapor.   It could thus be approximated from relations for liquid metals or gases in tubes; also [83] by $h = 6100(\Delta P/\Delta x)^{0.445}(D)^{1/3}$ Btu/(hr)(ft^2)(°F) for D in., x ft, and P psi. Poorly wetted tubes fluctuate unpredictably in temperature between wetted and unwetted (hotter) operation.

Type 18.   Forced convection saturated film boiling, as for type 15 boiling, has been studied for upward flow of nonaqueous liquids past horizontal cylinders of $\frac{3}{8}$- to $\frac{5}{8}$-inch diameter [74].   It was found that at low liquid velocity $[u/(gD)^{1/2} < 1]$, the same relation for h_{con} [Eq. (9-49e)] as for natural convection is valid, and $\bar{h} = \bar{h}_{con} + \frac{3}{4} h_r$.   For $u/(gD)^{1/2} > 1$,

$$\bar{h}_{con} = 2.7 \frac{uk_V\rho_V\lambda(1 + 0.4\,\Delta t\,C/\lambda)^{1/2}}{D\,\Delta t} \tag{9-52g}$$

and $\bar{h} = \bar{h}_{con} + \frac{7}{8} h_r$.   The local h_{con} and q/A are considerably larger along the leading edge than farther around.

In natural-convection nucleate boiling at a given q/A, burnout of the heating surface generally occurs when $(q/A)_c$ is reached and film boiling starts.   Only high-melting surfaces like platinum and graphite, which can stand the temperature necessary to radiate a large part of the $(q/A)_c$, generally avoid burnout.   In forced convection, however, the vapor film is thinner and can conduct more heat, particularly at high pressures. Thus burnout in film boiling is more readily avoided.   For instance, for hydrocarbon jet fuels [76] above 25 fps and 300 psia, stable-film boiling can be obtained above $(q/A)_c$ in 347 stainless tubes.   At $(q/A)_c$, Δt_{sat} jumps some 200 to 300°F but then stabilizes there.   For further increases in the film-boiling region, $\log q/A$ increases about three times as fast as $\log \Delta t_{sat}$, until the burnout wall temperature is reached.

In another case, the film-boiling plot of $\log q/A$ vs. Δt for hydrocarbon mixtures has, after a nucleate boiling jump in q/A, fallen back on the extension of the nonboiling curve.   This has been attributed [77] to the presence of higher boiling constituents.   If so, addition of such a substance to a reactor coolant may provide a method of protecting or warning against overheating.

9-20. Homogeneous Boiling. Saturated homogeneous boiling (type 19) would be expected in a homogeneous reactor or in a heterogeneous reactor using a liquid fuel solution if the fuel solution reached the boiling point plus the necessary small superheat to nucleate bubbles. Actually, because gaseous fission and decomposition products increase the vapor pressure and because fission-fragment recoil forms nuclei, boiling would no doubt occur sooner, probably with negligible superheat. It does not seem likely that local *subcooled* boiling with bubble collapse could be obtained in a homogeneous fuel solution, or even in aqueous suspensions of fuel-element compounds.

Boiling of an aqueous fuel solution may eventually become a desirable or feasible normal operating condition in power reactors. In any case, however, it offers a valuable safety feature for homogeneous reactors—a negative coefficient of substantially infinity. Accordingly it has been studied primarily from the standpoint of speed of response to a sudden heat pulse.

1. *Superheat to Initiate Boiling.* Available theories predict higher super-heat than is actually observed [65] before bubbling of pure liquid occurs in a container. For water at 100, 400, and 1000 psia the maximum superheat required was, respectively, 85, 35, and 10°F, the average superheat required being 25 per cent lower. Theory yields 275, 175, and 80°F, respectively, for pure liquid. In another study [158] 40°F super-heat was sufficient at 15 psia with an exponential power rise period of 0.017 sec, corresponding to a time delay of only 0.005 sec.

2. *Steady-state Boiling.* The higher the heating density H, the higher the necessary superheat, since more new nuclei need to develop into vapor bubbles per unit of time, and the lower the density of the vapor-liquid mixture. Circulation rate and mixture density along each assumed vertical streamline can be roughly estimated by trial and error from a reasonable slip velocity and a method similar to that outlined under type 10 boiling. In pool homogeneous boiling the friction is negligible and the buoyancy pressure ΔP_B would all go to acceleration of the fluids. The "slip" velocity in feet per second for water up to moderate pressures [91] roughly equals R_V, the fraction of vapor by volume.

3. *Rate of Bubble Growth* [65]. Bubble growth is steady for any one bubble at about 0.02 to 0.09 fps radially, which checks available theory. Rate of collapse of bubbles of substable size has also been studied.

4. *Rate of Density Change.* The decrease in density with time due to an increase in the heating density H follows available theory [85] for bubble initiation and growth. A sudden jump from low power to about 4 to 30 kw/liter for water boiling at atmospheric pressure requires a further super-

heat of about 1°C before the volume starts to increase, or about 0.3°C if gas is present. At 135 psia some 2°C is required either way, gas having little effect. The volume then increases at about 50 per cent per sec [85]. This delay of 0.1 to 0.3 sec decreases in higher power pulses but does not disappear.

9-21. Boiling in Nuclear Reactors. Originally it was generally assumed that saturated boiling could not be permitted in nuclear reactors [90]. This assumption was due primarily to fear of the flow instability obtainable in parallel coolant channels, which would immediately decrease the flow rate until boiling dry and burnout occurred in any channel in which boiling might commence (see Sec. 8-43). Other adverse possibilities were burnout due to film boiling even without flow instability (see Fig. 9-9) and burnout due to scale formation, as occurs in fuel-fired boilers. It was also anticipated that jumps in k, or reactivity ρ, and thus in power level, might take place, and possibly nuclear runaway, if the bubbles collapsed as a result of a pressure or cold-liquid surge. Scramming might also occur with increased turbine load owing to reactor-pressure falloff and coolant flashing just when more power is needed. Thermal fatigue of the fuel elements might also occur as a result of local oscillation of coolant density, and thus of cooling power and surface temperature. Local thermal-neutron flux and therefore heating density fluctuations would also occur because of the boiling, particularly with H_2O on account of its short neutron-diffusion length, and might also cause thermal fatigue. While subcooled boiling is much less unstable than saturated boiling, in that the increase in pressure drop and in volume are much smaller, some fluctuation would occur, and it was considered too close to saturated boiling for safety.

Accordingly the first liquid-cooled reactors have been designed so that no boiling could occur, by employing a high enough coolant pressure that the surface of the fuel elements never should reach the saturation temperature of the coolant. The design procedure of Table 9-11 is usually employed to estimate the maximum possible fuel-element surface temperature, or surface "hot spot." The operating pressure of the coolant system is then set as much above the saturation pressure of the hot spot as desired.* One actual water-cooled reactor [104] is operated at the saturation pressure corresponding to 96°F above the estimated maximum hot-spot temperature, which is equivalent to the hot-spot's reaching saturation temperature at double the rated load. Another reactor is pressurized to 75°F above the hot spot. There is added protection because t_w must exceed t_{sat} to cause boiling (Sec. 9-19). These high safety margins are intended to allow ample time for automatic controls

---

* Actually, subcooled boiling would initiate slightly downstream of the hot spot because of the rising coolant temperature with downstream distance.

actuated by high power, short period, high coolant temperature, or low flow to scram the reactor before boiling starts, in the event of a power surge or coolant flow failure.

The following benefits are obtainable if the fuel-element and core designs can tolerate incipient or slight localized boiling at infrequent peak temperatures:

1. Lower operating pressure at the same coolant temperature and flow, and thus smaller, cheaper, and lighter construction of the complete coolant circuit.

2. Lower coolant flow rate at the same pressure and inlet temperature, and thus smaller pipes and pumps and less coolant.

3. Lower coolant pressure drop for the same core and piping, and thus less pumping power, and somewhat smaller pump and motor.

4. Higher coolant outlet temperature at the same pressure and metal temperature, and thus higher net power output, or else a smaller heat exchanger.

These benefits will increase if localized subcooled boiling can be permitted at the hot spots during normal continuous operation, and will be greater yet if more general subcooled boiling, or localized saturated boiling, is feasible. For instance, the Pressurized Water Reactor (PWR) is planned to operate at $\bar{t}_2 = 542°F$ (Prob. 9-27), and 2000 psi (636°F saturation), to positively prevent boiling at hot spots under worst conditions. But the saturation pressure for 542°F is only 980 psi. If no margin of safety were required and localized boiling were permitted at hot spots, an intermediate pressure would suffice; possibly 1400 psi if slight subcooled boiling was permissible, and 1200 psi if saturated boiling at the outlet. The situation is favored also by the Δt_{sat} required before boiling can initiate, and by $\bar{t}$ at the hot spots being lower than $\bar{t}_2$.

When a reliable indicator of localized subcooled or saturated boiling becomes available, an additional benefit will become possible:

5. The initiation of boiling anywhere in the reactor could be used to activate a routine operating control or a scram circuit.

If extensive boiling is permissible in a power reactor, so that power can be removed as vapor, the first four benefits are yet further enhanced. For instance, a water-boiling reactor to replace PWR would operate at 500 psi and might have a 500°F hot spot instead of 2000 psi and 600°F, respectively [90]. Additional advantages also appear:

6. Alternatively, the saving of the temperature drop across the heat exchanger can be put into higher over-all efficiency of the power cycle.

7. One coolant circuit and its associated equipment is saved.

8. The external holdup or inventory of primary coolant is decreased by the use of vapor in the outgoing line, smaller flow rates in all lines, and no heat exchanger. This also means less external shielding.

9. The vapor bubbles may serve as an automatic control mechanism

in normal operation (see Fig. 9-18a). Either an H_2O or a D_2O boiling reactor can be designed so that it is self-regulating (as opposed to auto-catalytic—see Fig. 9-18b). Boiling increases (a) resonance absorption by expulsion of moderator (favorable to control in natural-U reactors);

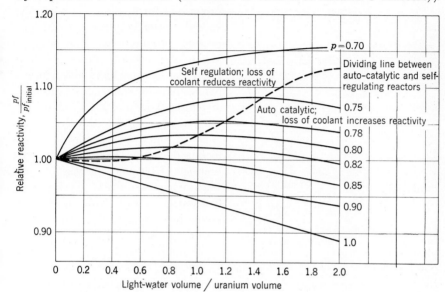

FIG. 9-18a. Effect of distributed steam voids on two types of boiling reactors. p is the resonance-escape probability. (From S. A. Untermyer, Nucleonics, July, 1954.)

(b) thermal utilization by expulsion of neutron absorber (unfavorable in H_2O boilers); and (c) neutron leakage by expulsion of moderator (favorable in small reactors). Some possible self-regulating boiling reactor types are as follows [90]:

Fuel	Moderator	Coolant	Self-regulation requirements
Natural uranium.............	D_2O	D_2O	None
Natural uranium..............	D_2O	H_2O	Limit on amount of coolant—see Fig. 9-18b
Natural uranium..............	Graphite	H_2O	Limit on amount of coolant—see Fig. 9-18b
Natural uranium..............	Graphite	D_2O	None
Partially enriched uranium.....	H_2O	H_2O	Limit on water-uranium ratio
Natural uranium–enriched control rods...............	H_2O	H_2O	Limit on water-uranium ratio
Fully enriched...............	H_2O	H_2O	Core must be small

Careful study of each specific reactor design seems necessary before it can be established that boiling is permissible as a normal general or as a localized condition. *Flow stability* will ordinarily be absolutely necessary i.e., that the flow not progressively avoid the boiling channel or hot spot enough to permit burnout. For roughly equiaxed channel sections such as empty pipes, inlet nozzles by the method of Sec. 8-43 could be employed for downward flow, and for upward flow if needed. Upward flow is preferable for stability, and except at high outlet velocities should not require nozzles (i.e., Fig. 8-17c).

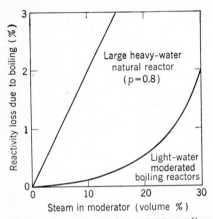

Fig. 9-18b. Effect of light-water cooling on reactivity of natural-uranium reactors with graphite or heavy-water moderator. (*From S. A. Untermyer, Nucleonics, July, 1954.*)

When the coolant flows in thin annuli, or in large ducts with a number of small fuel elements, or through an entire core in a single channel, inlet nozzles will be of little assistance in maintaining flow over a localized area that enters into boiling. However, inertia of the stream, redistribution of the heat dissipation away from a region going into film boiling, turbulent crossflow, and other effects may make boiling stable.

Actually, *thermal stability* (avoidance of burnout) without flow stability *may* be adequate for avoiding the need to design to absolutely prevent boiling. Thin, high-melting fuel elements in high-pressure, high-velocity water may not necessarily burn out whenever film boiling (see type 18 boiling) is reached.

Finally, *nuclear stability* would have to be assured in spite of the irregularities of boiling. For reactors boiling only at hot spots, H_2O might be preferred because the short neutron-diffusion length requires only a few mils of steam film to decrease the neutron flux and locally protect the surface. If there is no need to protect in this way against burnout, reactors producing net steam might be more stable with D_2O than with H_2O, because small temporary liquid-poor regions would have less effect on neutron flux.

The mass of coolant (considered here as moderator) displaced per unit of heat by boiling is $(V_V - V_L)/V_L\lambda$. For water,

Pressure, psia	14.7	100	500	1000	1500	2000
$(V_V - V_L)/V_L\lambda$, lb/Btu	1.65	0.28	0.061	0.030	0.0194	0.0136

Pressure, psia	2500	2800	3000	3100	3200
$(V_V - V_L)/V_L\lambda$, lb/Btu	0.0099	0.0080	0.0067	0.0060	0.0046

Thus the higher the pressure the greater the nuclear stability. From this standpoint it might also be desirable to design a boiling reactor so that it is neither significantly self-regulating nor auto-catalytic, i.e., on or only slightly above the dashed line of Fig. 9-18a. If the coolant is only a small part of the total moderator it also will have less effect on the neutron flux. Control circuits would have to prevent major fluctuations in the whole reactor, such as sudden pressure drops which might cause coolant flashing generally and localized burnout, or sudden pressure increases or drops which might cause power surges according to whether the reactor was self-regulating or autocatalytic respectively. An outlet nozzle might help in these cases. It is evident that only thorough analysis and tests for each specific reactor design and set of operating conditions will determine the feasibility of localized or net boiling. It seems likely, however, that boiling can be employed under proper conditions [173], both net boiling for direct production of steam, and localized subcooled or even saturated boiling at hot spots in "nonboiling" reactors.*

In a simple boiling reactor and turbogenerator (operating say at 500 psia) an increase in power demand causes a drop in boiling pressure, with consequent flashing of coolant. The increase in vapor volume decreases the reactivity, and the generated power diminishes until the controls bring the system to steady operation at the desired higher power.† This difficulty could be overcome by pressurizing the reactor coolant (say to 700 psia) so that boiling did not occur in the core, then flashing the coolant outside the reactor to produce steam at a lower pressure (i.e., 500 psia). This has the disadvantage of requiring a recirculating pump for the coolant in the reactor, as well as needing an appreciably higher coolant pressure to yield the same steam pressure. An interesting compromise is the General Electric dual-cycle reactor [170]. Here part of the steam is produced by direct boiling in the reactor (i.e., at 600 psia), and additional lower-pressure (i.e., 350-psia) steam is obtained by flashing additional coolant. The colder recirculated inlet water causes boiling to start farther up in the core, and less steam voids, reactivity decreases, and instability result. To compensate for the smaller expansion, thus smaller natural-circulation driving force, chimneys above the boiling-coolant channels can be added.

* The former possibility is, of course, being investigated by the Experimental Boiling Water Reactor (EBWR) and other boiling-reactor experiments, and the latter by progressively closer design and testing of pressurized reactors toward incipient boiling at hot spots. Some recent pressurized-water-cooled reactor designs are at incipient local boiling at the hot spot, when all pertinent hot-channel factors (Sec. 9-27) are applied.

† It was first thought that this instability might limit such reactors to some 10 kw/liter (20 per cent steam voids and 3 per cent reactivity decrease), too low to be attractive for central station operation. However, Borax-III has operated satisfactorily at 300 psig up to 16 thermal megawatts [188], corresponding to 28 kw/liter and only 2 per cent reactivity decrease. The thermal power fluctuated slightly during boiling at constant pressure, but the average power level could be increased smoothly by slowly withdrawing the control rods. The effect of faster load or pressure changes on the reactivity was slow enough so that manual control sufficed.

9-22. Contact Resistance at Liquid-metal–Solid-metal Interface. The high heat-transfer coefficients obtained with liquid metals are unfortunately not so correlatable as those with nonmetals. Figure 9-6 shows that a spread of 2:1 among experimental and theoretical results is frequent, and greater disagreement is not uncommon. Low results have been attributed to thermal resistance of gas films adsorbed on the pipe, but adsorbed films can scarcely exceed a monomolecular thickness and would have negligible thermal resistance. They have also been attributed to seal gas from the expansion chamber dispersed in the liquid metal [33, 187], which occurs at an agitated surface. Another possible cause is small absolute errors in wall or liquid temperatures which cause a high per cent error in the low Δt and thus high h values prevalent. Poor "wetting" has also been blamed; high surface tension could cause a nonwetting liquid to bridge over cavities in a rough wall. However, this explanation does not seem adequate, since absolute pressure has been reported to have no effect, and since nonwetting metals obey the usual streamline and turbulent friction pressure-drop relations.* Oxide layers are removed by hot alkali metals, as is shown in stainless-steel pipes by a sudden drop in electrical contact resistance to liquid Na at 200 to 350°C, depending on conditions and time. However, thin oxide layers can scarcely account for the large thermal-resistance variations observed.

Static tests [101] of Hg-Fe and Hg-Ni interfaces from 20 to 70°C without sodium present have shown thermal contact resistances $(1/h)$ up to 0.007°F/(hr)(ft²)(Btu) without wetting agent present and down to 0.00025 with it. These former exceed the observed flowing total thermal resistances, so evidently thermal contact resistance decreases as velocity increases. The simultaneous electrical contact resistances were much higher than would be expected if Eq. (9-20) applied, and they varied much more with wetting. It is concluded that electron flow at interfaces is through "active centers" and that most of the heat is conducted by atomic vibration through the intermediate area. On the other hand, higher-temperature tests with mercury between double-tube walls have shown negligible contact resistance. From a practical standpoint, it is desirable with liquid metals to design to avoid gas entrainment, to add a suitable wetting agent if possible when poor wetting is obtained, and to employ a reasonably conservative coefficient, such as 50 to 80 per cent of Eqs. (9-44i) to (9-44p) for forced convection in pipes. Provision should be made for skimming off, filtering, or dissolving oxides or other impurities if possible, "cold-trapping" Na_2O in Na, and otherwise keeping the metal clean [29].

9-23. Thermal Radiation. Thermal radiation causes heat to transfer from a hotter to a colder body without conduction or convection by intervening molecules. The Btu per hour radiated by a wall of A ft², emissiv-

---

* But nonwetting *plus* entrained gas decreases h considerably (see p. 396).

ity ϵ (relative to a "black" or maximum emitter) and $T°R$ is

$$q' = 1.73 \times 10^{-9}\epsilon A(T)^4 \qquad (9\text{-}53)$$

The net thermal radiation between two concentric or parallel surfaces is

$$q'_{1-2} = \frac{1.73 \times 10^{-9}A_1(T_1^4 - T_2^4)}{1/\epsilon_1 + (A_1/A_2)(1/\epsilon_2 - 1)} = (h_r)_1 A_1(T_1 - T_2) \qquad (9\text{-}54)$$

where A_1 is the smaller surface in square feet. If the surfaces are equal, A_1/A_2 disappears, as in Eq. (9-49g). The equivalent h, h_r in Eq. (9-54), generally will be found to be under 50 Btu/(hr)(ft²)(°F) for nuclear-reactor conditions. The radiant-heat transfer is thus usually negligible compared to the heat transferred by convection to a liquid coolant. Furthermore, most experimental correlations of convective h already include some radiation, not having been corrected for it.

With gaseous coolants the convective h is much lower, but again radiation is unimportant, this time because the absorptivity (emissivity) of the gas is small or zero (see App. H). The only apparent circumstance in which radiation directly to the coolant is significant in a reactor is in film boiling [Eq. (9-49g)]. In addition, with a gaseous coolant the fuel element can radiate to a cooler outer tube if there is one. This decreases somewhat the heat that must go directly to the coolant by convection, and thus it decreases the fuel-element temperature somewhat. The procedures here outlined are only approximate but will indicate whether any given thermal-radiation problem in a reactor merits more accurate analysis.

STEADY-STATE THERMAL DESIGN OF REACTORS

9-24. Temperature Rise of the Coolant. Although power reactors will have high q/A and thus high transverse temperature gradients in the coolant streams, the coolant channels will generally be thin. Thus for heat balances it will usually be adequate to consider the coolant temperature uniform at $\bar{t}$ at any given cross section of a stream. If the axial heat flow in the solid metal and coolant due to the coolant temperature rise is computed, it will generally be found negligible compared to the total power. Usually also the heat transferred from one channel to another in a direction normal to coolant flow is found to be negligible. Thus all the heat q generated per unit time in the "cell" of fuel, coolant, and moderator nearest to each coolant channel, from coolant inlet at x_1 to position x, is assumed to have gone into the coolant stream by position x.

The total heat dq/dx transferred to the coolant per unit length in any channel can be obtained by integrating Eq. (9-1) over the appropriate cross section of fuel cooled by the channel and multiplying by a factor to include the additional heating by γ absorption, moderation, and radioactive disintegration after neutron capture. The total heat q transferred from x_1 to x is then the integral $\int_{x_1}^{x} (dq/dx)\, dx$.

It is usually more convenient to work with the total desired reactor heat rate q_t and the ratios $\bar{H}/H_i$ and H/H_i of average and local H to the maximum, H_i (see Sec. 9-3). The total heat rate q_N for a given channel is

$$q_N = \frac{q_t}{N_c}\left(\frac{H}{\bar{H}}\right)_n \tag{9-55}$$

where N_c is the total number of channels and $(H/\bar{H})_n$ is the local to average ratio for the position of the channel in the plane of the reactor normal to the coolant flow. The local dq/dx at position x along the channel, for $x = 0$ at the midplane so that $x_1 = -x_2$, is

$$\frac{dq}{dx} = \frac{q_N}{2x_2}\left(\frac{H}{\bar{H}}\right)_x \tag{9-56}$$

The total q at x for the channel is thus

$$q = \frac{q_t}{2x_2 N_o}\left(\frac{H}{\bar{H}}\right)_n \int_{x_1}^{x}\left(\frac{H}{\bar{H}}\right)_x dx = \frac{q_N}{2x_2}\int_{x_1}^{x}\left(\frac{H}{\bar{H}}\right)_x dx \tag{9-57}$$

The only practically important orientation for coolant channels is normal flow between parallel inlet and outlet planes, or through a "slab." For homogeneous fuel loading the cosine distribution obtained can be integrated analytically (see Table 9-2). For "roof-topped" (see Sec. 9-35) or other irregular distributions a numerical or graphical integration is necessary for Eq. (9-57).

If the coolant undergoes a significant change in density and velocity while traversing the reactor, such as in boiling at low pressures, the method of Sec. 8-41 is used to obtain the relation between $\bar{t}$ and x from that between q and x.

If the pressure is high and the coolant expands significantly, as with supercritical or subcooled water at high temperatures, but velocities remain low, Eq. (8-20) is used in the form

$$h' = h_1' + \frac{q}{w} \tag{9-58}$$

If the pressure is substantially constant, $\bar{t}$ is obtainable directly from the calculated enthalpy h' and a table of h' vs. t at the given pressure. If the pressure decreases with x, P is calculated by Eq. (8-29) and t obtained from the table at the corresponding values of h and P.

For small expansions and moderate temperature rises the specific heat C as well as velocity remain substantially constant. This is usually the case with liquid metals, with subcooled water under about 500°F, etc. In this case $\bar{t}$ is directly obtainable as a function of q:

$$\bar{t} = \bar{t}_1 + \frac{q}{wC} = \bar{t}_1 + (\bar{t}_2 - \bar{t}_1)\frac{q}{q_N} \qquad (9\text{-}59)$$

For the cosine distribution with reflector (case 1, Table 9-2),

$$\frac{H}{\bar{H}} = \frac{x'_1 \cos x'}{\sin x'_1}$$

and Eqs. (9-57) and (9-59) yield

$$\bar{t} = \bar{t}_1 + \frac{\bar{t}_2 - \bar{t}_1}{2}\left(1 + \frac{\sin x'}{\sin x'_2}\right) = \bar{t}_1 + \frac{q_N}{2wC}\left(1 + \frac{\sin x'}{\sin x'_2}\right) \qquad (9\text{-}60)$$

where $x' = \pi x/2X_e$. Without reflector, $X_e = x_2$.

9-25. Control of Fuel-element Temperatures by the Coolant. The normal or local average wall temperature t_w at any position x is obtained by adding to $\bar{t}$ the Δt from the wall to the fluid at that position. If q/A and the coolant velocity are uniform over the heated perimeter P' at each position along a given channel or stream, as is usually substantially true, t_w will also be uniform and will be given by

$$t_w = \bar{t} + \frac{q}{A}\frac{1}{h} = \bar{t} + \frac{q_N}{2x_2hP'}\left(\frac{H}{\bar{H}}\right)_x \qquad (9\text{-}61)$$

The normal maximum temperature t_0 of the fuel element at x, which is at its axis or midplane, is obtained by adding to Eq. (9-61) the temperature drop Δt_e across the fuel element from Eqs. (9-26a), (9-37a), etc. Expressing $t_0 - t_w$ as equal to q/A times the ratio $\Delta t_e/(q/A)_1$, from the above equations

$$t_0 = \bar{t} + \frac{q}{A}\left[\frac{1}{h} + \frac{\Delta t_e}{(q/A)_1}\right] = \bar{t} + \frac{q_N}{2x_2P'}\left(\frac{H}{\bar{H}}\right)_x\left[\frac{1}{h} + \frac{\Delta t_e}{(q/A)_1}\right] \qquad (9\text{-}62)$$

A plot of t_w or t_0 against x for a symmetrical bare reactor will always show a maximum t_w between $x = 0$ and $x = x_2$ and a maximum t_0 between $x = 0$ and x at $(t_w)_{max}$. A reactor with reflector may show the maximum t_w either at x_2, or before x_2 (see Fig. 9-19).

For a cosine distribution with reflector, substituting for $\bar{t}$ by Eq. (9-60) and for $H/\bar{H}$ from Table 9-2,

$$t_0 = \bar{t}_1 + \frac{q_N}{2}\left\{\frac{1}{wC}\left(1 + \frac{\sin x'}{\sin x'_2}\right) + \frac{\pi}{2X_eP}\left[\frac{1}{h} + \frac{\Delta t_e}{(q/A)_1}\right]\frac{\cos x'}{\sin x'_2}\right\} \qquad (9\text{-}63)$$

The position x_{max} of maximum temperature is obtained by setting $dt_0/dx' = 0$, which yields

$$\frac{\pi x_{max}}{2X_e} = x'_{max} = \tan^{-1} \frac{2X_e P}{\pi w C \left[\dfrac{1}{h} + \dfrac{\Delta t_e}{(q/A)_1} \right]} \tag{9-64}$$

Substituting x'_{max} from Eq. (9-64) for x' in Eq. (9-63) and simplifying yields

$$q_N = \frac{2wC(t_{0,max} - \bar{t}_1)}{1 + 1/(\sin x'_2 \sin x'_{max})}$$

$$= \frac{2wC(t_{0,max} - \bar{t}_1)}{1 + \dfrac{1}{\sin x'_2} \left\{ 1 + \left(\dfrac{\pi w C}{2X_e P'} \right)^2 \left[\dfrac{1}{h} + \dfrac{\Delta t_e}{(q/A)_1} \right]^2 \right\}^{\frac{1}{2}}} \tag{9-65}$$

For t_w instead of t_0 the same equations (9-63), (9-64), and (9-65) may be employed by deleting the term $\Delta t_e/(q/A)_1$. A typical temperature distribution is shown in Fig. 9-19.

It is evident from Eq. (9-65) that if $\bar{t}_1$ and w have been fixed and the maximum satisfactory $t_{0,max}$ is set by the uranium phase change, strength, or other reasons, the maximum permissible power q_N of the channel is fixed. Similarly, if the maximum permissible t_w is set by corrosion, incidence of boiling, or other factors, the permissible q_N is fixed.* Again, if q_N, $\bar{t}_1$, and maximum t_w or t_0 are fixed,† the necessary w can be computed. However, w appears twice in Eq. (9-65) and also affects h. It must thus be found by repeated trials.

It is seen that all the temperatures in a nuclear reactor depend directly on the coolant inlet temperature and flow rate and the power. In a nuclear power plant the reactor coolant inlet temperature $\bar{t}_1$ may of course be kept constant by suitable instrumentation and controls, or it might be allowed to drop as low as possible, or to swing between limits, with changes in the power generated, main-condenser-coolant temperature, and other variables. In the latter case the temperature would have to be traced from the condenser-coolant inlet to the outlet, through the condenser, and through any intermediate boilers and heat exchangers before $\bar{t}_1$ could be estimated.

9-26. Thermal Design of a Gas-cooled Reactor. A number of the principles outlined in Chaps. 8 and 9 are illustrated in Prob. 9-19, which deals with the thermal design of a large gas-cooled research and isotope-production reactor. It is assumed that the dimensions and loading have

---

* Maximum t_0 is generally limiting in liquid-metal-cooled reactors, and maximum t_w in water-cooled reactors.

† As in computing coolant distribution among the channels of a reactor to obtain constant t_w or t_0 instead of $\bar{t}_2$.

already been specified by the reactor physics group (as per Chap. 6) and that rough preliminary estimates have yielded a set of specific coolant conditions as being near the optimum. It is desired to carry out a more detailed calculation of the coolant and fuel-element local temperatures at these particular conditions to check their suitability more precisely. It is under such circumstances that most thermal design calculations are normally carried out.

Prob. 9-19. As indicated above, this problem deals with heat transfer and coolant flow in an air-cooled high-power research reactor. The reactor, employing normal, natural uranium as the fuel and graphite as the moderator, is being considered for a rated thermal power of 30,000 kw. The core is to be a cylinder of graphite 21.5 ft in diameter and 30 ft in length, and in addition a reflector is to be employed such that the neutron flux at the surface of the cylinder of graphite is 25 per cent of its maximum value. [This is somewhat similar to the BNL reactor (App. N-1).] The fuel is in rods encased in thin aluminum tubes, with an OD of 1 in., and provided with four equidistant radial aluminum fins, $\frac{1}{2}$ in. long radially and 0.02 in. thick. The fuel is equally distributed throughout the reactor, in channels 2 in. in ID and on an 8-in.-square lattice. Assume that 90 per cent of the total heat is generated uniformly throughout the uranium and the other 10 per cent uniformly throughout the graphite.

It is being considered to cool this assumed reactor with air (assume dry, to be conservative) blown through the channels at such a rate that the outlet temperature is 500°F for all channels, with an inlet temperature of 100°F. Outlet pressure is atmospheric.

The following quantities are of interest:

a. Total Btu per hour dissipated by the fuel slugs of the center channel.

b. Btu per hour per foot dissipated by the central channel slugs at 0, 25, 50, 75, and 100 per cent of their length.

c. Temperature of the air and average local heat-transfer coefficient at above five positions.

d. Maximum temperature of the uranium at each above position.

e. Maximum temperature in the graphite at each above position.

f. Approximate kilowatts required to blow the air through the reactor against the pressure drop of the central channel, assuming 70 per cent over-all efficiency. No flow controlling nozzles are used in the central channel.

g. Temperatures and pumping power if the reactor is split in two with the air entering at the split (as in the Brookhaven reactor).

h. More exact calculation of pumping power, by finite increments, for the unsplit reactor.

Solution (see Table 9-9 and Fig. 9-19)

a.   Pile cross section per channel $= (\frac{2}{3})^2 = 0.444$ ft^2

$$\text{Approx no. of channels} = \frac{(21.5)^2 \pi/4}{0.444} = 816$$

(assume all are to be filled, even though unnecessary to reach criticality)
Average heat generation per channel:

$$\tilde{q}_N = \frac{q_t}{N_c} = 30,000 \times \frac{3413}{816} = 125,300 \text{ Btu/hr}$$

Radial variation of q_N [by Table 9-2 and Eq. (9-55)]:

$$\left(\frac{q_N}{q_{N,i}}\right)_n = \left(\frac{H}{H_i}\right)_n = J_0 \frac{2.4048r}{R_e} = J_0(r')$$

$J_0(r_1') = 0.25$. By a J_0 table or App. L, $r_1' = 1.955$ and $R_e = 13.22$ ft. **Then**

$$\left(\frac{\bar{q}_N}{q_{N,i}}\right)_n = \left(\frac{\bar{H}}{H_i}\right)_n = 1 - \frac{(1.955/2)^2}{2} + \frac{(1.955/2)^4}{12} - \cdots$$
$$= 0.592$$

Thus

$$(q_N)_{max} = q_{N,i} = \frac{125,300}{0.592} = 211,440 \text{ Btu/hr}$$

The fuel elements dissipate 90 per cent of this, or 190,300 Btu/hr, the moderator dissipating the balance, 21,140 Btu/hr.

b. Lengthwise variation of heat dissipation [by Table 9-2 and Eq. (9-56)]:

$$\frac{dq_N/dx}{(dq_N/dx)_i} = \left(\frac{H}{\bar{H}}\right)_x = \cos\frac{\pi x}{2X_e} = \cos x'$$
$$\cos x_1' = 0.25$$

By a cosine table $x_1' = 1.3173$ radians, and $X_e = 17.9$ ft. **Then**

$$\frac{dq_N/dx}{dq_N/dx} = \left(\frac{H}{H_i}\right)_x = \frac{x_1' \cos x'}{\sin x_1'} = 1.362 \cos x'$$
$$= 1.362 \cos\frac{1.3173\,x}{15} = 1.362 \cos\frac{x}{11.39}$$

Thus, for a central channel,

$$\frac{dq_N}{dx} = \frac{211,440}{30}\,1.362 \cos\frac{x}{11.39} = 9400 \cos\frac{x}{11.39} \qquad \text{Btu/(hr)(ft)}$$

c. Simplified calculation of air temperature by Eq. (9-60) (neglecting expansion):

$$i = 100 + \frac{400}{2}\left[1 + \frac{\sin(x/11.39)}{\sin(15/11.39)}\right] = 300 + 206.7 \sin\frac{x}{11.39} \qquad °F$$

Calculation of h from i [Eq. (9-44a) or curve ABC of Fig. 9-5 is used.   The inner-wall version of Eq. (9-44e) is not conservative and may be inapplicable because of the fins]:

From Eq. (9-59):

$$w = \frac{211,440}{400 \times 0.237}$$
$$= 2232 \text{ lb/hr}$$

Free-flow cross section $= 0.01607$ ft^2
Flow wall perimeter $= 1.107$ ft

$$D_e = \frac{4 \times 0.01607}{1.107} = 0.058 \text{ ft}$$

$$G = \frac{2232}{0.01607 \times 3600} = 38.6 \text{ lb/(sec)(ft}^2)$$

$$Re = \frac{0.058 \times 38.6}{0.0672\mu} = \frac{33}{\mu} \qquad \text{for } \mu \text{ in poises}$$

From Fig. 9-5:

h (on aluminum tube and fins) $= CGj\,(Pr)^{-2/3} = 33,000j\,(Pr)^{-2/3}$ Btu/(hr)(ft^2)(°F)

d. Fins:

$$x_f = 0.51 \text{ in.} \qquad \frac{P'}{S} = \frac{2}{\text{thickness}} = 1200 \text{ ft}^{-1}$$
$$k \text{ of aluminum} = 122 \text{ Btu/(hr)(ft)(°F)}$$

Efficiency, from Fig. 9-25 or Prob. 9-21:

$$\text{Average } x_f(hP'/kS)^{1/2} = 1.29$$

$$\text{Average } E_f = \frac{\tanh 1.29}{1.29} = 0.67$$

Δt_f across uranium-air film [by Eq. (9-91)] based on average E_f:

$$\Delta t_f = \frac{0.9}{h}\frac{dq_N/dx}{\pi D_b + 0.67 \times 8 \times x_f} = \frac{1.87}{h}\frac{dq_N}{dx} \qquad °F$$

Δt_e across the uranium [by Eq. (9-33) for simplicity and to be slightly conservative; Eq. (9-37) should be more accurate]:

$$k \text{ of uranium} = 19 \text{ Btu/(hr)(ft)(°F)}$$

$$\Delta t_e = \frac{1}{4\pi k}\left(\frac{dq}{dx}\right)_e = 0.00377\frac{dq_N}{dx}$$

e. $\bar{H}$ in the graphite (assumed uniform for simplicity and to be conservative):

$$\frac{0.1}{(64 - \pi)/144}\frac{dq_N}{dx} = 0.236\frac{dq_N}{dx} \qquad \text{Btu/(hr)(ft}^3)$$

k of graphite is taken as 87 Btu/(hr)(ft)(°F). Assuming that the graphite is a con-

TABLE 9-9. TEMPERATURE DISTRIBUTION ALONG A CENTRAL COOLANT CHANNEL
IN UNSPLIT REACTOR, BY APPROXIMATE CALCULATION

	Axial distance x, ft				
	-15	-7.5	0	$+7.5$	$+15$
dq_N/dx, Btu/(hr)(ft).......	2,160	6,840	8,650	6,840	2,160
$\bar{t}$ of air, °F...............	100	173.6	300	426.4	500
μ, centipoises.............	0.0190	0.0207	0.0238	0.0265	0.0279
Reynolds No., Re..........	173,600	159,500	138,600	124,600	118,300
j........................	0.0021	0.00215	0.0022	0.00225	0.0023
Prandtl No., Pr...........	0.696	0.685	0.671	0.660	0.653
h, Btu/(hr)(ft²)(°F)........	88.0	90.8	94.4	97.0	100.7
Δt_f, slug to air, °F........	50.5	155.1	188.5	145.0	44.2
Δt_e across uranium, °F......	9.0	28.6	36.2	28.6	9.0
t_{max} in uranium, °F........	159.5	357.3	524.7	600.0	553.2
Δt across graphite, °F......	0.4	1.4	1.7	1.4	0.4
Δt, graphite to air, °F......	5.2	16.0	19.4	14.9	4.6
t_{max} in graphite, °F........	105.6	191.0	321.1	442.7	505.0
q/N from slug by radiation, Btu/(hr)(ft)............	....	....	10.8	10.0	
f_F for smooth pipe........	0.0040	0.0041	0.0042	0.0043	0.0044

centric cylindrical shell of the same inner diameter ($R_i = 1$ in.) and cross section (R_0 comes out 4.52 in.), by Eq. (9-34) applied to inner cooling:

$$\Delta t_{graphite} = \frac{\bar{H}}{4k}\left[R_i^2 - R_0^2\left(1 - 2\ln\frac{R_0}{R_i}\right)\right] = 0.000197\frac{dq_N}{dx}$$

$$\Delta t_{graphite/air} = \frac{\Delta t_{U/air}(0.1/0.9)(\pi + 0.67 \times 4)}{2\pi}$$

$$= 0.103\,\Delta t_{U/air} \qquad (\text{since } h_{graphite/air} = h_{U/air} \text{ approximately})$$

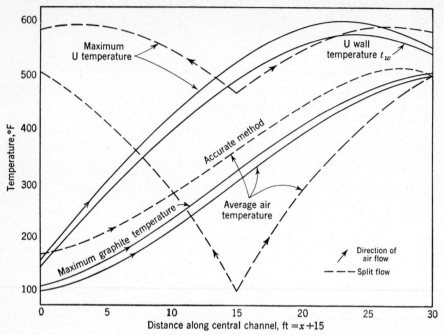

FIG. 9-19. Temperatures in large air-cooled research reactor.

Check radiation [Eq. (9-54) and App. H]: $\epsilon_{Al} = 0.1^{\pm}$; $\epsilon_{Gr} = 0.75^{\pm}$; neglecting the fins, $(q/A)_{\text{slugs}}$ radiated to graphite $\approx 17.3 \times 10^6 (T_{Al}{}^4 - T_{Gr}{}^4)\epsilon_{Al}$. The radiation is 0.2 per cent of $(q/A)_{\text{slugs}}$ and 2 per cent of $(q/A)_{Gr}$, so it does not affect either temperature appreciably.

f. Approximate calculation of ΔP [Eq. (8-34)]:

$$(144)^2[P_1{}^2 - (14.7)^2] = \frac{2 \times 1545 \times (38.6)^2 \times 760}{28.97 \times 32.2}$$
$$\left(\ln \frac{960 \, P_1}{560 \times 14.7} + \frac{2 \times .0042 \times 30}{0.058} \right)$$

$$P_1{}^2 = 1000 + 415 \log \frac{P_1}{8.58} \qquad P_1 = 35.4 \text{ psia}$$

Velocity check:

$$\bar{u}_2 = 930 \text{ fps} = 61 \text{ per cent of sonic velocity*}$$
$$\bar{u}_1 = 225 \text{ fps} \qquad \rho_1 = 0.171 \text{ lb/ft}^3$$
$$\text{Approximately } \Delta P_V = \frac{0.171(225)^2}{64.4} \times 144 = 0.94 \text{ psi}$$
$$\text{Total } P_{\text{inlet}} = 36.4 \text{ psia}$$

* This velocity is not entirely unreasonable. However, the walls and therefore all solids concerned will be further heated to the extent of the stagnation temperature rise times the temperature-recovery factor. The former by Eq. (8-20) is 72°F and the latter is $\text{Pr}^{1/3}$ or about 0.9. All solid temperatures at the outlet should thus be raised approximately 65°F. This correction falls to 4°F at the inlet.

Total pounds of air per hr $= 30,000 \times 3413(400 \times 0.237) = 1,080,000$

Assume compression averages isothermal at 100°F.

$$\text{Air ft-lb} = PV \ln P_1/P_2 = \left(35.4 \times 144 \times \frac{1,080,000}{0.171}\right) \ln \frac{36.4}{14.7}$$

$$= 2.91 \times 10^{10} \text{ per hour}$$

$$\text{Motor kw} = 15,700$$

g. If the reactor had been split, with air admission at the midplane, by the same method as part *f*:

$$P_1 = 19.4 \text{ psia} \qquad \text{and} \qquad \text{motor kw} = 4800$$

(The saving in pumping power is not so great a percentage with a gas as with a liquid coolant in this case of fixed outlet pressure.)

h. More exact check of air temperature and pressure distribution, by finite increments of length.   Accept same G, $\bar{t}_2$, and P_2 as in parts *c* and *f*, and work backward in 12 increments of 2.5 ft to check t_1 and P_1.   Set $dz = dW = 0$.   Use V in cubic feet per pound.

Total energy balance [Eq. (8-20) or (8-62)] in degrees Rankine:

$$T_{m-1} = T_m - \frac{\Delta q}{wC_r} + \frac{G^2(V_m{}^2 - V_{m-1}^2)}{2g_cJC_P} = T_m - \Delta\left(\frac{q}{wC_P}\right) + 0.25\bar{V}(\Delta V)$$

Mechanical energy balance [Eq. (8-21)] in pounds per square inch:

$$P_{m-1} = P_m + \frac{G^2}{\alpha g_c}(V_m - V_{m-1}) + \frac{2\bar{f}_F G^2 \Delta N \bar{V}}{g_c D_e}$$

$$= P_m + 0.348\,\Delta V + 27.7\,\bar{f}_F\bar{V}$$

Perfect gas law, in above units (actual PVT data could also be used):

$$V_{m-1} = \frac{RT_{m-1}}{MP_{m-1}} = 0.37\frac{T_{m-1}}{P_{m-1}}$$

Procedure: Given T_m, P_m, V_m, $\bar{f}_{F,m/m-1}$ and $\Delta(q/wC_P)_{m/m-1}$, assume V_{m-1}, compute ΔV and $\bar{V}$, then T_{m-1}, P_{m-1}, and V_{m-1} in succession by above equations.   If V_{m-1} comes out appreciably (say over 1 per cent) different from what was assumed, repeat with the new V_{m-1}.   When the V_{m-1} values check, start the next interval.   The final check values are given in Table 9-10.

Evidently the actual inlet pressure is some 3 psi higher than computed under *f*, and the inlet temperature some 73°F higher.   To match $\bar{t}_1$, $\bar{t}_2$, and P_2, as given, one or more new calculations at lower w would be necessary until $\bar{t}_1$ were exactly hit or could be interpolated with adequate accuracy.

It is noted in Fig. 9-19 that the maximum air temperature is only some 15° higher by the finite-increments method than by the approximate method.   However, the maximum metal and graphite temperatures are some 50° higher, because in that region the air temperature is that much higher by the accurate method.   Again, with

$\bar{t}_1 = 100°F$ the true maximum metal temperature would be 600°F, with w somewhat under 2232 in a central channel. Clearly, for reliable metal temperature and stress calculations the accurate method by small increments is necessary if the coolant velocity changes greatly.

TABLE 9-10. TEMPERATURE AND PRESSURE DISTRIBUTION IN UNSPLIT REACTOR, COMPUTED IN TWELVE FINITE INCREMENTS

Ft from outlet, $x + 15$	Ft from midplane, x	$\sin (x/11.39)$	$\Delta \sin (x/11.39)$	$\Delta \left(\dfrac{q}{wC_P}\right)$	$T,$ °R	f_F	$V,$ ft³/lb	ΔV	$p,$ psi
0	−15	−0.9682			960		24.15		14.70
			0.0777	16.05		0.0044		5.39	
2.5	−12.5	−0.8905			972.8		18.76		19.19
			0.1206	24.91		0.0044		2.82	
5	−10	−0.7699			960.2		15.94		22.28
			0.1575	32.53		0.0044		1.99	
7.5	−7.5	−0.6124			935.1		13.95		24.80
			0.1870	38.63		0.0044		1.55	
10	−5	−0.4254			901.5		12.40		26.91
			0.2075	42.86		0.0043		1.30	
12.5	−2.5	−0.2179			862.5		11.10		28.76
			0.2179	45.02		0.0043		1.12	
15	0	0			820.4		9.98		30.41
			0.2179	45.02		0.0042		0.95	
17.5	2.5	0.2179			777.7		9.03		31.87
			0.2075	42.86		0.0042		0.81	
20	5	0.4254			736.5		8.22		33.15
			0.1870	38.63		0.0042		0.68	
22.5	7.5	0.6124			699.3		7.54		34.31
			0.1575	32.53		0.0041		0.55	
25	10	0.7699			667.8		6.99		35.32
			0.1206	24.91		0.0041		0.42	
27.5	12.5	0.8905			643.6		6.57		36.23
			0.0777	16.05		0.0041		0.31	
30	15	0.9682		0	628.1		6.26		37.06
Inlet header	....	....	....		633.0	......		....	39.24

9-27. Steady-state Design by Hot-channel Factors. By Secs. 9-24 and 9-25 theoretical or "average" values of $\bar{t}$, $t_{w,max}$, $t_{0,max}$, and q_N at a given w, q_t, and t_1 can be computed, based on known average values of the local dimensions, physical properties, fuel concentration, neutron-flux density, etc. These average values are important in the design and operation of a reactor. However, when a reactor is being designed for maximum performance, as will in general be true for power reactors, it is even more important not to exceed the permitted temperatures *anywhere*. For instance, the several coolant channels nearest the axis of a reactor will all have identical values of $t_{0,\text{max}}$ by Eq. (9-65) if all other factors in the equation are (substantially) equal. In actual fact, however, there are bound to be appreciable differences among them in q_N, in w, and thus in $t_{0,\text{max}}$ and $t_{w,\text{max}}$. It is evidently necessary to estimate safety factors that will allow for these differences if it must be assured that nowhere does t_w or t_0 exceed the set maximum. For nuclear reactors, safety factors are known as "hot-channel" or "hot-spot" factors.

TABLE 9-11. TYPICAL HOT-CHANNEL FACTORS

Uncertainty in	Coolant temperature rise, $\bar{t} - \bar{t}_1$	Film temperature difference, Δt_f	Fuel temperature difference, Δt_e
Neutron flux............................	1.20	1.20	1.20
Fuel channel concentration..	1.10	1.20	1.20
Eccentricity and diameter.................	1.40	1.25	1.10
Flow distribution........................	1.02	1.02	
Fuel-element warpage....................	1.04	1.03	
Total product*........................	1.95	1.90	1.60
Film coefficient, h.........................	....	1.20	
Fuel-element thermal conductivity, k........	....	....	1.05
Fuel-element thickness...................	....	....	1.05
Total product........................	$1.95 = F_c$	$2.28 = F_f$	$1.76 = F_e$

Maximum possible coolant temperature $= \bar{t}_1 + 1.95(\bar{t} - \bar{t}_1)_\text{calc}$
Maximum possible surface hot spot $= \bar{t}_1 + 1.95(\bar{t} - \bar{t}_1)_\text{calc} + 2.28(\Delta t_f)_\text{calc}$
Maximum possible metal hot spot $= \bar{t}_1 + 1.95(\bar{t} - \bar{t}_1)_\text{calc} + 2.28(\Delta t_f)_\text{calc} + 1.76(\Delta t_e)_\text{calc}$
Maximum possible metal hot spot for Prob. 9-19 (calculated at $x = +7.5$ ft)
$\qquad = 100 + 1.95 \times 326.4 + 2.28 \times 145 + 1.76 \times 28.6 = 1117°F$
* S. McLain, ANL-5424, 1955.

The hot-channel factor for a given source of inaccuracy is a multiplier applied to the normal design value of a temperature differential to yield the maximum possible value of that differential. Such factors must be determined by careful analysis of manufacturing and inspection procedures and tolerances, the reliability of the physical properties and design relations used, methods of reactor control* and fuel-element routing in the reactor, tests on models, critical assemblies or "mock-up"

---

* For instance, the neutron-flux density varies much more with poison-control rods than with reflector or fuel solution level control (see Sec. 6-52).

reactors, etc. [186].   Table 9-11 lists typical values of these factors for the three significant temperature differentials in computing the hottest spot. These or similar factors apply at all positions in the reactor, but the values of most importance are usually those at the positions of maximum t_w and t_0, found in Sec. 9-25.   It is seen that the maximum possible t_0 in this case is about twice as much above the coolant inlet temperature $\bar{t}_1$ as the normal design hot spot.*

To design with hot-channel factors, in Eq. (9-60) through (9-65) w should be replaced by w/F_c, h by h/F_f, and Δt_e by $F_e \Delta t_e$.   Equations (9-64) and (9-65) become, respectively,

$$\frac{\pi x_{\max}}{2X_e} = x'''_{\max} = \tan^{-1} \frac{2X_e P' F_c}{\pi w C \left[\dfrac{F_f}{h} + \dfrac{F_e \Delta t_e}{(q/A)_1} \right]} \qquad (9\text{-}66)$$

$$q_N = \frac{2wC(t_{0,\max} - \bar{t}_1)}{F_c + \dfrac{F_c}{\sin x'_2 \sin x'''_{\max}}} \qquad (9\text{-}67)$$

Evidently, q_N is decreased by the ratio $1/F_c$, and if F_c, F_f, and F_e are equal, $x_{\max}$ is not changed.   As before, $t_{0,\max}$ can be replaced by t_w by setting $\Delta t_e = 0$.

It is seen in Table 9-11 that all the factors are assumed to apply simultaneously at the hottest spot.   It is true that some of them are interrelated, such as neutron flux, average fuel concentration, and warping of the fuel element as a result of high burnup.   However, this method of design is bound to be overconservative if each hot-channel factor is the maximum possible effect of the corresponding uncertainty.

A sounder procedure would be to determine or estimate the fractional standard deviation σ' for each independent uncertainty,† combine them to find the total σ of the limiting temperature, then design by the normal equations (without hot-channel factors) to an agreed-on number of σ's below the limiting temperature [105].   The probability of a hot spot's exceeding the limiting temperature is 16 per cent if the highest design temperature is σ lower, 2.3 per cent if 2σ lower, 0.13 per cent if 3σ lower, etc.

To illustrate this method, Table 9-11 has been recalculated in Table

---

* In later power reactors, considerably smaller F values are perforce being employed in order to obtain reasonable designs.   These are justifiable by smaller tolerances, greater experience and optimism, and a qualitative application of the method of Table 9-12.

† σ' is the root-mean-square fractional deviation from the mean due to each uncertainty.

9-12 on the following assumptions:

1. The first two uncertainties relate to heat generation and are cumulative (conservative).

2. The next three uncertainties relate to local flow and are cumulative (conservative).

3. The fuel-concentration factor is per unit length rather than volume, so the fuel-element-thickness factor only enters to the first power.

4. All the hot-channel factors correspond to 2σ (uncertain).

5. 3σ is adequate for hot-spot design (uncertain, as time, thermal fatigue, warning and hazards of failure, and other factors enter to problematical or arbitrary extents).

TABLE 9-12. STATISTICAL DETERMINATION OF HOT-SPOT TEMPERATURE
FOR PROB. 9-19

Temperature difference	Uncertainty in				
	Neutron flux and fuel concentration	Local flow w (all causes)	Film coefficient h	Fuel conductivity k	Fuel-element thickness
Coolant rise Δt_c (normal = 326.4°F):					
Hot-channel factor	1.32 (combined)	1.48 (combined)	1	1	1
Fractional σ'	0.16	0.24	0	0	0
$\sigma = \sigma'\Delta t_c$	52.2	78.3	0	0	0
Film difference Δt_f (normal = 145°F):					
Hot-channel factor	1.44 (combined)	1.31 (combined)	1.20	1	1
Fractional σ'	0.22	0.155	0.10	0	0
$\sigma = \sigma'\Delta t_f$	31.9	22.5	14.5	0	0
Fuel difference Δt_e (normal = 28.6°F):					
Hot-channel factor	1.44 (combined)	1.10	1	1.05	1.05
Fractional σ'	0.22	0.05	0	0.025	0.025
$\sigma = \sigma'\Delta t_e$	6.3	1.4	0	0.7	0.7
$\Sigma(\sigma) = \sigma$ due to each uncertainty	90.4	102.2	14.5	0.7	0.7
$[\Sigma(\sigma)]^2$	8,173	10,445	210.0	0.5	0.5

$\Sigma[\Sigma(\sigma)]^2 = 18,829$; total σ of $t_0 = (18,829)^{1/2} = 137°F$; $3\sigma = 411°F$.
Maximum hot-spot temperature $= 600 + 411 = 1011°F$.

The calculation has been based on the $+7.5$-ft position in Table 9-9, which is near the nominal metal hot spot. For simplicity, the "maximum possible" metal-hot-spot temperature has been calculated* from the normal-design value of 600°F, rather than the more common alternate procedure of computing the normal-design hot-spot temperature from the maximum allowable hot spot. It is seen that, in spite of the con-

* Bottom line, Table 9-11.

servativeness of the calculation in Table 9-12, a lower temperature deviation is obtained. This is in large part because the smaller F's and σ's have less effect by the statistical calculation. Further refinement in the statistical method is possible. For instance, the analysis might better be applied to Eq. (9-65) than to Eq. (9-63). However, this is probably not justified at the present level of experience. Furthermore, transient conditions are generally found to be more severe than steady-state conditions (see Secs. 9-45 to 9-47), so that further refinement of steady-state thermal design is not usually indicated.

9-28. Heat Generation in Flowing Liquids and Slurries. Homogeneous reactors employing solutions of fuel plus moderator, such as the HRE, generate fission heat internally when the solution is in the enlarged vessel which constitutes the core. Reactors employing fuel slurries instead of solutions, or dissolved or suspended fuel with separate moderator, such as the externally cooled LMFR, are similar cases. In these reactors there is usually no separate coolant in the core. Thus all the heat generated appears as temperature rise of the fuel fluid. This does not mean, however, that the outlet temperature t_2 is uniform even across any one channel. Apart from the radial variation of H (Sec. 9-3), small for small channels and long neutron-diffusion lengths, and vice versa, there is a variation due to the nonuniformity of coolant velocity u and therefore residence time in the core. The consequent temperature nonuniformity is continually being decreased by radial molecular heat conduction in laminar flow and much more effectively by eddy conduction (Sec. 8-29) in turbulent flow. However, radial temperature differentials may be of importance because of their effect on average density of the stream, because of thermal shock of surfaces on which the stream may impinge, etc. Accordingly, their evaluation is necessary in optimizing the design of such a reactor.

Volume heating of fuel solutions also occurs outside the reactor core, because of the radioactive disintegration of fission products. Heating of both coolant and fuel solutions also takes place within the core as a result of γ absorption, fast-neutron slowing down, and neutron reactions. These other sources of heat within the core can usually (conservatively) be neglected (since they are much smaller than the fission-product plus β energy), being included with the fission heat generation. However, they can be important in evaluating behavior of the reactor in the event of coolant flow failure.

If it is necessary to consider the axial or radial variation in H, the calculation must generally be made by finite increments (Sec. 9-39). In the case of uniform H, general mathematical solutions have been worked out for long tubes [99,100]. This case gives the highest temperature differentials; thus it is somewhat conservative for design purposes.

TABLE 9-13. VALUES OF $\dfrac{HL^2}{k(t - t_w)}$ FOR LIQUID WITH UNIFORM HEATING DENSITY

Wall conditions	Velocity distribution	Re	Pr	Value of t in $t - t_w$					
				Wide flat duct (L = wall spacing)			Round pipe (L = diameter)		
				t_{axial}*	$t_{\text{flow mean}}$*	$t_{\text{space mean}}$	t_{axial}†	$t_{\text{flow mean}}$†	$t_{\text{space mean}}$
Streamline									
Isothermal	Parabolic			8	10	12	16	24	32
Isothermal	Uniform			8	12	12	16	32	32
Adiabatic	Parabolic			−32	−46.7	−15	−32	−64	96
Adiabatic	Uniform			∞	∞	∞	∞	∞	∞
One wall isothermal and one adiabatic‡	Parabolic			2	2⅔	3			
One wall isothermal and one adiabatic‡	Uniform			2	3	3			
Turbulent									
Adiabatic		5,000	0.01	67	100	....	103	216	....
Adiabatic		10,000	0.01	94	182	....	143	296	....
Adiabatic		100,000	0.01	230	364	....	400	920	....
Adiabatic		1,000,000	0.01	10,000	1,700	....	2,200	4,200	....
Adiabatic		10,000	0.1	....	290	....	....	785	....
Adiabatic		100,000	0.1	....	2,000	....	....	5,750	....
Adiabatic		10,000	1	....	820	....	....	2,860	....
Adiabatic		100,000	1	....	12,900	....	....	40,000	....
Adiabatic		10,000	10	....	3,330	....	....	13,000	....
Adiabatic		100,000	10	....	85,000	....	....	300,000	....

* H. F. Poppendiek and L. D. Palmer, ORNL-1701, 1954.
† H. F. Poppendiek and L. D. Palmer, ORNL-1395, 1952; *Chem. Eng. Progr., Symposium Ser.*, no. 11, 1954.
‡ t_w is at the isothermal wall, and t_{axial} at the adiabatic wall.

447

From Eq. (9-15), the differential equation for a wide flat duct is

$$u \frac{\partial t}{\partial x} = \frac{\partial}{\partial y}\left[(\alpha + \epsilon_H)\frac{\partial t}{\partial y}\right] + \frac{H}{\rho C} \qquad (9\text{-}68)$$

where x is measured along the length and y along the thickness. For streamline flow $\epsilon_H = 0$ and u is obtained from Table 8-5. Equation (9-68) has been integrated for the conservative case of long ducts to give the transverse temperature distribution for the boundary conditions that both walls (at $y = 0$ and $y = L$) are at t_w and that the fraction of the heat generated that remains in the stream is F' ($F' = 0$ for isothermal walls and 1 for adiabatic walls). Substituting the position of the mid-plane gives the axial temperature t_{axial}. Integrating to the midplane gives the space mean temperature of the stream, or on weighting by the velocity, the flow mean temperature

$$t_{\text{axial}} - t_w = \frac{HL^2}{4k}\left(\frac{1}{2} - \frac{5F'}{8}\right) \qquad t_{\text{flow mean}} - t_w = \frac{HL^2}{4k}\left(\frac{2}{5} - \frac{17F'}{35}\right) \quad (9\text{-}69)$$

The integrations have also been carried out for turbulent flow, taking ϵ_H as equal to ϵ_M in Sec. 8-29 and using a generalized velocity distribution similar to that in Sec. 8-28. Results for the principal conditions for flat plates and for tubes are given in Table 9-13.

Experiments are lacking on laminar free convection in vertical fuel channels with uniform H, but theoretical analyses are available. They give conservatively high values of the maximum (axis-to-wall) temperature difference Δt_F across a horizontal section of the liquid fuel. For fuel sealed within channels of high height–breadth ratio Z/D, with an external coolant flowing upward which rises by Δt_c in height Z [157]:

$\dfrac{\Delta t_c \beta g D^4 C \rho^2}{Z k \mu}$	1	10	10^2	10^3	10^4	10^5
$\dfrac{D}{k\,\Delta t_F}\dfrac{q}{A}$, (flat channel).......	4	4	4.08	4.55	7.7	13.5
$\dfrac{D}{k\,\Delta t_F}\dfrac{q}{A}$, (pipe).............	4	4	4.02	4.4	7.85	25.5

9-29. Steady-state Thermal Circulation. Because of its simplicity and dependability, thermal circulation or gravity convection of coolant or liquid fuel in a loop is frequently considered for the primary method of heat removal from a nuclear reactor. When forced circulation of the coolant is the normal method, thermal circulation is still of importance; by locating the heat exchanger higher than the reactor, partial flow is maintained in the event of pump failure.

The pertinent driving heads in a thermal-circulation loop are shown in Fig. 9-20. Usually there is little, if any, temperature change along the exterior piping, and it is convenient to express the buoyancy pressure ΔP_B over the loop in terms of the (average) temperature difference between the hot and cold legs and the effective driving head Z_e [Z in Eq. (8-54)].* Z_e evidently equals the minimum head Z_0 plus the effective values of Z_h and of Z_c. If Z_0 exceeds some five times Z_h and Z_c, one-dimensional flow and

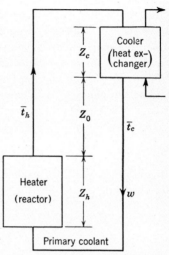

$$Z_e = Z_0 + \frac{Z_h + Z_c}{2}$$

may be assumed without sensible error. This case was considered in Sec. 8-38 for unsteady flow. In steady flow the denominator of the integral in Eq. (8-55) equals zero, and the dimensionless solution for w is obtainable from Table

Fig. 9-20. Thermal-circulation driving heads.

8-10 or Fig. 8-15 at $\theta = \infty$. For convenience, the following explicit solutions for w and $t_h - t_c$ are given: For streamline flow in a pipe loop,

$$w = \left(\frac{\pi}{128} \frac{\beta \rho^2 g Z_e D^4 q}{\mu N_e C}\right)^{1/2} \qquad \bar{t}_h - \bar{t}_c = \left(\frac{128}{\pi} \frac{\mu N_e q}{\beta \rho^2 g Z_e D^4 C}\right)^{1/2} \qquad (9\text{-}70)$$

For streamline flow in other cross sections, employing the ratio $\bar{u}\mu N/\Delta P\, g_c$ from Table 8-5,

$$w = \left(\frac{\bar{u}\mu N}{\Delta P\, g_c} \frac{\beta \rho^2 g Z_e S q}{\mu N_e C}\right)^{1/2} \qquad \bar{t}_h - \bar{t}_c = \left(\frac{\Delta P\, g_c}{\bar{u}\mu N} \frac{\mu N_e q}{\beta \rho^2 g Z_e S C}\right)^{1/2} \qquad (9\text{-}71)$$

For turbulent flow,

$$w = \left(\frac{\beta \rho^2 g Z_e D_e S^2 q}{2 N_e f_F C}\right)^{1/3} \qquad \bar{t}_h - \bar{t}_i = \left(\frac{2 N_e f_F q^2}{\beta \rho^2 g Z_e D_e S^2 C^2}\right)^{1/3} \qquad (9\text{-}72a)$$

or, employing Eq. (8-33) for f_F,

$$w = \left(\frac{\beta \rho^2 g Z_e D_e^{1.2} S^{1.8} q}{0.092 N_e C \mu^{0.2}}\right)^{0.357} \qquad \bar{t}_h - \bar{t}_c = \left(\frac{0.092 N_e \mu^{0.2} q^{1.8}}{\beta \rho^2 g Z_e D_e^{1.2} S^{1.8} C^{1.8}}\right)^{0.357} \qquad (9\text{-}72b)$$

* In a liquid-fuel reactor with external circulation, the external volume is minimized for economic and safety reasons, and the external heat generation may be negligible as a contribution to buoyancy. However, if the fuel volume in the pipes roughly equals or exceeds the volume in the core, it would be desirable to compute the local H and l around the loop (see second footnote on p. 369) and integrate per Eq. (8-19).

Turbulent flow starts when $\beta\rho^2 g Z_e D_e{}^4 q/2N_e\mu^3 CS$ reaches the critical value of $\mathrm{Re}^3 f_F$, approximately 10^8.

If Z_0 is relatively small, zero, or negative, more accurate values of effective Z_h and Z_c are needed. When the heated and/or cooled channels have $(\bar{t}_h - \bar{t}_c)/\Delta t_f$ exceeding about 1, corresponding roughly to N/D_e over 100, the space-mean or buoyancy temperature is usually close enough to the flow-mean temperature $\bar{t}$ that the latter can be employed, either for surface- or for volume-heated fluid.

The buoyancy effectiveness E_b of Z_h or Z_c is

$$E_b = \frac{\bar{t}_m - \bar{t}_{\text{in}}}{\bar{t}_{\text{out}} - \bar{t}_{\text{in}}} = \frac{1}{Z(\bar{t}_{\text{out}} - \bar{t}_{\text{in}})} \int_0^Z (\bar{t} - t_{\text{in}})\, dz \tag{9-73}$$

If ρ is not linear in t, each $\bar{t}$ should be replaced by ρ. In Fig. 9-21, E_b for the boiler is $(A_r + A_c)/(A_r + A_e + A_b)$, and for the reactor

$$\frac{A_b + A_c}{A_b + A_c + A_r}$$

E_b for a nuclear reactor with the usual symmetrical distribution of H along Z_h is evidently $\frac{1}{2}$. For any vertical heat exchanger with substantially constant over-all heat-transfer coefficient U and sink temperature (as in Fig. 9-21),* or substantially constant h and t_w,

$$E_b = \frac{\Delta t_1 - \Delta t_{lm}\dagger}{\Delta t_1 - \Delta t_2} = \frac{1}{1 - \Delta t_2/\Delta t_1} + \frac{1}{\ln(\Delta t_2/\Delta t_1)} \tag{9-74}$$

Typical values of E_b for the case of Fig. 9-21 ($Z_0 = 0$, $Z_h = -Z_c$) follow:

$\Delta t_2/\Delta t_1$	0.1	0.2	0.3333	0.5	0.6667
E_b for boiler	0.6768	0.6325	0.5889	0.5573	0.5333
E_b for loop	0.1768	0.1325	0.0889	0.0573	0.0333

The net Z_e for the loop is

$$\begin{aligned} Z_{e,\text{loop}} = ZE_{b,\text{loop}} &= Z_c E_{b,\text{boiler}} + Z_h E_{b,\text{reactor}} \\ &= Z_c(E_{b,\text{boiler}} - E_{b,\text{reactor}}) = Z(E_{b,\text{boiler}} - 0.5) \end{aligned} \tag{9-75}$$

If a uniform heating density H exists in the fluid simultaneously that heat is being transferred, the differential equation

$$wC\, d\bar{t} = HS\, dz - UP'(\bar{t} - t')\, dz$$

---

* The actual requirement is for single-pass flow with (UA/C) linear in z.

† See Eq. (9-90).

gives for $\bar{t}_m$ in Eq. (9-73)

$$\bar{t}_m = t' + \frac{HS}{UP'} + \frac{\Delta t_{\text{in}} - \Delta t_{\text{out}}}{\ln \dfrac{HS/UP' - \Delta t_{\text{in}}}{HS/UP' - \Delta t_{\text{out}}}} \tag{9-76}$$

where t' = constant temperature of wall or other fluid

U = film or over-all heat-transfer coefficient

$\Delta t = \bar{t} - t'$ at inlet or outlet

Equation (9-76) holds for heat transfer in either direction. It is seen to simplify to Eq. (9-74) if $H = 0$.

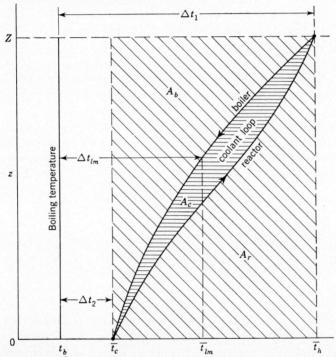

FIG. 9-21. Thermal-circulation loop between vertical tubes of the same height and elevation in a reflected reactor and in a boiler.

For a single-pass heat exchanger between a primary and a secondary coolant with constant specific heats and U, but neither stream at constant temperature, the mean temperature $\bar{t}_m$ for Eq. (9-73) may be approximated by heat balances plus the assumption that the two mean temperatures differ by Δt_{lm} and are at the same position along the channel. For either stream and for either parallel or counter flow,

$$\bar{t}_m = \frac{\bar{t}_{\text{I}} \bar{t}'_{\text{II}} - \bar{t}_{\text{II}} \bar{t}'_{\text{I}}}{(\bar{t}_{\text{I}} - \bar{t}'_{\text{I}}) - (\bar{t}_{\text{II}} - \bar{t}'_{\text{II}})} + \frac{\bar{t}_{\text{I}} - \bar{t}_{\text{II}}}{\ln \left[(\bar{t}_{\text{I}} - \bar{t}'_{\text{I}})/(\bar{t}_{\text{II}} - \bar{t}'_{\text{II}}) \right]} \tag{9-77}$$

where $\bar{t}$ and $\bar{t}'$ are for the two streams and I and II the two ends of the exchanger.

More complicated cases can also be worked out analytically, particularly by the Laplace transform, such as the mean temperature of a vertical stream on each side of a conducting wall, when heat is being generated in or adjacent to and removed by both streams as well as transferred between them. This case applies, for instance, to fluid-fuel reactors with an internal coolant, and to thermal shields cooled with thermal circulation of an enclosed coolant and external forced convection of a second coolant stream (which could be the primary coolant stream of the reactor [141]). In general, however, these and more complicated cases will need to be worked out by finite-difference calculations (see Sec. 9-39), so that integrated relations for buoyancy or heat-transfer mean temperatures $\bar{t}_m$ are not particularly helpful.

In the design of a thermal circulation loop or a complete plant, assuming one-dimensional flow, it is usually necessary to write the pertinent flow, heat-balance, and heat-transfer equations, combine equations to eliminate as many of the unknowns as conveniently possible, and proceed by trial and error to find the one or more desired solutions. For instance, for the loop in Fig. 9-21 the following equations are pertinent: Eq. (9-72) for w, the heat balance $q = wC(\bar{t}_h - \bar{t}_c)$, the heat-transfer relation

$$q = \frac{UA(\bar{t}_h - \bar{t}_c)}{\ln[(\bar{t}_h - t_b)/(\bar{t}_c - t_b)]}$$

Eq. (9-75) for Z_e, Fig. 8-10 for f_F as a function of Reynolds number $wD_e/S\mu$, and an empirical or theoretical relation for U as a function of w and other necessary knowns and unknowns in the problem. If it is desired to find $\bar{t}_h$ for a given q, t_b, geometry, and fluid the six unknowns would be $\bar{t}_h$, $\bar{t}_c$, E_b, w, f_F, and U. In this case w could be assumed, f_F and U calculated approximately (assuming they depend only slightly on average fluid temperature), $\bar{t}_h$ and $\bar{t}_c$ obtained by simultaneously solving the heat-balance and the heat-transfer equation, then Z_e computed, and finally w checked by Eq. (9-72). If not a close check, one or more values of w would be used until the trial w and that by Eq. (9-72) agreed, or w could be graphically found by plotting the ratio of assumed to calculated w against one of them and interpolating to the ratio $= 1$. Generalized solutions of the above relations for a specific case may be convenient in predicting operation under many conditions (see Prob. 9-20).

Prob. 9-20. Figure 9-22 is a generalized graphical solution [107] of thermal circulation in a vertical cylindrical homogeneous reactor containing a tall internal concentric nonconducting thin cylindrical baffle. The solution circulates upward in the baffle, then down between the baffle and the outer cylindrical wall of the reactor, which is maintained uniformly at t_w. The cross section of the whole reactor is S and that within the baffle is S_c. The outer cooling area is A and the heating density H is assumed uniform, both inside and outside the baffle. Friction within the baffle is neglected, and Z_e is given by Eqs. (9-73), (9-75), and (9-76). λ is $St/2f_F$, and from Eqs. (8-33) and (9-44a) is taken as $Pr^{-2/3}$ for nonmetals in turbulent flow, and roughly also for other flows. The solution for a given core is at the intersection of the appropriate curve from each family. Note that h must be assumed, then checked by computing G in the annulus from $\bar{t}_h - \bar{t}_c$. Also, if the outer perimeter P_0' is extended for better heat transfer by corrugations, etc., from the circle πD_0, it is recommended to divide λ by $(1 + \pi D_0/P_0')$.

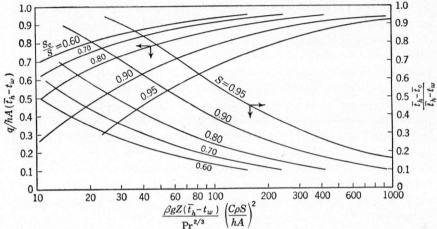

FIG. 9-22. Generalized solution of natural convection in a cylindrical homogeneous liquid reactor with internal cylindrical baffle and constant outer cylindrical wall temperature.

a. What power can be dissipated by a homogeneous aqueous reactor 35 in. in both diameter D_0 and height Z with the outer cylindrical wall maintained at 140°F if its perimeter is increased 75 per cent by corrugations and the maximum temperature permitted is 170°F? S_c/S is 0.9, and the physical properties of water are applicable. [Using the assumption that $h = 100$ Btu/(hr)(ft²)(°F) due to free gravity convection (Sec. 9-12), q is obtained as 35 kw in Ref. 107.]

b. In order to double the power, what change would need to be made separately in each of the following: $\bar{t}_h$, t_w, P_0', D_0, Z?

More detailed two-dimensional theoretical analyses have been carried out for laminar free convection with heat generation between parallel plates and within closed cylinders [108], and in single-pipe loops [109]. However, small variations in dimensions, operating conditions, physical properties, etc., may cause large changes in temperature and velocity distributions as well as difficulties in the analysis, and general solutions are impractical.* Thus, for short Z/D and simple specific shapes,

---

* This is particularly true if turbulence sets in, which will be at lower Re than in isothermal flow. In a 0.7-in. ID cold water loop with Gr (based on $\bar{t} - t_w$) about 10^5 and Pr = 6.5, (Re)$_c$ was about 100 [109].

numerical or analogue analysis is usually preferable. However, in all cases of low Z/D, particularly where geometries are complicated or turbulence is suspected, experimental tests under the exact conditions are needed for reliable results.

9-30. Comparison of Coolants. Many factors are involved in the selection of coolants, particularly primary coolants. These factors include cost, nuclear properties, hazards, and physical properties over the desired ranges of conditions. For instance, cost is an important argument against D_2O, and safety is a reason that pressurized gaseous CO_2 is attractive.

In designing a nuclear power plant, any coolant should properly be considered only in an optimized over-all design. If only a few factors are rigidly fixed, such as output and fuel, it is impossible to solve conveniently for the optimum values of all the other factors and thence the minimum cost. However, considerable guidance can be obtained by computing important derived quantities, pumping power in particular for power reactors. For a rough comparison of coolants, the ratio of their pumping powers for any desired fixed conditions is helpful. For a more thorough analysis, one or more of the conditions can be optimized for each coolant before computing the pumping power.

Employing Eq. (8-29) for ΔP and Eq. (8-33) for f_F, the pumping power in turbulent flow for substantially nonexpanding fluid is

$$P = \Delta P \frac{w}{\rho} = \frac{0.092 N_e}{g_c D_e{}^{1.2} S^{1.8}} \left[\frac{\mu^{0.2}}{\rho^2} \right] w^{2.8} \tag{9-78}$$

For streamline flow, using $(\bar{u}\mu N/\Delta P\, g_c)$ from Table 8-5

$$P = \frac{\Delta P\, g_c}{\bar{u}\mu N} \frac{N}{g_c S} \left[\frac{\mu}{\rho^2} \right] w^2 \tag{9-79}$$

Coolant Temperature Rise Δt_c. If the coolant-film temperature difference $\Delta t_f \ll \Delta t_c$, as might particularly be true for large reactors with long or small tubes and low q/A, the simplest basis of comparison is the pumping power required to give the same Δt_c. Replacing w in Eq. (9-78) by $q/(C\,\Delta t_c)$, for turbulent flow,

$$P = \frac{0.092}{g_c} \frac{N_e q^{2.8}}{D_e{}^{1.2} S^{1.8} (\Delta t_c)^{2.8}} \left[\frac{\mu^{0.2}}{\rho^2 C^{2.8}} \right] \tag{9-80}$$

Similarly, for streamline flow,

$$P = \frac{\Delta P\, g_c}{\bar{u}\mu N} \frac{N_e q^2}{g_c S (\Delta t_c)^2} \left[\frac{\mu}{\rho^2 C^2} \right] \tag{9-81}$$

Evidently, the smaller the physical-property groups in brackets in Eqs.

(9-80) and (9-81), the better the coolant from this standpoint. Values for typical coolants are listed in Table 9-14. Water is by far the best at temperatures it (and the organics) can reach, and fused salts and metals above that. If in a comparison of coolants all are not in the same (streamline or turbulent) regime, P should be computed by the appropriate equation for each.

Coolant-film Temperature Difference Δt_f. If $\Delta t_f \gg \Delta t_c$, as might be approached in an enriched power reactor, the pumping power for turbulent flow of nonmetals to obtain the same Δt_f and thus h by Eq. (9-44a) is

$$P = \frac{49{,}850 N_e S h^{3.5}}{g_c D_e^{0.5}} \left[\frac{\mu^{1.835}}{\rho^2 C^{1.167} k^{2.333}} \right] \tag{9-82}$$

For liquid metals in turbulent flow in tubes, Eq. (9-44i) yields

$$P = \frac{49{,}850 N_e S h^{3.5}}{g_c D_e^{0.5}} \left[\frac{0.803 \mu^{0.2}}{\rho^2 C^{2.8} k^{0.7}} \left(1 - \frac{4.8}{Nu} \right)^{3.5} \right] \tag{9-83}$$

Similar relations can be derived for other channel shapes. A plot of h against pumping power per unit of area P/A also is convenient for comparing coolants on the constant Δt_f basis for a specific geometry [116].

For streamline flow in round pipes, Eqs. (9-79) and (9-45b) (increased by the ratio 4.36:3.66 from Table 9-6 for constant q/A) yield

$$P = \frac{5.66 N^3 D^2 h^6}{g_c} \left[\frac{\mu}{\rho^2 C^2 k^4} \right] \tag{9-84}$$

In streamline flow this comparison is less appropriate, since Δt_f and h are more dependent on D than on w and constant D would be inappropriate to compare fluids at constant Δt_f and D. Table 9-14 shows that sodium and then the other liquid metals are by far the best in this respect.

Hot-spot Temperature. When neither Δt_c nor Δt_f predominates, both should be considered. One comparison has been made [112] based on the same Δt_c, Δt_f, and total volume of cylindrical channels V_c, as well as channel length $2x_2$, and q. Different D, ΔP, and w are required for each coolant. For turbulent flow of nonmetals,

$$P = \frac{4}{J g_c} \frac{q^3 x_2^2}{V_c^2 (\Delta t_c)^2 \Delta t_f} \left[\frac{\mu}{\rho^2 C^2 k} Pr^{-0.4} \right] \tag{9-85}$$

For streamline flow [113],

$$P = \frac{7.34}{J g_c} \frac{q^3 x_2^2}{V_c^2 (\Delta t_c)^2 \Delta t_f} \left[\frac{\mu}{\rho^2 C^2 k} \right] \tag{9-86}$$

Dividing Eq. (9-86) by Eq. (9-85) yields $P_{\text{streamline}}/P_{\text{turbulent}} = 1.835 \, Pr^{0.4}$, which suggests that, if optimum design of a reactor yields coolant flow

$$\text{TABLE 9-14. COOLANT PHYSICAL-PROPERTY GROUPS*}$$

Circulation: Flow: Constant: Equation: Coolant type:	Forced Turbulent Δ_e (9-80) Any	Forced Streamline Δ_e (9-81) Any	Forced Turbulent Δ_f (9-82) Nonmetal	Forced Turbulent Δ_f (9-83) Metal	Forced Streamline Δ_f (9-84) Any	Forced Turbulent $\Delta_e + \Delta_f$ (9-85) Nonmetal	Forced Streamline $\Delta_e + \Delta_f$ (9-86) Any	Thermal Streamline Δ_e (9-70) Any	Thermal Turbulent Δ_e (9-72b) Any
H_2O (saturated at 300°C)	0.188	0.000922	19.79	......	3.44×10^8	0.721	0.721	0.753	5.03
Organic liquids:									
Dowtherm (300°C)	1.582	0.01230	11,670	......	1.40×10^{12}	17.86	40.20	2.45	10.75
Circo XXX oil (300°C)	1.847	0.0339	130,800	......	8.97×10^{12}	36.2	136.8	5.51	14.10
Fused salts:									
NaOH (350°C)	0.875	0.0416	2,160	......	1.258×10^9	7.11	17.35	9.12	14.35
HTS (300°C)	2.95	0.0788	2.05×10^4	......	1.039×10^{11}	31.5	84.50	8.14	16.3
KCl-LiCl (400°C)	4.54	0.1552	625	......	1.042×10^8	17.79	25.0	11.83	19.4
Liquid metals:									
Na (300°C)	11.22	0.0479	......	1.188	43.6	2.02	0.262	7.30	28.9
Sn (300°C)	25.2	0.0927	......	4.73	2,210	6.94	1.151	7.54	31.1
Hg (300°C)	35.4	0.0532	......	13.89	85,100	11.78	1.898	3.08	22.6
Pb (500°C)	41.9	0.1278	......	13.61	68,400	16.91	3.45	6.38	29.6
Bi (300°C)	53.6	0.1413	......	15.39	37,300	18.32	3.21	2.83	39.8
K (300°C)	56.1	0.1178	......	8.92	1,110	10.18	1.160	8.51	41.2
Gases:									
H_2 (300°C, 10 atm abs)	2.77×10^4	61.60	2.07×10^6	......	2.24×10^4	999×10^4	850×10^4	351	580
CO_2 (300°C, 10 atm abs)	0.990×10^5	44.90	3.38×10^7	......	5.55×10^7	550×10^5	474×10^5	82.4	358
He (300°C, 10 atm abs)	2.48×10^5	454.0	2.84×10^7	......	5.97×10^{15}	9.86×10^5	8.66×10^5	524	825
Air (300°C, 10 atm abs)	2.49×10^5	119.7	7.56×10^7	......	8.96×10^{17}	1.299×10^6	1.111×10^6	132.4	496

* The smaller the group, the better the coolant in that particular column. Units are centimeters, grams, seconds, and calories.

near the critical Re, it may be desirable to assure turbulent flow with nonmetallic coolants and streamline flow with metals. From Table 9-14 water is by far the best in turbulent flow, but in streamline flow sodium is better.

Comparison based on simultaneous constant Δt_c and Δt_f is unnecessarily restricted, since the permissible $\Delta t_c + \Delta t_f$ or $\Delta t_c + \Delta t_f + \Delta t_e$ is the important consideration. Comparison on this basis for a cosine distribution would be made by computing w in Eq. (9-67) by trial and error for each coolant, then substituting each w into Eq. (9-78) or (9-79). For this or any other flux distribution the relative ranking of the fluids will be intermediate between that for constant Δt_c and for constant Δt_f; thus either water or sodium or NaOH might be best, depending on the design.

Thermal Circulation Δt_c. By Eq. (9-70), the efficiency of a coolant to dissipate heat with low coolant temperature rise is inversely proportional to the group $(\mu/\beta\rho^2 C)^{1/2}$. If turbulent flow is attained, by Eq. (9-72b) the pertinent group is $(\mu^{0.2}/\beta\rho^2 C^{1.8})^{0.357}$. These groups are also given in Table 9-14 and show in this respect the moderate superiority of water. However, taking Δt_f into account, the liquid metals would generally be best, and certainly such would be the case at the high temperatures that might be reached in emergency cooling by this method.

General Evaluations. Considerations apart from those so far discussed cannot readily be made quantitative. However, a number of qualitative evaluations of coolants for various reactor designs have necessarily been made. For instance, Parkins [114] concludes that liquid metals are preferable for (nonboiling) heterogeneous power reactors because of the high temperatures and low decomposition obtainable. Those recommended are Na, Na + K, Pb, and Pb + 2.5 per cent Mg. Shimazaki [115], from cost estimates for complete power plants using graphite-moderated solid-fuel reactors, finds Pb + 2.5 per cent Mg, Bi + 44.5 per cent Pb, Bi, Na, and HTS salt in that order of decreasing economy, though all are satisfactory. From the standpoint of current know-how, Na was believed best.

9-31. Comparison of Heat-transfer-surface Arrangements. Different shapes or arrangements of strip, plate, gauze, or porous fuel elements or of other heat-transfer surfaces can be compared, as coolants were in Sec. 9-30. The first step is to obtain the pressure-drop characteristics of the surfaces. To correct for any variation among test and/or use conditions, the results should be correlated as f_F vs. Re. As per Eqs. (9-78) and (9-80), and since $w = S_t \mu \, \mathrm{Re}/D_e = q/C \, \Delta t_c$,

$$\mathrm{P} = \frac{2f_F N w^3}{g_c \rho^2 D_e S_t^2} = \frac{2f_F N w^4}{\mathrm{Re} \, g_c \rho^2 S_t^3 \mu} = \frac{2f_F \, (\mathrm{Re})^2 \, \mu^2 N q}{g_c \rho^2 \, \Delta t_c \, D_e^3} \tag{9-87}$$

In this section, the total cross section of the heat-transfer arrangement

S_t is used, including both free flow space and solid, so that S will be the same for all core designs fitting the same volume. For a two-fluid heat exchanger the two sides are considered separately.

Comparison at Constant Δt_c. If $\Delta t_c \gg \Delta t_f$, the pressure-drop comparison of different heat-transfer surfaces can for simplicity assume Δt_c the same for all. An important case is that of some one arrangement (a single curve of f_F vs. Re), being considered for various operating conditions. If only the scale D_e is varied, P by Eq. (9-87) is proportional to f_F/D_e and thus to f_F/Re. Since f_F/Re always decreases when Re is increased, the highest Re or largest D_e gives the lowest P. Actually D_e will be limited by Δt_f and Δt_e becoming important as the cooling surface decreases and by the hot spot's overheating.

If S_t is not necessarily fixed but the other quantities are, P by Eq. (9-87) is proportional to f_F/S_t^2, and thus to $f_F Re^2$. Since f_F always varies much more slowly than Re, the smaller Re, the smaller P. Actually S_t will be limited by volume, weight, cost, etc.

If different surface roughnesses or finishes are being compared, the smoothest (lowest f_F) will, of course, have the lowest P at the desired w.

If different shapes, having different values of D_e and different curves of f_F vs. Re, are to be compared for use at the same N, w, ρ, and S_t, as might be the case if the reactor power, size, and Δt_c were fixed, Eq. (9-87) indicates that P will be proportional in each case to f_F/Re at the appropriate Re (a different Re according to D_e for each shape). To get a direct comparison of P it is convenient to plot the appropriate values of f_F/Re as ordinates at the *same* abscissa, so that the lower curve immediately designates the better shape from the pressure-drop standpoint. This can be accomplished by plotting f_F/Re for each of the shapes at the same conditions against the Re for some one of them, say Re_1, which equals Re $(D_e)_1/D_e$ for each other shape.

Comparison at Constant Δt_f. As considered in Sec. 9-30, nuclear reactors may operate at high q/A and large D_e so that $\Delta t_f \gg \Delta t_c$. This condition is more common in compact heat exchangers, which do not have the critical-size requirement of reactors, and for liquids of high Pr.

Information on the average heat-transfer coefficient is necessary. To correct for variations between test and use conditions, the surface coefficient $\bar{h}$ is usually correlated as j against Re. $\bar{h}$ is the local average coefficient, divided by the average fin efficiency if fins are involved (see Sec. 9-33). j is defined by

$$j \, Pr^{-\frac{2}{3}} = \overline{St} = \frac{\bar{h}}{CG} = \frac{\bar{h}S_t}{Cw} = \frac{qD_e}{4N \, \Delta t_f \, Cw} \tag{9-88}$$

Many data on j and f vs. Re are available [26, 118] Secs. 8-49 and 9-13, etc. Minor manufacturing deviations such as rough edges may affect these characteristics considerably, however, and tests are desirable in any important design. To compare different fuel-element designs (different D_e with different w required to yield the same Δt_f), D_e and w in Eq. (9-87) may be eliminated by Eq. (9-88) and by Re $= D_e w/S_t \mu$, yielding

$$P = \frac{f_F \, \mathrm{Re}}{j^2} \frac{q^2}{8 g_c S_t N (\Delta t_f)^2} \frac{\mu}{\rho^2 C^2} \mathrm{Pr}^{4/3} \tag{9-89}$$

The ratio $f_F \, \mathrm{Re}/j^2$ for different designs should be plotted against the same abscissa at the same operating point, say Re_1 for one of them. Obtaining w from Eq. (9-88) shows that $\mathrm{Re}_1 = \mathrm{Re}(j/j_1)[(D_e)_1/D_e]^2$. Thus the design which yields the highest $f_F \, \mathrm{Re}/j^2$ at the desired Re_1 is the best from this standpoint (smallest P for a given h).

Plots of h for a specific fluid and conditions [or of jG, which equals $(h/C)\mathrm{Pr}^{2/3}$] vs. P/A (which equals $f_F G^3/2 g_c \rho^2$) have been used [26] for comparing different heat-exchanger surfaces as to pumping power, the surface which has a higher curve being considered more efficient. These comparisons apply if A is fixed, as in comparing different surface finishes.*

For a nuclear-reactor core, or any heat exchanger for which the volume is the important specification, the proper comparison for pumping power vs. Δt_f with a given coolant and conditions is a plot of $ha = q/(S_t N \Delta t_f)$ vs. $Pa/A = P/S_t N$, where a is the heat-transfer surface per unit volume [113].† If the amount of fuel for the reactor core has been fixed in the nuclear design, the ha comparison is the correct one if the fuel elements can be manufactured with any desired fuel concentration. If an available fuel surface concentration must be used, A is fixed and the h vs. P/A comparison is correct. For the general case of the optimum core design for a given weight, amount of fuel, amount of poison, total cost, any combination of these, etc., the test values of h and of P/A for each core or arrangement should both be multiplied by the ratio of the weight, fuel, poison, cost, etc., or any weighted combination thereof, per unit of heat-transfer area. The design with the highest modified h coordinate values is the most efficient in that respect.

Hot-spot Temperature. The previous methods for evaluating heat-transfer surfaces at constant Δt_c and/or constant Δt_f are generally adequate for heat exchangers; temperatures are lower than in reactors, and Δt_e is usually smaller because of lower q/A and frequently thinner and better-conducting walls. However, for fuel elements it will generally

---

* The ratio $2j/f_F$, or $2h/CGf_F$ for gases, has been designated as the "efficiency" of heat-transfer surfaces. It approaches unity for smooth surfaces and is considerably less for rough surfaces. However, for constant Δt_f this comparison is excessively strict against rough surfaces.

† Maximum ha with flat fins is obtained [121] with fin spacing approximately equal to one boundary-layer thickness by the equations in Sec. 8-27.

be found that such evaluations made for constant Δt_c or Δt_f are not entirely applicable. The best shapes from a heat-transfer and pumping-power standpoint may not permit a large enough amount of surface to provide the desired amount of fuel with the fuel sandwich or rod selected. For instance, long straight surfaces seem poorer than transverse wires and strips, because the latter have higher h and can employ lower veloci-ties and smaller area to give a desired maximum Δt_f. However, this means also less fuel-element surface and mass and a looser structure. In most instances in reactor cores Δt_c and Δt_e also are high enough that they would be significantly increased. Thus when neither Δt_c or Δt_f pre-dominates, the hot-spot temperature should be evaluated for different fuel elements as in Sec. 9-30 for different coolants.

9-32. Mean Temperature Difference in Heat Exchangers. If the change in temperature of both fluids is small compared to the temperature difference between them, as in a condenser-boiler, Δt averaged over the

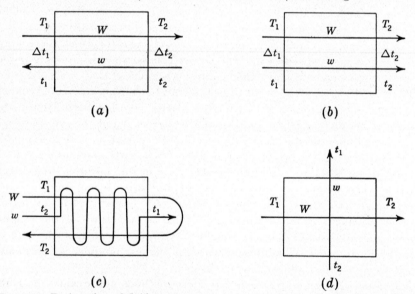

FIG. 9-23. Designation of fluid temperatures in heat exchangers. (a) Counterflow; (b) parallel flow; (c) two-pass flow on one side and many-pass (mixed one-pass) flow on the other; (d) crossflow.

heat-transfer surface, Δt_m, is substantially equal to the arithmetic mean $\Delta t_{am} = [(T_1 + T_2) - (t_1 + t_2)]/2$ in Fig. 9-23. If U (see Sec. 9-34) and the C values are constant, parallel or counter flow or one side at constant temperature requires the "log-mean" Δt:

$$\Delta t_m = \frac{(T_1 - t_1) - (T_2 - t_2)}{\ln\left[(T_1 - t_1)/(T_2 - t_2)\right]} = \frac{\Delta t_1 - \Delta t_2}{\ln\left(\Delta t_1/\Delta t_2\right)} = \Delta t_{lm} \qquad (9\text{-}90)$$

However, the ratio of terminal Δt values may exceed 1.4 before Δt_{am} introduces 1 per cent error.

For all other flow arrangements Δt_m is obtainable by multiplying Δt_{lm} computed as if the flow were countercurrent [Eq. (9-90) as per Fig. 9-23c and d] by an efficiency factor F [26]. F is given in Fig. 9-24 for the principal cases in nuclear power. Case a is of interest for primary to secondary liquid-coolant heat transfer in moderate units and case b for liquids in large units. Case c is useful in large or compact heat exchang-

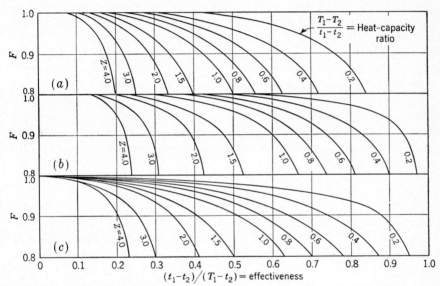

Fig. 9-24. Mean tempeature difference correction factor for heat exchangers. (a) One shell pass, transversely mixed, and 2, 4, 6, etc., tube passes. (b) Two shell passes, each transversely mixed, and 4, 8, 12, etc., tube passes in counterflow. (c) Crossflow, one unmixed pass on each side.

ers handling two gases, since it provides ready access for large flow volumes; the unmixed case is conservative but appropriate because the channel length-width ratio can scarcely exceed about 2, and there are usually vanes parallel to the flow that minimize cross mixing [118]. It is seldom practical to use more than one pass of crossflow because of the large duct volume required; counterflow would become preferable in spite of more difficult construction and access.

Values of F lower than about 0.8 should be avoided in precise designs because F then becomes very sensitive to the temperatures and flow rates.

9-33. Efficiency of Heat-transfer Surface. If a fluid film conductance hA is impractically low for the desired or available h and A, A can be increased (and with pins, short strip fins, etc., h is increased as well) by attaching additional surface that projects into the coolant. This method is particularly employed with gases because of their low h, or in compact designs such as aircraft heat exchangers. Unfortunately, Δt_f decreases on proceeding into the stream because of temperature drop in the fin, yielding a "fin efficiency" E_f less than unity.

Research reactors with low q/A, such as the X-10 reactor, do not need extended surface because Δt_{fb} from the base surface to the coolant stream is low enough without it. In high-power gas-cooled research production or power reactors, fins are found worthwhile, particularly if the fins also fill a need as fuel-element spacing or centering vanes. In general, fins are not so useful in liquid-cooled reactors and heat exchangers, since if they are long enough to be helpful E_f is low.

Theoretical analyses of fins assume—although it is not the case—that the local h is uniform and the fluid transversely mixed.* However, correlations are becoming available [i.e., Eq. (9-47g) and Refs. 26, 117, 118] of average local h, for various types of fins, obtained by correcting via Eq. (9-91) the observed effective h_{eff} by the theoretical E_f of the fins. Average local values of h from these correlations may be used for similar types of fins with considerable confidence by substituting back again into Eq. (9-91)

$$q = h_{eff}\,\Delta t_{fb}(A_b + A_f) = h\,\Delta t_{fb}(A_b + E_f A_f) \qquad (9\text{-}91)$$

For relatively thin fins, in which one-dimensional heat conduction can be assumed, the differential equation for extended surface uniformly arranged along a flat base or around a cylindrical base is [26], in terms of local fin temperature t and fluid temperature $\bar{t}$,

$$S_f \frac{d^2 t}{dx^2} + \frac{dt}{dx}\frac{dS_f}{dx} = \frac{h(t - \bar{t})P'}{k} \qquad (9\text{-}92)$$

S_f is the total cross section and P' the total perimeter (both sides) of the fin at distance x from its base. This has been integrated by Douglass's general solution, applicable to most fin designs [11]. Special shapes can be handled by other methods (see Secs. 9-39 to 9-43).

Values of E_f are given in Fig. 9-25 for types of fins and ranges of interest in nuclear power.

Prob. 9-21. (a) For the fins in Sec. 9-26, check by Fig. 9-25 the efficiency there calculated. (b) If E_f for a straight fin of uniform cross section S_f and cooled perimeter P is given by the expression

$$E_f = \frac{\tanh\left[(hP'/kS_f)^{\frac{1}{2}}x_f\right]}{(hP'/kS_f)^{\frac{1}{2}}x_f} = \frac{\tanh mx_f}{mx_f}$$

compute the terminal E_{term} at the outermost end of the fin, and the net effective or

---

* The decrease in efficiency due to nonmixing transversely to the coolant stream and parallel to flat fins is negligible under most practical conditions in nuclear power [122].

incremental E_{incr} for an increment of length at the outermost end x_f of the fins on a central fuel slug in Sec. 9-26.

Partial Solution. $E_{term} = (\cosh\ mx_f)^{-1}$; $E_{incr} = 1 - (\tanh\ mx_f)^2$.

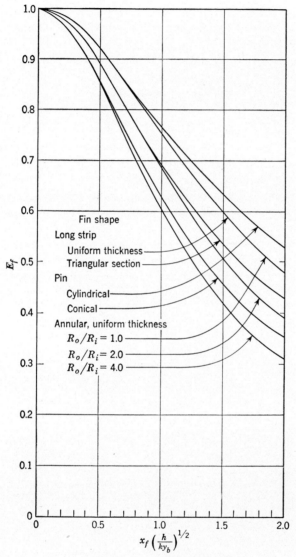

FIG. 9-25. Efficiency of extended surface.

In reactors and in light heat exchangers it is important to use efficiently the metal in extended surface to minimize neutron losses, weight, cost, etc. Experimental comparisons of $h_{eff}a$ vs. Pa/A at rated operating conditions can be used for different fin designs, similarly as in Sec. 9-31.

If there is a certain amount of heat per unit length of fuel element q/N to be dissipated at a given Δt_{fb}, and no limitation on the number and dimensions of the fins, it is readily seen that the more numerous (and narrower and thinner) the fins the smaller is the mass of fins required. Normally, however, the number of fins is first selected and thus $(q/N)_f$ fixed for each fin.

For flow parallel to a long or flat heat-transfer surface, straight fins in the direction of flow might be preferred. The optimum fin cross section is parabolic [11], having the profile

$$\frac{y}{y_b} = \left(\frac{x}{x_f} - 1\right)^2 \tag{9-93}$$

where y = local half-thickness at width x

y_b = local half-thickness at base

x_f = total width (maximum distance out from base)

This fin has concave faces and a sharp weak edge, so it is impractical. A triangular cross section is almost as efficient and more convenient but also comes to a weak sharp edge. A rectangular cross section is the most convenient mechanically but is appreciably less efficient. Dimensions of the optimum fins of these three shapes are given in Table 9-15. Any area at the outer edge (as for the rectangular fin) should in computing E_f be split between the two sides and added to x_f.

TABLE 9-15. DIMENSIONS AND EFFICIENCIES OF THIN STRAIGHT FINS OF
MINIMUM MASS*

Item	Dimension ratio	Fin cross section		
		Optimum	Triangle	Rectangle
1	$\dfrac{x_f h\,\Delta t_{fb}}{(q/N)_f}$	2	1.684	1.596
2	$\dfrac{y_b hk\,\Delta t_{fb}{}^2}{(q/N)_f{}^2}$	2	1.655	1.264
3	$\dfrac{y_b k}{x_f{}^2 h}$	0.5	0.583	0.496
4	$\dfrac{(V/N)_f h^2 k\,\Delta t_{fb}{}^3}{(q/N)_f{}^3}$	1.333	1.393	2.019
5	$\dfrac{(V/N)_f k}{x_f{}^3 h}$	0.1667	0.292	0.493
6	E_f	0.500	0.593	0.627
7	E_{term}	0	0.277	0.457

* q, V_f, and y_b are for one side (half fin) only.

If $(q/N)_f$ and Δt_{fb} are given, the necessary dimensions for the desired shape are obtainable from lines 1 and 2 of Table 9-15 and the necessary cross section (normal to the length and thickness and parallel to the width) or weight from line 4. For the optimum (parabolic) fin, Eq. (9-93) then gives the profile. If x_f is fixed (such as the radial distance between a cylindrical fuel element and its "process tube"), y_b and S_f are obtained from lines 3 and 5, and the $(q/N)_f$ for a given Δt_{fb} from line 1. If handling strength and ease of fabrication are considered, as well as mass, the best practical fin cross section for gas-cooled slugs is probably a truncated flat-sided wedge with y_b some two to three times y at the tip (cf. Fig. M-1). In most commercial extended-surface installations, E_f is 0.9 or higher. From Table 9-15 lower E_f evidently is preferable in nuclear or mobile installations.

For flow normal to cylinders, circular fins are preferable. With liquid coolants in forced convection (high h), the optimum r_f/r_b approaches unity, where r_b is the outer radius of the cylinder, and Table 9-15 is adequate. With lower h more detailed analysis is necessary. The fin of minimum mass for a given $\Delta t_{fb}/q_f$ and outer radius r_f has the profile [11]

$$y = r_f{}^2 \frac{h}{k} \left[\frac{1}{3}\left(\frac{r}{r_f}\right)^2 - \frac{1}{2}\frac{r}{r_f} + \frac{1}{6}\frac{r_f}{r} \right] \tag{9-94}$$

For any given conditions the unknown quantity in Eq. (9-95) is first found:

$$q_f = \frac{\pi}{3} h\, \Delta t_{fb} r_b{}^2 \left(\frac{r_f}{r_b} + 2\right)\left(\frac{r_f}{r_b} - 1\right) \tag{9-95}$$

and y_b and the half volume V_f per fin are then obtained by

$$y_b = \frac{h r_b{}^2}{6k}\left(\frac{r_f}{r_b} + 2\right)\left(\frac{r_f}{r_b} - 1\right)^2 \qquad V_f = \frac{\pi h}{6k} r_b{}^4 \left(\frac{r_f}{r_b} + 1\right)\left(\frac{r_f}{r_b} - 1\right)^3 \tag{9-96}$$

For a circular fin of uniform half-thickness y, the unknown quantities for the optimum fin are obtained by interpolation in the following table [11]. The optimum E_f is close to 0.4.

$q_f/\pi h r_b{}^2\,\Delta t_{fb}$	2	2.5	3	4	5	6	8	10	12	∞	
$r_b(h/ky)^{1/2}$	1.15	0.94	0.79	0.65	0.55	0.46	0.36	0.30	0.26	0	(9-97)
$r_f(h/ky)^{1/2}$	2.35	2.15	2.00	1.83	1.69	1.57	1.41	1.28	1.20	0	
r_f/r_b	2.04	2.29	2.53	2.82	3.07	3.41	3.92	4.27	4.62	∞	

The optimum circular fin of triangular cross section is close to the optimum of Eqs. (9-94) to (9-96), and its dimensions can be estimated by interpolating between the dimensions of the optimum and the uniform fin [Eq. (9-97)] proportionally to the corresponding dimensions in Table 9-15.

For pins or spines of constant cross section S_n and perimeter P', Table 9-15, lines 1 to 3, for rectangular cross sections are valid on substituting

S_n/P' for y_b.   An analysis has been made [11] of the optimum conical pins, taking into account the mass of the surface which they cover. E_f ranges from 0.3 to 0.95 depending on conditions.

9-34. Heat-exchanger Design.   The preceding principles form the basis for the thermal design of heat exchangers.   This subject will not be discussed in detail, as it is well covered elsewhere [26, 119] and has no major specialized nuclear aspects.

In general, heat-exchanger design consists of obtaining trial-and-error solutions of Eq. (9-98) and selecting the most economical one, including pumping cost, equipment fixed and maintenance charges, and fixed charges on the contained liquid if D_2O, a fuel solution, or other expensive liquid is involved.

$$\frac{\Delta t_m}{q} = \frac{F}{q} \frac{\Delta t_1 - \Delta t_2}{\ln (\Delta t_1/\Delta t_2)} = \frac{1}{A_i} \left[\left(\frac{1}{h_o} + r_o' \right) \frac{A_i}{A_o} + \left(\frac{D_o - D_i}{2\bar{k}_w} \right) \frac{A_i}{A_w} + \frac{1}{h_i} + r_i' \right]$$

$$= \frac{1}{U_n A_n} \quad (9\text{-}98)$$

F is obtained from Sec. 9-32 and h_o and h_i from Secs. 9-11 to 9-19, as appropriate.   The "over-all coefficient of heat transfer" U is a convenient term to describe the thermal resistance; n, usually i or o, should be stated.   The r' values are "fouling factors," surface thermal resistances from experience that are added on account of dirt or scale films or nonwetting by liquid metals (see Sec. 9-22).   Standard [120] and typical other values of r' in hours per square foot per degree Fahrenheit per Btu for $\bar{u} > 3$ fps are as follows:

0–0.0005—nonwetting liquid metals in forced convection and inorganic films formed with fused salts

0.0005—distilled water;* cold sea water; cold boiler feed water; organic vapors

0.001—cold city or spray pond water; clean river and lake water; hot sea water and boiler feed water; exhaust steam; clean oils and other organic liquids

0.002—air; hot city, well, and lake water; cold average river water

0.003 and upward—dirty river water and oils

A_o and A_i should each equal the corresponding $A_b + E_f A_f$ (see Sec. 9-33), if there are fins.   The design problem usually consists primarily of finding the A values for the heat exchanger, for selected tube diameters D_i and D_o.   For this purpose each A can be replaced by $\pi N_t D$, where N_t the total length of all tubes becomes the unknown.

---

* Probably conservative for highly purified recirculating water, as in MTR, SR, NRX, and STR (see App. N).   In designing the MTR 0.00004 was used [104].

Prob. 9-22. Show that the maximum temperature of primary coolant of specific heat C and flow w circulating through a boiler of area A with a constant over-all coefficient U and steam-side temperature t is $t + (q/wC)/(1 - e^{-UA/wC})$.

The sequence of steps for a given design might include the following:

1. Select reasonable velocities, or estimate optimum velocities,* for both sides of the heat exchanger (see Sec. 8-35).

2. Select a reasonable commercial tube size (usually $\frac{3}{8}$ to 1 in. OD),† or a type of heat-exchanger core, if not tubular.

3. Compute number of tubes n in parallel and tube spacing and shell diameter D_s, or size of nontubular core, to obtain the "inside" velocities under step 1 with countercurrent flow.

4. Compute h and select r' for both sides of the heat exchanger [i.e., Eqs. (9-44a) and (9-44h)].

5. Taking $F = 1$ make a preliminary calculation of A_i by Eq. (9-98) and then N/n, the tube length.

6. If N/n is reasonable the thermal design is complete, unless it is desired to repeat steps 1 and 2, then subsequent steps, on the basis of the approximate design.

7. If N/n is unreasonably long for one pass (N/n > the available space or tube lengths (usually 20 ft), or N/nD_s is uneconomically large), list the feasible pass arrangements for F (from the specified terminal temperatures) > 0.8 (see Sec. 9-32) and select an economical pass arrangement. Design shell-side baffling within each shell-side pass if necessary to obtain reasonable h_o or shell-side temperature uniformity in each pass. Then repeat all previous steps using the new value of F until a final consistent design is obtained.

8. If N/n is uneconomically short, use countercurrent flow but higher velocities, so that tubes are fewer and longer. Repeat all previous steps as needed to obtain a final consistent design.

If the specified performance must be assured, apply a safety factor of about 10 to 30+ per cent (Sec. 9-27) by increasing the tube length per pass (not the tubes per pass).

In a close optimum design it might be expected that no safety factors

---

* Simplified equations are available [26, 123] for estimating the optimum mass velocities in heat exchangers. These are not so far applicable with liquid metal or with significant wall resistance. If a valuable liquid such as D_2O or a fuel solution is to be handled, its fixed charges are proportional to those for the heat-transfer surface, and the two may be added to give the total fixed charges per unit of heat-transfer area for the optimizing formula.

† If D_2O or a fuel solution (the latter because of radioactivity and danger as well as cost) is being handled, the optimum tube size and spacing are generally the smallest commercial values, $\frac{1}{4}$ in. OD and $\frac{3}{8}$ in. between centers.

should be employed, particularly if underperformance can be corrected by slightly changed temperatures or circulation rates, and/or if unexpected overperformance is wasted by the use of flow throttling valves, partial bypassing of heat exchangers, etc., both of which are usually the case. However, a given per cent capacity increase usually only costs about half as much in per cent cost increase, so that liberal safety factors are normally applied to all heat transfer and fluid flow equipment, nuclear and otherwise. Furthermore, if unexpected overperformance is *not* wasted (i.e., improved heat transfer yields lower temperatures and longer life, or less pumping energy) optimum performance is not jeopardized, since costs near optimum design are relatively constant over appreciable variations in operating conditions.

The completed design is for certain selected fluid velocities, tube size, etc. Any heat exchanger that will have high fixed or pumping charges compared to designing cost should be completely designed for several values of each independent variable. The total annual fixed and pumping cost, plus the value of heat recovered, if this is not constant, should be computed for each design, and the most economical design selected or graphically interpolated. Because of the relative constancy of the total cost near its minimum, the design selected can be shifted as desired toward higher future capacity, lower investment, lower space or weight, etc., without much penalty on total cost.

The mechanical design of heat exchangers for nuclear power plants may have several distinct features to minimize loss of expensive or radioactive liquids or danger of explosive mixing of alkali metal and water. Greater joint tightness than usual may be desired, as can be achieved by thicker tube sheets and more grooves, or by welded joints. To maintain such tightness a single straight-tube pass would be avoided, and small-diameter bowed tubes, 180° return bends, or 90° bends with side outlet in the shell ("hockey-stick" tubes) would be employed. Greater insurance against tube failure is obtained by using tight-fitting double-walled tubes with a few small intermediate flow channels and two tube sheets, with the intervening thin space filled with He under higher pressure and connected to a leakage flow indicator.*

9-35. Thermal Design of Reactors. So many combinations exist as to fuel, fuel concentration, moderator, coolant, type of power cycle, etc. (Chap. 14), and so many problems in the optimizing of the design, that it is not possible to set down a rigid sequence of steps to yield a complete optimum design. Thus the power, fuel, moderator, coolant, etc., are usually selected arbitrarily after general discussions of all factors, including the rest of the plant. Other main aspects of the normal thermal

---

* Mercury between concentric smooth tubes has been used, but it has higher thermal resistance and its vapor is poisonous.

design are optimized if they are important; otherwise they are fixed from best judgment or experience or approximate estimates or correlations of optimum design. Finally, all remaining quantities, such as local values of coolant flow, temperature, and thermal stress throughout the reactor, under all conditions of normal operation and every conceivable eventuality, are investigated analytically or empirically to make certain that no serious condition exists, or until it is corrected. The whole field of reactor design should be considered wide open, and all possible ingenuity and originality brought to bear on it.

Though the development of liquid fuel reactors is not so far advanced as that of solid fuel reactors in the large sizes, their thermal design is generally simpler. Pertinent previous sections of this chapter can be directly applied (i.e., Secs. 9-11 to 9-14, 9-29, etc.).

Solid-fuel-element thermal design generally depends on the permissible hot-spot temperature. Reasonable limits for continuous service [104] are (1) surface temperature t_w in contact with coolant: Al in H_2O or D_2O, 240°C; Zr-U alloy in H_2O or D_2O, 340°C, in Na 500°C; stainless steel in Na, 650°C; any metal in polyphenyls, 440°C; (2) metal temperature t_0: U, 610°C (maybe the 668°C transformation). To minimize hot-spot temperature the neutron-flux distribution may be flattened radially by poisons* or by lighter fuel loading toward the center of the reactor.† In addition the peak flux may be shifted toward the coolant inlet end, a procedure commonly known as "roof-topping" (this occurs automatically when the control rods enter at the coolant outlet end); or two-pass flow may be used, with the coolant first going through the central channels [Pacific-Bechtel design Chap. 8, Ref. 72)] or to the center of all channels [BNL reactor (Fig. 9-19)].

A minimum cladding of the fuel of about 0.005 in. is necessary to keep fission products out of the coolant. Actually about 0.02 in. of Al [94] or somewhat less Zr (or stainless steel in intermediate and fast reactors) is desirable for mechanical reasons. To minimize Δt_e through the cladding and Δt_f it is desirable to have a large fuel-element area A. However, this increases the volume of the cladding metal, thus also of the reactor, the neutron losses, and the fuel-element cost. Accordingly it may be

---

* "Burnable" poisons such as B_4C in proportional amounts (see Sec. 6-52) can largely make up for faster fuel depletion at the center of a core and will decrease the frequency with which fuel elements must be replaced, with more uniformity and less mechanical difficulty than additional control rods. This is particularly important in minimizing shutdowns in high-pressure power reactors, i.e., PWR uses poison in its enriched fuel elements (see App. N-1).

† In large reactors with many control rods, "trimming" the rods to yield equal coolant outlet temperatures might increase the average power per channel by 15 per cent for the same maximum hot-spot temperature.

desirable to have the minimum feasible A, which is obtained by using high coolant velocities and reaching the maximum permissible hot-spot temperature. For economy and speed in refueling, large fuel elements or subassemblies of fuel elements are desirable. For economy in fabrication the fuel elements should be simple and preferably all of one type, such as in Fig. 9-26 [124]. The ratio of fuel to moderator and the spacing or "clumping" of the fuel in thermal reactors are set by the methods of Chap. 6, but within those requirements the thermal design has broad leeway.

To minimize fuel investment the specific power or heating density H should be high, but not so high that the reactor will be shut down too frequently for refueling or protection.* If B is the per cent burned up† per month of the original fuel present, the rate of heat generation q in Btu per hour per gram of fuel based on 192 Mev recovered per fission is approximately $860B$. At the surface of the solid cylinder of ρ_f lb of fuel per cubic foot and diameter D ft, $q/A = 98,000 B \rho_f D$ Btu/(hr)(ft²). At each surface of a sandwich containing ρ_f' lb of fuel per square foot of the sandwich $q/A = 196,000 B \rho_f'$. Thus, knowing q/A, from the necessary cladding and fuel thickness ρ_f' can be computed for any desired B. The specific power in kilowatts per gram of fuel is $0.252B$.

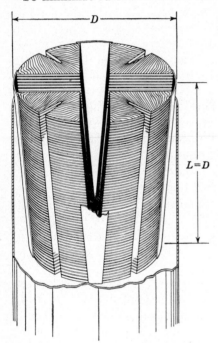

FIG. 9-26. Reactor core with special pancake fuel elements. Coolant flow is down the center and out between the pancakes.

The following equations for simple cases are convenient for computing the thickness L of a flat-plate fuel element or diameter D of a cylinder. They are based on a permitted hot spot $\Delta t_f + \Delta t_e = \Delta t_{fe}$, and given values of h at the surface, k of the fuel element (assumed all fuel or cladding), and H, all in Btu, degrees

---

* The current moderate fixed charges for fuel (App. K) yield lower optimum specific powers than originally considered.

† Since in U²³⁵ nonfission captures of thermal neutrons amount to 16 per cent of the total captures, the amount *fissioned* per month is 84 per cent of B; 1.1 g of U²³⁵ fissioned or 1.3 g consumed will yield 1 Mw-day of heat.

Fahrenheit, hours, and feet. H for the whole fuel-element volume is $392,000B\rho_f$.

1. Flat plate, cooled both sides, negligible cladding:

$$L = \frac{4k}{h}\left[\left(1 + \frac{2h^2\,\Delta t_{fe}}{kH}\right)^{\frac{1}{2}} - 1\right]$$ (9-99)

2. Flat plate, cooled both sides, "all" cladding:

$$L = \frac{2k}{h}\left[\left(1 + \frac{4h^2\,\Delta t_{fe}}{kH}\right)^{\frac{1}{2}} - 1\right]$$ (9-100)

3. Solid cylinder, negligible cladding:

$$D = \frac{4k}{h}\left[\left(1 + \frac{4h^2\,\Delta t_{fe}}{kH}\right)^{\frac{1}{2}} - 1\right]$$ (9-101)

Nonuniform fuel distributions may be desirable to increase the power, the average coolant-outlet temperature, the breeding gain, the burnup, and/or the interval between refuelings or to decrease the critical mass. This has been described [125] for a fast reactor with uniform q per passage and for a thermal reactor either with uniform specific power or uniform H.

Over-all methods have been developed for selecting the optimum temperatures in a power reactor plus the associated heat exchangers [126], and for evaluating alternative reactor designs [127]. In general, it is desirable to carry through to completion a number of rough designs over a reasonable range of each important variable, so as to know how important it may be to adhere closely to the optimum design found. The design is then carried out thoroughly for the conditions selected.

As illustrations of the scores of possibilities that should be looked into in the final detailed thermal and flow-design analysis, might be mentioned flow reversal and overheating in low-power outer tubes with a nonboiling coolant in downflow or a boiling coolant in upflow, instability of flow in parallel channels with boiling (Sec. 8-43) or with downflow (Sec. 8-37), local decrease in flow and overheating if fuel sandwiches or rods are flexible or bent and touch each other, local increased H if a control rod is suddenly withdrawn leaving a low Xe region, adequacy of control-rod cooling in all positions, cooling of spent fuel elements during and after removal (Sec. 9-4), correct distribution of flow through any shield-cooling channels, flow dead spaces with low h under or behind fuel-element supports, differential effects of thermal expansion during normal operation, overheating, or with any possible flow mismatches, existence of harmful vibration due to the flow, need of flow baffles to mix a stream and minimize thermal shock, etc. The "mechanical design" of reactors [128] embodies the constructional details for achieving satisfactory flow, thermal, and stress conditions, as well as convenient and dependable operation, control, refueling, and maintenance.

SPECIAL STEADY- AND UNSTEADY-STATE CALCULATION AND TEST METHODS

9-36. Analytical Unsteady-state Thermal Analysis.   The analytical solution of Eqs. 9-13 through 9-17 has been extensively developed [4] for zero or constant heating density H, constant physical properties, simple shapes or initial temperature distributions, and/or constant coolant h and $\bar{t}$.   These idealized solutions do not closely fit most unsteady-state reactor thermal problems.   However, they are frequently adapted to solve actual problems with adequate accuracy.   Generally the maximum temperature of the solid vs. time (to estimate incipient mechanical failure), the maximum surface temperature vs. time (to estimate incipient coolant boiling), and the average temperature of the solid (to estimate thermal stress or heat storage) are the results sought.

The principal solutions with nuclear-power applications are as follows:

1. Uniform initial temperature t_1.   Surface or coolant temperature suddenly changes to t_2. $H = 0$.
   a. Maximum, minimum, and average temperature of an infinite slab are obtainable as a function of time θ from Gurney-Lurie [26], Groeber [11], Hottell [26], or Bachmann [130] charts for constant h (Fig. 9-27).
   b. Same for an infinite cylinder [11, 26, 130] (Fig. 9-28).
   c. Same for a solid sphere [11, 26, 130].
   d. For an infinite bar, a parallelepiped, or a finite cylinder, the local or the over-all mean value of $(t - t_1)/(t_2 - t_1)$ is obtainable by multiplying together the values for the separate two or three dimensions (as in Table 9-2 for neutron flux).
   e. Temperature in a semi-infinite solid [4, 135] (Fig. 9-29).
2. Uniform initial temperature t_1.   Surface temperature changes linearly with time. $H = 0$.
   a. Temperature in a semi-infinite solid [26, 135] (Fig. 9-29).
   b. Temperature change Δt in an infinite slab at distance x from the heated face can be computed by the "method of reflections" [135], by summing terms $\Delta t'$ for a semi-infinite solid (Fig. 9-29):

$$\Delta t_x = \Delta t'_x + \Delta t'_{2R-x} - \Delta t'_{2R+x} - \Delta t'_{4R-x} + \Delta t'_{4R+x} + \Delta t'_{6R-x} - \Delta t'_{6R+x} - \cdots$$

   c. The maximum (at $\theta = \infty$) temperature difference across an infinite

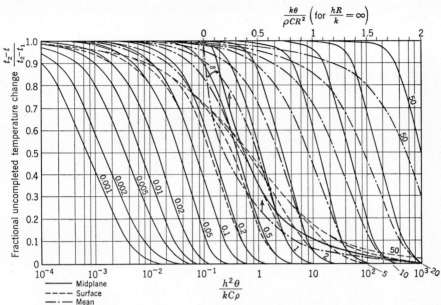

FIG. 9-27. Temperature change for an infinite slab of thickness $2R$, initially at t_1, with a sudden fluid temperature change to t_2 at both faces. Parameter $= hR/k$.

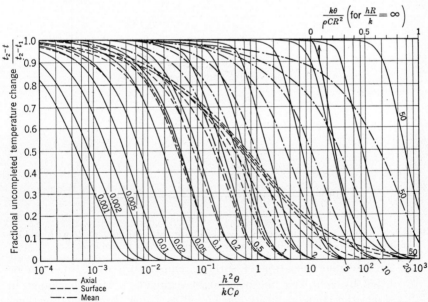

FIG. 9-28. Temperature change for an infinitely long cylinder of radius R, initially at t_1, with a sudden fluid temperature change to t_2 over the surface. Parameter $= hR/k$.

slab (obtainable by Eq. 9-23 since $dt/d\theta$ will be uniform):

$$= (t_{\max} - t_{\min})_\infty = \frac{R^2}{2\alpha}\frac{dt}{d\theta}$$

d. For a solid infinite cylinder, from Eq. (9-33):

$$(t_{\max} - t_{\min})_\infty = \frac{R^2}{4\alpha}\frac{dt}{d\theta}$$

e. For cases *c* and *d* at shorter θ [136], see Fig. 9-29.

f. For finite shapes, case 1*d* above applies.

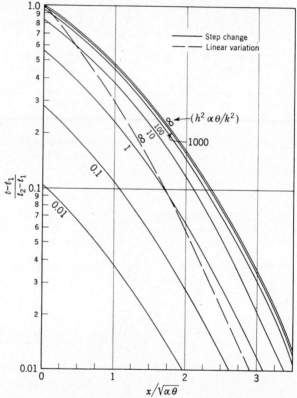

Fig. 9-29. Temperature change in a semi-infinite solid initially at t_1 with sudden (solid line) or linear (dashed line) change of ambient temperature to t_2.

g. An irregular variation of surface temperature with time can be fitted by a broken line of straight segments. At each change in slope, the change can be considered as a new $dt/d\theta$ starting then. The temperature changes at any location due to all previous straight segments are algebraically added [135].

3. A definite q/A is forced into the body, initially at t_1. $H = 0$.

    a. At $\theta = 0$, a constant q/A starts to enter one face of an infinite slab of thickness R, the other being adiabatic [131]. The curves for $hR/k = 0$ in Fig. 9-30 give t_{max} and t_{min}; $t_{av} = t_1 + \theta q/AC\rho R$.

    b. Same as case a, but q/A increases linearly [131] with θ from $q/A = 0$ at $\theta = 0$ (Fig. 9-30), or decreases linearly with θ [131].

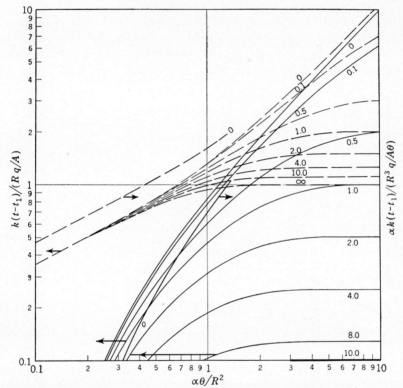

Fig. 9-30. Heating of an infinite slab of thickness R initially at t_1 by a sudden constant q/A at one face (left-hand ordinate), or by q/A increasing linearly from zero with time θ (right-hand ordinate). Parameter $= hR/k$, using h at the cooled face. Coolant remains at t_1. Cooled face, solid; heated face, dashed.

    c. Same as case a, but the unheated face is in contact at constant h with a coolant maintained at t_1 [132]. Figure 9-30 gives t_{max} and t_{min} as a function of hR/k.

    d. Same as case a, but the unheated face is in contact with a body of known heat capacity with h and $k = \infty$ [133].

    e. At $\theta = 0$ a constant q/A starts to enter the inner face of a concentric cylindrical shell [134].

    f. An irregular variation of q/A with time can be handled as under case $2g$ above. For instance, an unheated "equalizing period" is analyzed by continuing the initial heating, later starting an additional equal negative heating period, and adding the two temperature changes anywhere algebraically.

g. For an infinite bar, a parallelepiped, or a finite cylinder, the total local or mean temperature change is obtainable by adding the changes computed separately for each dimension.

4. The heating density throughout a body at t_1 suddenly changes from 0 to a constant value H.   Heisler [183] gives curves for the temperature distribution throughout infinite slabs, infinite solid cylinders, and solid spheres for the cases of constant surface temperature, and constant coolant temperature and h.

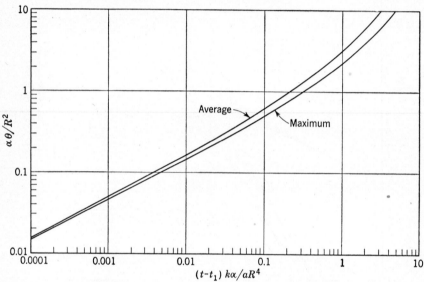

Fig. 9-31. Maximum and average temperature t for an infinite slab of thickness R with initial temperature t_1, one face maintained at t_1, and heating density $H = a\theta$.

5. The heating density H throughout the body varies with time.

a. For an infinite slab of thickness R, initial temperature t_1, cooled face kept at t_1, and linear uniform heating $H = a\theta$.   Letting $\alpha(\pi/2R)^2 = B$,

$$t = t_1 + \frac{2a}{\pi C \rho B^2} \sum_{n=1}^{\infty} \left[\frac{1 - (-1)^n}{n^5} \left(e^{-n^2 B \theta} + n^2 B \theta - 1 \right) \sin \frac{n\pi x}{2R} \right] \quad (9\text{-}102)$$

$t_{\max}$ is at $x = R$, and on integrating to obtain t_{av}, $\sin n\pi x/2R$ is replaced by $\dfrac{2[1 - \cos (n\pi/2)]}{n\pi}$.   Figure 9-31 covers the useful ranges.

b. Case a, except that $H = H_0 e^{b\theta}$.

$$t = t_1 + \frac{2H_0\theta}{\pi C\rho} \sum_{n=1}^{\infty} \left[\frac{1 - (-1)^n}{n} \frac{e^{b\theta} - e^{n^2 B\theta}}{b\theta + n^2 B\theta} \sin \frac{n\pi x}{2R} \right] \qquad (9\text{-}103)$$

$t_{\max}$ and t_{av} are obtained as in case a.

c. Same as case b, but cladding and stationary coolant layers without heat generation are added [166].

The above solutions cover many reactor problems.* Where heating or temperature rise is stated, cooling or temperature fall can be read. Also, although uniform initial temperature t_1 is stated, the temperature change calculated can be added to a steady-state temperature distribution originally present (see Secs. 9-7 and 9-8) provided the heat source remains (if not, method 3f is also employed).

Prob. 9-23. Assume a metal fuel element has specific heat c, total "lumped" mass m, and average temperature t_1. It is dissipating power q_1 through thermal conductance UA to a coolant stream of mass flow rate W and specific heat C. The coolant-inlet temperature is T_i and outlet temperature $T_{0,1}$. The mass of coolant in the reactor is M, and its average temperature T is the arithmetic mean of T_i and T_0. At time $\theta = 0$ the power jumps to q_2. Assuming the heat exchanger and secondary coolant are adequate to keep T_i substantially constant, compute metal t and coolant T as functions of θ.

Solution. The equations are

$$q_1 = UA(t_1 - T_1) = 2WC(T_1 - T_i)$$

$$q_2 = mc\frac{dt}{d\theta} + UA(t - T) = mc\frac{dt}{d\theta} + 2WC(T - T_i) + MC\frac{dT}{d\theta}$$

1. Approximate solution: Neglecting the last term and setting $1/UA + 1/2WC = B'$, there is obtained

$$\frac{t - T_i}{t_1 - T_i} = \frac{T - T_i}{T_1 - T_i} = \frac{T_0 - T_i}{T_{0,1} - T_i} = \frac{q_2}{q_1} - \left(\frac{q_2}{q_1} - 1\right) e^{-\theta/mcB'}$$

2. Exact solution: Including the last term and setting $E = UA/MC + 2W/M + UA/mc$, $F = 2WUA/Mmc$, $2\alpha = -E - (E^2 - 4F)^{1/2}$, and $2\beta = -E + (E^2 - 4F)^{1/2}$, there is obtained [137]

$$\frac{T - T_i}{T_1 - T_i} = \frac{q_2}{q_1} - \left(\frac{q_2}{q_1} - 1\right) \left(\frac{e^{\beta\theta}}{\beta} - \frac{e^{\alpha\theta}}{\alpha}\right) \left(\frac{1}{\beta} - \frac{1}{\alpha}\right)^{-1}$$

$$\frac{t - T_i}{T_1 - T_i} = \frac{q_2}{q_1} - \left(\frac{q_2}{q_1} - 1\right) \left(\frac{e^{\beta\theta}}{\beta} + e^{\beta\theta} - \frac{e^{\alpha\theta}}{\alpha} - e^{\alpha\theta}\right) \left(\frac{1}{\beta} - \frac{1}{\alpha}\right)^{-1} \qquad (9\text{-}104)$$

9-37. Approximate Analytical Steady-state Conduction Methods (No Heat Generation).
Two- and three-dimensional steady-state heat conduction without generation between isothermal or substantially iso-

---

* For example, Ref. 135 combines methods 2b and 2g to obtain the temperature distribution and maximum thermal stress in a plate subjected to a complex temperature cycle of a reactor coolant.

thermal surfaces in other than the simple geometries (flat slabs and concentric cylindrical and spherical shells) can be approximated by Langmuir's method [113]. The body is imagined subdivided by insulating surfaces reaching from one to the other isothermal surface and cutting up the body into narrow conducting prisms, truncated wedges, or truncated pyramids, for which the total heat conducted is calculated.* Repeating with other arbitrary arrangements of insulating surfaces, the arrangement which gives the largest q is the most accurate. The process is then repeated, but using many infinitely conducting surfaces intermediate between the isothermal surfaces; the arrangement of conducting surfaces which gives the smallest q is the most accurate. The average of the most accurate q by both methods is off by less than half their difference, usually sufficient accuracy for heat conduction through irregular thermal insulation, shields, and structural elements.

Potential theory is useful for estimating temperature drops from cylindrical or spherical heat sources to flat cooled surfaces in fuel elements, or from flat heated surfaces to coolant channels. The same relations of Sec. 8-19 for incompressible flow through porous bodies apply, substituting t for P, q for Q, and k for K'''.

Two-dimensional steady-state heat-conduction problems involving only isothermal and adiabatic boundaries are readily approximated by trial-and-error drawing of the "flow net," as in Sec. 8-20. The heat flowing across each curvilinear square q_{sq} equals $kb \, \Delta t_{sq}$, which equals $kb \, \Delta t_{total}/n$. Thus for a solid of the shape of the dashed area in Fig. 8-7, the total heat flowing between H and L would from Fig. 8-8 be $8kb \, \Delta t_{total}/22$.

9-38. Special Methods and Devices. In view of the high cost and the desired infallibility of nuclear reactors and even reactor experiments, it is customary to carry the preliminary design through all the anticipated steady-state operating conditions for every significant component, for separate subsystems, and for the whole reactor-power-plant system. The response to all possible intentional or accidental fluctuations or transients is then determined, to locate high-temperature and -stress regions, show up instabilities, fix permissible operating schedules, and specify control and warning devices.

Previous sections of this chapter suffice for the direct steady-state thermal design of most components. They also yield the data and simultaneous equations for the direct analytical or trial-and-error steady-state design of complete systems. Problems with varying physical properties, heat-transfer coefficients, heating densities, coolant flows and temperatures, etc., require special devices or methods to solve the systems of simultaneous equations describing the operations, physical properties, and boundary conditions.

* Usually by assuming shapes of differential width and integrating.

One general method consists of writing the differential equations as finite difference equations, using average values of the varying coefficients (see Secs. 9-39 and 9-40). This permits the direct computation of the change in each temperature, flow, etc., over a short finite increment of time. At the end of the increment, the coefficients are changed as needed and the effect of the next increment computed. The whole time interval is eventually covered in finite increments short enough to keep errors as small as desired. These methods are generally the best for analyzing one to a few reactor transient problems, or for maximum accuracy. This process may be carried out manually, but nuclear-power applications are usually so complex as to require an automatic device. Such a device is known as a "digital computer" and ranges from general-purpose mechanical "desk calculators" to electrical devices such as ENIAC, CPC, and UNIVAC, also special-purpose digital reactor simulators.

The other general method, analogue computers, employs a physical phenomenon or mechanism with the same governing equations, the desired quantities being read on scales, meters, etc. Thermal, mechanical, electrical, hydraulic, and diffusional processes are all applicable (see Secs. 9-41 to 9-44). These devices are frequently the best for analyzing steady-state reactor problems and for studying many variations of a problem. Once set up, they give results quickly with reasonable accuracy and economy [138]. Electronic differential analyzers are particularly useful for studying the transient response of complete systems. Reactor, coolant system, heat exchanger, pumps, and turbine, etc., reasonably subdivided, are each represented by computer units, the effect on the whole system due to sudden changes in the key flows and temperatures, etc., being readily recorded [139].

9-39. Numerical Design Methods. These methods subdivide the volume of each solid and fluid by a rectangular or a radial and circumferential grid, as fine as necessary for the desired accuracy, and assume that the mass of each subdivision is concentrated in a point, or "node" [140]. The thermal resistance between adjacent nodes is that between the sections of the subdivisions that pass through the nodes normal to the "bar" joining the nodes, namely, the distance between the nodes divided by k times the cross section (the average cross section, if it varies). In the interior of a body, nodes are placed at the centers of the subdivisions. Along surfaces half-sized subdivisions are generally used, with the nodes at locations of known or desired temperature.* The heat balance for each node of unknown temperature is then written in terms of the data and of the surrounding unknown temperatures, and the equations solved simultaneously or consecutively for the unknown temperatures. Figure

* Interpolation tables are available [146] which give temperatures at other locations conveniently. Otherwise simple graphical interpolation or extrapolation is used.

9-32 yields the following heat-rate components for node 0 in the directions shown on a two-dimensional rectangular grid. If conditions are not uniform along z, it should also be subdivided. In all cases Δx is in the direction between the two nodes under consideration, and Δy (and Δz) normal to that direction. In a square (or cubical) lattice they are, of course, equal and interchangeable.

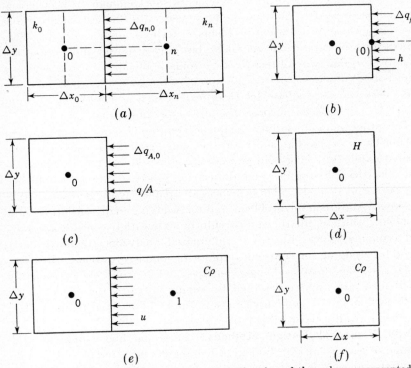

(a) (b)

(c) (d)

(e) (f)

Fig. 9-32. Analysis by finite increments of the heating of the volume represented by node O. (a) Conduction; (b) convection; (c) forced surface heating; (d) volume heating; (e) fluid transport; (f) accumulation.

Conduction. Heat conducted from adjacent node n at distance Δx is at a rate

$$\Delta q_{n,0} = \frac{(t_n - t_0)\,\Delta z\,\Delta y\,2}{\Delta x_n/k_n + \Delta x_0/k_0} = K_{n,0}\,\Delta t_{n,0}$$

Convection and Radiation. Heating rate by convection from fluid at t_f passing a wall node is $\Delta q_{f,0} = (t_f - t_0)h\,\Delta z\,\Delta y = K_{f,0}\,\Delta t_{f,0}$. For a central node the conduction resistance is added:

$$\Delta q_{f,0} = \frac{\Delta t_{f,0}\,\Delta z\,\Delta y}{1/h + \Delta x/2k_0} = K_{f,0}\,\Delta t_{f,0}$$

Convective h by Secs. 9-11 to 9-19 and radiation h by Eq. 9-54 are directly additive if the same $\Delta t_{f,0}$ applies.

Forced Surface Heating. Heating rate at a given q/A is

$$\Delta q_{A,0} = \Delta z\, \Delta y\, \frac{q}{A} = q'$$

Volume Heating. Heating rate at a given H is $\Delta q_v = H\, \Delta z\, \Delta x\, \Delta y = H'$.

Transport. Heating rate by a fluid flowing into a subdivision is $\Delta q_{T,0} = (t_T - t_0)u\rho C\, \Delta z\, \Delta y = C\, \Delta w\, \Delta t_T = C_T'\, \Delta t_T$. The outgoing stream is not of concern; thus this term changes on reversal of flow.

Accumulation. Heat accumulated by node 0 is at a rate

$$\Delta q_0 = (t_0' - t_0)\rho C\, \Delta z\, \Delta x\, \frac{\Delta y}{\Delta \theta} = \frac{C_0'\, \Delta t_0}{\Delta \theta}$$

where Δt_0 is the rise in temperature during time increment $\Delta \theta$.

The over-all balance of heat rates for node 0 is

$$\Delta q_0 = \Delta q_v + \Sigma(\Delta q_{n,0} + \Delta q_{f,0} + \Delta q_{A,0} + \Delta q_{T,0}) \qquad (9\text{-}105)$$

Transient Problems. In these problems all the temperatures and other quantities in Eq. (9-105) are known, except t_0' or Δt_0 in the accumulation term. Accordingly the equation can be written for each node in the problem, and each t' at the end of $\Delta \theta$ directly computed. The process is then repeated using the t' values in place of the t values, and t'' is computed for each node at $2\,\Delta\theta$. This "iteration" process is continued until the desired total time interval is covered. For convenience, Eq. (9-105) is converted to

$$t_0' = (q' + H')\frac{\Delta\theta}{C_0'} + \left\{ 1 - \left[\frac{\Delta\theta}{C_0'}(\Sigma K_{n,0} + K_{f,0} + C_T') \right] \right\} t_0$$

$$+ \frac{\Delta\theta}{C_0'}\Sigma(K_{n,0}t_n) + \frac{\Delta\theta}{C_0'}K_{f,0}t_f + \frac{\Delta\theta}{C_0'}C_T't_T$$

$$= B + B_0 t_0 + \Sigma B_n t_n + B_f t_f + B_T t_T \qquad (9\text{-}106)$$

The B coefficients defined above can be computed for each node, and the calculation of Eq. (9-106) carried out on a digital computer. If temperatures or flow vary sufficiently, the B values can be modified during the process.

The number of times that Eq. (9-106) must be employed to solve a given problem is proportional to $(\Delta\theta\, \Delta x\, \Delta y\, \Delta z)^{-1}$. Large increments favor speed and economy, and small increments favor accuracy. With a high-speed digital computer available, a single computation with a relatively fine grid is preferable; with hand calculation the use of two or three coarser grids and extrapolation to zero node spacing is preferable. If

$\Delta\theta$ is selected too large, it will be found that successive values of t_0, t_0', t_0'', etc., oscillate more and more widely. To prevent this "instability," B_0 should not be negative at any node; thus $\Delta\theta$ should not exceed

$$\frac{C_0'}{\Sigma K_{n,0} + K_{f,0} + C_T'}$$

anywhere.

The "truncation error" due to the finite size of the time increments can be estimated by repeating the calculation with smaller increments; for any node 0 it cannot exceed a maximum of $\theta_{\text{total}}(\Delta t_0' - \Delta t_0)/(2\ \Delta\theta)$, where the Δt_0 values are the most different successive increments [141].

Figure 9-33 shows a grid suitable for calculation by UNIVAC of SIR reactor transients [141]. The data included all dimensions, q_v for each node, the coolant-inlet temperature, the ΔP from inlet to outlet coolant plenums, and the metal temperatures where so marked. Some or all of these quantities in different analyses might vary with time. Thermal buoyancy was appreciable compared to friction (see Sec. 8-37). Accordingly, after t' was computed for each node the flow rates through channels A to E were modified in the next $\Delta\theta$ so as to yield the same total ΔP for each channel, which also when added to the inlet orifice ΔP_0 equaled the given over-all ΔP. The same process was independently carried out for the system of thermal circulation channels,* F to I, with the simultaneous conditions of total upflow equaling total downflow, and the same (but not constant) total ΔP across each channel. It was found that $\Delta\theta$ had to be much smaller to obtain stable computed flows than to obtain stable computed temperatures; so as to save time the temperature-change computations were not made as frequently as flow-change

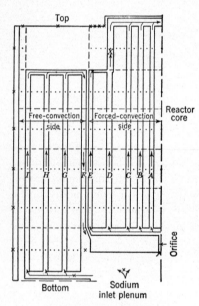

- Node of unknown temperature
- × Node of known temperature
- ——→ Assumed direction of coolant flow
- ----- Zero temperature-gradient boundary

FIG. 9-33. Nodes selected for transient analysis of SIR thermal shield. [S. K. Hellman et al., Trans. ASME, **78**:1155–1161 (1956).]

* This construction removes the smaller outer heat generation adequately without diverting the main coolant stream and without the many small orifices this would require.

computations [Eq. (8-55a), page 328].   Simpler one-dimensional [142] and two-dimensional [137] problems with constant flow have also been described.

Equation (9-106) simplifies considerably under some conditions.   In particular, for a node in a square subdivision, completely surrounded by two, four, or six identical equidistant nodes, $K_{n,0}$ equals $k \, \Delta z$.   If there is no heat convection or transport and the maximum stable $\Delta\theta_{\max}$ is employed, where

$$\underset{\substack{\text{(One-dimen-}\\ \text{sional)}}}{} \qquad \underset{\substack{\text{(Two-dimen-}\\ \text{sional)}}}{} \qquad \underset{\substack{\text{(Three-dimen-}\\ \text{sional)}}}{}$$

$$\Delta\theta_{\max} = \frac{(\Delta x)^2}{2\alpha}, \qquad \frac{(\Delta x)^2}{4\alpha}, \qquad \frac{(\Delta x)^2}{6\alpha} \qquad (9\text{-}107)$$

then B_0, B_f, and B_T all equal 0.   Under these conditions Eq. (9-106) yields

$$t_0' = (t_n)_{\mathrm{av}} \qquad (9\text{-}108)$$

Thus, merely averaging the surrounding node temperatures t_n gives the central one $\Delta\theta_{\max}$ later.   Equation (9-107) is thus the most frequently used $\Delta\theta$, on account of speed.   A somewhat more conservative procedure consists of averaging t_0 in equally with each of the m values of t_n.   In this case

$$t_0' = \frac{t_0 + \sum_1^m t_n}{m+1} \qquad (9\text{-}109)$$

and

$$\Delta\theta = \frac{(\Delta x)^2}{3\alpha}, \qquad \frac{(\Delta x)^2}{5\alpha}, \qquad \frac{(\Delta x)^2}{7\alpha} \qquad (9\text{-}110)$$

Heavier weighting of t_0 is possible [26, 140], though usually unnecessary and thus undesirable.   If an elementary volume is not rectangular but is fitted with nodes in a rectangular lattice, the arithmetic average cross section of the path between any two nodes is used, for simplicity.   Thus, for a node on a surface at 45° to the lattice, the average width for conduction to an interior node is $0.75 \, \Delta y$, with adequate accuracy.   The length employed is the full Δx.

Cylindrical and spherical surfaces can be fitted as closely as desired with a reasonably fine rectangular grid of nodes.   Local or over-all triangular lattices may fit the boundaries best, as with hexagonal coolant channels.   For a bar joining nodes a and c of triangle abc,

$$K_{a,c} = \frac{k \, \Delta z}{2 \tan \beta''}$$

for that one side only of the bar, where β'' (must $\leq 90°$) is the included angle at node b [140]. $K_{a,c}$ for the other side of the bar, if any, must be added to yield the total $K_{a,c}$.   Alternatively, cylindrical or spherical grids can be employed, Eq. (9-106) still being applicable.   Equation (9-108) would not be applicable, but a variation of Eq. (9-109), weighting t_0 more the larger the radius of the node, can be employed.   A constant $K_{n,0}$ of $(k\,\Delta z)$ may be obtained with a cylindrical grid if the angular spacing $\Delta\phi$ is small (say under $\frac{1}{3}$ radian) and set equal to $\ln\,[(r+\Delta r)/r]$, which yields equal radial and circumferential resistances between adjacent nodes.   For a cylindrical shell of radii r_1 and r_2 with p subdivisions, $\Delta\phi = [\ln\,(r_2/r_1)]/p$.   The ratio of successive radii $= (r_2/r_1)^{1/p}$.   The best procedure is to place nodes at r_1 and r_2, decide on p, and compute the corresponding $\Delta\phi$.   If $\Delta\phi$ is too large for accuracy, a larger p is employed.*

Equation (9-106) is "explicit" in that t' is the only unknown, and can be directly computed from the given data.   It is evidently equally logical to take the heat accumulation rate Δq_0 from the *previous* $\Delta\theta$ as from the *subsequent* $\Delta\theta$.   If Δq_0 is set $= C_0' \,(t_0 - 't_0)$, where $'t_0$ is the temperature of node 0 at $\theta - \Delta\theta$, the only changes in Eq. (9-106) are replacement of t_0' by $-'t_0$ on the left side, and of 1 by -1 on the right. The resulting "implicit" equation has the advantage that much larger $\Delta\theta$ values and fewer steps may be employed without encountering instability in the computed temperatures (or, alternatively, truncation error is less).   However, it is unsuitable for following a transient, because the several temperatures which are varying with time (except $'t_0$) are all unknown in advancing from $\theta - \Delta\theta$ to θ, θ to $\theta + \Delta\theta$, etc.   The obvious method of utilizing Eq. (9-106) numerically is to write it for each of the nodes (unknown temperatures) at θ and solve the many equations, each with several unknown temperatures, simultaneously at each θ.   This is normally prohibitive, but the general case of one-dimensional conduction with perpendicular heat transport, as with flow in channels [144] and of two- and three-dimensional conduction [154] can be solved conveniently on electrical resistive analogues.   Also, the 10 simultaneous equations for one-dimensional conduction with 10 equal increments have been solved [145] to yield a table of multipliers which bypasses the specific simultaneous solution at each $\Delta\theta$.

An unknown *steady-state* temperature distribution or heat flow is obtainable with these numerical methods by carrying out the transient calculation at the given conditions until no further significant change with time occurs.   While this generally requires many calculations, it may be the best method if the transient problem has already been set up on a high-speed computer.

---

* Taking the geometric mean of adjacent radii as the intervening boundary, the above simple formula gives $\Delta\phi$ approximately 1 per cent low at $\Delta\phi = 20°$ and 5 per cent low at 45°.   A more accurate relation is $\Delta\phi = [(r_b/r_a)^{1/2} - (r_a/r_b)^{1/2}] \ln\,(r_b/r_a)$.

At steady state, $t_0 = t_0'$ and Eq. (9-106) may be written

$$t_0 = \frac{q' + H'}{\Sigma K_{n,0} + K_{f,0} + C_T'} + \sum \frac{K_{n,0}t_n}{\Sigma K_{n,0} + K_{f,0} + C_T'} + \frac{K_{f,0}t_f}{\Sigma K_{n,0} + K_{f,0} + C_T'} + \frac{C_T't_T}{\Sigma K_{n,0} + K_{f,0} + C_T'} \qquad (9\text{-}111)$$

One such equation is obtained for each unknown temperature. A direct method to obtain the steady-state temperatures is to solve all the equations simultaneously. Though seldom used, this method is at times practical with a high-speed computer.

With other methods, to save time it is generally desirable to estimate the final temperature distribution closely in advance, then proceed to check and improve it. The usual iteration procedure consists of calculating in turn each node temperature along a prescribed route by Eq. (9-111) from the surrounding adjacent pertinent temperatures and changing the t_0 on record to the new value. The previous node temperatures should be employed when the next node is being balanced. For the frequent case of an interior node on a square (or cubic) lattice and no heat generation, convection, or transport , t_0 by Eq. (9-111) is the arithmetic mean of the surrounding t_n values [similar to Eq. (9-108)]. A skeleton outline of the body, showing the nodes, is a convenient form of record sheet.

A more rapid solution is obtained by "relaxation," or balancing (or

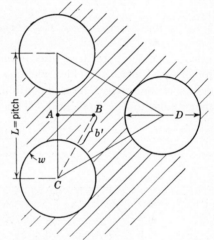

Fig. 9-34. Closest and furthest midplane points, A and B, respectively, in equilateral triangular lattice of parallel circular coolant channels. $b' = $ maximum ligament.

even overbalancing) each time the most unbalanced node. The temperature unbalance of each node is recorded, as well as its temperature. When a node has been balanced by changing its temperature the corresponding change in the balance or unbalance of each neighboring node is added to *its* previous unbalance. The node with the largest unbalance is always balanced next [140]. Table 9-16 was obtained in this way for the case of Fig. 9-34. Figure 9-35 shows the boundary fits obtainable for a specific case with coarse rectangular and cylindrical "square" grids.

For maximum speed it is desirable to solve the problem first on a coarse

TABLE 9-16. MAXIMUM TEMPERATURES AND HEAT VELOCITY IN MATRIX
WITH UNIFORM HEATING DENSITY H, COOLED BY PARALLEL CHANNELS AT
CONSTANT WALL TEMPERATURE t_w IN AN EQUILATERAL TRIANGULAR LATTICE*

	$\dfrac{L}{D}$ (see Fig. 9-34)				
	1.1	1.2	1.3	1.4	1.5
$\dfrac{(t_B - t_w)k}{HL^2}$	0.55	0.94	1.37	1.81	2.33
$\dfrac{t_B - t_w}{t_A - t_w}$	5.00	2.26	1.63	1.41	1.31
$\left(\dfrac{(q/A)_{\max}}{(q/A)_{\mathrm{av}}}\right)_w$ ..	2.07	1.20	1.17	1.16	1.16

* P. B. Richards, *Chem. Eng. Progr., Symposium Ser.*, no. 11, p. 127, 1954.

grid. The temperatures thus found around each square are averaged
[or inserted in Eq. (9-111)], yielding a fairly reliable temperature for the
center of the square. A grid containing twice as many points is obtained,
diagonally to the first grid. This finer grid of temperatures is then

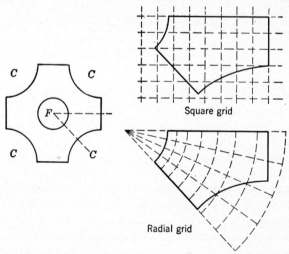

FIG. 9-35. Square grids for thermal analysis of LMFR fuel element (see Fig. M-3).

improved by iteration or relaxation, and a still finer one drawn and
improved as before, if desired [140]. This procedure is faster than going
at once to the finest grid, since it provides reliable estimates of the
temperatures, and fewer circuits of computation about the shape are

needed. It is also more accurate, as additional values of the desired temperatures and heat flows are provided at coarser grid sizes, permitting a graphical extrapolation to zero grid spacing.

An alternative to the use of coarse grids before finer grids consists of balancing or relaxing blocks of four, six, nine, etc., nodes at the same time [140]. The temperature unbalance found between the block and its adjacent nodes is applied to each node in the block in proportion to the number of adjacent nonblock nodes it has.* After the blocks as units are fairly well balanced, attention is turned to the individual nodes.

9-40. Graphical Methods. For one-dimensional cases, such as thin or flat fuel elements or shields, Eq. (9-108) or (9-109) can be readily carried out graphically by the Binder or Schmidt method [26]. To obtain t_b' by averaging t_a and t_c as per Eq. (9-108), a straight line is drawn

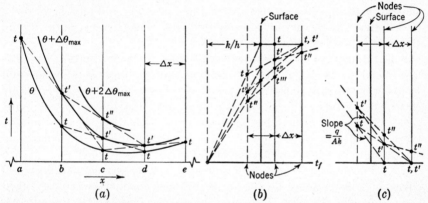

$$(a) \qquad\qquad\qquad (b) \qquad\qquad\qquad (c)$$

FIG. 9-36. Details of the Binder or Schmidt graphical method for one-dimensional heat conduction in a slab. (a) Central region with $H = 0$; (b) surface with convection cooling; (c) surface with known heat velocity.

between t_a and t_c in Fig. 9-36a to intersect the position of node b, and $\Delta\theta = (\Delta x)^2/2\alpha$. If the surface temperature is held constant or varies in a known manner with time, one node would be placed at the surface. If the surface temperature depends on coolant temperature t_f and coefficient h, or a known heat velocity q/A, it is more convenient to place the surface between an inner "actual" and an outer "imaginary" node, as in Fig. 9-36b and c. If heat generation occurs, the additional temperature rise $t' - t = H\,\Delta\theta/\rho C = H\,(\Delta x)^2/2k$ is added to each heated node at each time increment $\Delta\theta$. t_f, h, q/A, H, and even k at the surface may be varied as time or temperatures advance. If two materials A and

---

* Thus, if a block of $3 \times 3 = 9$ nodes, with conduction only, totals $t°$ warmer than the surrounding 12 nodes, $t°/6$ should be removed from each corner node, $t°/12$ from each midside node, and nothing from the central node.

B (i.e., fuel and cladding) are adjacent, the ratio $\Delta x_A/\Delta x_B$ may be set $= \sqrt{\alpha_A/\alpha_B}$ so as to maintain $\Delta\theta$ by Eq. (9-107) the same. If this gives inconvenient spacings, Δx may be kept constant and the design completed by weighting t_0 other than zero in one of the materials and using Eq. (9-106) for that material [26].

The solid or hollow cylinder and sphere with radial temperature and geometrical symmetry are also best treated as one-dimensional cases for either numerical or graphical solution. For the latter the same methods of Fig. 9-36 are employed, except that equally spaced nodes at r_a and r_b on the radius of a cylinder are spaced on the working plot [143] distances proportional to $(r_a + r_b)^{-1}$. For a sphere the spacing is as $(r_a + r_b)^{-2}$.

Comparable to Eq. (9-109) and (9-110), smaller $\Delta\theta$ than $\Delta\theta_{max}$ can also be used graphically (method of Nessi and Nisolle [11]). However, the added effort is probably better expended into decreasing Δx. In general, if graphical analysis is carried out carefully and on a large scale, it can give satisfactory results for one-dimensional cases more rapidly than the other methods.

9-41. Homogeneous Electrical Analogues (Steady State). As was mentioned in Sec. 8-21 for streamline flow, in heat-conduction and -convection design problems thermal resistance may be simulated by electrical resistance and temperature and heat flow by voltage and current. Merits of the electrical analogue over calculation, or performing a thermal experiment, are greater speed and economy, particularly in steady state or when various similar cases are to be covered.

Two-dimensional steady-state problems with isothermal and insulated edges and one or more homogeneous thermal conductors can be solved most conveniently with Teledeltos paper (see Sec. 8-21), though soft metal foil, brass shim stock, trays of conducting solution, and semi-conducting plastic sheeting have been used. Teledeltos grades available are L (approximately 1000 to 2000 ohms per square) and H (approximately 10,000 to 20,000 ohms per square). Crosswise resistance is some 12 per cent higher than lengthwise resistance along the rolled sheet; this can be approximately corrected for by a corresponding difference in the x and y scales. The resistance R may be increased almost isotropically up to fivefold by punching out holes on a reasonably fine grid. Equally spaced circles as follows are required (D/S is the ratio of diameter to center spacing):

Perforated R/unperforated R..	1	1.25	1.5	2	3	4	5	∞
D/S (90° grid)...............	0	0.39	0.51	0.65	0.79	0.86	0.90	1
D/S (60° grid; D/L in Fig. 9-34)	0	0.34	0.48	0.62	0.76	0.82	0.86	1

For conduction problems the shape is cut to scale out of the paper, allowing an extra $\frac{1}{4}$ in. at the isothermal edges to connect a good conductor.
Du Pont 4922 silver lacquer is generally used; it has a lengthwise resistance of only several ohms per foot for a thin $\frac{1}{8}$-in. strip, and also negligible contact resistance (a fine copper wire can be incorporated to
minimize the lengthwise resistance). Clean brass bars clamped against
the paper, which is backed with rubber, is more convenient for making
changes; contact resistance is about $\frac{1}{8}$ in. of the paper, negligible in large
analogue sizes.

Coolant film resistances are representable by additional paper slit into
strips normal to the surface (strips of finite width, of course, depart
from the true homogeneous analogue). To keep solid and film resistances

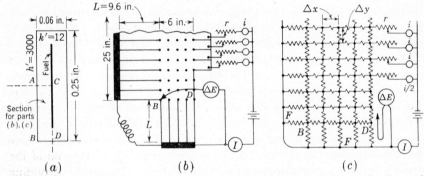

FIG. 9-37. Electrical analogues for two-dimensional steady-state heat conduction and
convection study of a long thin-strip stainless-clad fuel element. (a) Actual size and
thermal data; (b) possible homogeneous analogue arrangement [central dots apply to
(c)]; (c) possible lumped analogue arrangement [nodes are at dots in (b), and $\Delta x = \Delta y$].

in proportion, the strip length L is set equal to Sk/h, where S is the
geometrical scale ratio of the analogue. For instance, in Fig. 9-37, an
analogue for a very thin strip of U^{235} with stainless cladding,

$$ L = \left(\frac{6}{0.03}\right)\left(\frac{12}{3000}\right) \text{ft} = 9.6 \text{ in.} $$

If L is unreasonably long, additional resistors may be connected to each
strip, or some of the strips may be cut off at the base and the length of the
rest shortened to give the same total resistance of the strips. Strips
should be curled with string or small rubber bands to prevent contact
at their edges. If different sheets of paper are attached together, a
narrow strip of silver lacquer is used as glue. If the line of glue is nonisothermal, it should be frequently cut across with a razor to prevent
conduction along it.

Unsteady-state problems require electrical capacity (see Sec. 9-42).

Uniformly distributed capacity could in principle be obtained with a large grounded plate covered with thin dielectric and a conducting sheet on top; actually uniformity would be poor, and the size would be very large or time very short. Large capacitors can be attached at small points, but the electrical connection resistance is high; large electrodes decrease the resistance of the main sheet. Thus "lumped" or "network" or "passive-network" analogues (Sec. 9-42) are preferable for unsteady state.

Heat generation also requires electrodes as current lead-in. Here connection resistance is not important, though points at which temperature (voltage) is sought should be roughly equidistant from two or more lead-ins rather than near one. In Fig. 9-37b the fuel has been subdivided by short strips. Each rheostat r is adjusted until the currents i are distributed in proportion to the given heat-generation distribution (uniform for a small uniform strip fuel element such as this). If the resistances r are relatively high the settings will not need trial-and-error readjustments to match the heat generation. Equal currents without repeated adjustment can also be obtained by using a constant-current ballast tube at rated current, or a "beam power" (6SJ7 or 6V6) vacuum tube, which is grid-controllable over a 10-fold range of current, roughly.

Points A and C are the hottest surface and interior points, and B is the coldest. To obtain the temperature rise Δt of any such points above the coolant temperature, the voltage difference ΔE above the end of the strips is read by a potentiometer or high-resistance voltmeter. Dividing the electrical (Ohm's) law into the thermal law for any given shape

$$\frac{\Delta t}{\Delta E} = \frac{q}{I}\frac{R_t}{R_e} = \frac{q}{I}\frac{1}{kbR'_e} \qquad (9\text{-}112)$$

where R'_e is the ohms of a square of the paper used and b is the thickness (length of strip fuel element) of the section in Fig. 9-37b generating heat q. If the thickness of the U^{235} were significant, a larger analogue could be made with the fuel region of the proper resistance per square, and a two-dimensional array of current inputs over it.

For two- and particularly three-dimensional problems, acidified strong $CuSO_4$ solution can be employed. The container has the desired shape to scale, and isothermal edges or surfaces are sheet copper electrodes. Use of alternating current avoids electrode polarization. Solid electrolytes that can be cut and melted, such as 15 parts each of gelatin and glycerine in 100 parts of salt water, are also used. Fluid film resistances are more difficult to apply with these materials.

9-42. Lumped Component Electrical Analogues (Steady and Unsteady State). Replacing the conducting paper of region $ABCD$ in Fig. 9-37 by a network of resistors is a further departure from the homogeneous thermal problem. However, this is outweighted by practical advantages in many cases, such as greater speed of setup if a resistor board is available, greater accuracy of resistors than sheets, and greater convenience for unsteady-state problems. Portions of the problem are usually "lumped" in any case, as in Fig. 9-37b, and added "lumping" is not detrimental unless fine local temperature distribution in this region is desired.

A resistor board for general use may have clips to hold cartridge resistors, or a linear or tapered rheostat permanently at each resistor position. Up to three decades are also available in single plug-in resistor assemblies (Goodyear). Many nodes are provided, only the desired ones for any problem being connected. For single problems the network can be put together more quickly and economically by soldering the proper resistors together and mounting them on plywood. Good linear wire-wound rheostats (General Radio) are everywhere within 1 per cent of their scale value. In fixed resistors the lathe-cut carbon-film type (Aerovox) gives a good combination of cost, stability, and uniformity ($\pm$ 1 per cent).

In setting up a lumped-component analogue with constant fluid temperatures (i.e., small changes in stream temperatures, or for a cross section normal to the stream or streams) the general methods of Sec. 9-39, illustrated in Figs. 8-9 and 9-37c, are employed for the resistances. For convenience in obtaining surface temperatures, nodes are usually located along all surfaces. As many of the nodes as possible inside the conduction region are usually placed on a square grid, so the same size resistor can be used between them.

If each central resistor in the square grid in Fig. 9-37c is arbitrarily selected as 100 ohms, each edge resistor, between surface nodes, is 200 ohms. To maintain the proportion

$$\left(\frac{R_t}{R_e}\right)_{solid} = \left(\frac{R_t'}{R_e'}\right)_{square} = \left(\frac{R_t}{R_e}\right)_{film}$$

each full surface film resistor equals $R_e'k/h\,\Delta x$, or

$$\frac{100 \times 12}{3000 \times \dfrac{0.0075}{12}} = 640 \text{ ohms}$$

in this case. At corners B and D, 1280-ohm resistors are used, except that the two at B can evidently be replaced by one 640-ohm resistor. Equation (9-112) gives any desired Δt, as before.

For unsteady-state analyses, a condenser* is attached to each node with an electrical capacity proportional to the thermal capacity repre-

---

* A purely resistive network [154] can instead be used and stepwise subdivision of time (see Sec. 9-39), but the solution is more time-consuming.

sented by the node. Thus, the capacitor for a central node would be proportional to $bC\rho\,\Delta x\,\Delta y$, or $bC\rho(\Delta x)^2$ with a square grid. A side node has half this capacity, and a corner node (B and D) one-quarter. The other terminal of each capacitor is connected to the reference voltage, wire F in this case. Capacitors up to 20 μf are frequently employed. The dielectric must be good to keep self-discharge low over the time of operation; leakage time constants RC of 20,000 and 100,000 ohm-farads (seconds) at 25°C—still higher at lower temperatures—are available (i.e., Sprague). Mylar dielectric (Good-All) yields even higher RC.

The scale of resistances and capacitors may be independently selected, but they then control the ratio of thermal time to equivalent time in the electrical analog. The time-scale ratio is obtainable by dividing the heat-conduction equation [Eq. (9-14) with $H = 0$] by the comparable relation $dE/d\theta = \nabla^2 E/C''_e R''_e$, where C''_e and R''_e are the farads and ohms per unit length. The product $C''_e R''_e$ can be computed for any cross section (such as one transverse increment of the grid), because C_e and R_e vary inversely. They can also be C'''_e and R'''_e for one lengthwise increment of the grid if the distance variables in $\nabla^2 E$ are changed to grid steps. Thus

$$\frac{\alpha\theta_t}{(\Delta x)^2} = \frac{\theta_e}{C'''_e(R'''_e)} \tag{9-113}$$

For an estimated desired range of θ_t, θ_e can be calculated. If θ_e is impractically short or long, Δx, C'''_e, and/or R'''_e may be altered as desired or convenient. The capability of such an unsteady-state analogue to subdivide space is proportional to the number n of the largest capacitors $(C'''_e)_{\max}$ available. Its capability to subdivide time is, for practical purposes, proportional to $n(C'''_e)_{\max}(R'''_e)_{\max}/(\theta_e)_{\min}$, where $(R'''_e)_{\max}$ is the highest resistor size available allowing sufficient sensitivity and $(\theta_c)_{\min}$ is the minimum reliable response time of the electrical recorder, or indicator plus observer.*

Study of a transient in heat generation or in coolant temperature may be illustrated by the case of Fig. 9-37. Considered as a slab,

$$\frac{hR}{k} = \frac{3000 \times 0.03}{12 \times 12} = 0.625$$

From Fig. 9-27 $h^2\theta/kC\rho = 0.010$ for 90 per cent of the surface temperature change uncompleted and 1.7 for 10 per cent of the midplane change uncompleted, or a maximum reasonable range of θ_t of 0.0036 to 0.60 sec.

---

* In large complicated shapes the temperature may be desired at many points, for thermal-stress calculations. This was accomplished for the SIR reloading and shielding plug (which goes above Fig. 9-33) with a reasonable number of condensers by first setting nodes over the whole object up to the available number of condensers and finding their temperature history. Then subdivisions were set up in turn with finer grids and the boundary temperatures made to vary as previously found.

Taking C_e''' as 20 μF and R_e''' as 100 ohms yields $\theta_e = 0.0082$ to 1.37 sec. While this range would be feasible with fast instrumentation, it would be simpler to increase R_e''' to say 10^5 to 10^6 ohms, and θ_e proportionally. Voltage could be increased as desired. Rheostats r might need adjustment during the experiment to keep the currents constant. Indicating or recording instruments should be of high enough resistance not to alter the currents appreciably.

Lumped-component electrical analogue circuits can also be devised for steady-state and even transient analysis of complete systems, such as a reactor, pump, heat exchanger, turbine, and condenser. Such methods have not yet been developed in general terms.

9-43. Mass Transfer and Mass-transfer Analogues. Mass transfer is the passage of molecules of a dissolved or vaporized constituent from one phase to another, by diffusion and convection in the direction toward equilibrium. Normally the phases are insoluble or only slightly soluble in each other, and the substance being transferred may be the only, the major, or a minor constituent of one of the phases. Nuclear applications include the dissolving of fuel or fertile elements by liquid-fuel or breeder-blanket solutions, the removal of converted fertile elements, residual fuel, or fission products by evaporation, sparging, or extraction of their solution by a second liquid or a solid, the dissolving or differential thermal solubility corrosion of pipes and vessels, etc. In fact, a principal advantage of liquid-fuel and homogeneous reactors is that by mass transfer the fission-product concentration can be kept low for safety and neutron economy, and the fuel, fertile, and converted elements can be maintained at their most economical concentrations.

Most nuclear mass-transfer processes employ at least one dilute solution. This makes mass transfer analogous to heat transfer with constant properties, since the molecular and eddy diffusion in the laminar boundary layer and core are comparable to the corresponding conductions and the physical properties of the dilute phase or phases (the main resistance to mass transfer) are substantially constant at the values for the pure phase or phases. Theoretical approaches can be based on velocity distribution such as Eqs. (8-36), (8-37), and (8-38), Eq. (8-43) or (8-44) to yield ϵ, and finally integration of Eq. (8-42). Agreement is good at the same Re between the analogous heat- and mass-transfer expressions over the whole practical range of Schmidt number Sc (see Table 8-7) or 0.5 to 3000. Deissler has obtained theoretically [150], for flow in a smooth pipe with Sc above about 50,

$$\mathrm{St}_H = 0.079 f_F^{1/2}\,\mathrm{Pr}^{-3/4} \qquad \mathrm{St}_M = 0.079 f_F^{1/2}\,\mathrm{Sc}^{-3/4} \qquad (9\text{-}114)$$

which is adequate for diffusion-controlled thermal corrosion of pipes by most liquid metals and fused salts. Also, Chilton and Colburn's empirical analysis has shown that j_D and j_H [see Table 8-7 and Eq. (9-117)] are roughly equal for any given shape and Re value [151]. Thus both analyses indicate that Eqs. (9-44) and (9-47) for nonmetals would hold for mass transfer in metals or nonmetals on replacing Nu by Sh and Pr by Sc. At Re above about 10,000, j_H is also roughly equal to $\frac{1}{2}f_F$. Thus the average mass-transfer coefficient F' can be estimated from pressure drop as well as from heat-transfer data.

If D_V is not available it can be estimated by various methods. For liquid nonelectrolytes, such as metals dissolved in other liquid metals or fused salts, the Stokes-Einstein equation

$$D_V = \frac{RT}{3\pi\mu N_A D_a} \tag{9-115}$$

is suitable, where μ is the viscosity of the solute, N_A is Avogadro's number, and D_a is the atomic diameter of the solute, which may be computed as the diameter of a sphere from $M/\rho N_A = \pi D_a^3/6$, where M is the molecular weight. For conservative design Eq. (9-115) should be used for desired mass-transfer processes such as fuel solution or fission-product extraction, and twice Eq. (9-115) for undesirable processes such as corrosion by solution [148]. For electrolytes Eq. (9-115) does not hold, since D_a is increased by attached solvent molecules. For gases the Gilliland correlation [151] is suitable.

Laminar mass transfer is evidently analogous to laminar heat transfer. Equations (9-45) can be applied by further substituting $D_V\rho$ for k/C in Gz and Gz'. Also free-convection equations (9-46) apply; $\beta\Delta t$ should be computed [148] as $\Delta\rho/\bar{\rho}$.

In all cases, the total rate of mass transfer w_M is obtained by

$$w_M = F'A\,(\Delta c)_m = \frac{F'A}{R'T}\,(\Delta p) \tag{9-116}$$

where A = surface area through which mass transfer occurs

$(\Delta c)_m$ = average concentration difference causing mass transfer, mass per unit volume

F' = film mass-transfer coefficient, velocity units, i.e.,
$\text{lb}/(\text{sec} \times \text{ft}^2 \times \text{lb}/\text{ft}^3) = \text{fps}$

For a volatile solute, partial pressure p is more convenient. If one phase flows parallel or countercurrent to the other, or has a constant concentration at its surface, the log mean Δc [as per Eq. (9-90)] would apply. In other cases the mean Δc integrated over the area would be used.

Prob. 9-24. Show that the steady-state rate of solution w_M of an isothermal hot pipe and precipitation on an isothermal cold pipe in a loop is

$$\frac{Q(c_h^* - c_c^*)\left(1 - e^{-\frac{F_h'A_h}{Q}}\right)\left(1 - e^{-\frac{F_c'A_c}{Q}}\right)}{1 - e^{-\frac{F_h'A_h - F_c'A_c}{Q}}}$$

where c^* = solubility of pipe metal

$\qquad F'$ = mass-transfer coefficient

$\qquad A$ = pipe area

$\qquad Q$ = volumetric flow rate of the liquid

Intermediate nonisothermal portions of the loop are neglected, and h and c stand for hot and cold.

Solution. Equate w_M by Eq. (9-116) for each surface, using $(\Delta c)_{\text{log mean}}$ in terms of c_h and c_c of the stream, to $Q(c_h - c_c)$, then eliminate c_h and c_c.

Because of the similarity in the correlations for heat and mass transfer, the latter can be utilized as an analogue to yield information on the former. In particular, heat-transfer coefficients for fuel elements or other objects of special shape can be obtained by casting or machining the object of a material which will vaporize or dissolve slowly. The object, with all pertinent surroundings to scale, is placed in a nonrecirculating stream of an appropriate fluid at a temperature yielding a suitable mass-transfer rate, and at a velocity yielding the desired Re.

The average coefficient* $\bar{h}$ for the test fluid, or, more importantly, for the reactor fluid, if it is a different fluid, can be computed from the loss in weight of the object, or from the gain in concentration in the mixed stream. Even more simply, the ratio between the local h (based on approach temperature) at different positions, or between h and $\bar{h}$, is the ratio of the rate of vaporization or solution at those positions, as measured mechanically. For gases at 0 to about 50°C naphthalene is reliable and convenient [11]; such a study was made of the STR fuel-element assembly [24]. Camphor, acetamide, and other solids that sublime are also suitable [106]. For cold water as the fluid, β-naphthol, benzoic acid, and cinnamic acid are useful [151]. Heat-transfer information for liquid metals is not obtainable in this way, since the electronic contribution is absent in mass transfer and Sc values are not possible as low as the Pr values of liquid metals.

The Chilton-Colburn relation yields

$$j = \frac{F'}{u'} (Sc')^{\frac{2}{3}} = \frac{h}{C\rho u} (Pr)^{\frac{2}{3}}$$

$$\frac{h}{F'} = \frac{C\rho u}{u'} \left(\frac{Sc'}{Pr}\right)^{\frac{2}{3}} = C\rho' \frac{\mu}{\mu'} \frac{D'}{D} \left(\frac{Sc'}{Pr}\right)^{\frac{2}{3}} \qquad (9\text{-}117)$$

The mass-transfer quantities in Eq. (9-117) are primed, and the heat-transfer quantities are not primed. For air and naphthalene $Sc' = 2.5$.

---

* These mass- and heat-transfer coefficients must be based on analogous driving forces. The difference between the surface and the approach concentrations or temperatures is the most convenient, but the surface to local-mean driving force is common.

Expressing T' in degrees Rankine, evaporation rate w' in mils (0.001 in.) of naphthalene per hour, and naphthalene partial pressure driving force $\Delta p'$ in millimeters of Hg, $F' = 0.02574w'T'/\Delta p'$ ft/hr. Using fresh air and basing F' (and h) on inlet-stream concentration (or temperature), $\Delta p'$ is the vapor tension p'_s of naphthalene in millimeters, where

$$\log p'_s = 11.450 - \frac{6712.7}{T'}$$

If true local F' and h are desired, $\Delta p'$ is set equal to $p'_s - p'_b$, where p'_b is the local average partial pressure in the stream as calculated from upstream total vaporization rate. p'_b in millimeters is given by the ideal-gas law as

$$\frac{553.6W'T'}{Q'M'}$$

where W' = upstream rate of evaporation, lb/sec

$\quad\ Q'$ = gas flow rate, cfs

$\quad\ M'$ = molecular weight (128.2 for naphthalene)

$\quad\ T'$ = temperature, °R

9-44. Similitude Tests. For new and complicated shapes or assemblies of fuel elements, coolant channels, etc., heat-transfer coefficients and temperature and flow distributions can be approximately estimated by methods in this and the previous chapter. For best results, full-scale tests at working conditions are indicated. However, in many cases this would be quite difficult, expensive, and/or time-consuming. But fluid flow or heat-transfer characteristics for any given shape have regularly been expressed as relations among a certain number of dimensionless ratios (see also Sec. 8-22). Thus, if it is possible in a test to control all but one of the dimensionless ratios to the working condition values, the remaining ratio in the test will also have the same value as it would at working conditions. The desired ratio can be computed from the test data, and the unknown quantity therein computed for the working conditions.

The deviations from working conditions that are of interest are size, velocity, and fluid properties (different fluid and/or temperature). The following cases are most encountered. Arbitrarily, the primed quantities will refer to the test and unprimed to the working conditions. D is any characteristic length.

Flow Distribution, Type of Flow, Pressure Drop. In a scale model at the same Re as the working unit, the values of $N/D, f_F, C_D$, and any other such size and flow dimensionless ratios are the same as in the working unit. Then from Eq. (8-65) the ratio $\Delta P/\Delta P'$ for streamline or turbulent flow in a pipe, and also for flow through an orifice, is obtained as $(w/w')^2(\rho'/\rho)(D'/D)^4$. Since the ratio $\Delta P/\Delta P'$ is the same for all such standard cases, it is necessary and sufficient also for complex shapes to keep Re the same in the scale model as in the working unit to obtain similar flows.

$$\text{Re} = \frac{D\bar{u}\rho}{\mu} = \frac{D'\bar{u}'\rho'}{\mu'} \qquad \frac{Q\rho}{D\mu} = \frac{Q'\rho'}{D'\mu'} \tag{9-118}$$

ΔP can then be computed from $\Delta P'$ by the above ratio. Alternatively, the f vs. Re curve found with one liquid holds for all other liquids.

For example, the pressure vs. sodium flow rate curve of a SIR fuel-element assembly was obtained as f vs. Re much more conveniently with water than with sodium. A full-scale model was used, and (as a precaution, though not apparently required) actual velocity $\bar{u}$ by employing hot water so that $\mu'/\rho' = \mu/\rho$. The necessary water temperatures for sodium at 600 and 900°F are 171.5 and 211°F, respectively. Similarly, flow distribution in the STR core at high Re was studied with a full-scale model and air (instead of hot water). Also, a 1.5 scale model of the PWR thermal shield [190] was tested with air.*

Turbulent h. In general, to obtain the correct Nu or St under any conditions, the problem is to obtain in the test the same Re and Pr as in the working unit. However, for nonmetallic fluids in turbulent flow it is experimentally well established [Eq. (9-44a)] that

$$j_H = \text{St Pr}^{2/3} = \phi(\text{Re})$$

Thus only Re need be kept the same [Eq. (9-118)], in which case

$$\frac{h}{h'} = \left(\frac{C}{C'}\right)^{1/3} \frac{\bar{u}}{\bar{u}'} \frac{\rho}{\rho'} \left(\frac{\mu'}{\mu} \frac{k}{k'}\right)^{2/3} \tag{9-119}$$

For instance, average or local h may be desired over a fuel element too small for the installation of thermocouples and heating elements; in that case, a model several times larger than the working unit can be tested. A large model may also be used to permit testing at lower than actual working velocity. Most frequently tests at lower temperatures or with more convenient fluids are desired; the product $D'\bar{u}'$ is set at $D\bar{u}\rho\mu'/\rho'\mu$.

With metallic fluids, Eqs. (9-44i) to (9-44p) indicate that it is adequate to keep Pe the same for Nu to be the same in the test as in actual opera-

---

* A quarter-size model of the PWR core yielded useful flow-distribution data with air [190], even though only one sixty-seventh of the PWR Re was obtained. However, for accurate results the same Re is needed. This is difficult to achieve in a full-scale or smaller model because of the high pressures and/or velocities and powers required. For instance, CO_2 at 1200 psi has been used (to test the LSR core design). Much denser gases, particularly Freon C-318 (C_4F_8, $M' = 200$) or UF_6 ($M' = 352$) or very cold air, might perhaps be employed at atmospheric pressure.

tion. Because of the low values of Pr for liquid metals, no other fluids can replace them in similitude tests, however.

Streamline h. Equations (9-45) show that Nu in simple streamline flow is the same if Gz is the same. Since in a scale model N'/D' is the same as N/D,

$$\frac{C\bar{u}\rho D}{k} = \frac{C'\bar{u}'\rho'D'}{k'} \quad \text{and} \quad \frac{h}{h'} = \frac{D'}{D}\frac{k}{k'} \tag{9-120}$$

Conduction and Convection. If conduction is to be simulated as well as turbulent convection, it is necessary for h and k_s/D to be proportional in the test and the working conditions, where k_s is for the solid, or every one of several solids. Thus, besides Eq. (9-118) there is the requirement that $\mathrm{Nu}_s = hD'/k'_s = hD/k_s$. As before, Eq. (9-119) will hold for nonmetals. Eliminating $\bar{u}/\bar{u}'$ and h/h' among the three equations,

$$\frac{k'_s}{k_s} = \left(\frac{\mu'C'k'^2}{\mu Ck^2}\right)^{\frac{1}{3}} \tag{9-121}$$

Thus to simulate conduction and convection in a scale model of an irregular fuel element or assembly it is first necessary to pick the solid, fluid, and test temperature range such that Eq. (9-120) holds. A convenient procedure is to plot the left-hand side of Eq. (9-121) against temperature for each solid under consideration and the right-hand side for each fluid. Every solid-fluid intersection yields a feasible pair of materials and test temperature.

When the test temperature and materials have been selected, any D' and $\bar{u}'$ can be used which obey Eq. (9-118). Actual temperature differences Δt from fluid to solid at any position are obtained from the corresponding $\Delta t'$ by

$$\frac{\Delta t}{\Delta t'} = \frac{qk'_sD'}{q'k_sD} \tag{9-122}$$

Forced Plus Free Convection of Nonmetals. In a forced-convection film coefficient, free convection may be significant, as for a fuel element during low flow or pump failure (see Sec. 9-14). Setting the Gr × Pr products equal and solving with Eqs. (9-118), (9-119), and (9-122) gives the requirement for q'

$$\frac{q'}{q} = \frac{\beta}{\beta'}\left(\frac{\rho}{\rho'}\right)^2\left(\frac{\mu'}{\mu}\right)^{\frac{2}{3}}\left(\frac{k'}{k}\right)^{\frac{1}{3}}\left(\frac{C}{C'}\right)^{\frac{4}{3}} \tag{9-123}$$

D' and $\bar{u}'$ must satisfy Eq. (9-118), and any Δt can be obtained from the corresponding $\Delta t'$ by

$$\frac{\Delta t}{\Delta t'} = \frac{q}{q'} \left(\frac{D'}{D}\right)^3 \left(\frac{C\mu}{C'\mu'}\right)^{1/3} \left(\frac{k}{k'}\right)^{2/3} \tag{9-124}$$

Unsteady State. If steady-state conditions are properly simulated for the cooling of a solid as above described, the unsteady state is also simulated if $\theta\alpha_s/D^2$ is the same. The unsteady-state fractional temperature change at θ is thus the same as that in the test at $\theta' = \theta\alpha_s D'^2/\alpha_s' D^2$.

TRANSIENT THERMAL DESIGN OF REACTORS

9-45. Normal Transients. When the design of power reactors is analyzed in detail, it is found that transients in the operating conditions are more dangerous than steady-state operation at the maximum rating. This occurs because even a drop in temperature from steady operation produces thermal shock stresses, which are superimposed on the steady-state thermal and mechanical stresses (see Chap. 11).

A "normal" transient is one which occurs in the normal operation of the reactor. It should have been anticipated and analyzed in designing the reactor, and it should take place without damage to the reactor. The mildest transients are the fluctuations of neutron flux and thermal power, coolant flow, and/or temperature, caused by the automatic controls in the process of keeping the reactor in steady operation. In steady operation the reactivity is usually within a few cents (hundredths of the delayed-neutron reactivity) of delayed criticality, and the period is of the order of minutes. Though these fluctuations are mild, they are continuous and should be examined and controlled to ensure that they do not cause thermal fatigue in the fuel or structural elements. Besides the general transients there are localized transients to be analyzed, such as heating transients in the fuel elements near the end of a control rod in motion. In particular, the withdrawal of a rod in solid fuel exposes fuel low in xenon to a higher flux, with temporary overheating. Most reactors have an over-all or steady-state negative temperature coefficient of reactivity (see Sec. 6-47). In a power increase the fuel element or solution heats up first; this creates a "prompt" negative coefficient. Subsequently the moderator and/or reflector expand, yielding a "delayed" negative coefficient (neither refers to prompt or delayed neutrons). With a negative coefficient the steady-state control transients are a minimum or negligible; with a positive temperature coefficient they are frequent and steeper.

The most intense transients normally required of any given proposed reactor should be decided on, and the temperature and stress distribution found as functions of time by appropriate methods of Chaps. 6, 8, 9, 11, and 12. If the temperatures or stresses are too high the design must be appropriately modified. For a reactor without specific transient requirements, the fastest permissible start-up, shutdown, etc., should be computed for guidance in the operations. As far as possible, these calculated results should of course later be checked by actual incremental tests after construction.

For a military reactor the most rapid rise possible from cruising to full power is important, so it is necessary to find out how rapidly this can be done. If the reactor is too limited in this characteristic, as an alternative to redesign the plant might, when desired, be operated steadily at full reactor power with a special heat exchanger for discarding excess heat when using less than full power. For a central-station reactor the ability to respond to reasonable load change without damage (see Chap. 13) is necessary. Moderately fast start-up and shutdown are also desirable, to save time and expense in these operations.

Automatic scram circuits are provided in case the reactor starts to "run away" (see Chap. 12). Controls require up to about 1 sec to operate, which must be taken into account in setting the period and power scram levels. When a scram is called for (usually when the power has risen to about 150 per cent of rated, or the period has dropped to about 2 sec) the neutron flux is decreased at maximum speed. However, coolant flow should generally be decreased according to a schedule which minimizes the temperature and stress transients. For instance, the flow may be controlled to a constant outlet temperature, which automatically takes care of the delayed fission-product heat (Fig. 9-1) as well as regulating thermal stresses. In general, nonboiling reactors are controlled so that boiling should never occur, since then the flow distribution would deteriorate (see Sec. 9-21).

Control and safety considerations may actually dictate the choice of reactor type. In particular, D_2O-moderated reactors have the longest neutron lifetime, about 10^{-3} sec, allowing more time for control purposes. Also, natural or only slightly enriched U has a much lower ratio of $H/C\rho$ than U^{235}; thus temperature rise during a power surge is much slower. In any case, the usual procedure is to calculate for successive finite increments of time the distribution of neutron flux, then of heat generation, and finally of temperature and stress throughout the fuel-element cross section under consideration. If no unsafe temperature or stress is reached at any time in the transient, it is a safe operational transient.* The AEC standardized reactor safety inspections [164] include start-ups after long and short shutdowns, power changes, and

---

* For calculation of safe and unsafe power transients in homogeneous reactors with thermal expansion (HRE, etc.), see Ref. 155. For HRE, at 480°F and 1000 psi, reactivity decrease per degree Fahrenheit = 0.0005.

scrams by the high-power and short-period circuits. Actual maximum safe levels of operation are best determined experimentally [174].

9-46. Accidents. A departure from anticipated normal steady or transient operation due to an improper manipulation or test, or failure or incorrect performance of some portion or component of the reactor, controls, or cooling system is generally designated an "incident" or "accident." In addition to the normal transients, all accidents that conceivably might occur should be followed analytically by finite time increments or any other appropriate way, to ensure that there will be no damage to the core, or else to set the limits of the accidents at which damage or personnel exposure might occur and devise adequate controls.

Typical accidents that have been hypothesized for various reactors include the following:

1. A maintained power jump to 200 per cent rating (assuming the warning circuit at 110 per cent of rating and the scram circuit at 150 per cent fail to operate).

2. Sudden or linear arbitrary increases in multiplication constant above unity (assuming the power and period-limiting circuits fail), particularly at start-up.

3. Malfunction of the control system so that the control rods are withdrawn at their maximum speed regardless of power and period.

4. Sudden plugging or boiling dry of one channel, with falloff in heat generation, when the coolant is a moderator.

5. Sudden plugging or boiling dry of one channel when the coolant is a poison, the reactivity thereby increasing.

6. Sudden failure of all pump power without pump jamming, followed by a normal automatic scram.

7. Sudden failure of all pump power but without scram, due to non-operation of the scram circuit.

8. Sudden failure and jamming of the pump.

9. Sudden failure of all auxiliary and standby power.

10. Sudden drop in inlet water temperature to a boiling reactor, or to a high-temperature pressurized reactor, increasing the reactivity.

In the usual approximate procedure for accident analysis [111], two stages are assumed. In the first, the assumed mechanism of the accident increases the reactivity above critical, and all negative effects of the accident on the reactivity are neglected. In the second stage the reactivity drops below critical because of the negative effects (thermal expansion or ejection of the fuel and/or moderator), and the mechanism that caused the accident is neglected. If there are several methods of reactivity falloff, each can be considered separately; the least severe accident found is still conservatively severe.

In the detailed analysis of any transient, plots of the pertinent variables should be prepared. For instance, for a high-pressure nonboiling, forced-convection, water-cooled power reactor, it is postulated that the coolant-pump and the control-rod drives suddenly fail. A pressurized receiver keeps the water pressure, and therefore the boiling point, constant. Proceeding by short time increments, the falloff in coolant flow is computed by Sec. 8-38 and plotted in Fig. 9-38. For each time increment

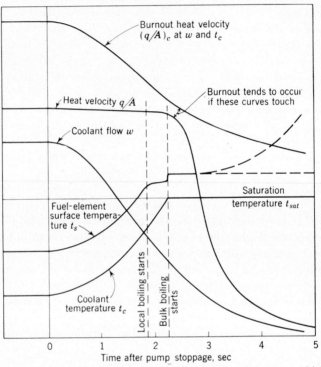

FIG. 9-38. Hot-point conditions during a hypothetical accident to a high-pressure water-cooled power reactor; coolant pump stops and control rods are not inserted. This type of graph is known as a "flow-coastdown" curve.

the heat-transfer coefficient is computed by Sec. 9-11 or 9-13; then the coolant and fuel-element surface temperature distributions along the coolant channels are computed by Sec. 9-39. When local boiling and bulk boiling initiate, flow decreases faster, as pressure drop is higher than otherwise (Sec. 8-41). When bulk boiling of a moderating coolant commences, strong demoderation occurs and the reactivity falls off rapidly. After the flow and temperature curves have been computed, that of burnout heat velocity $(q/A)_{bo}$ is obtained [normally $(q/A)_c$, from Sec. 9-18 or 9-19]. If at any position and time $(q/A) > (q/A)_{bo}$, burnout and an accident may occur.

Many accidents are safely terminated by some "prompt" phenomenon having a strong negative temperature coefficient, even if in the accident prompt criticality is exceeded. In homogeneous aqueous reactors this may be thermal expansion of the fuel solution, or radiolytic gas formation and/or boiling. In water-cooled heterogeneous reactors boiling is the main such mechanism. Several mils of steam on water-moderated, closely spaced MTR-type plates will drop reactivity from prompt critical to subcritical. The water may then condense, making the reactor super-critical again, the process ("chugging" [167]) repeating itself regularly at roughly 1-sec intervals until some other decrease in reactivity is applied.

In a rapid nuclear runaway the period T_1' for an e-fold increase in power is by Eq. (6-122) approximately $l^*/(\rho' - \beta')$, where l^* is the neutron mean lifetime, ρ' the reactivity, and β' the fraction of delayed neutrons (0.00755 by Table 6-6 for U^{235}). As a rough indicator [152], a reactor is inherently safe if the period T_h' for heat dissipation from the fuel to the coolant

$$T_h' = W'C\left(\frac{1}{h} + \sum \frac{\Delta x}{k}\right) \tag{9-125}$$

is smaller than T_1', and if also some mechanism (usually boiling or control-rod insertion) for decreasing ρ' develops reasonably promptly. W' is the mass of fuel per unit cooling surface, C its specific heat (a high tempera-ture average of 0.038 for U), and the $\Delta x/k$ values are the fuel and cladding thermal resistances. For an MTR-type fuel element with aluminum-clad sandwiches 1.5 mm thick, 0.078 lb of U per square foot, and

$$h = 1600 \text{ Btu/(hr)(ft}^2)(°F)$$

$T_h' = 0.007$ sec. In cold water $l^* = $ approximately 0.00006 sec and setting $T_1' = T_h'$ a maximum safe ρ' is obtained as 0.016. In D_2O and with wider sandwich spacing (i.e., the CP-5 reactor), $l^* = $ approximately 0.001 sec and $\rho' = 0.15$. More detailed calculations and tests confirm these rough values [172].

If the demoderation or poisoning action develops very rapidly, higher supercritical reactivity can be tolerated without damage, since the reac-tivity falls off sooner and the exponential rise is nipped off earlier. For instance, detailed calculations show that stainless-clad BSR fuel elements with $3/16$-in. sandwich spacing at atmospheric pressure safely limit a sudden Δk of 1.9 per cent in room temperature H_2O, and of 3.8 per cent in water at 212°F.

This point is also illustrated by the following temperatures calculated for a 1.5-mm-thick MTR-type sandwich with a sudden increase above rated power on a 0.01-sec period:

	Initial temperature, °C*	
	90	20
Midplane temperature t_{max} when surface temperature $t_{sur} = 100°C$, °C.	103	126
Estimated inertia pressure P_i opposing steam formation, atm.	0.7	3
Saturation temperature t_{sat} at P_i atm gauge pressure, °C. . .	115	143
t_{max} when t_{sur} reaches t_{sat}, °C.	122	183
t_{max} 0.01 sec later if exponential heating continued, °C.	177	440

* Assumed uniform, for simplicity.

The effect of steam in decreasing the reactivity of the core is designated the (negative) "reactivity of the steam." Calculations as per Chap. 6 for 1.5-mm MTR-type sandwiches in natural water yield the following uniformly distributed per cent steam by volume required for the indicated $\Delta\rho'$:

Plate spacing, mm:	8.6		13.0	
Approximate mean neutron life l^*, sec:	6×10^{-5}		10^{-4}	
Average water temperature, °F:	68	200	68	200
Reactivity decrease $-\Delta\rho'$	% of volume that is vapor			
0	0	0	0	0
0.01	4.5	5.6	8.8	7.8
0.02	8.2	9.5	15.1	14.2
0.03	11.9	13.4	19.9	19.7
0.04	15.5	17.3		

Over this range about 0.006 to 0.018 in. of steam on each surface suffices to return from prompt to delayed criticality.* If ρ' were suddenly increased from zero, the power would rise exponentially until vaporization started, then more gradually. It would at peak power overshoot the volume of steam which brings ρ' back to zero, because of sensible heat

* Actually, steam would be generated sooner and more copiously in the center of the core than elsewhere. Thus a given volume of steam might be twice as effective as if uniformly distributed.

in the fuel element, but would finally settle back to this value at the corresponding power by Fig. 9-14 or equivalent.*

The specific-heat effect causes a wide variation in time available for control operation, as will be seen in Table 9-17. Evidently, highly

TABLE 9-17. ADIABATIC HEATING OF FUEL ELEMENTS AT RATED POWER

	BNL	NRX	PWR	MTR
Rated power, thermal kw..............	28,000	40,000	200,000	30,000
Per cent U^{235} in total U..............	0.7	0.7	1.5 ± *	90
Total U, kg.........................	54,000	9,500	9,000	4.4
U^{235}, kg...........................	310	67	135	4
Kw/kg U^{235} (average)................	70	600	1,600	7,000
Average burnup of U^{235} per month, B, %	0.28	2.4	6.3	28
Kw/kg total U (average)..............	0.5	4.2	22	6,300
Adiabatic temperature rise at rated power, °F/sec, if the heating were localized in:				
Total U.........................	5.6	47	125	72,000
U alloy.............................				3,000
Total fuel element.................	5. ±	40 ±	40 ±	1,000 ±

* Uniform loading proposed for second core.

enriched fuel elements are likely to be too hazardous for commercial power reactors, which should have a high fuel burnup rate B to optimize total charges.

Dilution of the fuel with U^{238} or with an alloying element such as Al or Zr, or to some extent use of thicker Al cladding, decrease the danger by

* Booth [172] shows that the temperature t_y in a coolant stream at a distance y from an exponentially heated plate is given by

$$\ln \frac{t_{sur} - t_i}{t_y - t_i} = \frac{y}{(T'_1 \alpha_c)^{1/2}}$$

Boiling starts when t_{sur} is sufficiently superheated to form a vapor bubble. For water, $t_{sur,b}$ at initiation of boiling is obtainable from

$$(t_{sur,b} - t_{sat})^{4/3} = 0.1147 \frac{t_{sur,b} - t_i}{(T'_1)^{1/2}}$$

employing units of seconds and degrees Fahrenheit. The maximum surface temperature $t_{sur,max}$ for a Borax-type reactor in an excursion terminated by boiling was correlated satisfactorily by

$$\ln (t_{sur,max} - t_{sat}) = 0.445(t_{sur,b} - t_{sat}) + 5.82 + \frac{0.715}{(T'_1)^{1/2}}$$

increasing the heat capacity and slowing the rate of heating. In the case of a coolant-failure accident, this is important after scram as well as before. From Fig. 9-1 it is found for 1 week of operation that in the first 13 sec after complete scram the same energy is given off by the fission products as by the complete reactor in 1 sec of normal operation. Subsequent successive periods of 13.7, 14.3, and 15 sec, etc., each generate the same amount of heat (Table 9-18). As a compromise, it is frequently specified that a reactor coolant continue to circulate fairly strongly for at

TABLE 9-18. SECONDS AT FULL POWER REQUIRED TO EQUAL DELAYED-HEAT GENERATION FOR STATED TIMES AFTER SCRAM

Total seconds of delayed heating	13	27	41	56	100	1000	10,000	100,000
Total equivalent seconds at full power	1	2	3	4	6.8	50	300	1,100

least 15 sec after a pump power failure and that adequate cooling be available thereafter by natural circulation. With EM pumps a stand-by storage battery can be employed to maintain partial flow temporarily, and with rotary pumps a flywheel on the pump shaft.

9-47. Catastrophes. A nuclear accident in which the core is permanently damaged and fuel and radioactive fission products are disseminated into the coolant and/or the atmosphere may be designated a "catastrophe."

The conceivable external causes of catastrophes are earthquakes and nonnuclear collisions and explosions. Reactor shields and containers are so rigid, however, that damage from this direction is very improbable. Furthermore, it is much more likely that any such accident would separate the fuel and/or remove the moderator, making the assembly subcritical rather than supercritical.

One internal cause of reactor damage is fuel-element corrosion or melting by operation at too high a (constant) power or temperature. This would require repair or discard of the reactor but would generally involve no personnel hazard, because the initial damage would be localized, fairly gradual, and soon noticed.

Another possible cause of catastrophe is total coolant loss, on account of the fission-product heat generation, even though the reactor scrams correctly. Furthermore, this catastrophe could follow another type that damaged the reactor, then stopped itself by expelling the moderator. It is obvious from Tables 9-18 and 9-17 that in several seconds the MTR fuel elements would volatilize if the tank drained dry. With the other cores listed in Table 9-17, minutes or hours would be available for repairs or the arranging of temporary cooling.

By far the most important possible cause of catastrophes is an exponential power surge at a reactivity above prompt critical (see Sec. 9-46). Figure 9-39 shows the excess reactivity required beyond start-up criticality for a reactor to overcome fission-product build-up after various times. Further reactivity is needed to override additional xenon and samarium built up during a shutdown (Sec. 6-48), also to obtain reasonable start-up speed. In research and testing reactors extra reactivity is added to counteract neutron-absorbing experiments. Evidently, much more supercritical reactivity than that of the delayed neutrons alone (Sec. 6-43) must be available in practice for most reactors. It is thus usually possible by human error and selected equipment failures for a

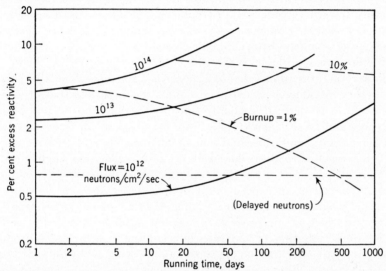

FIG. 9-39. Excess reactivity needed due to fuel depletion and fission-product growth. (*From M. Benedict, Chem. Eng. Progr., February, 1955, p. 14.*)

reactor to get above prompt criticality by enough that the period of T_1' (Sec. 9-46) of power rise is so short that the intrinsic negative coefficients have insufficient time in which to stabilize or shut down the reactor before damage. It is even more dangerous to have an initial net reactivity increase, or autocatalytic effect, when a power surge occurs. This can occur when boiling out of the coolant increases ρ' more than the prompt effects decrease it, as with H_2O-cooled natural-U reactors using graphite (Hanford) or D_2O (NRX) moderator.

Fortunately, within milliseconds (about one to two periods) after the instant a nuclear accident has damaged the core, the reactor will have become subcritical, on account of the accident if for no other reason. Furthermore, the neutron flux and the cooling effectiveness vary throughout the core; it is generally concluded from analysis of hypothetical catastrophes that no more than roughly 1 per cent of the radioactivity can be disseminated. However, this is still extremely great for a high-power heterogeneous reactor with high burnup, amounting to over 1000 lethal doses per kilowatt [152]. If the products remain within a tight enclosure (the favored "containment" policy), no external hazard should develop.

In the analysis of possible catastrophes the complete event should be followed, preferably by finite increments. Inertia, friction, ultimate strength, shock, and other aspects may enter into the manner and time of reduction of the reactivity. Well above prompt criticality the delayed neutrons can be neglected. Autocatalytic effects would contribute considerably toward causing a catastrophe, but if one occurs these effects would not generally increase its violence significantly, since they are usually small compared to the major changes in ρ', and since they occur at different times throughout the core [111].

The several catastrophes that have occurred* are not technically instructive, as they would have been expected under the conditions that occurred. The lesson to be derived is mainly that in experimental reactors proper awareness and good "bookkeeping" are necessary at all times as to the actual reactivity and the condition of all controls and fuel elements. The NRX reactor [153] was damaged in a test in which the cooling water to about 10 per cent of the fuel rods had been intentionally set low. Sudden removal of an experimental sample raised ρ' more than anticipated, subsequent boiling of the cooling H_2O in several tubes raised it further, and lack of air pressure prevented automatic insertion of the control rods. Melting of the fuel-rod cladding caused fission-product radioactivity to enter the cooling water, and melting of the outer H_2O-air and air-D_2O tube walls permitted the radioactive H_2O and D_2O to flood the air header room in the basement. The H_2O flow was shut down as fast as possible with time, but it had to be maintained to some extent to prevent the U from again overheating (Table 9-17). The basement water was pumped out to available tanks, then through a temporary pipeline to a sand-and-clay disposal area, which effectively adsorbed 10,000 curies. Fortunately for both the radioactivity disseminated and the subsequent heating, there had been very little burnup beforehand.

Another catastrophe occurred when a critical assembly with negligible cooling capacity was unwittingly brought gradually to over twice the reactivity of the delayed neutrons. The shortest period was 0.01 sec, demoderation occurred by boiling and decomposition of solid polyvinyl moderator, and substantially all the energy of the burst occurred in 0.05 sec.

EBR-1 was damaged during manual power-surge tests without coolant flow to investigate low-power instability observed with a core containing

---

* There have also been several shutdowns at Hanford due to coolant channel plugging [189].

fuel rods which were not rigidly supported throughout all their length. The automatic short-period scram circuit was not employed in the test, and when shutdown was called for, the regulating rods were started in, owing to a misunderstanding, instead of the (faster) shim rods [182]. When the automatic high-power scram occurred, the lack of coolant flow made the melting of some uranium unavoidable. The power surge itself was a slow one, starting with $T_1' = 60$ sec at a power of a few watts and decreasing to <5 sec at 200 kw. This positive coefficient of reactivity and the low-power instability were presumably due to an inward bowing of the fuel rods, because of higher heating density and thermal expansion at the inner sides of the rods. When bowing was prevented by fins long enough radially to touch each other, this instability was not observed.

Finally, the Borax-1 core in cold water was largely melted [167] when a reactivity near 4 per cent ($T_1' = 0.0026$ sec) was suddenly applied. It may be concluded that most thin fuel elements cannot safely stop power excursion periods of less than about 0.003 sec at atmospheric pressure, or even longer periods at higher pressures. Fuel elements with more fuel or cladding would be damaged by yet longer periods.

The following are general problems for this chapter:

Prob. 9-25. A BSR core (see App. N-1) using stainless steel instead of aluminum for the cladding has been stated not to be damaged by the delayed-heat generation if during long-time operation at 100 kw the water suddenly drained off (i.e., due to earthquake rupture of the pool) and only natural circulation of room air remained. Check this result by computing the maximum temperature to be expected in the core.

Prob. 9-26. The Oak Ridge X-10 reactor employs 1.75-in.-square channels on an 8-in. horizontal and vertical square lattice (see App. N-1). The channel diagonals are also horizontal and vertical. Making the (conservative) assumption that the neutron and γ heating of the graphite is uniform in the "cell" surrounding a given channel, find the ratio $Nk \, \Delta t/q$ by the relaxation method for the maximum Δt across the graphite.

Prob. 9-27. The water velocity through the seed assemblies (see Fig. M-4 and App. N-2) of the PWR is 20 fps and through the blanket assemblies 9.8 fps [175]. Ninety-six of the seed assemblies are employed, and 97 of the blanket assemblies. The maximum heat velocities are at about one-fourth of the fuel-element length from the inlet. They are 382,000 Btu/(hr)(ft^2) in the seed assemblies and 240,000 in the blanket, the average values being 112,000 and 65,000, respectively. Sixty per cent of the heat is generated in the blanket and 40 per cent in the seed. The total hot-channel products for the seed and blanket are, respectively, $F_c = 1.40$ and 1.25; $F_f = 2.03$ and 1.69; $F_e = 1.31$ and 1.06. Check as may be possible data above and in App. N-2 against each other. Show that the water and the fuel-element surfaces never reach the boiling point of 636°F. Estimate the maximum seed- and blanket-element surface and axial temperatures, and the maximum per cent per month burnup in each.

Prob. 9-28. (a) Assuming the BNL reactor $h = 50$ Btu/(hr)(ft^2)(°F), compute the average fin efficiency for the cross section given in Fig. M-1, in terms of the temperature at the base of the fin, by any suitable method. Compare your result with the equation in Prob. 9-21, and with Fig. 9-25 for the average thickness of the BNL fin. (b) From data in App. N-1 and Fig. M-1, and assuming the He-filled crack is

0.001 in. thick, compute by relaxation (Sec. 9-39) the temperature difference at the center of the reactor between the fuel element axis and l of the air.

Prob. 9-29. Harrer et al. [176] give the following data for the Experimental Boiling Water Reactor (EBWR): At 25 kw/1 of coolant H_2O and 20 thermal Mw, the required core is 4 ft in diameter and height. Cladding is Zircaloy-2.

a. Check the H_2O-U volume ratio for the natural-U elements and for the whole core (given as 2.5 and 3.6).

b. Check the mean and maximum q/A [given as 40,000 and 150,000 Btu/(hr)(ft²)].

c. Check the steam flow (1,000 lb/min.), average steam voids in heated channels (20 per cent), and exit steam quality (2.56 per cent), considering that the water is in natural convection between fuel plates in the open tank.

d. Feed water at 110°F enters at the top of the annulus 1.5 ft wide which serves as the recirculating water downpath. Subcooling at the heated channel inlet (bottom of the annulus, after mixing) is 11°F. Check the recirculating water-steam ratio.

e. Check where boiling starts (stated at one-third up the fuel elements) and burn-out q/A [stated 600,000 Btu/(hr)(ft²)(°F)].

f. Check the following stated maximum temperatures: fuel centerline 561°F, fuel surface 508°F, and the respective average temperatures 518 and 495°F.

Prob. 9-30. Zinn [177] gives the following data for the CP-5 reactor: At 4000 kw, a practical minimum of 16 fuel elements (see Fig. M-2 and App. N-1), each containing 140 g of U^{235}, is needed. Each fuel element has 12 plates and 9100 cm². Average q/A is 6.5, and maximum is about 9.8 cal/(cm²)(sec). Average water velocity between the plates is 1.6 m/sec and $\bar{h}$ is 0.2 cal/(cm²)(sec)(°C). Thus $(\Delta t_f)_{max} = 49$°C. $l_2 - l_1$ is 12.6°C at 1200 gpm. The midpoint of the (vertical) elements is under roughly 3 ft static head of D_2O above atmospheric. With an inlet l_1 of 40°C, film boiling is just avoided.

a. From the dimensions, power, and coolant rate, check all other possible quantities given. Also compute the maximum metal temperature.

b. Average fuel burnup is 15 per cent and maximum 30 per cent before replacement. However, to achieve this burnup five more fuel elements are gradually added. Check the frequency of core replacement (once a year stated).

Prob. 9-31. a. Weinberg et al. [178] report that the water velocity between the MTR fuel plates averages 10 m/sec, yielding $h = 0.9$ cal/(cm²)(sec)(°C). Check this by Eq. 9-44a (see Fig. M-2 and App. N-1).

b. They also report an average Δt_f of 27°C and water rise of 6.1°C. Check how many fuel elements this corresponds to at 30 Mw.

Prob. 9-32. Lichtenberger et al. [179] describe EBR. The core and inner blanket are cooled by 292 gpm of NaK entering at 228°C and leaving at an average of 316°C.

a. On a certain day, the core generated 960 kw and had an average q/A of 209,000 Btu/(hr)(ft²)(°F) at the fuel-element surface. The NaK cross section was 0.1008 ft² and velocity 6.5 fps. The 0.384-in. OD fuel slugs are in NaK inside a 0.02-in. wall, 0.448-in. OD, 347 stainless-steel tubes located on 0.494-in. hexagonal centers. Compute the maximum U temperature (357°C stated) and the NaK outlet temperature.

b. On the same day, the inner blanket generated 196 kw and had NaK at 1.77 fps flowing through 0.368 ft². Check the average surface q/A [11,900 Btu/(hr)(ft²) stated] and the outlet temperature.

c. The outer blanket was cooled by 5800 cfm of air entering at 3°C and 0 psig and leaving at 91°C. Compute the heat generated (213 kw stated). The air flows through 60 roughly equally spaced 1-in. OD annuli in the outer blanket. If the U is nowhere above 200°C, what is the ID of the annuli? Discuss the conclusion that cooling of the outer blanket is the power limitation of this reactor. How would you correct this?

d. The primary NaK coolant transmits its heat to a secondary NaK coolant to control radioactivity. The primary coolant flows in two passes inside 102 0.75-in. OD, 16-gauge hairpin-bend tubes with 495 ft² of outside heat-transfer area. The tubes are in a 17-in. OD shell in which the secondary NaK flows countercurrent (also in two passes). The secondary NaK is 280 gpm, and Δt_m is 19°F. Compute the over-all U_o. Check U_o by available correlations in Sec. 9-11.

e. The water flows countercurrent to the secondary NaK, entering the preheaters at 214°F, going to the boilers and boiling at 446°F, and leaving the superheaters at 529°F and 405 psia. The water is inside the tubes, which are all 9 ft 6.75 in. in effective length, and 2 in. in ID, and have successive wall thicknesses of $\frac{3}{32}$ in. of nickel, $\frac{1}{8}$ in. of copper, and $\frac{3}{32}$ in. of nickel, all well bonded. For NaK entering the super-heater at 583°F and leaving the preheater at 419°F, check the following values of U from the data given, and by available relations in this chapter:

	Shell, iron pipe size, in.	Number of units	Connection	Total A_i, ft²	Over-all U	NaK velocity, fps
Superheaters........	5	4	Series	20	150	6.15
Boilers.............	3	18	Parallel	90	702	2.51
Preheaters.........	5	9	Series	45	311	6.15

Prob. 9-33. The steam generator in HRE-1 transfers 1000 kw through 112 horizontal stainless return-bend tubes of $\frac{1}{4}$-in. OD and 0.049 in. wall, which average 110 in. of effective length. The fuel solution enters at 482°F and 1000 psia and leaves at 407°F. The rate is 100 gpm, and the properties are reasonably close to those of water. The shell is $26\frac{1}{4}$-in. ID, and the water boils therein by natural convection. Predict (*a*) the obtainable steam pressure (design value 200 psia); (*b*) the fuel-solution pressure drop (design value 25 psi).

Prob. 9-34. Figures 9-35 and M-3*b* show one proposed cross section for the graphite core elements of an internally cooled LMFR. The central hole *F* is filled with the noncirculating fuel, a $\frac{1}{2}$ per cent solution of U²³³ in bismuth. The channels *C*, formed between four adjacent core elements, carry the coolant, pure bismuth, in upflow at 2.63 fps. The thermal conductivity *k* of the graphite may be taken at 30 Btu/(hr)(ft)(°F), a low value reached with old graphite at the high temperatures involved, averaging 500°C.

a. Assuming the inner and outer graphite surfaces to be isothermal at t_i and t_0, respectively, compute the average value of $t_i - t_0$ for a cylindrical reactor 9.45 by 9.45 ft in size at 300 Mw. Use "flux plotting" to obtain the resistance of the graphite. Assume the conservative case that all the heat is generated in the fuel (design value 164°F given by O. E. Dwyer [180]).

b. Repeat *a* by "relaxation," approximating the cross section with either a rectangular or a radial grid (Fig. 9-35).

c. Compute the average contact temperature drop at the fuel-graphite interface from the assumed contact conductance [180] of 10,000 Btu/(hr)(ft²)(°F).

d. Compute the bismuth-film heat-transfer coefficient and the average temperature drop across the bismuth film (check with design value of 81°F).

e. The sum of the Δt's computed in steps *c*, *d*, and either *a* or *b* is the total Δt from the surface of the fuel to the coolant stream, assuming the graphite surfaces isothermal, or infinitely conducting. This resulting Δt must be somewhat too low. Compute the combined Δt by relaxation using the appropriate h and nodes that lie exactly on the surface of the graphite, and compare with the previous result.

f. Assuming that the fuel solution does not move and that the heating density H in it is uniform, compute the temperature drop from the axis of the fuel solution to its periphery (check with design value 232°F).

g. Find Δt across the fuel solution from the most appropriate correlation if it undergoes natural convection.

h. If the simplification is not made that all the heat is generated in the fuel, roughly what fraction do you expect would be generated in the graphite and what fraction in the coolant, and by what per cent would the over-all Δt be decreased?

NOMENCLATURE

A = surface area

a = surface area per particle; heat-transfer surface per unit volume

C = specific heat; also electrical capacitance

C_P = specific heat at constant pressure

C_V = specific heat at constant volume

c = concentration

c^* = solubility

D = diameter

D' = diameter, in.

D_e = equivalent diameter = $4 \dfrac{\text{cross section}}{\text{wetted perimeter}}$

D_V = diffusivity

E = effectiveness; efficiency; also electrical potential

F = efficiency factor

F' = mass-transfer coefficient

f_F = Fanning friction factor

G = mass velocity

g = gravitational acceleration

g_c = ratio between units of F and of ma

H = total heating density; also height above a datum

h = local film heat-transfer coefficient

I, i = electrical current

I_0, J_0, K_0 = zero-order Bessel functions (see App. L)

J = radiation intensity; also mechanical equivalent of heat

k = thermal conductivity; also multiplication constant

L = average diffusion length; also dimension in general

M = molecular weight

m = mass

N = number of atoms per unit volume; also length

n = neutron-flux density (also frequently designated nv); also number in general

P = pressure

P' = perimeter

P_R = reduced pressure = $\dfrac{P}{P_{\text{critical}}}$

p = per cent gas by volume; partial pressure

Q = quantity of heat; also volumetric flow rate

q = rate of heat flow

R = fraction by volume; also thickness; also resistance in general

r = radius

r' = fouling factor

r_e = electrical resistivity

S = cross-section area

s = surface tension

T = absolute temperature

t = thermometric temperature

U = over-all heat-transfer coefficient

u = velocity

V = volume; specific volume

w = mass rate of flow

x = fraction by weight; also vapor quality; also distance along a channel

x' = fraction by volume

y = distance transverse to a flow, or away from a channel wall

z, Z = height above a horizontal datum

α = thermal diffusivity $= \dfrac{k}{C\rho}$

β = volumetric coefficient of thermal expansion

Γ = mass flow rate per unit of width

ϵ = eddy diffusivity; also emissivity

θ = time

λ = decay constant; latent heat of vaporization

μ = molecular viscosity

μ_B = γ-ray attenuation coefficient

ν = kinematic viscosity

ρ = density

σ = microscopic cross section; also standard deviation

σ_a = absorption cross section

ψ = wetting angle

Gr, Gz, j, Nu, Pe, Pr, Re, Sc, Sh, St = dimensionless ratios (see Table 8-7)

REFERENCES

1. *Chem. Eng. News*, **30**:660 (1952).
2. Way, K., and E. P. Wigner: *Phys. Rev.*, **73**:1318 (1948); **70**:115 (1946).
3. Hougen, J. O., and E. L. Piret: *Chem. Eng. Progr.*, **47**:295–304 (1951).
4. Carslaw, H. S., and J. C. Jaeger: "Conduction of Heat in Solids," Oxford University Press, New York, 1947.
5. "NBS-NACA Thermal Tables," National Bureau of Standards, 1954. Hilsenrath, J., and Y. S. Touloukian, *Trans. ASME*, **76**:967–985 (1954).
6. Bromley, L. A.: UCRL-1852.
7. Lindsay, A. L., and L. A. Bromley: *Ind. Eng. Chem.*, **42**:1508–1511 (1950).
8. Keyes, F. G.: *Trans. ASME*, **62**:589–596 (1951).
9. Keyes, F. G.: *Trans. ASME*, **76**:809–816 (1954).
10. Lenoir, J. M., W. A. Funk, and E. W. Comings: *Chem. Eng. Progr.*, **49**:542 (1953).
11. Jakob, M.: "Heat Transfer," vol. 1, John Wiley & Sons, Inc., New York, 1949.
12. Deissler, R. G., and C. S. Eian: NACA RME52C05 and E53G03.

13. Coryell, C. D., and N. Sugarman: "Radiochemical Studies: The Fission Products," NNES, div. IV, vol. 9, bk. 1, p. 349, McGraw-Hill Book Company, Inc., New York, 1951.

14. Lane, J. A.: Proceedings of the University Reactors Conference, Oak Ridge Institute for Nuclear Science, AECU-2900, 1954.

15. Deissler, R. G.: *Trans. ASME*, **76**:73–85 (1954).

16. Verschoor, H., and S. Stemerding: "General Discussion on Heat Transfer," pp. 201–203, Institution of Mechanical Engineers, London, 1951.

17. Perry, J. H.: "Chemical Engineers' Handbook," 3d ed., McGraw-Hill Book Company, Inc., 1950.

18. Aladev, I. T.: *Bull. U.S.S.R. Acad. Sci., Tech. Sci. Div.*, no. 11, pp. 1669–1681, 1951.

19. Dow, W. M., and M. Jakob: *Chem. Eng. Progr.*, **47**:637 (1951).

20. Vreedenberg, H. A.: "General Discussion on Heat Transfer," pp. 373–377, Institution of Mechanical Engineers, London, 1951.

21. Leva, M.: *Ind. Eng. Chem.*, **39**:857 (1947).

22. Thomson, A. S. T., et al.: "General Discussion on Heat Transfer," pp. 177–180, Institution of Mechanical Engineers, London, 1951.

23. Monrad, C. C., and J. F. Pelton: *Trans. AIChE*, **38**:593 (1942).

24. DeBortoli, R. A., et al.: *Nuclear Sci. Eng.*, **1**:239–251 (1956).

25. Norris, R. H., and D. D. Streid: *Trans. ASME*, **36**:525 (1940).

26. McAdams, W. H.: "Heat Transmission," 3d ed., McGraw-Hill Book Company, Inc., New York, 1954.

27. Eckert, E. R. G.: "Introduction to the Transfer of Heat and Mass," McGraw-Hill Book Company, Inc., New York, 1950.

28. Kettenring, K. N., et al.: *Chem. Eng. Progr.*, **46**:139–145 (1950).

29. "Liquid Metals Handbook" (2d ed., R. N. Lyon, ed.; 3d ed., C. B. Jackson, ed.), Government Printing Office, Washington, D.C., 1956.

30. Poppendiek, H. F., and W. B. Harrison: *Chem. Eng. Progr., Symposium Ser.*, no. 9, p. 93, 1954.

31. Trefethen, L. M.: NP-1788, 1950.

32. Untermeyer, S., and Weills, J. T.: AECD-3454, 1952.

33. MacDonald, W. C., and R. C. Quittenton: *Chem. Eng. Progr., Symposium Ser.* no. 9, pp. 59–67, 1954.

34. Wilke, C. R., and O. A. Hougen: *Trans. AIChE*, **41**: 445–451 (1945).

35. Dwyer, O. E., et al.: Proceedings of the 1953 Conference on Nuclear Engineering, University of California (Berkeley), p. F-73, 1953; *Ind. Eng. Chem.*, **48**:1836 (1956).

36. BNL-289 (S-21), p. 36, 1954.

37. Brooks, R. D., and S. K. Friedlander: "Reactor Handbook," vol. 2, pp. 277–286, McGraw-Hill Book Company, Inc., New York, 1955.

38. Hyman, S. C., C. F. Bonilla, and S. W. Ehrlich: NYO-564, 1951.

39. Jakob, M., and P. C. Gupta: *Chem. Eng. Progr., Symposium Ser.*, no. 9, p. 15, 1954.

40. Kyte, J. R., et al.: *Chem. Engr. Progr.*, **49**:653 (1953).

41. Rogers, F. T., et al: *J. Appl. Phys.*, **22**:1476 (1951).

42. Schwartz, H.: NAA-SR-40, 1949.

43. Isakoff, L., and F. W. DeVries: M. S. thesis in chemical engineering, Columbia University, 1951.

44. Eckert, R. G., and A. J. Diaguila: *Trans. ASME*, **76**:497–504 (1954).

45. Sigel, L. A.: M.S. thesis in chemical engineering, Johns Hopkins University, 1948.

46. Seban, R. A.: *Trans. ASME*, **76**:299–303 (1954).

47. Misra, B.: Ph.D. thesis in chemical engineering, Columbia University, 1955; *Chem. Eng. Progr.*, *Symposium Ser.*, 1957.
48. Carpenter, F. G., and A. P. Colburn: "General Discussion on Heat Transfer," pp. 20–26, Institution of Mechanical Engineers, London, 1951.
49. Coppage, J. E.: Ph.D. thesis in mechanical engineering, Stanford University, 1952.
50. Tidball, R. A.: *Chem. Eng. Progr.*, *Symposium Ser.*, no. 5, 1953.
51. Donohue, D. A.: *Ind. Eng. Chem.*, **41**:2499–2511 (1949).
52. Cichelli, M. T., and C. F. Bonilla: *Trans. AIChE*, **41**:755–787 (1945).
53. Robinson, D. B., and D. L. Katz: *Chem. Eng. Progr.*, **47**:317–324 (1951).
54. Rohsenow, W. H.: *Trans. ASME*, **74**:969–975 (1952).
55. Grady, J. J.: M.S. thesis in chemical engineering, Columbia University, 1950.
56. Bonilla, C. F., and C. W. Perry: *Trans. AIChE*, **37**:685–705 (1941).
57. Kutateladze, S. S.: *Otdelenie Tekhnicheskikh Nauk*, no. 4, pp. 529–536, 1951.
58. Bromley, L. A.: *Chem. Eng. Progr.*, **46**:221–227 (1950).
59. Hayworth, H. C., and N. Nicholaus: M.S. thesis in chemical engineering, Columbia University, 1951.
60. Banchero, J. T., et al.: *Chem. Eng. Progr.*, *Symposium Ser.*, no. 17, p. 21, 1955.
61. Paschkis, V., and G. Stolz: Quenching Program, Columbia University, Mechanical Engineering Department, Mar. 1, 1955.
62. Hatch, L. P.: *Am. Scientist*, **41**:410–421 (1953).
63. Sanders, A.: M.S. thesis in chemical engineering, Columbia University, 1954.
64. Pigford, R. L.: *Chem. Eng. Progr.*, *Symposium Ser.*, no. 17, pp. 79–92, 1955.
65. Martin, W., et al.: AECU-2169, 1952.
66. Pike, F. R., P. D. Miller, Jr., and K. O. Beatty, Jr.: American Institute of Chemical Engineers Heat Transfer Symposium, Preprint 2, St. Louis, 1953.
67. McAdams, W. H., et al.: *Ind. Eng. Chem.*, **41**:1945 (1949).
68. Clark, J. A., and W. M. Rohsenow: *Trans. ASME*, **76**:553 (1954).
69. Jens, W. H., and P. A. Lottes: ANL-4915, 1952.
70. Gunther, F. C.: *Trans. ASME*, **73**:115 (1951).
71. Buchberg, H., et al.: "Proceedings of the Heat Transfer and Fluid Mechanics Institute," pp. 177–191, Stanford University, 1951.
72. Jeffery, R. W.: B. S. thesis in mechanical engineering, Massachusetts Institute of Technology, 1952.
73. Motte, E. I., and L. A. Bromley: UCRL-2511, 1954.
74. Bromley, L. A., et al.: UCRL-1894, 1952.
75. Moscicki, I., and J. Broder: *Roczniki Chem.*, **6**:321 (1926).
76. Beighley, C. M., and L. E. Dean: *Jet Propulsion*, May-June, 1954, pp. 180–186.
77. Jens, W. H.: *Mech. Eng.*, **76**:981–986 (1954).
78. Roy, G. M., and C. A. Pursel: General Electric Company, August, 1954.
79. Nakagawa, Y., and T. Yoshida: *Chem. Eng. (Japan)*, **16**:74–82, 104–110 (1952).
80. Styuschchin, N. G., and L. S. Sterman: *J. Tech. Phys. (U.S.S.R.)*, **21**:448 (1951).
81. Piret, E. L., and H. S. Isbin: *Chem. Eng. Progr.*, **50**:305–311 (1954).
82. Lyon, R. E., et al.: American Institute of Chemical Engineers, Heat Transfer Symposium, Preprint 6, St. Louis, 1953.
83. Nerad, A. J., and A. Howard: unpublished reports, General Electric Company, 1936.
84. Carter, J. C.: ANL-4766, 1952.
85. Greenfield, M. L., et al.: Report 54-77, Department of Engineering, University of California at Los Angeles, 1954.
86. Gessler, J. L.: M.S. thesis in chemical engineering, Johns Hopkins University, 1948.

87. Goldmann, K.: NDA Report 10-68, 1953.
88. Armacost, W. H., et al.: *Trans. ASME*, **76**:715–748 (1954).
89. Mumm, J. F., at ANL; reported at BNL, 1954.
90. Untermyer, S.: *Nucleonics*, **12**(7):43–47 (July, 1954).
91. Bailey, R. V., at Tulane University.
92. Bernath, L.: Du Pont Memo DPW-54-526, 1954; *Chem. Eng. Progr., Symposium Ser.* (Preprint 8, Louisville Meeting, 1955).
93. Cook, W. H., at ANL; reported at BNL, 1954.
94. Untermyer, S.: *Nucleonics*, **12**(1):12–15 (January, 1954).
95. Jones, C. D., and D. J. Masson: ASME Paper 54-A-147, 1954.
96. Rohsenow, W. M., et al.: ASME Papers 54-A-144 and 54-A-145, 1954.
97. Kays, W. M.: ASME Paper 54-A-151, 1954.
98. Lietzke, A. F., at NACA; reported at BNL, 1954.
99. Poppendiek, H. F., and L. D. Palmer: ORNL-1395, 1952; *Chem. Eng. Progr., Symposium Ser.*, no. 11, 1954.
100. Poppendiek, H. F., and L. D. Palmer: ORNL-1701, 1954.
101. Bonilla, C. F., and S. J. Wang: NYO-3091, 1952.
102. Addoms, J. N.: Sc.D. thesis in chemical engineering, Massachusetts Institute of Technology, 1948.
103. Lane, J. A., and S. McLain: *Chem. Eng. Progr.*, **49**:287–293 (1953).
104. McLain, S.: ANL-5424, 1955.
105. Buckley, P. S.: *Chem. Eng.*, **57**:112–114 (September, 1950).
106. Plewes, A. C. et al.: *Chem. Eng. Progr.*, **50**:77–80 (1954).
107. Schwartz, H.: NAA-SR-40, 1949.
108. Ostrach, S.: *NACA TN*-3141, 1954.
     Woodrow, J.: AERE E/R 1267, 1953.
     Wordsworth, D. V.: AERE E/R 1270, 1953.
109. Hamilton, D. C., et al.: ORNL-1624, 1954.
110. Richards, P. B.: *Chem. Eng. Progr., Symposium Ser.*, no. 11, p. 127, 1954.
111. Hurwitz, H., Jr.: *Nucleonics*, **12**(3):57–61 (March, 1954).
112. Goodman, C., et al.: "The Science and Engineering of Nuclear Power," vol. II, pp. 120–176, Addison-Wesley Publishing Company, Cambridge, Mass., 1949.
113. Bonilla, C. F.: AEC Report M-4476, 1949.
114. Parkins, W. E.: F55-71, "Proceedings of the 1953 Conference on Nuclear Engineering," University of California (Berkeley), 1953.
115. Shimazaki, T. T.: *Chem. Eng. Progr., Symposium Ser.*, no. 12, pp. 113–119, 1954.
116. Carberry, J. J.: *Chem. Eng.*, **60**:225–227 (June, 1953).
117. Norris, R. H., and W. A. Spofford: *Trans. ASME*, **64**:489–496 (1942).
118. Kays, W. M., et al.: "Gas Turbine Plant Heat Exchangers," ASME, 1951.
119. Kern, D. Q.: "Process Heat Transfer," McGraw-Hill Book Company, Inc., New York, 1950.
120. "Standards of Tubular Exchanger Manufacturers Association," 3d ed., New York, 1952.
121. Gardner, K. A.: *Trans. ASME*, **67**:621–631 (1945).
122. Dunn, W. E., and C. F. Bonilla: *Ind. Eng. Chem.*, **40**:1101–1105 (1948).
123. Cichelli, M. T., and M. S. Brinn: ASME Paper 54-A-125, 1954.
124. Grebe, J. J., and A. W. Hanson: *Chem. Eng. Progr., Symposium Ser.*, no. 13, pp. 200–209, 1954.
125. Goertzel, G., and W. A. Loeb: *Chem. Eng. Progr., Symposium Ser.*, no. 12, pp. 82–87, 1954.
126. Robbins, C. H.: *Chem. Eng. Progr., Symposium Ser.*, no. 12, pp. 181–200, 1954.

127. Miles, F. T., and I. Kaplan: *Chem. Eng. Progr., Symposium Ser.*, no. 11, pp. 159–175, 1954.
128. Stahl, C. R.: *Mech. Eng.*, **76**:978–980 (1954).
129. McGoff, M. J., and J. W. Mausteller: MSA Technical Report 32, 1954.
130. Bachmann, H.: "Tafeln über Abkühlungsvorgange einfacher Körper," Springer, Berlin, 1938.
131. Newman, A. B.: *Trans. AIChE*, **30**:598–612 (1934).
132. Newman, A. B., and L. Green: *Trans. Electrochem. Soc.*, **66**:345–358 (1934).
133. Beatty, K. O., et al.: *Ind. Eng. Chem.*, **42**:1527–1532 (1950).
134. Rubin, T.: AEC Report Y-F-10-28, 1950.
135. Fritz, R. J.: *Trans. ASME*, **76**:913–921 (1954).
136. Russell, T. F.: First Report, pp. 149–187, Alloy Steels Research Committee (England).
137. Monson, H. O., J. P. Silvers, and P. A. Lottes: "Heat Transfer Problems in Nuclear Reactors," ASEE Conference on Nuclear Physics in Engineering, September, 1954.
138. Korn, G. A., and T. M. Korn: "Electronic Analog Computers," McGraw-Hill Book Company, Inc., New York, 1952.
139. Moise, J. C.: *Chem. Eng. Progr., Symposium Ser.*, no. 12, pp. 96–106, 1954.
140. Dusinberre, G. M.: "Numerical Analysis of Heat Flow," McGraw-Hill Book Company, Inc., New York, 1949.
141. Hellman, S. K., et al.: *Trans. ASME*, **78**:1155–1161 (1956).
142. Bonilla, C. F.: *J. Eng. Educ.*, **45**:328–332 (1954).
143. Patton, T. C.: *Ind. Eng. Chem.*, **36**:990 (1944).
144. Pardo, J. F.: Ph.D. dissertation in chemical engineering, Columbia University, 1954.
145. Leppert, G.: *J. Am. Soc. Naval Engrs.*, **65**: 741–752 (1953).
146. Wu, C. H.: *NACA TN* 2214, 1950.
147. Lubarsky, B., and S. J. Kaufman: *NACA TN* 3336, 1955.
148. Bonilla, C. F.: Paper 122, vol. 9, p. 331, UN Conference on Atomic Energy, Geneva, 1955.
149. Durham, F. P.: *Nucleonics*, **13**(5):42–46 (May, 1955).
150. Deissler, R. G.: *NACA TN* 3145, 1954.
151. Sherwood, T. K., and R. L. Pigford: "Absorption and Extraction," McGraw-Hill Book Company, Inc., New York, 1952.
152. Benedict, M.: *Chem. Eng. Progr.*, **51**:53–56 (1955).
153. Hatfield, G. W.: *Mech. Eng.*, February, 1955, pp. 124–126.
154. Liebmann, G.: ASME Paper 55-SA-15, 1955.
155. Kasten, P. R.: "Homogeneous Reactor Safety," ORNL, 1955.
156. Douglas, W. J. M., and S. W. Churchill: *Chem. Eng. Progr., Symposium Ser.* (Preprint 16, Louisville Meeting, 1955).
157. Hamilton, D. C., et al.: ORNL-1769, 1954.
158. Rosenthal, M. W.: ORNL, 1955.
159. Lottes, P. A.: "Proceedings of the 1955 Conference on Nuclear Engineering," p. A-1, University of California at Los Angeles.
160. Ellion, M. E.: Memo 20-88, Jet Propulsion Laboratory, California Institute of Technology, 1954.
161. Pramuk, F. S., and J. W. Westwater: *Chem. Eng. Progr., Symposium Ser.* (Louisville Meeting, 1955).
162. Rohsenow, W., and P. Griffith: *Chem. Eng. Progr., Symposium Ser.* (Louisville Meeting, 1955).

163. Eckert, E. R. G., and G. M. Low: *NACA TN* 2401, 1951.
164. Graham, R. H.: *Nucleonics*, **13**(3): 25–27 (March, 1955).
165. Weatherhead, R. T.: ANL, 1955.
166. "Reactor Handbook," vol. 2, "Engineering," AECD-3646, 1955.
167. Dietrich, J. R.: Paper 481, vol. 13, p. 88, UN Conference on Atomic Energy, Geneva, 1955.
168. Stein, R. P., and W. Begell: Department of Chemical Engineering, Columbia University, 1955.
169. McGoff, M. J., and J. W. Mausteller: MSA Memo Report 87, 1955.
170. Untermyer, S. A.: *Nucleonics*, **13**(7): 34–35 (July, 1955).
171. Miller, P., et al.: Preprint 47, Nuclear Engineering and Science Congress, Cleveland, 1955.
172. Booth, M.: M.S. thesis in chemical engineering, Massachusetts Institute of Technology, 1955.
173. Untermyer, S. A.: Preprint 25, Nuclear Engineering and Science Congress, Cleveland, 1955.
174. Graham, R. H., and D. G. Boyer: TID-8009, 1956.
175. Simpson, J. W., et al.: Paper 815, vol. 3, p. 211, UN Conference on Atomic Energy, Geneva, 1955.
176. Harrer, J. M., et al.: Paper 497, vol. 3, p. 250, UN Conference on Atomic Energy Geneva, 1955.
177. Zinn, W. H.: Paper 861, vol. 2, p. 456, UN Conference on Atomic Energy, Geneva, 1955.
178. Weinberg, A. M., et al.: Paper 490, vol. 2, p. 402, UN Conference on Atomic Energy, Geneva, 1955.
179. Lichtenberger, H. V., et al.: Paper 813, vol. 3, p. 345, UN Conference on Atomic Energy, Geneva, 1955.
180. Dwyer, O. E.: *Chem. Eng. Progr., Symposium Ser.*, no. 11, pp. 75–91, 1954.
181. Bonilla, C. F., et al.: NYO-7638, 1956.
182. Atomic Energy Commission: Twentieth Semi-Annual Report, pp. 45–47, July, 1956.
183. Heisler, M. P.: *Trans. ASME*, **78**:1187-1192 (1956).
184. Farbar, L., and M. J. Morley: University of California (Berkeley), 1956.
185. Stein, R. P.: AEC Reactor Heat Transfer Conference, New York, Nov. 1, 1956.
186. LeTourneau, B. W., and R. E. Grimble: *Nuclear Sci. Eng.*, **1**:359–369 (1956).
187. Rickard, C. K., et al.: AEC Reactor Heat Transfer Conference, New York, Nov. 1, 1956.
188. Zinn, W. H., et al.: *Nuclear Sci. Eng.*, **1**:420–437 (1956).
189. Graham, R. H.: *Nucleonics*, **13**(10): 42–44 (October, 1954).
190. Hazard, H. R., and J. M. Allen: BMI-1141, 1956.

CHAPTER 10

METALLURGY OF URANIUM AND URANIUM ALLOYS

By George L. Kehl

10-1. Introduction. The operation of a nuclear reactor requires an appropriate fissionable material as a fuel, and uranium is at present by far the most important one. Natural uranium is composed of 0.712 per cent U^{235}, a trace of U^{234}, and the remainder U^{238}. Uranium 238 does not undergo fissioning by thermal neutrons, which from an industrial viewpoint is the most important fissioning reaction. Thus the isotope U^{235}, which does, is separated by appropriate techniques (see Chap. 1). However, through suitable nuclear reactions U^{238} produces the nuclide Pu^{239} and Th^{232} produces U^{233}, which are fissionable by slow neutrons.

10-2. Sources of Uranium. Approximately 0.0004 per cent of the earth's crust is natural uranium, it being more plentiful than silver, cadmium, bismuth, or mercury. Briefly, uranium occurs in the form of a variety of minerals in igneous rocks, and to a lesser extent in sandstone and other sedimentary formations. In the United States, the principal sources of uranium are carnotite ores (mined in the past solely for the vanadium present), other oxidized ore bodies, and primary pitchblende. Successful attempts have been made in recovering the uranium from phosphate ores processed principally for fertilizing use.

10-3. Preparation for Use. Owing to the inherent difficulties in obtaining uranium by direct reduction of the oxide, e.g., with carbon, the tetrafluoride is generally employed as the starting material. From this, metallic uranium is obtained either by "bomb" reduction with sodium, calcium, or magnesium to produce what is known as "biscuit" or "derby" metal, or by electrolysis of fused salts. High-purity uranium has been successfully produced in moderate quantities by the latter method at the Argonne National Laboratory. The process consists essentially of carefully melting and liquating electrolytic crystals of uranium deposited from a fused salt electrolyte containing uranium tetrafluoride and a eutectic mixture of lithium and potassium chlorides.

The biscuit metal as removed from the reduction bomb is particularly rough on the top surface and may contain a relatively high percentage of volatile materials depending upon whether calcium or magnesium is used as the reductant. Refining and consolidation of the uranium biscuits is accomplished in one of several ways, depending upon the ultimate use of the material. Generally the process is one of induction melting in vacuo (usually below about a 500-μ pressure), within an inert atmosphere, or under a suitable flux or slag. Arc melting in an inert atmosphere, such as argon or helium, is particularly advantageous in the melting and manufacture of alloys. Typical impurities are given in Table 10-1.

TABLE 10-1. TYPICAL ANALYSIS OF URANIUM METAL

Element	"Biscuit form," ppm	"Biscuit form" after remelting, ppm	High-purity uranium after melting, ppm
Al	<20	10	5
Ba	<10		
C	30–50	25	18
Ca	<10	<20	<20
Cu	7	1	1
Fe	20	18	3
H$_2$	3		
K	<10	<10	<10
Mg	15	5	<0.5
Mn	8	3	<0.5
Mo	<20	<20	<20
N$_2$	....	....	10
Nb	<50	....	<5
O$_2$	4	....	<10
P	<10	<10	<20
Pb	2	<1	<1
Si	....	20	<10
Sr	<10		
Te	<10		
V	<10		
Y	<15		
Zn	<20	<20	<20
Zr	<10		

10.4. Fabrication Methods. Uranium in ingot form may be fabricated into useful shapes by most of the conventional metallurgical forming methods—forging, swaging, rolling, extruding, drawing, etc.—and by powder-metallurgy techniques. The former operations with the exception of extruding may be carried out either cold or hot (preferably below 600°C). If uranium is hot-worked, suitable precautions must be taken to prevent its contamination and reaction with air at elevated temperatures. This may be accomplished by the use of salt or oil baths or inert atmospheres during heating for forming, and during subsequent annealing if this operation is included.

Fabrication of uranium shapes may be accomplished by either welding or brazing, although the methods are not completely satisfactory. Special precautions are required to prevent oxidation of the uranium. Most of the commercial brazing materials alloy with uranium to form brittle intermetallic compounds at the brazed joint. This circumstance is undesirable with regard to a metallurgically sound joint, and may be prevented to some degree by first electroplating the uranium with silver or nickel.

Shaping uranium by conventional machining methods is relatively easy. Owing, however, to the pyrophoric and toxic characteristics of uranium, it is essential that a suitable lubricant be employed in the machining operations. If the machining is done dry, adequate exhausting facilities should be provided to protect the operator from inhalation of air-borne debris.

10-5. Protection from Corrosion.   Uranium is in general very reactive chemically.   In both liquid and solid metallic form it reacts readily with the atmosphere, water, melting-crucible refractories, and a large variety of organic and inorganic reagents.   Because of the relatively poor corrosion resistance of the solid metal, considerable attention has been given to protective coatings.

Provided the surface of the uranium is adequately cleansed—a requirement that has presented some difficulty—the metal may be electroplated with any one of several common metals, namely, nickel, copper, chromium, iron, tin, brass, silver, and gold.

Alternate to electroplating, protective coatings of other metals may be applied to uranium by roll cladding methods, or, as in the case of fuel elements, the uranium rods may be jacketed in aluminum cans.   Usually, the rods are bonded to the inside surfaces of the cans by immersing in an aluminum-silicon alloy.

10-6. Physical- and Mechanical-Property Characteristics.   The handling of massive uranium metal constitutes a mild health hazard from the standpoint of body radiation and toxic effects attending inhalation and ingestion of air-borne debris (see Chap. 5).   The toxicity is largely related to the high density of uranium (19.1 g/cm^3) and attendant heavy-metal poisoning potentialities.

Uranium metal exhibits three allotropic modifications in the solid state. The α phase, which is stable between ordinary temperatures and 660°C, is orthorhombic and contains four atoms per unit cell.   Between 660 and 760°C, the β phase is stable and is crystallographically a complicated σ-phase-type structure containing 30 atoms per unit cell.   The γ phase is stable from 760°C to the melting point (1133°C) and is body-centered cubic (2 atoms per unit cell). The various allotropic forms of uranium have quite different properties. The α phase is reasonably soft and plastic—particularly at temperatures above 300°C—and because of this many fabrication operations are best conducted at temperatures below 650°C. The β phase is relatively brittle, whereas the γ phase is extremely soft (comparable to lead at ordinary temperatures) and to a degree unsuitable for successful forming under most conditions.

Uranium metal is strongly anisotropic; this is reflected in the electrical and thermal conductivities and other physical properties (see thermal-expansion data in Table 10-2). The mechanical properties are thus

TABLE 10-2. PHYSICAL CONSTANTS OF URANIUM METAL

Density:
  α phase* (25°)................................ 19.04 g/cm³
  β phase (extrapolated to 20°C)................... 18.77 g/cm³
  γ phase* (extrapolated to 25°C)................. 18.86 g/cm³
  Wrought metal................................ 18.5–19.0 g/cm³
Melting point.................................... 1132 ± 1°C
Boiling point.................................... 3900(?)°C
Heat of fusion (estimated)........................ 4.7 kcal/mole
Heat of vaporization............................. 106.7 ± 0.1 kcal/mole
Specific heat at 27°C............................ 6.649 cal/(mole)(°C)
Coefficient of thermal expansion (α phase, X-ray data) per °C:

Direction parallel to crystallographic axis	25–125°C	25–650°C
a........................	21.7×10^{-6}	36.7×10^{-6}
b........................	-1.5×10^{-6}	-9.3×10^{-6}
c........................	23.2×10^{-6}	34.2×10^{-6}
Volume coefficient..........	43.4×10^{-6}	61.6×10^{-6}

Electrical resistivity at 27°C...... 20–40 μohm-cm
Thermal conductivity at 20°C.... 0.060 cal/(sec)(cm)(°C)
Lattice constants, at 25°C:
  α (orthorhombic).............. $a_0 = 2.8541 \pm .003$A, $b_0 = 5.8692 \pm .0015$A
                                       $c_0 = 4.9563 \pm .0004$A
  β (tetragonal)................. $a_0 = b_0 = 10.590 \pm 0.001$A, $c_0 = 5.634 \pm 0.001$A
  γ (body-centered cubic)........ $a_0 = 3.467$A
Allotropic transformation temperatures:

	Upon cooling, °C	Upon heating, °C
$\alpha \to \beta$.........	...	663
$\beta \to \gamma$.........	...	764
$\gamma \to \beta$.........	762	
$\beta \to \alpha$.........	660	

* Calculated.

similarly directionally dependent and vary with the texture, which is dependent upon fabrication and the thermal-treatment history of the metal.

This anisotropic effect is well illustrated by the fact that the yield strength of uranium, when determined in a direction parallel to the rolling direction in "as rolled" sheet, is of the order of 135,000 psi, compared to 104,000 psi when determined 45° to the rolling direction. Also, anisotropy is manifested by the distortion and dimensional instability of uranium when thermally cycled within the temperature limits of the stable α phase.

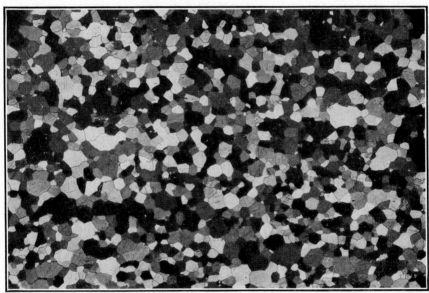

Fig. 10-1. Photomicrograph of high-purity α uranium, 100×, with polarized light, metal extruded, cold-rolled 50 to 75 per cent, and recrystallized at 550°C. (*Courtesy of H. H. Chiswick, ANL.*)

Uranium at ordinary temperatures is also optically anisotropic owing to the orthorhombic structure of the α phase. This circumstance, coupled with the difficulty of satisfactorily etching metallographic specimens, is responsible for the widespread use of polarized light in microscopic examination of uranium and uranium alloys. The structure of α uranium, as observed with polarized light, is shown in Fig. 10-1.

Some typical mechanical properties of uranium, as influenced by selected heat treatments, are given in Table 10-3.

10-7. Alloys of Uranium. Owing to the allotropic modifications existent in uranium, and the differences in solubility of elements in the various allotropic forms, uranium is quite amenable to alloying. Addi-

TABLE 10-3. TYPICAL ROOM-TEMPERATURE MECHANICAL PROPERTIES
OF URANIUM

Heat treatment	Yield stress (0.2% offset), lb/in.²	Ultimate tensile strength, lb/in.²	Elongation, %
As cast..........................	28,000	56,000	4.0
Rolled (1020°F)..................	31,000	96,000	11.0
Rolled (570°F):			
α annealed (1110°F)*..............	43,300	111,100	6.8
β annealed (1290°F)..............	24,550	63,800	8.5
Rolled (1110°F):			
α annealed (1110°F)..............	26,000	88,400	13.5
β annealed (1290°F)..............	24,600	61,700	6.0
Rolled (α):			
γ annealed (1560°F)..............	26,000	57,000	5.0

* Annealed in an inert atmosphere.

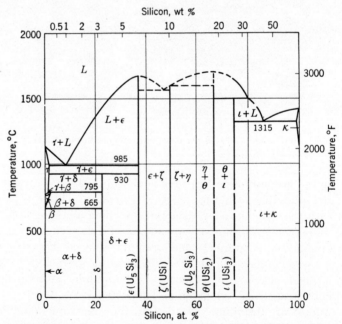

FIG. 10-2. Equilibrium diagram of the system U-Si.   (*H. A. Saller and F. A. Rough*, BMI-1000, 1955.)

tions of one or more appropriate alloying elements followed by suitable heat treatment, will improve mechanical properties at ordinary and elevated temperatures, alter some physical properties, improve corrosion resistance (particularly with solid solution-type alloys) and dimensional stability under irradiation, and improve fabricating characteristics.

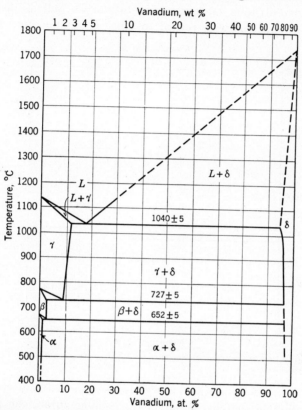

FIG. 10-3. Equilibrium diagram of the system U-V.   (*H. A. Saller and F. A. Rough,* BMI-1000, 1955.)

Similarly to the more common alloy systems, binary uranium alloys fall into several standard types. Owing to the partial covalent-type binding in α uranium, solid solubility is usually very restricted in the α phase.

Uranium alloy systems may exhibit one or more intermetallic compound formations, as illustrated in Fig. 10-2,* the U-Si system. This is typical of the binary uranium alloy systems with beryllium, bismuth,

---

* The constitutional diagrams given here are taken from H. A. Saller and F. A. Rough [1], which should be consulted for full details on these and other U and Th systems.

lead, tin, copper, iron, and oxygen. Other systems, however, do not exhibit compound formation; rather they are characterized by eutectic and eutectoid formations as illustrated in the U-V system (Fig. 10-3).

Alloying is advantageous in the case of highly enriched fuel as a means of bringing about its dilution. The selection of an alloy addition for this purpose is based upon nuclear considerations as well as upon metallurgical considerations with regard to fabrication and other treatment

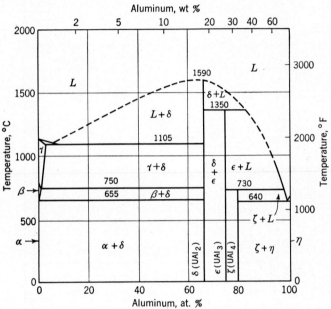

Fig. 10-4. Equilibrium diagram of the system U-Al. (*H. A. Saller and F. A. Rough,* BMI-1000, 1955.)

that may be imposed. Heat removed is also of prime importance, since the lower the fuel content the larger the cooling surface can be and the higher the thermal conductivity. Because of the low thermal-neutron cross sections of aluminum, beryllium, and zirconium (0.22, 0.009, and 0.18 barns per atom respectively), these elements (Figs. 10-4 to 10-6) have received considerable attention as diluents for thermal reactors, aluminum being interesting because of its cheapness, availability, and thermal conductivity. For fast reactors, Fe-U and Cr-U have been proposed, but unfortunately they do not depress the melting point of uranium very much (Figs. 10-7 and 10-8). For liquid-metal-fuel reactors, Bi-U and Bi-Pb-U are available, though the solubility of U is not high at moderate temperatures (Fig. 10-9).

In the case of natural or slightly enriched fuels cooling is less of a problem, but protection of the fuel against accidental corrosion by the

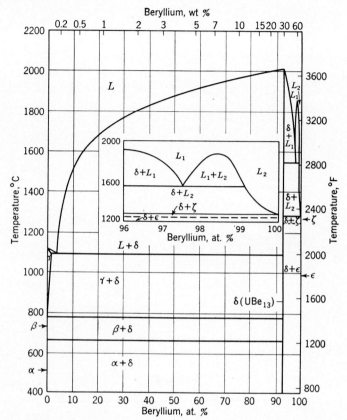

FIG. 10-5. Equilibrium diagram of the system U-Be.   (*H. A. Saller and F. A. Rough,* BMI-1000, 1955.)

coolant is desirable. The water-corrosion resistance of U is greatly improved by Mo (Fig. 10-10), 3 to 6 per cent of Nb (Fig. 10-11), 2.5 per cent of Si, and 1.5 to 5 per cent of Nb + Zr (Chap. 9, Ref. 102).

10-8. Radiation Damage. The changes in properties of a solid material arising from irradiation, or the amount of damage so induced, will depend upon the kind of radiant energy involved and the kind of atomic bonding predominant in the solid. In principle, the property changes are associated with one or more of the following irradiation effects: (1) creation of vacant lattice sites through collision of fission fragments or energetic neutrons with atoms of the solid; (2) introduction of interstitial atoms through displacement of atoms from their equilibrium positions in the lattice; (3) thermal spiking, i.e., creation of lattice oscillations in the wake of fission fragments or knocked-on atoms; (4) extensive ionization of the irradiated solid; and (5) creation of a different element. The role

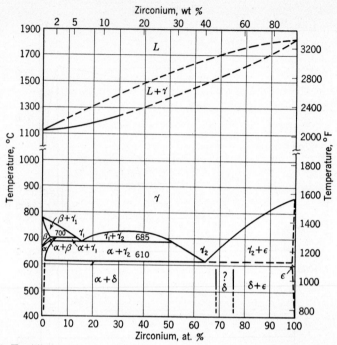

FIG. 10-6. Equilibrium diagram of the system U-Zr. (*H. A. Saller and F. A. Rough,* BMI-1000, 1955.)

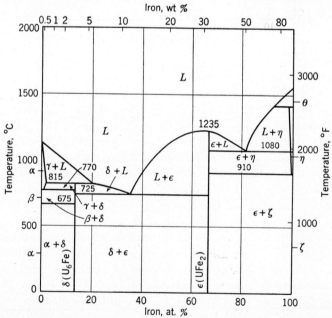

FIG. 10-7. Equilibrium diagram of the system U-Fe. (*H. A. Saller and F. A. Rough,* BMI-1000, 1955.)

of fission-fragment damage to solids and the attendant introduction of fission products is confined mainly to reactor fuel material. The range of fission fragments is only of the order of a few microns. However, the damage that can be produced in fuel may be extensive, as illustrated in Fig. 10-12.

Ionically bonded solids are readily damaged by most types of radiation —β, γ, neutron, and other heavy-particle radiation—through ionization effects. This mode of damage consists in the displacing of electrons to

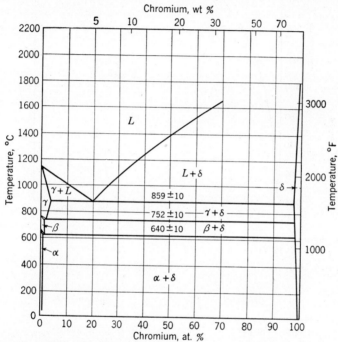

FIG. 10-8. Equilibrium diagram of the system U-Cr.   (*H. A. Saller and F. A. Rough,* BMI-1000, 1955.)

form free ions, usually without change in lattice structure. This damage is generally only temporary, because with time in normal surroundings the displaced electrons will migrate back to form the original configuration. With some solids, such as ordinary and optical glasses, the displaced electrons may become entrapped at sites of lattice imperfections to form so-called "F centers." Such an agglomeration of electrons has little effect on the mechanical properties of the glass, but anisotropy is introduced regarding light transmission that is manifested by change in color of the glass. The F centers can be partially dispersed by subjecting the colored glass to intense white illumination, or by annealing at moderately elevated temperatures. For use in cave or "hot" laboratory cells,

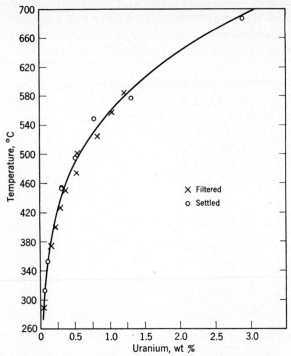

FIG. 10-9. Solubility of U in molten Bi. [*R. J. Teitel, D. H. Gurinsky, and J. S. Bryner, Nucleonics*, **12**(7):14 (*July*, 1954); *courtesy of Nucleonics.*]

TABLE 10-4. BREAKDOWN EXPOSURE OF SOME SOLID INSULATING MATERIALS*

Material	Breakdown exposure, $neutrons/cm^2$
Polyethylene, $(C_2H_4)_n$	$> 10^{19}$
Teflon, $(C_2F_4)_n$	5×10^{18}
Silicone rubber	4×10^{18}
Rubber	1.3×10^{18}
Formvar (polyvinyl formal)	$> 2 \times 10^{18}$
Kel-F $(C_2ClF_3)_n$	10^{18}

* D. S. Billington, "Radiation Damage in Reactor Materials," P/744, UN Conference on Atomic Energy, Geneva, 1955.

nonbrowning glass is available. Such glass contains high-purity cerium which, presumably, undergoes a valence change when the glass is irradiated, and F centers do not form, or if they do, they are immediately dispersed.

Plastics, elastomers, oils, organic compounds, etc., are covalently bonded, and in most cases have bonding energies of only a few electron volts. When irradiated, such materials quite readily suffer bond rupture with consequent formation of free radicals and new compounds. Plastics, for example, may swell, shrink, harden, or soften under irradiation and therefore, along with many other organic materials, are inappropriate for use in reactor construction.   The breakdown exposures of several organic materials of construction are given in Table 10-4.

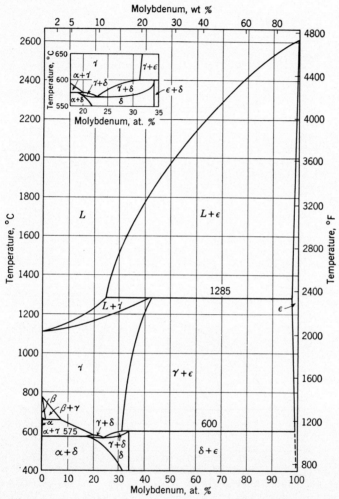

FIG. 10-10. Equilibrium diagram of the system U-Mo.   (*H. A. Saller and F. A. Rough, BMI-1000, 1955.*)

Some organic materials, chiefly those with conjugated double bonds and benzenoid rings, have relatively high bond strengths and are therefore more resistant to radiation damage than those referred to above.   One of the most resistant organic compounds is diphenyl, and because of this circumstance it is employed as coolant and moderator in the OMRE.

Owing to the nature of the metallic bond, ionizing radiation has no effect on metals and alloys except to raise the temperature above ambient during irradiation. However, heavy-particle radiation of sufficient energy, such as fission fragments in the case of fuel material and fast

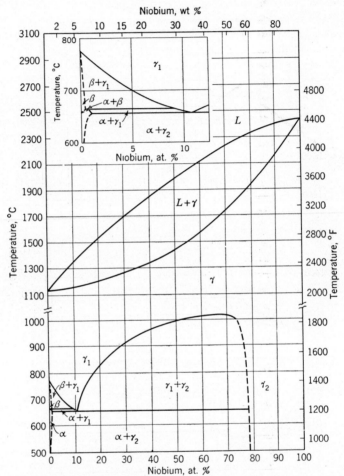

Fig. 10-11. Equilibrium diagram of the system U-Nb. The monotectoid has been subsequently reestablished at ANL at 8 ± 1 wt % Nb and 634 ± 2°C. (*H. A. Saller and F. A. Rough, BMI-1000, 1955.*)

neutrons* or high-speed heavy accelerator particles in the more general case, is the major cause of radiation damage. Although imperfectly

---

* Slow or thermal neutrons, having energies of some 0.023 ev, are not sufficiently energetic to produce damage in metals. To displace an atom of copper from its normal lattice position requires a displacement energy of about 25 ev; for iron, 27 ev; and for nickel, 35 ev.

understood, the attendant changes in properties of a metal or alloy are attributable to the creation of vacancies attending the elastic collision of a neutron, for example, with ions in equilibrium positions in the lattice, the knocked-on atoms eventually occupying interstitial lattice sites. Also contributing are the stresses introduced along the lines of disarranged atoms comprising the path of thermal spikes. The general effect of these structural changes is manifested in changes in the properties of the metal or alloy after irradiation, in a manner similar to changes brought about by solution hardening, and in part by work hardening.

(a)

(b)

Fig. 10-12. Physical damage induced in uranium fuel due to neutron irradiation (fission). (a) Before appropriate metallurgical treatment; (b) after metallurgical treatment. [W. H. Zinn, Nucleonics, 10(9):8 (September, 1952); courtesy of Nucleonics.]

Radiation effects reach a saturation limit, but generally at a magnitude of property changes below those accomplished by solution or work hardening. This saturation limit is further dependent upon other characteristics of the metal. In general, radiation effects are least pronounced in close-packed structures and in materials possessing a low degree of anisotropy.

In the general case, irradiation of metals and alloys produces an increase in hardness and in yield and tensile strengths with an accompanying decrease in ductility, impact strength, and creep rate. Such physical properties as electrical and thermal conductivities are decreased slightly, and the density is decreased very slightly. The magnitude of changes in properties depends in part upon the initial structural condition of the material. The percentage change is greater for metals in the soft or

TABLE 10-5. EFFECT OF IRRADIATION ON HARDNESS OF SOME METALS
AND ALLOYS

| Material | Condition | Irradiation data | | | Rockwell hardness | | |
| | | Calculated nvt | | Tem-pera-ture, °F | | | |
		Slow	Fast		Scale	Initial	Final
Plain carbon steel, SAE-1042	Annealed	1×10^{19}	....	450–500	C	21–23	27–29
	Hardened	....	....	....	C	45–46	46–48
Plain carbon steel, SAE-1095	Annealed	1×10^{19}	....	450–500	C	31–32	33
	Hardened	....	....	....	C	53–55	54–55
Stainless steel, type 304....	Annealed	1.2×10^{19}	....	400–500*	B	72–76	72–76
Stainless steel, type 442....	Annealed	3×10^{19}	....	540*	A	53–57	59–61
Nickel, type A	Annealed	1×10^{19}	....	400–500*	F	64–78	82–91
Monel metal..	....	5×10^{19}	4×10^{19}	70–140	B	81	95
Inconel.......	....	5×10^{19}	4×10^{19}	70–140	B	88	100

TABLE 10-6. EFFECT OF IRRADIATION ON TENSILE PROPERTIES OF SOME
METALS AND ALLOYS

| Material | Irradiation data | | | Ultimate tensile strength, psi | | Yield strength, 0.2 % offset, psi | | Elongation, 2 in., % | |
| | Calculated nvt | | Tem-pera-ture, °F | | | | | | |
	Slow	Fast		Initial	Final	Initial	Final	Initial	Final
Stainless steel, type 304.....	1×10^{19}	....	400–500	92,500	95,500	44,000	48,500	70.8	41.0
Stainless steel, type 316.....	1×10^{19}	....	400–500	90,000	89,500	35,000	36,500	70.0	66.5
Stainless steel, type 316.....	3.7×10^{20}	5.1×10^{20}	70–80	79,500	151,000				
Nickel........	1×10^{19}	....	400–500*	67,500	69,500	22,000	28,000	45	33.5
Monel metal....	4×10^{19}	....	540*	86,000	91,500	......	......	31	29
Inconel.........	4×10^{19}	....	540*	108,500	113,000	......	......	33	29

* In flowing water.

fully annealed condition compared with the cold-worked or thermally
hardened condition.

Typical property changes are illustrated in Tables 10-5 to 10-7 [5].

Many of the property changes in metals arising from irradiation may
be restored by annealing at elevated temperatures. Usually, the anneal-

TABLE 10-7. EFFECT OF IRRADIATION ON DENSITY AND ELECTRICAL RESISTIVITY OF SOME METALS AND ALLOYS

Material	Irradiation data			Average density, g/cm³		Electrical resistivity, µohm-cm	
	Calculated nvt		Temperature, °F	Pre-irradiation	Difference, postirradiation	Initial	Final
	Slow	Fast					
Stainless steel, type 304..	1×10^{19}	....	400–500*	.....	....	80.2 at 28°C	80.4 at 27.8°C
Stainless steel, type 316..	1×10^{19}	....	400–500*	.....	....	77.0 at 28.4°C	78.01 at 28.4°C
	2×10^{20}	3.5×10^{19}	70–140	7.995	-0.005 ± 0.002		
Nickel, type A..........	1×10^{19}	....	400–500*	.....	....	9.8 at 27°C	9.63 at 27.2°C
	2×10^{20}	3.5×10^{19}	70–140	8.894	-0.006 ± 0.002		
Monel metal..........	2×10^{20}	3.5×10^{19}	70–140	8.836	-0.004 ± 0.003		

* In flowing water.

ing temperature for this purpose is lower than that required for restoration of properties after work hardening. Because of the former circumstance, materials irradiated at elevated temperatures exhibit less damage than when irradiated at ordinary or subnormal temperatures. Irreversible damage, such as accumulation of fission products in fuel, creation of new elements, physical cracking or dimensional changes, etc., are unaffected by annealing.

REFERENCES

1. Saller, H. A., and F. A. Rough: BMI-1000, 1955.
2. Teitel, R. J., D. H. Gurinsky, and J. S. Bryner: *Nucleonics*, **12**(7): 14 (July, 1954).
3. Bopp, C. D., and O. Sisman: *Nucleonics*, **13**(7): 28 (July, 1955).
4. Foote, F.: Physical Metallurgy of Uranium and Its Alloys, vol. 9, UN Conference on Atomic Energy, Geneva, 1955.
5. Sutton, C. R., and D. O. Leeser: *Chem. Eng. Progr.*, *Symposium Ser.*, no. 12, pp. 208–221, 1954.

BIBLIOGRAPHY

Radiation Damage

Seitz, F., and J. S. Koehler: The Theory of Lattice Displacements Produced during Irradiation, vol. 7, UN Conference on Atomic Energy, Geneva, 1955.
Billington, D. S.: Radiation Damage in Reactor Materials, vol. 7, UN Conference on Atomic Energy, Geneva, 1955.
Paine, S. H., and J. H. Kittell: Irradiation Effects in Uranium and Its Alloys, vol. 7, UN Conference on Atomic Energy, Geneva, 1955.
Wilson, J. C., and D. S. Billington: Effect of Nuclear Radiation on Structural Materials, Nuclear Science and Engineering Congress, American Institute of Chemical Engineers, 1955.
Siegel, S.: Radiation Damage as a Metallurgical Research Technique, "Modern Research Techniques in Physical Metallurgy," American Society for Metals, Cleveland, 1953.
Calkins, V. P.: Radiation Damage to Nonmetallic Materials, *Chem. Eng. Progr.*, *Symposium Ser.*, no. 12, pp. 28–42, 1954.
Friedemann, H. C.: "Effect of Irradiation on Solids—A Bibliography," Nuclear Engineering Division, Penn-Texas Corp., New York, 1956.

Metallurgy of Beryllium

Stacy, J. T.: Beryllium and Its Alloys, "The Reactor Handbook," vol. 3, McGraw-Hill Book Company, Inc., New York, 1955.
White, D. W., and J. E. Burke: "The Metal Beryllium," American Society for Metals, Cleveland, 1955.

Metallurgy of Zirconium

Dayton, R. W.: Zirconium and Its Alloys, in "The Reactor Handbook," vol. 3, McGraw-Hill Book Company, Inc., New York, 1955.

Hawkins, R. R.: "Zirconium," American Society for Metals, Cleveland, 1953.

Lustman, B., and F. Kerze, Jr. (eds.): "The Metallurgy of Zirconium," McGraw-Hill Book Company, Inc., New York, 1955.

"Zirconium—A Bibliography of the Unclassified Literature," TID-3010, Technical Information Service, AEC, Oak Ridge, Tenn.

Metallurgy of Thorium

Keeler, J. R.: Thorium and Its Alloys, in "The Reactor Handbook," vol. 3, McGraw-Hill Book Company, Inc., New York, 1955.

Carlson, O. N., et al.: The Metallurgy of Thorium and Its Alloys, vol. 9, UN Conference on Atomic Energy, Geneva, 1955.

Metallurgy of Plutonium

Jette, E. R., and A. S. Coffinberry: Plutonium and Its Alloys, in "The Reactor Handbook," vol. 3, McGraw-Hill Book Company, Inc., New York, 1955.

Coffinberry, A. S., and F. H. Ellinger: The Intermetallic Compounds of Plutonium, vol. 9, UN Conference on Atomic Energy, Geneva, 1955.

CHAPTER 11

THERMAL-STRESS ANALYSIS AND MECHANICAL DESIGN

By Alfred M. Freudenthal

GENERAL CONSIDERATIONS

11-1. General Aspects of the Problem of Mechanical Strength in Nuclear Reactors. At all power levels of operation of a nuclear reactor the mechanical strength of the materials used in its construction is of vital importance. It is generally true with respect to any apparatus, equipment, or plant that no matter how far advanced and how well developed the principles on which its operation is based, actual and efficient construction is possible only if and when materials of adequate mechanical performance are available. The most serious limitations in the development of equipment for optimal performance are frequently those imposed by the mechanical properties of the materials used in its construction. In the case of nuclear power reactors the most stringent of these limitations are due to the need for neutron economy, in conjunction with the requirements concerning the mechanical properties at the desired high operating temperatures. Very few of the conventional engineering metals are acceptable in the construction of nuclear reactors because of their high specific absorption of slow neutrons. The development of structural materials with good high-temperature mechanical properties as well as low neutron-capture cross section is needed before the multiple purpose or power-breeder reactor can be fully economical. The stringent requirements with respect to neutron losses in breeder reactors are difficult to reconcile with the practical design requirements of power reactors and with the use of such structural materials and heat-transfer fluids as are so far available or economical.

The most significant difference in the stress analysis and mechanical design of nuclear power reactors and of more conventional heat-transfer equipment is the emergence of the thermal stresses as the dominant design feature, particularly with respect to the solid fuel, the cladding, and the circulation and heat-exchanger system, in all of which severe temperature gradients exist normal to the direction of coolant flow.

538

While thermal stresses may also be significant in the design of the structural parts, the controls, and the shielding, where thermal stresses are due, in general, to temperature gradients in the direction of coolant flow, they usually affect, but no longer govern, the design of these parts, which is determined primarily by the load stresses. In general, mechanical design problems in the field of reactors require a more thorough knowledge of the behavior of materials and of the interaction between the stresses developed in the reactor parts, their design for adequate strength, and the selection of the material than does the design of conventional equipment or structures. At the present stage of development in which the possible reactor materials are either rare or costly or difficult to form, the principal criterion for their selection for the various reactor parts is their over-all nuclear and mechanical performance in service rather than their cost.

11-2. Materials of Reactor Construction. While there is little amplitude in the selection of fuel, since any reactor must contain either one or the other of the fissionable isotopes of uranium or plutonium, the investigation of the thermal and mechanical properties of the possible fuel elements is of considerable importance, particularly in combination with the effective moderators.

In general, uranium metal can be reasonably well rolled, forged, and drawn into any number of shapes; it can also be machined using proper lubricants and coolants to prevent burning of the metal. The main fabrication problem is its chemical reaction with the atmosphere, which makes welding or brazing rather difficult, since it requires a vacuum or a protective atmosphere. It has a very poor corrosion resistance in most media. Hence protective cladding is desirable, for which a number of metals are available.

Thorium, a secondary reactor fuel, also fabricates very well, since it is softer than uranium. Because of its high ductility it can be easily forged, rolled, drawn, extruded, and readily machined. Its corrosion resistance, though poor, is better than that of uranium; its weldability, however, poorer. Since, like uranium, thorium is neither corrosion-resistant nor adequately strong, other metals must be used for cladding and support.

The hardness and tensile strength of uranium can be appreciably raised by cold-working; considerable differences exist, therefore, in the properties of the annealed and the cold-worked metal. Uranium alloys containing small amounts of aluminum, zirconium, molybdenum, or chromium in solid solution show improved strength and corrosion resistance. Uranium ceramics, such as uranium oxide (urania) in solid form or dispersed in a matrix of stainless steel, zirconium, or aluminum, are also used as fuel elements.

In comparison with the anisotropic uranium, the properties of thorium are isotropic. Alloys and intermetallic compounds of thorium are being developed that are appreciably stronger than pure thorium.

The main objective in fuel-element design is to produce an optimal combination of life and specific power. The life of the fuel element is determined by operating temperature, heat flow, corrosion, and inter-diffusion of fuel and cladding, as well as by the tolerable burnup, which is of the order of 2 per cent. While massive metallic uranium clad in aluminum or stainless steel has been predominantly used in existing reactors, more recent designs take advantage of the fact that finely divided fuel elements have lower thermal stresses and provide more heat-transfer surface per unit of core volume and improved heat-transfer rates, particularly in water- and gas-cooled reactors. Since ease of fabrication and low cost are desirable features, elements of the simplest geometries consistent with required performance and adequate mechanical stability should be selected, such as parallel flat, curved, or corrugated thin plates, thin-walled tubes, slender rods, helically twisted strips, or small spheres.

Effective neutron moderators are limited to elements of low atomic number and weight. Production difficulties and nuclear, thermal, and mechanical performance of the reactor limit the choice of solid moderators in the order of increasing cost to carbon (graphite), beryllium oxide (beryllia), and beryllium. The principal advantages of graphite are its low cost, its abundance, the well-established techniques of purification and fabrication, and its good mechanical properties. Both beryllia and beryllium are preferable on the basis of their nuclear properties but require considerable development with respect to commercial production and fabrication.

Beryllium, which has a high melting point, low capture cross section, and good moderating properties was recognized very early in reactor development as a very promising reactor material. Its main disadvantage, so far, is its toxicity and its brittleness, which make the fabrication of beryllium structural elements difficult and costly. In spite of this a number of shapes have been fabricated. Some of the difficulties have more recently been overcome by the production of more ductile beryllium alloys. Welding or brazing of beryllium is possible, though complicated by the rapid oxidation of the metal in the atmosphere; machining is difficult. Its corrosion resistance is better than that of uranium and thorium, but still unsatisfactory. Processes of electroplating with atmospherically more stable metals such as copper, nickel, iron, or silver are therefore being developed.

Because of its considerable strength at elevated temperatures beryllium is used not only as a moderating material but also as a structural and cladding material within the reactor core.   Three other metals are in the same category for such use because of a similarly low capture cross section for thermal neutrons (<0.5 barn per atom): aluminum, magnesium, and zirconium.

Of these, aluminum and magnesium are well-established structural metals, the usefulness of which in power reactors is, however, limited by their low melting point and the consequent rapid loss of strength with increasing temperature.   In designs requiring better performance at moderately elevated temperatures, either metal can be alloyed with other metals, mostly at some expense to neutron economy.   Thus copper-bearing aluminum alloys and alloys of magnesium with aluminum, zinc, and zirconium might be used.

Zirconium, with its relatively high melting point and thermal stability, good corrosion resistance, and satisfactory mechanical properties is, at present, a preferred structural and cladding material.   In a highly purified state it is very ductile and can be easily fabricated by any conventional fabrication method in a wide variety of shapes.   Its machining properties resemble those of aluminum and it can be easily welded in a protective atmosphere (helium).   Brazing is difficult.   It oxidizes to some extent, but can be protected by electroplating.   Alloys of zirconium with various metals such as copper or magnesium may become of considerable importance as reactor materials.

Many more structural and cladding materials become available when a higher capture cross section can be tolerated, as in a power reactor with enriched fuel.   Table 11-1 shows the principal metals grouped according to their capture cross sections for thermal neutrons.

TABLE 11-1. CAPTURE CROSS SECTIONS OF STRUCTURAL METALS FOR THERMAL NEUTRONS AT ENERGY LEVEL 0.025 EV

Group A, <0.5 barn		Group B, 0.5–5 barns		Group C, >5 barns	
		Mo	2.4		
C	0.0045	Fe	2.5	Ti	5.8
Be	0.009	Cr	2.9	Mn	12.6
Mg	0.059	18-8 stainless	2.9	W	19.2
Si	0.13	Cu	3.6	Ta	21.3
Pb	0.17	Inconel X	4.1	Co	34.8
Zr	0.18	Monel	4.2	B	750
Al*	0.22	Ni	4.5	Cd	2400
		V	4.8		

* Structural aluminum alloys contain small percentages of Mn, by which their cross section is almost doubled.

All structural metals that are at present used in nuclear reactors have neutron-capture cross sections larger than that of beryllium.   Even aluminum, which is one of the less objectionable metals from the nuclear point of view and has therefore been used in reactors from the very beginning of reactor development, has a capture cross section more than twenty times larger than that of beryllium.

In general, it can be said that, at operating temperatures beyond those of aluminum, magnesium, and zirconium and its alloys (700 to 800°F), iron-base alloys, such as the austenitic stainless steels 321 and 347, are used in the temperature range up to 1500°F; nickel-base alloys, such as Inconel and Hastalloy, in the range from 1500 to 1800°F; and cobalt and mixed-base alloys in the range from 1600 to 2200°F. Because of their strength at elevated temperatures and their good corrosion and oxidation resistance, the austenitic stainless steels are widely used in reactors for cladding, piping, and auxiliary reactor components and for construction of heat exchangers. The higher neutron absorption as compared with zirconium is compensated by the superior strength properties. While in reactor design the cost of the material is usually a secondary consideration, it should, however, be noted that the costs of the nickel-base and cobalt-base alloys are about four and eight times, respectively, that of stainless steel. The high neutron-capture cross section of titanium has restricted its use in thermal reactors.

Recently developed materials for service at very high temperatures include metal alloys produced from powder, such as sintered aluminum; ceramics, such as oxides of beryllium, magnesium, aluminum, and zirconium; carbides of beryllium, titanium, and silicon; and silicides, particularly of molybdenum. The strength, at 900°F, of wrought sintered aluminum parts is far superior to that of strong cast or wrought alloys, which permits the utilization of aluminum at temperatures conventionally considered as excessive for aluminum alloys. In general, metals and compounds prepared by powder metallurgy may find increasing use in reactor construction. The ceramics have melting points of the order of 2000 to 2800°C, but very poor ductility and, therefore, relatively poor resistance to transient thermal stresses ("thermal shock" and "thermal fatigue"). To improve this performance and to achieve both high strength at high temperatures and adequate ductility and thermal conductivity to provide resistance to thermal shock, compounds of ceramics and metals ("cermets," "ceramals," and "intermetallics") are being studied; of these silicon carbide or zirconium boride with various metal (Mo, Co, W, Ni) and alloy additions are real "cermets," with structures made up of a brittle (ceramic) phase and a ductile (metal) phase, while molybdenum disilicide and chromium carbide are combinations of intermetallic compounds. The "cermets" show, in general, better resistance to thermal shock and thermal fatigue, the intermetallics better creep and fracture strength at temperatures as high as 2000°F.

Because of their low ductility both types of material are highly notch-sensitive, especially at room temperature, and present serious problems in the design of structural connections.   Moreover their impact resistance is inferior to that of conventional high-temperature alloys.

The possibility of constructing reactors that will operate at temperatures in excess of 800 to 1000°C will most probably depend on the further development of high-temperature materials and of their fabrication processes.

The development of structural materials which meet the requirement of low capture cross section is, in itself, a task of major magnitude.   It is safe to assume that the production of any structural material that is not now commercially used involves, in general, considerable difficulty. The task is further complicated by the fact that materials undergo changes in their physical properties when subjected to neutron bombardment and to secondary radiation of the intensities existing in a nuclear reactor (radiation damage; see Chap. 10).   The practical information available in this respect is still limited, particularly since, for design purposes, it is not sufficient to study the radiation damage of pure metals alone.   As most engineering materials are alloys, the specific properties of which frequently depend on one of the minor alloying elements, the particular alloys themselves and the effect of nuclear radiation on each of the alloying elements must be investigated, as well as the effect of minor alloying elements on the nuclear properties of the various alloys.   Thus, for instance, the common alloying elements boron, cobalt, and manganese have such high capture cross sections that the presence of minute amounts of those elements appreciably affects the capture cross section of the alloy.   Significant properties of the principal reactor materials are summarized in Table 11-2.

The principal property of the materials for controls is a high cross section for the capture of thermal neutrons.   Adequate strength, ductility for fabrication, and corrosion resistance to the reactor coolant are also required.   Boron, cadmium, and hafnium meet most of these requirements.   Boron and cadmium have been widely used; hafnium, with mechanical properties resembling those of zirconium, has not yet been available in sufficient quantities.   Some of the rare earths have very high cross sections and might be developed for use as control materials.

The reactor operates at a maximum temperature T_R, taking in the cooling fluid at a temperature T_i and discharging it at an exit temperature T_e.   Its efficiency of operation depends on the temperature T_e which, of course, can be higher the higher T_R is.   The temperature T_R is bound to be significantly higher than any temperature encountered in subsequent equipment; thus the reactor is usually the critical unit in the complete plant.   The power level of the reactor varies with the load as well

$$\text{TABLE 11-2. PRINCIPAL PHYSICAL PROPERTIES OF SEVERAL REACTOR MATERIALS*}$$

Property	Temperature	Fuels		Graphite	Structural materials					
		Uranium	Thorium		Aluminum	Beryllium	Magnesium†	Zirconium	Molybdenum	Stainless 18-8
Density ρ, g/cm³	Room	19.1	11.7	2.2	2.7	1.85	1.74†	6.5	10.2	7.9
Melting point, °C	......	1133	1690	3700	660	1350	650	1850	2620	1420
Coefficient of linear expansion α, $\times 10^6$ per °C	Room	22	11.2	2–3‡	23.9	12.4	26	4–10‡	5.5	16.7
Thermal conductivity k, cal/(cm)(sec)(°C)	Room	0.06	0.09	0.3	0.53	0.38	0.35	0.05	0.53	0.04
Elastic modulus E, $\times 10^{-6}$ psi	Room	30	10	1.0	10	42	6.5	12	45	30
Poisson ratio ν, $\times 10^{-3}$ psi	Room	0.23	0.26	0.20–0.33	0.3	0.03	0.35	0.33	0.31	0.38
Yield stress σ_0, $\times 10^{-3}$ psi	Room	25	27	......	15–21	33	15–33	18	57–85§	45§
	600°F	18	12		1.5	15	5	10		30§
Tensile strength σ_t, $\times 10^{-3}$ psi	Room	90	38		16–24§	45–80§	25–45§	35–85§	100–250§	90§
	600°F	32	22		2.5	40	8	16–45§	60–80	65§
Elongation at fracture in 2 in., %	Room	14	40		35	2	5–15§	16–31§	5–20§	55§
	600°F	43	38		90	17	23	60		40§
Thermal-neutron capture cross section σ_a, barns/atom	Room	......	......	0.005	0.215	0.009	0.059	0.18	2.4	2.9

* See also App. F.
† Characteristic alloy.
‡ Depending on orientation.
§ Depending on heat-treatment and cold-working.

as periodically about any controlled level as a result of minor fluctuation of the neutron flux (see Chap. 12), and the operating temperature of the reactor at every point as well as the exit temperature of the coolant are therefore subject to similar variation. The heat generated in the fuel and moderator produces temperature gradients and thus thermal stresses transverse to the direction of flow of the coolant; the temperature gradients along the coolant circulation tubes give rise to thermal stresses in the direction of the coolant flow. All thermal stresses are necessarily subject to fluctuations because of the periodically fluctuating power and temperature levels, and particularly because of the major transient changes in conditions at start-up, shut down, or change in power level.

The heat removal from a reactor depends primarily on mass flow. Since the heat capacity and the density of the coolant are fixed for the selected coolant, the reactor designer must determine the optimal cross-sectional areas for coolant flow, optimal velocity, and temperature rise. Obviously, this temperature rise will be larger the smaller the reactor, thus the smaller the flow rate.   A large temperature rise necessarily leads to steep thermal gradients which, in turn, cause high thermal stresses. These stresses will be especially severe under the transient conditions of changing power level during which the metal adjacent to the coolant changes temperature rapidly, while the main part of the reactor core follows with considerable delay.

Because of the severity of the temperature gradients one of the main tasks of the designer is to minimize the resulting thermal stresses.   This is usually attempted in two different ways: by selecting structural materials of optimal performance with respect to thermal stresses (see Sec. 11-19), or by making the individual elements small or flexible, and using hinged or flexible joints.   The strength of the joints must, however, be sufficient to sustain vibration stresses and mechanical shock, wherever necessary.

Limitations on the power output and efficiency of nuclear reactors therefore result from the limitations on the thermal stresses due to temperature gradients that can be safely sustained by the various parts of the reactor, as well as on the maximum level of operating temperatures in relation to the melting temperature of the materials used in the reactor. The higher the operating temperature in relation to the melting point, the lower the strength of the metal and the more important and rapid the changes in its mechanical performance due to the corrosive attack of the environment and to changes in the microstructure.   Both phenomena affect ductility and strength of the material, the first by intergranular cracking and loss of effective metal as a result of chemical reaction on the surface, the second by embrittlement, strain-aging, and other effects of precipitation and phase change.

For use in nuclear reactors, therefore, materials must meet the following requirements:

1. Low capture cross section for thermal neutrons.

2. High melting point T_m and adequate mechanical strength and ductility at the operating temperature level and stress. Unfavorable nuclear properties might be compensated by reduced volume of the material as a result of its high strength.

3. High resistance to radiation damage.

4. High resistance to corrosion, as well as high stability of the microstructure at elevated temperatures.

Moreover, such materials must be suitable for use in the actual construction of mechanically stressed parts in general and of heat-transfer equipment in particular: they must have ductility adequate for fabrication of structural shapes, high elastic modulus and low creep rates to limit the total deformation, and high thermal conductivity to ensure effective heat transfer.

It has, in general, not been easy to develop structural materials for service at high temperatures, even without the additional nuclear requirements. These requirements severely limit the selection of structural materials out of the already limited number of materials that have been developed for general high-temperature service. On the other hand, the desire for compactness of the reactor to ensure neutron economy as well as small volume and weight, particularly in mobile reactors, requires high rates of heat transfer. It leads, therefore, to high thermal and mechanical stresses. Moreover, the difficulty, if not impossibility, of repairs inside an operating reactor and the serious consequences of even minor mechanical failures make it imperative to design the reactor with the greatest possible safety against mechanical failure of any of its components. These conflicting requirements put a heavy burden on the technologist developing materials for nuclear reactors as well as on the designer using them.

11-3. Stress and Strength Analysis. Design of the components of a nuclear reactor for mechanical strength involves the analysis of the stresses within the reactor parts produced by loads, temperatures, and temperature-gradients, and their comparison with the strength of the parts at the critical locations at which the stresses attain their highest intensity. Stresses and strength characteristics are both functions of temperature and of time and should, therefore, be evaluated in terms of their temperature-time histories.

Stress analysis and strength analysis must be based on certain abstractions and idealizations of the behavior of the real reactor parts. For stress analysis this idealization or simplification refers to the operational requirements imposed upon the various reactor parts, to the design features required by functional considerations, and to the stress-strain-time relations introduced as representing the deformational response of the material to applied forces under specific conditions of temperature.   For the strength analysis it refers to the modes of failure and the failure criterion under the specific conditions of stress and temperature imposed by the operational requirements on reactor parts and connections of parts of various geometrical configuration.

The stresses in the various parts of a reactor are due to one or a combination of the following loading conditions:

1. Mechanical forces due to gravity, imposed service loads, and reactions in the supports and connections due to the action of the forces and to the restraints of the system.

2. Primary coolant pressure, usually assumed as 15 to 25 per cent in excess of the normal steady-state operating pressure, or 5 to 10 per cent in excess of the maximum surge pressure.

3. Thermal gradients associated with steady-state heat flow.

4. Thermal gradients due to heat generation in fuel elements and to γ radiation in other parts.

5. Transient thermal gradients resulting from normal operation (power changes, including warm-up and shutdown), as well as from emergency operation (rapid power change, emergency shutdown, and emergency cooling following loss of pumping power).

6. Thermal reaction forces due to restrained expansion or contraction, as well as to interaction of materials with different coefficients of thermal expansion.

7. Mechanical vibration and shock, the latter particularly under emergency conditions (accidents, natural forces, military action).

Although during normal operation the system is subject to relatively narrow cyclic variations of pressure and temperature and heat losses along the surface, these conditions may in general be considered to be steady-state conditions.   The start-up and shutdown cycles, however, both normal and emergency, must be considered as repeated transient thermal stressing, the number of repetitions being an inverse function of the severity of the thermal stresses.   Thus with a total of less than 10^5 transient temperature cycles, not more than 10^4 will probably exceed in severity the normal cycle associated with power generation, and of these less than 10^2 might be assumed to represent conditions of emergency cooling with loss of pressure.

Conventional stress analysis is based on the assumption of a time-independent, linear, and instantaneously reversible deformational response of the material (Hooke's law).   This assumption leads to the linear differential equations of the classical theory of elasticity, the solutions of which describe the field of stresses and small strains in terms of the applied loads and external conditions. Time effects appear only as the result of time-dependent loading and transient temperatures. The criterion of failure that is generally associated with this analysis is a function of the invariants of the stress tensor, a critical value of which is assumed to define failure. The failure condition is usually based on the true yield-point stress or on a designated limit of elastic behavior, expressed in terms of the second invariant of this tensor (Mises yield condition), and is thus defined as impending deviation from elastic behavior.

The inadequacy of the assumptions underlying the above procedure with respect to the design of structural parts for service at elevated temperatures has long been appreciated by designers. From the earliest observations of the behavior of metals at elevated temperatures, the existence of a time-dependent irreversible component of the deformation was in evidence. The importance of this phenomenon of "creep" with respect to the performance of structures at elevated temperatures is now generally recognized, and the development of materials for high-temperature service has been guided mainly by the requirement to reduce the creep rate at the expected temperature and loading conditions to acceptable limits, determined by total deformation and period of service. Such a failure criterion is no longer time-independent but is defined in terms of a critical stress producing a creep rate which, in the course of the expected service life of the designed part, will lead to a limiting total permanent deformation compatible with the safe operation of the structure. An interrelation is therefore established among stress level, anticipated service life, and creep rate, as a result of which the failure criterion becomes a function of the service life; since the critical deformation can be attained in a short time under a high stress intensity, or in a long time under a low stress intensity, the designer has to specify the desired service life. The effect of time increases with increasing temperature of operation, partly because of the disproportionately rapid increase of creep rate with temperature, and partly because of the increasing importance of the decrease with time of the fracture strength of the creeping material ("creep rupture"). The failure criterion under these conditions must therefore be expressed in the *dual* terms of (1) a stress producing a limiting creep deformation, and (2) a stress causing fracture, both defined with respect to the anticipated service life of the designed structure in terms of an invariant of the stress tensor. The lower of these two stresses represents the design criterion of failure.

The recognition of the importance of the deviation from elastic behavior of real materials at elevated temperatures and the consideration of creep and creep rupture as the dominant characteristics of high-temperature design have, however, so far been limited to the definition of a failure criterion or failure condition, and thus to the strength analysis of structural elements.   In stress analysis, however, particularly the analysis of thermal stresses, the assumption of linear elasticity has generally been maintained in spite of its obvious conflict with the assumptions of the strength analysis.   The principal reason for this reluctance to replace the assumption of linear elasticity by a more realistic idealization is the mathematical difficulty of solving the predominantly nonlinear differential equations or the linear equations with variable coefficients resulting from the introduction of stress-strain relations containing inelastic strain components.

The assumption of elasticity of the structural materials at the high operating temperatures desirable for nuclear power reactors can be justified only as a very rough approximation.   It can easily be shown that the introduction of time-dependent inelastic strain components representing the observed creep, while only moderately affecting the stresses due to the applied loads, reduces the level of thermal stresses very significantly.   Since the relative importance of load stresses in comparison to the thermal stresses in structural parts operating at elevated temperatures decreases with increasing operating temperatures, creep is of considerable practical importance in the analysis of thermal stresses in nuclear reactors.   A rational evaluation of the level and distribution of thermal stresses requires, therefore, the introduction of an idealization of the mechanical response of real materials at elevated temperatures that includes both the elastic and the creep component of the deformation.   This procedure leads to differential equations for the stresses which can be solved with a reasonable effort.   The simplest such idealization is the linear viscoelastic response combining the linear elastic (Hookean) relation between stress and strain with the linear viscous (Newtonian) relation between stress and strain rate to produce a material that responds nearly elastically to rapidly applied loads but creeps under loads of longer duration (Maxwell body) [1].   The temperature sensitivity of creep should be considered by introducing the temperature-dependence of the coefficient of viscosity.   In comparison the much less pronounced temperature-dependence of the elastic constants may usually be neglected.

Such a simple idealization is, however, not quite adequate, since for real materials the relation between creep rate and applied stress is usually nonlinear and the creep rate itself is a function of time.   Moreover, a definite temperature-dependent yield stress frequently exists, which would require a more complex idealization.   The consideration, even in the simplest form, of all significant types of mechanical response at elevated temperature results in a combined viscoelastic-viscoplastic type of material with temperature-dependent parameters, leading to nonlinear differential equations of considerable complexity.   For specific boundary-value problems solutions may be obtained only by elaborate numerical procedures. In the few attempts to solve simple inelastic thermal-stress problems, linear viscoelastic [2] and ideal plastic materials [3] have, so far, been considered.

At present, high-temperature design is based almost exclusively on elastic solutions of thermal stress fields. As long as effective methods for the solution of the differential equations for thermal-stress fields in inelastic bodies have not been developed, the elastic thermal stresses are regarded as the best available approximation to the real thermal stresses, since the latter will, in general, be substantially lower. The elastic thermal stresses under a gradually increasing temperature difference represent an upper limit of the real stresses. Their use in design will, in all cases in which the temperature change is not reversed, result in an increase of the safety of the designed structure beyond its specified nominal level.

However, the neglect of temperature reversal and of the effects of the temperature history on the momentary level of thermal stresses may result in differences in sign as well as in intensity between the computed elastic and the actual stresses. Temperatures varying between a zero reference level and a maximum level may thus be accompanied by thermal stresses *alternating* between tension and compression, while the elastic solution would indicate only a change between zero and a maximum, parallel to the temperature change.

Moreover, the lack of differentiation in elastic stress analysis between the *sources* of stress, i.e., between stresses due to loads and stresses due to restraints of deformation, produces a distorted picture of the relative effects of the two types of stress. To obtain the critical design stresses in elastic analysis, "load stresses" and "deformation stresses" are simply added. Such superposition disregards the fact that load stresses in real materials are considerably less sensitive to time and temperature than the highly time-sensitive deformation stresses [4]. Thus, for instance, the simple addition of elastic hoop stresses due to fluid pressure to the elastic thermal stresses in a coolant-carrying tube inside a heat exchanger may produce a completely unrealistic picture of the real maximum stresses after a certain period of operation.

While, from the point of view of structural safety, the use in design of the elastic thermal stresses may under certain simple conditions be admissible, economy makes it desirable to use stresses that are nearer to reality, particularly in nuclear reactors, in which the excess volume of unnecessary materials and the neutron losses they cause are infinitely more important than the cost of overconservative design. Since the capture cross section of most structural materials except carbon and beryllium is high, their volume inside the reactor must be a minimum. To minimize the investment per unit output, high power and high heat generation are desirable. This means high temperatures, thermal stresses, and coolant velocity, and therefore high coolant maximum pressure. Such increase of pressure in turn requires thicker tube walls, and thus a larger volume of objectionable materials within the reactor.

Thus the balance among the nuclear aspect, the heat-transfer aspect, and the mechanical-strength aspect of the design of a power reactor is extremely delicate and requires very careful analysis. Any solution can necessarily be only a compromise among the conflicting requirements. The evaluation of the thermal and operating stresses, and the rational selection of adequate values of strength to withstand these stresses and of the safety factor to be used represents a problem of considerable complexity. Satisfactory solutions have not yet been developed, even for conventional heat-transfer equipment, because of the difficulties of evaluating actual thermal stresses under the complex conditions of the high and somewhat fluctuating temperatures of steady operation, and of the transient temperature of the starting and shutdown periods, as well as of determining the actual strength of the material under the operating conditions.

11-4. The Heat-transfer Problem in Relation to Stress Analysis. The thermal stresses occurring in the various parts of a nuclear reactor are the results of the temperatures and temperature gradients produced by the transfer of heat. Since economical operation depends on high heat generation, the thermal stresses will necessarily increase with efficiency of design. Their analysis requires the knowledge of the temperature distribution and its variation with time, due to changes in sources of heat and in power level during operation, and during start-up and shutdown.

The higher the maximum temperature reached by the coolant, the more mechanical and electrical power can be produced. Thus, a prime goal of strength analysis is to determine, in conjunction with considerations of heat transfer, fluid flow, and the power cycle, the maximum possible and the optimum outlet temperatures of the coolant. Such analysis is, however, necessary all along the travel of the coolant, since the maximum temperature gradients frequently do not occur at the maximum coolant temperature.

The differential equations governing heat-transfer problems and methods for the computation of temperature distributions in reactor elements of simple geometrical shapes have been presented in Chap. 9. With respect to those equations and their solutions, it should be realized that the pronounced time- and temperature-dependence of the deformation of reactor materials at high operating temperatures prevents, in general, the build-up to their full elastic level of the thermal stresses associated with steady-state temperature distributions. The higher the operating temperature, the more rapidly will the intensity of those stresses decay toward zero. Severe thermal stresses can therefore be produced only as the result of the transient temperature distributions associated with routine or emergency start-up or shutdown, or changes in power level. In solving the heat-transfer equations for these conditions, it is necessary to consider the surface resistance due to the existence of a boundary layer and expressed by the surface heat-transfer coefficient (Chap. 9), as well as the temperature-dependence of the conductivity.

The neglect of surface resistance in the heat-transfer equations is equivalent to the unjustified assumption that the surface temperature itself can be controlled, rather than the temperature of the heat-transfer medium (coolant). The surface resistance produces a delay in the occurrence of the maximum surface temperature, and the maximum temperature difference between surface and interior is reduced. The thermal stresses caused by this difference can therefore not build up to the level associated with zero surface resistance, and since the maximum stress is also delayed with respect to the change of temperature of the heat-transfer medium, it is further reduced by the creep of the material during the delay period. Neglect of the fluid heat-transfer resistance leads therefore to overconservative design [5].

An increase of the thermal conductivity with temperature reduces the transient thermal stresses developing during heating of the coolant but increases the stresses due to cooling of the ambient. Use of the initial conductivity underestimates the transient thermal stresses during heating and overestimates them during cooling of the ambient if the thermal conductivity decreases with temperature. The variation with temperature is given in App. F.

THEORY OF ELASTIC STRESS

11-5. Thermal Stresses. Thermal stresses are caused by (1) the resistance of the structural element or of the continuum to nonuniform expansion or contraction, due to temperature differences and gradients; (2) a resistance, along a boundary of a structure or a continuous body, restricting free displacement of this boundary due to uniform or nonuniform heating or cooling.

Thermal stresses therefore require either temperature gradients producing nonuniform volume changes incompatible with the conditions of continuity of the system ("compatibility conditions"), or restrictions of the free over-all expansion or contraction of the system by fixed or by force-resisting boundaries; frequently both conditions exist simultaneously. The source of the thermal stresses is therefore not an external force, the intensity of which is independent of the deformational response of the system, but an internal or "resistance" force, the intensity of which is determined by this response. Hence thermal stresses are "deformation stresses." The time- and temperature-sensitivity of the stress-strain relation on which their intensity depends will necessarily be reflected in the time- and temperature-variations of the thermal stresses. The computed intensity of these stresses is thus determined by the assumptions introduced concerning the stress-strain–strain-rate relation of the material.

Conventional thermal-stress analysis is based on the assumption of time- and temperature-independent linear elasticity expressed by a linear stress-strain relation with temperature independent parameters. It has already been pointed out that at the same time the inelasticity of real materials at elevated temperatures is recognized by the introduction of a limiting rate of creep and of a time- and temperature-dependent fracture stress as criteria of design and that this contradiction between the bases of the stress analysis and of the strength analysis results in unrealistic design procedures. However, the simplicity of the assumptions of linear elasticity has made the solution of thermal-stress problems possible. In spite of its increasingly recognized lack of reality, particularly at very high temperatures, elastic thermal-stress analysis will, for some time still, remain the standard procedure.

Since the temperature distribution must be known before the thermal-stress problem can be set up, the solution of the heat-transfer equations under the imposed boundary conditions and conditions of heat generation must precede the solution of the thermal-stress problem. The temperature distribution is introduced as the "load functions" into the differential equations of the thermal-stress problem. Because of the complexity of the solutions for all but the simplest problems, and particularly for "transient" problems, integration of the thermal-stress equations for specific temperature distributions and boundary conditions, and computation of numerical values of the stress components represent, in general, a computational effort of considerable magnitude, requiring a thorough familiarity with the mathematical theory of elasticity and with the general analytical and numerical methods that have been developed for the integration of elastic boundary-value problems [6]. Only a general outline of the procedures for the determination of the thermal stresses can be presented in this chapter. They will be illustrated by problems of particular simplicity, such as the circular cylindrical rod, the thick-walled tube, the sphere, and the long narrow plate. Fortunately, these simple elements are of considerable importance in reactor design, as illustrated by Fig. 11-1 in which they are shown schematically as reactor elements.

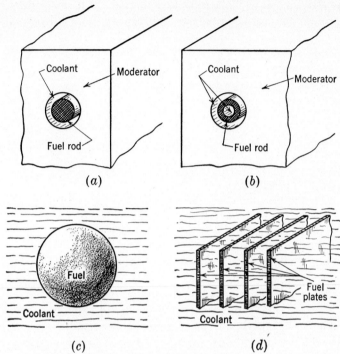

FIG. 11-1. Simplified reactor elements: (*a*) solid fuel rod; (*b*) tubular fuel rod; (*c*) homogeneous fuel, graphite sphere; (*d*) cladded fuel plates.

11-6. Equilibrium and Compatibility Conditions. Considering a volume element $(dx_1 \, dx_2 \, dx_3)$ in equilibrium in a rectangular coordinate system (x_1, x_2, x_3) (Fig. 11-2) the six stress components σ_{11}, $\sigma_{12} = \sigma_{21}$, $\sigma_{13} = \sigma_{31}$, σ_{22}, $\sigma_{23} = \sigma_{32}$, σ_{33} acting on the surfaces of the element must fulfill the force-equilibrium conditions (the first subscript refers to the direction of the normal to the plane on which the stress component acts, the second subscript to the direction of the stress component):

$$\frac{\partial \sigma_{11}}{\partial x_1} + \frac{\partial \sigma_{21}}{\partial x_2} + \frac{\partial \sigma_{31}}{\partial x_3} + X_1 = 0$$

$$\frac{\partial \sigma_{12}}{\partial x_1} + \frac{\partial \sigma_{22}}{\partial x_2} + \frac{\partial \sigma_{32}}{\partial x_3} + X_2 = 0 \qquad (11\text{-}1)$$

$$\frac{\partial \sigma_{13}}{\partial x_1} + \frac{\partial \sigma_{23}}{\partial x_2} + \frac{\partial \sigma_{33}}{\partial x_3} + X_3 = 0$$

where X_1, X_2, X_3 are the components of a body force X. The equalities $\sigma_{12} = \sigma_{21}$, $\sigma_{13} = \sigma_{31}$, and $\sigma_{23} = \sigma_{32}$ follow from the condition of moment equilibrium.

In spherical coordinates with central symmetry the equilibrium equations between the stress components σ_{rr}, $\sigma_{r\vartheta} = \sigma_{\vartheta r}$, and $\sigma_{\vartheta\vartheta}$ become

$$\frac{\partial \sigma_{rr}}{\partial r} + \frac{1}{r}\frac{\partial \sigma_{r\vartheta}}{\partial \vartheta} + \frac{1}{r}(\sigma_{rr} - \sigma_{\vartheta\vartheta}) + X_r = 0$$

$$\frac{\partial \sigma_{r\vartheta}}{\partial r} + \frac{1}{r}\frac{\partial \sigma_{\vartheta\vartheta}}{\partial \vartheta} + \frac{2}{r}\sigma_{r\vartheta} + X_\vartheta = 0$$

(11-1a)

For cylindrical symmetry with coordinates (r,z) the equilibrium relations between the stress components σ_{rr}, $\sigma_{rz} = \sigma_{zr}$, $\sigma_{\vartheta\vartheta}$, and σ_{zz} are

$$\frac{\partial \sigma_{rr}}{\partial r} + \frac{\partial \sigma_{rz}}{\partial z} + \frac{1}{r}(\sigma_{rr} - \sigma_{\vartheta\vartheta}) + X_r = 0$$

$$\frac{\partial \sigma_{rz}}{\partial r} + \frac{\partial \sigma_{zz}}{\partial z} + \frac{1}{r}\sigma_{rz} + X_z = 0$$

(11-1b)

Considering a small displacement of the corner (x_1,x_2,x_3) of the volume

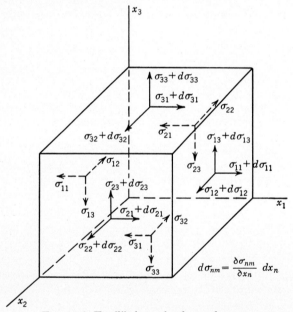

FIG. 11-2. Equilibrium of volume element.

element with the components (u_1,u_2,u_3) in the direction of the coordinate axes, the displacement in the same directions of the adjacent corners of the strained element can be expressed by $\left(u_1 + \dfrac{\partial u_1}{\partial x_1}dx_1\right)$, $\left(u_2 + \dfrac{\partial u_2}{\partial x_2}dx_2\right)$, $\left(u_3 + \dfrac{\partial u_3}{\partial x_3}dx_3\right)$, while the angular distortions of the element change the initially right angles of the cube by $\left(\dfrac{\partial u_1}{\partial x_2} + \dfrac{\partial u_2}{\partial x_1}\right)$, $\left(\dfrac{\partial u_1}{\partial x_3} + \dfrac{\partial u_3}{\partial x_1}\right)$, $\left(\dfrac{\partial u_2}{\partial x_3} + \dfrac{\partial u_3}{\partial x_2}\right)$. Using the notations ϵ_{11}, ϵ_{22}, ϵ_{33} for the normal strain components and ϵ_{12}, ϵ_{13}, ϵ_{23} for the shear-strain components, these infinitesimal components can be related to the displacement components by the equations

$$\epsilon_{11} = \frac{\partial u_1}{\partial x_1} \qquad \epsilon_{22} = \frac{\partial u_2}{\partial x_2} \qquad \epsilon_{33} = \frac{\partial u_3}{\partial x_3}$$

$$\epsilon_{12} = \frac{1}{2}\left(\frac{\partial u_1}{\partial x_2} + \frac{\partial u_2}{\partial x_1}\right) \qquad \epsilon_{13} = \frac{1}{2}\left(\frac{\partial u_1}{\partial x_3} + \frac{\partial u_3}{\partial x_1}\right) \qquad (11\text{-}2)$$

$$\epsilon_{23} = \frac{1}{2}\left(\frac{\partial u_2}{\partial x_3} + \frac{\partial u_3}{\partial x_2}\right)$$

It is easily shown by double differentiation of the strain components with respect to two of the three directions of the coordinate axes that, for Eq. (11-2) to be compatible with the existence in the continuum of a continuous displacement u with the components (u_1, u_2, u_3), the following "compatibility" conditions must be fulfilled:

$$\frac{\partial^2 \epsilon_{11}}{\partial x_2\,\partial x_3} = \frac{\partial}{\partial x_1}\left(-\frac{\partial \epsilon_{23}}{\partial x_1} + \frac{\partial \epsilon_{31}}{\partial x_2} + \frac{\partial \epsilon_{12}}{\partial x_3}\right) \qquad 2\frac{\partial^2 \epsilon_{12}}{\partial x_1\,\partial x_2} = \frac{\partial^2 \epsilon_{11}}{\partial x_2{}^2} + \frac{\partial^2 \epsilon_{22}}{\partial x_1{}^2}$$

$$\frac{\partial^2 \epsilon_{22}}{\partial x_3\,\partial x_1} = \frac{\partial}{\partial x_2}\left(-\frac{\partial \epsilon_{31}}{\partial x_2} + \frac{\partial \epsilon_{12}}{\partial x_3} + \frac{\partial \epsilon_{23}}{\partial x_1}\right) \qquad 2\frac{\partial^2 \epsilon_{23}}{\partial x_2\,\partial x_3} = \frac{\partial^2 \epsilon_{22}}{\partial x_3{}^2} + \frac{\partial^2 \epsilon_{33}}{\partial x_2{}^2} \qquad (11\text{-}3)$$

$$\frac{\partial^2 \epsilon_{33}}{\partial x_1\,\partial x_2} = \frac{\partial}{\partial x_3}\left(-\frac{\partial \epsilon_{12}}{\partial x_3} + \frac{\partial \epsilon_{23}}{\partial x_1} + \frac{\partial \epsilon_{31}}{\partial x_2}\right) \qquad 2\frac{\partial^2 \epsilon_{31}}{\partial x_3\,\partial x_1} = \frac{\partial^2 \epsilon_{33}}{\partial x_1{}^2} + \frac{\partial^2 \epsilon_{11}}{\partial x_3{}^2}$$

When each volume element in the unrestrained continuum is subjected to a temperature change* $T(x_1, x_2, x_3)$ the normal strain components are $\epsilon_{11} = \epsilon_{22} = \epsilon_{33} = \alpha T$, where α is the coefficient of linear thermal expansion, while the shear-strain components $\epsilon_{12} = \epsilon_{23} = \epsilon_{31} = 0$, provided such expansion is compatible with the continuity of the medium. The coefficient of thermal expansion increases with temperature. Stress analysis with constant initial value of α therefore underestimates the thermal stresses due to temperature increase. Introducing these strain components into the first three compatibility conditions, three differential equations for the temperature distributions $T(x_1, x_2, x_3)$ are obtained, which produce stress-free deformations in the unrestrained continuum:

$$\frac{\partial^2 T}{\partial x_1{}^2} + \frac{\partial^2 T}{\partial x_2{}^2} = 0 \qquad \frac{\partial^2 T}{\partial x_2{}^2} + \frac{\partial^2 T}{\partial x_3{}^2} = 0 \qquad \frac{\partial^2 T}{\partial x_3{}^2} + \frac{\partial^2 T}{\partial x_1{}^2} = 0 \quad (11\text{-}4)$$

* Throughout Chap. 11 the symbol T generally represents the difference between the local temperature and the temperature at some designated reference surface, rather than the actual local temperature.

Two conclusions can immediately be drawn from these equations concerning simply connected regions:

1. No thermal stresses result from linear temperature distributions (uniform parallel heat flow), such as $T = a_1x_1 + a_2x_2 + a_3x_3$ or $ax_1x_2x_3$, provided no restraints are imposed along boundaries of the system.

2. Since Eqs. (11-4) are identical with Eq. (9-14) for $\partial T/\partial t = 0$ and $H = 0$, it follows that for conditions of steady-state two-dimensional heat flow without heat sources or sinks thermal stresses can occur only as a result of imposed boundary conditions, not as a result of temperature gradients in the systems.

These conclusions are independent of the stress-strain relations assumed, since they have been derived from the compatibility conditions alone.

The relations between the strain and the displacement components in spherical coordinates with central symmetry are

$$\epsilon_{rr} = \frac{\partial u_r}{\partial r} \qquad \epsilon_{\vartheta\vartheta} = \frac{1}{r}\frac{\partial u_\vartheta}{\partial \vartheta} + \frac{u_r}{r} \qquad \epsilon_{r\vartheta} = \frac{1}{2r}\left(\frac{\partial u_r}{\partial \vartheta} - u_\vartheta + r\frac{\partial u_\vartheta}{\partial r}\right) \quad (11\text{-}2a)$$

while for cylindrical symmetry

$$\epsilon_{rr} = \frac{\partial u_r}{\partial r} \qquad \epsilon_{\vartheta\vartheta} = \frac{u_r}{r} \qquad \epsilon_{zz} = \frac{\partial u_z}{\partial z} \qquad \epsilon_{rz} = \frac{1}{2}\left(\frac{\partial u_r}{\partial z} + \frac{\partial u_z}{\partial r}\right) \quad (11\text{-}2b)$$

11-7. Stress-strain Relations. The stress-strain relations in the form suitable for thermal-stress analysis are obtained from the consideration that the total strain components are composed of the strain components due to the stresses and the normal strain components αT resulting from a temperature change $T(x_1,x_2,x_3)$. With the constants E, ν, G, and K denoting respectively the elastic modulus, Poisson's ratio, the shear modulus, and the bulk modulus, these relations are

$$\epsilon_{11} = \frac{1}{E}[\sigma_{11} - \nu(\sigma_{22} + \sigma_{33}) + E\alpha T] = \frac{1}{2G}(\sigma_{11} - p) + \frac{p}{K} + \alpha T$$

$$\epsilon_{22} = \frac{1}{E}[\sigma_{22} - \nu(\sigma_{33} + \sigma_{11}) + E\alpha T] = \frac{1}{2G}(\sigma_{22} - p) + \frac{p}{K} + \alpha T$$

$$\epsilon_{33} = \frac{1}{E}[\sigma_{33} - \nu(\sigma_{11} + \sigma_{22}) + E\alpha T] = \frac{1}{2G}(\sigma_{33} - p) + \frac{p}{K} + \alpha T$$

$$\epsilon_{12} = \frac{\sigma_{12}}{2G} \qquad \epsilon_{13} = \frac{\sigma_{13}}{2G} \qquad \epsilon_{23} = \frac{\sigma_{23}}{2G}$$

(11-5)

where $p = \frac{1}{3}(\sigma_{11} + \sigma_{22} + \sigma_{33})$. The effect of external restraints on the thermal stresses in a volume element can be rapidly estimated by using the first three equations (11-5) in conjunction with the assumption that for this element the strain components due to restrained temperature expansion can be expressed as linear functions of the temperature change

$\epsilon_{ii} = c_{ii}\alpha T$, where the subscript $i = 1, 2, 3$. Hence, from Eqs. (11-5),

$$2G(c_{ii} - 1)\alpha T = \sigma_{ii} + \frac{2G - K}{K}\, p \qquad (11\text{-}6)$$

Adding the three equations (11-6), the relation is obtained

$$K \left(\sum_1^3 c_{ii} - 3 \right) \alpha T = 3p \qquad (11\text{-}7)$$

Combining Eqs. (11-7) and (11-6), the stress components are

$$\sigma_{ii} = 2G\alpha T \left(c_{ii} - \tfrac{1}{3} \sum_1^3 c_{ii} \right) + K\alpha T \left(\tfrac{1}{3} \sum_1^3 c_{ii} - 1 \right) \qquad (11\text{-}8)$$

A condition of "no restraint" is defined by $c_{ii} = 1$ and $\sigma_{ii} = 0$; full restraint by

$$c_{ii} = 0 \qquad \text{and} \qquad \sigma_{ii} = -K\alpha T = -\frac{E\alpha T}{1 - 2\nu} \qquad (11\text{-}9)$$

Equation (11-9) therefore specifies the highest intensity of elastic thermal stress that can be produced by a temperature increase T, when the actual restraints on any volume element produced by the compatibility and boundary conditions are replaced by the maximum possible restraints.

Extending the above considerations to the continuum made up of an infinite number of volume elements expanding as a result of the temperature increase T, continuity would require the application of a hydrostatic pressure $p = -\tfrac{1}{3}\Sigma\sigma_{ii} = K\alpha T$ to each element. Such a pressure distribution could be produced by the application of a combination of body forces and surface forces which evidently should satisfy both the equilibrium and the boundary conditions. Introducing $\sigma_{ii} = -p = -K\alpha T$ into Eqs. (11-1), and considering that all shear stresses are zero, the components of the required force are obtained:

$$X_1 = K\alpha \frac{\partial T}{\partial x_1} \qquad X_2 = K\alpha \frac{\partial T}{\partial x_2} \qquad X_3 = K\alpha \frac{\partial T}{\partial x_3} \qquad (11\text{-}10)$$

Moreover, the pressure $K\alpha T$ is to be applied to the surface to fulfill the boundary conditions. Assuming complete continuity of the volume elements, the elastic thermal stresses are obtained by superimposing on the hydrostatic pressure $(-K\alpha T)$ the elastic stresses produced by a body force with the components $-X_1$, $-X_2$, $-X_3$ and by a normal surface tension $K\alpha T$.

11-8. Two-dimensional Problems. Many practical problems can be so formulated that the temperature distribution and either the displacements or the stresses in the direction of the coordinate axis x_3 or z

can be assumed as constant or vanishing, leading respectively to problems of plane strain ($\partial \epsilon_{33}/\partial x_3 = $ const) or plane stress ($\sigma_{33} = 0$).

In problems of plane strain the temperature T does not depend on x_3, so that $\partial T/\partial x_3 = 0$. Hence $X_3 = 0$ and the normal surface tension $K\alpha T$ has to be applied to the boundary in the (x_1, x_2) plane only. The equilibrium conditions (11-1) are therefore

$$\frac{\partial \sigma_{11}}{\partial x_1} + \frac{\partial \sigma_{12}}{\partial x_2} - K\alpha \frac{\partial T}{\partial x_1} = 0$$

$$\frac{\partial \sigma_{12}}{\partial x_1} + \frac{\partial \sigma_{22}}{\partial x_2} - K\alpha \frac{\partial T}{\partial x_2} = 0$$

(11-11)

Of the compatibility equations all but one are satisfied; the second equation (11-3) can be expressed in terms of the stress components, by introducing the stress-strain relations for plane strain

$$E\epsilon_{33} = \sigma_{33} - \nu(\sigma_{11} + \sigma_{22}) + E\alpha T$$

(11-5a)

and

$$\epsilon_{11} = \frac{1}{E}[(1 - \nu^2)\sigma_{11} - \nu(1 + \nu)\sigma_{22}] + (1 + \nu)\alpha T - \nu\epsilon_{33}$$

$$\epsilon_{22} = \frac{1}{E}[(1 - \nu^2)\sigma_{22} - \nu(1 + \nu)\sigma_{11}] + (1 + \nu)\alpha T - \nu\epsilon_{33}$$

(11-5b)

$$\epsilon_{12} = \frac{1}{2G}\sigma_{12} = \frac{(1 + \nu)}{E}\sigma_{12}$$

into the compatibility condition

$$\frac{\partial^2 \epsilon_{11}}{\partial x_2^2} + \frac{\partial^2 \epsilon_{22}}{\partial x_1^2} = 2\frac{\partial^2 \epsilon_{12}}{\partial x_1 \, \partial x_2}$$

(11-3a)

When the first of Eqs. (11-11) is differentiated with respect to x_1, the second with respect to x_2, and the equations are added, the following relation is obtained:

$$\frac{\partial^2 \sigma_{11}}{\partial x_1^2} + \frac{\partial^2 \sigma_{22}}{\partial x_2^2} = K\alpha \left(\frac{\partial^2 T}{\partial x_1^2} + \frac{\partial^2 T}{\partial x_2^2}\right) - 2\frac{\partial^2 \sigma_{12}}{\partial x_1 \, \partial x_2}$$

(11-12)

Considering this equation and the stress-strain relations (11-5b), the compatibility condition (11-3a) is transformed into

$$\left(\frac{\partial^2}{\partial x_1^2} + \frac{\partial^2}{\partial x_2^2}\right)(\sigma_{11} + \sigma_{22}) = \frac{2\nu K\alpha}{1 - \nu}\left(\frac{\partial^2 T}{\partial x_1^2} + \frac{\partial^2 T}{\partial x_2^2}\right)$$

(11-13)

Equations (11-11) and (11-13) are usually solved by the introduction of a stress function ϕ defined by the following relations:

$$\frac{\partial^2 \phi}{\partial x_2^2} = \sigma_{11} - K\alpha T \qquad \frac{\partial^2 \phi}{\partial x_1^2} = \sigma_{22} - K\alpha T \qquad \frac{\partial^2 \phi}{\partial x_1 \, \partial x_2} = -\sigma_{12}$$

(11-14)

Substituting these expressions into Eq. (11-13), the differential equation is obtained:

$$\nabla^4 \phi = \frac{\partial^4 \phi}{\partial x_1{}^4} + 2 \frac{\partial^4 \phi}{\partial x_1{}^2 \partial x_2{}^2} + \frac{\partial^4 \phi}{\partial x_2{}^4} = - \frac{\alpha E}{1 - \nu} \nabla^2 T \qquad (11\text{-}15)$$

since $(1 - 2\nu)K = E$. In cylindrical coordinates with rotational symmetry,

$$\nabla^4 \phi = \left(\frac{\partial^2}{\partial r^2} + \frac{1}{r} \frac{\partial}{\partial r} + \frac{\partial^2}{\partial z^2} \right) \left(\frac{\partial^2 \phi}{\partial r^2} + \frac{1}{r} \frac{\partial \phi}{\partial r} + \frac{\partial^2 \phi}{\partial z^2} \right) = - \frac{\alpha E}{1 - \nu} \nabla^2 T \quad (11\text{-}15a)$$

The third component of stress, according to Eq. (11-5a), is

$$\sigma_{33} = \nu(\sigma_{11} + \sigma_{22}) + E\epsilon_{33} - E\alpha T \qquad (11\text{-}16)$$

or, considering Eqs. (11-14),

$$\sigma_{33} = \nu\nabla^2 \phi - K\alpha T(1 - 4\nu) + \epsilon_{33} E \qquad (11\text{-}17)$$

where

$$\epsilon_{33} = a_1 x_1 + a_2 x_2 + a_3 \qquad (11\text{-}18)$$

because of $\partial \epsilon_{33}/\partial x_3 = 0$.

The absence of applied external forces requires that the total force in the direction x_3, as well as the components in the directions x_1 and x_2 of the bending moment due to the stresses σ_{33}, be zero; therefore

$$\iint \sigma_{33} \, dx_1 \, dx_2 = 0 \qquad \iint \sigma_{33} x_1 \, dx_1 \, dx_2 = 0 \qquad \iint \sigma_{33} x_2 \, dx_1 \, dx_2 = 0 \quad (11\text{-}19)$$

Substituting Eqs. (11-16) and (11-18) into (11-19), and selecting the center of gravity of the section as the origin of the coordinates, and the principal axes of inertia as the coordinate axes

$$\iint x_1 \, dx_1 \, dx_2 = \iint x_2 \, dx_1 \, dx_2 = \iint x_1 x_2 \, dx_1 \, dx_2 = 0$$

while

$$\iint x_1{}^2 \, dx_1 \, dx_2 = I_1 \qquad \text{and} \qquad \iint x_2{}^2 \, dx_1 \, dx_2 = I_2$$

The constants are obtained by solving the three simultaneous equations derived from Eq. (11-19):

$$a_1 = - \frac{1}{EI_1} \iint x_1 [\nu(\sigma_{11} + \sigma_{22}) - E\alpha T] \, dx_1 \, dx_2$$

$$a_2 = - \frac{1}{EI_2} \iint x_2 [\nu(\sigma_{11} + \sigma_{22}) - E\alpha T] \, dx_1 \, dx_2 \qquad (11\text{-}20)$$

$$a_3 = - \frac{1}{EA} \iint [\nu(\sigma_{11} + \sigma_{22}) - E\alpha T] \, dx_1 \, dx_2$$

In the case of plane stress defined by $\partial T/\partial x_3 = 0$ and $\sigma_{33} = 0$ the stress-strain relations (11-5) become

$$\epsilon_{11} = \frac{1}{E} (\sigma_{11} - \nu\sigma_{22}) + \alpha T$$

$$\epsilon_{22} = \frac{1}{E} (\sigma_{22} - \nu\sigma_{11}) + \alpha T \qquad (11\text{-}5c)$$

$$\epsilon_{12} = \frac{\sigma_{12}}{2G}$$

The same considerations as in the case of plane strain, using the above stress-strain relations instead of Eqs. (11-5b) and (11-12), lead to the differential equation for the stress function

$$\nabla^4\phi = \frac{\partial^4\phi}{\partial x_1{}^4} + 2\frac{\partial^4\phi}{\partial x_1{}^2\,\partial x_2{}^2} + \frac{\partial^4\phi}{\partial x_2{}^4}$$

$$= -\alpha E\left(\frac{\partial^2 T}{\partial x_1{}^2} + \frac{\partial^2 T}{\partial x_2{}^2}\right) = -\alpha E\nabla^2 T \qquad (11\text{-}21)$$

The stress function that represents the solution of the plane strain or plane stress problem must satisfy the respective differential equation and produce the normal tension $K\alpha T$ along the boundary.

Considering that, according to Eq. (9-14),

$$\nabla^2 T = \frac{1}{a^2}\frac{\partial T}{\partial t} - \frac{H}{k}$$

the differential equations of the stress function for plane strain

$$\nabla^4\phi = -\frac{\alpha E}{(1-\nu)}\left(\frac{1}{a^2}\frac{\partial T}{\partial t} - \frac{H}{k}\right) \qquad (11\text{-}15b)$$

and for plane stress

$$\nabla^4\phi = -\alpha E\left(\frac{1}{a^2}\frac{\partial T}{\partial t} - \frac{H}{k}\right) \qquad (11\text{-}21a)$$

For steady heat flow ($\partial T/\partial t = 0$) and no heat generation ($H = 0$) both equations are transformed into the biharmonic equation

$$\nabla^4\phi = 0 \qquad (11\text{-}22)$$

in which case the thermal stresses are due to the restraints on the temperature displacements imposed along the boundaries only, not to temperature gradients.

The boundary conditions along a force-free boundary C result from the condition that the components in the x_1 and x_2 directions of the resultant stress over any length ds of the boundaries are zero:

$$\sigma_{11}\frac{\partial x_2}{\partial s} - \sigma_{12}\frac{\partial x_1}{\partial s} = 0 \qquad \text{and} \qquad \sigma_{12}\frac{\partial x_2}{\partial s} - \sigma_{22}\frac{\partial x_1}{\partial s} = 0 \quad (11\text{-}23)$$

Introducing Eq. (11-14) into Eq. (11-23) and integrating along the boundary with the integration constants a_1, a_2, and a_3, the condition is obtained for the value of ϕ_c along a force-free boundary C:

$$\phi_c = a_1 x_1 + a_2 x_2 + a_3 \qquad (11\text{-}23a)$$

Analytical as well as numerical methods for the solution of Eq. (11-22) for specific boundary conditions have been well established in the mathematical theory of elasticity; for the solution of particular problems the respective references should be consulted [6].

Unless the boundaries of the problem are of relatively simple geometrical shape, numerical procedures will usually represent the best approach to the problem. In this approach the partial differentials of Eq. (11-22) are replaced by finite differences, and the transformed finite-difference equation for the function ϕ is established for every nodal point of a regular network of adequate shape covering the surface and containing its boundaries [7]. For the simplest case of a square network with interval h and central difference expansion of the differentials (Fig. 11-3),

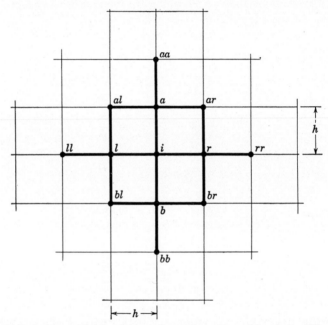

FIG. 11-3. Square grid for solution of equation $\nabla^4\phi = 0$ by finite differences.

Eq. (11-22) is transformed for each nodal point i into the set of linear equations:

$$h^4\nabla^4\phi_i = \phi_{aa} + \phi_{bb} + \phi_{rr} + \phi_{ll} + 2(\phi_{al} + \phi_{ar} + \phi_{bl} + \phi_{br})$$
$$- 8(\phi_a + \phi_b + \phi_r + \phi_l) + 20\phi_i \quad (11\text{-}22a)$$

The problem is thus reduced to the solution of a system of m simultaneous linear equations for the values of ϕ_m at the m nodal points [8].

11-9. Three-dimensional Problems. An approach similar to that of the stress function for the plane problem permits the solution of three-dimensional problems. If a function ψ of the coordinates and of time is defined in terms of the displacements

$$u_1 = \frac{\partial \psi}{\partial x_1} \qquad u_2 = \frac{\partial \psi}{\partial x_2} \qquad u_3 = \frac{\partial \psi}{\partial x_3} \qquad (11\text{-}24)$$

the volumetric strain

$$e = (\epsilon_{11} + \epsilon_{22} + \epsilon_{33}) = \nabla^2 \psi \qquad (11\text{-}25)$$

because of the relation between strain and displacement specified by Eqs. (11-2). If Eqs. (11-5) are solved for the stress components, the following relations are obtained:

$$\begin{aligned}
\sigma_{11} &= \lambda e + 2G\epsilon_{11} - E\alpha T & \sigma_{12} &= 2G\epsilon_{12} \\
\sigma_{22} &= \lambda e + 2G\epsilon_{22} - E\alpha T & \sigma_{13} &= 2G\epsilon_{13} \qquad (11\text{-}5d) \\
\sigma_{33} &= \lambda e + 2G\epsilon_{33} - E\alpha T & \sigma_{23} &= 2G\epsilon_{23}
\end{aligned}$$

where for the sake of convenience the Lamé constant is introduced

$$\lambda = \frac{E}{(1 + \nu)(1 - 2\nu)} = \frac{2\nu G}{1 - 2\nu} = \frac{3K\nu}{1 + \nu}$$

Expressing the components of strain in Eq. (11-5d) by Eqs. (11-2) and (11-24), and substituting the resulting expressions for the stress components into the equilibrium equations (11-1), the equations are obtained

$$\begin{aligned}
\frac{\partial}{\partial x_1} \nabla^2 \psi &= \frac{1 + \nu}{1 - \nu} \alpha \frac{\partial T}{\partial x_1} \\
\frac{\partial}{\partial x_2} \nabla^2 \psi &= \frac{1 + \nu}{1 - \nu} \alpha \frac{\partial T}{\partial x_2} \qquad (11\text{-}26) \\
\frac{\partial}{\partial x_3} \nabla^2 \psi &= \frac{1 + \nu}{1 - \nu} \alpha \frac{\partial T}{\partial x_3}
\end{aligned}$$

which are satisfied by any function ψ that represents a solution of the potential equation,

$$\nabla^2 \psi = \frac{1 + \nu}{1 - \nu} \alpha T \qquad (11\text{-}27)$$

The use of this approach with the aid of methods developed in the theory of potentials is particularly expedient for the solution of problems of heating or cooling of localized regions within an infinite or semi-infinite medium, in which stresses and displacements tend toward zero at large distances from the heat source [9]. If the temperature varies with time, the substitution of Eq. (9-14) into Eq. (11-27) after differentiation with respect to time gives the relation, for a thermal diffusivity a,

$$\nabla^2 \frac{\partial \psi}{\partial t} = \frac{1 + \nu}{1 - \nu} \alpha a^2 \left(\nabla^2 T + \frac{H}{k} \right) \qquad (11\text{-}28)$$

With $H = 0$ for problems without heat generation, therefore,

$$\frac{\partial \psi}{\partial t} = \frac{1 + \nu}{1 - \nu} \alpha a^2 T \tag{11-28a}$$

which is an easily integrable differential equation for the function ψ. While the introduction of stress functions ϕ or functions ψ represents a general approach to the solution of elastic thermal-stress problems, solutions of problems involving very simple shapes, such as spheres, cylinders, cylindrical tubes, and flat plates can be obtained by direct integration of either the equilibrium equations or the compatibility conditions. It has already been pointed out (Sec. 11-2) that simple shapes are of considerable importance in reactor design. These cases are now considered.

ELASTIC THERMAL STRESSES IN SIMPLE SHAPES

11-10. Sphere with Polar-symmetrical Temperature Distribution. Of the two equilibrium equations (11-1a), the only equation remaining is

$$\frac{d\sigma_{rr}}{dr} + \frac{2}{r} (\sigma_{rr} - \sigma_{\vartheta\vartheta}) = 0 \tag{11-29}$$

The compatibility condition follows immediately from the definition of the components of strain $\epsilon_{rr} = du/dr$ and $\epsilon_{\vartheta\vartheta} = u/r$, so that

$$\frac{d}{dr} \epsilon_{\vartheta\vartheta} + \frac{1}{r} (\epsilon_{\vartheta\vartheta} - \epsilon_{rr}) = 0 \tag{11-30}$$

The stress-strain relations (11-5) with $\sigma_{11} = \sigma_{rr}$, $\sigma_{22} = \sigma_{33} = \sigma_{\vartheta\vartheta}$, and shear stresses vanishing because of the symmetry, are

$$\epsilon_{rr} = \frac{1}{E} (\sigma_{rr} - 2\nu\sigma_{\vartheta\vartheta}) + \alpha T$$

$$\epsilon_{\vartheta\vartheta} = \frac{1}{E} [\sigma_{\vartheta\vartheta} - \nu(\sigma_{rr} + \sigma_{\vartheta\vartheta})] + \alpha T \tag{11-31}$$

Substituting Eqs. (11-31) into the compatibility condition and considering the equilibrium equation (11-29), the following differential equation for the stress component σ_{rr} is obtained:

$$\frac{d^2\sigma_{rr}}{dr^2} + \frac{4}{r} \frac{d\sigma_{rr}}{dr} = \frac{1}{r^4} \frac{d}{dr} \left(r^4 \frac{d\sigma_{rr}}{dr} \right) = - \frac{2\alpha E}{1 - \nu} \frac{1}{r} \frac{dT}{dr} \tag{11-32}$$

the solution of which, for a solid sphere of radius R, with the boundary conditions $\sigma_{rr} = 0$ for $r = R$ and $u = 0$ for $r = 0$, is

$$\sigma_{rr} = \frac{2\alpha E}{1 - \nu} \left(\frac{1}{R^3} \int_0^R Tr^2\, dr - \frac{1}{r^3} \int_0^r Tr^2\, dr \right) \tag{11-33}$$

The tangential stress component follows from Eq. (11-29):

$$\sigma_{\vartheta\vartheta} = \frac{\alpha E}{1 - \nu} \left(\frac{2}{R^3} \int_0^R Tr^2\, dr + \frac{1}{r^3} \int_0^r Tr^2\, dr - T \right) \tag{11-34}$$

For a hollow sphere of outer radius R and inner radius r_0, with the boundary conditions $\sigma_{rr} = 0$ for $r = R$ and $r = r_0$,

$$\sigma_{rr} = \frac{2\alpha E}{1 - \nu} \left[\frac{r^3 - r_0{}^3}{(R^3 - r_0{}^3)r^3} \int_{r_0}^R Tr^2\, dr - \frac{1}{r^3} \int_{r_0}^r Tr^2\, dr \right]$$

$$\sigma_{\vartheta\vartheta} = \frac{2\alpha E}{1 - \nu} \left[\frac{2r^3 + r_0{}^3}{2(R^3 - r_0{}^3)r^3} \int_{r_0}^R Tr^2\, dr + \frac{1}{2r^3} \int_{r_0}^r Tr^2\, dr - \frac{1}{2}\, T \right] \tag{11-35}$$

The strain components are obtained by introducing Eqs. (11-35) into Eqs. (11-31), and the displacements by substituting the results into the definition of strain.

The steady-state solution for the temperature difference between interior and surface of a sphere with uniform heat generation at rate H [Eq. (9-39)] is

$$T = \frac{H}{6k}\, (R^2 - r^2)$$

Introducing this equation into Eqs. (11-33) and (11-34), the thermal-stress components are obtained

$$\sigma_{rr} = - \frac{\alpha E H}{15k(1 - \nu)}\, (R^2 - r^2)$$

$$\sigma_{\vartheta\vartheta} = \frac{\alpha E H}{15k(1 - \nu)}\, (2r^2 - R^2) \tag{11-36}$$

with maximum intensity of $\sigma_{\vartheta\vartheta}$ at the surface $r = R$. The solution for a hollow sphere of inner radius r_0 cooled at outer radius R is

$$T = \frac{H}{3k} \left[\frac{1}{2}\, (R^2 - r^2) + r_0{}^3 \left(\frac{1}{R} - \frac{1}{r} \right) \right]$$

The thermal-stress components are obtained by substituting this equation into Eqs. (11-35).

For steady heat flow between the temperature $T = T_0$ at $r = r_0$ and $T = 0$ at $r = R$, without heat generation, the temperature distribution is

$$T = T_0 \frac{r_0}{R - r_0} \left(\frac{R}{r} - 1 \right)$$

Substituting this expression into Eqs. (11-35), the stress components are

$$
\begin{aligned}
\sigma_{rr} &= \frac{E\alpha T_0}{1-\nu} \frac{r_0 R}{R^3 - r_0^3} \left[R + r_0 - \frac{1}{r}(R^2 + Rr_0 + r_0^2) + \frac{R^2 r_0^2}{r^3} \right] \\
\sigma_{\vartheta\vartheta} &= \frac{E\alpha T_0}{1-\nu} \frac{r_0 R}{R^3 - r_0^3} \left[R + r_0 - \frac{1}{2r}(R^2 + Rr_0 + r_0^2) - \frac{R^2 r_0^2}{2r^3} \right]
\end{aligned}
\tag{11-37}
$$

For $r = r_0$ and $r = R$ the stress $\sigma_{rr} = 0$; it reaches its maximum for $r^2(R^2 + Rr_0 + r_0^2) = 3R^2 r_0^2$. The stress $\sigma_{\vartheta\vartheta}$ increases with r, reaching its highest intensity for $r = R$. For a relatively thin shell $R = r_0(1 + m)$, where m is small. Introducing this expression into Eqs. (11-37) and neglecting higher powers of m, the approximations for the stress components are obtained:

$$
\sigma_{rr} = 0 \qquad (\sigma_{\vartheta\vartheta})_{r=r_0} = -\frac{E\alpha T_0}{2(1-\nu)} \left(1 + \frac{2}{3}m\right)
$$
$$
(\sigma_{\vartheta\vartheta})_{r=R} = \frac{E\alpha T_0}{2(1-\nu)} \left(1 - \frac{2}{3}m\right)
\tag{11-38}
$$

Equations (11-37) and (11-38) also hold for negative values of T_0 (heat flowing inward), the stresses changing in sign but not in magnitude.

11-11. Long Cylinder with Axially Symmetrical Temperature Distribution (Plane Strain). The two equilibrium equations (11-1b) can be written in the simplified form:

$$
\frac{d\sigma_{rr}}{dr} + \frac{1}{r}(\sigma_{rr} - \sigma_{\vartheta\vartheta}) = 0 \qquad \text{and} \qquad \frac{d\sigma_{zz}}{dz} = 0 \tag{11-39}
$$

The definition of the components of strain

$$
\epsilon_{rr} = \frac{du_r}{dr} \qquad \epsilon_{\vartheta\vartheta} = \frac{u_r}{r} \qquad \epsilon_{zz} = \frac{du_z}{dz} \tag{11-40}
$$

produces the compatibility condition

$$
\frac{d}{dr}\,\epsilon_{\vartheta\vartheta} + \frac{1}{r}(\epsilon_{\vartheta\vartheta} - \epsilon_{rr}) = 0 \tag{11-40a}
$$

The stress-strain relations are identical with Eqs. (11-5), in which the subscripts 11, 22, and 33 are replaced respectively by rr, $\vartheta\vartheta$ and zz. As in the case of spherical symmetry, the shear stresses are zero.

For conditions of plane strain ($\epsilon_{zz} = \text{const}$), the third of the equations (11-5) expresses σ_{zz} in terms of the other two stress components. Assuming $\epsilon_{zz} = 0$,

$$
\sigma_{zz} = \nu(\sigma_{rr} + \sigma_{\vartheta\vartheta}) - E\alpha T \tag{11-41}
$$

the remaining stress-strain relations can therefore be written

$$\epsilon_{rr} = \frac{1 - \nu^2}{E} \left(\sigma_{rr} - \frac{\nu}{1 - \nu} \sigma_{\vartheta\vartheta} \right) + (1 + \nu)\alpha T$$

$$\epsilon_{\vartheta\vartheta} = \frac{1 - \nu^2}{E} \left(\sigma_{\vartheta\vartheta} - \frac{\nu}{1 - \nu} \sigma_{rr} \right) + (1 + \nu)\alpha T$$

(11-42)

Substituting Eqs. (11-42) into the compatibility condition (11-40a), and considering the first equilibrium equation (11-39) the differential equation for the stress-component σ_{rr} is obtained

$$\frac{d^2\sigma_{rr}}{dr^2} + \frac{3}{r}\frac{d\sigma_{rr}}{dr} = \frac{1}{r^3}\frac{d}{dr}\left(r^3 \frac{d\sigma_{rr}}{dr} \right) = -\frac{\alpha E}{1 - \nu}\frac{1}{r}\frac{dT}{dr}$$

(11-43)

The solution of Eq. (11-43) for a solid cylinder of radius R with the boundary conditions $\sigma_{rr} = 0$ for $r = R$ and $u = 0$ for $r = 0$ is

$$\sigma_{rr} = \frac{\alpha E}{1 - \nu}\left(\frac{1}{R^2} \int_0^R Tr\, dr - \frac{1}{r^2} \int_0^r Tr\, dr \right)$$

(11-44a)

and from Eq. (11-39)

$$\sigma_{\vartheta\vartheta} = \frac{\alpha E}{1 - \nu}\left(\frac{1}{R^2} \int_0^R Tr\, dr + \frac{1}{r^2} \int_0^r Tr\, dr - T \right)$$

(11-44b)

while from Eq. (11-41)

$$\sigma_{zz} = \frac{\alpha E}{1 - \nu}\left(\frac{2}{R^2} \int_0^R Tr\, dr - T \right)$$

(11-44c)

For a thick-walled hollow cylinder, with the boundary conditions $\sigma_{rr} = 0$ for $r = R$ and $r = r_0$,

$$\sigma_{rr} = \frac{\alpha E}{1 - \nu}\frac{1}{r^2}\left(\frac{r^2 - r_0^2}{R^2 - r_0^2} \int_{r_0}^R Tr\, dr - \int_{r_0}^r Tr\, dr \right)$$

$$\sigma_{\vartheta\vartheta} = \frac{\alpha E}{1 - \nu}\frac{1}{r^2}\left(\frac{r^2 - r_0^2}{R^2 - r_0^2} \int_{r_0}^R Tr\, dr + \int_{r_0}^r Tr\, dr - Tr^2 \right)$$

(11-45)

$$\sigma_{zz} = -\frac{\alpha E}{1 - \nu}\left(T - \frac{2}{R^2 - r_0^2} \int_{r_0}^R Tr\, dr \right)$$

The equation for σ_{zz} has been obtained by superimposing on the solution according to Eq. (11-41) a self-equilibrating uniform axial stress which eliminates the axial force resulting from the condition $\epsilon_{zz} = 0$.

The steady-state solution for the temperature difference T between the interior and the surface of a solid cylinder with uniform heat generation per unit volume at rate H [Eq. (9-33)] is

$$T = \frac{H}{4k}(R^2 - r^2)$$

Substitution of this equation into the stress equations (11-44) gives the stress components for a solid cylinder:

$$\sigma_{rr} = \frac{\alpha E H}{16k(1-\nu)}(r^2 - R^2)$$

$$\sigma_{\vartheta\vartheta} = \frac{\alpha E H}{16k(1-\nu)}(3r^2 - R^2) \qquad (11\text{-}46)$$

$$\sigma_{zz} = \frac{\alpha E H}{16k(1-\nu)}(4r^2 - 2R^2)$$

The maximum stresses occur at $r = R$.

For a hollow cylinder in which heat is generated at a rate H and removed from the inner surface of the tube at temperature T_c while the external surface is insulated, so that the boundary conditions are $dT/dr = 0$ for $r = R$ and $T = T_c$ for $r = r_0$, the solution of the heat-transfer equation is

$$T_0 = T - T_c = \frac{H}{2k}\left(R^2 \ln\frac{r}{r_0} - \frac{r^2 - r_0^2}{2}\right)$$

The expressions for the stress components are obtained by substituting this equation into the general stress equations (11-45):

$$\sigma_{rr} = \frac{\alpha E H}{16k(1-\nu)r^2}\left[\frac{r^2 - r_0^2}{R^2 - r_0^2}\left(4R^4 \ln\frac{R}{r_0} - 3R^4 + 4r_0^2 R^2 - r_0^4\right)\right.$$
$$\left. + 2R^2 r^2\left(1 - 2\ln\frac{r}{r_0}\right) - 2r_0^2(R^2 + r^2) + r^4 + r_0^4\right]$$

$$\sigma_{\vartheta\vartheta} = \frac{\alpha E H}{16k(1-\nu)r^2}\left[\frac{r^2 + r_0^2}{R^2 - r_0^2}\left(4R^4 \ln\frac{R}{r_0} - 3R^4 + 4r_0^2 R^2 - r_0^4\right)\right.$$
$$\left. - 2R^2 r^2\left(1 + 2\ln\frac{r}{r_0}\right) + 2r_0^2(R^2 - r^2) + 3r^4 - r_0^4\right] \qquad (11\text{-}47)$$

$$\sigma_{zz} = \frac{\alpha E H}{16k(1-\nu)}\left[8R^2 \ln\frac{r}{r_0} - 4r^2 + 4r_0^2\right.$$
$$\left. - \frac{2}{R^2 - r_0^2}\left(4R^4 \ln\frac{R}{r_0} - 3R^4 + 4r_0^2 R^2 - r_0^4\right)\right]$$

If heat is removed only from the outer surface at temperature T_c while the inner surface is insulated, the boundary conditions are $dT/dr = 0$ for $r = r_0$ and $T = T_c$ for $r = R$. The solution of the heat-transfer equation is

$$T_0 = T - T_c = \frac{H}{2k}\left(\frac{R^2 - r^2}{2} - r_0^2 \ln\frac{R}{r}\right)$$

The expressions for the stress components are obtained by substitution into Eq. (11-44):

$$\sigma_{rr} = \frac{\alpha E H}{16k(1-\nu)r^2} \left[\frac{r^2 - r_0{}^2}{R^2 - r_0{}^2} \left(R^4 - 4r_0{}^2 R^2 + 3r_0{}^4 + 4r_0{}^4 \ln \frac{R}{r_0} \right) \right. $$
$$\left. - \left(2r^2 R^2 - r^4 - 4r_0{}^2 r^2 \ln \frac{R}{r} - 2r_0{}^2 r^2 - 2r_0{}^2 R^2 + 3r_0{}^4 + 8r_0{}^4 \ln \frac{R}{r_0} \right) \right]$$

$$\sigma_{\vartheta\vartheta} = \frac{\alpha E H}{16k(1-\nu)r^2} \left[\frac{r^2 + r_0{}^2}{R^2 - r_0{}^2} \left(R^4 - 4r_0{}^2 R^2 + 3r_0{}^4 + 4r_0{}^4 \ln \frac{R}{r_0} \right) \right.$$
$$- \left(2r^2 R^2 - r^4 - 4r_0{}^2 r^2 \ln \frac{R}{r} - 2r_0{}^2 r^2 \right.$$
$$\left. - 2r_0{}^2 R^2 + 3r_0{}^4 + 4r_0{}^4 \ln \frac{R}{r_0} \right) \quad (11\text{-}47a)$$
$$\left. - r^2 \left(4R^2 - 4r^2 - 8r_0{}^2 \ln \frac{R}{r} \right) \right]$$

$$\sigma_{zz} = - \frac{\alpha E H}{16k(1-\nu)} \left[4R^2 - 4r^2 - 8r_0{}^2 \ln \frac{R}{r} \right.$$
$$\left. - \frac{2}{R^2 - r_0{}^2} \left(R^4 - 4r_0{}^2 R^2 + 3r_0{}^4 + 4r_0{}^4 \ln \frac{R}{r} \right) \right]$$

When both surfaces are cooled, the maximum temperature difference and the maximum thermal tensile stress are, respectively,

$$T_0 = \frac{q/N}{k} f_1 \quad \text{and} \quad \sigma_{\vartheta\vartheta,\text{max}} = \frac{q/N}{k} \frac{E\alpha}{1-\nu} f_2 \quad (11\text{-}47b)$$

where q/N is the heat generated per unit length and f_1 and f_2 are coefficients tabulated as functions of r_0/R in Table 11-3.

TABLE 11-3. COEFFICIENTS OF EQS. (11-47b)*

r_0/R	f_1	f_2
1.00	0.000	
0.90	0.004	0.00138
0.80	0.009	0.00285
0.70	0.015	0.00460
0.60	0.023	0.00655
0.50	0.035	0.00885
0.40	0.050	0.01155
0.30	0.070	0.01450

* Courtesy of Stuart McLain, ANL.

For a hollow cylinder without heat generation, with a temperature difference T_0 between the higher inner and lower outer surface temperatures, the steady-state temperature difference T between the temperature at a distance r from the center and the outer surface temperature is

$$T = T_0 \frac{\ln (R/r)}{\ln (R/r_0)}$$

and the thermal stresses, according to Eq. (11-44), are

$$\sigma_{rr} = \frac{E\alpha T_0}{2(1 - \nu) \ln (R/r_0)} \left[-\ln \frac{R}{r} - \frac{r_0^2}{R^2 - r_0^2} \left(1 - \frac{R^2}{r^2}\right) \ln \frac{R}{r_0} \right]$$

$$\sigma_{\vartheta\vartheta} = \frac{E\alpha T_0}{2(1 - \nu) \ln (R/r_0)} \left[1 - \ln \frac{R}{r} \right.$$

$$\left. - \frac{r_0^2}{R^2 - r_0^2} \left(1 + \frac{R^2}{r^2}\right) \ln \frac{R}{r_0} \right] \quad (11\text{-}48)$$

$$\sigma_{zz} = \frac{E\alpha T_0}{2(1 - \nu) \ln (R/r_0)} \left(1 - 2 \ln \frac{R}{r} - \frac{2r_0^2}{R^2 - r_0^2} \ln \frac{R}{r_0} \right)$$

For positive T_0 the stress σ_{rr} is compressive throughout, becoming zero at the surfaces as required by the boundary conditions, while $\sigma_{\vartheta\vartheta}$ and σ_{zz} are compressive near the inner surface and tensile near the outer surface. The largest values, which occur at the inner and the outer surfaces, respectively, are obtained by introducing $r = r_0$:

$$(\sigma_{\vartheta\vartheta})_{r_0} = (\sigma_{zz})_{r_0} = \frac{E\alpha T_0}{2(1 - \nu) \ln (R/r_0)} \left(1 - \frac{2R^2}{R^2 - r_0^2} \ln \frac{R}{r_0} \right) \quad (11\text{-}49)$$

and $r = R$:

$$(\sigma_{\vartheta\vartheta})_R = (\sigma_{zz})_R = \frac{E\alpha T_0}{2(1 - \nu) \ln (R/r_0)} \left(1 - \frac{2r_0^2}{R^2 - r_0^2} \ln \frac{R}{r_0} \right) \quad (11\text{-}50)$$

For small wall thickness $R/r_0 = 1 + m$, and

$$\ln \frac{R}{r_0} = m - \frac{m^2}{2} + \frac{m^3}{3} - \cdots$$

The surface stress values for $r = r_0$, if m is small, are

$$(\sigma_{\vartheta\vartheta})_{r_0} = (\sigma_{zz})_{r_0} = -\frac{E\alpha T_0}{2(1 - \nu)} \left(1 + \frac{m}{3} \right) \quad (11\text{-}49a)$$

and for $r = R$:

$$(\sigma_{\vartheta\vartheta})_R = (\sigma_{zz})_R = \frac{E\alpha T_0}{2(1 - \nu)} \left(1 - \frac{m}{3} \right) \quad (11\text{-}50a)$$

For tubes of moderate wall thickness the bending stresses at a free end are about 25 per cent higher than the stresses computed above, but decrease very rapidly toward the above values with the distance from the ends [10].   Equations (11-48) to (11-50) also hold for negative values of T_0 (heat flowing inward).

If the temperature is constant through the thickness δ of the wall but varies along the length of the thin-walled cylindrical shell according to the function $T = T(z)$, and if it is assumed that the shell is divided into rings of thickness dz, the radial expansion of the rings due to the temperature is $\alpha R T(z)$. This expansion can be suppressed by an external pressure [see Eq. (11-75)] $p_e = \alpha E \delta T(z)/R$ which produces circumferential stresses $\sigma'_{\vartheta\vartheta} = -p_e R/\delta = -E\alpha T(z)$. The total thermal stresses are obtained by freeing the shell surface of the external load p_e, applying the load $-p_e$. The stresses produced by this load are obtained by integrating the differential equation of the radial displacement w:

$$\frac{d^4w}{dz^4} + 4\beta^4 w = -\frac{p_e}{D} \tag{11-51}$$

where $D = E\delta^3/12(1-\nu)$ and $\beta^4 = 3(1-\nu^2)/R^2\delta^2$. The bending moments are

$$M_z = -D\frac{d^2w}{dz^2} \quad \text{and} \quad M_\vartheta = \nu M_z \tag{11-52}$$

and the stresses are

$$\sigma''_{zz} = \frac{6M_z}{\delta^2} \quad \text{and} \quad \sigma''_{\vartheta\vartheta} = \frac{6M_\vartheta}{\delta^2} \tag{11-53}$$

The total temperature stresses are σ''_{zz} and $\sigma'_{\vartheta\vartheta} + \sigma''_{\vartheta\vartheta}$.

A linear temperature change between T_1 and T_0 over a length c of a cylindrical shell according to

$$T(z) = \frac{(T_1 - T_0)z}{c} + T_0 \tag{11-54}$$

produces no thermal stresses, since $\sigma'_{\vartheta\vartheta} + \sigma''_{\vartheta\vartheta} = 0$ and $\sigma''_{zz} = 0$. If, however, the end that is at temperature T_0 is fixed to a shell which is at uniform temperature T_0, the break in the angle of the generatrix of magnitude $\psi = \alpha R(T_1 - T_0)/c$ must be removed by the application of a bending moment M_{oz} on both ends of the discontinuity (assumed to be far from the cylinder ends)

$$M_{oz} = \frac{1}{2}\psi\beta D = \frac{\beta D\alpha R(T_1 - T_0)}{2c} \tag{11-55}$$

producing a maximum longitudinal thermal stress at that section of $(\sigma_{zz})_{\max} = 6M_{oz}/\delta^2$ [10]. This stress decays very rapidly with the distance from the point of application of M_{oz}. The "length of attenuation" λ, which represents the distance at which the longitudinal thermal stress has decayed to about $1/e$ of its maximum, is approximately $\lambda = 0.78 \sqrt{R\delta}$; for a thin-walled shell with $\delta \leq 0.2R$, $\lambda \leq 0.35R$. At a distance of roughly $2.5\lambda \leq 0.88R$ the longitudinal stresses have practically vanished.

11-12. Flat Plate of Unit Width, Thickness $2d$, and Length L. Considering a temperature distribution $T(x_2)$ varying with the relative distance x_2/d from the center line, but independent of the location along the axis x_1 and uniform over the width, the free expansion or contraction parallel to the length L of a thin plate can be completely suppressed by applying to each element the longitudinal compressive stress $\sigma'_{11} = -\alpha E T(x_2)$, while $\sigma'_{22} = \sigma'_{33} = 0$. If the plate is free from external forces and moments, the thermal stresses are obtained by superimposing on the stresses σ'_{11} the stresses due to self-equilibrating axial forces and moments balancing the force F' and bending moment M' resulting from the stresses σ'_{11} at the ends of the plate. Since

$$F' = \int_{-d}^{+d} \sigma'_{11}\, dx_2 = -\int_{-d}^{+d} E\alpha T(x_2)\, dx_2$$
$$M' = \int_{-d}^{+d} \sigma'_{11} x_2\, dx_2 = -\int_{-d}^{+d} E\alpha T(x_2) x_2\, dx_2 \qquad (11\text{-}56)$$

and since the stresses in the plate resulting from the forces $-F'$ and the moment $-M'$ under the assumption of simple beam theory are

$$\sigma''_{11} = -\frac{F'}{2d} \mp \frac{12M'}{(2d)^3}\, x_2 \qquad (11\text{-}57)$$

the total temperature stresses are

$$\sigma_{11} = \sigma'_{11} + \sigma''_{11} = -E\alpha T(x_2) + \frac{1}{2d}\int_{-d}^{+d} E\alpha T(x_2)\, dx_2$$
$$\pm \frac{3x_2}{2d^3}\int_{-d}^{+d} x_2{}^2 E\alpha T(x_2)\, dx_2 \qquad (11\text{-}58)$$

In the case of a thick plate the complete suppression of the expansion or contraction in both the directions x_1 and x_3 requires, according to Eqs. (11-8), the application of the stresses $\sigma'_{11} = \sigma'_{33} = -\alpha E T/(1 - \nu)$. Hence the total temperature stresses $\sigma_{11} = \sigma_{33}$ can be obtained from Eq. (11-58) by multiplying all three terms on its right side by $1/(1 - \nu)$.

The temperature distribution for a long and wide slab of thickness $2d$ in which heat is generated at a uniform rate H and removed symmetrically at both surfaces [Eq. (9-23)] is

$$T - T_s = \frac{H}{2k}\, (d^2 - x_2{}^2)$$

with $T_0 = (T - T_s)_{max} = H/2k$, where T_s is the surface temperature at $x_2 = \pm d$. Introducing this equation into Eq. (11-58) multiplied by $1/(1 - \nu)$, and considering that the temperature distribution in the plate is symmetrical, the thermal-stress distribution

$$\sigma_{11} = \frac{\alpha E H}{2k(1 - \nu)}\left(x_2{}^2 - \frac{1}{3}d^2\right) \tag{11-59}$$

is obtained. The maximum (tensile) stress $\sigma_{11,max} = \alpha E H d^2/3k(1 - \nu)$ occurs at the surface $x_2 = \pm d$ and is proportional to the difference between the mean temperature and the surface temperature.

If heat is generated at the center of the slab only, the temperature distribution is linear with the difference between center and surface $T_0 = qd^2/kA$, as per Eq. (9-22). The maximum (tensile) stress at the surface $x_2 = \pm d$ is therefore

$$\sigma_{11,max} = \frac{\alpha E q d^2}{2k(1 - \nu)A} \tag{11-59a}$$

Since the maximum temperature difference between the surface and the center of a square bar of cross section ($2d \times 2d$) generating heat uniformly at rate H is $T_0 = Hd^2/3k$, the maximum (surface) stress is approximately

$$\sigma_{11,max} \doteq \frac{5}{8}\frac{\alpha E H d^2}{3k(1 - \nu)} \tag{11-59b}$$

If the heat generation in a slab is nonsymmetrical and the surface temperatures therefore unequal, the maximum thermal stress will occur at the surface with the lower temperature, for which the difference between the mean and the surface temperature is largest. By adjusting the cooling on both surfaces so as to equalize the surface temperature, the thermal stresses can be reduced to their minimum level.

Thus, for instance [11], in the case of heat generated in a plate of thickness d by γ radiation in accordance with $H = H_0 e^{-\mu_B x}$, where x denotes the distance from the inner wall which is cooled to zero reference temperature while the outer wall is perfectly insulated, the temperature at any point in the wall is

$$T(x) = \frac{H_0(1 - e^{-\mu_B x} - \mu_B x e^{-bd})}{k\mu_B{}^2} \tag{11-60}$$

The stress at any point in the wall therefore is

$$\sigma_{11}(x) = \frac{\alpha E H_0}{2(1 - \nu)k\mu_B{}^2}$$
$$\left[2e^{-\mu_B x} - \frac{2}{\mu_B d} + \left(\frac{2}{\mu_B d} - \mu_B d + 2\mu_B x\right)e^{-\mu_B d}\right] \tag{11-61}$$

The maximum stress at the inner wall is

$$\sigma_{11,msx} = \frac{\alpha E H_0}{2(1 - \nu)k\mu_B{}^2}\left[\left(\mu_B d + \frac{2}{\mu_B d}\right)e^{-\mu_B d} + 2 - \frac{2}{\mu_B d}\right] \quad (11\text{-}61a)$$

If the outer wall is cooled to a known temperature T_a, the temperature at any point in the wall is

$$T(x) = T_a\frac{x}{d} + \frac{H_0}{k\mu_B{}^2}\left[1 - \frac{x}{d}(1 - e^{-\mu_B d}) - e^{-\mu_B x}\right] \quad (11\text{-}60a)$$

and the stress is therefore

$$\sigma_{11}(x) = \frac{\alpha E}{2(1 - \nu)}\left\{T_a\left(1 - \frac{2x}{d}\right)\right.$$
$$\left. + \frac{H_0}{k\mu_B{}^2}\left[\left(1 + \frac{2}{\mu_B d} - \frac{2x}{d}\right)(e^{-\mu_B d} - 1) + 2e^{-\mu_B x}\right]\right\} \quad (11\text{-}62)$$

The curvature that develops in a thermally stressed plate may be computed by standard methods [6].

11-13. Stresses in Spheres and Cylinders Subject to Symmetrical Radial Forces. In spherical or cylindrical elements, coolant pressure produces polar symmetrical or axially symmetrical stress fields that are determined by the equilibrium equations, the compatibility conditions, the stress-strain relations, and the boundary conditions through which the coolant pressure is introduced.

For the hollow sphere with outer radius R and inner radius r_0 under the action of external and internal hydrostatic pressures p_a and p respectively, the homogeneous differential equation (11-32) with $dT/dr = 0$ must therefore be integrated for the boundary conditions

$$\sigma_{rr} = -p_a \text{ for } r = R \qquad \text{and} \qquad \sigma_{rr} = -p \text{ for } r = r_0$$

From the solution of the homogeneous equation (11-32)

$$\sigma_{rr} = C_1 + \frac{C_2}{r^3}$$

the conditions are transformed into the equation for the integration constants C_1 and C_2:

$$C_1 + \frac{C_2}{R^3} = -p_a \qquad \text{and} \qquad C_1 + \frac{C_2}{r_0{}^3} = -p$$

or, with $\rho_0 = r_0/R$ and $\rho = r_0/r$,

$$\sigma_{rr} = -\frac{p}{1 - \rho_0{}^3}\left[\rho^3 - \rho_0{}^3 + \frac{p_a}{p}(1 - \rho^3)\right] \quad (11\text{-}63)$$

Substituting Eq. (11-63) into Eq. (11-29), the tangential stress component

$$\sigma_{\vartheta\vartheta} = \frac{p}{1 - \rho_0{}^3}\left[\left(\frac{1}{2}\rho^3 + \rho_0{}^3\right) - \frac{p_a}{p}\left(1 - \frac{1}{2}\rho^3\right)\right] \qquad (11\text{-}64)$$

The maximum stresses in a hollow sphere under zero external pressure and internal pressure p at $r = r_0$ (or $\rho = 1$) are

$$(\sigma_{rr})_{max} = -p \qquad \text{and} \qquad (\sigma_{\vartheta\vartheta})_{max} = p\left(\frac{1}{2} + \rho_0{}^3\right)(1 - \rho_0{}^3) \quad (11\text{-}65)$$

For a thin spherical shell under internal pressure with $R = r_0(1 + m)$, where m is small in relation to one, so that $\rho_0 \doteq (1 - m)$, the maximum tangential stress is

$$\sigma_{\vartheta\vartheta} = \frac{p(0.5 - m)}{m} \doteq \frac{p}{2m} = \frac{pr_0}{2\delta} \qquad (11\text{-}64a)$$

where δ is the thickness of the shell.

The strains are obtained by substituting the expressions for the stress component into Eq. (11-31); the displacements follow from Eq. (11-30). Thus, for instance, for a spherical cavity in the elastic medium under internal pressure for which

$$\sigma_{rr} = -p\rho^3 \qquad \text{and} \qquad \sigma_{\vartheta\vartheta} = p\rho^{3/2} \qquad (11\text{-}66)$$

the radial strain

$$\epsilon_{rr} = -\frac{p\rho^3(1 + \nu)}{E} = -\frac{pr_0{}^3}{2Gr^3} \qquad (11\text{-}67)$$

and the radial displacement

$$(u_r)_{r_0} = \int_{r_0}^{\infty} \epsilon_{rr}\, dr = -\frac{pr_0{}^3}{2G}\int_{r_0}^{\infty}\frac{dr}{r^3} = \frac{pr_0}{4G} \qquad (11\text{-}68)$$

For an infinitely long thick-walled cylinder with outer radius R and inner radius r_0 under the action of external and internal hydrostatic pressures p_a and p respectively, the radial stress σ_{rr} is obtained by integrating the homogeneous differential equation (11-43) for the boundary conditions (11-60). Integration of Eq. (11-43) with $dT/dr = 0$ gives

$$\sigma_{rr} = C_1 + \frac{C_2}{r^2} \qquad (11\text{-}69)$$

The integration constants are determined by the boundary conditions

$$C_1 + \frac{C_2}{R^2} = -p_a \qquad \text{and} \qquad C_1 + \frac{C_2}{r_0{}^2} = -p \qquad (11\text{-}70)$$

Hence, with $\rho_0 = r_0/R$ and $\rho = r_0/r$,

$$\sigma_{rr} = -\frac{p}{1 - \rho_0{}^2}\left[(\rho^2 - \rho_0{}^2) + \frac{p_a}{p}(1 - \rho^2)\right] \qquad (11\text{-}71)$$

Introducing Eq. (11-71) into Eq. (11-39), the tangential stress component is

$$\sigma_{\vartheta\vartheta} = \frac{p}{1 - \rho_0{}^2}\left[(\rho^2 + \rho_0{}^2) - \frac{p_a}{p}(1 + \rho^2)\right] \tag{11-72}$$

From Eq. (11-41)

$$\sigma_{zz} = \frac{2\nu p}{1 - \rho_0{}^2}\left(\rho_0{}^2 - \frac{p_a}{p}\right) = \text{const} \tag{11-73}$$

The maximum stresses at $r = r_0$ or $\rho = 1$ of a thick-walled cylinder under internal pressure p with $p_a = 0$ are therefore

$$(\sigma_{rr})_{\max} = -p \quad \text{and} \quad (\sigma_{\vartheta\vartheta})_{\max} = \frac{p(1 + \rho_0{}^2)}{1 - \rho_0{}^2} \tag{11-74}$$

For a thin-walled tube under internal pressure with $R = r_0(1 + m)$, where m is small in relation to one, so that $\rho_0 \doteq (1 - m)$, the maximum tangential stress at a large distance from the ends is

$$(\sigma_{\vartheta\vartheta})_{\max} = \frac{p(1 - m)}{m} \doteq \frac{p}{m} = \frac{pr_0}{\delta} \tag{11-75}$$

where δ is the wall thickness of the tube. Near the ends additional stresses arise due to the end conditions. If the end of the shell is fixed, bending stresses are introduced by the condition that along the end section both the radial deflection w and the slope dw/dz must vanish. If the homogeneous equation (11-51) is solved for these boundary conditions, the following expressions are obtained for the bending moment M_{oz} and the shear force Q_{oz} at $z = 0$ [10]:

$$M_{oz} = \frac{p}{2\beta^2} \quad \text{and} \quad Q_{oz} = -\frac{p}{\beta} \tag{11-76}$$

For a freely rotating edge

$$M_{oz} = 0 \quad \text{and} \quad Q_{oz} = -\frac{p}{2\beta} \tag{11-77}$$

The stresses due to the end conditions decay very rapidly.

The strains are obtained by substituting the equations for the stress components into Eq. (11-42). The displacements follow from Eq. (11-40). Thus, for a long cylindrical cavity under internal pressure for which the stresses in the surrounding medium are

$$\sigma_{rr} = -p\rho^2 \quad \text{and} \quad \sigma_{\vartheta\vartheta} = p\rho^2 \tag{11-78}$$

the radial strain is

$$\epsilon_{rr} = -p\rho^2(1 - \nu^2)\frac{1 + \nu/(1 - \nu)}{E} = -\frac{pr_0^2}{2Gr^2} \tag{11-79}$$

and the radial displacement is

$$(u_r)_{r_0} = \int_{r_0}^{\infty} \epsilon_{rr}\,dr = -\frac{pr_0^2}{2G}\int_{r_0}^{\infty}\frac{dr}{r^2} = \frac{pr_0}{2G} \tag{11-80}$$

11-14. Thermal Stresses in the Infinite Elastic Medium Due to Spherical and Cylindrical Inclusion.   Spherical and cylindrical inclusions produce thermal stresses in the surrounding medium if either their temperature or their coefficient of expansion differs from that of the medium.

For a spherical cavity of radius r_0 under a hydrostatic pressure p in an infinite elastic medium, the displacement of any point on the spherical surface $u_{r_1} = (u_r)_{r_0}$ is given by Eq. (11-68). For a cylindrical cavity the length of which is large in relation to its radius r_0 (plane strain), the surface displacement $(u_r)_{r_0}$ due to p is given by Eq. (11-80).

The displacements of any point on the surface of a solid sphere and an infinitely long solid cylinder of bulk modulus K subject to a hydrostatic pressure p are, respectively,

$$u_{r_2} = \frac{pr_0}{3K} \tag{11-81}$$

for the sphere and

$$u_{r_2} = \frac{2pr_0(1 + \nu)}{9K} \tag{11-82}$$

for the cylinder.

A temperature difference ΔT of a spherical or cylindrical inclusion of coefficient of thermal expansion α would produce a displacement of the points of the surface of both sphere and cylinder

$$u_{r_3}' = \alpha\,\Delta T\,r_0 \tag{11-83}$$

if such displacement were not restrained by the medium. A joint temperature change T of medium and inclusion the coefficient of thermal expansion of which differs by $\Delta\alpha$ would produce a relative displacement of the surfaces of inclusion and medium by

$$u_{r_3}'' = \Delta\alpha\,Tr_0 \tag{11-84}$$

if such displacement were free.

The interaction of medium with shear modulus G_0 and inclusion of bulk modulus K requires that, at the joint surface $r = r_0$,

$$u_{r_1} = u_{r_3} - u_{r_2} \tag{11-85}$$

Hence with $u_{r_3} = u'_{r_3} + u''_{r_3}$ the pressure on the sphere is

$$p = \frac{12G_0K(\alpha\,\Delta T + T\,\Delta\alpha)}{3K + 4G_0} \tag{11-86}$$

and on the cylinder

$$p = \frac{18G_0K(\alpha\,\Delta T + T\,\Delta\alpha)}{9K + 4G_0(1 + \nu)} \tag{11-87}$$

The maximum tangential tension in the medium at the surface of the cavity is equal to $\frac{1}{2}p$ for the sphere and to p for the cylinder.

THEORY OF INELASTIC STRESS

11-15. Types of Inelastic Behavior. The deformational response of structural metals at elevated temperatures is a combination of elastic, quasiviscous, and plastic (work-hardening) components, with the plastic component dominating at the lower temperature levels, the quasiviscous component at the higher levels. The three mechanical constants associated with these components of deformation—the elastic (shear) modulus, the coefficient of viscosity, and the yield stress—are temperature-dependent. Volume changes in isotropic materials are essentially elastic; the inelastic response is associated with the shear or "deviator" components of stress and strain, not with the total stress and strain components [14]. See, for instance, Eq. (11-90).

Of the mechanical constants, the temperature variation of the elastic modulus is relatively unimportant, particularly in relation to the very rapid change with temperature of the coefficient of viscosity. It is also quite likely that the apparent slowness in the reduction of the modulus of elasticity with increasing temperature is, at least partly, due to the difficulty of observing the true elastic response of a material in the presence of creep and that, therefore, the elastic modulus is, in fact, even less temperature-dependent than is suggested by test results obtained in conventional experiments.

The coefficient of viscosity is the most temperature-sensitive of the mechanical constants. The relation

$$\eta = \frac{\eta_0 \exp\,(Q/RT^*)}{\exp\,(Q/RT_0^*)} = \eta_0 \exp\left[\frac{Q}{RT_0^*}\left(\frac{T_0^*}{T^*} - 1\right)\right] \tag{11-88}$$

or

$$\ln\frac{\eta}{\eta_0} = \frac{Q}{RT_0^*}\left(\frac{T_0^*}{T^*} - 1\right) = \frac{Q}{RT_0^*}\frac{\mp T/T_0^*}{1 \pm T/T_0^*}$$

is generally used [12] to describe the change of this coefficient from its value η_0 at the absolute (Kelvin) temperature T_0^* to its value η at the temperature $T^* = T_0^* \pm T$. For temperature changes that are small with respect to T_0^* Eq. (11-88) can be approximated by

$$\ln\left(\frac{\eta}{\eta_0}\right) = \frac{Q}{RT_0^*}\left[\pm\frac{T}{T_0^*} + \left(\frac{T}{T_0^*}\right)^2\right] \doteq \frac{Q}{RT_0^*}\frac{T}{T_0^*} \qquad (11\text{-}88a)$$

For metals the ratio Q/R [Q in calories per mole and $R = 2$ cal/(mol)(°C)] is of the order of magnitude of 10^4 to 10^5 (e.g., for zirconium $Q \approx 78,000$ cal/mole, for magnesium alloys 50,000 to 60,000 cal/mole, for steel about 80,000 to 100,000 cal/mole, for beryllium 65,000 cal/mole, and for graphite nearly 120,000 cal/mole [18b]).

The change of the yield stress with temperature varies with the material. It is very slow for high-temperature alloys such as Inconel, moderate for titanium alloys, but rather rapid for the more conventional structural materials such as aluminum 7075, SAE 4340 steel, or stainless steel. A relatively simple expression by which the observed change can be roughly approximated is [13]

$$\sigma_0 = \bar{\sigma}_0\left[1 - \left(\frac{2T}{T_m}\right)^2\right] \qquad \text{for } T < 0.5T_m \qquad (11\text{-}89)$$

where $\bar{\sigma}_0$ is the yield stress at the reference temperature $T = 0$, and T_m is the melting temperature.

The simplest idealization of the mechanical behavior of structural materials at elevated temperatures is the assumption of viscoelastic response. The viscous component of the deformation is usually nonlinear. This complicates the solution of even simple viscoelastic stress problems to such an extent that it appears expedient, particularly in a general discussion of the effect of creep in design, to consider in first approximation a linear viscoelastic response. This response can be assumed to represent roughly the behavior of metals either at relatively low stresses or at very high temperatures; at stresses in the vicinity of the yield limit and within the range of moderately elevated temperatures the nonlinearity of the viscous response must be considered.

The simplest linear viscoelastic relation between the deviatoric components of stress and strain for a material with creep (Maxwell body) has the general form (dot denotes time derivative)

$$2G(\dot{\epsilon}_{ij} - \tfrac{1}{3}\dot{e}\delta_{ij}) = \left(\frac{\partial}{\partial t} + \frac{1}{\tau}\right)(\sigma_{ij} - p\delta_{ij}) \qquad (11\text{-}90)$$

where the relaxation time $\tau = \eta/G$; the subscripts i, j take, consecutively, the values 1, 2, 3; and $\delta_{ij} = 0$ for $i \neq j$ and $\delta_{ij} = 1$ for $i = j$. The volumetric strain component $e = (\epsilon_{11} + \epsilon_{22} + \epsilon_{33})$ and the hydrostatic stress $p = (\sigma_{11} + \sigma_{22} + \sigma_{33})/3$. Because of the similarity of form of the equations of elasticity (11-5) for $K = \infty$ (incompressible medium) with

Eq. (11-90), the distribution of load stresses in the incompressible linear viscoelastic material is the same as in the linear elastic material under identical loading conditions.

Nonlinear viscous equations of different forms have been proposed. The simplest form is that of a power function

$$\dot{e}_r = \left(\frac{s_r}{\lambda_r}\right)^n \tag{11-91}$$

where λ_r is a parameter of dimension stress $\times$ (time)$^{1/n}$ and s_r and e_r denote Hencky's stress and strain "intensities" [14]:

$$s_r = \frac{\sqrt{2}}{2} \left[(\sigma_{11} - \sigma_{22})^2 + (\sigma_{22} - \sigma_{33})^2 + (\sigma_{33} - \sigma_{11})^2 + 6(\sigma_{12}{}^2 + \sigma_{23}{}^2 + \sigma_{31}{}^2) \right]^{1/2} \tag{11-92}$$

$$e_r = \frac{\sqrt{2}}{3} \left[(\dot{\epsilon}_{11} - \dot{\epsilon}_{22})^2 + (\dot{\epsilon}_{22} - \dot{\epsilon}_{33})^2 + (\dot{\epsilon}_{33} - \dot{\epsilon}_{11})^2 + \tfrac{3}{2}(\dot{\epsilon}_{12}{}^2 + \dot{\epsilon}_{23}{}^2 + \dot{\epsilon}_{31}{}^2) \right]^{1/2}$$

Because of the desired symmetry of the stress-strain-rate relation, n should be an odd integer, usually 5 or 7.

For stresses at the yield point, the three-dimensional yield condition

$$s_r = \sigma_0 \tag{11-93}$$

determines admissible states of stress in the plastic range of the material.

In the linear viscoplastic material the general relation between the components of the stress deviator and the strain-rate deviator ($s_r > \sigma_0$) is

$$2\eta(\dot{\epsilon}_{ij} - \tfrac{1}{3}\dot{e}\delta_{ij}) = \left(1 - \frac{\sigma_0}{s_r}\right)(\sigma_{ij} - p\delta_{ij}) \tag{11-94}$$

For $s_r \leqq \sigma_0$ the viscoplastic strain rate vanishes. The behavior is then governed by the viscoelastic relation (11-90), the simplifying assumption for establishing Eq. (11-94) being that the viscoelastic strain rates are negligibly small in relation to the viscoplastic strain rates. If this assumption cannot be made, Eqs. (11-90) and (11-94) must be combined for $s_r > \sigma_0$. The relation between e and s has for all materials the form $\dot{e} = \dot{p}/K$.

For inelastic thermal stresses in infinite plates the above equations can be simplified by the assumptions $\sigma_{11} = \sigma_{22} = \sigma$, $\sigma_{33} = 0$ and therefore $\epsilon_{11} = \epsilon_{22} = \epsilon$. Under this condition the following relations are obtained:

Linear viscoelastic material:

$$\frac{\partial}{\partial t}\epsilon = \left(\frac{\partial}{\partial t} + \frac{1}{\tau}\right)\frac{\sigma}{6G} + \frac{2}{3K}\frac{\partial \sigma}{\partial t} + \frac{\partial}{\partial t}(\alpha T)$$

$$= \left(\frac{1 - \nu}{1 + \nu}\frac{\partial}{\partial t} + \frac{1}{3\tau}\right)\frac{\sigma}{2G} + \frac{\partial}{\partial t}(\alpha T) \tag{11-90a}$$

Nonlinear viscous material:

$$\frac{\partial}{\partial t}\,\epsilon = \left(\frac{\sigma}{\lambda_r}\right)^{n} + \frac{2}{3K}\frac{\partial \sigma}{\partial t} + \frac{\partial}{\partial t}\,(\alpha T) \qquad (11\text{-}91a)$$

Nonlinear viscoelastic material:

$$\frac{\partial}{\partial t}\,\epsilon = \frac{1-\nu}{1+\nu}\frac{1}{2G}\frac{\partial \sigma}{\partial t} + \left(\frac{\sigma}{\lambda_r}\right)^{n} + \frac{\partial}{\partial t}\,(\alpha T) \qquad (11\text{-}91b)$$

Ideal plastic material:

$$\sigma = \sigma_0 \qquad (11\text{-}93a)$$

Linear viscoplastic-material beyond the yield point:

$$\frac{\partial}{\partial t}\epsilon = (\sigma - \sigma_0)\frac{1}{6\eta_1} + \frac{2}{3K}\frac{\partial \sigma}{\partial t} + \frac{\partial}{\partial t}\,(\alpha T) \qquad (11\text{-}94a)$$

Linear viscoelastic-viscoplastic material beyond the yield point:

$$\frac{\partial}{\partial t}\,\epsilon = \left(\frac{1-\nu}{1+\nu}\frac{\partial}{\partial t} + \frac{1}{3\tau}\right)\frac{\sigma}{2G} + \frac{1}{6\eta_1}\,(\sigma - \sigma_0) + \frac{\partial}{\partial t}\,(\alpha T) \qquad (11\text{-}95)$$

The above equations are somewhat simplified if incompressibility of the material can be assumed ($K \to \infty$ and $\nu \to \frac{1}{2}$).

The effect of inelastic behavior on the level of thermal stresses will now be considered for the special case of the flat plate.

11-16. Surface Stresses in a Plate Subject to Rapidly Changing Ambient Temperatures. *Instantaneous Change of Temperature (Thermal Shock).* Consider an infinite slab of thickness d, or a curved slab the radius of curvature of which is large in relation to its thickness, with one face exposed to a rapidly changing ambient temperature and the other face insulated. The temperature-distribution problem is one-dimensional and equivalent to that of a plate of thickness $2d$ exposed at both faces to symmetrically changing ambient temperature. If the plate is not restrained along the faces, the thermal stresses at any distance x_2 from the center plane will be the difference between the stresses suppressing the local thermal expansion due to $T(x_2)$ and the uniformly distributed stresses due to the forces at infinity reestablishing the force-free edges (see Sec. 11-12). For the linear elastic or viscoelastic plate with temperature-independent properties, in which the thermal stress intensity is directly proportional to the temperature, this stress difference is proportional to the difference $T_a - T$ between the local temperature $T(x_2)$ and the average temperature T_a. For nonlinear behavior the stresses must be computed by setting $d\epsilon/dt = 0$, and substracting from the resulting stress $\sigma_T(x_2)$ the uniform stress $\sigma_{T,0} = \dfrac{1}{d} \displaystyle\int_0^d \sigma_T \, dx_2$ produced by the force which reestablishes the free boundary. The maximum stress $\sigma = \sigma_{T,0} - \sigma_T$ (compression for increasing, tension for decreasing, ambient temperature) will usually be at the surface. Consideration of the variation with temperature of α and k may, however, shift the maximum stress toward the interior of the plate.

In order to compute the surface stress the full temperature distribution over the thickness of the plate must be known. The solutions of the transient heat-transfer problem giving the distribution in terms of an infinite series

$$\frac{T}{T_0} = 1 - \sum_0^\infty \phi_k(t)\psi_k(x_2) \tag{11-96}$$

is well known [15]; the time functions are negative exponentials

$$\phi_k(t) = \exp\left(-\frac{a_k^2 t}{\vartheta}\right)$$

$\vartheta = cd^2/k$ is a time parameter and a_k are the roots of the transcendental equation $a_k \tan a_k = \beta$, where $\beta = hd/k$ is the Nusselt number employing the surface heat-transfer coefficient.

To compute the stresses σ_T and $\sigma_{T,0}$ in nonlinear inelastic materials by combining Eq. (11-96) with any of the nonlinear equations (11-90) to (11-95), numerical methods must be applied. In a first approximation it may, however, be assumed that in both linear and slightly nonlinear materials the stress $\sigma = \sigma_{T,0} - \sigma_T$ is proportional to the temperature difference $T_a - T$. In this case, therefore, the critical difference, as a function of t/ϑ for different values of the parameter β, obtained from a numerical evaluation of Eq. (11-96) for the surface at which $T = T_s$, determines the surface stress. If the peak values of $T_a - T_s$ and the dimensionless time of their occurrence t_0/ϑ, taken from a plot of $T_a - T_s$ vs. t/ϑ, can be related to the parameter by fitting simple functions to the points $[(T_a - T_s)_{\max}, \beta]$ and $[(t_0/\vartheta), \beta]$, respectively, and if the general shape of the plotted functions can be reproduced by a relatively simple, easily integrable function, the effect of the inelasticity of the material on the thermal stresses can be roughly analyzed.

On the basis of published numerical evaluations of Eq. (11-96) for various parameters [16], it appears that a function of the type

$$\left(\frac{y}{y_0}\right)^n = x \exp(1 - x) \tag{11-97}$$

where $y = (T_a - T_s)/T_0$, $y_0 = y_{max}$ and $x = t/t_0$ reproduces the shape of the theoretical functions fairly well. By curve fitting, the following approximate relations, valid for $5 < \beta < 25$, have been obtained:

$$y_0 = \frac{1}{1 + \sqrt{5/\beta}} \qquad \frac{t_0}{\vartheta} = \frac{0.18}{\sqrt{\beta}} \qquad n = 4 \tag{11-98}$$

Introducing Eq. (11-98) into Eq. (11-97), the equation for the transient change $y(t/\vartheta)$ is

$$y = \frac{T_a - T_s}{T_0} = -A(\beta) \left(\frac{t}{\vartheta}\right)^{\frac14} \exp\left[-\frac{a(\beta)t}{\vartheta}\right] \tag{11-99}$$

where

$$A = \frac{1.96 \sqrt[8]{\beta}}{1 + \sqrt{5/\beta}} \qquad \text{and} \qquad a = 1.39 \sqrt{\beta}\ (5 < \beta < 25) \tag{11-100}$$

The effect of the inelasticity of the material on the thermal stresses under conditions of thermal shock can now be discussed by introducing the difference between the rate of deformation due to T and the rate of deformation due to the negative force restoring the free edge and roughly proportional to T_a

$$\frac{\partial \epsilon(\sigma_{T,0})}{\partial t} - \frac{\partial}{\partial t}(\alpha T) \doteq \frac{\partial}{\partial t}(T_a - T)$$

into the inelasticity equations. Thus, for instance, for the simplest case of linear viscoelastic response the differential equation for the stress at the surface σ_s takes the form

$$\left(\frac{\partial}{\partial t} + \frac{1 + \nu}{1 - \nu}\frac{1}{3\tau}\right)\sigma_s = -2G\alpha\frac{1 + \nu}{1 - \nu}\frac{\partial}{\partial t}(T_a - T_s) \tag{11-101}$$

Assuming a temperature-independent relaxation time,

$$\tau_1 = 3\tau \frac{1 - \nu}{1 + \nu}$$

and introducing $T_a - T_s$ according to Eq. (11-99), the solution of (11-101) can be written in the form of a series:

$$\begin{aligned} \sigma_s = -2G\alpha T_0 \frac{1 + \nu}{1 - \nu} A e^{-at/\vartheta} \left(\frac{t}{\vartheta}\right)^{\frac14} &\left[1 - \frac{1}{5}\left(\frac{4\vartheta}{\tau_1}\right)\left(\frac{t}{\vartheta}\right)\right. \\ &+ \frac{1}{5 \times 9}\left(\frac{4\vartheta}{\tau_1}\right)^2\left(1 - \frac{\tau_1 a}{\vartheta}\right)\left(\frac{t}{\vartheta}\right)^2 \\ &\left. - \frac{1}{5 \times 9 \times 13}\left(\frac{4\vartheta}{\tau_1}\right)^3\left(1 - \frac{\tau_1 a}{\vartheta}\right)^2\left(\frac{t}{\vartheta}\right)^3 + \cdots\right] \end{aligned} \tag{11-102}$$

or

$$\sigma_s = (\sigma_s)_{\text{elastic}} \left[1 - \frac{1}{5}\left(\frac{4t}{\tau_1}\right) - \frac{1}{5 \times 9}\left(\frac{\tau_1}{4t_0} - 1\right)\left(\frac{4t}{\tau_1}\right)^2 \right.$$
$$- \frac{1}{5 \times 9 \times 13}\left(\frac{\tau_1}{4t_0} - 1\right)^2\left(\frac{4t}{\tau_1}\right)^3$$
$$\left. - \frac{1}{5 \times 9 \times 13 \times 17}\left(\frac{\tau_1}{4t_0} - 1\right)^3\left(\frac{4t}{\tau_1}\right)^4 \cdots \right] \quad (11\text{-}102a)$$

where $(\sigma_s)_{\text{elastic}}$ is the expression in front of the bracket in Eq. (11-102). For the maximum temperature difference at $t = t_0$ the surface stress is

$$\bar{\sigma}_s = (\sigma_{s_{\max}})_{\text{elastic}} \left\{ 1 - \frac{4t_0}{5\tau_1}\left[1 + \frac{1}{9}\left(\frac{4t_0}{\tau_1} - 1\right) + \frac{1}{9 \times 13}\left(\frac{4t_0}{\tau_1} - 1\right)^2 \right.\right.$$
$$\left.\left. + \frac{1}{9 \times 13 \times 17}\left(\frac{4t_0}{\tau_1} - 1\right)^3 - \cdots \right] \right\} \quad (11\text{-}102b)$$

If, for instance, $t_0 = \frac{1}{2}\tau_1$ the stress $\bar{\sigma}_s = 0.56(\sigma_{s_{\max}})_{\text{elastic}}$; no thermal stresses are built up when the stress relaxation is as fast as or faster than the increase in the difference $T_a - T_s$.

Even for delay times t_0 that are so short in relation to the relaxation time of the material that almost the full elastic stress level can be built up, these stresses are only momentary, and they decrease at a faster rate than the temperature difference $T_a - T_s$ itself. As $T_a - T_s$ decreases the stresses may change sign before relaxing but will disappear before the temperature gradients disappear. Since the maximum stress itself is only of short duration, the critical design-strength values of the material are the very short-time strength values, representing the stresses that can be safely sustained during minutes or even fractions of minutes. These stresses are significantly higher than even the $\frac{1}{2}$-hr values which are usually considered to represent short-time strength. Figure 11-4 [3b] illustrates the variation of the viscoelastic stresses with time.

When the response of the material is pronouncedly nonlinear as, for instance, in the case of an elastic-plastic plate, a plastic region will spread from the surface as soon as $\sigma_s > \sigma_0$ and extend as far inward as the momentary elastic stress $\sigma_s > \sigma_0$. Subsequent reduction of the difference $T_a - T_s$ will produce elastic stresses of opposite sign proportional to this reduction and thus residual stresses $\sigma_{res} = \sigma_0 - (\sigma_s)_{\text{elastic}}$. In the course of a rapid change of ambient temperature of such magnitude that the elastic stresses at the peak difference $T_a - T_s$ exceed the yield point over a finite depth from the surface, the subsequent reduction of $T_a - T_s$ due to heat flow [Eq. (11-99)] may produce severe elastic stresses of opposite sign and, therefore, substantial residual stresses, even without reversal of the ambient temperature. If the residual stresses are not relaxed they will increase the elastic stresses resulting from such reversal before and at the peak difference $-(T_a - T_s)$.

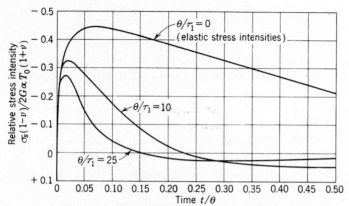

FIG. 11-4. Relative transient stress intensity $\sigma_s(1 - \nu)/2G\alpha T_0(1 + \nu)$ for a visco-elastic flat plate (thermal shock, $\beta = 5$).

Whenever the yield stress is attained in a viscoelastic-plastic plate, the stress will remain at this level only for a short period. During this time the viscoelastic stress distribution is changed and residual stresses are built up as soon as the stresses decrease below the yield point; these residual stresses are then eliminated by relaxation.

Gradual Heating. Conditions of gradual heating are much less severe than those of thermal shock. Assuming that the mean temperature increase of the surface is governed by the relation

$$T(t) = T_0(1 - e^{-\gamma t}) \tag{11-103}$$

and that the transient temperature difference between surface and center of the plate produced by a temperature increase dT at the surface may be approximated by Eq. (11-99), the total temperature difference is obtained by integration:

$$\frac{T_a - T_s}{T_0} = -A\gamma \int_0^t \left(\frac{t}{\vartheta}\right)^{\frac{1}{4}} e^{-(a+\gamma\vartheta)(t/\vartheta)} \, dt \tag{11-104}$$

where the initial temperature is taken as zero, T_0 is the total temperature rise, and γ is an inverse time parameter. Introducing this expression into Eq. (11-101), the differential equation for the viscoelastic stress at the surface is obtained:

$$\left(\frac{\partial}{\partial t} + \frac{1}{\tau_1}\right)\sigma_s = -2G\alpha T_0 \frac{1 + \nu}{1 - \nu} A \left(\frac{t}{\vartheta}\right)^{\frac{1}{4}} e^{-(a+\gamma\vartheta)(t/\vartheta)} \tag{11-105}$$

the solution of which is

$$\sigma_s = -2G\alpha T_0 \frac{1+\nu}{1-\nu} A\gamma t \left(\frac{t}{\vartheta}\right)^{\frac{1}{4}} e^{-(a+\gamma\vartheta)(t/\vartheta)} \left[1 + \frac{4}{9} c \left(\frac{t}{\vartheta}\right)\right.$$
$$\left. + \frac{4^2}{9 \times 13} c^2 \left(\frac{t}{\vartheta}\right)^2 + \frac{4^3}{9 \times 13 \times 17} c^3 \left(\frac{t}{\vartheta}\right)^3 + \cdots \right] \quad (11\text{-}106)$$

where $c = a + \gamma\vartheta - \vartheta/\tau_1$. The maximum surface stress is obtained by solving the equation $d\sigma_s/dt = 0$; neglecting terms in t/ϑ of higher than first order, the time t_0 at which this stress is reached is, in first approximation, $t_0/\vartheta = \frac{9}{4}(a + \gamma\vartheta + \vartheta/\tau_1)$. Hence $ct_0/\vartheta = 9(1 - \kappa)/4(1 + \kappa)$ where $\kappa = \vartheta/\tau_1(a + \gamma\vartheta)$. For elastic response $\tau_1 \to \infty$ and therefore $\kappa \to 0$; the viscoelastic stress reduction depends on the value of $0 < \kappa < 1$, in which the effects of thermal coefficients of the material, surface heat-transfer resistance, rate of heating and relaxation time are combined. For $\kappa = 0.5$ the maximum viscoelastic surface stress is about 0.42 of the elastic stress.

11-17. Surface Stresses in Plates Subject to Linear Temperature Gradients. When the two surfaces of a flat plate or of a thin curved plate of thickness d are kept at temperatures T_e and T_i, a steady heat flow and a linear temperature gradient are established after a certain time. The difference between the change of length of the colder and the hotter surface produces a spherical curvature of the unrestrained plate:

$$\frac{1}{\rho} = \frac{\alpha(T_e - T_i)}{d} = \frac{\alpha T}{d} \quad (11\text{-}107)$$

If the edges of the flat plate are perfectly restrained against rotation or if the plate is part of a large circular cylinder or sphere, a constant bending moment $M_x = M_y = M$ is induced which restores the initial curvature.

Since for the elastic plate the relation between deflection w, bending moment M, and radius of curvature ρ is

$$w_{xx} = w_{yy} = \frac{-M}{D(1+\nu)} = -\frac{1}{\rho} \quad (11\text{-}108)$$

where $\qquad D = \frac{Ed^3}{12(1-\nu)^2} \qquad w_{xx} = \frac{\partial^2 w}{\partial x_1{}^2} \qquad w_{yy} = \frac{\partial^2 w}{\partial x_2{}^2}$

the temperature moment is

$$M = \frac{\alpha T \, D(1+\nu)}{d} \quad (11\text{-}109)$$

and the maximum elastic fiber stresses are

$$\sigma = \pm \frac{\alpha E T}{2(1-\nu)} \quad (11\text{-}110)$$

For the linear viscoelastic plate, according to Eq. (11-90a),

$$\dot{w}_{xx} = \dot{w}_{yy} = -\left(\frac{1-\nu}{1+\nu}\frac{\partial}{\partial t} + \frac{1}{3\tau}\right)\frac{6M}{Gd^3} = -\frac{d}{dt}\frac{1}{\rho} \qquad (11\text{-}111)$$

since at the surface $\epsilon_{11} = \epsilon_{22} = \mp dw_{xx}/2 = \mp dw_{yy}/2$ and $\sigma = 6M/d^2$. Hence the temperature moment is obtained from the differential equation

$$\left(\frac{1-\nu}{1+\nu}\frac{\partial}{\partial t} + \frac{1}{3\tau}\right)M = -\frac{\alpha\dot{T}D_v}{d} \qquad \text{where } D_v = \frac{Gd^3}{6} \quad (11\text{-}112)$$

For constant T, and $\dot{T} = 0$, the solution of this equation describes the relaxation of an initially induced temperature moment M_0, which proceeds exponentially with the relaxation time $\tau_1 = 3\tau(1-\nu)/(1+\nu)$. If the temperature difference T increases slowly so that the temperature gradient can be considered to remain invariably linear, the temperature moment is obtained as a solution of Eq. (11-112) for a given time function $T(t)$. For $T(t)$ according to Eq. (11-103),

$$M(t) = \frac{\alpha T_0 D_v \tau_1 \gamma (e^{-\gamma t} - e^{-t/\tau_1})}{d(1 - \gamma\tau_1)} \qquad (11\text{-}113)$$

The moment thus increases from $M = 0$ at $t = 0$ to a maximum, which is attained at $t = \tau_1 \ln \tau_1\gamma/(\tau_1\gamma - 1)$. This maximum is considerably smaller than the elastic moment if τ_1 and $1/\gamma$ are of the same order of magnitude; for $1/\gamma = 0.5\tau_1$ the viscoelastic temperature moment does not attain one-half the elastic moment.

For a linear increase of the temperature difference $T = T_0 t/t_0$

$$M(t) = \frac{\alpha T_0 D_v \tau_1 (1 - e^{-t/\tau_1})}{t_0} \qquad (11\text{-}114)$$

If, from a time $t = t_1$ the difference remains constant, the moment $M(t_1)$ relaxes exponentially with relaxation time τ_1. Therefore, a decrease of the temperature difference after a time which is sufficiently long to permit a substantial relaxation of the temperature moment caused by the preceding increase of the temperature difference, will produce thermal moments of opposite direction, as this increase is equivalent to an increase of negative sign with respect to the new stress-relaxed condition. Hence, in viscoelastic plates sufficiently long cycles of temperature difference between zero and a maximum will be accompanied by almost complete moment and stress reversals, leading to fatigue failures (Fig. 11-5).

Consideration of the temperature dependence of τ_1 would require the introduction of a variable viscosity over the thickness of the plate according to Eq. (11-88a). The resulting stress distribution is nonlinear, since the viscous strain rate $\dot{\epsilon} = (d/dt)(x_2/\rho) = -x_2\dot{w}_{xx} = \sigma/\lambda(T)$ and therefore

$\sigma = -\ddot{w}_{xx}x_2 f(x_2)$ where $f(x_2)$ is obtained by introducing a linear temperature distribution into Eq. (11-90a). With increasing temperature difference the nonlinearity becomes more pronounced; the neutral axis moves toward the colder surface in which the fiber stress therefore increases much faster than at the hotter surface.

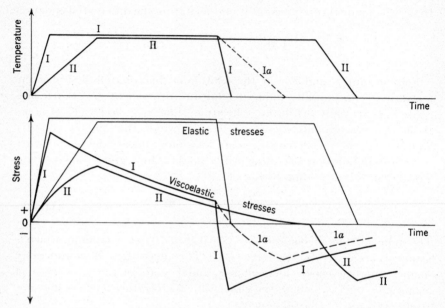

FIG. 11-5. Schematic comparison between elastic and linear viscoelastic thermal stresses due to temperature cycles I and II.

The temperature-dependence of the viscosity not only affects the distribution and peak value of the thermal stresses but changes the distribution of the load stresses in the presence of a temperature gradient, since the more rapidly creeping fibers near the hotter surface transfer their stresses gradually toward the colder surface. Thus, for instance, the initially uniform stress in a thin elastic or linear viscoelastic spherical shell under internal pressure and constant temperature [Eq. (11-64a)] will, in the viscoelastic shell, be changed by a temperature gradient across the thickness of the shell. Assuming a steady linear temperature difference of $T = 30°C$ and a mean temperature in degrees Kelvin of $T^* = 900°K$ or roughly $1100°F$, a creep rate of $\dot{\epsilon} = 10^{-7}/hr$ at a stress of 20,000 psi which, in a linear viscoelastic material with a temperature-independent elastic modulus $E = 3 \times 10^7$ psi, is associated with a relaxation time τ_0 of ~ 6660 hr, a value of $Q/R = 5 \times 10^4$ and the temperature-dependence of the coefficient of viscosity and, therefore, of the relaxation time according to Eq. (11-88a), the ratio of the stresses in the colder and the hotter surface, obtained by introducing $\dot{\epsilon}$ and $\tau(x_2)$ into

Eq. (11-90a) and solving for σ, is $e^{0.83}/e^{-0.83} = e^{1.56} = 5.5$.   The maximum fiber stress in the colder surface $\sigma_{\vartheta\vartheta} \doteq pr_0/\delta$ represents an increase by 100 per cent over the uniform elastic stress of $pr_0/2\delta$.   For nonlinear viscosity, according to Eq. (11-91) the ratio between the stresses in the cold and the hot external fibers is roughly $e^{\alpha/n}$, where $\alpha = QT/RT^{*2}$ and n is an odd integer, $n > 3$ [17].

For an incompressible nonlinear viscous material according to Eq. (11-91a) the differential equation for the thermal moment is obtained by expressing ϵ and σ in terms of ρ and M and introducing ρ according to Eq. (11-107).   Hence

$$M^n = -K_v{}^n \frac{d}{dt}\left(\frac{1}{\rho}\right) = \frac{-\alpha \dot{T} K_v{}^n}{d} \tag{11-115}$$

where
$$K_v = nd^{(2n+1)} \frac{\lambda_r}{2(2n+1)} \tag{11-116}$$

Therefore for a nonlinear viscoelastic plate following Eq. (11-91b) the differential equation of the temperature moment M is

$$\frac{1-\nu}{1+\nu} \frac{\partial}{\partial t}\left(\frac{M}{D}\right)_v + \left(\frac{M}{K_v}\right)^n = -\frac{\alpha \dot{T}}{d} \tag{11-117}$$

or, with $c_n = \dfrac{1+\nu}{1-\nu} \dfrac{D_v}{K_v{}^n}$,

$$\dot{M} + c_n M^n = -\frac{1+\nu}{1-\alpha} \alpha \dot{T} D_v \tag{11-117a}$$

For a stationary temperature difference $T = $ const and $\dot{T} = 0$, the relaxation of an initially induced temperature moment M_0 thus proceeds more slowly than in the linear material (Fig. 11-6), according to the equation

$$\left(\frac{M}{M_0}\right)^{n-1} = \frac{1}{1 + (n-1)M_0{}^{n-1}c_n t} \qquad n > 1 \tag{11-118}$$

For a linear increase of the temperature difference $T = T_0 t/t_0$ the equation

$$\dot{M} + c_n M^n = -\frac{1+\nu}{1-\nu} \frac{\alpha T_0 D_v}{t_0} = \text{const} \tag{11-119}$$

is directly integrable.   For less simple temperature histories numerical integration is required.

In an elastic-plastic plate the yield limit is reached in the surface fibers when σ according to Eq. (11-110) reaches σ_0.   The critical temperature difference $T_{0,c}$ becomes

$$T_{0,c} = \frac{2\sigma_0(1-\nu)}{\alpha E} \tag{11-120}$$

If the temperature-dependence of the yield point according to Eq. (11-89) is considered, the critical temperature T_c in terms of the critical temperature $T_{0,c}$ is obtained from the equation

$$T_c = T_{0,c}\left[1 - 4\left(\frac{T_c}{T_m}\right)^2\right] < T_{0,c} < 0.5T_m \qquad (11\text{-}120a)$$

For aluminum with $\sigma_0/E = 2 \times 10^{-3}$, $\alpha \doteq 2 \times 10^{-5}$ per °C and $\nu = 0.35$ the critical temperature difference $T_{0,c} = 130°C$; with $T_m = 660°C$ the critical temperature difference corrected for the temperature-dependent yield point becomes $T_c \doteq 113°C$.

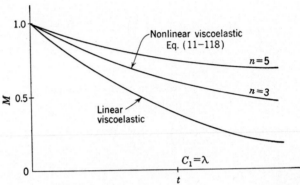

FIG. 11-6. Comparison between relaxation curves of linear and nonlinear viscoelastic materials.

For $T > T_c$ the plate under a constant moment is in the elastic state over a depth $z = \pm \xi(d/2)$ from the middle plane, and in the plastic state between $z = \pm \xi(d/2)$ and $z = \pm d/2$. Since the curvature of the plate is constant and determined by the maximum strain ϵ_0 in the elastic core

$$\frac{1}{\rho} = -w_{xx} = -w_{yy} = \frac{2\epsilon_0}{\xi d} = \frac{2\sigma_0(1 - \nu)}{E\xi d} = \frac{\alpha T_c}{\xi d} \qquad (11\text{-}121)$$

Employing the relation between the bending moment M and the depth of the elastic core in an elastic-plastic plate [1, p. 422],

$$M = M_{\text{flow}}(1 - \tfrac{1}{3}\xi^2) \qquad (11\text{-}122)$$

where the "flow moment" $M_{\text{flow}} = \sigma_0 d^2/4$ denotes the theoretical ultimate plastic carrying capacity of the plate, the relation between ξ and T obtained from Eqs. (11-107) and (11-121),

$$\xi = \frac{T_c}{T} \leq 1.0 \qquad (11\text{-}123)$$

yields as the relation between temperature difference and temperature moment in the elastic-plastic plate:

$$\frac{M}{M_{\text{flow}}} = 1 - \frac{1}{3}\left(\frac{T_c}{T}\right)^2 \tag{11-124}$$

where $\frac{2}{3}M_{\text{flow}} \leqq M \leqq M_{\text{flow}}$.

A reduction of the temperature difference is accompanied by a reversed elastic moment according to Eq. (11-109), with a temperature difference $-T$. Hence for $T > T_{0,c}$ a residual bending moment M_{res} will remain in the plate after complete removal of the temperature difference:

$$M_{res} = M_{\text{flow}}\left[1 - \frac{1}{3}\left(\frac{T_{0,c}}{T}\right)^2 - \frac{2}{3}\frac{T}{T_c}\right] \tag{11-125}$$

The residual fiber stress is

$$\sigma_{res} = \sigma_0\left(1 - \frac{T}{T_c}\right) \tag{11-126}$$

For $T > 2T_{0,c}$, $M_{res} > \frac{5}{12}M_{\text{flow}}$, and $\sigma_{res} > \sigma_0$. Hence the removal of the temperature difference is accompanied by plastic deformation in the opposite direction. Repetition of temperature cycles in excess of $2T_{0,c}$ will therefore gradually lead to failure by "alternating plasticity" or "thermal fatigue" [16, 18].

In a heat-generating plate with elastic thermal stresses according to Eq. (11-59), the yield stress is reached at the surfaces whenever

$$H = \frac{3k(1 - \nu)\sigma_0}{\alpha E d^2} = H_0$$

For $H > H_0$ plastic regions spread inward until a new plastic region starts to form at the center, spreading outward. If the plate generates heat at the center with elastic thermal stresses according to Eq. (11-59a), plastic regions form simultaneously at the center and the surfaces.

11-18. Linear Viscoelastic Thermal Stresses under Conditions of Spherical and Cylindrical Symmetry. The thermal-stress equations for the linear viscoelastic (Maxwell) sphere are easily derived by combining the equilibrium and the compatibility conditions (11-29) and (11-30) with the linear viscoelastic deviatoric stress-strain relations of the Maxwell body, Eq. (11-90), written for the directions r and ϑ:

$$2G(\dot{\epsilon}_{rr} - \tfrac{1}{3}\dot{e}) = (\dot{\sigma}_{rr} - p) + \frac{\sigma_{rr} - p}{\tau}$$

$$2G(\dot{\epsilon}_{\vartheta\vartheta} - \tfrac{1}{3}\dot{e}) = (\dot{\sigma}_{\vartheta\vartheta} - p) + \frac{\sigma_{\vartheta\vartheta} - p}{\tau} \tag{11-127}$$

where $\qquad p = \frac{\sigma_{rr} + 2\sigma_{\vartheta\vartheta}}{3} \qquad$ and $\qquad e = \epsilon_{rr} + 2\epsilon_{\vartheta\vartheta} \tag{11-128}$

The compressibility relation is

$$K\dot{e} = p + K\alpha\dot{T} \tag{11-129}$$

The resulting differential equation for the radial stress component is

$$\left(\frac{\partial}{\partial t} + \frac{1+\nu}{1-\nu}\frac{1}{3\tau}\right)\sigma_{rr} = -\frac{2E\alpha}{1-\nu}\left(\frac{1}{r^3}\int r^2\dot{T}\,dr\right) - \frac{C_1}{r^3} + C_2 \tag{11-130}$$

where the integration constants C_1 and C_2 are determined by the boundary and initial conditions.

Equation (11-130) is valid for constant coefficients G and τ. Variability of G and η without change of τ does not change the character of the equation and, considering the limited range of the variability of G, has relatively little effect on the stresses. The variability of the relaxation time, on the other hand, changes the character of the differential equation by making its coefficient variable. The effect of such variation on both load and temperature stress is considerable [19].

The form of Eq. (11-130) suggests that, in a linear viscoelastic (Maxwell) medium with constant τ, temperature stresses are built up only during transient temperature states $\dot{T}(r,t) \neq 0$; steady-state distributions $\dot{T}(r,t) = 0$ produce no thermal stresses. An initial state of thermal stresses existing at $t = 0$ will relax exponentially with a relaxation time $3\tau(1-\nu)/(1+\nu)$. For a Maxwell medium with temperature-dependent relaxation time the form of the relaxation equation

$$\dot{\sigma}_{rr} + \frac{1+\nu}{1-\nu}\frac{\exp\,(Q/RT_0^*)}{3\tau_0}\exp\left(-\frac{Q}{RT^*}\right)\sigma_{rr} = 0 \tag{11-131}$$

indicates that the stress relaxation at any point is still exponential but that, since the relaxation times vary with the radius r because

$$T^* = T^*(r)$$

the stress distribution changes with time.

Relatively simple solutions of the heat-transfer problem are obtained for the solid sphere of radius R suddenly cooled from a uniform constant initial temperature T at constant surface temperature $T_s < T$, heated with linearly increasing surface temperature, or generating heat at a constant rate. In all these cases the solution has the form

$$T(r,t) = \phi_0(r) + \sum_1^\infty \phi_n(r)e^{-k\alpha_n^2 t} \tag{11-132}$$

where $\phi_0(r)$ represents the steady-state solution and $\alpha_n = n\pi/R$. Solutions of the form of Eq. (11-132) are obtained if the boundary-layer heat-transfer resistance is neglected, and the temperature of the surface of the sphere is controlled rather than that of the surrounding medium.

From Eq. (11-132) it follows that

$$\dot{T}(r,t) = -k \sum_1^\infty \alpha_n^2\, \phi_n(r) e^{-k\alpha_n^2 t} \tag{11-133}$$

Considering that the boundary conditions for a solid sphere are $\sigma_{rr} = 0$ at $r = R$ and σ_{rr} finite at $r = 0$, the integration constants are

$$C_1 = 0 \qquad C_2 = \frac{2E\alpha}{1-\nu} \frac{1}{R^3} \int_0^R r^2 \dot{T}\, dr \tag{11-134}$$

Introducing the abbreviation

$$F_n(r) = \frac{1}{r^3} \int_0^r r^2\, \phi_n(r)\, dr - \frac{1}{R^3} \int_0^R r^2\, \phi_n(r)\, dr \tag{11-135}$$

the right-hand side of Eq. (11-130) becomes

$$\frac{2E\alpha}{1-\nu} \sum_1^\infty k\alpha_n^2 F_n(r) e^{-k\alpha_n^2 t}$$

leading, for the initial and end conditions $\sigma_{rr} = 0$ at $t = 0$ and $t = \infty$, to the solution

$$\sigma_{rr} = \frac{6E\alpha\tau}{1+\nu} \sum_1^\infty \frac{k\alpha_n^2}{3k(1-\nu)\tau\alpha_n^2/(1+\nu) - 1} F_n(r) \left[\exp\left(-k\alpha_n^2 t\right) \right.$$
$$\left. - \exp\left(-\frac{1+\nu}{1-\nu} \frac{t}{3\tau}\right) \right] \tag{11-136}$$

containing only terms with $k\alpha_n^2 < (1+\nu)/3\tau(1-\nu)$.

Equation (11-136) indicates a simultaneous exponential build-up and relaxation of the linear viscoelastic thermal stresses. The stress distribution is that of the elastic sphere. The stress intensity due to each term, however, depends on the difference of two exponential functions—the larger the difference the higher the transient stress intensity. The upper limit of this intensity is that of the elastic body with $\tau \to \infty$; the lower limit is zero and is attained for $k\alpha_n^2 = (1+\nu)/[3\tau(1-\nu)]$.

Yielding of the sphere over a region bounded by the surface $r = \rho'$ will occur if the momentary viscoelastic stresses at this surface fulfill the yield condition which, because of the spherical symmetry, can be written in the form

$$\sigma_{rr} - \sigma_{\vartheta\vartheta} = \pm\sigma_0 \tag{11-137}$$

where σ_0 denotes the uniaxial yield stress. Substituting Eq. (11-137) into Eq. (11-29), the radial stress in the plastic region of the sphere is obtained:

$$\sigma_{rp} = \pm 2\sigma_0 \ln r + C \qquad (11\text{-}138)$$

The integration constant C is determined by the condition that at the free surface of this region $r = a$, $\sigma_{rp} = 0$. Hence

$$\sigma_{rp} = \pm 2\sigma_0 \ln \frac{r}{a}$$

and

$$\sigma_{\vartheta p} = \pm \sigma_0 \left(2 \ln \frac{r}{a} - 1 \right) \qquad (11\text{-}139)$$

Since the radial stress must be continuous across the viscoelastic-plastic boundary $r = \rho'$, the momentary location of this boundary can be determined from the condition that $\sigma_\rho = \sigma_{rp}$ where $\sigma_\rho = (\sigma_{rr})_{r=\rho}$ from Eq. (11-130), or

$$\sigma_\rho = \pm 2\sigma_0 \ln \frac{\rho'}{a} \qquad (11\text{-}140)$$

While the stress components in the plastic part of the sphere must fulfill the time-independent yield condition (11-137), they are themselves time-dependent because of the time-dependence of ρ'.

Reversal of the temperature change $T(r,t)$ after part of the sphere has become plastic will produce residual stresses that can be obtained by superimposing upon the transient viscoelastic-plastic stresses due to $T(r,t)$ a transient viscoelastic stress field caused by $-\dot{T}(r,t)$. The residual stresses will relax exponentially.

For the cylinder with axially symmetrical temperature distribution, equilibrium equation (11-39), and compatibility condition (11-40a), the stress-strain relations (11-91) remain valid with

$$p = \frac{\sigma_{rr} + \sigma_{\vartheta\vartheta} + \sigma_{zz}}{3} \qquad \text{and} \qquad e = \epsilon_{rr} + \epsilon_{\vartheta\vartheta} + \epsilon_{zz} \qquad (11\text{-}141)$$

where the subscript z refers to components in the direction of the cylinder axis.

The stresses can be obtained by direct integration only for the thin circular disk (plane stress with $\sigma_{zz} = 0$). Introducing the auxiliary variables

$$\sigma_{rr} + \sigma_{\vartheta\vartheta} = 3p \qquad \text{and} \qquad \sigma_{rr} - \sigma_{\vartheta\vartheta} = f \qquad (11\text{-}142)$$

and substituting them into the Eqs. (11-90), (11-39), and (11-40), the following equations are obtained:

$$\left[\left(1 + \frac{G}{3K}\right)\frac{\partial}{\partial t} + \frac{1}{\tau}\right]\left(\frac{\partial f}{\partial r} + \frac{2f}{r}\right) = 3G\alpha\frac{\partial \dot{T}}{\partial r} \qquad (11\text{-}143)$$

and
$$\frac{\partial f}{\partial r} + \frac{2f}{r} = -2\frac{\partial p}{\partial r} \qquad (11\text{-}144)$$

Hence the equations for the auxiliary variables are

$$3\left[\frac{\partial}{\partial t} + \frac{3}{2}(1 + \nu)\tau\right]\frac{\partial p}{\partial r} = -E\alpha\frac{\partial T}{\partial r} \qquad (11\text{-}145)$$

and
$$f = -r\frac{\partial p}{\partial r} + \frac{F(t)}{r^2} \qquad (11\text{-}146)$$

where $F(t)$ is an integration function to be determined from the initial and boundary conditions.

For the long cylinder (plane strain) the combination of the equilibrium equations and the compatibility condition with the linear viscoelastic stress-strain relations, the compressibility equation, and the plane strain condition $\dot{\epsilon}_{zz} = 0$ produces two simultaneous partial differential equations for σ_{rr} and σ_{zz} that can be separated only by the assumption of incompressibility. In this case, the equation for σ_{rr} for the Maxwell cylinder can be directly written on the basis of the elastic viscoelastic analogy [20]:

$$\left(\frac{\partial}{\partial t} + \frac{1}{\tau}\right)\frac{\partial}{\partial r}\left(r^3\frac{\partial \sigma_{rr}}{\partial r}\right) = -6G\alpha r^2\frac{\partial \dot{T}}{\partial r} \qquad (11\text{-}147)$$

while the component σ_{zz} is obtained in a similar way from

$$\left(\frac{\partial}{\partial t} + \frac{1}{\tau}\right)\left(\sigma_{zz} - \sigma_{rr} - \frac{1}{2}r\frac{\partial \sigma_{rr}}{\partial r}\right) = -3G\alpha\dot{T} \qquad (11\text{-}148)$$

after Eq. (11-147) has been solved.

ASPECTS OF DESIGN

11-19. Selection of Structural Reactor Materials. The selection of stress-carrying materials for nuclear reactors must be guided by the nuclear requirements in conjunction with the requirements for a reasonably low level of thermal stresses at as high a fracture strength and as low a creep rate as can be obtained at the level of operating temperatures, considering the expected service life of the reactor.

Since, according to Sec. 11-10, the level of elastic thermal stresses under partial restraint due to a temperature change ΔT or heat generation at a rate H is proportional to $E\alpha\,\Delta T/(1 - \nu)$ or $E\alpha H/k(1 - \nu)$, the ratio $E\alpha/k(1 - \nu) = Ev_1$ represents a quality figure for the selection of materials; it indicates the intensity of thermal stresses produced by a given temperature difference or a given rate of heat generation. The coefficients E, ν, α, and k and, therefore, also G and K are temperature-sensitive (see App. F). Because of the temperature-sensitivity of all pertinent coefficients, the level of thermal stresses not only will depend on the temperature difference or the rate of heat generation but will be significantly affected by the absolute level of operating temperatures.

The elastic modulus E decreases with increasing temperature, while the coefficient of thermal expansion and the Poisson ratio increase, so that $1 - \nu$ decreases. Since for many metals used for high-temperature service the thermal conductivity generally increases with temperature, the counteracting tendencies have a somewhat compensating effect on the thermal stress level. Hence, in rough approximation, the room-temperature value (Ev_1) can be used as a general comparative measure of the intensity of thermal stresses produced in various materials by a given temperature difference or rate of heat generation.

The thermal stresses must be safely resisted by the reactor components during the expected temperature history and service life t_L. The fracture strength $\sigma_F(t_L)$ under the actual conditions of temperature and stress represents, therefore, a second quality figure of the material. Under the simplest conditions of uniaxial stress the relation of fracture strength at service life t_L vs. absolute temperature can be roughly approximated by the relation

$$\sigma_F(t_L) = \sigma_{F0}(t_L) \exp c \, \frac{T_0 - T}{T_m - T} = \sigma_{F0}v_2 \qquad (11\text{-}149)$$

where σ_{F0} denotes the fracture stress at reference temperature T_0. For metals and ceramic materials the coefficient c varies between 0.6 and 2.4; an average value of $c = 1.5$ has been used to compute the factor v_2, in order to illustrate the decay of strength with temperature. Some "quality ratio" $Ev_1/\sigma_{F0}v_2$ is therefore an expression of the high-temperature performance of a material with respect to both the level of elastic thermal stresses produced and the strength available to resist the stresses; the smaller this "quality ratio," the better the high-temperature performance of the material.

In Table 11-4 high-temperature quality ratios have been computed for several representative materials and for an operating temperature level of 450°C and subsequently multiplied by their slow-neutron-capture cross section per unit volume $N\sigma_a$. The best reactor materials are, obviously, those for which some combination of both figures is lowest, representing the best compromise between nuclear and structural requirements. These materials have been marked by asterisks.

Material	σ_a, barns/atom	ρ, g/cm³	W_a, g/mole	N, atoms/cm³ × 10^{-24}	$N\sigma_a$, cm^{-1}	T_m, °C	α, cm × 10^6 cm-°C	E, 10^6 psi	ν	k, cal/(°C)(cm)(sec)	σ_{F0}, 10^3 psi	v_1	v_2	$Ev_1/\sigma_{F0}v_2$ × 10^{-5}	$N\sigma_a Ev_1/\sigma_{F0}v_2$
Metals:															
Be	0.009	1.85	9.02	0.121	0.0011	1350	12.4	40		0.38	25	49	0.487	1.60	176*
C (graphite)	0.005	2.22	12.01	0.081	0.0004	3980	3.0	1.5		0.06	3.5	75	0.882	0.39	16*
Mg	0.059	1.74	24.32	0.043	0.0025	650	26.0	6.5		0.38	32	103	0.041	5.12	1,280
Al	0.215	2.70	26.97	0.060	0.0129	660	23.9	10		0.53	15	68	0.045	10.10	13,030
Ti	5.6	4.54	47.90	0.056	0.314	1820	8.5	17		0.41	75	32	0.625	1.15	3,610
Fe (steel)	2.4	7.87	55.85	0.085	0.203	1535	12.0	29		0.12	50	150	0.552	1.58	32,100
Ni	4.5	8.9	58.69	0.091	0.408	1450	7.4	30	0.33†	0.22	50	50	0.525	0.57	23,300
Monel		8.85	.....	0.087	0.360	1350	14.0	26		0.06	70	350	0.489	2.66	96,000
Co-Cr-W‡		9.3	.....	0.083	2.840	2330	12.0	30		0.19	105	940	0.712	0.38	108,000
Cu	3.6	8.96	63.54	0.085	0.306	1080	16.5	16		0.94	30	26	0.361	0.39	11,900
Zr	0.18	6.50	91.22	0.043	0.0078	1900	5.5	12		0.05	40	164	0.641	0.77	600*
Mo	2.4	10.20	95.95	0.064	0.153	2620	4.9	50		0.35	160	210	0.741	0.09	1,380
W	19.2	19.20	183.92	0.063	1.210	3400	4.3	50		0.48	150	13	0.803	0.05	6,050
Ceramic compounds:															
Al₂O₃		3.80	.....	0.112	0.010	2020	8.3	55		0.01	47	1035	0.664	18.30	18,300
BeO		2.80	.....	0.134	0.001	2540	10.6	45		0.20	21	66	0.733	0.59	59*
MgO		3.60	.....	0.110	0.003	2800	14.7	12	0.20†	0.02	3	920	0.760	48.50	14,550
ThO₂		9.69	.....	0.066	0.132	3000	10.2	21		0.02	18	640	0.779	9.58	126,500
ZrO₂		5.72	.....	0.084	0.010	2700	7.7	36		0.01	30	960	0.750	15.40	15,400
Be₂C		2.40	.....	0.147	0.001	2100§	10.8	42		0.03	25	450	0.677	11.25	1,125

* Optimal quality figure for service at 450°C.
† Average.
‡ Co, 55; Cr, 35; W, 10.
§ Decomposes.

11-20. Procedures of Design. After selection of the materials most suitable for the various reactor parts on the basis of the general considerations embodied in Table 11-4, and computation of the thermal stresses and load stresses for preliminary assumed dimensions of those parts, their dimensions must be redetermined more precisely by the following procedures:

1. By comparing the computed stresses in the critical sections with the fracture strength of the respective material under a similar state of stress at the operating temperature and the expected service life, as obtained from creep-rupture tests (design for strength)

2. By estimating, from the results of creep tests, the total permanent deformation at the end of the expected service life which would be produced under a similar state of stress at the operating temperature by the computed stresses, and comparing it with the limiting admissible deformation (design for creep)

If the computed stress is sufficiently below the observed creep-rupture stress, and the total permanent deformation resulting from it is smaller than the specified limiting total creep value, the dimensions of the designed parts are satisfactory. The margin of safety which is to be introduced in this comparison depends on the reliability, completeness, and reproducibility of the test results; on the close correlation between test and service conditions with respect to state of stress, operating temperatures, and time; and on the reliability and accuracy of the results of the stress analysis. It also reflects the adequacy of the design assumptions concerning operating temperatures and conditions, as well as anticipated service life. The better the actual operating conditions are approximated in the assumptions of the stress analysis and the materials tests, and the more reliable the basic design assumptions, the smaller can the factor of safety properly be and the higher, therefore, the working (design) stresses to be selected.

Alternatively, for given (necessary) dimensions of the reactor components the highest working temperature or the highest rate of heat generation may have to be determined on the basis of a reasonable working stress at the operating temperature and anticipated life.

The working stresses are usually based on the assumption that the so-called "gross structural discontinuities," representing major changes in contour, such as all openings and primary junctions between various reactor parts, are considered in the stress analysis either rigorously or by the aid of a semiempirical stress-concentration factor, while the "local structural discontinuities," representing local stress concentrations, mainly in the connections, are to be covered by the safety factor.

For parts operating at relatively low temperatures (up to about 700°F for low-alloy steel) the effect of creep and of creep fracture is usually neglected and the working stresses are specified as fractions of the maximum "static" yield stress (in either shear or tension) or, alternatively, as fractions of the so-called "ultimate tensile strength." In general, up to 60 per cent of yield stress is permitted for primary load stresses under operating conditions and up to 100 per cent of yield stress for the combination of peak primary load stresses and steady-state temperature stresses. The latter limit already implies the recognition of the fact that the computed elastic thermal stresses are partly relieved by yielding, so that the sum of the elastic load stresses and steady-state thermal stresses need not be limited to values below the yield stress.

The concept of safety in high-temperature design is more complex than in conventional design, since time appears as an additional variable. A relatively small reduction in stress increases the useful life of the structure very considerably because of the roughly straight-line negative relation between strength and logarithm of time under load. A conventional safety factor of 2 or 3 therefore appears overconservative, since its application would be equivalent to design for practically indefinite creep life. On the other hand, underestimation of the stresses under operating conditions might be extremely dangerous, since the actual useful life could thus be reduced to a fraction of the design life of the structure. An additional safety-reducing effect is the increasing statistical dispersion with temperature of the observations of creep life at an applied stress. Thus, for instance, creep-rupture tests of super alloys at 1500°F have shown observed times to fracture at constant stress of 25,000 psi from 50 to 250 hr. Statistical extrapolation for design purposes from these observations suggests that a probable range of variation of at least 25 to 500 hr would have to be considered in estimating the useful life at this stress. [21].

Results of creep tests are originally plotted as creep strain vs. time diagrams for various applied stress levels at a certain temperature (Fig. 11-7). These diagrams usually indicate three definite stages of creep: (1) an initial relatively short stage of "primary" creep proceeding at a decreasing rate, (2) a relatively long period of "secondary" creep at constant rate, followed by (3) a short period of "third stage" creep proceeding at an increasing rate, terminated by fracture. Design for creep deformation is usually based on the constant "secondary" creep rate, and design for creep strength on the time to initiation of third-stage creep rather than on time to actual fracture, since third-stage creep is, in fact, already a process of progressive deterioration of strength, or fracture.

From the original creep diagrams the relations can be derived and plotted between time to fracture, or time to specified creep strain, at various stress levels and temperatures. For creep fracture these relations can be combined into a single "master curve" for each material, valid over a wide range of temperature. The combined effect of absolute temperature T^* and time t (in hours) to fracture or to a specified percentage of permanent strain at any constant stress level σ is fairly well expressed by the combined parameter

$$P = T^*(C + \log t) = F(\sigma) \tag{11-150}$$

where for various metals $17 < C < 23$, with $C = 20$ as the best average value. Figure 11-8 shows a typical conventional design diagram for one temperature level. Figure 11-9 shows "master curves" according to Eq. (11-150) for several high-temperature alloys [22]. The function $F(\sigma)$ is of the form $F(\sigma) = A - B(\log \sigma)^n$, with $0 < n < 1$.

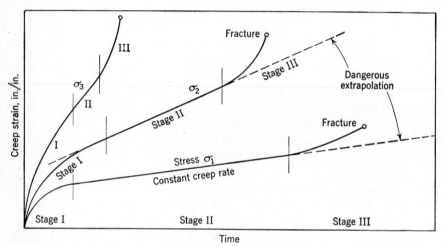

FIG. 11-7. Schematic creep diagrams at constant temperature for $\sigma_1 < \sigma_2 < \sigma_3$.

To be useful in reactor design the test conditions should preferably include a specified intensity of nuclear radiation, as the creep strength and creep values are affected by time, temperature, and radiation damage. Since these effects are not simply additive and therefore cannot be obtained by combining the results of conventional creep-rupture tests at constant temperature with creep-rupture tests at certain intensities of nuclear radiation, tests under the various desired combinations of temperature and intensity of nuclear radiation should actually be performed. Slow neutrons do not seem to affect reactor structural materials seriously even at room temperature, and the effect decreases with increasing temperature (see Table 10-6, also [23]). At 10^{20} neutrons/cm² the effects seem to approach maximum values; however, 10^{24} neutrons/cm² integrated flux may be desirable in power reactors. In general, irradia-

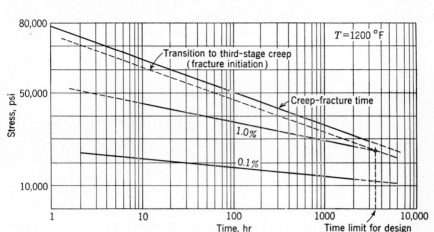

FIG. 11-8. Schematic creep-design diagram for given temperature (fracture stress and stress to specified strain as functions of time).

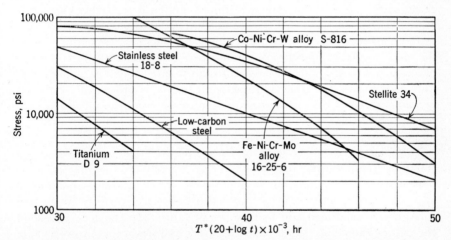

FIG. 11-9. Master creep-fracture curves for representative alloys. [*Symposium on Effect of Cyclic Heating, etc., ASTM Special Technical Publication* 165, 1954; F. R. Larson and J. Miller, *Trans. ASME,* **74**:765 (1952); E. E. Baldwin, *ASME Preprint* 54-A-231, 1954.]

tion raises the yield strength and increases hardness and annealing temperature.

From the stress-time diagrams at constant temperature and intensity of radiation, other diagrams needed for particular design problems can easily be derived, such as sets of creep-strength vs. time diagrams for different temperatures at given radiation levels, creep-strength vs. temperature diagrams for specified rupture times, and rupture-time vs.

temperature diagrams for specified levels of stress. Entering the pertinent diagram with the specified operating temperature, radiation intensity and service life, the respective stress levels causing rupture or producing a specified limiting strain are obtained and can be compared with the computed stresses under critical conditions; the ratios of the ormer and the latter represent, respectively, the safety factors of the designs for strength and for total creep deformation.

An observed linearity of the log-log plot of stress vs. time or vs. the combined temperature-time parameter [Eq. (11-150) with $n = 1$] is a characteristic feature only within a limited range and cannot generally be extrapolated to a longer period than that for which the straight-line relations have been established by tests. For certain combinations of stress, time, and temperature very distinct curves replace the straight-line relations (see Fig. 11-9). A strength analysis based on a strength-time diagram extrapolated from short-time creep-rupture tests is likely to overestimate the long-time creep strength of the material and thus to endanger the life of the designed structure. The frequently observed rather sharp break in the strength-time diagrams at relatively long times is assumedly associated with a metallurgical instability producing changes in the deformation and fracture mechanism, such as a transition from a ductile, transgranular type of fracture to a predominantly brittle intergranular fracture. This may result from the intensification, under a certain combination of time and temperature, of a process of oxidation or embrittlement due to changes in the microstructure. The effect of nuclear radiation on this transition may also be significant, although no published information is available.

The relations between creep or creep-fracture stress vs. time or vs. a combined temperature-time parameter are usually obtained from simple tension tests. Only a limited number of systematic two-dimensional creep and creep-rupture tests have been performed, mainly by testing tubes under axial load and hydrostatic pressure. In the general case of three-dimensional stress the uniaxial stress and strain (or strain rate) are replaced by the "invariant" stress and strain "intensities" (or strain-rate intensity) as defined by Eq. (11-92). Explicitly, in terms of the principal stress and strain components,

$$s_r = \frac{1}{\sqrt{2}} [(\sigma_1 - \sigma_2)^2 + (\sigma_2 - \sigma_3)^2 + (\sigma_3 - \sigma_1)^2]^{1/2}$$

$$\text{and} \qquad e_r = \frac{\sqrt{2}}{3} [(\epsilon_1 - \epsilon_2)^2 + (\epsilon_2 - \epsilon_3)^2 + (\epsilon_3 - \epsilon_1)^2]^{1/2}$$

(11-151)

Since for uniaxial stress with $\sigma_1 = \sigma$, $\sigma_2 = \sigma_3 = 0$ the stress intensity $s_r = \sigma$, the observed relations of uniaxial creep or creep-fracture stress vs. temperature and time can, in first approximation, be extended to two- and three-dimensional design problems, by considering them directly as s_r versus $T^*(C + \log t)$ relations.

In the design procedures so far discussed, it has been tacitly assumed that both the stresses and the operating temperatures are continuously sustained at constant levels during the service life of the reactor. Actual operating conditions of nuclear reactors, as well as of any other heat exchanger, however, are considerably more complex. Because of the periodic shutdown and starting periods required for the removal of fission products, as well as because of the fluctuations in power and temperature during continuous operations, considerable variation in the thermal stress level must be expected during the service life of the reactor. It has also been shown that the stress relaxation due to viscoelastic response of metals at the operating-temperature levels produces transient thermal stresses of relatively short duration rather than sustained thermal stresses, as assumed on the basis of elastic stress analysis. Moreover, it causes stress reversal during shutdowns and gradually transforms the initial stress pulsation around the operating temperature level due to the fluctuation of the power level into a stress reversal around zero mean stress [2]. Therefore, the resistance to fracture under repeated thermal stress application and stress reversal, which is usually referred to as "thermal-fatigue strength" may become a more significant design characteristic of reactor materials than the resistance to fracture under sustained stress ("creep strength"). And even where the resistance to sustained stress remains the relevant design criterion, the creep strength under the relatively short duration of the transient thermal stresses rather than the long-time creep strength may have to be considered in selecting the dimensions of reactor parts and in evaluating the safety of the design for thermal stresses.

However, very little is known at present about the resistance and deformation properties of structural materials under repeated application of thermal and mechanical stress cycles at power-reactor temperatures. When the number of cycles is large ($> 10^5$) and the alternating strain amplitudes remain essentially within the elastic range, the number N of cycles to fracture can be expressed as an inverse function of the stress amplitude of the type observed in conventional mechanical-fatigue tests, although it seems necessary to warn against the direct use of the results of mechanical-fatigue tests. The relatively low frequency of the cycle in the reactor, particularly the thermal cycle, will significantly reduce the fatigue lives, probably by as much as 60 per cent, in comparison with the lives observed under conditions of conventional fatigue testing at the same temperature. When the number of cycles is small ($\ll 10^5$) and the alternating strain amplitudes are beyond the elastic range, conditions of alternating plasticity rather than of fatigue prevail. The number N of cycles to failure must be expressed as an inverse function of the inelastic strain amplitude per cycle $\pm \epsilon_p$ of the general form $N^k \epsilon_p = \text{const.}$ For stainless steel 347 the relation $\sqrt{N} \epsilon_p = 0.36$ has been proposed on the basis of a limited number of tests [16, 18, 22], but good thermal-fatigue data for reactor design are very scarce. For different metals, different values of the exponent k and of the constant must be expected. As in the case of comparatively long lives ($> 10^5$) the thermally produced strain cycle appears to be more damaging than a mechanical cycle of identical alternating strain amplitude under the same steady temperature level. It has been suggested, again on the basis of limited test results, that the reduction of life may be to as low as one-third or one-fourth of the life under mechanically induced cyclic strain.

It is rather difficult to correlate results of conventional creep and creep-fracture tests, obtained at constant temperature levels in air, with the strength and deformation of reactor parts under actual operating conditions at varying temperature levels and in different environments. The information so far available concerning the strength and deformation of metals under conditions of creep, thermal fatigue, and thermal shock in different environments (hydrogen, liquid metals, or fused salts) is hardly adequate for design purposes. In reactors with alkali metal coolants the effect of caustic embrittlement [24] may be quite serious, even outside the reactor itself.

In order to consider the effects of intermittent heating and stressing, the following are the simplest possible design assumptions and have frequently been proposed:

1. Simple summation of creep increments over the entire stress-temperature history based on observed creep rates under constant stress and constant temperature.

2. Linear accumulation of creep damage over time to fracture at constant stress and constant temperature, and absence of stress-and-temperature interaction. This leads, for creep rupture under an arbitrary stress-temperature history, to the linear damage rule $\sum\limits_i t_{imn}/t_{mn} = 1.0$,

where t_{imn} are the actual times at stress σ_m and (absolute) temperature T_n, and t_{mn} are the rupture times under these conditions obtained from Eq. (11-150).

It appears, however, on the basis of the scarce experimental evidence [25], that both creep and creep damage are accelerated by intermittent stressing and heating as compared with steady conditions, probably owing to the effects of primary creep, recrystallization, microthermal stresses, and oxidation or embrittlement. Hence the times to attain a specified creep strain or to produce rupture may be only between 0.5 and 0.3 of those computed on the basis of assumption 1 or 2.

Even the relatively simple problem of the influence of time at constant temperature on the notch effect in structural metals is still waiting for a solution useful in design. Theoretical considerations and a few tests suggest that plastic relief of the stress concentrations at the root of a notch will decrease with increasing temperature and time, as a result of the increasing significance of the viscous deformation mechanism. Therefore, the ratio of strength of the notched and unnotched section may decrease with increasing temperature, but no useful design information has been developed.

A further difficulty in design of reactor parts is the adequate combination of load stresses and temperature stresses. Simple superposition would be justified only for perfect elasticity of the structural material. Even in linear viscoelastic materials, for which superposition is possible, the transient character of the temperature stresses makes such superposition meaningless. In nonlinear viscoelastic, plastic, or viscoplastic materials temperature stresses and load stresses are interdependent and are therefore both affected by load and temperature history. Thus, for instance, the thermal stresses produced by a linear temperature difference in an elastic-plastic tube under internal pressure that has been plastically strained by this pressure over part of its thickness will depend on the depth of the plastic zone at the time of occurrence of the temperature difference.

Combined effects of variable and steady stress components σ_a and σ_m, respectively, are at present considered on the basis of the empirical linear interaction criterion defining failure

$$\frac{\sigma_a}{\sigma_e} + \frac{\sigma_m}{\sigma_u} = 1$$

where σ_e is the fatigue strength for complete reversal for a specified number of stress cycles and σ_u is the steady stress which alone would produce failure. It should be noted, however, that, if the variable stress is caused by temperature variation, σ_e must not be confused with the fatigue strength obtained in conventional constant-temperature fatigue tests, nor must σ_u be confused with the conventional ultimate strength. Thus, while the linear relation may have the merit of simplicity and is probably on the safe side, the specification of σ_e and σ_u is rather problematic. Creep-rupture or yield strength (whichever is smaller) might be a reasonable value of σ_u when the results of thermal cycling tests can be expressed in terms of a stress σ_e. Otherwise σ_e will have to be replaced by the temperature strain cycle $\epsilon_p \pm (\alpha T - \epsilon_0)$, where ϵ_0 represents the elastic limit strain.

It is now generally assumed that both thermal fatigue and thermal shock increase the severity of the load effect, particularly under conditions when welds, sharp notches, abrupt changes in section, or other stress concentrations are present, or when metals with different coefficients of expansion and conductivity are joined metallurgically (by welding, casting, etc.) or mechanically (by riveting, shrink-fitting, etc.). However, no reliable and reproducible test results exist to support this assumption and to provide numerical values for design.

Since temperature changes and fluctuations in the presence of loads, as well as notches and stress concentrations, form part of the actual operating and design conditions of reactors and other heat exchangers, even an elaborate procedure of stress analysis and design for strength and deformation of a nuclear reactor can, at present, provide no more than a crude estimate of the safety limits of the design. Actual performance tests are therefore necessary as yet for precise results.

Prob. 11-1. A gas-cooled reactor employing He at 10 atm abs as a coolant is in a 44-ft-diameter steel sphere. The mean coolant temperature is (a) 390°F, (b) 700°F. Determine the wall thickness of the (elastic) sphere, which is perfectly insulated on the outside, for low-alloy steel with permissible working stresses of 12,000 psi at 390°F and 9000 psi at 700°F for the stresses due to pressure alone. As an alternative, assume that the thickness is determined under the condition that only 60 per cent of the above permissible stresses be utilized for resistance to pressure. What is the minimum permissible outside surface temperature of an imperfectly insulated sphere in cases a and b?

Prob. 11-2. A stainless-steel pipe carries 3,500,000 lb/hr of D_2O at an average pressure of 800 psi and a temperature of 440°F. Determine the ID and wall thickness of the elastic pipe under the assumption of an average coolant velocity of 25 fps and maximum possible temperature difference across the wall thickness of 40°F, with a permissible stress of 18,000 psi.

Prob. 11-3. A stainless-steel tube of $\frac{5}{8}$-in. ID transfers heat across its wall at a maximum rate of 250,000 Btu/(hr)(ft²). The fluid temperature inside the tube is 350°F, the inner boundary conductance is 5000 Btu/(ft²)(hr)(°F), the difference between inside and outside pressure is 200 psi. Determine the stresses in the elastic tube as a function of the wall thickness.

Prob. 11-4. A solid cylindrical U fuel rod steadily and uniformly generates 280,000 Btu/hr per square foot of surface. The maximum U temperature is 1000°F. Determine the maximum elastic thermal stress and the change of diameter as a function of the rod diameter.

Prob. 11-5. Determine with respect to the elastic thermal stresses a safe diameter of a solid spherical U fuel element steadily and uniformly generating heat at a rate of 10^6 Btu/hr per square foot of surface. The maximum surface temperature of the sphere is 800°F. Determine the specific change in diameter of this sphere as a function of the rate of heat generation.

Prob. 11-6. Determine the maximum steady temperature difference across the wall that an elastic tube of 1.0 in. OD and $\frac{3}{4}$ in. ID can withstand if the fluid pressure in the tube is 50 atm and the tube is made of (a) aluminum with a working stress of 5000 psi, (b) carbon steel with a working stress of 10,000 psi, (c) stainless steel with a working stress of 18,000 psi, (d) zirconium with a working stress of 14,000 psi.

Prob. 11-7. A uniform U fuel plate of 0.10-in. thickness with 0.01-in.-thick stainless-steel cladding plates generates heat at a uniform rate of 10^6 Btu/(ft²)(hr). The surface temperature of the U plate is 800°F. Calculate the elastic thermal stresses in the fuel plate and in the cladding plates under the assumption of joint expansion of fuel plate and cladding. (HINT: Calculate thermal stresses in fuel plate due to temperature distribution, thermal stresses in cladding plates due to temperature distribution, and stresses of interaction due to unequal expansion of fuel plate and cladding plates, and superimpose.)

Prob. 11-8. Determine the maximum thermal stress under optimal cooling conditions in a cylindrical nuclear-reactor shell of 6-ft diameter constructed of stainless steel 5 in. thick and heated by γ rays. The γ-ray heating rate per unit volume is $H(x) = H_0 \exp(-\mu_B x)$ where H_0 is the maximum effective heat-source strength per unit volume, μ_B the mean absorption coefficient for γ-ray energy, and x the distance from the inner shell surface. Assume the following values for the relevant parameters: $k = 18$ Btu/(hr)(ft)(°F), $E = 28 \times 10^6$ psi, $\alpha = 9 \times 10^{-6}/$°F, $\nu = 0.3$, $\mu_B = 0.75/$in., $H_0 = 60$ Btu/(hr)(in.³). (HINT: Determine temperature distribution through wall, and maximum temperature difference associated with equal temperature at both surfaces.)

Prob. 11-9. Assuming that the creep-rupture strength vs. time and temperature relation of 18-8 stainless steel is given by the equation

$$T^*(20 + \log t) = 4 \times 10^4 (4.9 - \log \sigma)^{0.3}$$

valid for $t < 10^6$ hr, determine the required wall thickness of an elastic steel tube of $\frac{3}{4}$-in. ID subject to an internal pressure of 60 atm and (a) a steady temperature difference of 40°F, (b) a steady heat transfer at a rate of 200,000 Btu/(hr)(ft²), for an expected service life of (i) 1000 hr, (ii) 10,000 hr, (iii) 100,000 hr, at a temperature level of (i) 600°F, (ii) 1500°F, with a safety factor of 2.0 with respect to stress. What is the associated safety factor with respect to service life? (T^* is the absolute temperature in degrees Kelvin.)

Prob. 11-10. An uninsulated cylindrical reactor shell of carbon steel of 6-ft diameter and 4-in. wall thickness is subjected to a steady linear temperature difference across its wall of 20°F due to heat losses. If, at the operating temperature, the steel has a yield limit of 25,000 psi, determine the stresses in the shell following the application of thermal insulation by which the temperature difference is reduced to 2°F. What linear temperature difference would be required to produce yielding over a thickness of $\frac{1}{8}$ in.?

Prob. 11-11. If the shell referred to in Prob. 10 were made of a linear viscoelastic material with an elastic modulus of 3×10^7 and a (constant) relaxation time of 1000 hr, calculate the maximum thermal stress in the shell if during the starting period the linear temperature difference increases at a rate of 200°F/hr to the constant difference at operating level of 20°F. Determine the thermal stress immediately after a shutdown operation in which the temperature difference is reduced to 0°F at a rate of 90°F/hr, assuming that the shutdown occurs after (a) 1000 hr, (b) 10,000 hr of continuous operation.

NOMENCLATURE

a = parameter
A = area (cross section), parameter
c = constant, heat capacity
d = (half-)thickness, differential operator

D = plate rigidity

e, e_{ij} = volumetric strain, deviatoric strain component

E = elastic modulus

f, F = function

G = shear modulus

H = heating density

k = thermal conductivity

K = bulk modulus

l = length

m = thickness ratio, index

M = moment

n = index

p = hydrostatic pressure

Q = activation energy

r = radius

r_0, R = inner, outer radius

s_{ij} = deviatoric stress component

t, t_{mn} = time, time at conditions σ_m, T_n

T, T_0 = temperature difference with respect to reference location, temperature

T^* = absolute temperature

u, u_i = displacement, displacement component

w = deflection

x, x_i = distance, coordinate

X_i = force component

z = coordinate

α = coefficient of thermal expansion

β, γ = parameters

δ = thickness

ϵ, ϵ_{ij} = strain, strain component

η = coefficient of shear viscosity

θ = time parameter

Θ = angle

λ = coefficient of viscous traction, length

ν = Poisson's ratio

ξ = dimensionless ordinate

ρ, ρ_0 = radius ratio

ρ = radius of elastic-plastic boundary

σ, σ_{ij} = stress, stress component

τ = relaxation time

ϕ = stress function

ψ = displacement function

REFERENCES

1. Freudenthal, A. M.: "Inelastic Behavior of Engineering Materials and Structures," p. 205, John Wiley & Sons, Inc., New York, 1950.
2. Freudenthal, A. M.: "Von Mises Memorial Volume," Academic Press, Inc., New York, 1954.
3. a. Freudenthal, A. M.: *J. Appl. Phys.*, **25**:1110 (1954).
   b. Freudenthal, A. M.: *J. Aeronaut. Sci.*, **21**:772 (1954).

4. Freudenthal, A. M.: chap. III, Structural Engineering Aspects, in "Rheology of Building Materials," Interscience Publishers, Inc., New York, 1954.
5. Heisler, M. P.: *Trans. ASME*, **76**:920 (1954).
6. Timoshenko, S., and J. N. Goodier: "Theory of Elasticity," 2d ed., pp. 421ff., McGraw-Hill Book Company, Inc., New York, 1951.
7. Salvadori, M., and M. Baron: "Numerical Methods in Engineering," pp. 45ff., Prentice-Hall, New York, 1952.
8. Holms, A. G.: *NACA Rept.* 1059, 1952.
9. Goodier, J. N.: *Phil. Mag.*, **23**:23 (1937).
10. Timoshenko, S.: "Theory of Plates and Shells," pp. 398–427, McGraw-Hill Book Company, New York, 1940.
11. Durham, E. P.: ASME Paper 54-A-126, 1954.
12. Houwink, R.: "Elasticity, Plasticity and Structure of Matter," pp. 38ff., Cambridge University Press, New York, 1937.
13. Heimerl, G. H., and P. J. Hughes: *NACA TN* 2975, Washington, 1953.
14. Ref. 1, p. 257.
15. Cheng, C. M.: *J. Am. Rocket Soc.*, **21**:147 (1951).
16. Manson, S. S.: *NACA TN* 2933, Washington, 1953.
17. Freudenthal, A. M.: *Trans. N.Y. Acad. Sci.*, vol. 19, no. 4, 1957.
18. *a.* Coffin, L. F.: *Trans. ASME*, **76**:931 (1954).
    b. Symposium on Effect of Cyclic Heating, etc., ASTM Special Technical Publication 165, p. 31, 1954.
19. Ref. 2, p. 259.
20. Alfrey, T.: *Quart. Appl. Math.*, **2**:113 (1944).
21. Freudenthal, A. M.: *Trans. Am. Soc. Civil Engrs.*, **121**:1337 (1956).
22. Larson, F. R., and J. Miller: *Trans. ASME*, **74**:765 (1952).
    Baldwin, E. E.: ASME Preprint 54-A-231, 1954.
23. Sutton, C. R., and D. O. Leeser: "Nuclear Engineering," pt. II, p. 208, American Institute of Chemical Engineers Symposium, 1954.
24. Sines, G., and E. C. McLean: *Mech. Eng.*, **78**:1105 (1956).
25. Ref. 18*b*, pp. 65, 75.

CHAPTER 12

INSTRUMENTATION AND CONTROL

By John W. Hoopes, Jr.

GENERAL CONSIDERATIONS

12-1. Introduction. The instrumentation and control of a nuclear reactor must ensure its safe and efficient operation as a research instrument, a chemical producer, or a power producer. As more and more flexibility in performance is demanded, more accurate and faster-acting control is required. The designer of a control system must consider the nuclear and thermal characteristics of the reactor assembly and the coolant loop; the design principles involved, however, are similar to those encountered in the analysis of any other process.

To the designer of the control system, the reactor is a source of heat, produced at a rate substantially proportional to the average neutron flux. The response of a reactor to its controls, or to a change in external conditions, depends on the design of the reactor as outlined in Chap. 6; i.e., moderator, arrangement of fuel, energy range of neutrons utilized, etc. Operating power levels must be limited by the ability of the heat-removal system to keep temperatures in the reactor below those which might cause mechanical damage. The maximum rates of change of power permissible from thermal considerations depend on the resistance of the structure to thermal stresses, upon the temperature coefficients of reactivity, and upon the thermal lags in the reactor and its cooling system (see Sec. 9-45). Specifications for control action should include the allowable watt-seconds overload resulting from a power increase. Unfortunately any such specifications can be only estimated, using the principles outlined in Chaps. 8 to 11.

The importance of the instrumentation may be seen by the fact that the control system may represent an investment of about $30,000 for a $61,000 research reactor [1] or $1.5 million for a $33 million sodium-cooled power reactor [2].

12-2. Scope of Instrumentation and Control. The principal circuits of information and energy are sketched schematically in Fig. 12-1. It is seen that, in addition to a control loop for maintaining the flux or power output at a desired level, there is a safety and maintenance system. Dangerously high neutron flux, or a rate of change of flux which would in a short time lead to dangerous levels, will actuate the rods. Other monitoring devices give warning in case of a plugged fuel channel, pump or control power failure, or sudden appearance of greater-than-normal radioactivity in the coolant or space around the reactor. As a typical example, the Brookhaven reactor is protected by an interlocking safety network by which any one of about 50 causes will call for a shutdown or

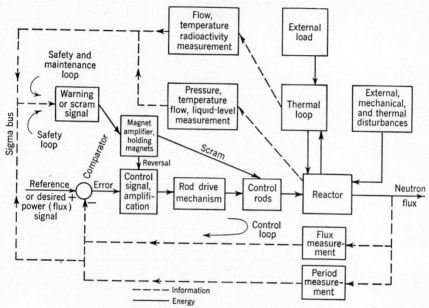

FIG. 12-1. Circuits of information and control.

prevent a start-up [3]. Other typical causes besides those given above are [3, 4] (1) loss of oil pressure on bearings of control-rod motors; (2) loss of power on trip circuits; (3) loss of instrument air pressure; (4) broken wires in safety circuits, e.g., failure of all thermocouples or flow indicators in any flow channel; (5) failure of auxiliary power supply.

If an operating difficulty is not sufficient to involve immediate danger to the reactor or its control system, or will not impair the proper functioning of the control system, an audible or visual warning signal may be sufficient.

12-3. Measuring Elements. Table 12-1 lists some of the measurements which must be made and indicates some of the commonly used methods of measurement. Except for the nuclear quantities, the variables measured and instruments used are those common to ordinary

TABLE 12-1

Measured quantity	Method of measurement	Application
Nuclear:		
Neutron flux:		
a. Low level	Pulse discrimination, using U^{235} fission chamber or BF_3 proportional counter	Inside reactor for start-up power level; monitor coolant for slug failure; shield leakage
b. High level......	Compensated ionization current chambers, using B^{10}, fed to d-c amplifier or galvanometer; boron thermopile, to potentiometer	Inside reactor, normal operation, indication of power level, flux; safety circuit, triggered; "backup" flux measurement
Pile period:		
a. Low level.......	Counter, signal fed to log count-rate meter	Inside reactor, for start-up
b. High level......	Flux signal fed to log n amplifier, differentiator; flux and rate of change of flux fed to potentiometer connected as ratio meter*	Inside reactor, start-up and power changes; safety shutdown, triggered
Gamma flux.........	Geiger tube; ionization chambers; scintillation counters	Outside shield, coolant stream, for maintenance and safety
Nonnuclear:		
Temperature........	Thermocouple to potentiometer; resistance thermometer in Wheatstone bridge; bimetal element position; liquid- or gas-filled bulb and capillary to Bourdon gauge	Coolant temperature inlet, and outlet each channel; fuel temperature, few locations; control-rod temperature; control-rod coolant temperature; moderater temperature; shield temperature
Pressure or pressure difference	Tubing to manometer or Bourdon gauge; Bourdon or magnetic transducer to potentiometer or voltmeter; strain gauge in Wheatstone bridge to potentiometer; diaphragm or bellows, transmitted air pressure to indicator or recorder	Coolant pressure, each channel; moderator pressure; gas space pressure; flow rate (see below); liquid level (see below)
Flow rate............	Pressure drop across orifice, venturi tube, or other constriction; pressure developed by impact or pitot tube; rotameter (constant head, varying constriction); electromagnetic flow meter (for metals or salt solutions); magnetic rotor to revolution counter	Coolant flow, total; coolant flow, each channel; control-rod coolant flow; gas space flow; make-up moderator or coolant flow
Liquid level	Pressure difference between top and bottom of tank, due to head of liquid; buoyancy or position of float	Coolant liquid level; moderator liquid level; fuel liquid level; leakage of coolant or moderator
Position..............	Lever arm; differential gears; selsyn generator and receiver	Reactor distortion; control-rod position; spent-fuel position after ejection
Composition (physical methods)	Electrolytic conductivity; pH (acidity); specific gravity	Composition of coolant Composition of moderator Composition of fuel if solution

* J. Weill, *Nucleonics,* **11**(3):36–39 (1953).

chemical-process or power-plant control.   The techniques are standard, with due regard for several special factors [5, 6]:

1. The instruments must be extremely dependable because of the exceptional safety requirements on the one hand and great inconvenience and cost of unnecessary shutdowns on the other.   Duplication of instruments of crucial importance is common.   To avoid false control action (especially emergency shutdown or "scram"), several signals may be fed into a coincidence circuit, which requires that several instruments agree before action is taken.   In start-up, when reactor periods can be dangerously short, the output of several period meters can be fed into an "auctioneer circuit," which initiates action according to the highest (i.e., shortest period) reading.

2. Wherever possible, the sensing elements should be designed and located to allow maintenance and replacement without resort to remote control.   For example, ion chambers and thermocouples can be inserted into the interior of the reactor by means of long thimbles extending at least through the thermal shield; the whole assembly can be removed during shutdown periods for replacement.   If the element must be located so that replacement is difficult or impossible, duplication is wise. Elements in "hot" areas are equipped with crane loops and quick-disconnect electrical connections for easy replacement.

3. Since the calibration of elements such as thermocouples and strain gauges may be affected by high neutron fluxes, these should be placed in regions of low flux near the edge of the reactor if possible.   Otherwise they must be recalibrated at intervals, or empirical corrections for drift must be applied.

4. Connections must be absolutely leaktight.   These must be no possibility of the transmittal of radioactivity to points outside the shield by means of manometer leads, air lines of pneumatic control systems, etc. Electrical transmission of information is ideal in this respect.

12-4. Control Elements.   The reactivity of a nuclear reactor can be controlled by several methods:

1. Adding or removing fuel or moderator.   This method is most easily used with homogeneous or slurry fuel reactors or those moderated with heavy water.   Burnup, xenon poisoning on shutdown, or temperature effects can be compensated for by controlling the fluid level.   Emergency shutdown can be accomplished by dumping if other shutdown methods are inoperable because of distortion of the reactor structure due to severe conditions—earthquakes, bombing, etc.

2. Changing the composition of the fuel or moderator.   Usually this is considered a shutdown measure, as the process—at least in the case of liquids—is difficult to reverse.   For example, neutron-absorbing poisons such as boron (borate) solutions can be introduced.   Conceivably

"reversible" control could be obtained by using ion exchange or similar means to remove the poison.

3. Changing the shape of the reactor.

4. Dividing the reactor structure into several parts which can be moved relative to each other; for example, using a movable reflector [8]. Methods 3 and 4 are of most use in reactors utilizing "intermediate" or "fast" neutrons, with highly enriched fuel and high volumetric rates of heat generation.

5. Changing the temperature of the reactor by changing the flow or temperature of coolant or by varying the ambient temperature. Usually the temperature coefficient of reactivity is used as a stabilizing influence to improve the response to other types of control action or to load changes when the reactor is delivering power to a steam plant.

6. Inserting solid neutron-absorbing material into the core or reflector. Rotatable drums with poison on one side and moderator in the other have been used, but movable rods are used in most of the thermal reactors.

To avoid excessive thermal stress and to provide safety and flexibility, several rods are usually used; those serving a given function are moved in unison. Some, called "shim" rods, are used to make coarse adjustments of reactivity, usually by manual control. Fine adjustments are made by the "regulating" rods, which may be actuated manually or automatically. Safety rods, if used specifically as such, are normally kept withdrawn unless shutdown is required. Hollow rods may be used for irradiation of special materials.

A fully inserted regulating rod is useless for safety or shutdown action; completely removed, it cannot be used to increase power. It is obvious, then, that a reactor should be designed with at least sufficient excess reactivity that compensation for burnup and Xe^{135} poisoning after shutdown is obtained by the shim rods alone. Table 12-2 lists the losses which should be taken into account for a typical research reactor. High-flux power reactors must provide for greater reactivity changes because of greater Xe^{135} poisoning, temperature changes, and burnup.

The rods are inserted vertically through the top shield or horizontally through the walls. They must be located geometrically so as to take into account several factors, which can be expressed qualitatively:

1. If several rods are too close together, the more rods are inserted the lower the flux becomes, locally, and the less effective a subsequent rod will be. This is the "shadowing" effect.

2. Insertion of a rod near the center of a reactor tends to raise the flux near the edges if the total power is kept constant. This increases the effectiveness of rods near the edges, which normally would be operating in a region of very low flux.

3. Judicious insertion of rods near the center can produce a more nearly uniform flux distribution; such a distribution may result in lower thermal stresses in the structure.   In production reactors, more nearly equal production rates are obtainable for all fuel channels.

In high-flux reactors, provision must be made for cooling the rods to remove the heat of absorption by circulating a coolant—often a gas such as air or helium—through the thimble in which the rod moves.

TABLE 12-2. TYPICAL REACTIVITY ALLOWED FOR CONTROLS UNDER
NORMAL CONDITIONS*

Operational requirements (regulating rods)†	0.003
Temperature effects	0.00375
Fuel depletion and miscellaneous fission products	0.0015
Xenon poisoning	0.0045
Required for operation	0.01275
Allowance for experiments (isotope production, etc.)	0.02
Total for shim rods (minimum)	0.03275

* Individual losses taken from T. E. Cole [Design of a Control System for a Low-cost Research Reactor, *Nucleonics*, **11**(2):32–37 (1953)] for the light-water-moderated "swimming-pool" reactor, or BSR.

† The maximum change is kept [K. Way and E. P. Wigner, *Phys. Rev.*, **73**:1318 (1948)] below the delayed-neutron effect (Table 6-5) so that in the event of mis-operation of the regulating rods (such as malfunctioning of a servo control) the reactor cannot attain dangerously short periods.

12-5. Actuating Mechanisms.   It is usually desired to drive control rods at rates somewhere within the range of 0.001 to 0.1 per cent of reactivity change per second.   The lower limit is generally adequate for research reactors; in fact, it is desirable for keeping the neutron flux constant.   Military propulsion reactors use the highest rates, near the safe limit of the reactor.   Civilian power reactors have intermediate needs.

Direct-current motors are suitable up to about 20 hp in size.   Constant field excitation is used.   The armature voltage can be provided by a variable autotransformer and a rectifier, yielding a speed range adjustable over a ratio of about 10:1.   Starting time constants [10] are of the order of 0.10 to 0.01 sec.   With armature voltage control, dynamic braking is obtained, with accurate positioning resulting.

With servo control, the armature voltage is taken from a Thyratron output stage [12] or from a d-c generator driven by an a-c induction motor.   If larger powers are needed, as for moving a heavy reflector, the servo output can be boosted by interposing a d-c motor generator (Ward Leonard system) or an amplidyne generator [13] before the d-c motor. Three-phase servo motors which save 40 per cent of the weight of a d-c system have also been developed. For continuous servo control of lightweight regulating rods, two-phase servo motors can be used. These are reversible induction motors with special construction to give low inertia, ability to stall without overheating, and an approximately linear torque-speed curve at constant excitation. Maximum power is about 100 watts.

Standard single- and three-phase 60-cycle a-c motors can be used to drive the shim and safety rods at a constant speed, which can be set by a variable-speed gear drive if desired.

Hydraulic or pneumatic drive of control rods is feasible. It is found more often on safety rods. Since neutrons and γ radiation cause deterioration of oils, hydraulic systems must in general be located outside the shield. Control rods and gravity-actuated safety rods are frequently held up by magnets excited by a "magnet amplifier." This receives its input from the safety instrumentation and provides "trigger" release for rapid shutdown if the period is less than, say, 1 sec. A less immediately dangerous period such as 5 sec may be cause for insertion of regulating and shim rods by their drive motors.

The position of the rods can be measured within 0.0005 in. or so by counting the revolutions of the driving motor or gears. A selsyn generator on the motor shaft may be connected to a selsyn receiver and a turn indicator on the control panel [10, 11, 35].

12-6. Reactor Operation: Start-up Using Period Information. Even with all control and safety rods inserted, the effective reactor multiplication factor k_{eff} is of the order of 0.90 to 0.95; the steady-state neutron concentration is

$$n_0 = \frac{Sl^*}{1 - k_{eff}} \qquad (12\text{-}1)$$

where S represents the contribution of any actual neutron source plus effects of cosmic radiation, residual radioactivity from previous operation, etc., and l^* is the neutron lifetime.

If k_{eff} is increased to unity, the neutron concentration originally at n_0, corresponding to 1 to 10^3 counts per second, will rise slowly at a constant rate until theoretically it becomes infinite. Actually the contribution of the source becomes negligible as the rate of fuel fission becomes high so that a finite flux results. Because it would take a very long time for n/n_0 to approach a useful value of 10^{10} or higher under these conditions, for practical reasons the reactor is brought to power in a supercritical state. If flux-level information alone is available, the approach to criticality, supercriticality, and final power level consists of a cautious series of step changes in position of the control rods. Where period information is available, however, the process can be speeded up considerably.

For about four decades of neutron intensity above the source level (unless the reactor has been only very recently shut down) the flux must be measured by a logarithmic count-rate meter;* if its output is plotted on semilog paper, the period can be obtained by taking the slope of the graph. Alternately, the output can be differentiated electrically by an RC network to give a meter reading which is proportional to the reciprocal of the period. Such a counter is inserted near the center of the reactor; its usable range can be increased by withdrawing it into the shield as the flux increases.

To start up some reactors, the safety rods are withdrawn and locked into position if the count-rate meters give a reading. Although this may increase the multiplication constant to perhaps 0.99 or even greater, Eq. (12-1) shows that the flux level is still negligible. The control rods can be withdrawn at any desired speed, with the stipulation that at about the time the reactor becomes slightly supercritical (which unfortunately cannot be indicated by control-rod position, because the k_{eff} corresponding to a given rod position is affected by temperature, poisoning, burnup, etc.) the period will be long enough to allow the reactor to be brought smoothly and safely into the desired power.

The power at criticality is still rather low; solution of the subcritical neutron equations for various rates of change of k_{eff} show that n/n_0 is of the order of 10^2 regardless of the rate of withdrawal of the rods. As the flux rises, however, the capacity of the counter to resolve pulses is soon exceeded. Power information is then obtained from ionization chambers, usually located near the edge of the active zone or in the shield.† Use of chambers compensated for γ radiation, which will indicate a flux over a range of 100 times as large as that measurable by uncompensated ones, plus locating one or more chambers near the core of the reactor allow accurate flux measurement over the full range of power.

The period at criticality depends on the rate at which the rods are being withdrawn, and it keeps decreasing as the k_{eff} keeps rising above unity [38]. As soon as period information is available from amplifiers connected to the ion chambers, the rod movement can be controlled to keep the period constant at the minimum desired for the rise to full power—say 30 sec, as for the MTR. Once this stable period is attained, the rods need be moved merely to compensate for the change in temperature of the reactor as it heats up.

---

* This has the advantage of allowing measurement over a wide range of counting rates without scale switching.

† They are not located in the core because the saturation flux at which most ionization chambers become nonlinear because of ion recombination is of the order of 10^8 to 10^{10} neutrons/(cm²)(sec). This may be only 10^{-3} to 10^{-5} of the flux in the active section of a high-flux reactor.

12-7. Approach to Full Power; Start-up from Idling Power or Soon after Shutdown. Actual power-level information is needed only in the last two or three decades below operating power level. The flux can be measured roughly over a range of 10^6 by means of a logarithmic amplifier. For the final approach to the desired power level, a linear indication or record of the ion-chamber current is desirable. A stable d-c amplifier feeding a recorder, or a galvanometer which reads chamber current directly, can be used to measure power level within 1 per cent over a limited range, which can be extended to about 10^5 by the use of suitable

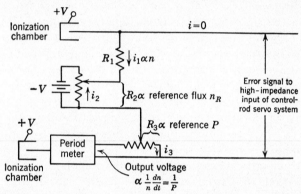

FIG. 12-2. Schematic circuit for shifting automatically from period to flux level control.

shunts. As the flux comes within, say, 1 per cent of the desired value, the actuation of the control rods must be based on the difference between the reference (desired) flux and the actual flux rather than on period. One circuit for changing reference signals from period to flux is given by Schultz [27].

Another simple system for doing this [37] is schematically shown in Fig. 12-2.

If R_1 is large compared to R_2 and R_3, then $i_1 \ll i_2$ and i_3, the error voltage $e(t)$ transmitted to the servo system is approximately

$$e(t) = i_1 R_1 - i_2 R_2 + i_3 R_3$$
$$= k_1 n - k_2 n_R + \frac{k_3}{n}\frac{dn}{dt}$$

At low fluxes, the error is determined by the value of $(1/n)dn/dt$ or $1/T_R$. As n becomes larger, $(1/n)dn/dt$ must decrease correspondingly for a given error.

Figure 12-3 sketches qualitatively the variations of flux and control-rod position during a startup [37]. Sources of information are indicated. In starting up soon after shutdown, chamber information is used—i.e., one starts higher on Fig. 12-3.

The detailed design of a control system by which a reactor can be started up and then controlled at a given power level is beyond the scope of this chapter. However, it is of interest to discuss some of the principles involved in combining the system to be controlled, the proper controller, and other components to give a desired performance.

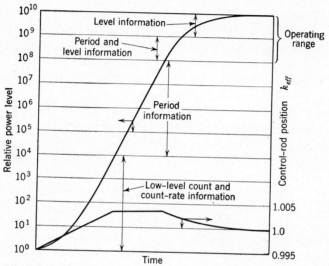

FIG. 12-3. Main sources of information for control and control-rod position during start-up.

12-8. Automatic versus Manual Control.

Automatic control is worthwhile only if the advantages over manual control justify the added investment. In most of the process industries the criteria for justification are economic; in atomic-energy installations, safety and military considerations may overbalance the economic ones. However, arguments in favor of automatic control are similar for all industries: (1) higher production rates; (2) control to closer tolerances, with more uniform output and fewer shutdowns; (3) saving of skilled labor; (4) performance of tasks which would be difficult or impossible under manual control.

It goes almost without saying that familiar process variables, such as the flow, pressure, and temperature of coolant, are regulated automatically in a nuclear plant wherever possible. Pneumatic or electrical control systems for these quantities have been developed by many manufacturers. On the other hand, few of the existing reactors *require* automatic control of neutron flux. The need for such control will increase, however, as more rigid specifications are placed on performance of reactors.

For example, manual control* of a production reactor can give satisfactory results. Such reactors operate continuously for weeks or months at the highest possible neutron flux consistent with the ability of the coolant to keep the temperatures of the fuel elements at safe levels. The only adjustment of the control rods normally required is an occasional withdrawal to counteract the poisoning effect of the increasing fission products. Shutdowns and start-ups for charging and maintenance are infrequent and can be carried out slowly under manual control. Extra supervision can be made available at these times if necessary. However, the average power level and corresponding production rate is always slightly below the maximum because of the intermittent nature of the adjustments. Automatic control, with more nearly continuous withdrawal of the rods, would give production rates more closely approaching the maximum by smoothing out the "sawteeth."

The primary purpose of the research reactor is usually to provide a steady neutron flux at specified points for special purposes, such as isotope production, materials testing, or the determination of nuclear properties. The design of such a system is complicated by the fact that the presence of experimental apparatus may change the characteristics of the reactor, control rods, or measuring systems. Since start-ups and shutdowns may be weekly or even daily, it may be economical to equip the reactor with instruments such that (1) it may be brought safely to the desired power automatically as quickly as thermal limitations permit, (2) start-up and shutdown may be carried out by operators of "technician" skill. The MTR can be brought up to power under automatic controls with periods as short as four times the minimum safe period; this is 40 per cent of the minimum period allowed under manual control.

Power reactors serve primarily as a source of heat in response to an external demand. In stationary-plant service, sudden demand changes would be small; larger changes usually would be periodic and predictable. If the power reactor is to be operated in conjunction with a fuel-fired or a hydroelectric plant, the nuclear plant will normally be run at full power, and fuel or water will be saved if the combined capacity is not needed. In mobile duty in aircraft or naval vessels there may be sudden, large, unpredictable changes in power requirements. In either case, however, the reactor power must correspond closely to that demanded; discrepancies should be slight enough and of short enough duration to be absorbed by the rest of the generating system without influencing the delivered power. It would be quite possible for an operator to shut

---

* By "manual control" is meant that a human operator turns the knob which sets the control rods in motion. The way that the rods follow the knob setting is determined by the rod positioning system, whose design requires a knowledge of the principles treated here.

down the plant inadvertently while attempting to follow a "step" change in demand; an automatic system can be designed so that this could not happen. In mobile reactors, especially those utilizing fast or intermediate-energy neutrons, it may be desirable or necessary that possible reactor periods be so short that safe or accurate manual control is difficult. It is obvious that automatic control of power-producing reactors is almost a necessity.

12-9. The Control Loop. A nuclear reactor and its associated control mechanisms make up a "closed-loop" or "feedback" system. The essential characteristic of such a system is that action of the controlling mechanism is initiated by deviation of the controlled variable, which in this case is the neutron flux, from the desired value.

Such a feedback control system is shown schematically in Fig. 12-4a. A reference signal, $r(t)$, a function of time which may be a voltage corresponding to the power demand, is fed into the "comparator" or "error"-measuring device.* Here it is compared with the actual value of the controlled variable $c(t)$ (suitably converted to a voltage $b(t)$ by the measuring elements and amplifiers); the difference between the two signals is designated the "error" $e(t)$. The error signal produces an action of the control elements intended to reduce the error to zero or to as small a value as possible with the control mechanism used. If manual control is used, the operator determines the action ("completes the loop").

This method of operation contrasts with that of an "open-loop" control system, in which the power demand signal alone is responsible for the position of the control rods. Sketched in Fig. 12-4b, such a controller depends for its accuracy upon the stability of the original calibration and unchanging characteristics of the controlled system. It does not know how good a job it is doing, nor can it do anything to improve a bad job. The dead-weight pressure regulator on certain types of pressure cookers is a common example of open-loop control.

In addition to being affected less by changing controller or process characteristics, closed-loop control puts less stringent requirements on controller sensitivity and "dead zone" in its response to changes in input. This results from the fact that small percentage changes in the controlled variable produce large percentage changes in the error, if the error is relatively small. By the use of closed-loop or feedback control, it is thus possible to achieve very close control over long periods of time with equipment built to reasonable tolerances and stability. A drawback is an increased tendency for the system to oscillate or "hunt." This tendency must be examined and allowed for in the design of the control system.

* The function $R(s)$, shown also in the figure, will be discussed later.

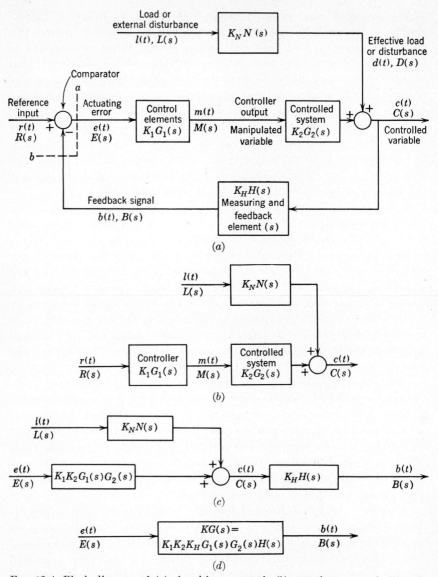

FIG. 12-4. Block diagram of (a) closed-loop control; (b) open-loop control; (c) open loop, equivalent of a; (d) open loop, no external load or disturbance.

12-10. Design Methods. To design a feedback control system which is to give a specified performance, one must be able to describe the behavior of each of the components shown in Fig. 12-4a. In a theoretical analysis the first step is to write the differential equation of the complete system (or its transformation according to rules of operational calculus).

The equation is then solved to show the behavior of the controlled variable or the error when the system is subjected to certain disturbances.

Alternatively, an experimental loop can be made up of actual components, together with a simulated controlled system, such as an electronic pile simulator [14, 15, 39]. The performance of the loop is then studied, and the loop is modified until the response is acceptable. Interchangeable components ideally suited for such "breadboard" synthesis are available commercially.

The two most commonly used methods of analysis are characterized by the type of disturbance imposed on the system:

1. "Transient" analysis subjects the system to a step or impulse change in reference signal or load. Values of the controlled variable or the error are expressed as functions of time as the controller attempts to reduce the error.

2. "Steady-state" analysis considers the system in the form of an open loop as shown in Fig. 12-4c, to which a sinusoidal error signal is continuously supplied. The performance of the system is described by relative amplitudes and phase angles of the controlled variable and the error signal as functions of the frequency of oscillation.

The transient-analysis approach gives in general a more realistic and easily visualized idea of the performance of the system in actual service. On the other hand, with complicated systems it may be difficult to write the higher-order differential equation in terms of the properties of the components, tedious to solve it and interpret the solution, and more difficult to determine what parameter to change in order to obtain the desired response characteristics.

The sinusoidal steady-state method applies techniques used in electrical-network or feedback-amplifier design. Its chief advantage is the ease with which the equation for a complex system can be built up from the "transfer functions" of each component. The effect of changing the characteristics of a given component is readily seen. The chief difficulty lies in correlating the performance of the system under sinusoidal excitation with its performance under transient conditions encountered in practice. The bases for correlating in common use are mainly empirical ones.

Either procedure of theoretical analysis is at all simple only when the components of the system are "linear"; i.e., when the behavior of the components can be expressed as linear differential equations with constant coefficients. When system elements are not linear over wide ranges of the variables, they may often be considered so for small changes. Best results with nonlinear systems come from experimental study of a simulated system made up of mechanical, hydraulic, pneumatic, or electrical analogues.

CHARACTERISTICS OF SYSTEM COMPONENTS

12-11. Description by Differential Equation. Many of the building blocks of a feedback control system can be visualized in fairly simple terms as resistances to the flow of energy or material and as storage reservoirs. In addition there are devices whose primary function is to transform the system variables. For example, an ideal amplifier multiplies; a synchro transmitter-control transformer combination changes angular difference into electrical potential. These "transformers" have resistance and storage characteristics which may be negligible under the conditions of normal operation but which can easily become important in high-speed control.

The process of analyzing the characteristics of the components of a system may be made clearer by the several examples.

Single-capacity Systems (*Simple Time Lag*). Of obvious interest in the control of nuclear reactors might be a mechanical positioning device by which an angular position θ_0 is obtained in response to some reference position θ_i. The device may be visualized as a control drum with poison on one side, as in the SIR, mounted at the end of a long shaft supported by bearings, with a crank at the input end outside the shield. Its angular position determines the pile reactivity. Alternatively, if the shaft were being used to raise and lower a vertical control rod through a rack-and-pinion arrangement, the weight of the rod might produce a torque Γ_0 on the output end. If it is assumed that the inertia effects are small, the net torque produced by twisting the input end with a torque Γ_i must be balanced by the bearing friction, which is proportional to the angular velocity:*

$$\Gamma_i + \Gamma_0 = f \frac{d\theta}{dt} \qquad (12\text{-}2)$$

The output angle will not correspond to the input position because of the elasticity of the shaft:

$$\theta_i - \theta_0 = \frac{1}{k} \Gamma_i \qquad (12\text{-}3)$$

where k is the coefficient of elasticity for the device.

A balance of torques, since inertia terms are absent, gives

$$\frac{f}{k} \frac{d\theta_0}{dt} + \theta_0 = \theta_i + \frac{1}{k} \Gamma_0 \qquad (12\text{-}4)$$

which is the differential equation from which the behavior of the positioner can be deduced.

* Γ_0 and Γ_i may be considered positive when they are in the direction of increasing θ.

A thermocouple, weighing W pounds, immersed in an air stream, is an example of a simple thermal system. Any difference in temperature between the air at T_i and the couple at an average temperature T produces a flow of heat at a rate dQ_0/dt. The flow is opposed by the thermal resistance R_{th} of the fluid film, commonly expressed [16] by the reciprocal of the product of the surface area A_0 and the "heat-transfer coefficient" h_0:

$$T_i - T = R_{th}\frac{dQ_0}{dt} = \frac{1}{h_0 A_0}\frac{dQ_0}{dt} \tag{12-5}$$

The temperature of the thermocouple will change from its original value T_0 as it stores or gives up heat:

$$T - T_0 = \frac{1}{W c_p}\int_{T_0}^{T}\frac{dQ_0}{dt}\,dt \tag{12-6}$$

where c_p is the specific heat of the metal. Differentiation gives the rate of change of T due to heat storage; if dQ_0/dt is eliminated by means of Eq. (12-5),

$$\frac{dT_0}{dt} = (T_i - T_0)\frac{h_0 A_0}{W c_p} \tag{12-7}$$

or

$$\frac{W c_p}{h_0 A_0}\frac{dT_0}{dt} + T_0 = T_i = \tau\frac{dT_0}{dt} + T_0 \tag{12-8}$$

The analogous electrical circuit is the resistance-capacitance network shown in Table 12-3(9). It is open-circuited at the output end. The flow of charge q_i through the resistance R requires a driving force $V_i - V_0$.

$$V_i - V_0 = R\frac{dq_i}{dt} \tag{12-9}$$

The charge is stored in the condenser, originally at potential V:

$$\frac{1}{C}\int\frac{dq_i}{dt}\,dt = V_0 - V \tag{12-10}$$

Differentiating Eq. (12-10) and eliminating the current dq_i/dt using Eq. (12-9) the differential equation of the combination is obtained:

$$RC\frac{dV_0}{dt} + V_0 = V_i \tag{12-11}$$

The first-order differential equation obtained in each case is characteristic of "single-capacity" devices. These have one energy or material storage unit, and the rate of transfer must be proportional to the first power of the potential difference. The coefficient of the derivative has the dimensions of time in each example and is known as the "time constant" τ. Table 12-3 makes evident the analogous quantities in the mechanical, electrical, and pneumatic terminology; mechanical, pneumatic, hydraulic, or chemical process control problems can be often solved by means of more readily constructed electrical analogues. The equations in the table are written in terms of the differential operator d/dt; d/dt applied to V gives dV/dt, etc.

TABLE 12-3. CONTROL-SYSTEM BUILDING BLOCKS

Component	Diagram	Differential equation*
1. Amplifier (a-c or d-c)		$V_2 = KV_1$
2. Synchro transmitter and control transformer		$V_2 = K(\theta_2 - \theta_1) = Ke$
3. Proportional controller		$m = \mu e$
4. Rate generator (tachometer)		$V_2 = K \dfrac{d\theta_1}{dt}$
5. Ideal proportional and derivative (lead) controller		$V_2 = \mu \left(RC \dfrac{d}{dt} + 1 \right) e$
6. Practical lead network		$\left(\dfrac{R_1}{R + R_1} RC \dfrac{d}{dt} + 1 \right) V_2 = \dfrac{R_1}{R + R_1} \left(RC \dfrac{d}{dt} + 1 \right) V_1$

626

7. Ideal proportional and integral controller (phase lag)	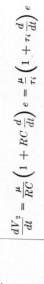 $\mu k = 1$	$\frac{dV_2}{dt} = \frac{\mu}{RC}\left(1 + RC\frac{d}{dt}\right)e = \frac{\mu}{\tau_i}\left(1 + \tau_i\frac{d}{dt}\right)e$
8. Practical phase lag network (to reduce steady-state error)		$\left[1 + (R + R_1)C\frac{d}{dt}\right]V_2 = \left(1 + R_1C\frac{d}{dt}\right)V_1$
9. Single-capacity system (simple time lag)	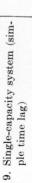 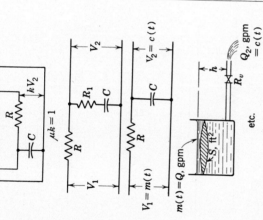 etc.	$\left(\tau\frac{d}{dt} + 1\right)c(t) = m(t)$ $\tau = RC$ $\tau = R_vS$
10. Two-capacity system (non-interacting)	 etc.	$\left(\tau_1\frac{d}{dt} + 1\right)\left(\tau_2\frac{d}{dt} + 1\right)c(t) = m(t)$ $\tau_1 = R_1C_1$ $\tau_2 = R_2C_2$

* Zero initial conditions.

TABLE 12-3. CONTROL-SYSTEM BUILDING BLOCKS (Continued)

Component	Diagram	Differential equation*
11. Rotating elastic mass (shaft positioner)		$\left(J\dfrac{d^2}{dt^2} + f\dfrac{d}{dt} + k \right)\theta = k\theta_R$
12. LRC stage		$\left(L\dfrac{d^2}{dt^2} + R\dfrac{d}{dt} + \dfrac{1}{C} \right)V_2 = \dfrac{1}{C}V_1$
13. D-c motor, shunt field control		$\left(J\dfrac{d^2}{dt^2} + f\dfrac{d}{dt} \right)\theta = ki_f$
14. Two-phase motor		$\left[J\dfrac{d^2}{dt^2} - \left(\dfrac{\partial\Gamma}{\partial n}\right)_v \dfrac{d}{dt} \right]\theta = \left(\dfrac{\partial n}{\partial V}\right)_n V_1$ where $\left(\dfrac{\partial\Gamma}{\partial n}\right)_v$ and $\left(\dfrac{\partial n}{\partial V}\right)_n$ are obtained from the motor torque-speed curves
15. D-c motor, armature voltage control		$\left(L_a J\dfrac{d^3}{dt^3} + JR_a\dfrac{d^2}{dt^2} + K_1 K_2\dfrac{d}{dt} \right)\theta = K_2 V_a$

628

Prob. 12-1. Estimate the time constant τ for a small thermocouple consisting of a spherical head of metal, having the properties of copper, 1 mm in diameter and submerged in an air-coolant stream moving such that the heat-transfer coefficient between gas and solid is 100 Btu/(hr)(ft²)(°F).

Solution. Neglecting the effect of the lead wires, the mass of the bead is $\rho \pi d^3/6$ and its area is πD^2. The density ρ is 557 lb/ft³ and $c_p = 0.093$ Btu/(lb)(°F) at the gas temperature. For a spherical particle τ in Eq. (12-8) equals $Dc_p\rho/6h_0$, which for the example given is 2.82×10^{-4} hr or 1.02 sec.

Multiple-storage-element Systems. If the inertia of the positioner becomes significant, a second-order differential equation is required to describe its properties. The torque balance [Eq. (12-3)], must include an equivalent torque $J(d^2\theta/dt^2)$ required to accelerate the rod, where J is the moment of inertia of the system. Thus Eq. (12-4) becomes

$$\frac{J}{k} \frac{d^2\theta_0}{dt^2} + \frac{f}{k} \frac{d\theta_0}{dt} + \theta_0 = \theta_i + \frac{1}{k} \Gamma_0 \qquad (12\text{-}12)$$

The equivalent electrical circuit is the inductance-resistance-capacity network shown in Table 12-3(12). A voltage drop of

$$L \frac{di}{dt} = L \frac{d^2q}{dt^2} = LC \frac{d^2V_0}{dt^2}$$

must be included.

The equation for the inductive circuit is

$$LC \frac{d^2V_0}{dt^2} + RC \frac{dV_0}{dt} + V_0 = V_i \qquad (12\text{-}13)$$

A second-order equation is also obtained when two single-capacity devices are connected in series, as in the electrical example of Table 12-3 (10). In this case, the voltage output V_1 of the first stage is the voltage input of the second; part of the entering charge q_i is used to build up V_1 in the condenser C_1.

The resistance equations are

$$V_i - V_0 = \frac{dq_i}{dt} R_1 + \frac{dq_0}{dt} R_2 \qquad (12\text{-}14)$$

$$V_1 - V_0 = \frac{dq_0}{dt} R_2 \qquad (12\text{-}15)$$

while the storage of energy is given by

$$C_1V_1 = \int \frac{dq_i}{dt} dt - \int \frac{dq_0}{dt} dt \qquad (12\text{-}16)$$

$$C_2V_0 = \int \frac{dq_0}{dt} dt \qquad (12\text{-}17)$$

E_i, q_i, and q_0 can be eliminated to give the equation for the two-capacity system:

$$(R_1C_1)(R_2C_2) \frac{d^2V_0}{dt^2} + (R_1C_1 + R_1C_2 + R_2C_2) \frac{dV_0}{dt} + V_0 = V_i \qquad (12\text{-}18)$$

The coefficients of the derivatives in Eq. (12-18) include not only the individual time constants $\tau_1 = R_1C_1$ and $\tau_2 = R_2C_2$ but also a term R_1C_2 containing properties from both parts of the system. This is characteristic of "interacting" systems, where the response of the first stage is influenced by that of the following stages. Such an interacting network must be treated as a single unit.

If the amount of charge taken by the second stage is negligible (i.e., if C_2 becomes very small*), a "noninteracting" system results. This is true even if the time constants τ_1 and τ_2 are kept the same as before by increasing R_2. If the term R_1C_2 is neglected, the equation for a two-capacity, noninteracting RC system is obtained:

$$(R_1C_1)(R_2C_2)\frac{d^2V_0}{dt^2} + (R_1C_1 + R_2C_2)\frac{dV_0}{dt} + V_0 = V_i \quad (12\text{-}19)$$

Equation 12-19 contains only τ_1 and τ_2. If the derivatives are written in terms of the operators d/dt and d^2/dt^2, the equation for n-capacity, noninteracting systems becomes obvious:

$$\left[\tau_1\tau_2\frac{d^2}{dt^2} + (\tau_1 + \tau_2)\frac{d}{dt} + 1\right]c(t) = m(t) \quad (12\text{-}20)$$

and, for n RC combinations in cascade,

$$\left[\left(\tau_1\frac{d}{dt} + 1\right)\left(\tau_2\frac{d}{dt} + 1\right) \cdots \left(\tau_n\frac{d}{dt} + 1\right) \cdots\right]c(t) = m(t) \quad (12\text{-}21)$$

$c(t)$ and $m(t)$ refer to the output and input variables, respectively. It should be noted that, although some of the devices listed in Table 12-3 contain several sources of resistance, inductance, and capacitance, they may behave like systems of lower order if some of the time constants are relatively small. For example, the time constant of a measuring element such as a thermocouple can be negligible compared to that of the self-balancing potentiometer to which it is connected. If the thermocouple is placed in a protecting well, the reverse may be true.

12-12. Differential Equation of Nuclear Reactor. It has been shown in Chap. 6 that the rate of change of neutron-flux concentration n at a given point in the reactor can be expressed by equations of the form

$$\frac{dn}{dt} = \frac{k_{eff} - 1}{l^*}n - \frac{\Sigma_i\beta_i}{l^*}n + \Sigma_i\lambda_ic_i + S \quad (12\text{-}22)$$

$$\frac{dc_i}{dt} = \frac{\beta_i}{l^*}n - \Sigma_i\lambda_ic_i \quad (12\text{-}23)$$

A fraction β_i of the total neutrons produced per fission comes from fission products of the ith kind, which have the respective decay constant λ_i.

* A cathode-coupled vacuum tube can be a much more practical isolating device [17].

The first two terms on the right-hand side of Eq. (12-22) represent the net rate of production of prompt neutrons; the third term is the contribution of the delayed neutrons coming from the fission products of concentration c_i. S is the effect of the source or residual radioactivity and is important only at the very low powers encountered during start-up or shutdown. l^* is the effective lifetime of a neutron.

Since the input variable k_{eff} and controlled or output variable n appear as a product, the reactor equation is nonlinear if k_{eff} varies.

Single Group of Delayed Neutrons.   For the purposes of this chapter the fission products of interest with U^{235} can be considered as one group producing the fraction $\beta = \Sigma_i \beta_i = 0.00755$ of the total neutrons and having a mean decay constant $\lambda = \beta/[\Sigma_i(\beta_i/\lambda_i)] = 0.081$ sec^{-1}.   Equations (12-22) and (12-23) become

$$\frac{dn}{dt} = \frac{k_{eff} - 1 - \beta}{l^*} n + \lambda c + S \tag{12-24}$$

$$\frac{dc}{dt} = \frac{\beta}{l^*} n - \lambda c \tag{12-25}$$

"Linearized" Reactor Equation.   Feedback-control techniques can be applied in a fairly straightforward manner for small excursions of power and values of k_{eff} near unity if the reactor behavior can be described in terms of a single group of delayed neutrons.   Under conditions such that the neutron concentration and fission-product concentrations depart from their equilibrium values n_0 and c_0 by small quantities δn and δc, respectively, Eq. (12-24) and (12-25) become

$$\frac{d(\delta n)}{dt} = \frac{n_0}{l^*} \delta k + \frac{1}{l^*} \delta k\, \delta n - \frac{\beta}{l^*} \delta n + \lambda\, \delta c \tag{12-26}$$

$$\frac{d(\delta c)}{dt} = \frac{\beta}{l^*} n_0 + \frac{\beta}{l^*} \delta n - \lambda c_0 - \lambda\, \delta c \tag{12-27}$$

where $\delta k = k_{eff} - 1$.   Since at equilibrium

$$\frac{\beta}{l^*} n_0 - \lambda c_0 = 0 \tag{12-28}$$

Eq. (12-27) can be rewritten as

$$\delta c = \frac{\beta/l^*}{d/dt + \lambda}\, \delta n \tag{12-28a}$$

If δn is small enough that the product $\delta k\, \delta n$ can be neglected, elimination of δc gives

$$\frac{d}{dt}\left(\frac{d}{dt} + \frac{\beta + \lambda l^*}{l^*}\right) \delta n = \frac{\lambda n_0}{l^*} \delta k + \frac{n_0}{l^*} \frac{d(\delta k)}{dt} \tag{12-29}$$

The reactor equation can be expressed in terms of $\ln n$ or the period τ_R:

$$\frac{d}{dt}\left(\frac{d}{dt} + \frac{\beta + \lambda l^*}{l^*}\right) \ln n = \frac{\lambda}{l^*}\,\delta k + \frac{1}{l^*}\frac{d(\delta k)}{dt} \qquad (12\text{-}30)$$

$$\frac{d(1/\tau_R)}{dt} + \frac{\beta + \lambda l^*}{l^*}\frac{1}{\tau_R} = \frac{\lambda}{l^*}\,\delta k + \frac{1}{l^*}\frac{d(\delta k)}{dt} \qquad (12\text{-}31)$$

Thus n or its logarithm varies as a second-order equation, while the reciprocal of the period is describable in terms of a first-order relationship.

The kinetic equations for reactors with circulating fuel, such as the proposed Brookhaven U-Bi reactor, must be modified to take into account the fact that some of the delayed-neutron emitters are being swept away before they have a chance to decompose. The result is that the delayed-neutron contribution to n is warped in the direction of flow of fuel, relative to its distribution if the fuel were stationary. The situation can be treated by an "effective" β in the noncirculating equations given here [36].

12-13. Transient Response of Simple Systems. One of the most useful indications of a system's performance characteristics is the response to a "step" change in an input variable. This sort of disturbance may occur when the control-rod position or the control point of a closed loop (e.g., the desired value of the flux) is suddenly altered.* The effect of a sudden sustained change in the external supply or demand is also of importance in the regulation of thermal, hydraulic, or power-conversion parts of a nuclear plant.

The transient response can be found by solving the differential equation. The actual path of the output variable as it approaches its new value is the sum of the "transient," or "force-free" solution (the complimentary function) plus the particular solution determined by the form of input to the system or "driving function" as given by the right-hand side of the equations shown.

Single-capacity Systems. The "force-free" part of the first-order equations typified by Eq. (12-4) is

$$\tau \frac{d\theta_0}{dt} + \theta_0 = 0 \qquad (12\text{-}32)$$

Where $\tau = f/k$. The force-free or complementary solution is obtained by separating variables and integrating to give

$$\theta_{0,comp} = A e^{-t/\tau} \qquad (12\text{-}33)$$

---

* Note the similarity between Eq. (12-29) and that for the motor-driven positioner of Prob. 12-2. The response of the reactor to a "step" change in δk would be the same as if the (error) input to the positioner were held at a fixed value.

If the right-hand side of Eq. (12-4) has a constant value $\theta_i + \Gamma_0/k$ for $t > 0$, a particular integral is the steady-state solution $\theta_{0,ss}$, which is obtained by letting the derivative term equal zero:

$$\theta_{0,ss} = \theta_i + \frac{1}{k}\Gamma_0 \qquad (12\text{-}34)$$

The general solution is then

$$\theta = Ae^{-t/\tau} + \theta_{0,ss} \qquad (12\text{-}35)$$

Substitution of the initial condition $\theta_0 = \theta_{00}$ at $t = 0$ gives a value for A. The final equation can be written in the form

$$\frac{\theta_0 - \theta_{00}}{\theta_{0,ss} - \theta_{00}} = 1 - e^{-t/\tau} \qquad (12\text{-}36)$$

which is plotted in Fig. 12-5. Differentiation of the equation will show that the initial rate of change of $(\theta_0 - \theta_{00})/(\theta_{0,ss} - \theta_{00})$ is $1/\tau$; in 4τ sec

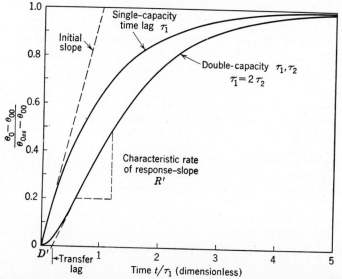

Fig. 12-5. Response of one- and two-capacity systems to "step" change in reference variable. Positioner without inertia taken as example of single-capacity system.

θ_0 has changed $(1 - e^{-4})$ or 98 per cent of the way to its new value. Obviously then, giving the crank end of the positioner a sudden twist results in a damped change in the position of the other end. Both input and output ends eventually turn through the same angle. Equation (12-34) indicates that a sudden change of load at the other end will result in a similar type of response, but output position will be different from before at the same crank position.

Prob. 12-2. A shim rod of a small reactor is driven by a d-c motor through a gear box and cable hoist. The motor armature voltage is fixed, and the operator controls the speed manually by changing the input voltage to an amplifier which excites the field. The motor torque can be assumed to be 0.25 lb-ft per volt input, and the effective inertia of rotor plus control rod is 0.03 slug-ft². The frictional damping is 1 lb-ft/(radian/sec). Because of the gear box, the control rod produces a negligible steady-state torque on the motor. One thousand revolutions of the motor corresponds to 1 in. of rod travel. If the amplifier input voltage is suddenly changed from zero to 10 volts, how fast will the rod be moving at steady state, and how long will it take for it to approach this speed within 2 per cent? The rod is originally at rest.

Solution. The differential equation of the rod drive is obtained by a torque balance:

$$J \frac{d^2\theta}{dt^2} + f \frac{d\theta}{dt} = KV_i$$

where $K = 0.25$ is the gain of the amplifier and motor field in terms of the input voltage V_i. In terms of rod speed,

$$\frac{dx}{dt} = \dot{x} = \frac{2\pi}{1000} \frac{d\theta}{dt} \quad \text{ips}$$

$$\frac{1000}{2\pi} J \frac{d\dot{x}}{dt} + \frac{1000}{2\pi} f\dot{x} = KV_i$$

which is a first-order equation in $\dot{x}$. Substituting numerical values gives

$$0.03 \frac{d\dot{x}}{dt} + \dot{x} = 0.00157 V_i$$

For $V_i = 10$ at $t > 0$, the steady-state rod speed, found by setting $d\dot{x}/dt = 0$, is 0.0157 in./sec. The coefficient of $d\dot{x}/dt$ in the last equation is the time constant τ; steady state will be reached within 2 per cent by 4τ or 0.12 sec.

Multicapacity Systems. The response of an "inductive" system such as that described by Eq. (12-12) can be found in the same manner. The value of the output variable at infinite time is again given by Eq. (12-34).

The transient solution is found by solving the left-hand side, which may be written as

$$\left(\frac{d^2}{dt^2} + \frac{f}{J} \frac{d}{dt} + \frac{k}{J} \right) \theta_0 = 0 \tag{12-37}$$

Equation (12-37) may be considered as a quadratic in the operator d/dt, having the roots r_1 and r_2.

$$\left(\frac{d}{dt} - r_1 \right) \left(\frac{d}{dt} - r_2 \right) \theta_0 = 0 \tag{12-38}$$

where
$$r_1, r_2 = -\frac{1}{2} \frac{f}{J} \pm \sqrt{\left(\frac{f}{J} \right)^2 - 4 \frac{k}{J}} \tag{12-39}$$

If $(f/J)^2 > 4(k/J)$, both roots will be real and negative. For this "overdamped" condition, θ_0 will change exponentially to its final value:

$$\theta_0 = \theta_{0,ss} + A_1 e^{+r_1 t} + A_2 e^{+r_2 t} \tag{12-40}$$

If it is assumed that the shaft is initially at rest and in equilibrium, substitution of the two initial conditions $\theta_0 = \theta_{00}$ and $d\theta_0/dt = 0$ at $t = 0$ will allow the evaluation of A_1 and A_2.

$$A_1 = -(\theta_{00} - \theta_{0,ss}) \frac{r_2}{r_1 - r_2}$$

$$A_2 = (\theta_{00} - \theta_{0,ss}) \frac{r_1}{r_1 - r_2}$$

The equation for the overdamped response is then

$$\frac{\theta_0 - \theta_{00}}{\theta_{0,ss} - \theta_{00}} = 1 - \frac{1}{r_1 - r_2} (r_1 e^{-r_2 t} - r_2 e^{-r_1 t}) \tag{12-41}$$

The rate at which θ_0 approaches $\theta_{0,ss}$ may be increased by decreasing the frictional damping f or by increasing the shaft stiffness k. Eventually a point is reached where $(f/J)^2 = 4k/J$, or $f = 2\sqrt{kJ}$. The mechanism is then "critically damped," and any less damping will result in an oscillating approach to $\theta_{0,ss}$.

At the critically damped condition, the two roots of the characteristic equation (12-37) are equal (the square-root term becomes zero):

$$r = r_1 = r_2 = -\frac{1}{2}\frac{f}{J} \tag{12-42}$$

The solution of Eq. (12-34) is now

$$\theta_0 = \theta_{0,ss} + A_1 e^{-rt} + A_2 t e^{-rt} \tag{12-43}$$

If the constants A_1 and A_2 are found for the same initial conditions as above, the equation for the critically damped response is obtained:

$$\frac{\theta_0 - \theta_{00}}{\theta_{0,ss} - \theta_{00}} = 1 - e^{-rt} - rt e^{-rt} \tag{12-44}$$

A damped oscillating approach occurs when $(f/J)^2 < 4k/J$, or $f < 2\sqrt{kJ}$. The roots r_1 and r_2 are complex:

$$r_1, r_2 = -\frac{1}{2}\frac{f}{J} \pm \frac{j}{2}\sqrt{4\frac{k}{J} - \left(\frac{f}{J}\right)^2} \tag{12-45}$$

where $j = \sqrt{-1}$. The general solution is

$$\theta_0 = \theta_{0,ss} + e^{-\frac{1}{2}(f/J)t}(A_1 e^{j\omega t} + A_2 e^{-j\omega t}) \tag{12-46}$$

where

$$\omega = \frac{1}{2}\sqrt{4\frac{k}{J} - \left(\frac{f}{J}\right)^2} = \sqrt{\frac{k}{J} - \left(\frac{f}{2J}\right)^2} \tag{12-47}$$

The exponentials in $j\omega t$ can be defined as

$$e^{j\omega t} = \cos \omega t + j \sin \omega t \tag{12-48}$$
$$e^{-j\omega t} = \cos \omega t - j \sin \omega t \tag{12-49}$$

so that Eq. (12-46) becomes

$$\theta_0 = \theta_{0,ss} + e^{-\frac{1}{2}(f/J)t}[(A_1 + A_2)\cos \omega t + (A_1 - A_2)j \sin \omega t] \tag{12-50}$$

A_1 and A_2 are complex conjugate numbers of the form $x \pm jy$. Therefore, Eq. (12-50) can be expressed in terms of two constants $A_3 = A_1 + A_2$ and $A_4 = (A_1 - A_2)j$, whose values can be found from the initial conditions. The final solution is

$$\frac{\theta_0 - \theta_{00}}{\theta_{0,ss} - \theta_{00}} = 1 - e^{-\frac{1}{2}(f/J)t}\left[\cos\omega t + \frac{\frac{1}{2}f/J}{\omega}\sin\omega t\right] \tag{12-51}$$

Equation (12-51) can be written to show more clearly that the "underdamped" response is a damped oscillation of frequency ω:

$$\frac{\theta_0 - \theta_{00}}{\theta_{0,ss} - \theta_{00}} = 1 - e^{-\frac{1}{2}(f/J)t}\left[\sqrt{1 + \left(\frac{f}{2J\omega}\right)^2}\sin(\omega t + \psi)\right] \tag{12-52}$$

where

$$\psi = \tan^{-1}\frac{2J\omega}{f} \tag{12-53}$$

If the damping is reduced to zero, the system will oscillate indefinitely at its "natural" frequency $\omega_n = \sqrt{k/J}$. Brown and Campbell [18] have shown that it may be convenient to write the equations dimensionlessly in terms of ω_n and the ratio $\zeta = f/f_c$ of the actual damping f to the critical damping $f_c = 2\sqrt{kJ}$. The actual frequency of oscillation ω is equal to $\omega_n\sqrt{1 - \zeta^2}$, and the damping factor $f/2J$ becomes $\zeta\omega_n$. Thus Eq. (12-12) becomes

$$\left(\frac{d^2}{dt^2} + 2\zeta\omega_n\frac{d}{dt} + \omega_n{}^2\right)\theta_0 = \omega_n{}^2\theta_i + \frac{\Gamma_0}{J} \tag{12-54}$$

and the roots r_1, r_2 are

$$r_1, r_2 = -\zeta\omega_n \pm f\omega\sqrt{1 - \zeta^2} \tag{12-55}$$

Equations (12-52) and (12-53) are changed to

$$\frac{\theta_0 - \theta_{00}}{\theta_{0,ss} - \theta_{00}} = 1 - e^{-\zeta\omega_n t}\left[\frac{1}{\sqrt{1 - \zeta^2}}\cos(\omega_n\sqrt{1 - \zeta^2}\,t + \psi)\right] \tag{12-56}$$

$$\psi = \tan^{-1}\frac{\sqrt{1 - \zeta^2}}{\zeta} \tag{12-57}$$

Figure 12-6 shows the response of such a system to a step input for several degrees of damping.

The noninteracting two-capacity system exemplified by Eq. (12-20) always has real, negative roots; consequently it is not oscillatory by itself. A typical curve is given in Fig. 12-5. The primary effect of the added capacity is to introduce an initial "transfer lag" into the response curve and to lower its maximum rate of rise. The product of these two characteristics of the curve is often used to characterize the controllability of process systems whose differential equations cannot be written exactly. The larger the products, the more difficult is the control.

In general, the differential equation of a linear system will be in the form

$$\left(a_n\frac{d^n}{dt^n} + a_{n-1}\frac{d^{n-1}}{dt^{n-1}} + \cdots a_0\right)c(t) = b_m\left(\frac{d^m}{dt^m} + \cdots b_0\right)m(t) \tag{12-58}$$

The response to any input $m(t)$ will be made up of the sum of (1) exponentials of the form $Ae^{\pm t/T_i}$ for every real root of the characteristic equation; (2) oscillatory terms of the form $Be^{\alpha t}\cos(\omega t + \psi)$ for every complex root; and (3) a particular solution which is determined by the "driving function" $m(t)$. T_i, α, and ω are made up of system properties; A, B, and ψ are evaluated from the initial conditions.

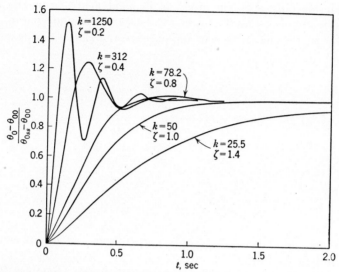

FIG. 12-6. Response of positioner with inertia $J = 2$ slug-ft^2, $f = 10$ ft-lb/(radian/sec) plotted against time.

Prob. 12-3. Plot on the same graph the transient response $(c - c_0)/(c_{ss} - c_0)$ to a step input of the following systems: (a) simple single-capacity time lag of $\tau = 1$ sec; (b) three-capacity "noninteracting" system $\tau_1 = \tau_2 = \tau_3 = \frac{1}{3}$ sec; (c) three-capacity "noninteracting" system $\tau_1 = 10\tau_2 = 10\tau_3 = 1$ sec.

Self-regulation. Systems which contain the a_0 term and no roots with positive real parts show a finite steady state c_{ss} in response to a step change in m. They are known as "self-regulating" systems and are typified by those mentioned thus far in the discussion. The d-c motor with field control, in Table 12-3, is not self-regulating with respect to shaft angle, but Prob. 12-2 showed it to be so with respect to speed.

12-14. Response of Nuclear Reactor to Step Input. If a reactor is operating in a steady state corresponding to a neutron concentration n_0 such that the contribution of the source is negligible, the variation in n following a sudden sustained change of k_{eff} from its initial value of unity can be found by solving the simultaneous equations (12-22) and (12-23). Substitution of Eq. (12-23) into Eq. (12-22) gives

$$\left[\frac{d}{dt} - \frac{k_{eff} - 1 - \sum_i \beta_i}{l^*} - \sum_{i=1}^{i=k} \frac{\lambda_i \beta_i}{l^* \left(\frac{d}{dt} + \lambda_i \right)} \right] n = 0 \qquad (12\text{-}59)$$

The response n/n_0 is the sum of $k + 1$ exponentials in time corresponding to the $n + 1$ roots of Eq. (12-59). The solution has been treated in detail elsewhere [19, 20, 38].

Considering only a single group of delayed neutrons, if the change in multiplication constant $\delta k = (k_{eff} - 1)$ is small enough that $\beta - \delta k + \lambda l^* \gg 2\lambda l^* \, \delta k$, the change in neutron concentration can be approximated by

$$\frac{n}{n_0} = A_1 e^{r_1 t} + A_2 e^{r_2 t} \tag{12-60}$$

where
$$r_1 \approx \frac{\lambda \, \delta k}{\beta - \delta k + \lambda l^*} \tag{12-61}$$

$$r_2 \approx -\frac{\beta - \delta k + \lambda l^*}{l^*} - \frac{\lambda \, \delta k}{\beta - \delta k + \lambda l^*} \tag{12-62}$$

$$A_1 \approx \frac{\beta + \lambda l^*}{\beta - \delta k + \lambda l^* + (\lambda l^* \, \delta k)/(\beta - \delta k + \lambda l^*)} \tag{12-63}$$

$$A_2 \approx \frac{\delta k(\beta - \delta k)}{(\beta - \delta k + \lambda l^*)^2 + 2\lambda l^* \, \delta k} \tag{12-64}$$

The constants A_1 and A_2 can be found from the initial conditions $n = n_0$ and $dn/dt = (k/l^*) \, n_0$ at $t = 0$.

For a typical natural-uranium reactor such as the Argonne CP-2 [21], l^* is approximately equal to the value 1.25×10^{-3} sec. For reactivities of about 0.003 or less, the terms $\lambda \, \delta k$ and $\lambda l^* \, \delta k$ are small enough that the response can be taken to be

$$\frac{n}{n_0} = \frac{\beta}{\beta - \delta k} e^{\frac{\lambda \delta k}{\beta - \delta k} t} - \frac{\delta k}{\beta - \delta k} e^{-\frac{\beta - \delta k}{l^*} t} \tag{12-65}$$

which is plotted in Fig. 12-7 for $\delta k = 0.0025$ and $\delta k = 0.00025$. There is a very rapid initial change due to the steady-state neutron concentration, corresponding to a "period" $T_R = \dfrac{n}{dn/dt} = \dfrac{l^*}{\delta k}$.† The periods at zero time were 0.5 and 5 sec, respectively. After about $4l^*/(\beta - \delta k)$ sec, the power is changing at a rate determined by the slow neutrons alone, with a stable period of $(\beta - \delta k)/(\lambda \, \delta k)$ sec. For $\delta k = 0.0025$, this period is 25 sec; for decreasing power at $\delta k = -0.0025$, the stable period is -49 sec. It is seen that the delayed neutrons make control of a thermal reactor much easier than if all the neutrons were "prompt."

The reactor as described by the equations above is evidently not a "self-regulating" system; however, this of course does not mean that it is uncontrollable.

† The initial transient is similar to that which would occur if there were no delayed neutrons.

If the "linearized" reactor equation (12-29) is integrated for a step change in δk and the same initial conditions, and if λl^* is neglected in comparison to β, the response is found to be

$$\frac{n}{n_0} \approx \frac{\delta k}{\beta} (1 - e^{-\beta t/l^*}) + 1 + \frac{\lambda \, \delta k}{\beta} t \qquad (12\text{-}66)$$

Equation (12-66) indicates an eventual linear increase in flux with time; except for the constant term $\delta k/\beta$, it is equivalent to Eq. (12-65) when δk is small compared to β and if only the first two terms of the expansion of the positive exponential are considered.

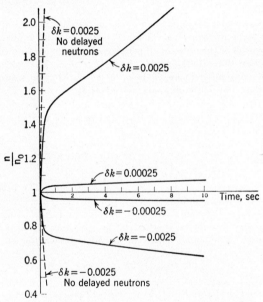

FIG. 12-7. Reactor response to step input in δk.

12-15. Effect of Temperature Coefficient on Response.

A negative temperature coefficient of reactivity can offer protection against power overloads and can give a "self-regulating" characteristic to reactor performance. For example, a finite power increase may be obtained merely by pulling out the control rods an amount corresponding to δk_0 at the original temperature. If δk_0 is small enough to avoid fuel overheating or undesired vaporization of coolant or moderator immediately after the reactivity is increased, the power level and temperature level will rise until δk_0 is counteracted. The flexibility of such a method of control is limited by the rates of heat transfer from fuel to coolant and from coolant to moderator and by the heat capacity of the reactor. Because the temperature effect will tend to cancel the effect of control rod motion, a "scram" or other rapid decrease in power requires more rapid rod insertion than would be needed to obtain the same dn/dt on withdrawal to increase power [20].

The thermal effect can be treated analytically by considering the reactor neutron equations simultaneously with those relating the average reactor temperature to power level [22]. Since the temperature in a large heterogeneous reactor varies locally from fuel to coolant, rises parallel to the coolant flow, and falls off perpendicular to the cooling channels much as does the neutron concentration, an "average" temperature must be chosen. For small or homogeneous reactors, the problem is simpler.

As an example, consider a small, "fast" reactor, of mass W and mean specific heat c_p. It is cooled by immersion in a bath held at a constant temperature T_{cm}. For changes in power level of the order of 10 per cent or less the neutron concentration can be assumed to obey the equation $d(\delta n)/dt = (\delta k/l^*)n_0$. The excess multiplication factor δk may be considered to be made up of a contribution from the control-rod motion δk_i plus that due to the temperature effect $\delta k_T = \alpha(T - T_0)$, where T is the average reactor temperature, T_0 the temperature under the steady-state conditions at $t < 0$, when the flux is equivalent to n_0, and α is the temperature coefficient of reactivity. The neutron equation is then

$$\frac{d\,\delta n}{dt} = \left[\frac{\delta k_i}{l^*} + \frac{\alpha}{l^*}\,(T - T_0)\right]n_0 \qquad (12\text{-}67)$$

The reactor generates heat at a rate $Kn = K(n_0 + \delta n)$. The heat is removed to the coolant at a rate proportional to the over-all heat transfer coefficient U from reactor to coolant, the area A available for heat transfer, and the driving force $T = T_{cm}$. Any excess will accumulate in the reactor at a rate $Wc_p(dT/dt)$, appearing as a rise in reactor temperature. The heat balance can therefore be written as

$$Wc_p\frac{dT}{dt} = K(n_0 + \delta n) - UA(T - T_{cm}) \qquad (12\text{-}68)$$

If T is eliminated between Eq. (12-67) and (12-68), a second-order equation is obtained:

$$\left(\frac{d^2}{dt^2} + \frac{UA}{Wc_p}\frac{d}{dt} - \frac{\alpha}{l^*}\frac{Kn_0}{Wc_p}\right)\delta n = \frac{UAn_0}{l^*Wc_p}\,\delta k_i + \frac{n_0}{l^*}\frac{d\,\delta k_i}{dt} - \frac{\alpha}{l^*}\frac{UAn_0}{Wc_p}\,(T_0 - T_{cm})$$
$$+ \frac{\alpha Kn_0{}^2}{l^*Wc_p} \qquad (12\text{-}69)$$

The exponential response to a step increase in δk_i is in terms of the roots

$$r_1, r_2 = -\frac{UA}{2Wc_p} \pm \frac{1}{2}\sqrt{\left(\frac{UA}{Wc_p}\right)^2 + \frac{4k\alpha n_0}{Wc_p l^*}} \qquad (12\text{-}70)$$

Any positive temperature coefficient will result in a positive r_1, or an increase of flux to infinity. If α is a small negative number, the neutron concentration will level off in an overdamped manner at a new steady-state value

$$\delta n_{ss} = -\frac{UA}{K}\frac{\delta k_i}{\alpha} = +\frac{UA}{K}\,(T_{ss} - T_0) \qquad (12\text{-}71)$$

If α is more negative than $\alpha_c = -(UA)^2 l^*/4KWc_p n_0$, the approach to n_{ss} will be oscillatory. The type of response can be influenced by the power level or by merely changing the degree of agitation of the coolant (and hence U).

The thermal effects obviously introduce nonlinearities in the response which complicate the theoretical design of the control system and which are not easily built into a simulator. Often the heat of a heterogeneous reactor is removed by a circulating coolant, which in turn transfers it to an external load such as a boiler (an essentially constant temperature sink) or another circulating loop. The latter case is used as an isolating loop when the primary coolant may be corrosive as well as radioactive, e.g., sodium. Each set of heat-transfer surfaces will contribute a time constant τ to the equations which will determine the reactor temperature. In addition there will be a series of "transport lags," due to the time it takes for coolant to travel from reactor to boiler, for example. These may be approximated by additional time constants (usually two or three will suffice; the effect was shown in Fig. 12.5) in the equation. The resulting set of simultaneous equations is probably best solved using a complete process analogue [22a, 39].

12-16. Response to Steadily Changing Input. If a single-capacity system such as an ionization chamber is exposed to a neutron concentration* n increasing at a steady rate $dn/dt = n'$, the output voltage V across the grid resistance of the amplifier can be found by solving the differential equation

$$\left(\tau \frac{d}{dt} + 1\right) V = n = n't \qquad (12\text{-}72)$$

where τ is of the order of 0.01 sec [21].

V_{ss}, the particular solution, will be of the form $V_{ss} = a + bt$; if this solution is substituted into Eq. (12-72), a and b can be evaluated:†

$$\tau b + a + bt = n't$$

which is satisfied if $b = n'$ and $a = -\tau b = -\tau n$.

The complete solution for $V = 0$ at $t = 0$, is

$$V = \tau n' e^{-t/\tau} + n'(t - \tau) \qquad (12\text{-}73)$$

The steady-state output is linear with time and lags the input by τ sec.

Prob. 12-4. A thermopile with a time constant $\tau = 10$ sec is being used to monitor power in a research reactor. If the reactor has been operating at a steady neutron

---

* Neutron flux Φ is more appropriate than concentration n, but if the effective neutron velocity is constant (see Sec. 6-6), n may be used.

† This method of obtaining the particular solution is known as the "method of undetermined coefficients" [23].

flux n_0 and the control rods are withdrawn suddenly, the flux may be assumed to increase exponentially in accordance with the stable period $T_R = 20$ sec and

$$n = n_0 e^{t/T_R}$$

Show that (a) the thermopile will be in error by a constant *percentage* of the actual flux during the time of power change; (b) the error in the indicated reactor *period* will be constant and equal to 10 sec.

The underdamped positioner described by Eq. (12-12), when turned at a constant rate $d\theta_i/dt = \omega_i$ will eventually settle down so that its output end rotates at the same angular velocity as the input but lagging behind at an angle of $\Gamma_0/k + (f/k)\omega_i$. The error under such conditions is known as the "steady-state follow-up error" and will be discussed further in Sec. 12-37.

12-17. Reactor Response to Steady Change in Reactivity. No control rod of finite mass can be moved so fast as to produce a "step" change in reactivity; so it is of interest to estimate the response if the multiplication factor is changed at a constant rate $d\delta k/dt$. This may be considered to be approximately the effect of withdrawing rods slowly at a constant speed.

If the reactivity changes slowly, Hurwitz [24] has shown that the neutron concentration at time t after the beginning of an arbitrary change in k_{eff} may be approximated by

$$\frac{n}{n_0} = \sqrt{\frac{1 + \Sigma \beta_i/l^* \lambda_i}{1 + \Sigma \dfrac{\beta_i/l^* \lambda_i}{(1 + p/\lambda_i)^2}}}\ e^{\int_0^t p(t')dt'} \tag{12-74}$$

where p is the algebraically largest root of Eq. (12-59) and $k = (k_{eff} - 1)$ is a function of time. For a single group of delayed neutrons,

$$\frac{n}{n_0} = \sqrt{\frac{1 + \beta/\lambda l^*}{1 + \dfrac{\beta/\lambda l^*}{(1 + p/\lambda)^2}}}\ e^{\int_0^t p(t')dt'} \tag{12-75}$$

If the multiplication factor changes at a constant rate, departing from its initial value of unity by less than about 0.003, $p(t')$ is given by

$$p(t') = \frac{\lambda \dfrac{d\,\delta k}{dt} t'}{l^* + \beta - \dfrac{d\,\delta k}{dt} t'} \tag{12-76}$$

$$\int_0^t p(t')dt' = \lambda t \left[\frac{l^*\lambda + \beta}{\dfrac{d\,\delta k}{dt} t} \ln \frac{1}{1 - \dfrac{(d\,\delta k/dt)t}{\lambda l^* + \beta}} - 1 \right] \tag{12-77}$$

The square-root term in Eq. (12-75) is so nearly constant that an average value may be used.

If k_{eff} of the reactor considered in the numerical example of the previous paragraph is changed from unity to 1.0025 in t sec, the neutron concentration at the end of that time is approximated by

$$\frac{n}{n_0} = \frac{1.0 + 1.46}{2} e^{0.0173t} \tag{12-78}$$

If $t = 10$ sec (probably a shorter time than the assumptions will allow), the power will have increased by 46 per cent by the time that the reactivity change has been made. If the rods are moved more slowly so that $t = 50$ sec, $n/n_0 = 2.82$ by the time the rods have completed their travel.

If the rods are immediately returned to their original positions at the same rate, the exponent in Eq. (12-78) is doubled, while the average square-root term remains the same. The net increase in power due to a 20-sec excursion to $\delta k = 0.0025$ and return is 73 per cent; if the excursion were to take 100 sec, the final power at $\delta k = 0$ would be 693 per cent of the original. Although the reactivity change considered is much larger than would usually occur from, say, a sudden change in coolant inlet temperature, it is obvious that the control system must be able to respond rapidly to correct accidental changes in k_{eff}.

12-18. Response of "Linearized Reactor" to Constant Rate of Change of δk. The response for small power changes and very small δk's can be found by integrating Eq. (12-29) for $d \, \delta k/dt = $ constant $= \delta k'$ and initial conditions $n = n_0$, $dn/dt = 0$ at $t = 0$. The change in neutron concentration can be approximated by a response which eventually becomes a parabolic function of time:

$$\frac{n}{n_0} = 1 - l^* \delta k'(1 - e^{-\beta/l^*t}) + \frac{\delta k'}{\beta} t + \frac{\lambda}{2} \frac{\delta k'}{\beta} t^2 \tag{12-79}$$

Prob. 12-5. The power of a thermal reactor is to be controlled by driving the rods at only one constant speed corresponding to $\delta k' = 10^{-6}$ sec^{-1}. Ideally, in the absence of temperature-coefficient compensation (cf. Sec. 12-15), if the power is to be changed by a certain amount, the motion of the rods must be reversed ($\delta k'$ changes sign) before the desired flux has been attained. As the final power is reached, the rods will then be back in their original position. Actually this is difficult to carry out; there may be an oscillating approach. If the neutron flux can be approximated by Eq. 12.29, with $l^* = 1.25 \times 10^{-3}$ sec, $\beta = 0.081$ sec^{-1}, and $\lambda = 0.00755$, for a 5 per cent power increase:

a. At what value of n/n_0 should the motion of the rods be reversed? (HINT: For the return of the rods, find an equation similar to Eq. (12-79) but use n/n_0 and dn/dt at the end of the rod withdrawal as initial conditions for the insertion.)

b. How long does the complete excursion take?

c. Plot the period as a function of time during the power change.

SYNTHESIS OF CONTROL SYSTEM

12-19. Addition of Controller—Proportional Controller Action. The performance of the closed loop is based on the equation of the entire loop, in which the input variable to the controlled system derives from the output value through the error-measuring device and the control action. The simplest controller is one which changes the input or manipulated variable by an amount proportional to the error. In the regulation of reactor coolant flow, for example, such control might be arranged so that there exists a given position of the coolant supply valve for every outlet temperature of the coolant. At some reactor power level there may be zero error between the actual outlet temperature and the desired constant value; if the power level changes, the outlet temperature will change enough to give a new coolant flow at some new steady state. The more sensitive the controller, the less the outlet temperature will have to vary, assuming that the system remains stable.

Inserted into the elastic shaft-positioning device of Eq. (12-12) between the handwheel position θ_R and the shaft, a proportional controller (actually consisting of a means to measure the output position θ_0, a comparator, and a lever system to move the input end of the shaft) can be built so as to allow zero error at one given θ_R and Γ_0, denoted by θ_{R0} and Γ_{00}. If zero error $e(t)$, zero acceleration, and zero angular velocity are the initial conditions,

$$\theta_{R0} = \theta_{00} = \theta_{i0} + \frac{1}{k}\Gamma_{00}$$

The proportional control action is defined by

$$\theta_i - \theta_{i0} = \mu(\theta_R - \theta_0) = \mu e(t) \tag{12-80}$$

If at time $t = 0$ the reference angle and load torque are suddenly changed to new constant values θ_R and Γ_0, the differential equation of the closed-loop system can be found by substituting the action of the controller for the input angle θ_i on the right of Eq. (12-12):

$$\frac{J}{k}\frac{d^2\theta}{dt^2} + \frac{f}{k}\frac{d\theta_0}{dt} + \theta_0 = \mu(\theta_R - \theta_0) + \theta_{00} + \frac{1}{k}(\Gamma_0 - \Gamma_{00}) \tag{12-81}$$

The steady-state output angle will be

$$\theta_{0,ss} = \frac{\mu}{1+\mu}\theta_R + \frac{\theta_{00}}{1+\mu} + \frac{\Gamma_0 - \Gamma_{00}}{k(1+\mu)} \tag{12-82}$$

corresponding to an error $\theta_R - \theta_{0,ss}$ of

$$e_{ss} = \frac{1}{1+\mu}(\theta_R - \theta_{R0}) - \frac{\Gamma_0 - \Gamma_{00}}{k(1+\mu)} \tag{12-83}$$

Comparison of Eq. (12-82) with Eq. (12-34) shows that addition of the controller decreases the error due to load changes by a factor $1/(1+\mu)$: however, the controlled system will not follow exactly a change in the set point or desired value θ_R.

The transient response will be the same as that of an open-loop system of increased stiffness $k(1 + \mu)$. Equations of the form of those in Sec. 12-13 will still describe the performance, except that, e.g., the "natural" undamped frequency of oscillation will be a function of the controller sensitivity μ:

$$\omega_n = \sqrt{\frac{k(1 + \mu)}{J}} \tag{12-84}$$

If the mechanical system is considerably overdamped, the nature of the response can be controlled by variation of μ, since

$$\zeta = \frac{f}{2\sqrt{kJ(1 + \mu)}} \tag{12-85}$$

If it is nearly critically damped without the controller, the controller must be of low sensitivity if the response is to be kept in the optimum, slightly underdamped range $0.4 < \zeta < 1.0$. The curves of Fig. 12-5 still apply, but the effects of changing k on ω_n, and ζ on the time scale of the graphs must be understood.

It is easy to show that addition of proportional control to a single-capacity system can never produce oscillations. Adding it to a two-capacity noninteracting system [Eq. (12-20)] will produce damped oscillations for all values of μ greater than $\mu_c = (\tau_1 - \tau_2)^2/4\tau_1\tau_2$.* A three-capacity system will be critically damped at $\mu_c = f(d/dt)_{\min}/a_0$, where $f(d/dt)_{\min}$ is the most negative minimum value of the characteristic equation $f(d/dt) = 0$ of the *system without controller*, found by setting $d[f(d/dt)]/d(d/dt)$ equal to zero, and a_0 is defined in Eq. (12-58). If the gain of the controller is increased beyond the limiting value

$$\mu_0 = \frac{a_2 a_1}{a_3} - 1 = \left(\frac{1}{\tau_1} + \frac{1}{\tau_2} + \frac{1}{\tau_3}\right)\left(\frac{1}{\tau_1\tau_2} + \frac{1}{\tau_1\tau_3} + \frac{1}{\tau_2\tau_3}\right)(\tau_1\tau_2\tau_3) - 1 \tag{12-86}$$

a positive real root will exist and the system will be unstable, since the response will include an exponential term with a positive exponent. It is of interest to note that the uncontrolled three-capacity noninteracting system is nonoscillatory when subjected to a step input change. The same is true for an n-capacity RC system.

Prob. 12-6. A control rod is being positioned by a d-c motor whose torque Γ is proportional to its field current i_f, which is merely the amplified error signal e between the desired position of the rod θ_R and the actual position θ. The motor plus control rod have an effective moment of inertia $J = 0.01$ lb$_F$-ft-sec^2, and the frictional damping coefficient f is 0.16 lb$_F$-ft-sec. (a) What must be the gain $K = \Gamma/e$ of the amplifier plus motor for the system to be critically damped? (b) Plot the control-rod position θ and the error e as functions of time following a "step" change in the set point θ_R from zero to $\pi/2$ (at $\theta_R = 0$, there is zero error), assuming critical damping and a gain equal to 25 times K_{crit}. (c) Repeat for a steady rotation of the set point at a constant rate $d\theta_R/dt = \pi/10$ rad/sec. (d) What are the maximum horsepower

* The fact that any control at all will produce oscillatory response if $\tau_1 = \tau_2$ indicates that multielement systems with equal time constants are the most difficult to control.

and torque of the motor? (HINT: For the differential equation of the system, consider Prob. 12-2.)

12-20. Generalized Criteria for Stability. For a closed-loop system to be stable, the characteristic equation of the complete system (including controller) must have no positive real roots or complex roots with positive real parts. If the characteristic equation is written as

$$a_0 \frac{d^n}{dt^n} + a_1 \frac{d^{n-1}}{dt^{n-1}} + \cdots a_{n-1} \frac{d}{dt} + a_n = 0 \qquad (12\text{-}87)$$

and the coefficients a_k do not have the same sign or if any coefficient is missing, the system will be unstable. The criterion of Routh [25,26] puts further restrictions on the coefficients: For a system to be stable, all the terms in the first column of the following array must have the same algebraic sign. If n is equal to 5, for example,

$$\begin{matrix} a_0 & a_2 & a_4 \\ a_1 & a_3 & a_5 \\ b_1 & b_2 & \\ c_1 & c_2 & \\ d_1 & & \\ e_1 & & \end{matrix} \qquad (12\text{-}88)$$

where

$$b_1 = \frac{a_1 a_2 - a_0 a_3}{a_1} \qquad\qquad b_2 = \frac{a_1 a_4 - a_0 a_5}{a_1}$$

$$c_1 = \frac{b_1 a_3 - a_1 b_2}{b_1} \qquad\qquad c_2 = \frac{b_1 a_5 - a_1 0}{b_1}$$

$$d_1 = \frac{c_1 b_2 - b_1 c_2}{c_1} \qquad\qquad (12\text{-}89)$$

$$e_1 = \frac{d_1 c_2 - c_1 0}{d_1}$$

12-21. Proportional plus Integral Action. If an integrating action is added to the controller, the steady-state error will be zero for any constant value of set point or load. The output of such a controller as applied to the shaft positioner will be

$$\theta_i - \theta_{i0} = \mu \left[\frac{1}{\tau_i} \int (\theta_R - \theta_0)\, dt + (\theta_R - \theta_0) \right] \qquad (12\text{-}90)$$

Differentiation of Eq. (12-90) to give (12-91) shows that the integration changes the output of the controller at a *rate* proportional to the error. The action is often called "reset action" in the process industries:

$$\frac{d\theta_i}{dt} = \frac{\mu}{\tau_i} (\theta_R - \theta_0) + \mu \left(\frac{d\theta_R}{dt} - \frac{d\theta_0}{dt} \right) \qquad (12\text{-}91)$$

If Eq. (12-12) is differentiated and Eq. (12-91) is substituted as the "force function," a third-order differential equation is obtained describing the mechanical positioner with proportional plus integral control under constant Γ_0:

$$J \frac{d^3\theta_0}{dt^3} + f \frac{d^2\theta_0}{dt^2} + k(1 + \mu) \frac{d\theta_0}{dt} + \frac{\mu k \theta_0}{\tau_i} = \frac{\mu k \theta_R}{\tau_i} + k\mu \frac{d\theta_R}{dt} \qquad (12\text{-}92)$$

According to Eq. (12-92), if the output comes to a steady value the error will be zero for any steady load torque; i.e., $\theta_{0,ss} = \theta_R$. However, two constants in the equation are functions of the controller properties. At certain values of the sensitivity μ and integral action "time" τ_i, the damping of the third-order system will vanish. The system will then "hunt" indefinitely, oscillating at a frequency ω_0.

These limiting conditions can be found by considering that the three roots of the left-hand side of Eq. (12-92) will be of the form

$$r_1, r_2 = \alpha_1 \pm j\beta \qquad (12\text{-}93)$$
$$r_3 = \alpha_2 \qquad (12\text{-}94)$$

corresponding to the general solution

$$\theta_0 = \theta_{0,ss} + A_1 e^{\alpha_2 t} + e^{\alpha_1 t}(A_2 e^{j\beta t} + A_3 e^{-j\beta t}) \qquad (12\text{-}95)$$

For undamped oscillations, $\alpha_1 = 0$. Substituting $\theta_0 = A e^{j\beta t}$ into the left side of Eq. (12-92) and equating respectively the real and imaginary terms gives the combinations of μ_0 and τ_{i0} that will produce hunting:

$$\frac{1 + \mu_0}{\mu_0} = \tau_{i0} = J/f \qquad (12\text{-}96)$$

$$\omega_0 = \sqrt{\frac{k(1 + \mu_0)}{J}} \qquad (12\text{-}97)$$

Equation 12-96 can be found directly by applying Routh's criterion (Sec. 12-20).

12-22. Rate Action. In "rate," "derivative," or "lead" action, the controller changes its output by an amount proportional to the rate at which the error is changing. Since this provides no means of determining the actual error, proportional or integral action must be used as well. The equation for proportional plus rate action is

$$\theta_i - \theta_{i0} = \rho \frac{d(\theta_R - \theta_0)}{dt} + \mu(\theta_R - \theta_0) \qquad (12\text{-}98)$$

If Eq. (12-98) is substituted as the force function of Eq. (12-12), it is seen that the addition of rate action increases the effective damping coefficient of the mechanical system to $f + \rho k$. The steady-state error $\theta_i - \theta_{0,ss}$ following step changes in θ_i or Γ_0 is the same as that for proportional control alone. However, the limiting sensitivity μ_{crit} for critical damping ($\zeta = 1$) is greater, since

$$\zeta = \frac{f\left(1 + \dfrac{k}{f}\rho\right)}{2\sqrt{kJ(1 + \mu)}} \qquad (12\text{-}99)$$

so that the errors due to load changes are smaller than when proportional control alone is used at the same damping ratio ζ. The greater permissible μ increases the "stiffness" of the system and ω_n, as is seen by Eq. (12-84), so that the time for approach to steady state is greatly decreased. The "follow-up" lag to a steady rotation of the input end of the shaft at a constant rate (see Sec. 12-16) $d\theta_i/dt$ was

$$\frac{f}{k}\frac{d\theta_i}{dt} = \frac{2\zeta}{\omega_n}\frac{d\theta_i}{dt}$$

for the uncontrolled system; since k is effectively increased faster than f the follow-up error is reduced.

The advantages of rate action are limited by the tendency of the control system to "saturate" when exposed to step disturbances or to noise in the error signal.

Prob. 12-7. A steel shaft $\frac{3}{4}$ in. in diameter, 20 ft long, held in a lubricated sleeve, is being used to position an irradiation sample mounted in a drum. The effective moment of inertia J reflected to the crank end is 1.24 lb$_F$-ft-sec^2 or slug-ft^2. The sleeve friction is such that a steady torque of 1 lb-ft is required to rotate the crank at 0.1 radian/sec; under the same condition the drum position lags the crank by 0.0077 radians. There is no external torque except friction. It is desired to speed up the response of the drum to a "step" motion of the crank end. If a proportional plus rate controller having a derivative control factor $\rho = 0.1$ sec is added, what may be the maximum sensitivity μ of the controller and by what factor will the response time be shortened? The damping factor ζ should be unchanged.

Solution. From the data given, it is seen that for the uncontrolled system,

$$J = 1.24 \text{ slug-ft}^2 \qquad f = 1/0.1 = 10 \text{ lb}_F\text{-ft-sec}$$

and $k = f\omega_i/\theta_{0,ss} - \theta_i = (10)(0.1)/0.0077 = 129.5$ lb$_F$-ft/radian; $\zeta = f/2\sqrt{kJ} = 0.395$ which is quite underdamped. $\omega_n = \sqrt{k/J} = 10.22$ radians/sec. Figure 12-6 indicates that the system, if subjected to a step input, will have approached the final steady-state position within 2 per cent in a time corresponding to $\omega_n t = 8.2$, or $8.2/10.22 = 0.8$ sec. With the controller of $\rho = 0.1$, substitution into Eq. (12-99) shows that a controller sensitivity of 4.25 will give a damping ratio of 0.395; this corresponds to a new natural frequency $\omega_n = \sqrt{k(1 + \mu)/J}$ of 23.5 radians/sec. The time for 98 per cent approach to a steady-state angle $\theta_{0,ss}$ is now $8.2/23.5 = 0.35$ sec, 43.8 per cent of the original response time. The steady-state follow-up error, if the reference crank is turned at $\omega_R = 0.1$ radian/sec, is given by $f\omega_R/k$ with the effective friction and stiffness factors substituted:

$$\theta_{0,ss} - \theta_R = \frac{(f + k\rho)\omega_R}{k(1 + \mu)} = \frac{22.95 \times 0.1}{129.5 \times 5.25} = 0.0338 \text{ radian}$$

12-23. Application of Control Action to Reactors.* *Proportional Control.* Assume that the correction of control rod position (and likewise the multiplication constant, to a first approximation) is made proportional

---

* For the sake of simplicity, only the "linearized" neutron equation (12-29) will be considered.

to the power or neutron-concentration error:

$$\delta k = \mu(n_R - n) = \mu(\delta n_R - \delta n) \tag{12-100}$$

If the reactor has been in steady operation at a flux corresponding to n_0, with zero error, and the reference signal n_R is changed suddenly to a new constant value $n_0 + \delta n_R$, Eqs. (12-100) and (12-29) can be combined to give

$$\left(\frac{d^2}{dt^2} + \frac{\beta + \lambda l^* + \mu}{l^*} \frac{d}{dt} + \frac{\mu \lambda n_0}{l^*}\right) \delta n = \mu \frac{\lambda n_0}{l^*} \delta n_R \tag{12-101}$$

The transient response is given in terms of the roots of the left-hand side of the equation; these are

$$r_1, r_2 = -\frac{\beta + \lambda l^* + \mu}{2l^*} \pm \frac{1}{2}\sqrt{\left(\frac{\beta + \lambda l^* + \mu}{l^*}\right)^2 - \frac{4\mu\lambda n_0}{l^*}} \tag{12-102}$$

The steady-state error is presumably zero.   Critical damping μ_c, given by

$$\mu_c = \frac{(\beta + \lambda l^* + \mu_c)^2}{4\lambda n_0 l^*} \tag{12-103}$$

is an impossibility with the nuclear constants used previously.   The system is always considerably overdamped.

The inconvenience of having the required sensitivity for a given response be inversely proportional to the power level can be circumvented [27] by the use of a comparator whose output at a given error is inversely proportional to n_R.   Alternatively [28], logarithmic measuring circuits can be used to give a corresponding controller action:

$$\delta k = \mu(\ln n_R - \ln n) \tag{12-104}$$

The response to a step change in power demand is given by the solution of

$$\left(\frac{d^2}{dt^2} + \frac{\beta + l^*\lambda}{l^*} \frac{d}{dt} + \frac{\lambda\mu}{l^*}\right) \ln n = \mu \frac{\lambda}{l^*} \ln n_R \tag{12-105}$$

Critical damping then corresponds to the condition

$$\mu_c = \frac{(\beta + \lambda l^*)^2}{4\lambda} \tag{12-106}$$

but is again unattainable in practice.

Unfortunately, thermal and poisoning effects will soon cause the equilibrium or zero-error position of the control rods to correspond to a k_{eff} differing from unity by, say, δk_d. The steady-state error will be $\delta k_d l^*/\mu$. The effect is analogous to that of an external torque on a positional servomechanism with proportional control alone.

Integral Control. The drift can be eliminated by driving the control rods at a speed which is proportional to the error. If logarithmic circuits are used,

$$\frac{d\,\delta k}{dt} = \frac{1}{\tau_i}\,(\ln n_R - \ln n) \tag{12-107}$$

This is equivalent to integral control, since

$$\delta k = \frac{1}{\tau_i}\int_0^t (\ln n_R - \ln n)\,dt \tag{12-108}$$

If Eqs. (12-29), (12-107), and (12-108) are combined and differentiated to eliminate the integral, the third-order differential equation of the system is obtained:

$$\left(\frac{d^3}{dt^3} + \frac{\beta + \lambda l^*}{l^*}\frac{d^2}{dt^2} + \frac{1}{\tau_i l^*}\frac{d}{dt} + \frac{\lambda}{\tau_i l^*}\right)\ln n = \frac{1}{\tau_i l^*}\left(\lambda + \frac{d}{dt}\right)\ln n_R \tag{12-109}$$

For $\beta = 0.00755$, $\lambda = 0.081$, and $l^* = 1.25 \times 10^{-3}$,

$$\left(\frac{d^3}{dt^3} + 6.12\frac{d^2}{dt^2} + \frac{800}{\tau_i}\frac{d}{dt} + \frac{64.8}{\tau_i}\right)\ln n = \frac{800}{\tau_i}\left(\frac{d}{dt} + 0.081\right)\ln n_R \tag{12-110}$$

The response to a step change in the demand neutron concentration n_R is determined by the roots of the left-hand side of the equation. Application of Routh's criterion shows that the system will always be stable. One root is always real and negative for any positive value of $1/\tau_i$; at "reset rates" $1/\tau_i$ less than about 0.0024, however, corresponding to critical damping, the other two roots become complex conjugate numbers, resulting in oscillatory approach to steady state.

Prob. 12-8. The neutron concentration in a reactor with no delayed neutrons can be assumed to obey the equation $dn/dt = n(\delta k/l^*)$ or $(d \ln n)/dt = (\delta k/l^*)$, where δk denotes the departure of the multiplication factor from unity. Show (a) that use of integral control alone as defined by Eq. (12-108) will result in an unstable system and (b) that addition as well of proportional control [defined by Eq. (12-104)] of sensitivity greater than a certain minimum will stabilize the reactor.

As higher performance is demanded of the control-rod mechanism, the tendency for instability arising from such factors as inertia and friction of the driving motors and rods, as well as reactor characteristics, must be considered. Treatment of the necessarily more complex equations by means of the transfer function-frequency response concept is simpler, though more empirical, and will be described in the following sections.

12-24. Steady-state Sinusoidal Response. If a system has three or more energy storage elements (or two storage elements plus integral control), calculation of its response by the methods in the preceding paragraphs becomes laborious. Even when the roots are found by approximation methods [29], synthesis to give a desired response becomes difficult because of the complex manner in which individual system properties enter into the over-all response.

The alternative approach has been to consider the steady-state response of the system, with its feedback loop opened, to a sinusoidal signal of constant amplitude A and frequency ω. The output will differ in magnitude and phase angle from the input; these differences determined for all frequencies can be used to predict the service performance of the system. The differential equation of each component of the system can be converted into an ordinary algebraic equation by use of operational methods such as the Laplace transformation and can be expressed as the "transfer function" of the component [40].

12-25. Laplace Transformations [30]. Because space is limited, the Laplace transformation can be considered here as merely a set of rules by which a function of time $f(t)$ can be converted into an equivalent "transform of $f(t)$," $\mathcal{L}\{f(t)\}$; $\mathcal{L}\{f(t)\} = F(s)$ is a function of the complex variable s. Table 12-4 gives a few typical relationships for obtaining $F(s)$.

TABLE 12-4. LAPLACE TRANSFORMATIONS OF $f(t)$*

	$f(t)$	$\mathcal{L}[f(t)] = F(s) = \int_0^\infty e^{-st}f(t)\,dt$
1	a	$\dfrac{a}{s}$
2	t	$\dfrac{1}{s^2}$
3	t^k (k = positive integer)	$\dfrac{k!}{s^{k+1}}$
4	e^{at}	$\dfrac{1}{s-a}$
5	te^{at}	$\dfrac{1}{(s-a)^2}$
6	$\dfrac{1}{a}\sin at$	$\dfrac{1}{s^2+a^2}$
7	$\cos at$	$\dfrac{s}{s^2+a^2}$
8	$af(t) + bg(t)$	$aF(s) + bG(s)$
9	$\dfrac{df(t)}{dt}$	$sF(s) - f(0)$
10	$\dfrac{d^nf(t)}{dt^n}$	$s^nF(s) - s^{n-1}f(0) - s^{n-2}\left(\dfrac{df(t)}{dt}\right)_{t=0+} - \cdots$
11	$\lim\limits_{t\to\infty} f(t)$	$\lim\limits_{s\to 0} sF(s)$

* A more complete list is given in R. V. Churchill, "Modern Operational Mathematics in Engineering," McGraw-Hill Book Company, Inc., New York, 1944.

As an example, the transform of Eq. (12-12) is found as follows:

$$\mathcal{L}\left\{\frac{J}{k}\frac{d^2\theta_0}{dt^2}\right\} = \frac{J}{k}\left\{s^2\Theta_0(s) - s\theta_{00} - \left(\frac{d\theta_0}{dt}\right)_{t=0}\right\} \tag{12-111}$$

$$\mathcal{L}\left\{\frac{f}{k}\frac{d\theta_0}{dt}\right\} = \frac{f}{k}\left\{s\Theta_0(s) - \theta_{00}\right\} \tag{12-112}$$

$$\mathcal{L}\{\theta_0\} = \Theta_0(s) \tag{12-113}$$

$$\mathcal{L}\{\theta_i\} = \Theta_i(s) = \frac{1}{s}\theta_i \text{ for a step input at } t = 0 \tag{12-114}$$

$$\mathcal{L}\left\{\frac{1}{k}\Gamma_0\right\} = \frac{1}{s}\left(\frac{1}{k}\Gamma_0\right) \text{ for constant } \Gamma_0, \text{ etc.} \tag{12-115}$$

The transformed Eq. (12-20) is then

$$\left\{\frac{J}{k}s^2 + \frac{f}{k}s + 1\right\}\Theta_0(s) = \Theta_i(s) + s\left(\frac{J}{k}\right)\theta_{00} + \frac{f}{k}\theta_{00} + \left(\frac{d\theta_0}{dt}\right)_{t=0}$$
$$+ \frac{1}{s}\frac{\Gamma_0}{k} \tag{12-116}$$

which can be written as

$$\Theta_0(s) = \frac{k\Theta_i(s)}{Js^2 + fs + k} + \frac{(1/s)\Gamma_0 + sJ\theta_{00} + f\theta_{00} + k(d\theta_0/dt)_{t=0}}{Js^2 + fs + k} \tag{12-117}$$

12-26. Transfer Function. The first term on the right of Eq. (12-117) gives the response if all initial conditions and external loads are zero; the second term contains all the initial conditions and external influences.* The "transfer function" may be defined as the ratio of the Laplace transform of the output to the Laplace transform of the input under zero initial and load conditions:

$$\frac{\Theta_0(s)}{\Theta_i(s)} = \frac{k/J}{s^2 + (f/J)s + k/J} = \frac{k/J}{(s - r_1)(s - r_2)}$$
$$= \frac{\omega_n^2}{s^2 + 2\zeta\omega_n s + \omega_n^2} = KG(s) \tag{12-118}$$

where $K = k/J$ or ω_n^2, the "gain" of the system, is a frequency-invariant property. $G(s)$ determines the response to inputs of different frequencies. The result is equivalent to substituting s, s^2, . . . for d/dt, d^2/dt^2, . . . in the differential equation.

The transfer function of a single-capacity positioner with negligible mass, described by Eq. (12-12), can be found similarly to be

$$\frac{\Theta_0(s)}{\Theta_i(s)} = \frac{1}{f/ks + 1} = \frac{1}{\tau s + 1} = \frac{1/\tau}{s + 1/\tau} = KG(s) \tag{12-119}$$

* If the equation is rearranged so that the inverse Laplace transform can be read from a table, the complete solution of the equation will be obtained directly.

12-27. Synthesis of System Equations.   If two or more *noninteracting* systems are connected in series, the over-all transfer function for the composite system is the product of those for the individual components:

$$\frac{C(s)}{M(s)} = (K_1K_2 \cdots K_n)[G_1(s)G_2(s) \cdots G_n(s)] \quad (12\text{-}120)$$

If a stage influences the output of the preceding one, the two stages must be considered as a single stage in evaluating the transfer function.

The performance of the generalized feedback control system shown in Fig. 12-4 can be written in terms of the transfer functions of the individual stages.   The equivalent open-loop assembly 12-4c obtained by breaking the loop of 1-24a at $a - b$ can be described by the relationships

$$E(s) = R(s) - B(s) \quad (12\text{-}121)$$
$$B(s) = K_HH(s)C(s)$$
$$= K_HH(s)[K_1K_2G_1(s)G_2(s) + K_NN(s)L(s)] \quad (12\text{-}122)$$

Eliminating $B(s)$ from Eqs. (12-121) and (12-122),

$$E(s) = \frac{1}{1 + K_1K_2K_HG_1(s)G_2(s)H(s)} R(s)$$
$$- \frac{K_NK_HN(s)H(s)}{1 + K_1K_2K_HG_1(s)G_2(s)H(s)} L(s) \quad (12\text{-}123)$$

The effects of changing set point and of external disturbances are seen to be additive.

The transform of the controlled variable is

$$C(s) = K_1K_2G_1(s)G_2(s)E(s) + K_NN(s)L(s) \quad (12\text{-}124)$$

which becomes, when Eq. (12-121) is substituted, the equation of the closed loop

$$C(s) = \frac{K_1K_2G_1(s)G_2(s)}{1 + K_1K_2K_HG_1(s)G_2(s)H(s)} R(s)$$
$$+ \frac{K_NN(s)}{1 + K_1K_2K_HG_1(s)G_2(s)H(s)} L(s) \quad (12\text{-}125)$$

$K_NN(s)$ can be found by writing the transformed differential equation of the system in the form of Eq. (12-125) and comparing terms.   For the example of Eq. (12-81),

$$K_NN(s) = \frac{1/\mu}{Js^2 + fs + k} \quad (12\text{-}126)$$

if there is no external load or disturbance, $L(s) = 0$.   If the output and reference signal are compared directly in the same units ("unity feedback"), $K_HH(s) = 1$ and $B(s) = C(s)$.

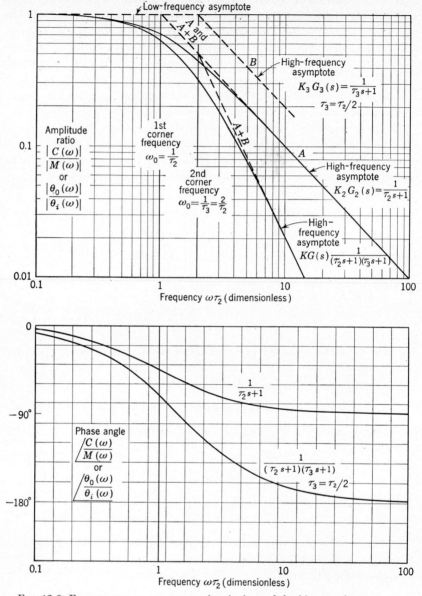

FIG. 12-8. Frequency-response spectra for single- and double-capacity systems.

12-28. Frequency Spectra.   The response of a single-capacity system to a sinusoidal input can be obtained by substituting $\theta_i = A \sin \omega t$ in the differential equation (12-12) and solving by "classical" methods, or by substituting $\mathcal{L}[\theta_i] = H_i(s) = A\omega/(s^2 + \omega^2)$ into Eq. (12-117) and rearranging the resulting equation so that Table 12-4 can be used in reverse. The complete solution is

$$\theta_0(t) = \frac{A\omega\tau}{1 + \tau^2\omega^2} e^{-t/\tau} + \frac{A}{\sqrt{1 + \tau^2\omega^2}} (\sin \omega t + \psi) \qquad (12\text{-}127)$$

$$\psi = \tan^{-1} \omega\tau \qquad (12\text{-}128)$$

After the transient exponential term has died out, the steady response is seen to be sinusoidal, of attenuated peak magnitude $A/\sqrt{1 + \tau^2\omega^2}$,

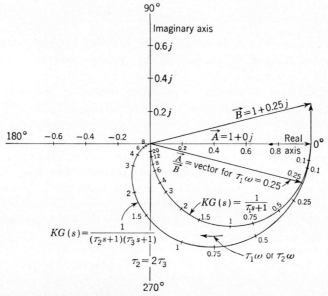

FIG. 12-9. Transfer locus (polar diagrams) for single- and double-capacity systems.

lagging by an angle $-\psi$ behind the input. Figure 12-8 shows the response as ratio of output to input and phase angle as functions of frequency. The data of Fig. 12-8 can be combined to give a single curve on polar coordinates, in which frequency is a parameter; the length of each vector is $\dfrac{|\theta_0|}{|\theta_i|} = \sqrt{1 + (\tau\omega)^2}$ and its angle is the phase angle. If the tips of all the vectors are connected, a semicircle results. By considering fictitious negative frequencies the circle can be completed.

The semicircle of Fig. 12-9 is called the "transfer locus" of the system. By considering the polar diagram as a diagram representing complex numbers, with real abscissa and imaginary ordinate, it can be derived quickly by substituting $s = j\omega$ into the transfer function:

$$\frac{\theta_0(\omega)}{\theta_i(\omega)} = \frac{1}{1 + \tau\omega j} = \frac{1 + 0\omega j}{1 + \tau\omega j} \tag{12-129}$$

Equation (12-129) represents the quotient of the two vectors $\vec{A}$ and $\vec{B}$ shown in Fig. 12-9. Both vectors have real (horizontal) components equal to unity; the divisor has an imaginary (vertical) component $\tau\omega j$, so that its vector length is $\sqrt{1 + (\tau\omega)^2}$ and its angle with the real axis is $\tan^{-1} \tau\omega$. Vectors are divided by subtracting their angles and dividing their magnitudes; therefore the quotient can be represented by a vector of magnitude $1/\sqrt{1 + (\tau\omega)^2}$ of phase angle $-\tan^{-1} \tau\omega$, written as

$$\frac{\theta_0(\omega)}{\theta_i(\omega)} = \left| \frac{1}{\sqrt{1 + (\tau\omega)^2}} \right| \underline{/-\tan^{-1} \tau\omega} \tag{12-130}$$

The transfer loci of more complicated transfer functions can be found in a similar manner by expressing the transfer function in terms of vectors on the complex plane and performing the indicated multiplications and divisions to obtain the resultant vector for a given frequency. The loci of some representative systems are given in Table 12-5.

The transfer locus of a series combination of components can be found by multiplying the curves for the components; i.e., by adding the phase angles and multiplying the vector lengths for each frequency.

12-29. Approximate Representation of Frequency Spectra. When the frequency response is plotted as in Fig. 12-8, in terms of the logarithms of the amplitude ratio and of the frequency, the resulting curve can be approximated by straight lines. For example, the amplitude of a single-capacity system, in logarithmic form, is

$$\log \left| \frac{\theta_0(\omega)}{\theta_i(\omega)} \right| = \log \left| \frac{1}{1 + \tau\omega j} \right| = -\frac{1}{2} \log \sqrt{1 + (\tau\omega)^2} \tag{12-131}$$

At small values of ω, the right-hand side of Eq. (12-131) approaches $\log 1^{-\frac{1}{2}} = \log 1$; at high frequencies, it is asymptotic to $-\log \tau\omega$. On a log-log plot, these asymptotes correspond respectively to a horizontal line through $|\theta_0(\omega)/\theta_i(\omega)| = 1$ and a line of slope -1 which intersects the low-frequency asymptote at the point $(|\theta_0(\omega)/\theta_i(\omega)| = 1, \tau\omega = 1)$. The frequency of intersection ω_0 is known as the "corner frequency"; it is that frequency at which the real and imaginary parts of the transfer function are equal.

It has become customary to express the amplitude ratio in decibels, such that

$$\left| \frac{\theta_0(\omega)}{\theta_i(\omega)} \right|_{db} = 20 \log \left| \frac{\theta_0(\omega)}{\theta_i(\omega)} \right| \tag{12-132}$$

This permits the use of an arithmetic ordinate scale and makes it easy to plot the phase angle on the same graph paper. The high-frequency asymptote of slope -1 on log-log paper has a slope of -20 db per decade (10-fold increase in frequency) on db-log paper.

TABLE 12-5. REPRESENTATIVE TRANSFER LOCI
Arrow on locus points in direction of increasing frequency

System	$KG(j\omega)$	Transfer locus on complex plane	Asymptote on plot of log $|KG(\omega j)|$ vs. log ω	Phase angle vs. log ω
1. Tachometer or rate controller	$\rho\omega j$			
2. Amplifier or proportional controller	μ			
3. Proportional plus rate controller	$\mu + \rho\omega j$			
4. Integral controller	$\dfrac{1}{\tau_i\omega j}$			
5. Single-capacity system (also practical lag network)	$\dfrac{1}{\tau\omega j + 1}$			
6. Shaft angle, d-c motor; simple servomechanism	$\dfrac{k/f}{(J/f\omega j + 1)\omega j}$			

Since multiplication corresponds to the addition of logarithms, the response of series combinations can be obtained by adding magnitude ordinate distances as well as angle ordinates for each frequency. Two single-capacity elements in series would approach asymptotically a slope of -2 (or -40 db per decade) at very high frequencies.

For example, if elements such as a thermocouple and a galvanometer, or two tanks of liquid [see Table 12-3 (9)] having transfer functions G_2 and G_3 are connected in series in a noninteracting manner, where $G_2(s) = 1/(1 + \tau_2 s)$ and $G_3(s) = 1/(1 + \tau_3 s)$

the transfer function of the combination is

$$G(s) = G_2(s)G_3(s) = \frac{1}{1 + \tau_2 s} \frac{1}{1 + \tau_3 s} \tag{12-133}$$

The steady-state ratio of output to input for sinusoidal excitation will be

$$\frac{C(\omega j)}{M(\omega j)} = G(\omega j) = \frac{1}{1 + \tau_2 \omega j} \frac{1}{1 + \tau_3 \omega j} \tag{12-134}$$

which can be approximated by the straight lines shown dashed in Fig. 12-8 for $\tau_2 = 1$, $\tau_3 = 0.5$ sec. The equivalent polar diagram is given in Fig. 12.9.

Prob. 12-9. A galvanometer will follow rapidly changing currents most satisfactorily if it is slightly underdamped. Assume that Eq. (12-54) can be used to describe the behavior of the galvanometer if $\Gamma_0 = 0$ and θ_i is taken as the current input. Plot the transfer function on polar coordinates and on logarithmic coordinates, using ω/ω_n as a dimensionless measure of the frequency for conditions of (a) critical damping $\zeta = 1$; (b) $\zeta = 0.8$. [HINT: Write the left-hand side of Eq. (12-54) in the form of Eq. (12-38).]

12-30. Effect of Controllers. Addition of a proportional controller of sensitivity μ (whose transfer function, from Table 12-4, is also seen to be μ) merely multiplies $G(s)$ by the constant μ. This is equivalent to moving the entire logarithmic response curve upward by 20 log μ decibels or, on a polar plot, to increasing the magnitude of the radial scale by a factor μ:

$$\frac{C(\omega j)}{E(\omega j)} = \frac{M(\omega j)}{E(\omega j)} \frac{C(\omega j)}{M(\omega j)} = \mu G(\omega j) \tag{12-135}$$

On the other hand, addition of integral control action changes the shape of the response curve. The transfer function $K_1 G_1(s)$ of the controller is $1/\tau_i s$ and has the response shown in Table 12-5(4).*

The approximate open-loop response of the system plus controller

$$\frac{C(\omega j)}{E(\omega j)} = \frac{1}{\tau_i \omega j} \frac{1}{1 + \tau_2 \omega j} \frac{1}{1 + \tau_3 \omega j} \tag{12-136}$$

is shown in Fig. 12-10 for $\tau_2 = 1$, $\tau_3 = 0.5$, $\tau_i = 3\tau_2$, and $\tau_i = \tau_2/3$. It is seen that the gain, infinite at zero frequency, decreases by 20 db per decade until the first corner frequency at $\omega \tau_2 = 1$. Attenuation increases at the rate of 40 db per decade until the second break point at $\omega \tau_3 = 1$, after which it reaches its final rate of -60 db per decade.

---

* The response is easily seen to be given by

$$\left| \frac{m(\omega)}{e(\omega)} \right| = \frac{1}{\sqrt{0 + (\tau_i \omega)^2}} = (\tau_i \omega)^{-1} \quad \text{or} \quad \left. \left| \frac{m(\omega)}{e(\omega)} \right| \right|_{db} = -20 \log \tau_i \omega$$

which is a line of slope -20 db per 10-fold increase in frequency, crossing the zero-decibel line at $\tau_i \omega = 1$ or $\omega = 1/\tau_i$. The phase angle is $\psi = \tan^{-1}(\omega/0) = -90°$.

The phase angle is $-114°$ at unity gain (zero decibels); the quantity $180° - (-\psi)$ at unity gain is often called the "phase margin," so that the phase margin for $\tau_i = 3\tau_2$ is $66°$. It is negative for $\tau_i = \tau_2/3$, which denotes an unstable system.

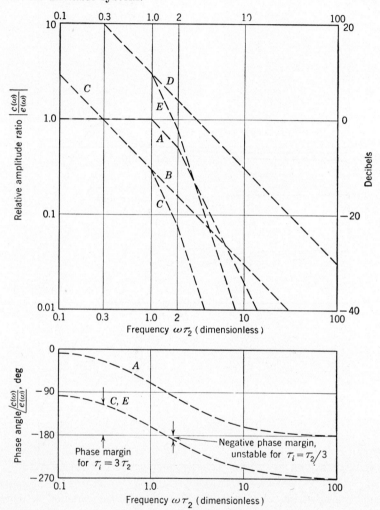

Fig. 12-10. Effect of integral control on two-capacity system. Asymptote A, system alone; B, integral controller, $\tau_i = 3\tau_2$; C, system plus controller of curve B; D, integral controller, $\tau_i = \frac{1}{3}\tau_2$; E, system plus controller of curve D.

12-31. Calculation of Closed-loop Response from Open-loop Frequency Spectrum.
The sinusoidal open-loop response given by

$$\frac{C(\omega j)}{E(\omega j)} = KG(\omega j) = K_1 K_2 \cdots G_1(\omega j)G_2(\omega j) \cdots \quad (12\text{-}137)$$

is related to the performance of the system as connected in a closed loop by Eqs. (12-123) and (12-125). Only the simplest case of unity feedback

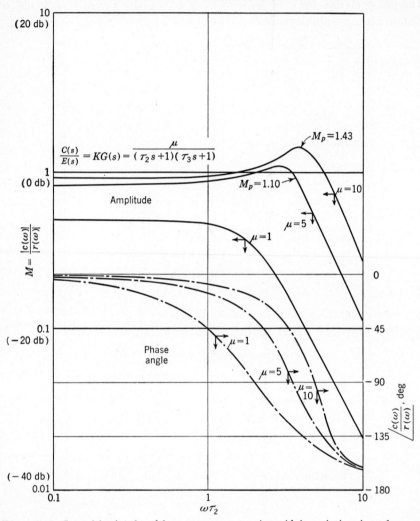

$$\frac{C(s)}{E(s)} = KG(s) = \frac{\mu}{(\tau_2 s +1)(\tau_3 s +1)}$$

FIG. 12-11. Logarithmic closed-loop response to sinusoidal variation in reference signal, proportional control of two-capacity system.

$(K_H H(s) = 1)$ will be treated here. If the external load is constant, the effect of set-point changes is given by

$$\frac{C(\omega j)}{R(\omega j)} = \frac{KG(\omega j)}{1 + KG(\omega j)} \tag{12-138}$$

$$\frac{E(\omega j)}{R(\omega j)} = \frac{1}{1 + KG(\omega j)} \tag{12-139}$$

Values of $|KG(\omega)|$ and $\underline{/KG(\omega)}$ for each frequency can be calculated or taken from plots of the form of Figs. 12-8 or 12-9. Substitution into

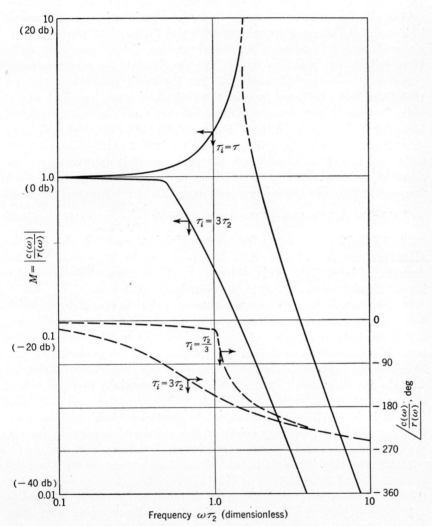

FIG. 12-12. Logarithmic closed-loop response to sinusoidal variation in reference signal, proportional plus integral control of two-capacity system.

Eqs. (12-123) and (12-125) will give the closed-loop response of the system to a sinusoidal variation in load $L(\omega j)$.

Figure 12-11 shows the closed-loop response of the two-capacity system of Fig. 12-8, with proportional control of several different sensitivities, while Fig. 12-12 gives the response with integral control added.

12-32. Correlation between Frequency Response and Service Performance.

The usual necessary criterion of satisfactory service performance is the suitable response of the controlled system to a step or constant rate of change of reference signal or load. Specifications are expressed in terms of the maximum transient error and the required recovery time, although external factors of noise and random disturbances may make these criteria not sufficient ones. The steady-state frequency-response curves can be evaluated similarly. For the easily predictable systems that have been discussed here, a comparison of some transient curves with the closed-loop frequency-response curve will indicate empirical bases for evaluating the latter. For example, both Figs. 12-6 and 12-11 show increased peaking with increased controller gain. The resonant frequency in Fig. 12-11 at which the peak occurs is approximately the same as the frequency of oscillations of the underdamped transient response at the same controller sensitivity. The service performance will probably be satisfactory if the resonant peak $\left|\dfrac{c(\omega)}{r(\omega)}\right|_{max} = M_p$ is about 1.3 to 1.5 and if it occurs at a frequency which is at least $3/t$, where t is the allowable time for a transient to disappear (say by 98 per cent) in the case of step input [31]. This allows for a certain amount of underdamping for a quick return of $c(t)$ to the neighborhood of $R(t)$.

If the changes in reference variable, or external disturbances which will be encountered in actual service, are known functions of time, and if the transfer function of the open-loop system plus controller has been determined by calculation or by experiment, the performance can be calculated (by methods beyond the scope of this chapter) and compared with the specifications. Process analogues are especially useful in giving a more exact idea of system response.

Since it is much more convenient to predict probable performance from the *open-loop* frequency behavior, it is of interest to consider further the effect of the open-loop characteristics on closed-loop performance.

12-33. Stability.

Instability of a closed-loop system occurs when the denominator of Eq. (12-123) or (12-125) vanishes; this occurs when the transfer function of the series loop (as sketched in Fig. 12-4d) is equal to $-1 = -1 + j0$.* Under this condition the feedback signal is equal in magnitude to the error and lags it by 180°. Subtraction in the comparator results in a reinforcement of the error and increasingly larger controlled action whose effect is eventually limited by "saturation" (i.e., nonlinearity) or destruction of the system.

---

* Multiloop systems, i.e., those with internal loops characteristic of the system or intentionally inserted to give desired response characteristics, can be stable when the main loop is closed even though the open loop may be unstable. Space does not allow their consideration.

Zero or negative "phase margin" corresponds to instability.

On a polar diagram of a simple open-loop system, closed-loop instability is indicated when the curve passes through the point $(-1, j0)$ or more

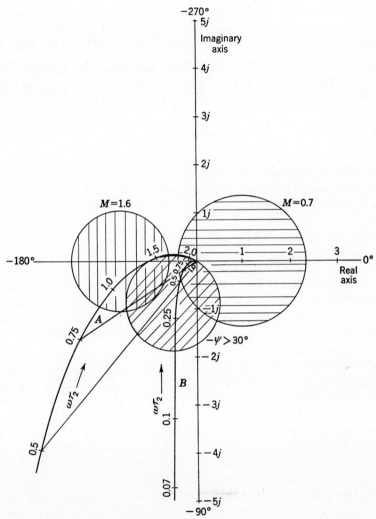

FIG. 12-13. Polar diagram of proportional plus integral control of two-capacity system showing M circles.   Curve A, $\tau_i = \frac{1}{3}\tau_2$; curve B, $\tau_i = 3\tau_2$.

generally by the encirclement of that point as the frequency varies from $-\infty$ to $+\infty$.*   By increasing the gain, for example, the system of Fig. 12-9 cannot be made unstable, while that of Fig. 12-13 can be.

* The general Cauchy-Nyquist criterion is treated more fully in servomechanism texts [32].

12-34. Degree of Underdamping. The system should not be violently oscillatory even though stable, such as in Fig. 12-6 for $\zeta = 0.4$, although slight underdamping is desirable for quick response.

The values of $M = \left| \dfrac{c(\omega)}{r(\omega)} \right|$ which may be expected can be found from the open-loop transfer locus $\dfrac{C(\omega j)}{E(\omega j)}$ by plotting lines of constant magnitude M on the same diagram. These may be calculated by substituting $x + jy$ for the transfer function of Eq. (12-138). The loci of constant M are found to be circles of radius $M/(M^2 - 1)$ and centers at

$$x_0 = \frac{M^2}{M^2 - 1} \qquad y_0 = 0$$

The response of the system given by Eq. (12-136) is plotted in Fig. 12-13 with the M contours superimposed. The areas for $M > 1.6$ are shaded vertically and would be entered if the gain were increased sufficiently.

12-35. Attenuation. Extraneous "noise" composed of frequency components higher than the frequency range over which the system is normally expected to respond should produce little effect. This means that the magnification or "control ratio" M should be approximately constant (say 0.7 to 1.6) for frequencies characteristic of the variations of $R(t)$ or load in service but should be attenuated rapidly at higher frequencies. The area corresponding to $M < 0.7$ is shaded with horizontal lines in Fig. 12-13.

For essentially steady conditions (low frequencies) small errors can be maintained by providing integration in the loop. This gives very high gain at the low frequencies of interest and attenuates noise well. Unfortunately, as is evident from the controller transfer locus [Table 12-5(4)], such a controller produces lagging phase angles.

12-36. Phase Lag. Usually the magnification M of a simple system such as the positioning mechanism can be held within the desired limits over the specified frequency range by suitable adjustment of the sensitivity (gain) of the controller. Unfortunately, a fairly uniform M obtained in this manner has a concomitant rapidly increasing phase lag of $C(\omega)$ relative to $R(\omega)$ with increasing frequency. In a single-loop system, specification of the magnitude spectrum determines the phase spectrum, and conversely.

If the tangent of the closed-loop phase angle ψ between the controlled variable and the reference variable is denoted by N, and if the transfer function of the system is considered as $x(\omega) + jy(\omega)$, it is found that loci of constant lag angle can be drawn on the transfer locus plot. They appear as circles of radius r and center at x_0, y_0, where

$r = (1/2N) \sqrt{N^2 + 1}$, $x_0 = -\frac{1}{2}$, $y_0 = \frac{1}{2}N$. An upper limit on the phase angle would correspond to another shaded "off-limits" area on the locus diagram. The diagonally shaded area in Fig. 12-13 indicates phase angles greater than 30°.

It is common to specify that the controlled variable shall not lag the reference variable by more than, say, 30° over the range of frequencies in which M lies between 0.7 and 1.6.

Either the polar diagram of $KG(j\omega)$ or the logarithmic plots of $|KG(\omega)|$ and $\underline{/KG(\omega)}$ can be used in synthesizing a system to give the specified performance. The system behavior is perhaps more easily visualized from the typical transfer loci shapes. However, graphical synthesis from building blocks is easier on the logarithmic plots. The principles involved are the same.

For example, a criterion often applied to the open-loop response when working with the logarithmic curves is that the controlled variable should lag the error signal by 120 to 150° when $|KG(\omega)| = 1$ (i.e., the phase margin should be of the order of 30 to 60°). In addition, the "gain margin," or value of $|KG(\omega)|$ at the frequency at which $\underline{/KG(\omega)} = -180°$, should be approximately 0.4 to 0.1 (corresponding to a value of $|KG(\omega)|$ of -8 to -20 db).

12-37. Low-frequency Response. The low-frequency end of the spectrum of the transfer function is perhaps the most important in determining service performance. The limit approached at $\omega = 0$ is the steady-state output for constant input. Closed-loop Eqs. (12-138) and (12-139) indicate zero steady-state error only for infinitely large values of $KG(j\omega)$. A very large value of the frequency invariant K would produce instability at higher frequencies in many cases; therefore $G(j\omega)$ must become infinite as $\omega \to 0$.

Prob. 12-10. Determine the steady-state (zero-frequency) error of the positioning device described by Eq. 12-81.

Solution

$$KG(j\omega) = \frac{\theta_0(j\omega)}{E(j\omega)} = \frac{\mu}{(J/k)(j\omega)^2 + (f/k)(j\omega) + 1} \tag{12-140}$$

Setting $\omega = 0$ shows that, for a constant input value given by $\theta_R(0) = \theta_R$,

$$\frac{\theta_{0,ss}}{e_{ss}} = \frac{\theta_{0,ss}}{\theta_R - \theta_{0,ss}} = \mu \qquad \text{or} \qquad e_{ss} = \frac{\theta_R}{1 + \mu} \tag{12-141}$$

which corresponds to Eqs. (12-83) and (12-84) with zero load and initial conditions.

The same answer can be obtained in a more general way by considering the "limiting value" relationship for the Laplace transform (Table 12-4):

$$f(\infty) = \lim_{s \to 0} sF(s) \tag{12-142}$$

$$E(s) = \frac{1}{1 + KG(s)} R(s) + \frac{K_N N(s)}{1 + KG(s)} L(s) \tag{12-143}$$

The steady-state error at infinite time can be found by applying Eq. (12-142):

$$e(\infty) = e_{ss} = \lim_{s \to 0} \left[\frac{s}{1 + KG(s)} R(s) + \frac{sK_N N(s)}{1 + KG(s)} L(s) \right] \qquad (12\text{-}144)$$

For a "step" reference signal of value r, $R(s) = r/s$. Applied to the positioner with proportional control, no load, and $\theta_{00} = 0$,

$$e_{ss} = \lim_{s \to 0} \left[\frac{(J/k)s^2 + (f/k)s + 1}{(J/k)s^2 + (f/k)s + 1 + \mu} \right] \theta_R = \frac{\theta_R}{1 + \mu} \qquad (12\text{-}145)$$

which is identical with Eq. (12-141).

Addition of integral control to the positioner gives the transfer function

$$KG(s) = \frac{\Theta_0(s)}{E(s)} = \frac{\mu(s + 1/\tau_i)}{s[(J/k)s^2 + (f/k)s + 1]} \qquad (12\text{-}146)$$

Substitution into Eq. (12-142) shows that there is zero steady-state error following a step change in θ_R.

If the reference signal is changed at a constant rate $d\theta_R/dt = \omega_R$,

$$\theta_R = \theta_{R0} + \omega_R t \qquad (12\text{-}147)$$

$$\Theta_R(s) = \frac{1}{s} \theta_{R0} + \frac{1}{s^2} \omega_R \qquad (12\text{-}148)$$

Substitution of Eqs. (12-146) and (12-148) into Eq. (12-142) shows that the system will not follow such a command exactly but will eventually come to a steady error $e_{ss} = \omega_R/\tau_i$. If another component providing integration is added to the loop, say $K_3 G_3(s) = 1/\tau_i's$, the positioner will eliminate this error. The resulting system, described by a quartic equation, is inherently less stable than before.

Systems having transfer functions from whose denominator s (or $j\omega$) can be factored possess "reset" properties for a step input command. Those from which s^2 [or $(j\omega)^2$] can be factored have the ability to follow inputs varying at a constant rate. If s^3 [or $(j\omega)^3$] can be factored, they can follow without error an input r which is changing with steady acceleration, i.e., $d^2r/dt^2 = a$, or

$$r = r_0 + at + \frac{a}{2} t^2$$

The low-frequency characteristics of such systems correspond on a logarithmic plot of $KG(j\omega)$ to limiting slopes of -1, -2, and -3 (-20, -40, and -60 db per decade), respectively. On a polar plot, the curves are asymptotic to the axes at $-90°$, $-180°$, and $-270°$, respectively, at $\omega = 0$.

12-38. Limitations of Gain Adjustment Alone. A simple servomechanism control-rod drive suitable for manual control of the reactor power would be a remote-controlled version of the drive which has been discussed in Prob. 12-2. It could consist basically of a comparator, which could be a synchro transmitter and control transformer pair or a pair of potentiometers in a Wheatstone-bridge circuit; an amplifier, which determines the gain of the system; and the drive motor plus rod. A first approximation of the transfer function of the open loop is given in Table 12-5(6). Since the amplifier and synchros offer additional time lags, a better approximation would be given by an equation of the form of Eq. (12-136) and the response curves of Figs. 12-10, 12-12, and 12-13. As applied to the positioner, τ_2 is the time constant J/f of the motor and control rod, τ_3 the effective time constant of the other components, and τ_i the reciprocal of the "velocity constant" or k/f of the mechanism.

Such a rod positioner gives zero error at any fixed set point or reference angle. Attempts to reduce the "follow-up" error by increasing gain (the equivalent of decreasing τ_i in Figs. 12-10, 12-12, and 12-13) results in lagging phase angles which may cause overpeaking (entry into the $M = 1.6$ circle of Fig. 12-13) and eventual instability, as in curve A. Various compensating devices must be resorted to if improved performance is desired.

12-39. Compensating Elements. An increase of the drag coefficient f will allow the use of a higher gain for the same follow-up error and M_p and will shorten the response time of the system by decreasing τ_2 or J/f. The effective f can be increased without increasing the frictional power losses by connecting a tachometer to the shaft of the motor with its output fed back to a point earlier in the loop, e.g., to the input terminals of the motor. This internal loop provides "kinetic damping."

By adding a "lead" compensating network such as that in Table 12-3 (6), which provides leading phase angles at high frequencies, the transfer locus of curve A of Fig. 12-13 can be warped out of the $M = 1.6$ circle while keeping the high gain in the low-frequency (steady-state) region.

By adding a "lag" network such as shown in Table 12-3 (8) or 12-5 (5), the low-frequency response of curve B of Fig. 12-13 can be improved by providing lagging phase angles and effectively higher gain at the lower frequencies.

Instead of the networks mentioned, internal feedback elements, consisting of simple RC combinations connected to the amplifier so as to feed back part of its output into its input, can be used to give lag or lead compensation. Internal feedback is also used to reduce nonlinearities or the effects of changing characteristics of the main parts of the loop.

12-40. Application of Frequency-response Technique to Reactor Control. For small variations of δk and flux, the transfer function of a reactor with a single group of delayed neutrons can be obtained by substituting s (or $j\omega$) for d/dt in Eq. (12-29):

$$KG(s) = \frac{N(s)}{\delta K(s)} = \frac{(n_0/l^*)(s + \lambda)}{s\left(s + \dfrac{\beta + \lambda l^*}{l^*}\right)} \qquad (12\text{-}149)$$

where $N(s)$, $\delta K(s)$ refer to the Laplace transforms of the neutron concentration and excess multiplication constant, respectively. In terms of the transform $LN(s)$ of $\ln n$,

$$KG(s) = \frac{LN(s)}{\delta K(s)} = \frac{\dfrac{\lambda}{\beta + \lambda l^*}\left(\dfrac{1}{\lambda} s + 1\right)}{s\left(\dfrac{l^*}{\beta + \lambda l^*} s + 1\right)} \qquad (12\text{-}150)$$

A more exact transfer function, considering five or six groups of delayed neutrons, has been derived as a basis for the design of an electric pile simulator [33] to be used in loop analysis.

$$\frac{LN(s)}{\delta K(s)} = \frac{1/l^*}{s\left(1 + \displaystyle\sum_1^6 \frac{\beta_i/l^*}{s + \lambda_i}\right)} \qquad (12\text{-}151)$$

which can be multiplied out, factored, and rewritten (for easy vector multiplication) in a form

$$KG(s) = \frac{(1/l^*)(s + \lambda_1)(s + \lambda_2) \cdots (s + \lambda_6)}{(s + \alpha_1)(s + \alpha_2) \cdots (s + \alpha_6)} \qquad (12\text{-}152)$$

Experimental transfer functions have been determined for the open loop by oscillating the control rods [21] or calculated from Eq. (12-125) from the response of a closed-loop servo system subjected to sinusoidal oscillations of n_R [28]. The single-group form of the equations offers a good approximation of the response; for the British Harwell reactor [28], empirically

$$\frac{LN(s)}{\delta K(s)} = \frac{30(5s + 1)}{s(0.2s + 1)} \qquad (12\text{-}153)$$

which can be compared to Eq. (12-150) after the values of β, l^*, and λ previously used have been inserted for d/dt:

$$\frac{LN(s)}{\delta K(s)} = \frac{10.6(12.35s + 1)}{s(0.163s + 1)} \qquad (12\text{-}154)$$

12-41. Addition of Controller. The transfer function of the control action exemplified by Eq. (12-108) is

$$\frac{\delta K(s)}{LE(s)} = K_2 G(s) = \frac{1}{\tau_i s} \qquad (12\text{-}155)$$

The transfer function of the combined system with the loop open is given by the product of Eq. (12-150) or its experimentally determined

equivalent and Eq. (12-155).   The polar diagram of the CP-2 reactor
[21] and the reactor plus a controller with two values of τ_i are sketched in
Fig. 12-14.   It is seen that the system is always stable within the limitations of the very small δk's considered.

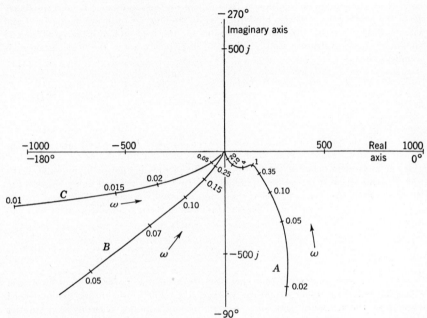

FIG. 12-14. Polar plot of transfer function of CP-2 reactor (data of Harrer) and reactor
with control rods driven at speeds proportional to ln n/n_R (or $\delta n/n_R$, since departures
from equilibrium are small).   Radial distance is measured in [ln (n/n_R)].   Curve A,
reactor alone; curve B, with integral control, $\tau_i = 10$ sec; curve C, with integral
control, $\tau_i = 100$ sec.

Prob. 12-11.   Solve Prob. 12-5 by considering the transfer functions and transfer
loci of the reactor plus the control actions.

Prob. 12-12.   Plot the transfer function of the reactor described by Eq. (12-154)
on the polar diagram and on logarithmic coordinates.   (a) Show that the reactor is
stable with proportional or integral control, and examine the stability limits with
proportional plus integral control action.   (b) Compare with similar plots for the
"linearized" reactor.   (HINT: For a first approximation, use the asymptotes of the
logarithmic plot.)

When the required speeds of control action are so rapid that the characteristics of the rod-drive mechanism must be considered, the transfer
function becomes much more complicated.   Since satisfactory performance is usually obtained only by using additional stabilizing networks,
the exact design of which is beyond the scope of this chapter, the over-all
transfer functions for the examples given in the next two sections will
not be plotted.

12-42. D-C Motor Rod Drive. A pertinent illustration of the additional complexity introduced by including the characteristics of the drive mechanism is a d-c motor operated at constant armature current, with its field voltage the amplified error signal. The rotor may be considered to develop a torque proportional to the air-gap flux, which in turn is proportional to the field current. The build-up of field current at a given applied voltage is determined by the inductance L_f and resistance R_f of the field circuit. Forces hindering the rotation of the rotor are its inertia J and viscous drag f. If the differential equations for the field and rotor are written and combined, the transfer function relating the shaft position θ to the field voltage v_f is found to be

$$\frac{\Theta(s)}{V_f(s)} = K_2 G_2(s) = \frac{K_f K_T}{(Js^2 + fs)(L_f s + R_f)}$$
$$= \frac{(K_f/R_f)K_T/f}{s[(J/f)s + 1][(L_f/R_f)s + 1]} \tag{12-156}$$

where K_f is the proportionality factor between flux and field current, and K_T is the proportionality factor between torque and flux.

If the gear ratio between the motor and control rods is high or if the mass of the rods is small (often it is possible to include the time lag of the rods along with that of the rotor, J/f), $\delta K(s)$ can be considered as approximately $K_D\Theta(s)$. If the field voltage is merely $K_E E(s)$, the open-loop transfer function of the reactor plus control rods and drive is

$$\frac{LN(s)}{LE(s)} = \frac{K_E(K_f/R_f)(K_T/f)K_D(\lambda/\beta + \lambda l^*)[(1/\lambda) s + 1]}{s^2[(J/f)s + 1][(L_f/R_f)s + 1][l^*s/(\beta + \lambda l^*) + 1]} \tag{12-157}$$

A self-braking action can be obtained by operating the motor with constant field excitation and controlled armature voltage V_a. The rotor torque is influenced directly by the armature back emf, so that the mechanical and electrical properties are interacting. The transfer function is of the form

$$\frac{\Theta(s)}{V_a(s)} = \frac{K_a}{s(L_a J s^2 + J R_a s + K_a K_b)} \tag{12-158}$$

where K_b = ratio of back emf to armature angular velocity

K_a = proportionality factor between armature current and torque

L_a = inductance of armature

R_a = resistance of armature

If L_a is small,

$$\frac{\Theta(s)}{V_a(s)} = \frac{K_a}{s(J R_a s + K_a K_b)} = \frac{1/K_b}{s[(J R_a/K_a K_b)s + 1]} \tag{12-159}$$

12-43. Hydraulic Rod Drive. A hydraulic transmission can be used [10, 28] as a high-performance rod drive which can develop a torque essentially independent of velocity and over a speed range of about 20:1. In principle, a mechanical linkage actuated by the amplified error signal controls the volumetric flow rate Q of hydraulic fluid; the flow rate determines the output shaft speed. Any change in shaft position is resisted by the inertia J of the output motor plus load and is diminished because of the compressibility $(1/B)$ of the oil and leakage in the pump and motor, denoted as L ft³/(sec)(psi) pressure developed by the pump. The transfer function of the hydraulic drive can be approximated [17] by

$$\frac{\delta K(s)}{E(s)} \propto \frac{\theta(s)}{X(s)} = \frac{d_p n}{s\left(\dfrac{VJ}{Bd_m}s^2 + \dfrac{LJ}{d_m}s + d_m\right)} = \frac{K_v}{s\left(\dfrac{s^2}{\omega_n{}^2} + \dfrac{2\zeta s}{\omega_n} + 1\right)} \qquad (12\text{-}160)$$

where $d_p n$ = maximum rate of oil flow (product of pump displacement and number of strokes per minute)

V = total volume of oil in the system

d_m = motor displacement

It is obvious that the transmission can be oscillatory or unstable with certain values of the constants; proper performance usually requires the use of internal stabilizing loops. Valve design becomes important in fast-responding drives [34].

A more accurate description of the characteristics of typical electrical rod-drive mechanisms should take into account the power source—Thyratrons, rectifier, magnetic amplifiers, motor generators, or amplidyne generators driven by the amplified and modified error signal. These are treated in more detailed analyses of control-system performance.

12-44. Conclusion. This chapter has covered some of the characteristics of the control system of a nuclear reactor and has attempted to show the basis for design of such systems. The mathematical relationships given describe only to a certain approximation the behavior of actual equipment. However, they can be used with greater or less refinement in the preliminary synthesis of system analogues or mock-ups from which a more exact knowledge of performance can be obtained. As has been pointed out, the availability of packaged components, reactor simulators, disturbance generators, and response analyzers has made such semiempirical design quite straightforward.

No attempt has been made to present circuit diagrams of specific pieces of equipment or control systems or to describe the mechanical details of control rod drives, ion chambers, self-balancing potentiometers, etc. Some of these have been treated in Chap. 4 and in many of the references and in most cases are available as standard commercial components.

The philosophy of design and operation of the control systems—with especial emphasis on the safety loop—has been given by Stephenson [35].

NOMENCLATURE

c_P = specific heat at constant pressure, Btu/(lb$_M$)(°F)

C = electrical capacitance, coulombs/volt

$c(t)$ = controlled variable, a function of time

$C(s)$ = a transformed function of the complex variable s

c = concentration of delayed-neutron emitters, atoms/cm^3

c_i = concentration of the ith kind of emitter

$e(t)$ = error, difference between desired value and controlled variable

f = damping or viscous drag coefficient, lb$_F$-ft/(radian/sec)

h = film heat-transfer coefficient, Btu/(hr)(ft^2)(°F)

i = electrical current, coulombs/sec

$j = \sqrt{-1}$

J = moment of inertia, slug-ft^2 = lb$_F$-ft-sec^2

k = stiffness, lb$_F$-ft/radian

k_0, k_{eff} = effective multiplication constant of nuclear reactor, dimensionless

l^* = mean effective lifetime of neutrons, sec

$m, m(t)$ = manipulated variable, a function of time

M = ratio of controlled to reference variable

n = average neutron concentration, neutrons/cm^3

Q = quantity of heat transferred, Btu

r = roots of an equation; reference variable

R = electrical resistance, ohms; reference variable

R_{th} = thermal resistance, °F/(Btu)(hr).

s = complex variable; $KG(s)$, $KH(s)$, etc., transfer functions, are functions of s

t = time, sec or hr

T = temperature, °F

T_R = reactor period, sec, $\dfrac{n}{dn/dt}$

U = over-all heat-transfer coefficient, Btu/(hr)(ft^2)(°F)

V = electrical potential, volts

W = mass of material, lb

α = reactor temperature coefficient, (°F)$^{-1}$

β = fraction of total neutrons produced which are delayed

β_i = denotes the fraction of the ith kind.

Γ = torque, lb$_F$-ft

δ = small increment

ζ = damping factor, dimensionless

$\Theta(s)$ = Laplace transform of angular position; a function of the complex variable s

$\theta, \theta(t)$ = angular position, a function of time

λ = average decay constant of fission products

λ_i = denotes the decay constant of the ith kind

μ = sensitivity of proportional controller

ρ = rate action proportionality factor

τ = time constant, sec

τ_i = integral controller time constant

ψ = phase angle, degrees or radians

ω = frequency, radians/sec

ω_n = natural frequency of oscillation, radians/sec

REFERENCES

1. Breazeale, W. M.: The "Swimming Pool"—A Low Cost Research Reactor, *Nucleonics*, **10**(11): 56–60 (1952).
2. "Reports to the U.S. Atomic Energy Commission on Nuclear Power Reactor Technology," p. 82, Government Printing Office, May, 1953.
3. Sheehan, T. V., and J. Weisman: personal communication.
4. Ref. 2, pp. 7, 24.
5. Cochran, D., and C. A. Hansen, Jr.: Some Instrumentation Requirements in an Atomic Power Plant, *Mech. Eng.*, **71**:808 (1949).

6. Roy, T. R. V.: Instrumentation Problems of Production Reactors, *Chem. Eng. Progr.*, **52**:233–237 (1956).

7. Weill, J.: *Nucleonics*, **11**(3):36–39 (1953).

8. Ref. 2, p. 78.

9. Cole, T. E.: Design of a Control System for a Low-cost Research Reactor, *Nucleonics*, **11**(2):32–37 (1953).

10. Harrer, J. M.: Reactor Operation, *Nucleonics*, **11**(6):35–40 (1953).

11. Baker, C. P., et al.: "Water Boiler," AECD-3063, Sept. 4, 1944.

12. Ahrendt, W. R., and J. F. Taplin: "Automatic Feedback Control," chap. 9, McGraw-Hill Book Company, Inc., New York, 1951.

13. Garr, D. E.: Electrical Positioning System of High Accuracy for Industrial Use, *Gen. Elec. Rev.*, **50**(7):17 (1947).

14. Bell, P. R., and H. A. Straus: "Electronic Pile Simulator," AECD-2764, declassified Dec. 20, 1949.

15. Pagels, W.: "A Portable Electronic Pile Kinetic Simulator," AECD-2941, Apr. 3, 1950.

16. Walker, W. H., W. K. Lewis, W. H. McAdams, and E. R. Gilliland: "Principles of Chemical Engineering," 3d ed., chap. IV, p. 106, McGraw-Hill Book Company, Inc., New York, 1937.

17. Brown, G. S., and D. P. Campbell: "Principles of Servomechanisms," chap. 5, John Wiley & Sons, Inc., New York, 1948.

18. Ref. 17, pp. 48–53.

19. Isbin, H. S., and J. W. Gorman: Applications of Pile-kinetic Equations, *Nucleonics*, **10**(11):68–71 (1952).

20. Moore, R. V.: The Control of a Thermal Neutron Reactor, *Proc. Inst. Elec. Engrs. (London)*, **100**(5):90–101, pt. 1 (1953).

21. Harrer, J. M., R. E. Boyar, and D. Krucoff: Transfer Function of Argonne CP-2 Reactor, *Nucleonics*, **10**(8):32–36 (1952).

22. *a.* Goodman, C.: "The Science and Engineering of Nuclear Power," vol. II, chap. 8, p. 100, Addison-Wesley Publishing Company, Cambridge, Mass., 1949.

    b. A. M. Weinberg: *Nucleonics*, **11**(5):18–20 (1953).

23. Sherwood, T. K., and C. E. Reed: "Applied Mathematics in Chemical Engineering," chap. III, p. 101, McGraw-Hill Book Company, Inc., New York, 1939.

24. Hurwitz, H., Jr.: Derivation and Integration of the Pile-kinetic Equations, *Nucleonics*, **5**(7):61–67 (1949).

25. Ref. 12, pp. 77–80.

26. Chestnut, H., and R. W. Mayer: "Servomechanisms and Regulating System Design," vol. 1, pp. 134–137, John Wiley & Sons, Inc., New York, 1951.

27. Schultz, M. A.: "Automatic Control of Power Reactors," AECD-3163, Nov. 6, 1950.

28. Bowen, J. H.: Automatic Control Characteristics of Thermal Neutron Reactors, *Proc. Inst. Elec. Engrs. (London)*. **100**(5):102, pt. 1 (1953).

29. Ref. 17, chap. 3.

30. Churchill, R. V.: "Modern Operational Mathematics in Engineering," McGraw-Hill Book Company, Inc., New York, 1944.

31. Hall, A. C.: Application of Circuit Theory to the Design of Servomechanisms, *J. Franklin Inst.*, **242**:279 (1946).

32. Ref. 17, chap. 6.

33. Franz, J. P.: "Pile Transfer Functions," AECD-3260, July 18, 1949.

34. Lee, S., and J. F. Blackburn: Contributions to Hydraulic Controls, *Trans. ASME*, **74**:1005, 1013 (1952).

35. Stephenson, R.: "Introduction to Nuclear Engineering," McGraw-Hill Book Company, Inc., New York, 1954.
36. Feck, J. A., Jr.: Kinetics of Circulating Reactors at Low Power, *Nucleonics*, **12**(10):52 (1954).
37. Taylor, D.: Trends in Nuclear Instrumentation, *Nucleonics*, **12**(10):12–19 (1954).
38. "Reactor Handbook," vol. I, chap. 1.6, AECD-3645, February, 1955.
39. Stone, J. J., and E. R. Mann: "Oak Ridge National Laboratory Reactor Controls Computor," ORNL-1632, Apr. 20, 1954.
40. O'Meara, F. E.: Reactor Transfer Functions, *Atomics*, **7**(2):49–51 (1956).

BIBLIOGRAPHY

A.E.C. Inspects Reactors for Safety, *Nucleonics*, **13**(3):25 (1955).
Bonnaure, P., P. Braffort, I. Pelchowitch, and J. Weill: Automatic Control of a Nuclear Reactor, *J. Nuclear Energy*, **1**(8:)24–38 (1954).
Glasstone, S.: "Principles of Nuclear Reactor Engineering," chaps. IV–VI, D. Van Nostrand Company, Inc., Princeton, N.J., 1955.
Schultz, M. A.: "Control of Nuclear Reactors and Power Plants," McGraw-Hill Book Company, Inc., New York, 1955.
Stewart, H., F. LaViolette, C. McClelland, G. Gavin, and T. Synder: Low-power Thermal Test Reactor for Nuclear Physics Research, *Nucleonics*, **11**(5):38 (1953).
Trimmer, J. D., and W. H. Jordan: Instrumentation and Control of Reactors, *Nucleonics*, **9**(4):60–68 (1951).

Other general references may be found in the books or articles referred to in Refs. 5, 9, 10, 12, 17, 20, 26 to 28, 35, and 40.

CHAPTER 13

POWER GENERATION

By Theodore Baumeister

13-1. Power Requirements [10, 16].   Much has been said about the tremendous potential of nuclear energy as a raw fuel for use in commercial power plants (see Table 13-1) [32].   This means that the substitution of the heat of the nuclear reaction for the chemical combustion of fossil fuel,

TABLE 13-1. WORLD RESERVES OF FOSSIL FUELS AND NUCLEAR ENERGY*

Fuel	World reserves	Source of data	Unit energy, Btu	Total energy, Btu
Fossil:				
Crude oil........	610×10^9 bbl	Weeks & Moulten	6.4×10^6	35×10^{17}
Natural gasoline	11.5×10^9 bbl	American Petroleum Institute	6.4×10^6	0.74×10^{17}
Shale oil........	620×10^9 bbl	Bureau of Mines	6.4×10^6	40×10^{17}
Natural gas.....	560×10^{12} ft³	American Gas Association	1000	6×10^{17}
Coal............	3482×10^9 tons	Bureau of Mines	$\left.\begin{array}{l} 27 \times 10^6 \\ 28 \times 10^6 \\ 9 \times 10^6 \end{array}\right\}$	722×10^{17}
Total.........	....	....	....	80×10^{18}
Nuclear:				
Uranium........	25×10^6 tons	AEC	3.5×10^{10}	1700×10^{18}†
Thorium........	1×10^6 tons	AEC	3.5×10^{10}	71×10^{18}
Total.........	....	....	....	1800×10^{18}

* W. L. Cisler, Electric Power Systems and Nuclear Power, *Edison Elec. Inst. Bull.*, **20**:289 (September, 1952); J. L. Schanz, Prospects for Nuclear Power, *Elec. Light and Power*, **32**:78–87 (December, 1954).
   † At 1:1 minimum breeding.

the energy of an elevated water supply, or the power of the wind must be reckoned with in the future evaluation of alternative power-generation methods [15, 24].   In order better to proceed with that evaluation it is essential to know the competitive position of present-day acceptable power plants.   These power plants should include stationary and transportation applications; electric utilities and industrials; and aircraft, marine, automotive, and railroad services.   The extent of the present power-plant establishment is represented by the data of Table 13-2

TABLE 13-2. ESTIMATED INSTALLED CAPACITY IN 1950 OF POWER PLANTS OF
THE UNITED STATES*

Type of plants	Millions of kw
Public-utility central stations	69
Industrial	21
Agricultural	46
Railroad motive power	83
Marine, civilian	22
Aircraft, civilian	17
Military establishment (1945)	750
Automotive	3370
Miscellaneous	10
Total	4388

* T. Baumeister, Steam and Electric Power—Its Past and Its Future, Centennial
of Engineering, *Combustion*, December, 1952; after J. A. Waring, Jr., in *Steelways*.

which shows the distribution of prime movers by type of service application. The aggregate capacity of all the power plants of the nation is
thus of the order of 6 billion hp or 4.5 billion kw. The internal-combustion engine dominates the entire field, as witness the automotive, air-

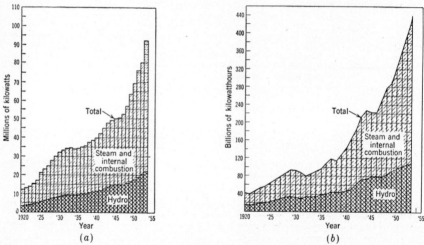

FIG. 13-1. Electric-utility capacity and generation by source of energy, United States,
1920–1953. (*Federal Power Commission.*)

craft, railroad, and military installations. Steam power dominates the
public-utility and industrial stationary applications, as is illustrated by
the data of Fig. 13-1. The central-station capacity of the country was
over 100 million kw in 1954, and the industrial plant was about one-third
of that, or 30 million kw. The output of the utilities reached 472 billion
kwhr in 1954. The bulk of this energy was generated by steam power.

Hydro power will become an ever-decreasing fraction of the total, as the 1952 estimate of the Federal Power Commission for the entire potential hydro capacity of the country is 106 million kw and 487 billion kwhr in an average water year. Only a fraction of this amount can be harnessed for generation purposes because of (1) other demands for water, (2) physical location, and (3) economics. Figure 13-2 shows an attempt to predict the electric-utility demands by the year 1970. A figure of 1000 billion kwhr by that date is not unreasonable and could easily be exceeded.

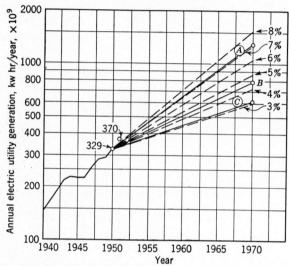

Fig. 13-2. Estimated annual electric-utility generation to the year 1970, United States. Broken lines show annual growth rates from 3 to 8 per cent as applied to the 1950 generation of 329 × 10⁹ kwhr. Solid lines show growth rates as related to population growth, as follows: (A) from the extrapolated per capita consumption trend of the last three decades; (B) from the same per capita consumption growth as prevailed during the 1940–1950 decade; (C) from the average per capita consumption of the last three decades. (T. Baumeister, Steam and Electric Power—Its Past and Its Future, Centennial of Engineering, Combustion, December, 1952.)

This will come in large measure from fossil fuels; a small amount will come from hydro and a conjectural amount (5 to 15 per cent) from atomic energy. The energy output of the utility plants averaged about 2500 kwhr per capita for the year 1951, and the industrial plants of the nation probably added another one-third to that figure. When all the transportation plants of Table 13-2 are added, as to output, the annual mechanical and electrical energy consumption of the country is of the order of 5000 to 6000 kwhr per capita. If the average load for the stationary plants is compared to the installed capacity, the resultant capacity factor is of the order of 50 to 60 per cent. On the other hand the annual capacity factor of the transportation plants is but a fraction

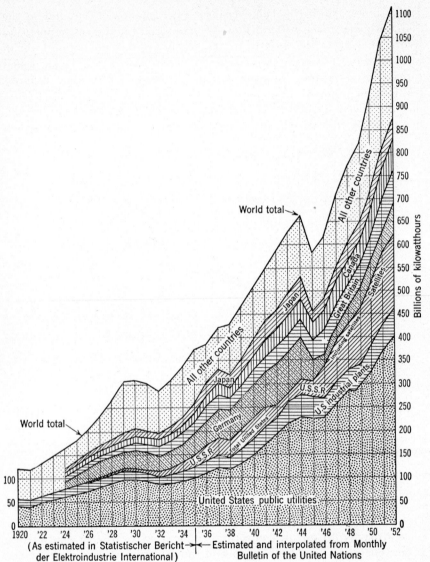

FIG. 13-3. World production of electricity. Generation by all agencies, including industry for its own use. (*Edison Electric Institute.*)

of this value, perhaps less than 1 or 2 per cent on passenger automobiles. A recent estimate of the electrical energy requirements of the world is shown in Fig. 13-3.

13-2. Performance of Established Power Plants and Systems. The competitive situation into which an atomic power plant must fit is represented by the data on performance of stationary and transportation

plants as given in summary form in Table 13-3. These data are not
exact in their definition, but they are included in order to demonstrate
the nature of the competitive problem. Any nuclear power plant which
is offered commercially must do better than existing or conventional

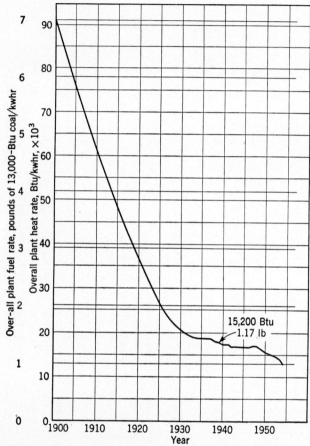

FIG. 13-4. Over-all thermal performance of electric-utility power plants, United
States, 1900–1951. (*Adapted from Federal Power Commission data.*)

plants. The data on the latter, as given in Table 13-3, should serve at
least to define the nature of the competitive limits. Figure 13-4 is useful
as an historical record because it shows the improvement in the fuel consumption and heat rate of the central stations of the United States since
1900. While average practice is now of the order of 15,000 Btu/kwhr,
the best steam plants of current design are delivering 1 kwhr for 9000 Btu.
It is this latter figure which is the standard of comparison for a nuclear
plant for stationary service.

TABLE 13-3. PERFORMANCE AND COST DATA OF SELECTED COMMERCIAL POWER PLANTS

Plant type	Plant size, kw	Unit size, kw	Equivalent heat rate, Btu/kwhr	Plant weight, lb/kw	Plant bulk, ft³/kw	Plant floor space, ft²/kw	Labor requirements, output, kwhr/man-hr	Investment, $/kw	Annual fixed-charge rate applicable to investment, %*
Central station:									
Steam, small	10,000–50,000	10,000–25,000	15,000	....	30–50	0.4–0.6	200–1000	200–300	12–15
Steam, large	200,000–500,000	75,000–200,000	10,000	....	20–30	0.2–0.3	3000–5000	100–200	12–15
Industrial by-product, steam	5000–10,000	2000–10,000	5,000	....	50–75	....	100–500	200–500	15–20
Stationary Diesel	2000–5000	500–1000	12,000	....	....	....	200–500	200–300	13–16
Hydroelectric	50,000–500,000	10,000–100,000	....	....	....	....	10,000–15,000	100–500	10–12
Diesel locomotive	2000–3000	1,000	12,000	100–200	2–3	0.3–0.6	200–500	100–150	20
Steam locomotive	2,500	2,500	50,000	100–200	2–3	0.3–0.6	100–400	50–100	20
Automobile	200	200	15,000	5–10	0.10	0.05	....	....	
Aircraft	1,000	1,000	12,000	1–3	0.05–0.1	0.02–0.03	....	75–125	
Motor ship	10,000	5,000	10,000	300–500	....	....	1000	200–250	12–20
Steamship	25,000	15,000	18,000	200–300	....	....	1000	175–250	12–20

* These data apply to business-managed enterprises. For government operation the rates are lower.

The nature of the variation in thermal performance of present-day power plants with type of service, size, application, cycle, and load is shown in Figs. 13-5 and 13-6. These are the heat-rate data which determine the fuel costs of existing plants. Figure 13-7 gives conversion

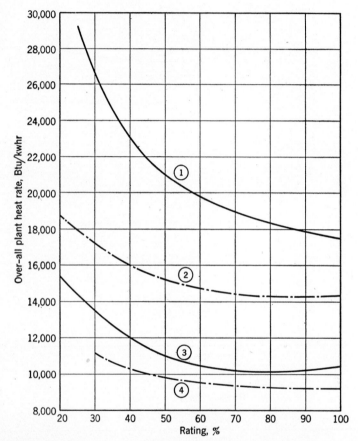

FIG. 13-5. Comparative heat rates—gas-turbine, steam, and diesel plants. (1) Gas-turbine plant, 3500 kw, 6:1 compression ratio, 1600°F, 50 per cent regenerator; (2) steam-turbine plant, 3500 kw, 650 psi, 825°F, 1 in. Hg absolute; (3) diesel plant, supercharged; (4) steam-turbine plant, 125,000 kw, 2000 psi, 1050°F/1000°F, 1 in. Hg absolute.

data from the usual fuel prices of dollars per ton, cents per gallon, dollars per 1000 ft³ to the more realistic figure of cents per million Btu. Table 13-4 demonstrates better the significance of some of these data. This table shows the competitive price situation which might obtain for representative fuels in various localities of the United States. Table 13-5 gives some equally important figures on the bulk and weight of fuel

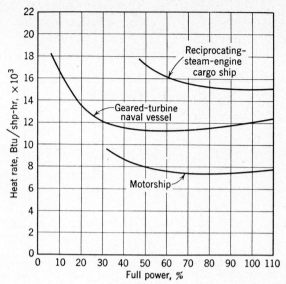

FIG. 13-6. Comparative thermal performance of seagoing ships.

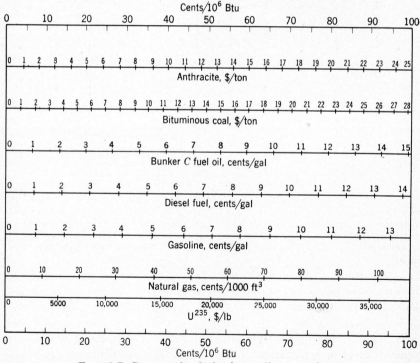

FIG. 13-7. Comparative fuel prices—alignment chart.

which are significant to transportation plants.   The labor requirements of representative power plants are shown in Table 13-3.

The competitive market for the atomic power plant, on the basis of costs, is reflected in the data of Tables 13-3 and 13-6 and Fig. 13-8. Table 13-3 shows the investment in power plant for several representative

TABLE 13-4. COMPARATIVE FUEL COSTS* IN SELECTED UNITED STATES MARKET AREAS

Fuel	East Coast		Middle West		Gulf		West Coast	
	A	B	A	B	A	B	A	B
Bituminous coal....	10	36	8	29	10	36	18	65
Anthracite.........	12	50	15	63	18	75	20	83
Bunker C fuel oil...	6	40	10	68	5	34	5	34
Diesel fuel.........	12	84	13	91	8	56	8	56
Gasoline..........	25	208	25	208	20	167	20	167
Natural gas.......	100	91	70	64	30	27	30	27

* Column A gives usual sales price basis of dollars per ton for coal; cents per gallon for oil; and cents per 1000 ft³ for gas.   Column B gives fuel price in cents per million Btu.

TABLE 13-5. WEIGHT AND SPACE REQUIREMENTS OF SELECTED FUELS

Fuel	High heating value, Btu/lb	Specific weight, lb/ft³	Bulk, ft³ per million Btu	Weight, lb per million Btu
Bituminous coal........	14,000	52	1.38	71
Anthracite............	12,000	57	1.47	83
Bunker C.............	18,500	60	0.90	54
Diesel fuel............	19,500	55	0.93	51
Gasoline..............	20,000	45	1.11	50
Natural gas...........	1,100* (NTP)	0.04 (NTP)	910 (NTP)	36

* Per cubic foot.

applications.   The range of cost is from \$50 to \$500 per kw.   The annual charges which are made against such investments are also included in this table and contain allowances for (1) return on investment or cost of money; (2) depreciation, both physical and functional; (3) property taxes and insurance.   These items are difficult to define, because certain activities are conducted under government subsidy while others are operated under the private-enterprise system.   All charges are considerably lower under government ownership, with lesser amounts charged for cost of money, depreciation, and property taxes. In some instances these items are actually made zero. Even allowing for such differences, there are still many further complications which arise with business-managed properties because of managerial and financial decisions on the proper accounting methods to be used by a business enterprise. Some activities are conducted on a pay-as-you-go basis. Others utilize debt financing. Some operations write the investment off over a long period of time, while others are short-lived by choice or by the nature of the product or service.

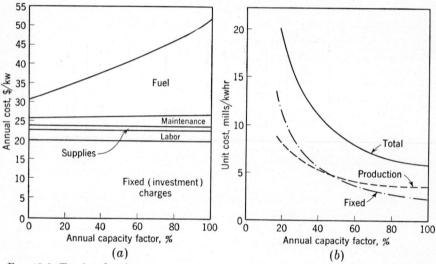

FIG. 13-8. Total and unit costs of power as a function of annual capacity factor, for representative conventional-fuel-fired, condensing, steam-electric power plant. (*a*) Total costs; (*b*) unit costs.

These matters are seldom subject to manipulation, change, or decision by engineers. They are essentially the prerogative and the duty of management on policy and procedures.

In the 1953 hearings before the Joint Committee on Atomic Energy, the Federal Power Commission supplied the following estimate on steam electric-power costs of electric utilities of the United States for the year 1952, as "the kwh unit costs for 27 major interconnected power pools generating over 60% of the total":

	Estimated cost, *mills per kwhr*
Fuel	3.04
Operation and maintenance	1.01
Estimated fixed charges	2.69
Total	6.74

TABLE 13-6. PERFORMANCE AND COSTS FOR SELECTED STATIONARY POWER PLANTS

Plant type	Steam central stations		Hydro		Diesel		Steam industrial	
	Large	Small	Large	Small	Large	Small	Large	Small
Plant capacity, net, kw	400,000	50,000	150,000	20,000	10,000	1,000	10,000	1,000
Capacity factor, per cent	80	60	75	60	75	70	90	75
Load factor, per cent	85	70	75	60	80	75	90	75
Annual energy generation, 10^6 kwhr	2,800	262	985	105	65.7	6.12	78.6	6.57
Investment, \$/kw	150	250	200	300	200	275	175	225
Fuel cost, cents per 10^6 Btu	20	30	......	......	50	75	30	40
Plant heat rate, Btu/kwhr	9,500	14,000	......	......	11,000	13,000	4,300	5,000
Annual fixed-charge rate, per cent	12	12	10	10	13	13	15	15
Fixed costs, mills/kwhr	2.57	5.71	3.05	5.72	3.96	5.82	3.34	5.13
Production costs:								
Fuel, mills/kwhr	1.90	4.20	......	......	5.50	9.75	1.29	2.00
Labor and supervision, mills/kwhr	0.15	1.40	0.30	0.50	0.90	5.00	0.30	1.00
Maintenance, mills/kwhr	0.15	1.40	0.15	0.25	0.80	2.00	0.20	0.60
Supplies and expense, mills/kwhr	0.05	0.30	0.10	0.10	0.30	0.50	0.10	0.20
Total production cost, mills/kwhr	2.25	7.30	0.55	0.85	7.50	17.25	1.89	3.80
Total cost of power, mills/kwhr	4.82	13.01	3.60	6.57	11.46	23.07	5.23	8.93

685

The cost of power is very sensitive to the loads carried, as is illustrated by Fig. 13-8. Recognizable components of cost, such as fixed charges, fuel, labor, maintenance, and supplies are there segregated for a representative stationary steam power plant. Some of these costs are substantially constant, regardless of load. Others are widely variable. The lines, as is shown in Fig. 13-8a, are not necessarily straight. They may be (1) concave upward, as with fuel, because of variations in efficiency; or (2) discontinuous, because of the cutting in or out of service of units in a multiunit plant or because of the addition of labor at certain points of increasing load. In any event, the cost equation for a power plant will recognize that there are (1) constant components of cost, and (2) variable components, i.e., constant or variable with load. The unit costs which obtain are the consequence of dividing the abscissa of Fig. 13-8a into the ordinate, with the result as shown in Fig. 13-8b. The unit cost curve is hyperbolic in nature and demonstrates the desirability of operating a power plant at high, rather than low, loads. Several typical cost equations follow:

ELECTRIC POWER

Case 1:

Annual cost, $/kw = 12.00 + 0.0018 × 8760 × capacity factor*

Case 2:

Annual cost, $/kw = 18.00 + 0.25 × 12 + 0.004 × 8760 × load factor*

Case 3:

Annual cost, $/kw = 15.00 + 0.003 × 8760 × capacity factor*

DISTRICT STEAM HEATING

Demand charge = $2 per pound of steam per hour demand
Energy charge = 90 cents per 1000 lb delivered

13-3. Loads and Load Curves. No power plant can be properly designed, specified, built, or operated without an exact knowledge of the loads and their variations. Some plants operate under the most extreme conditions of load fluctuation, while others, by comparison, operate with a fairly uniform, nonswinging load. An example of the former is a naval-vessel power plant, such as a destroyer or aircraft carrier, where the cruising power may be only 4 or 5 per cent of the full-speed power and where it may be necessary to swing from dead stop to full power, forward or astern, in a matter of seconds. An example of greater load constancy

* These factors are defined in Sec. 13-3.

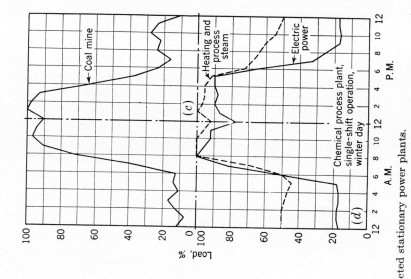

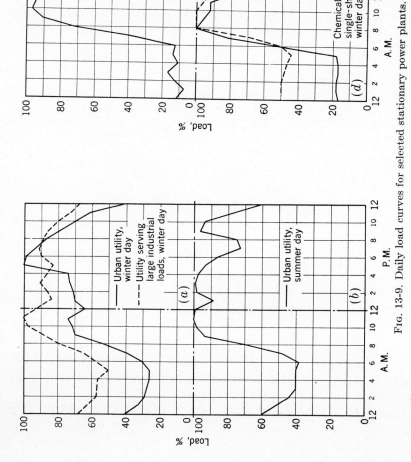

Fig. 13-9. Daily load curves for selected stationary power plants.

687

is a central station where troughs are from 20 to 30 per cent of peaks and where the scheduling of operations and the allocation of load to selected units can be made hours ahead of time. Because of the variations in loads represented by these two examples, coupled with the fact that there is no present practical way in which to store the output of a power plant, it is imperative that a never-ending collection of data be accumulated if a plant is to be adequately designed and economically operated. There is no substitute for adequate load data. Several tables and graphs are included here to give some idea of the load problem. Table 13-3 contains information on the size of power plants for different

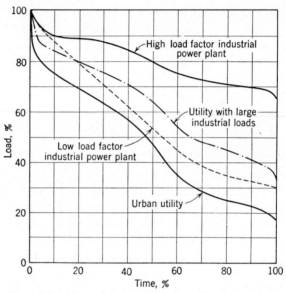

FIG. 13-10. Load-duration curves for selected stationary-power-plant services.

services. Details on load variations are shown in Fig. 13-9 as daily load curves for several stationary applications.

Load data can be replotted for greater convenience and utility in load and cost studies, unit scheduling, and plant design by rearrangement in the form of load-duration curves (Fig. 13-10). Here the loads no longer appear chronologically but are so accumulated as to show the extent or duration of any load. Load-duration curves are generally more useful when obtained over extended operating periods, such as a month or a year, and when plotted on the percentage basis. Percentages are particularly helpful in the analysis and comparison of performance among several plants. In such comparisons it is common practice to relate average loads to (1) peak loads and (2) rated capacities. The following

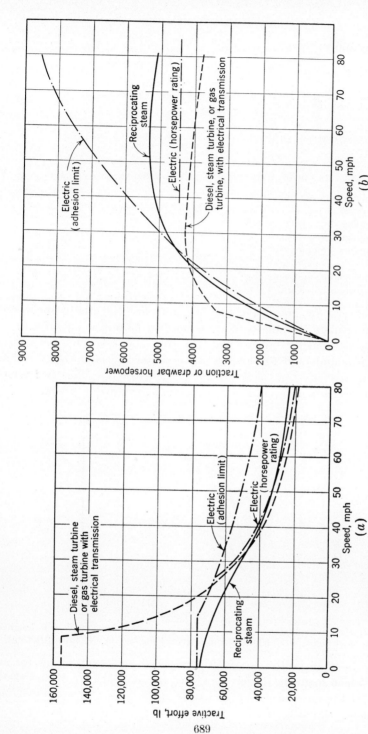

FIG. 13-11. (a) Tractive-effort curves for selected types of locomotives, 4000 to 5000 rated tractive horsepower. (b) Tractive or drawbar horsepower-speed curves for selected types of locomotives, 4000 to 5000 rated tractive horsepower.

689

definitions apply:

$$\text{Load factor, } \% = \frac{\text{average load for period}}{\text{peak load in period}} \times 100 \qquad (13\text{-}1)$$

$$\text{Capacity factor, } \% = \frac{\text{average load for period}}{\text{rated capacity of plant}} \times 100 \qquad (13\text{-}2)$$

Some representative values of power-plant capacity factor are given in Table 13-7.

TABLE 13-7. RANGE OF CAPACITY FACTORS FOR SELECTED POWER PLANTS

Public-utility systems, in general............................ 50–70
Chemical or metallurgical plant, three-shift operation.......... 80–90
Seagoing ships, long voyages............................... 70–80
Seagoing ships, short voyages.............................. 30–40
Airplanes, commercial..................................... 20–30
Private passenger cars..................................... 1–3
Main-line locomotives..................................... 30–40
Interurban buses and trucks............................... 5–10

Load curves for propulsive power plants are essentially the result of matching plant output to the resistance to motion. This resistance can be expressed as horsepower, but it is often more useful to show values of drag in pounds and to balance this resistance by a thrust, drawbar pull, or tractive effort of the power plant. The tractive effort and horse-power curves for a group of locomotives are shown in Fig. 13-11.

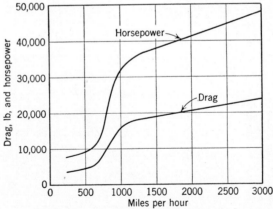

FIG. 13-12. Minimum drag and power required to drive a 100,000-lb well-designed airplane in level flight, at 30,000 ft altitude, at various speeds. (*Gordon McKay.*)

Figure 13-12 gives the horsepower required and drag curves for an airplane which might be driven by a nuclear power plant. Figure 13-13 shows speed-power curves of a marine installation. Power-available data can be related to these power-required curves where the excess is

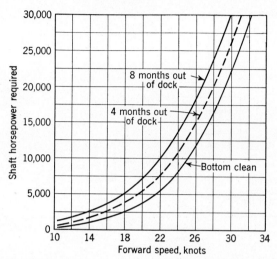

Fig. 13-13. Effect of fouling on the speed-power curve of a ship.

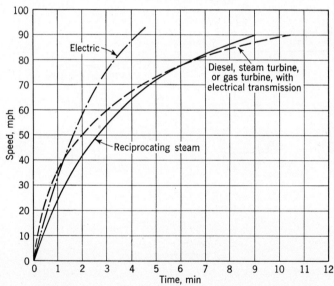

Fig. 13-14. Acceleration curves for selected types of locomotives, 4000 to 5000 rated tractive horsepower.

variously available for flexibility, maneuverability, acceleration, or rate of climb. Figure 13-14 shows some acceleration curves for locomotive drives.

The data which have been given here for power plants should not be construed as all-inclusive. They are in most cases oversimplified, but they should serve to show the conditions under which a power plant

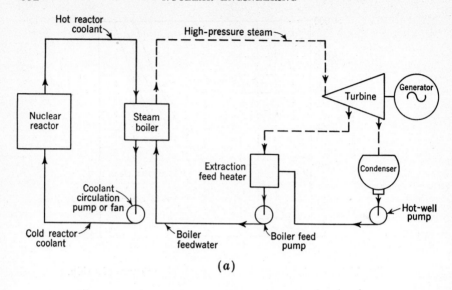

(a)

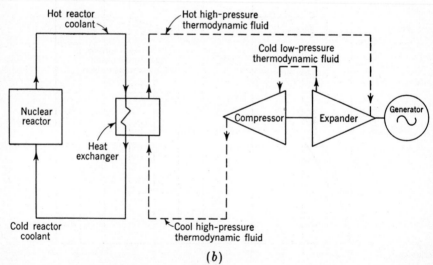

(b)

FIG. 13-15. Cycle diagrams for rudimentary nuclear power plants. (a) Steam-electric plant; (b) gas-turbine–electric plant.

may be expected to work. Existing plants have, by use, established certain performance and cost standards. These form the competitive base which any atomic power plant must better if that plant is to be considered a success.

13-4. Atomic-power-plant Cycles and Their Performance. While there have been many suggestions as to basic types of atomic-power-plant cycles, many of them, as yet, are fantastic, impractical, or uneconomic. In fact, there are only two cycles which warrant real consideration *at this time*—the gas-turbine cycle and the steam cycle. These cycles both presume the use of equipment most of which has been proved by experience. There is a minimum of substitution. That substitution is essentially to use a nuclear reaction as the source of heat instead of the chemical reaction of air with a conventional fuel. This means, as in Fig. 13-15a, the substitution of a nuclear reactor for the normal boiler furnace. The heat developed in the reactor is transferred by a coolant fluid to a waste-heat type of boiler. Beyond the boiler, all equipment is

TABLE 13-8. HEAT-RELEASE RATES FOR SELECTED OPERATIONS

	$Btu/(hr)(ft^3$ of space$)$
Open-feed water heater	50,000–100,000
Pulverized-coal-fired furnace	25,000–50,000
Stoker-fired furnace	50,000–100,000
Oil-fired, Navy-type, boiler furnace	1,000,000–2,000,000
Aircraft engine cylinder, combustion chamber	5,000,000
Candle flame	1,000,000
Flaming arc	800,000,000
Kitchen gas-range burner	500,000
Oxyacetylene flame	5,000,000
Surface combustion burner	2,000,000
Aircraft gas-turbine combustion chamber	4,000,000
V-2 rocket*	170,000,000
Brookhaven reactor*	6,000
Experimental breeder reactor*	23,500,000

* W. H. Zinn, Basic Problems on Central Station Nuclear Power, *Nucleonics*, **10**(9): 8–14 (September, 1952).

standard and subject to the usual variations in design as they are encountered in ordinary steam-power practice. Similarly in Fig. 13-15b, the combustion of fuel in a conventional gas turbine is replaced by a reactor or by a reactor and heat exchanger, all other apparatus conforming to the usual gas-turbine design practice. Whether the plant to be considered for atomic application is a steam or a gas-turbine plant, it is essential to recognize the cyclic potentialities and limitations—especially those set by thermodynamic considerations. Any standard engineering thermodynamics textbook gives information (1) for gas-turbine plants, on the influence of compression ratio, initial temperature, reheat, regeneration, and physical properties of the working fluid; and (2) for steam plants, on the influence of steam pressure, steam temperature, exhaust pressure or vacuum, reheat pressure and temperature, and regenerative feed heating on the basic performance cycles.

This present limitation of nuclear fuels to gas-turbine and steam cycles should not, however, detract from the high thermal efficiency and low heat rate which is theoretically obtainable with nuclear fuels. If nuclear reaction temperatures run to millions of degrees, then the Carnot cycle thermal efficiency would approach 100 per cent (heat rate = 3413 Btu/kwhr) with atmospheric temperature as that of heat rejection. Table 13-8 gives heat-release data for different heat sources.

13-5. Steam Plants. Figures 13-16 show the heat rate which can be expected with the ideal Rankine cycle as influenced by pressure, temperature, vacuum, and regenerative feed heating. The ideal Rankine cycle heat rate is found by the use of steam tables and the Mollier diagram

$$\text{Rankine cycle heat rate, Btu/kwhr} = \frac{3413}{\text{thermal efficiency}}$$

$$= 3413 \, \frac{h_{\text{in}} - h_{liq} - \text{FPW}}{h_{\text{in}} - h_{\text{out}} - \text{FPW}} \quad (13\text{-}3)$$

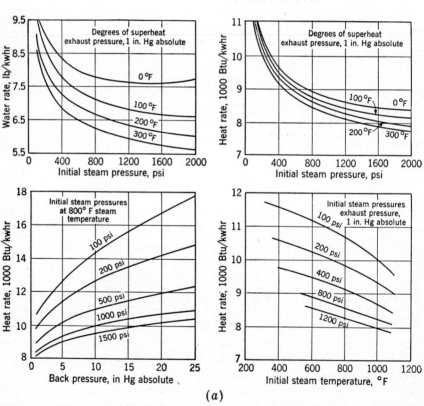

(a)

FIG. 13-16a. Curves showing influence of steam pressure, temperature, and vacuum on the heat rate and water rate of the ideal Rankine condensing steam cycle.

where h_{in} = enthalpy of steam at superheater outlet or turbine throttle, Btu/lb

h_{out} = enthalpy of steam at turbine exhaust (same entropy as h_{in}), Btu/lb

h_{liq} = enthalpy of saturated liquid at turbine exhaust conditions, Btu/lb

$$\text{FPW} = \text{feed pump work, Btu/lb} = \frac{\text{head on feed pump, ft}}{778} \qquad (13\text{-}4)$$

Unless steam pressures are high, the feed-pump work can be neglected in Eq. (13-3). The Rankine-cycle calculation is most conveniently carried out on a Mollier chart where the terms of Eq. (13-3) are as indicated in Fig. 13-17.

The preceding equations and statements summarize the method of computing the ideal thermal performance of steam plants. On any

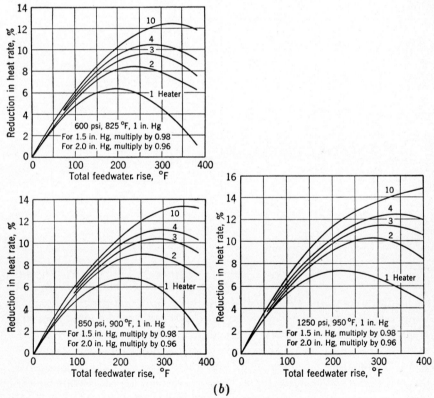

(b)

Fig. 13-16b. Curves showing influence of feed conditions and steam conditions on the heat rate of ideal Rankine regenerative steam cycle.

actual thermal power plant the heat rate is determined by metering the heat supplied during any selected period and dividing it by the metered generation during the same period, or

Actual over-all plant heat rate, Btu/kwhr

$$= \frac{\text{heat supplied during period, Btu}}{\text{electric generation during period, kwhr}} \quad (13\text{-}5)$$

The theoretical value can be reconciled with the actual performance by including all practical items such as turbine efficiency; generator efficiency; blowdown losses; make-up; power required to drive pumps, fans,

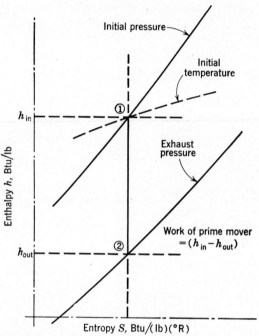

Fig. 13-17. Mollier diagram showing ideal Rankine-cycle expansion line.

and other auxiliaries; boiler efficiency; starting, stand-by, and radiation losses. Heat-balance diagrams, as prepared for power plants, contain such allowances and show heat quantities, fluid quantities, pressures, temperatures, enthalpies, efficiency, and power for each component element and for the assembly. A typical heat-balance diagram and its summary calculations are shown in Fig. 13-18 and Table 13-9.

An attempt has been made in Fig. 13-19 to give some comparative thermal-performance data for a wide assortment of real steam-power-plant cycle conditions. These data are applicable to unit sizes of the range 75,000 to 150,000 kw. No allowance is included for the boiler

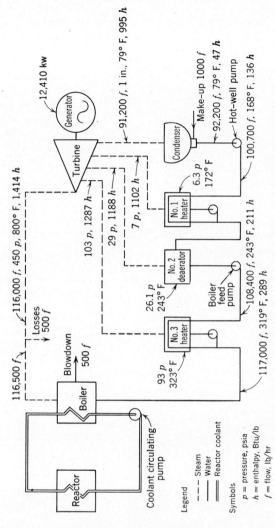

FIG. 13-18. Heat-balance diagram for a nuclear steam-electric power plant.

TABLE 13-9. HEAT-BALANCE CALCULATIONS FOR NUCLEAR STEAM-ELECTRIC
POWER-PLANT CYCLE ILLUSTRATED IN FIG. 13-18

Nominal rating of turbine-generator set.....................	10,000 kw
Throttle pressure...	450 psia
Throttle temperature.....................	800°F
Condenser:	
Steam flow...	91,200 lb/hr
Pressure..	1 in. Hg abs
Make-up..	1000 lb/hr
Hot well temperature..................................	79°F
Hot well enthalpy.....................................	47 Btu/lb
No. 1 extraction feed heater:	
Water flow entering...................................	92,200 lb/hr
Turbine shell pressure................................	7.0 psia
Pressure drop to heater...............................	0.7 psia
Heater shell pressure.................................	6.3 psia
Heater saturation temperature.........................	172°F
Terminal temperature difference	4°F
Water temperature out.................................	168°F
Water temperature in..................................	79°F
Water temperature rise................................	89°F
Heat required...	8.2×10^6 Btu/hr
Extraction enthalpy...................................	1102 Btu/lb
Drain enthalpy..	140 Btu/lb
Heat available..	962 Btu/lb
Extraction required...................................	8,500 lb/hr
No. 2 extraction feed heater and deaerator:	
Water flow entering...................................	100,700 lb/hr
Turbine shell pressure................................	29.0 psia
Pressure drop to heater...............................	2.9 psia
Heater shell pressure.................................	26.1 psia
Heater saturation temperature.........................	243°F
Terminal temperature difference.......................	0°F
Water temperature out.................................	243°F
Water enthalpy out....................................	211 Btu/lb
Water enthalpy in.....................................	136 Btu/lb
Enthalpy rise of water................................	75 Btu/lb
Heat required...	7.55×10^6 Btu/hr
Extraction enthalpy...................................	1188 Btu/lb
Drain enthalpy..	211 Btu/lb
Heat available..	977 Btu/lb
Extraction required...................................	7700 lb/hr
No. 3 extraction feed heater:	
Water flow entering...................................	108,400 lb/hr
Turbine shell pressure................................	103 psia
Pressure drop to heater...............................	10 psia
Heater shell pressure.................................	93 psia
Heater saturation temperature.........................	323°F
Terminal temperature difference.......................	4°F
Water temperature out.................................	319°F
Water enthalpy out....................................	289 Btu/lb

TABLE 13-9. HEAT-BALANCE CALCULATIONS FOR NUCLEAR STEAM-ELECTRIC POWER-PLANT CYCLE ILLUSTRATED IN FIG. 13-18 (*Continued*)

Water enthalpy in.................................... 211 Btu/lb
Enthalpy rise of water................................ 78 Btu/lb
Heat required.. 8.5 × 10⁶ Btu/hr
Extraction enthalpy.................................. 1287 Btu/lb
Drain enthalpy....................................... 293 Btu/lb
Heat available....................................... 994 Btu/lb
Extraction required.................................. 8600 lb/hr
Feed to waste-heat boiler............................ 117,000 lb/hr
Waste-heat boiler:
  Enthalpy of steam.................................. 1414 Btu/lb
  Enthalpy of feed................................... 289 Btu/lb
  Heat added to steam................................ 1125 Btu/lb
  Boiler blowdown.................................... 500 lb/hr
  Steam made... 116,500 lb/hr
  Heat added in boiler............................... 131 × 10⁶ Btu/hr
Turbine-generator performance:
  Steam output of boiler............................. 116,500 lb/hr
  Steam losses....................................... 500 lb/hr
  Turbine throttle flow.............................. 116,000 lb/hr
  Throttle enthalpy.................................. 1414 Btu/lb
  Exhaust enthalpy, including turbine leaving losses.......... 995 Btu/lb
  Internal work, from throttle to condenser.................. 419 Btu/lb
      No. 1 extraction point......................... 312 Btu/lb
      No. 2 extraction point......................... 226 Btu/lb
      No. 3 extraction point......................... 127 Btu/lb
  Internal generation by flow to condenser.................... 11,200 kw
      No. 1 extraction heater........................ 780 kw
      No. 2 extraction heater........................ 512 kw
      No. 3 extraction heater........................ 318 kw
  Total internal generation.......................... 12,810 kw
  Generator losses................................... 400 kw
  Load at generator terminals........................ 12,410 kw
Plant performance:
  Auxiliary power use in steam section of plant.............. 4%
  Auxiliary power use in steam section of plant.............. 500 kw
  Net plant output................................... 11,910 kw
  Plant realization ratio............................ 0.95
  Heat supplied to boiler............................ 138 × 10⁶ Btu/hr
  Net plant heat rate................................ 11,600 Btu/kwhr

room, its inefficiency and its auxiliary use, as in the conventional coal-fired steam plant. The data are therefore in such form as to be directly usable in calculations for nuclear steam-electric power plants. It is, of course, necessary to modify the heat rates as shown by appropriate allowances for auxiliary power use by the reactor plant, by the pumps, blowers, or fans circulating the coolant fluid, and by radiation losses to the ambient.

13-6. Gas-turbine Plants. The ideal thermal efficiency of the Brayton cycle (Fig. 13-20) is given by the relation

$$\text{Thermal efficiency} = 1 - \left(\frac{1}{R_v}\right)^{k-1} = 1 - \frac{T_a}{T_b} \qquad (13\text{-}6)$$

where R_v = volumetric ratio of compression, V_a/V_b
$\qquad k$ = ratio of specific heats, C_p/C_v
$\qquad T$ = absolute temperature

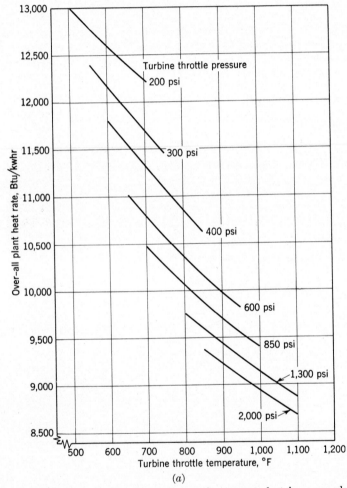

FIG. 13-19a. Estimated plant heat rates for nuclear steam-electric power plants as a function of steam pressure and temperature, nonreheat cycles. Includes allowances for auxiliary power beyond the boiler, make-up, and blowdown. Contains no allowance for power consumption of reactor or reactor auxiliaries.

If air is the working substance, with $k = 1.4$, the thermal efficiency and heat rate are as plotted in Fig. 13-21. These ideal values are much better than for steam (cf. Fig. 13-16). These better values constitute probably the principal inducement for the use of a gas turbine for nuclear power. Efforts to harness the cycle with a real compressor and a real expander, however, lead to quite different conclusions. If an allowance is made for actual compression efficiency and turbine efficiency, the ideal T-S diagram of Fig. 13-20 takes on the form of the real diagram of Fig. 13-22. The compression phase follows the path a-b', and the expansion phase follows the path c-d', both characterized by irreversibility and increasing entropy. The inefficiencies raise both the compressor and turbine exhaust temperatures. In Fig. 13-22, the areas account for the energy quantities as follows:

Compressor losses $= 1$
Turbine losses $= 3$
Heat rejected as heat $= 1 + 2 + 3$
Actual net work $= 4 - 1 - 3$
Heat supplied as heat $= 2 + 4$

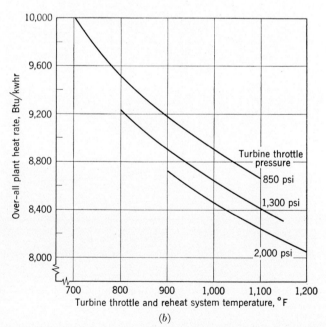

(b)

FIG. 13-19b. Estimated plant heat rates for nuclear steam-electric power plants as a function of steam pressure and temperature, reheat cycles. Includes allowances for auxiliary power beyond the boiler, make-up, and blowdown. Contains no allowance for power consumption of reactor or reactor auxiliaries.

It takes relatively little inefficiency in the compressor and expander to make the actual net work entirely disappear. The top allowable temperature T_c is fixed by fuel combustion or by reactor operation, and increasing inefficiency in the compressor means that T_b' approaches this

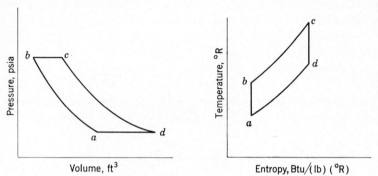

FIG. 13-20. Brayton cycle—ideal pressure-volume and temperature-entropy diagrams.

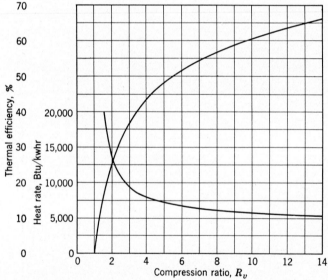

FIG. 13-21. Brayton cycle—thermal efficiency and heat rate as functions of compression ratio.

value of T_c. Less heat can be added, more fluid must be circulated, and compressor and expander grow prohibitively large. High internal efficiency and high throttle temperatures are basic requirements for a successful gas-turbine power plant. Some of these requirements are reflected in the data of Fig. 13-23 [31]. The conclusion from such curves is that a practical gas-turbine plant must have a high working fluid

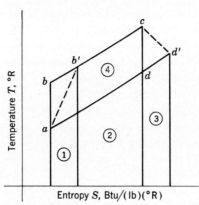

FIG. 13-22. Temperature-entropy diagram for a real Brayton cycle.

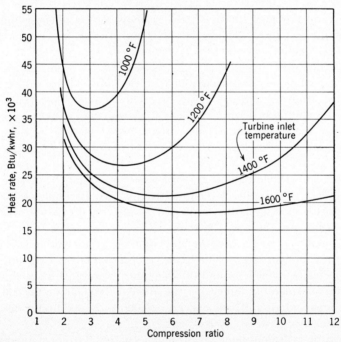

FIG. 13-23. Simple cycle, gas-turbine plant. Thermal performance as a function of compression ratio and turbine inlet temperature. Engine and compression efficiencies = 80 per cent. No pressure drops in system. 80°F inlet air temperature. Gas cycle with $C_p = 0.24$. No refrigeration.

temperature for the gas entering the expander. In numbers, this translates to a gas temperature of at least 1000°F and preferably over 1200°F. Conversely, any attempt to utilize a gas-turbine cycle with low temperatures is inherently unsound. The consequence is that with the present or near-future limits on reactor operating temperatures, the gas turbine is not a promising device for nuclear application. Those power services which require minimum weight or minimum bulk, and which can accept short life for the plant, may utilize gas-turbine principles provided suitable reactors can be built for operation at elevated temperatures. Otherwise the steam cycle, with flow diagrams as pictured in Fig. 13-15a or 13-18, is likely to dominate the nuclear-power field.

13-7. Structural and Functional Aspects of Nuclear Power Plants— Size. Many statements are made concerning the physical size of a nuclear power plant, especially the reactor. The energy equivalent of 1 lb of U^{235} as 1500 tons of coal presumably indicates that a nuclear power plant, with its attendant fuel, can be concentrated in the absolute minimum of weight and space. There is merit to this observation, because 1 lb of U^{235} would occupy 0.00086 ft^3, while 1500 tons of coal would occupy 58,000 ft^3. The weight ratio is 3×10^6 to 1, and the volume ratio is 6.8×10^7 to 1. These features are overwhelmingly in favor of nuclear fuel. There are, however, certain accompanying disadvantages which need scrutiny. Some enthusiasts feel that these numbers mean that a pellet of U^{235} would be ample to supply the needs of an automobile engine throughout its useful life. The mean free path of neutrons, however, is some 6 in., so that if a chain reaction is to be sustained a minimum critical size, much greater than a pellet, is essential (see Chap. 6). This critical-mass requirement thus fixes the present minimum size of a nuclear power plant. In addition to the limitations on the mass of fuel, there is the need for heavy shielding, described in Chap. 7. This shielding is of the order of 2 ft of lead, 10 ft of concrete, or 15 ft of water or diesel fuel. Even for aircraft application, where every effort is made to reduce bulk and weight, the estimated weight of a reactor is between 50 and 100 tons. When a reactor is to be used for propulsion of a seagoing ship, the limitation on weight and bulk is far less severe than on aircraft, locomotives, or automobiles. Machinery for naval service is far heavier and bulkier than for aircraft service. Power plants are considerably larger—of the order of 100,000 to 200,000 hp for capital ships and aircraft carriers, and 25,000 to 75,000 hp for destroyers and submarines.

This same reasoning can be extended to the stationary-power field, where generating units are apparently being built in ever-increasing sizes. Today, many units are in service which deliver 150,000 to 200,000 kw. Units of 475,000 kw are on order, and sizes in excess of 500,000 kw are seriously projected. For such sizes the critical mass of the nuclear power plant is negligible. This is the area where large units could be employed without handicap [8, 11, 29].

13-8. Contamination. A problem which is present in all fuel-burning power plants is that of the disposal of waste and refuse without pollution of the atmosphere or of the natural water supply. The problem is greatly magnified with nuclear fuels because of the radioactive elements present in the operation. No process—not even a nuclear process—can be carried on without losses and inefficiency. This means that there will ultimately be left, in the pile, contaminated spent fuel which will have to be removed and reprocessed for possible recovery of valuable constituents, with ultimate disposal of nonrecoverable residue. Likewise, fluids circulated through a reactor may be subject to activation by neutrons and/or deterioration by γ radiation (cf. Chaps. 3 and 7). The maintenance problems posed by radioactivity are unprecedented in the entire field of technology. Totally different solutions are employed for different ranges of half-life. No matter how well equipment is designed, or how well materials are selected, or how well the workmanship or construction is effected, or how carefully the plant is operated, there are bound to be breakdowns in service. There is thus the demand for maintenance either on the forced or the scheduled outage basis, or both. Even a simple problem, such as the repair of a gate valve in a pipeline, can become a trying experience which taxes the imagination and ingenuity of the most skillful designers and operators; it may even be an impossible task. Entire new techniques are called for with remote controls, indirect vision through mirror systems, and the manipulation of tools and instruments through complicated handling devices. The wastes from a nuclear plant cannot be casually disposed of, like the stack gases and ashes from a conventional coal-fired plant. They can be concentrated by evaporation, precipitation, or base exchange to decrease bulk, then stored in tanks until sufficiently inactive, encased in concrete and dumped at sea, or whatever other treatment may be adequate and most economical.

Likewise the hazard of an explosive condition in the reactor, the coolant, or the waste-heat system must be meticulously considered [28]. The potential hazard here is greatly magnified beyond the number of accidents which occur with conventional boilers. The hazard of scattering radioactive particles by an explosion is so severe that it is hardly rational to project a nuclear power plant, in the present stage of development, for installation in a congested, populated area unless the maximum possible explosion can be contained in a closed space. Otherwise it must be placed in a region of large acreage, where damage to surroundings from contamination would be minimized. A circular "exclusion circle" of the order of 1 acre per 5 thermal kw may be required, though improvements in safe design and low fission-product inventory may greatly decrease this area. If abandonment is necessary, the radioactive parts can be buried under a mound of earth and left to exhaust themselves— maybe for thousands of years. This offers a serious present-day limitation on stationary-power-plant applications. An illustration of the containment method for a small stationary installation (approximately 10,000 kw of output) is the housing of the SIR submarine prototype power plant at KAPL in a welded steel sphere. This sphere is 225 ft in diameter, with 1-in. walls, and designed to withstand a pressure of 100 psi. Such a solution is economically justifiable on an experimental plant for military service, but its cost could be prohibitive for a large-size stationary plant competing with conventional steam and hydro installations.

The offshore location of a nuclear power house has much to recommend it. While this proposal would introduce new problems and add new costs, the disposal of waste and the meeting of the contamination hazard could be effective. The submarine or the aircraft carrier would appear a natural place for the installation of an atomic power plant. Maintenance as well as contamination problems would be somewhat simplified. If the plant ultimately had to be abandoned, the radioactive parts, by suitable design of the hull, could conceivably be scuttled at sea without sacrifice of the entire ship. In a similar fashion, the airplane application has merit because the power plant alone, or the complete craft, could be sunk at sea.

In any attempt to use a nuclear power plant for commercial purposes this contamination hazard will have to be met by adequate insurance. Any accident which would lead to scattering of radioactivity entails the obligation of proper insurance. The wreck of a nuclear locomotive on a railroad right of way, an accident on a nuclear-powered ship in port, or the drift and subsequent fallout of a radioactive cloud at a remote distance constitute the type of contingency which will have to be faced. There are no statistical or actuarial data on nuclear operations; there are no insurance reserves accumulated or set aside. The extent of the risk is unknown, so that protection will have to be arranged not only through insurance companies and insurance underwriters but with the added support of government subsidy. The only reactor accidents from which real experience has been gained, so far, are (1) the fast-neutron reactor "Clementine" at Los Alamos in 1950, and (2) the NRX reactor at Chalk River, Ontario, in 1952 [20] (see Sec. 9-47).

13-9. Reactor Coolants and Cooling. The heat which is generated within the reactor must be transferred to a fluid or fluids which can serve for the thermodynamic conversion of heat into useful work. In conventional internal-combustion plants the operation is accomplished more simply and directly than in the conventional steam plant. With the former, the heat is generated directly in the thermodynamic fluid. With the latter, the heat is developed in the furnace gases, then transferred through the boiler surface to the thermodynamic fluid, steam. All serious proposals for nuclear power plants are akin to the steam plant, rather than to the internal-combustion plant, with respect to the working fluids. Two fluids are usually desired. One fluid brings the heat from within the reactor to the outside but becomes too radioactive from neutrons in the reactor to be used in the prime mover. It therefore delivers its heat to a second fluid in a heat exchanger, or waste-heat-type boiler. The heat is transferred without activation to the second, or thermodynamic, fluid for use in the prime mover. The first fluid may give off strong γ rays and require shielding. The γ rays, however, do not activate the second fluid, which thus may be distributed freely. This double-fluid circuit for the transfer of heat, with its additional temperature loss, limits the conversion efficiency from heat to work, because the temperatures (1100°F) in the most modern steam plants are higher than present feasible reactor coolant temperatures. The double coolant also adds substantially to plant cost, bulk, weight, loss of flexibility, maintenance problems, and operating hazards. However, if purified water can be boiled directly in the reactor, little shielding is necessary (see Sec. 7-10).

Many reactor coolants have been proposed and tested [23, 36]. All seem to offer some merits but many disadvantages. Gases generally offer favorable radioactive properties. But their low density, low specific heat, and high thermal resistance make for large flow passages and excessive heat-transfer surfaces and/or high pumping and temperature losses, both within the reactor and in the external heat-exchange system. Most gases are noncorrosive, so that materials specifications are broad and liberal. If the coolant system is operated at low pressure, the requisite metal thickness of the containing parts, for bursting strength, is a minimum. However, an economic balance will usually yield a high optimum pressure for gases (see Prob. 8-28b). The use of a closed-cycle gas turbine can be anticipated with a single fluid circuit if a gas, such as pure He, with zero cross section and a high-temperature reactor were available.

Liquids are often favored for reactor cooling because of their high density, high specific heat, and low thermal resistance (see Sec. 9-30). The passages and the surfaces in the reactor and in the heat exchanger will all be smaller with liquid, rather than gas, cooling. Water, on account of its favorable physical properties and prevalence, is a useful coolant. Its low boiling point, however, requires high pressure if vaporization is to be prevented in the reactor at the high temperatures thermodynamically desired, as well as those to be expected at "hot spots." Mercury, lithium, and fused salts or eutectic salt mixtures, such as chlorides and fluorides of alkali metals, metallic alloys, or eutectic mixtures such as NaK, Pb-Bi, etc., are all possible, though they tend to bring corrosion troubles. Some properties of liquid coolants are contained in Table 13-10. Such coolants will allow the reactor to be run at tempera-

TABLE 13-10. COMPARISON OF SOME HEAT-TRANSFER FLUIDS AT HIGH TEMPERATURE*

(Flowing at 10 fps in 1-in. ID pipe)

	Sodium	NaK (40% K)	HTS†	Dowtherm A‡	Mercury	Lead	Lead-bismuth eutectic	Lithium	Water
Melting point, °F	208	65	290	55	−37	622	257	367	32
Boiling point (14.7 psia), °F	1616	1518	.....	485	675	3170	3038	2403	212
Liquid density, g/cm³	0.9	0.9	1.7	.....	13	10	10	0.5	1
Specific heat, Btu/(lb)(°F)	0.33–0.30	0.28	0.373	0.63–0.69	0.033	0.034	0.035	1.05–1.22	1.005
Thermal conductivity, Btu/(hr)(°F)(ft²)	49–38	14–16	.....	0.105	5–9	8	5.3–6.5	24	0.406
Heat-transfer coefficient, Btu/(hr)(°F)(ft²)	6400	3500	1500	200–300	5700	4100	3700	5800	3000
Pumping power (water = 1)	0.925	0.925	1.92	0.925	13.1	11.5	11.5	0.5	1
Heat capacity (water = 1)	0.24	0.185	0.63	.....	0.42	0.38	0.30	0.48	1

* T. Trocki, Engineering Aspects of Liquid Metals for Heat Transfer, *Nucleonics* 10(1):28–32 (January, 1952).

† Heat-transfer salt of NaNO₂, NaNO₃, and KNO₃.

‡ Eutectic mixture of diphenyl and diphenyl oxide.

tures of the order of 1000°F. Steam generators and turbines of the 400- or 600-psi class then follow, and with such pressures and temperatures the nuclear power plant has utility in some applications.

13-10. Gas-heated Waste-heat Boilers and Heat Exchangers. If the reactor coolant is a gas, it will be circulated by fans through a duct system to a waste-heat boiler and thence in an open cycle to a stack, or in a closed cycle returned to the reactor. The open system would use air at atmospheric pressure as the heat-transfer fluid to be economical. If clean, it can safely be discharged a couple of hundred feet above the ground level without danger to the surroundings from radioactivity, as in the Brookhaven reactor. The boiler design would be similar to conventional waste-heat installations. Because of the low pressure the heat-transfer rates would be low. When coupled with low temperature differences and economical velocities, the required heat-transfer surfaces would be large. An approximate over-all heat-transfer equation is

$$U = 2 + 0.002G_{max} \tag{13-7}$$

where U is in Btu per hour per square foot per degree Fahrenheit and G is the gas mass velocity in pounds per hour per square foot at the minimum flow cross section between tubes. The most economical values of G are those for which the total annual operating cost (mainly power to the fans) plus fixed charges (mainly on the heat-transfer surface) are a minimum. From experience with conventional waste-heat installations it is doubtful whether G greater than 5000 to 10,000 can be justified. The cleanliness required of nuclear coolants permits the use of small passages and extended or finned surfaces, reducing the bulk, weight, and cost of the heat exchanger. However, maintenance imposes a unique problem, not present with conventional boilers. It must be possible to remove parts of the apparatus and replace them after the system has been suitably deactivated. Scavenging with air and washing with water or special decontaminating solutions will probably be necessary before the parts can be made accessible. These maintenance requirements and the other special conditions can lead to a design totally different from conventional ones. Figure 13-24 serves as an illustration of one solution.

13-11. Liquid-heated Equipment [23]. Boilers heated by a liquid heatant are more promising for nuclear power plants than those using gas. With a liquid the coefficient U can be improved 25- to 50-fold. The consequent reduction in heat-exchanger size, coupled with the improved heat transfer in the reactor itself and the smaller pipes and accessories for transporting the liquid, results in substantial savings in investment, weight, and bulk of the heat-recovery system. New problems due to radioactivity and chemical activity may arise and outmode conventional designs and materials. Double-tube construction for the

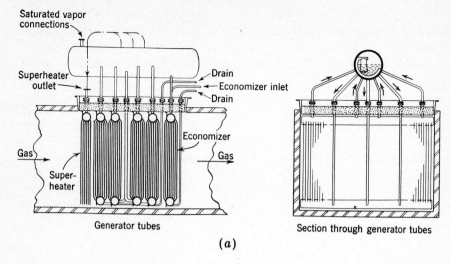

Saturated vapor connections

Superheater outlet

Drain
Economizer inlet
Drain

Gas

Economizer

Gas

Super-heater

Generator tubes

Section through generator tubes

(a)

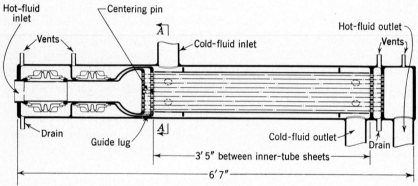

Hot–fluid inlet

Centering pin

Cold-fluid inlet

Vents

Hot-fluid outlet

Vents

Drain

Guide lug

Cold-fluid outlet

Drain

3′ 5″ between inner-tube sheets

6′ 7″

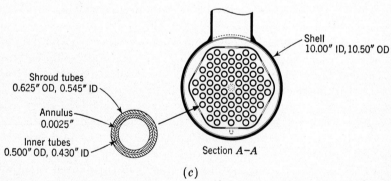

Shell
10.00″ ID, 10.50″ OD

Shroud tubes
0.625″ OD, 0.545″ ID

Annulus
0.0025″

Inner tubes
0.500″ OD, 0.430″ ID

Section *A–A*

(c)

Fɪɢ. 13-24. (a) Gas-heated steam generator. (*The Babcock and Wilcox Company.*) fluid. [*T. J. Trocki and D. B. Nelson, Report on Liquid Metal Heat Transfer and Steam Company; The Babcock and Wilcox Company.*] (c) Arrangement of intermediate metal heat exchanger, single tube. (*Foster Wheeler Corporation.*)

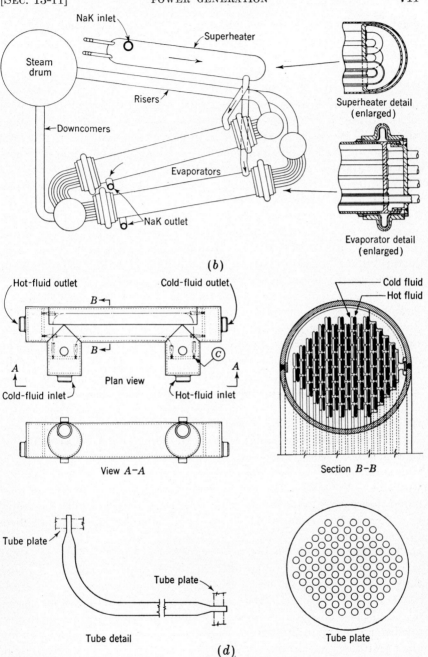

(b) Liquid-metal steam generator.   Double tube with mercury as the intermediate
Generation System for Nuclear Power, Mech. Eng., **75**:472–476 (1953); *General Electric*
liquid-metal heat exchanger. (*The Babcock and Wilcox Company.*)   (d) Liquid-

heat exchanger, with the heatant within the inner tube and the steam outside of the outer tube, is expensive but is useful to prevent explosion from contact between the two fluids on tube failure. Mercury has been employed (Fig. 13-24b)* with tubes having appreciable radial clearance, and helium with tubes that are shrunk or forced together, leaving only grooves for the helium [46]. Two tube sheets are needed at the ends of the tubes, with the intermediate fluid under a pressure higher than that of the two other fluids. Thus neither principal fluid can leak into the other on a single tube failure or even a nondestructive double failure, and indication of a leak is immediately given by a drop in the pressure of the intermediate fluid. Possible hazards to be avoided include explosive reactions of water with sodium, caustic embrittlement, and corrosion. Double-tube construction with stainless steel may run to $100 or $200 per ft², at least ten times the price which would be paid for waste-heat surface with gas as the heatant. However, such high prices seem to be justified by smaller surface requirement than with a gaseous heatant and the greater safety than with a single tube (see Chap. 9).

On the water side of the boilers there is need for the maintenance of clean surfaces to retain the rated steam-making capacity. There is, however, no danger from overheating, as with direct-fired boilers, and the circulation requirements are not so severe. Ample disengagement surface in the boiler drum, with suitable dry pipes, separators, and scrubbers, will reduce the solids carry-over below 1 ppm. This dust might become radioactive, and at one time it was thought that steam quality might have to be raised to an unprecedented 1 in 10⁹ parts. Currently this problem is believed to be less serious (Chap. 7).

13-12. Auxiliary Equipment. Every power plant is an assembly of certain major pieces of apparatus such as boilers and turbines, with a wide assortment of accessory or auxiliary equipment to make the assembly workable. No attempt is here made to discuss the usual auxiliaries found in a conventional steam plant, but some reference must be made to a few specific auxiliaries which take on peculiar features when applied to a nuclear power plant.

13-13. Fans [4]. If the reactor coolant is a gas, such as air, it is necessary to have a fan or blower to deliver the coolant through the reactor, the duct system, and the heat exchanger. If the operating temperature range of the gas is low and the volumes of gas to be circulated are large, in the interest of sustained reliability the capacity can be economically divided among multiple fans. Maintenance becomes possible without plant shutdown or curtailment. The casings, wheels, and driver should be designed for easy replacement without hazard from radioactivity. The optimum heat-transfer rates, and therefore pressure drop, are high,

* This construction may show high thermal resistance (Sec. 9-22).

but rugged, high-speed blowers, which develop 20 to 30 in. of water head, are not exceptional (see Sec. 8-49). Fan requirements can be visualized by considering a 30 thermal Mw reactor, operating with air at 300°F and a 200°F temperature rise (see Sec. 9-26). The removal would be

$$\frac{30,000 \times 3,413}{60} = 1710 \times 10^3 \text{ Btu/min}$$

and the requirement

$$\frac{1,710,000}{0.24 \times 200} \times 12.4 \times 760/492 = 680,000 \text{ cfm}$$

at 300°F and atmospheric pressure. If the head were 30 in. of water and the fan efficiency were 70 per cent, the fan power would be

$$\text{shp} = \frac{0.0001575 \times 30 \times 680,000}{0.7} = 4600 \text{ hp} = 3500 \text{ kw}$$

If the power plant had 20 per cent over-all thermal efficiency, it would deliver $0.2 \times 30,000 = 6000$ kw gross. Thus the auxiliary power for the fan system alone would be $3500/6000$, or 58 per cent of the gross output, and the net salable power output could not reach 2500 kw. Such figures demonstrate the need for higher reactor temperatures and larger allowable temperature rises of the coolant to reduce the auxiliary power consumption to an economic minimum.

13-14. Pumps. Liquid coolants go far toward reducing the auxiliary power requirements for the coolant circulating system. With water the problems of pumping are relatively easy, but for molten metals or molten salts, conventional styles of centrifugal, reciprocating, or rotary pumps are unsuitable [14]. High liquid density, corrosivity, and radioactivity impose severe problems on pump design. Leakproof pumps with drive through sealed diaphragms instead of through stuffing boxes (Sec. 8-49) are advantageous. Electromagnetic pumps (Sec. 8-50) are leakproof, foolproof, and effective with liquid metals. Hot, radioactive, dense, corrosive liquids can be handled with safety, but these pumps are not efficient. Figures 13-25 and 8-19 show performance results for three designs of EM pumps with sodium [9]. Peak efficiencies are less than 40 per cent. This figure, judged by usual pump performance standards, indicates the need for extensive development work to improve efficiency. Reliability, weight, bulk, and cost could also benefit from such developmental effort.

13-15. Ducts and Piping. The large volumes required with gas-cooled systems are best transported through concrete ducts. The piping systems for handling the same power in molten metals and salts are much smaller. The optimum sizes as computed by Sec. 8-35 correspond to linear velocities of the order of 25 fps, much higher than usual power-plant practice. The materials for pipe, joints, and packings employed in conventional practice are suitable with water. However, with the other liquid coolants leak-tested welded construction is necessary [40] to reduce leakage to a minimum. When couplings are needed in liquid-metal systems, stainless-steel ring seals are employed. Materials to contain liquid metals must not be corroded by the liquid metals [23] or even be perceptibly soluble because of thermal corrosion and radioactivity

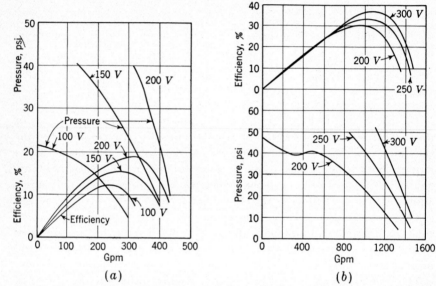

Fig. 13-25. Electromagnetic-pump performance curves. (a) Helical-flow induction pump at 25 cps handling sodium at 750°F. (b) Linear induction pump at 60 cps handling sodium at 700°F. [J. F. Cage, Jr., Electromagnetic Pumps for High Temperature Liquid Metal, Mech. Eng., **75**:467–471 (June, 1953).]

transfer to unshielded regions. With molten metals and salts the hazard of freezing the lines, reactor, heat exchangers, and pumps is avoided by draining before shutdown, or by steam or electric tracer lines during idleness. These tracer lines may be wrapped around the pipe, or they may be enclosed adjacent to the pipe in a common insulating casing. The necessary electric energy or steam consumption can be estimated by the methods of Chap. 9.

13-16. Fuel and Refuse Handling. One of the vital problems of the nuclear power plant is the removal of spent fuel from the reactor and its replacement by a fresh supply. With a homogeneous-fluid-fuel reactor the process can be continuous, but with a heterogeneous reactor optimum design generally yields a batch process, with shutdown before refueling.

The handling, disposal, treatment, and recovery of radioactive wastes imposes a technical and financial burden, including any value they may have, which is unmatched by any auxiliary equipment or process in conventional fuel-burning power plants. There must be accessibility for removal and replacement, but with complete protection of operating and maintenance personnel and the avoidance of contamination of the neighborhood. For some installations it is economical to abandon the plant after a certain life.

13-17. Control and Regulation. No power plant operates under a truly constant load. Some plants operate with loads that show only gradual and small changes, while others are subjected to the most violent swinging load conditions imaginable. Power plants operate to generate power at the instant of its use, having to meet rising loads without loss of speed or stability and falling loads without harm to plant or component parts from overheating, overspeeding, or excess pressure. These service conditions are severe in the case of the boiler room of a conventional plant, but they are greatly magnified with a reactor-waste-heat installation for an atomic plant. It is possible to obtain constant load operation by diverting the excess power above the instantaneous demand into storage-battery charging (for very small plants) or water elevating or by discharging it as heat. However, the former requires high investment and the latter yields low efficiency; consequently, these procedures are seldom justified.

Chapter 12 discusses the operation of control rods. Even if their motion could be made to respond instantaneously to load changes, there would still be problems, such as localized overheating. This calls for continuance of reactor coolant flow with a delayed adjustment to maintain the coolant outlet temperature at a desired value. If heat is to be discarded when the load suddenly drops, hot coolant can be diverted to a heat-wasting cooler before the boiler, or into extra condensers. Probably some nuclear power plants will be run base-loaded. This solution would seem highly desirable, since the investment is high and the fuel cheap. However, it would shift the responsibility for matching generation against the increased fluctuation in demand to a plant using some other source of raw energy. Hydro plants in particular are suitable because of their quick response, but they are not widely distributed. In any event, the satisfactory control and regulation of a reactor-heat-exchanger system for a nuclear power plant needs the maximum in skill and ingenuity.

13-18. Economics [1, 21, 25, 26, 29]. Many attempts have been made to put a price tag on nuclear power [22, 33, 45]. This is, of course, a fundamental step in any engineering activity. Almost all previous work in this field has been by government subsidy. Risk capital heretofore has been conspicuous by its absence. The 1954 amendments to the U.S. Atomic Energy Act (see Chap. 15) should encourage enterprise capital. The law was liberalized to the extent of permitting, under government license, private ownership and operation of equipment and facilities, but it retained government ownership of fissionable material, though such material may be licensed for private use [38, 39]. There is still a stringent legal monopoly of fissionable material by the Federal government. Any proposals for nuclear power must be made within the framework of that monopoly, the guarantee of a fair price for fuel and a continuation of its supply.

It must be remembered that there has never been one all-important method of power generation. Nuclear energy cannot be expected to displace all other sources of raw energy. It should be viewed as complementary to, rather than competitive with, existing methods of generation. There are areas, doubtless, where it will find real acceptance [18, 19]. In the first three sections of this chapter some standards of comparison are given for conventional plants. In any comparison of costs it should be remembered that competition will be with the most modern, economical, conventional plants and not with the averages of past practice.

Most variations in proposed nuclear plants stem from the different types of reactor which might be used [2]. Some reactors are designed primarily for generating heat and therefore power. Others have as their major purpose the breeding of Pu^{239} from U^{238} or U^{233} from Th^{232}, or the generation of radiation for research purposes. In the latter cases heat generation is secondary, but instead of dissipating the heat, as to a river at Hanford or to the air at Brookhaven, it is possible to construct a multipurpose plant in which the heat is treated as a recoverable by-product. The economics of such a proposal, like that for an industrial by-product steam power plant, is much more controversial than for a single-purpose power plant. The argument always hinges about the allocation of costs to the components. This allocation is not simple, and informed people do not agree as to methods, details, or conclusions. A "by-product" power plant may offer an advantage initially because some progress can be expected when the financial burden is not too heavy. On the other hand, if nuclear power is to become economically independent of subsidy it must ultimately stand on its own feet and deliver at less cost than alternative methods.

Much thinking has gone into this economic picture [27, 39, 41]. The present consensus is that further experimental and developmental work is needed before the nuclear plant will become a serious competitor for conventional methods of producing electric power. Excerpts follow from some of the pertinent economic analyses which have been made.

The first nuclear reactor to deliver heat to a power plant for the

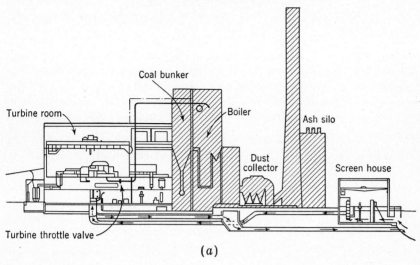

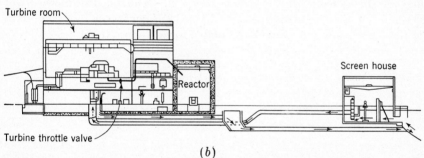

FIG. 13-26. (a) Cross section of the new St. Clair thermal electric power plant of the Detroit Edison Company. Shaded area might be replaced by nulcear heat power reactors. [*W. L. Cisler, Electric Power Systems and Nuclear Power, Edison Elec. Inst. Bull.*, **20**:289 *(September,* 1952).]

	Steam generation	Electric generation	Total plant
Total cost..........................	$48,293,000	$50,407,000	$98,700,000
Cost per kilowatt capability..........	77	81	158
Per cent of total plant...............	49	51	100
Per cent of total company investment.	17*	18*	35*

* These percentages represent the sum of the investments in steam-generation facilities, electric-generation facilities, and power plants, respectively, in the four power plants of The Detroit Edison Company, as of June 30, 1951. The new St. Clair Plant is not included.

(b) Present conception of how St. Clair power plant might look if nuclear heat-power reactors were substituted for coal-fired boilers. [*W. L. Cisler, Electric Power Systems and Nuclear Power, Edison Elec. Inst. Bull.*, **20**:289, *(September,* 1952).]

generation of electric power was the EBR, at Arco, Idaho. Cisler says [12] of this project:

Recent [1952] published information gives the cost of the experimental breeder reactor at Arco as $2,800,000, excluding the cost of the site. This installation apparently has a capacity of about 250 kw, or a unit cost of $11,200 per kw. It is recognized that this is an experimental unit but it is readily apparent that its cost is many times that which can be justified for practical power generation.

To be more realistic, the same author [13] estimates costs for a 625-Mw nuclear plant compared to a modern large electric-utility steam plant. Fig. 13-26 shows a conventional coal-burning plant and a nuclear power plant reduced to a common scale for the cross section. An allocation of costs among components is also shown. Of a total investment of $450 per kw on this utility system (Detroit Edison Co.), 35 per cent or $158 per kw is represented by the steam power plants. In turn, approximately one-half ($77) of this amount is for the steam-making section and one-half ($81) for the electric section. The nuclear power-plant designer, on these figures, thus actually has $77 per kw, or about one-sixth of the total system investment available for a competitive nuclear power plant, neglecting differences in operating cost. It is evident that even

TABLE 13-11. CALCULATION OF JUSTIFIED INVESTMENT IN A
NUCLEAR HEAT-POWER REACTOR AND APPURTENANCES*

	Annual operating cost, St. Clair thermal electric power plant†	Justified investment in reactor‡
Operating manpower	$ 180,000	$ 1,800,000
Operating and maintenance, materials, and supplies	307,000	3,070,000
Investment in thermal plant		48,293,000
Fixed charges on investment:		
Depreciation 3%		
Interest 5%		
Taxes and insurance 2%		
10%	4,829,300	
Fuel cost	8,855,000	88,550,000
Total	$14,171,300	$141,713,000
Investment per kilowatt capability	$ 77	$ 227
Cost per million Btu at turbine throttle	$ 0.577	$ 0.577

* W. L. Cisler, Electric Power Systems and Nuclear Power, *Edison Elec. Inst. Bull.*, **20**:289 (September, 1952).

† 625,000 kw capacity.

‡ No labor, maintenance, or fuel cost. Steam plant annual operating costs capitalized at 10 per cent.

if nuclear power becomes cheaper than fossil-fuel power no significant decrease in *delivered* energy cost can occur, since five-sixths of the investment is not affected.   Cisler's complete analysis, summarized in Table 13-11, employs an annual capitalization rate of 10 per cent, equivalent to a 33-year life, an allowable return of 5 per cent, and no corporate income taxes.   This rate is on the low side and tends toward government-financed operation rather than business-operated utility practice.   Coal is taken at $7.28 per ton, and nuclear-plant fuel, labor, and maintenance are assumed zero or negligible.   The conventional boiler-plant operating and fixed charges are comparable to a first cost of $227 per kw for the power reactor installation, the maximum possible competitive investment for a nuclear power plant under these conditions.   For a breeder reactor the cost of the fuel actually would be negligible, since U, quoted at $35 per lb [2] and yielding the energy of 2,600,000 times as much coal, corresponds to a coal price of 2.7 cents per ton.   If a more substantial fuel price must be borne by the nuclear power plant, the allowable investment will be reduced below the $227 figure.

Zinn has also made an analysis of economic feasibility [44] for a stationary nuclear power plant generating electric energy compared to a modern coal-burning plant with 200,000 kw units costing $200 per kw, one-half of this investment being for furnace and boiler, replaceable by the reactor. The economic results for three alternative types of reactor are shown in the cost data of Table 13-12.   Reactor 1 uses separated U^{235} as the fuel

TABLE 13-12. ECONOMICS OF REACTOR OPERATION FOR
NUCLEAR STEAM-ELECTRIC POWER PLANT*

Plant type as set by type of reactor operation	Unit costs investment credit per kilowatt of capacity		200,000-kw installation		
	For part of conventional coal plant displaced	For fuel savings (capitalized at 12 per cent)	Permitted cost of reactor and boiler	Permitted total cost of nuclear steam-electric power plant	Annual uranium fuel requirements, tons
1. Separated U^{235} is consumed; no fuel is regenerated.......	$100	−$204	....	....	77
2. Conversion to Pu^{239} or U^{233}; 1 % of fuel is consumed; no allowance for chemical processing................	$100	$128	$45,600,000	$65,600,000	27
3. Breeder reactor, fuel cost, including chemical processing is considered negligible..	$100	$204	$60,800,000	$80,800,000	0.27

* W. H. Zinn, Basic Problems in Central Station Nuclear Power, *Nucleonics*, **10**(9):8–14 (September, 1952).

TABLE 13-13. COMPARATIVE DATA ON INDUSTRY PROJECTS*

Reactor data	Commonwealth Edison–Public Service		Pacific Gas and Electric–Bechtel		Monsanto–Union Electric	
	Plant A	Plant B	Plant C	Plant D	Plant E	Plant F
Primary coolant..........	Helium	Heavy water	Light water	Liquid sodium	Liquid sodium	Liquid sodium
Moderator...............	Graphite	Heavy water	Heavy water	None	Graphite	Graphite
Fuel....................	Natural uranium slugs	Natural uranium	Natural uranium	Enriched uranium	Enriched uranium	Enriched uranium
Shielding...............	Cast iron, concrete	Steel-lined concrete	Concrete	Concrete	Concrete	Concrete
Control method..........	Boron steel rods	Boron steel rods	Movable plates	Blanket		
Plant capacities:						
Steam generation, lb/hr.....	906,000	3,210,000	....	1,600,000	3,000,000	4,800,000
Gross heat, Mw..........	350	1,064	500	500	1,000	3,000
Gross electrical, Mw......	61.7	225.5	106.2	154.6	220	579
Net electrical, Mw.......	46.7	211.5	100.6	1,453	210	554
Capital costs:						
Reactor and associated equipment	$11,500,000	$ 48,398,000	$24,350,000	$33,100,000	$26,000,000	
Total plant............	$40,000,000	$118,000,000	$41,000,000	$51,000,000		$61,000,000
Cost per installed gross kw capacity	$ 649	$ 492	$ 388	$ 330	$ 124	$ 110

* H. Bergen, Comparative Data Abstracted from AEC Report, *Consulting Eng.*, **2**:28–36 (July, 1953).

and consumes it without fuel regeneration; fuel cost is taken at $9000 per lb.   Reactor 2 uses natural U or Th with some conversion of U^{238} into Pu^{239} or Th^{232} into U^{233} so that 1 per cent of the nuclear fuel can be consumed in the operations; fuel cost is taken at $35 per lb.   Reactor 3 is a breeder reactor, with fuel price negligibly small.   The investment credits in Table 13-12 show that the first possibility is economically unsound, and the breeder reactor shows the highest allowable investment,

$$\$100 + 204 = \$304 \text{ per kw}$$

These figures are of the same order of magnitude as those shown for the previous example of Table 13-11, and present a picture for electric-power generation by nuclear fission.

The results of three industrial surveys [37] made under the aegis of the AEC were published in 1953 and given comparatively by Bergen [7] in Table 13-13.   The surveys were as follows:

1. Commonwealth Edison–Public Service studied a gas-cooled and a heavy-water-cooled reactor plant.

2. Pacific Gas and Electric–Bechtel studied a water-cooled thermal reactor plant and a sodium-cooled fast reactor plant.

3. Monsanto–Union Electric studied a sodium-cooled reactor plant for producing power and plutonium.

An analysis made by *Electric World* [17] on a 200,000-kw nuclear-electric plant with 16 per cent annual fixed charges and 7000 hr operation a year is summarized in Table 13-14.

TABLE 13-14. ESTIMATED COSTS ON 200,000-KW NUCLEAR ELECTRIC
POWER PLANT*
Financial Data

Item	Conventional plant	Nuclear plant
Construction cost:		
Utility......................................	$26,000,000	$44,000,000
AEC†..	....	16,000,000
Outlay, $/kw:		
Total....	130	300
By AEC†....................................	....	80
By utility...................................	....	220
Fixed charges, mills/kwhr....................	3.0	5.0
Labor and maintenance......................	0.5	1.0
Fuel...	3.0	0.2
Total energy cost, mills/kwhr.................	6.5	6.2

* Engineering Reference Sheet, *Elec. World*, Jan. 24, 1955.
† AEC allotment to research and development.

TABLE 13-14. ESTIMATED COSTS ON 200,000-KW NUCLEAR ELECTRIC
POWER PLANT (*Continued*)
Uranium Fuel Costs per Million Kilowatthours

Item	Pounds	$/lb	Mills/kwhr
Source material, natural uranium..............	85	20	1.7
Special material, enriched uranium.............	0.2	9,000	1.8
AEC charge for use of special material..........	....	......	0.1
Fuel fabrication cost....................	....	......	0.2
Freight (new and used fuel)...................	....	......	0.1
Separation expense...........................	....	......	0.5
Total.......................................	....	......	4.4
AEC credit for used fuel elements:			
Plutonium content........................	0.4	10,000	4.0
Depleted uranium	....	......	0.2
Total.......................................	....	......	4.2
Net fuel cost...............................	....	......	0.2

Uranium and Plutonium Quantities
per Million Kilowatthours

Fuel (U^{235}) consumption rate per million Btu‡.	0.000031 lb
Heat rate...	16,000 Btu/kwhr
Corresponding thermal efficiency............................	21.3%
U^{235} per million kwhr...................................	0.5 lb
40% from enriched uranium§...............................	0.2 lb
60% from natural uranium§...............................	0.3 lb

‡ Since 4,000 MWD irradiation leaves half the U^{235} in the natural uranium, the amount purchased for replenishment would be (with 0.007115 enrichment):

$$\frac{2 \times 0.3}{0.007115} = 85 \text{ lb per million kwhr}$$

With an 0.8 conversion ratio, the 0.5 of U^{235} consumed per million kilowatthours would result in 0.4 lb of plutonium.

§ Above based on 1.2 per cent average feed enrichment.

The AEC, under the 1954 amendments of the Atomic Energy Law, is sponsoring development programs for civilian power reactors in an effort to foster economical nuclear-electric power [2, 40, 42]. Because of the diversity of reactor types, each with its attendant advantages and disadvantages, the AEC has selected certain projects which appear to have promise. Essential data on the selected projects are summarized in Table 13-15. The only plant to be built full scale under the program

TABLE 13-15. AEC "FIVE-YEAR" DEVELOPMENT PROGRAM FOR CIVILIAN POWER REACTORS (1954–1959)*

Reactor concept	Name	Symbol	Designer and location	Scheduled completion date	Moderator	Coolant	Fuel	Experimental scale	Thermal capacity		Electrical output, kw	Cost, (000,000 omitted)	Date of full-scale† reality	Order of economic promise
									10^6 Btu/hr	Kw				
Pressurized water	Pressurized Water Reactor	PWR	Westinghouse; Shippingport, Pa.	1957	Ordinary water (H₂O)	Ordinary water (H₂O)	Slightly enriched uranium	Full	900	264,000	60,000	$85	1957	5
Boiling water	Experimental Boiling Water Reactor	EBWR	ANL; Chicago	1957	Ordinary water (H₂O)	Ordinary water (H₂O)	Natural and highly enriched uranium	Medium	68	20,000	5,000	$17	1960	3
Sodium graphite	Sodium Reactor Experiment	SRE	NAA; Santa Susana, Cal	1956	Graphite	Sodium	Slightly enriched uranium	Medium	68	20,000	None	$10	1959	4
Fast breeder	Experimental Breeder Reactor, No. 2	EBR-2	ANL; Chicago	1958	None	Sodium	Uranium-plutonium alloy	Medium	214	62,500	15,000	$40	1959	2
Homogeneous	Homogeneous Reactor Experiment, No. 2	HRE-2	ORNL; Tenn.	1956	Heavy water (D₂O)	Heavy water (D₂O)	Highly enriched uranium in UO₂SO₄ solution	Small	34	10,000	2,000	. . .	1962	1

* Largely from 17th and 18th Semi-annual Reports of AEC.
† See App. N-3 and Table 14-2 for further details.

723

is the PWR, which is least attractive from the viewpoint of anticipated ultimate economic acceptability but is closest to immediate reality. Figure 13-27 shows a schematic diagram of this power plant as it is to be built by the Westinghouse Electric–Duquesne Light group near Pittsburgh. It will operate with light water circulated through the reactor at approximately 2000 psi. Steam will be generated in the boilers at 600 psi and delivered saturated to the turbine throttle. The other plants receiving technical and financial support from the AEC, as given in Table 13-15, are on a relatively small experimental scale.

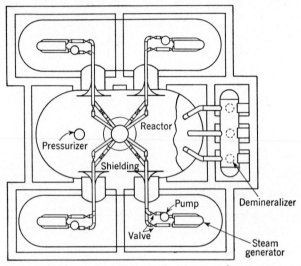

Fig. 13-27. Plan view of proposed layout of reactor and steam-generating section, PWR plant at Shippingport, Pa. Water pressure = 2000 psi, temperature = 600°F. Steam pressure = 600 psi, dry and saturated.

The joint effort of the AEC and the Department of Defense has lead to the Army Package Power Reactor (APPR) program. This calls for the development of power plants to supply heat and electricity in a range of a few hundred kilowatts to 40,000 kw. Air transportability of major components is required. A heterogeneous pressurized-water, enriched-uranium reactor, to deliver 10,000 kw of heat and producing 1200 kw electric output and 2500 kw steam for space heating, was chosen for the initial step and a contract let to American Locomotive Company for such a unit at a price* of $2,096,753.

The tenor of these examples can be pursued further by considering the total investment (Fig. 13-28) of a power plant employing a reactor delivering 10^9 Btu/hr to the boiler. The heat-rate data of Fig. 13-19a

* This highly competitive bid is generally believed to be less than half the actual cost.

are used with the net electric output of the order of 100,000 kw. The higher-pressure plants, by virtue of their better heat rates, have a larger net output. Selecting a constant temperature of 1000°F for the heatant leaving the reactor, the series of projected plants can be defined exactly. If various prices, from zero to $50,000,000, are used for the reactor and added to the estimated prices for the rest of the power plant, a series of curves, as plotted in Fig. 13-28, result which give investment as a function of steam pressure.

The minimum unit investment justifies the selection of higher steam pressures as the reactor cost rises. The improved heat rates at high

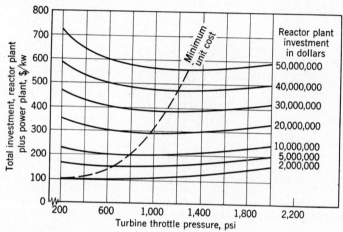

FIG. 13-28. Estimated investment for complete nuclear steam-electric power plants as a function of steam pressure. Data are based on regenerative, nonreheat, condensing steam cycles. Size range is of the order of 1 billion Btu/hr supplied as heat and 100,000 kw of usable generating capacity.

pressures spread the total investment over more salable generating capacity, so that with high, rather than low, prices for reactors it is desirable to select the maximum operating steam pressures and temperatures [6]. An analysis for reheat cycles can be made similarly by use of the data of Fig. 13-19b. Since the heatant can be brought close to the turbine, there is no inherent reason why multiple reheat cycles should not also be considered. The Liquid Metal Fuel Reactor (LMFR) [43] of Fig. 13-29 is an example of a power-plant design which attempts to take full advantage of high pressure, high temperature, and reheat.

These data on stationary-power applications are helpful in indicating possible avenues of development for nuclear plants [21]. The economic factor is so vital in any proposal to adapt nuclear fuels to electric-energy-generation systems that it imposes burdens in the present early stage of development [1]. A promising field, where many technical problems will

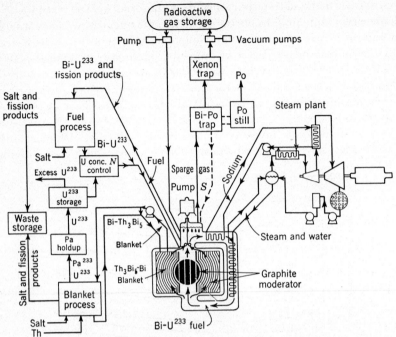

FIG. 13-29. Core, heat exchanger, steam power generator, shown schematically for LMFR. Core is a graphite-moderator structure with liquid-metal fuel flowing through it. Fuel is solution of fissionable U^{233} in molten bismuth. Shape and size (about 5 ft in diameter) allow chain reaction to occur. Fission energy raises fuel temperature. Fuel, pumped through heat exchangers, transfers its heat to liquid sodium, then boiling water; cooled fuel is pumped back into bottom of core. Steam flows to turbogenerator; sodium provides superheat, reheat. Thermal efficiency is over 35 per cent. Integral processing systems pass sparge gas through fuel to remove gaseous fission products and Po, remove solid fission products from fuel with fused salts, and remove fissionable material bred in thorium-containing blanket. Principal advantages are high specific power; high-temperature operation and thus high thermodynamic efficiency, but without high pressures in the reactor; good enough neutron economy to give some breeding with resulting low fuel costs; relatively simple construction; easy fuel handling by pumping; cheap chemical processing to remove fission products, integrated into reactor design; lack of radiation damage to liquid-metal fuel; unlimited burnup of fuel (two preceding items); removal of relatively fresh fission products in concentrated form. This may allow reduction of exclusion area. Fission products may be useful by-products, are at worst easily stored wastes. [*C. Williams and F. T. Miles, Liquid Metal Fuel Reactor Systems for Power, Nucleonics,* **12**(7): 11–13 (*July,* 1954).]

be solved and many lessons learned, is the military power plant. The naval vessel, the submarine, or the large airplane all can use the highly concentrated energy source of the nuclear reaction. There the price of fuel or the cost of the power plant is secondary to military or naval efficiency. The present extensive work in these applications will teach many lessons which can be used when the cloak of secrecy is removed.

Plants such as those for the submarines U.S.S. *Nautilus* and *Sea Wolf* will teach invaluable lessons [30]. These lessons must be supplemented by vigorous prosecution of research and development by industry and by government. The Five Year Reactor Development Program (Table 13-15) and the newly announced Power Demonstration Program of the AEC should give needed factual data and experience for the perfection of reliable, economical nuclear-electric plants. Many estimates have been made of the rapidity of this development, all within the framework of the strong ultimate prospects for nuclear power [34]. There are optimistic estimates which expect as much as 50 per cent of the new electric generating capacity being installed by 1975 in the United States to be nuclear [26]. On the less optimistic side, there are estimates which limit the nuclear expectation to 3 to 5 per cent of the total installed generating capacity of the nation by 1975. In any event there is little likelihood that an energy source which has over twenty-five times the reserves of all fossil fuel can long lie dormant and undeveloped [3, 11].

Prob. 13-1. The possibility of using all the Great Lakes and Long Island Sound as exclusion areas for nuclear power plants has been proposed. Roughly estimate the permissible kilowatts based on 5 kw/acre, and compare with a capacity estimate of 10^7 kw for the regions.

Prob. 13-2. It is being considered to raise by 100°F the reactor and boiler temperatures in Fig. 13-18, without changing the condenser pressure. Compute the net plant heat rate if the additional 100°F is put into (a) superheat at the same pressure, (b) saturation pressure at the same superheat. Results of detailed calculations can be checked by Figs. 13-16 and 13-19.

Prob. 13-3. Compute the effect on the thermal performance of the cycle in Fig. 13-18 if heater 3 is decreased in heat-transfer area (to save investment) so that the feed-water temperature drops to 310°F.

Prob. 13-4. A reheat cycle is to be substituted for the cycle of Fig. 13-18, using steam at 800 psi with initial and reheat temperatures equal to 700°F. What would you estimate the power output and heat rate to be?

Prob. 13-5. In Fig. 13-18 an open feed-water heater is to be substituted for the closed heater 1. What is the change in plant thermal performance, and what justifiable investment could be borne by the change if the heat at the reactor is worth 10 cents per 10^6 Btu, the annual load factor equals 70 per cent, and fixed costs on capital are as given in Table 13-14?

Prob. 13-6. On the basis of a 20,000-hp nuclear power plant in a ship, estimate the cruising radius at 20 knots if the consumable uranium is 1 oz. Reactor temperatures limit steam conditions to 600 psi and 700°F.

Prob. 13-7. Lay out a nuclear-reactor–gas-turbine power plant where the objective is minimum weight and bulk of plant. Gas temperatures are limited to 1100°F. What would you estimate the fissionable-fuel requirement to be for operation of a 10,000-hp plant for 24 hr?

Prob. 13-8. It has been assumed that a complete 500,000-kw breeder reactor and power plant can eventually be built with a physical-plant investment of $243 per kw and an additional $33 per kw for engineering, start-up, and miscellaneous charges. The annual fixed charges exclusive of amortization of the investment are taken at

12.25 per cent of the total capital requirement, and the annual operating cost at $19,400,000.   One gram of Pu is produced for each 7300 kwhr sold.   Show that Fig. 13-30 gives the period of amortization as a function of the selling price of Pu and energy.   (Note that such a plant would not be competitive with typical steam plants and published Pu production costs in the United States but could be competitive in remote locations where delivered fuel is more expensive or where Pu had no competition from separated U^{235}.)

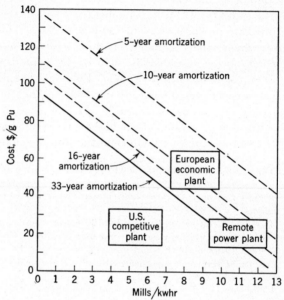

Fig. 13-30. Cost of power vs. value of plutonium for dual-purpose plants.   (*H. W. Huntley and R. G. Lorraine, General Electric Company*, 1953.)

REFERENCES

1. American Power Conference, Forum on the Industrial Use of Atomic Energy, Mar. 25, 1953.
2. Atomic Industrial Forum, Forum Report on Nuclear Reactor Development, July, 1954.
3. Baumeister, T.: A Power Engineer Looks at Nuclear Energy for Public Utilities, *New York Society of Security Analysts, Commercial and Financial Chronicle*, June 4, 1953.
4. Baumeister, T.: "Fans," McGraw-Hill Book Company, Inc., New York, 1935.
5. Baumeister, T.: Steam and Electric Power—Its Past and Its Future, Centennial of Engineering, *Combustion*, December, 1952.
6. Beldecos, N. A., and A. K. Smith: Comparative Performance of Turbine Generator Units in Saturated Steam Cycles, ASME Paper 54-SA-65, 1954.
7. Bergen, H.: Comparative Data Abstracted from AEC Report, *Consulting Eng.*, **2**:28–36 (July, 1953).
8. Brooks, R. D., and A. L. Rosenblatt: Nuclear Power Plants, *Mech. Eng.*, **75**: 363–368 (May, 1953).

9. Cage, J. F., Jr.: Electromagnetic Pumps for High Temperature Liquid Metal, *Mech. Eng.*, **75**:467–471 (June, 1953).

10. Chadwick, Sir J.: "Prospects for Atomic Power," University of Toronto Press, Toronto, 1954.

11. The Chase National Bank: "Peacetime Prospects of Atomic Energy," June, 1954.

12. Cisler, W. L.: Electric Power Systems and Nuclear Power, *Edison Elec. Inst. Bull.*, **20**:289 (September, 1952).

13. Cisler, W. L.: The Future of Atomic Energy in Industry, *SAE Journal*, **16**:73–74 (February, 1953).

14. Clark, P. M.: Mechanical Pumps for High Temperature Liquid Metal, *Mech. Eng.*, **75**:615–618 (1953).

15. Conant, J. B.: A Skeptical Chemist Looks into the Crystal Ball, *Chem. Eng. News*, **29**:3847–3849 (1951).

16. Dunning, J. R.: The Future of Atomic Energy, *Am. Scientist*, **37**:4 (1949), **38**:1 (1950).

17. Engineering Reference Sheet, *Elec. World*, Jan. 24, 1955.

18. Grebe, J. J.: Nuclear Power and Industry, *J. Franklin Inst.*, **255**:409–426 (May, 1953).

19. Hafstad, L. R.: Atomic Energy, When, Where, How Much? *Chem. Eng. News*, **30**:3808–3815 (1952).

20. Hatfield, G. W.: A Reactor Emergency, *Mech. Eng.*, **77**:124–136 (1955).

21. Irwin, J. M.: Industry Bids for Atomic Power, *Harvard Business Rev.*, July–August, 1953.

22. Isard, W., and V. Whitney: "Atomic Power: An Economic and Social Analysis," McGraw-Hill Book Company, Inc., New York, 1952.

23. "Liquid Metals Handbook," 2d ed. (R. N. Lyon, ed.), Government Printing Office, 1952.

24. Materials Policy Commission: "A Report to the President on Resources for Freedom," vol. 3, "The Outlook for Energy Sources," June, 1952.

25. Menke, J. R.: "Nuclear Fission as a Source of Power," AEC, MDDC-1104, 1947.

26. Morehouse, E. W., and T. Baumeister: How Will Atomic Power Affect the Electric Power Industry? *Land Economics*, **31**:93–107 (May, 1955).

27. National Industrial Conference Board: "Atomic Energy in Industry," Oct. 16–17, 1952.

28. National Industrial Conference Board: Atomic Energy Meeting, Oct. 29–30, 1953.

29. Power Reports on Atomic Energy Today, *Power*, July, 1953.

30. Roddis, L. H., and J. W. Simpson: The Nuclear Propulsion Plant of the U.S.S. Nautilus, *Soc. Naval Architects Marine Engrs., Trans.*, **62**:491–521 (1954).

31. Salisbury, J. K.: The Basic Gas Turbine Power Plant, *Mech. Eng.*, **66**:373–383 (1944).

32. Schanz, J. L.: Prospects for Nuclear Power, *Elec. Light and Power*, **32**:78–87 (December, 1954).

33. Schurr, S. H., and J. Marschak: "Economic Aspects of Atomic Power," Princeton University Press, Princeton, N.J., 1950.

34. Sporn, P.: Nuclear Power's Progress, *Edison Elec. Inst. Bull.*, **20**:411–416 (December, 1952).

35. Trocki, T.: Engineering Aspects of Liquid Metals for Heat Transfer, *Nucleonics*, **10**(1):28–32 (January, 1952).

36. Trocki, T. J., and D. B. Nelson: Report on Liquid Metal Heat Transfer and Steam Generation System for Nuclear Power, *Mech. Eng.*, **75**:472–476 (1953).

37. "Reports to the U.S. Atomic Energy Commission on Nuclear Power Reactor Technology," Government Printing Office, May, 1953.

38. U.S. Joint Congressional Committee, Atomic Power and Private Enterprise, 82d Cong., 2d Sess., December, 1952.
39. U.S. Joint Congressional Committee, Hearings on Atomic Power Development and Free Enterprise, June 24–July 31, 1953.
40. U.S. Joint Congressional Committee, Report of the Subcommittee on Research and Development on the Five Year Power Reactor Development Program, 83d Cong., March, 1954.
41. Ward, J. C., Jr.: "A Look at Atomic Energy from the Investment Viewpoint," New York Society of Security Analysts, Jan. 3, 1954.
42. Weinberg, A. M.: Power Reactors, *Scientific American*, **191**:33–39 (December, 1954).
43. Williams, C., and F. T. Miles: Liquid Metal Fuel Reactor Systems for Power, *Nucleonics*, **12**(7):11–13 (July, 1954).
44. Zinn, W. H.: Basic Problems in Central Station Nuclear Power, *Nucleonics*, **10**(9): 8–14 (September, 1952).
45. Davis, W. K.: Capital Investment Required for Nuclear Energy, Paper 477, UN Conference on Atomic Energy, Geneva, 1955.
46. King, E. C., and R. C. Andrews: American Institute of Chemical Engineers, Heat Transfer Symposium, Preprint 5, St. Louis, 1953.

CHAPTER 14

NUCLEAR-REACTOR TYPES

By John W. Landis

14-1. Methods of Classification. Nuclear reactors may be classified by application or by design. The first method is simple and straightforward but does not of course contribute much to description or delineation of the reactor's characteristics and may be quite misleading. The second method is complicated by the existence of many parameters but, if properly utilized, specifies quite thoroughly both the operational and the constructional features of the reactor.

The major actual and proposed uses for nuclear reactors are to produce heat for thermoelectric power, process or space heat, fissionable materials, radioisotopes, neutrons, and γ radiation and to drive submarines, surface ships, aircraft, and locomotives. Thus, the application classifications of reactors include:

1. Central-station power reactors—those which supply steam or gas to turbines for the generation of electricity

2. Package power reactors—small units for power production in remote or unusual locations

3. Process heat reactors—those which supply heat for industrial operations

4. Central heat reactors—those which are used for space heating

5. Production reactors—those which operate primarily to produce fissionable materials, radioisotopes, or other unique substances

6. Test reactors—those which are used to test the radiation stability of materials and equipment components for nuclear applications

7. Research reactors—those which are designed to produce neutrons and γ radiation for nuclear, chemical, or biological research or are used for instruction or demonstration (usually small, low-power machines)

8. Food-irradiation reactors—those which are designed specifically for the irradiation of perishables

9. Propulsion reactors—those which supply power to propel marine, land, or air craft

Reactors are sometimes built to serve more than one of these purposes simultaneously. An example is the C-pile at Hanford which not only produces plutonium but also supplies heat to the production facilities. Since it is impossible in one chapter to discuss all the potential uses for reactors, the following discussion will be confined to central-station power reactors.

Designwise, reactors may be classified in accordance with many variables, chief among which are the following:

Neutron energy	Coolant
Geometry	Moderator
Fuel cycle (fuel,	Structural material
fertile material, and	Method of control
fuel regeneration)	

Table 14-1 lists some of the possibilities in each of these categories. As Weinberg [33] has pointed out, assuming just the possibilities mentioned, there is a plethora of reactor types to be studied and "the central issue in reactor development is to trace out of this welter [of combinations] the dozen or so which are most likely to succeed."

The most basic distinctions from the physics standpoint are probably the first three listed. In a thermal reactor the majority of the fissions occur at a neutron energy equivalent to the temperature of the core, whereas in a fast reactor the majority of the fissions result from the absorption of fast neutrons. A homogeneous reactor is one in which the fuel, moderator, and coolant (or just fuel and coolant in the fast type) are mixed intimately, often in liquid form, while a heterogeneous reactor is one in which the components are separate, usually in a definite geometric arrangement. The differences among the three fuels and between the two fertile materials have been covered in Chaps. 1, 3, and 6. Finally, the ability to regenerate new fuel is of course vitally important in both the economic and the fuel-availability aspects of reactor operation.

The limiting or determining factors in reactor design, however, are more often than not the coolant, the moderator, and/or the structural material. In view of this fact and the fact that there are more alternatives in these categories, the most concise method of reactor classification is based upon these variables. This system has the additional advantage of fitting into current terminology.

Briefly, the system is to name first the coolant, then the moderator, and then the structural material; these are followed, if desired, by the atomic weights of the fuel (with enrichment in parentheses) and the fertile material. If the coolant also serves as the moderator, no moderator

TABLE 14-1. REACTOR DESIGN PARAMETERS

Neutron energy	Geometry	Fuel Cycle			Coolant	Moderator	Structural material	Method of control
		Fuel material	Fertile material	Fuel regeneration				
Thermal	Heterogeneous	U^{235}	None	None (burner-upper)	Pressurized H_2O	H_2O	Al	Rods
Intermediate	Homogeneous	U^{233}	Uranium	Less than consumed (converter)	Pressurized D_2O	D_2O	Zr	Burnable poison
Fast	1. Single region	Pu^{239}	Thorium	More than consumed (breeder)	Boiling H_2O	Graphite	Stainless steel	Soluble poison
	2. Double region				Boiling D_2O	Be	Carbon steel	Negative temperature coefficient
					Air	BeO	Ti	Fuel movement
					Nitrogen	Hydrides	Mg	Reflector movement
					CO_2	Organics	etc.	Moderator movement
					He	etc.		etc.
					Na			
					NaK			
					Diphenyl			
					Terphenyl			
					Pb			
					Bi			
					Fuel solutions			
					Fuel suspensions			
					etc.			

term is used.* The moderator term is omitted also in the case of the fast reactor. If the fuel serves as its own coolant, its state, such as "pressurized H_2O solution," is substituted for the coolant. If different structural materials are used, the fuel-element cladding (if any) is named first.

With this system, both the neutron energy and the geometry will be apparent from the components named in almost all cases. Further, a burner-upper will be designated by the omission of the second atomic weight, and in the majority of cases where fertile material is employed, even the beginning student in nuclear technology will be able to tell whether conversion or breeding will result.

The "method of control" parameter has been neglected in this system because no reactor conceived to date has been eliminated by inability of the designers to devise suitable control components.

Application of this nomenclature to existing reactors may help to clarify it for the reader. The reactor being built at Shippingport, Pa., by Westinghouse Electric Corporation for the U.S. Atomic Energy Commission and Duquesne Light Company in the AEC's Five Year Reactor Development Program (commonly called the Pressurized Water Reactor, or PWR) is a "pressurized-H_2O Zr-SS 235 (90%/0.7%) 238" type. The reactor conceived and developed by Brookhaven National Laboratory (known as the Liquid Metal Fuel Reactor, or LMFR) is a "U-Bi-alloy graphite croloy 235 (90%) 232" type. The Sodium Reactor Experiment being constructed by Atomics International for the AEC's Five Year Program is a "sodium graphite SS-Zr-SS 235 (2.8%) 238" type. Generalized categories are designated by use of only the coolant and moderator terms, substituting the word "fast" for the moderator in the case of the fast reactor.

14-2. Power-reactor Summary. Reactors of the following generalized types have been or are being considered for central-station power plants:

1. Pressurized H_2O cooled
   a. Pressurized H_2O moderated
   b. D_2O moderated
   c. Graphite moderated
2. Pressurized D_2O cooled
   a. Pressurized D_2O moderated
3. Boiling H_2O cooled
   a. Boiling H_2O moderated
   b. Graphite moderated
4. Boiling D_2O cooled
   a. Boiling D_2O moderated

* No confusion should result here, because natural water, heavy water, and organic compounds are the only substances commonly used in both capacities.

5. CO_2 or N_2 cooled
   a. Graphite moderated
   b. BeO moderated
6. He cooled
   a. Graphite moderated
   b. BeO moderated
7. Na cooled
   a. Graphite moderated
   b. Be or BeO moderated
   c. D_2O moderated
   d. Fast
8. NaK cooled
   a. Fast
9. Diphenyl cooled
   a. Diphenyl moderated
   b. Graphite moderated
10. Bismuth cooled
   a. Graphite moderated
11. Pressurized H_2O-UO_2SO_4 solution or slurry cooled
   a. Pressurized H_2O moderated
12. Pressurized D_2O-UO_2SO_4 solution or slurry cooled
   a. Pressurized D_2O moderated
13. Pressurized D_2O-UO_3-PuO_2 slurry cooled
   a. Pressurized D_2O moderated
14. Pressurized D_2O-UO_3-ThO_2 slurry cooled
   a. Pressurized D_2O moderated
15. Pressurized H_2O–uranium–phosphoric acid solution cooled
   a. Pressurized H_2O moderated
16. Molten U-Bi alloy cooled
   a. Graphite moderated
17. Fused uranium compound cooled
   a. Graphite moderated
   b. Fast

A list of most of the power reactors which have been operated, are under construction, or are planned, classified in the categories above, is given in Table 14-2 (see pages 764 to 768).

14-3. Descriptions of Prototypes. To provide more detailed information on the differences among the various types of power reactors, the salient features of the reactor concepts and experiments in many of the categories in Table 14-2 will be described. The discussion will not be definitive, but sufficient data will be presented to give the reader a picture of the complexity of the field. Some other details will be found in Apps. M and N.

Pressurized Water Reactor (PWR). The pressurized-H_2O (category 1) prototype is the Pressurized Water Reactor now under construction by Westinghouse Electric Corporation at Shippingport, Pa. A schematic diagram of the primary system of this plant is shown in Fig. 14-1. Figure 14-2 is an artist's concept of the completed plant, and Fig. 14-3

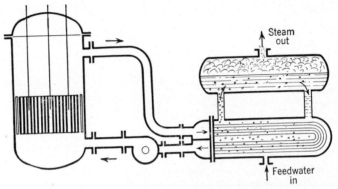

FIG. 14-1. Schematic diagram of PWR primary system.

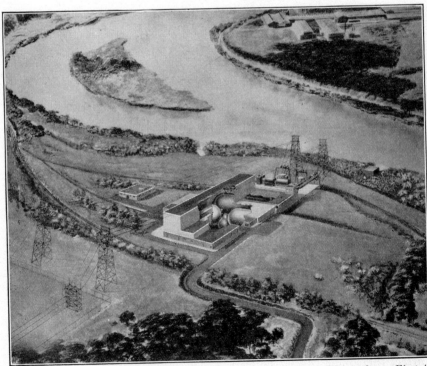

FIG. 14-2. Artist's concept of PWR plant at Shippingport, Pa. (*Westinghouse Electric Corporation.*)

illustrates how the reactor—in the center containment shell—is connected to the four heat exchangers, two in each of the two lateral containment shells. The cost of this project will be approximately $100 million.

The PWR will utilize natural uranium seeded with fully enriched uranium as fuel.* The core arrangement is shown in Fig. 14-4. The natural uranium elements are assemblies of Zircaloy-2 tubing filled with uranium oxide pellets. The enriched uranium elements are assemblies of U^{235}-zirconium-alloy plates clad with Zircaloy-2. Cross-shaped control rods move up and down within the enriched uranium elements. Drives are mounted on the reactor-vessel head.

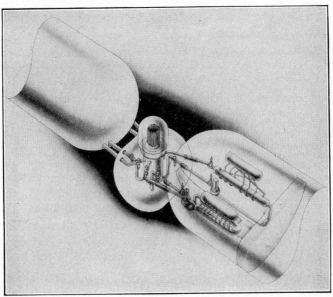

Fig. 14-3. Layout of PWR primary system. (*Westinghouse Electric Corporation.*)

The pressure in the primary loops will be 2000 psig. Flow through the reactor vessel (33 by 10½ ft) will be 50,400 gpm single-pass from bottom to top. The inlet temperature will be 508°F and the outlet temperature 542°F. The average temperature of the water in the heat exchangers will be 524°F.

Each of the heat exchangers is rated at 263×10^6 Btu/hr. Each will deliver 287,000 lb/hr of steam. At full load the steam will be 600 psia saturated. At no load the pressure will rise to 885 psia.

* It is anticipated that "seed cores" will provide more economical power by permitting the discard of the burned-up natural uranium, thus requiring the reprocessing of only the much smaller quantity of enriched fuel. However, a uniform slightly enriched core will probably be tested as the second loading of PWR.

Flow in each of the primary loops will be 16,800 gpm, 30 fps. Only three of the four loops will be required to supply the 861,000 lb/hr of steam necessary to produce the rated 60,000 kw net electrical output. In anticipation of better performance a 100,000-kw turbogenerator has been installed.

Canned rotor pumps are being used in the primary loops. Isolation valves near the reactor inlet and outlet in each loop will make it possible to isolate the loop for repairs without shutting down the reactor. Eight manually operated valves will act as backups for the main valves.

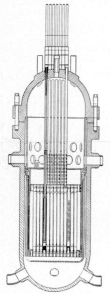

Pressure will be maintained by electric immersion-type heaters in a pressurizing vessel. Water purity will be maintained by a system of filters and ion exchangers.

Nuclear Power Demonstration Reactor (NPDR). The pressurized-D_2O (category 2) prototype is the Canadian Nuclear Power Demonstration Reactor to be located at Des Joachims, Canada. This reactor, which is being built by the Canadian General Electric Company, is scheduled to be completed in 1958. It will produce about 20 Mw of electric power and cost about $15 million.

The fuel for this reactor will be natural uranium in oxide form spiked with plutonium. The structural material will be zirconium. Two hundred fuel rods will comprise the core, which will be 10 ft high by 10 ft in diameter. Control will be accomplished by varying the amount of moderator in the core.

FIG. 14-4. PWR core arrangement. (*Westinghouse Electric Corporation.*)

The layout of this plant will be similar to that of the PWR, with the pressure of the primary loop at least 1200 psia and the outlet temperature of the D_2O coolant about 500°F.

Experimental Boiling Water Reactor (EBWR). The prototype of the boiling-H_2O category is the Experimental Boiling Water Reactor at Argonne National Laboratory, the second of the reactors in the U.S. AEC's Five Year Reactor Development Program. The main features of this reactor are pictured in Fig. 14-5.

EBWR is designed to use either H_2O or D_2O, but it is being operated first on H_2O. It is also designed for either natural or forced circulation, but the pumps for forced circulation have not yet been installed.

In the natural-circulation boiling heterogeneous reactor the core is placed near the bottom of an upright heavy cylindrical vessel and is covered to a depth of approximately two feet with chemically pure water. The heat generated in the core causes the water to boil and circulate through the upper portions of the vessel, out through a steam dryer, and thence directly to the turbine. The condensate is pumped back into the reactor vessel and around the cycle again.

EBWR produces about 5 Mw of electric power from a 4- by 5-ft cylindrical core of slightly enriched uranium plates. The fuel is clad with Zircaloy-2, and the control rods are made of hafnium. The steam

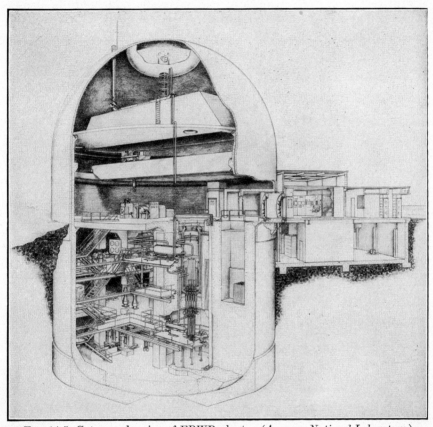

FIG. 14-5. Cutaway drawing of EBWR plant.   (*Argonne National Laboratory.*)

pressure in the plant is about 600 psig. This plant went into operation before PWR because of its much smaller size.

The total cost of the EBWR project will be about $17 million, including development work.

Calder Hall Reactors. The Calder Hall reactors were the first central-station power reactors to go "on the line." They are graphite-moderated natural-uranium reactors and utilize carbon dioxide as the coolant (category 5). The background technology derives from the British production reactors at Windscale. The plant at Calder Hall comprises two reactors, one on each side of a 92-Mw turbogenerator building. A recent photograph of the plant is shown in Fig. 14-6. A schematic diagram of the gas system is shown in Fig. 14-7.

The primary circuit consists of a large carbon-steel pressure vessel (71½ by 37 ft) constructed to extremely close tolerances, containing the 1000 tons of graphite core, and of four parallel coolant loops, each with a large carbon dioxide–steam heat exchanger and a 2000-hp d-c-driven blower. The pressure of the carbon dioxide gas is 100 psi. It leaves the

Fig. 14-6. View of one of the Calder Hall reactors after start-up. (*United Kingdom Atomic Energy Authority.*)

pressure vessel at 336°C and generates 210 psia steam superheated to 595°F and 63 psia steam superheated to 350°F in the heat exchangers.

The natural-uranium fuel elements are clad with Mg-Be-Al alloy. They are charged into the core through special guide tubes in the head of the pressure vessel. The skin temperature of these elements is maintained at about 400°C. The burnup expected is 3000 Mw-days/ton. The control rods are of boron steel clad with stainless steel; they also enter the core through the pressure-vessel head.

The internal parts of the primary circuit have been thoroughly cleansed to prevent build-up of radioactivity. An 8-ft-thick biological shield of heavy-aggregate concrete attenuates the neutron and γ radiation to tolerable levels.

Assuming an 80 per cent load factor and a total cost of about $60 million for the Calder Hall plant, the unit cost of the power delivered will be about 12 mills per kwhr.   No credit has been taken for plutonium in this calculation, but fixed charges are only about 7 per cent in England, compared with 15 per cent for private utilities in the United States.

Helium Graphite Reactor.   The prototype installation of category 6 will be the Gas Cooled Reactor Experiment, being designed by Aerojet

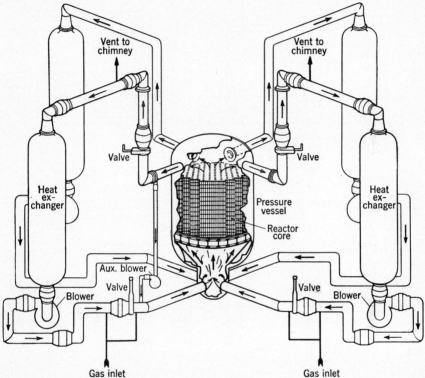

FIG. 14-7. Schematic diagram of gas system of Calder Hall–type reactor.   (*United Kingdom Atomic Energy Authority.*)

General.   As of this writing, however, the details of this plant have not been announced.

A reactor of this type proposed by Daniels [6] is worthy of some discussion.   The proposed plant arrangement and design conditions are shown in Fig. 14-8.   The core is composed of boron-free graphite blocks in vertical columns supported on a molybdenum grid held up by molybdenum or beryllium oxide pillars.   Four fuel holes and twelve coolant holes are carefully machined through each block.   The fuel holes are loaded with special UC_2-graphite cartridges, each about three inches long.

Twenty-two thousand of these cartridges are required to fill the 1000 fuel holes.   The UC_2 is 10 per cent enriched.

The primary circuit of this reactor is pressurized to 225 psi.   Helium, circulated by a canned-motor fan, enters the core at 820°F and exits at 1350°F.   It gives up its heat to air at 400 psi in a special "leakproof" heat

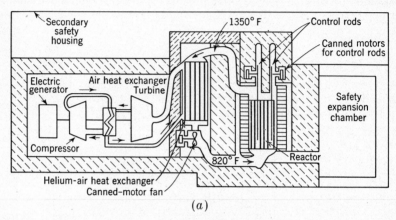

(a)

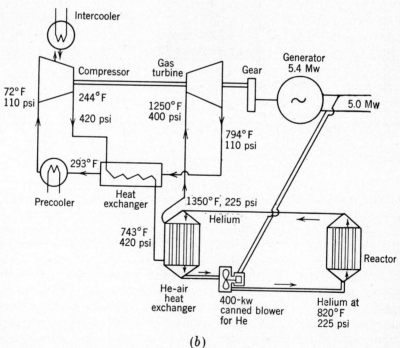

(b)

FIG. 14-8. (a) Proposed plant arrangement for closed-cycle gas-cooled reactor. (b) Flow diagram for (a). [From Nucleonics, 14(3): 35 (1956).]

exchanger.   The entire primary circuit is kept as nearly gastight as possible.

The control rods, which are driven by canned motors, are made of molybdenum impregnated with boron.   Concrete roof and underground walls, along with a safety expansion chamber, not only provide required shielding but will contain any radioactive dust or gas that might be released in an accident.

Sodium Reactor Experiment (SRE), Experimental Breeder Reactor No. 2 (EBR-2), and the Los Alamos Molten Plutonium Reactor Experiment (LAMPRE-1).   The sodium-cooled reactors are so divergent in characteristics that at least two prototypes should be discussed—the Sodium Reactor Experiment and the Experimental Breeder Reactor No. 2.   For completeness, the main features of LAMPRE-1 are also outlined.

The Sodium Reactor Experiment is another of the reactors being built under the aegis of the U.S. Atomic Energy Commission in the Five Year Reactor Development Program.   It is the first of the "sodium graphite" series, the main advantage of which is the ability to attain high temperatures in the thermal-neutron range without pressurization.

The prime contractor for the SRE is Atomics International.   Construction is almost complete, and operation should begin in 1957.   The rated output of the plant is 20,000 kw (heat) and Southern California Edison is planning to convert this heat to about 7500 kw of electric power.

The core of the SRE is a vertical array of hexagonal graphite prisms with center holes for the fuel, control, and safety elements.   The active volume is a right cylinder 6 ft high, 6 ft in diameter.   The fuel elements are clusters of seven stainless-steel tubes 6 ft long, separated from each other by spiral wire, filled with 0.750-in.-diameter slightly enriched (2.8 per cent) uranium slugs about six inches long.   The tubing walls are 10 mils thick and the annulus around the fuel is also 10 mils.   The annulus is filled with NaK.   The space above the slugs in the tubes is filled with helium.   About thirty-one fuel elements and about four control elements (B-Ni rods) are required.

The core is contained in a stainless-steel tank about nineteen feet high by eleven feet in diameter.   It is supported on a stainless-steel grid near the bottom of the tank.   The graphite prisms are clad with zirconium to prevent reaction with the sodium.   The sodium enters the bottom plenum chamber at about 500°F.   It then passes up through the axial tubes in the graphite, cooling the fuel elements, and into the upper plenum chamber where the average temperature is 960°F.   The rate of flow is 1200 gpm.   The core construction is illustrated in Fig. 14-9.

The primary sodium is activated by passage through the core.   To reduce the hazards associated with this phenomenon, a secondary sodium loop is employed to carry the heat into the steam generators. The steam generators are designed to generate 850°F steam at 800 psia.

A graphite reflector 2 ft thick surrounds the core of this reactor. The core tank in turn is ringed with a 5½-in.-thick thermal shield, and the entire assembly is contained in another tank the function of which is to hold any sodium which might leak from the core tank. There is a foot

Fig. 14-9. Diagram showing SRE core construction.   (*Atomics International, Division of North American Aviation, Inc.*)

of insulation around the outer tank, and around this the "cavity liner," which acts as a form for the concrete shield. Toluene is circulated in the cavity liner to remove the heat passed to that region. Toluene does not react with sodium and therefore is used to cool all parts of the system where there is danger of sodium leakage.

The main pumps will be modified hot-process pumps similar to the type used in refinery service. Valves will use frozen-sodium seals. Special handling coffins and cleaning facilities are required for fuel-element removal operations. Also, certain special equipment such as cold traps,

a sodium service system, a toluene system, and an inert-gas system is vital
to safe operation of the plant.

The total cost of the SRE will be about $14,500,000.

The Experimental Breeder Reactor No. 2 is the sequel to the Experimental Breeder Reactor No. 1, both fast reactors. It falls in a different
category because it will use sodium as the coolant instead of NaK. Also,
it will use about 25 per cent enriched uranium instead of fully enriched
uranium as its fuel. But in many other ways it will be quite similar to
EBR-1. Plutonium may be used both in EBR-2 and in the rebuilt
EBR-1 at some future date.

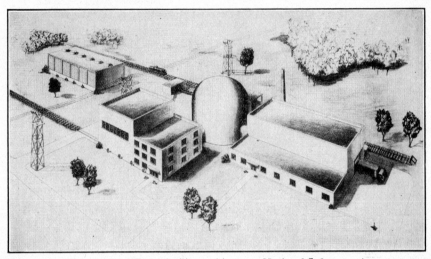

Fig. 14-10. EBR-2 facility. (*Argonne National Laboratory.*)

EBR-2 is the fourth reactor being developed in the U.S. AEC's Five
Year Program. The prime contractor is Argonne National Laboratory.
Construction is scheduled to begin in 1957 and operation in 1959. The
purpose of the project is "to demonstrate the engineering feasibility of a
fast power reactor system which will be economically competitive with
existing power sources" [29]. Total cost of the project will probably
run to $40 million.

The power rating of EBR-2 will be 62,500 kw (heat) and 20,000 kw
(electric). The plant will comprise the reactor, the primary sodium
cooling system, the secondary sodium cooling and steam-generation system, the turbogenerator station, and an integral fuel-reprocessing facility.
The layout of the plant will be approximately as shown in Fig. 14-10.

The entire primary system, including the reactor, the primary sodium
pumps and piping, the intermediate heat exchanger, and the fuel transfer
and storage system, will be submerged in sodium in a single vessel as

shown in Fig. 14-11.   This arrangement reduces the probability of loss
of reactor coolant essentially to zero, provides reliable shutdown cooling
in the event of loss of forced convection or other emergency, prevents
rapid changes in load demand from being reflected back to the reactor
as inlet temperature changes, permits unloading of the core immediately
after shutdown, maximizes the integrity of the radioactive sodium containment, permits some leakage in the primary piping and therefore the
use of slip joints to facilitate the removal of components, simplifies the
shielding, and simplifies the start-up process.   The chief disadvantages

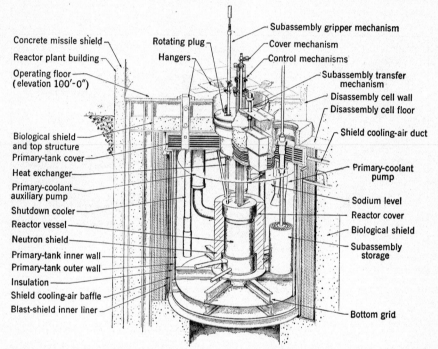

FIG. 14-11. EBR-2 primary system.   (*Argonne National Laboratory.*)

of this arrangement are that all reloading operations must be carried out
under molten Na, maintenance is difficult, and much heat is lost to the
large pool of sodium.   The sodium will circulate from the tank through
d-c electromagnetic pumps (rated preliminarily at 6000 gpm at 40 psi)
to the reactor inlet plenum chambers, then upward through the reactor
core to the common top plenum chamber, out through the piping to the
intermediate heat exchanger, and back to the tank.   The very large
current required by the pumps (250,000 amp) will be supplied by homo-
polar generators or special rectifiers.

As is shown in Fig. 14-12, the reactor core and blanket will consist of
a large number of close-packed hexagonal-prism subassemblies identical

in external dimensions.   The core subassemblies, containing 91 pin-type
fuel elements spaced on a triangular lattice in the core section and 19
pin-type fertile material elements in the upper and lower blanket sections,
will be arranged in the approximate shape of a cylinder.   The blanket
subassemblies just outside this cylinder will be similar in construction to
the blanket sections of the core subassemblies, i.e., they will contain
19 pin-type fertile material elements on triangular pitch.   The outer
blanket subassemblies will consist of six right-triangular prisms of fertile
material fitted longitudinally with coolant tubes.

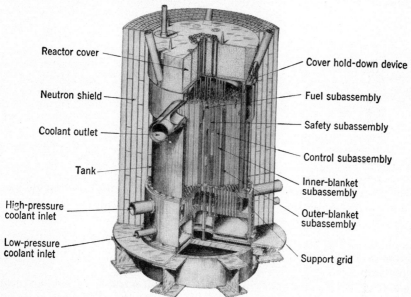

FIG. 14-12. EBR-2 core-and-blanket arrangement.   (*Argonne National Laboratory.*)

The fuel will be 25 per cent enriched-U metal alloyed with various
stabilizing materials; the blanket material will be natural U.   The
structural material will be stainless steel.   Sodium flow in the core and
central and inner blanket subassemblies will be around the stainless steel
tubes containing the pins.   Flow in the outer blanket subassemblies will
be through the tubes.

The control subassemblies will be modified core subassemblies and control will be accomplished by movement of the fuel in the control subassemblies into and out of the core region.

A special stainless-steel tank will contain, position, and support the
core and blanket subassemblies and direct the flow of sodium.   About
two feet of moderating and absorbing material will be wrapped around
the tank to attenuate the escaping neutron flux sufficiently so that pri-

mary system components will not be too highly activated. The complete reactor-and-shield assembly will be about 13 ft high by 10 ft in diameter.

The average core power density expected in this reactor is 1000 kw/liter. The average core heat flux expected is about 1,000,000 Btu/(ft²)(hr). Thus, reactor cooling will be a critical problem both during operation and after shutdown. Reactor loading and unloading operations will therefore be carried out under liquid sodium as was mentioned previously.

In normal operation the following conditions are expected in this system:

Reactor outlet temperature........................	900°F
Heat-exchanger primary outlet temperature...........	750°F
Primary circuit flow................................	5,000,000 lb/hr
Heat-exchanger secondary inlet temperature..........	580°F
Heat-exchanger secondary outlet temperature.........	880°F
Secondary circuit flow.............................	2,200,000 lb/hr
Feed-water temperature.............................	450°F
Steam temperature.................................	850°F
Steam pressure....................................	1,250 psig
Steam production..................................	200,000 lb/hr

The heat exchanger will be a counterflow, shell-and-tube, single-pass, sodium-to-sodium unit. The steam generators will be of the once-through type, sodium to water.

Dump tanks will be provided for storage of sodium removed from the primary tank. Oxide will be cleaned out of the system by means of cold traps.

LAMPRE-1 will be constructed to go critical in 1958. It will be fueled with a molten alloy of plutonium and iron near the eutectic composition (a third constituent may be added to give further fuel dilution) and will develop about 1000 kw (heat), which will be dumped to the atmosphere. Expected total cost is $3 million, including research and development and a power prototype.

To handle the high temperatures which will be developed in the core of this reactor, tantalum will probably have to be used as the structural material. The coolant, Na, will be carried through the semihomogeneous core in a tight-packed array of small-diameter tubes. A movable reflector will be used for control. The heat generated will probably be passed directly to the atmosphere in a Na-to-air heat exchanger.

Experimental Breeder Reactor No. 1 (EBR-1). Only one power-reactor experiment falls in category 8 (NaK-cooled). It is the Experimental Breeder Reactor No. 1, which was designed, built, and operated successfully at the National Reactor Testing Station in Idaho from 1951 to 1955, when a partial melting of the core in an accident forced a shutdown. This reactor not only proved the feasibility of the fast reactor and of the breeding process, but also was the first reactor to produce electric power. Cooled with NaK in both the primary and secondary circuits, it produced 1400 kw of heat power and 160 kw of electric power from a cylindrical core about 10 in. in diameter and 1 ft high.

The core of EBR-1 consisted of a tightly packed array of $\frac{3}{8}$-in. uranium rods (center section U^{235}, top and bottom sections U^{238}), each jacketed in stainless steel. There were spaces for 217 of these rods. Around the core were two annular blanket areas separated from the core and from each other by two stainless-steel walls. The inner wall served as a NaK flow baffle, directing the NaK downward through the inner blanket area and upward through the core. The outer wall was the reactor tank.

The inner-blanket elements were $\frac{3}{8}$-in. natural-uranium rods jacketed in stainless steel. The outer-blanket elements were keystone-shaped bricks of natural uranium (also jacketed in stainless steel) built up in the form of a cylinder with a curved bottom, completely enveloping all but the top of the reactor tank. The outer blanket was air-cooled.

Control of this reactor was achieved by movement of the outer blanket as a whole and of natural-uranium control rods within the blanket. About 48 kg of U^{235} constituted a critical mass of fuel.

The primary coolant (NaK) left the reactor at about 600°F, passed through a double-pass shell-and-tube heat exchanger, and returned to the reactor at about 442°F. The secondary coolant (NaK) left the heat exchanger at 583°F, came out of the superheater at 572°F, left the boiler at 455°F, and returned from the water preheater to the heat exchanger at 419°F. The steam conditions were 540°F and 400 psig.

The total cost of the EBR-1 project was $6 million, including laboratory work.

Organic Moderated Reactor Experiment (OMRE). This reactor, prototype of category 9, will be built at the National Reactor Testing Station in Idaho by Atomics International for the U.S. Atomic Energy Commission. Operation is scheduled to begin in 1957. The heat developed—16,000 kw—will be dissipated to the atmosphere. The total cost of the project is estimated to be $1,800,000.

The combination coolant and moderator, diphenyl, will be circulated through a heterogeneous core of slightly enriched uranium fuel elements clad in stainless steel. The system will be pressurized to 300 psig, and the reactor outlet temperature will be 530°F.

Bismuth Graphite Reactor (BGR). No definite plans have been laid as yet for construction of a category 10 reactor. Several utility study groups, however, have shown interest in this type. A full-scale power plant might have the following characteristics:

1. A reactor consisting of a vertical array of several thousand Croloy-2¼ tubes, each ½ to 1 in. in diameter and filled with sintered slightly enriched uranium oxide pellets, arranged in clusters within Croloy-2¼ coolant tubes which in turn are set in graphite prisms

2. A primary loop consisting of a relatively low-pressure reactor vessel, isolation valves, two gas-sealed centrifugal pumps, a bismuth cleanup system, a once-through steam generator, an expansion tank, a sump tank, auxiliary heaters, and piping

3. A secondary system consisting of the normal turbogenerator, boiler feed pump and condenser components, and special decontamination units

4. A fuel-element unloading and cleansing system capable of preventing carry-over of the radioactive coolant to the storage chambers or coffins

5. Control by a circular array of hydraulically driven boron-steel control rods

6. Temperature rise in the reactor from 550 to 950°F

7. Steam conditions of 850°F and 800 psia

8. A thick-walled gastight enclosure for the reactor and an adjacent thin-walled gastight enclosure for the remainder of the primary loop equipment

Plutonium produced in the U^{238} in this reactor would contribute significantly to the long-term reactivity of the core. Some additional conversion might be achieved in a blanket of natural uranium oxide elements around the core, but the cost of these elements would have to be low to make such conversion economical.

Homogeneous Reactor Experiment (HRE). The Homogeneous Reactor Experiment which was run at Oak Ridge National Laboratory from April, 1952, until early in 1954 was the prototype of category 11 reactors and in fact the first homogeneous power reactor of any type. It produced about 1000 kw of heat power and about 150 kw of electric power.

Essentially this reactor was a sphere of stainless steel 18 in. in diameter filled with a dilute solution of highly enriched uranyl sulfate (UO_2SO_4) in ordinary water which was continuously circulated from the sphere out through a U-tube heat exchanger and back at a rate of 100 gpm. The primary circuit was pressurized to 1000 psi to prevent the solution from boiling (by heating to 545°F in a pressurizer). This required a sphere wall thickness of $\frac{5}{16}$ in.

A reflector consisting of a 10-in. layer of heavy water, pressurized about the same as the fuel solution, was wrapped around the sphere. The heat generated in the heavy water (50 kw) was removed by circulating the water through a boiler feed-water preheater, maintaining the temperature in the reflector at 350°F.

At full power the rate of dissociation of the water in the uranyl sulfate solution was 10 cfm of gas at STP. The hydrogen and oxygen were converted back to water in flame and catalytic recombiners. A catalytic recombiner was used also in the heavy-water circuit. The recombination reactions produced slightly more than 45 kw (heat).

The temperature of the fuel solution at the reactor outlet was maintained at 482°F. The temperature drop through the heat exchanger under full power conditions was 72°F. Steam generated was 3000 lb/hr. The steam pressure was 200 psia.

HRE proved that mechanical control is unnecessary in a reactor of this type. The negative temperature coefficient of the fuel solution is sufficient to stabilize the power output. The safety measures employed in HRE included dumping of the reflector, dumping of the fuel, dilution of the fuel, and neutron-absorbing plates.

In consuming 0.002 lb of U^{235} per day, HRE produced about 20 cm^3 per day of fission-product gases. These were mixed in with the hydrogen and oxygen effluent and were extracted from the recombiner circuit by adsorption by activated charcoal. The radioactivity of the full-power fuel solution was about 30 curies/cm^3.

The entire core and reflector assembly of this reactor was housed in a forged-steel pressure vessel of 45 in. OD, which in turn was housed in a shield of barytes concrete block 7 ft thick.

The total cost of the HRE program from 1950 to 1953 was $12,800,000, including $1,100,000 for the actual construction of the HRE.

Homogeneous Reactor Test (HRT). The fifth reactor in the U.S. AEC's Five Year Reactor Development Program will be the Homogeneous Reactor Test, officially termed the Homogeneous Reactor Experiment No. 2. This is the first of the category 12 reactors. Again, highly enriched UO_2SO_4 solution will be the fuel, but the solvent will be heavy water instead of light water. The U^{235} concentration will be 9.6 g per kilogram of heavy water.

At first, this reactor will employ a simple reflector of heavy water. After several months of operation to determine the relative merits of continuous chemical separations processes and the reliability of the system as a whole, however, experiments will be performed with uranyl sulfate solutions, and possibly with thorium oxide slurries in the blanket.

HRT is laid out so that the fuel process chamber is on one side of the reactor and the blanket process chamber is on the other. Each comprises a gas separator, a steam generator, a canned motor pump, a pressurizer, a letdown valve, a letdown heat exchanger, a boiler feed pump, a dump valve, a dump tank, an evaporator, an off-gas system, and auxiliary equipment. An artist's concept of the plant is shown in Fig. 14-13. Figure 14-14 is a schematic flow sheet. The location of the major pieces of equipment is perhaps best shown in Fig. 14-15, a plan view of the reactor system.

The estimated cost of the homogeneous-reactor program at ORNL from July, 1953, to July, 1958, will be $38,800,000. This includes about $3,500,000 for the construction of HRT. HRT will be operated at a power level of 5000 to 10,000 kw (heat), producing 300 kw of electric power (limited by the size of the turbogenerator), and will require about 4 kg of U^{235}, compared with 3 kg for HRE, for a critical mass. Like HRE, control will be effected by the large negative temperature coefficient and by changing the uranium concentration.

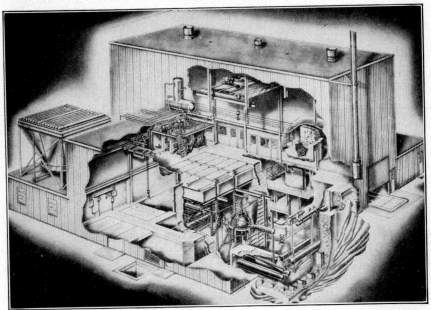

Fig. 14-13. Cutaway drawing of HRT plant. (*Oak Ridge National Laboratory.*)

The primary loop of HRT will be kept under a pressure of 2000 psi. This is more than 700 psi in excess of the vapor pressure at the rated reactor outlet temperature of 572°F. The fuel circulation rate will be 400 gpm, and the flow will be in at the bottom and out at the top of the core. The core tank is 32 in. in diameter and is made of Zircaloy-2. All other parts of the reactor, which is now completely constructed, are made of stainless steel, with some titanium at points of high turbulence such as the pump impellers and the gas separators.

Radiolytic gases will not be removed directly from the core, as they were in HRE, but rather from the core exit pipe. They will be recombined at low pressure by means of a platinized alumina catalyst. Even at full power, gas evolution will be small because of the presence of dissolved copper, which catalyzes internal recombination.   Fission-product gases, stripped from the core by the radiolytic gases, will be extracted from the recombination circuit by adsorption by activated charcoal.

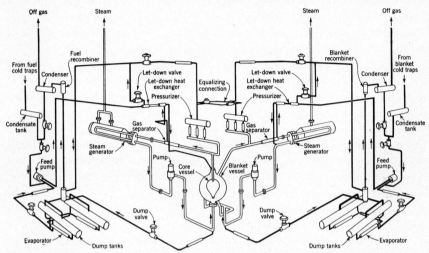

FIG. 14-14. Schematic flowsheet for HRT.   (*Oak Ridge National Laboratory.*)

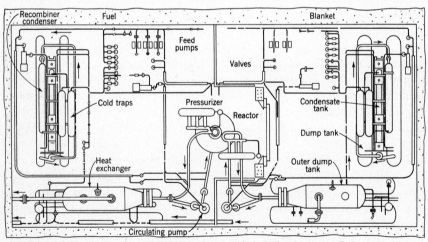

FIG. 14-15. Plan view of HRT system.   (*Oak Ridge National Laboratory.*)

Assuming a 5000-kw output, or a specific power of 17 kw/liter, the reactor inlet temperature will be 493°F and the temperature rise in the reactor will be 79°F.   Gases will be generated at 0.96 cfs.   The equivalent drop in temperature in the U-tube heat exchanger will produce 520 psia saturated steam.   Some of this will be fed to the 350-kva turbogenerator left over from HRE, and the remainder will be fed to an air-cooled steam condenser.

The blanket fluid, also pressurized to 2000 psi, will be circulated through its own heat exchanger and will develop 220 kw, or 0.14 kw/liter. The circulation rate will be 230 gpm, the inlet temperature 532°F, and the outlet temperature 540°F. Gases will be generated at 0.013 cfs. The blanket processes will be identical with the fuel-solution processes except for the amount of material handled.

The shielding for this installation is a large tank of ¾-in. steel plate (54 by 30½ by 25 ft) backed by concrete and earth and roofed with 5 ft of high-density concrete. The tank will be capable of withstanding 30 psi when sealed. The chemical processing pits are similarly enclosed.

A special wall between the reactor and the control room, consisting of two vertical ½-in. steel plates spaced 5½ ft apart and filled with high-density barytes gravel and water, will permit flexibility in the location of service piping and conduit. A neutron shield of barytes sand, colemanite, and water will be placed around the reactor proper.

All equipment in the HRT system has been designed and located to facilitate maintenance. Shield compartments will be flooded with water to protect workers manipulating the special long-handled maintenance tools.

Los Alamos Power Reactor Experiment No. 1 (LAPRE-1) and Los Alamos Power Reactor Experiment No. 2 (LAPRE-2). The Los Alamos Power Reactor Experiment No. 1, prototype of category 15 reactors, is a homogeneous forced-circulation reactor employing a dilute solution of highly enriched uranium and phosphoric acid in ordinary water as the fuel. The highly corrosive fuel requires a gold lining in the core tank and heat exchanger. These are in turn contained in a single gold-lined vessel 6 ft high by 15 in. in diameter. The circulating fuel is maintained at a pressure of 3900 psig and exits from the reactor at about 800°F, generating 2000 kw (heat), which goes to form high-pressure steam. Rods and temperature coefficient are used to control this reactor.

LAPRE-1 cost about $2,200,000. It went into operation at the end of 1956. About 8.4 kg of fuel are required for a critical mass.

LAPRE-2 is a natural-convection-cooled version of LAPRE-1, using a concentrated solution of highly enriched U in phosphoric acid as the fuel. It will develop about 1300 kw (heat) from a maximum fuel temperature of 800°F. The system will be kept under 1000 psig. This reactor is scheduled for operation in 1957. It will cost only $125,000.

Liquid Metal Fuel Reactor (LMFR). The prototype of category 16 reactors has not been designed as yet. It will be known as the Liquid Metal Fuel Reactor Experiment and will probably be rated at about 20,000 kw (heat). Nevertheless, based on a study by a group of industrial concerns led by The Babcock & Wilcox Company in 1955 [12], the following general comments can be made about a full-scale power plant of this type.

The core could well consist of an assemblage of graphite wedges in the shape of a right cylinder (about 5 by 5 ft) drilled with approximately 500 vertical holes (about 2 in. in diameter) through which the fuel alloy is passed. The fuel alloy would be uranium-bismuth, and the flow would be single-pass from bottom to top.

The core would be encased in a thin Croloy-$2\frac{1}{4}$ vessel and around this would be a blanket of graphite containing several hundred flow passages for a slurry of thorium bismuthide in bismuth. Located at the

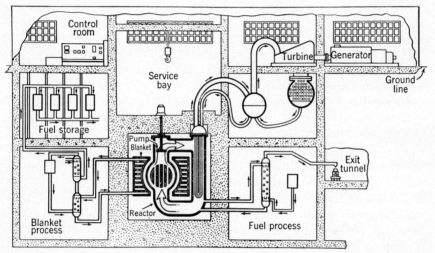

FIG. 14-16. Proposed arrangement for LMFR. (*Brookhaven National Laboratory.*)

top of the reactor would be a degassing vessel in which the liquid fuel is fountained to effect the removal of volatile fission products.

This possible arrangement is shown in Fig. 14-16.

In addition to the reactor proper there would be two primary loops— one for the fuel, the other for the blanket fluid. These loops would consist chiefly of isolation valves, electromagnetic pumps, and heat exchangers. A portion of the fluid in each loop would be continuously bypassed into an integrated chemical reprocessing system.

The heat exchangers would transfer the heat from the fuel alloy or the blanket slurry, as the case might be, to lead-bismuth eutectic. The lead-bismuth eutectic would be carried on the shell side in each case. Design pressure would be 350 psi on both sides, and the design temperatures would be 1025°F on the tube side and 1000°F on the shell side. Both shell and tubes would be U-shaped to permit use of fixed tube sheets without danger of differential expansion problems.

A total of 550,000 kw of heat power could be developed from a plant such as this. The design parameters would be as follows:

	Core	Blanket
Flow, gpm......................	36,000	3600
Inlet temperature, °F............	752	752
Outlet temperature, °F...........	1,022	1022
Heat output, Mw...............	500	50

The throttle steam conditions might then be 800 psig and 900°F.

Because of the low solubility of uranium in bismuth, fully enriched uranium would have to be used as the fuel in this plant. A practical concentration would be 600 ppm.

14-4. General Principles of Evaluation. The first principle in evaluating power reactors for a given application is to define clearly and concisely what that application is to be and how it will limit or otherwise affect the design of the reactor. After the application or function factors have been listed, one should then proceed in an orderly fashion through the categories in Table 14-1, taking them in the sequence dictated by the application requirements and/or the status of the particular branches of reactor technology involved. In view of the present rather embryonic state of the reactor art, it is recommended that the advantages and disadvantages of each of the reactors under consideration, particularly those which are pertinent to the key requirements for the installation, be tabulated and analyzed by reference to the literature before a decision is made.

Many general features of the broad classes of reactors must be taken into consideration in the early stages of such an evaluation. Some of these are listed below.

Neutron energy
  Thermal
    Heat removal ordinarily is not so difficult as in most fast reactors.
    Wide choice of coolants is available.
    High specific power can be achieved.
    Usually operates on any enrichment of fuel.
    Critical mass may be made small if desired.
    High percentage burnup is frequently attainable.
    Total fuel investment seldom is excessive.
    Reactivity is sensitive to build-up of fission products.
    High neutron-absorption cross sections prevent use of many good structural materials.
    Breeding is not possible except with U^{233}.

Power density for high-Z moderators is often low.

Moderator may be expensive.

Fast

No moderator may be employed.

Very high power density may be attained.

High breeding gain is possible.

Operation is not limited by xenon poisoning.

Fissions more U^{238} than thermal type.

Takes full advantage of the superior heat-removal capabilities of liquid-metal coolants.

Permits wide latitude in selection of materials of construction.

Is flexible with respect to breeding cycle.

Can be operated with small excess reactivity.

Can utilize thorium or depleted uranium in blanket.

May be designed to take advantage of increased nonfission capture of neutrons due to Doppler effect for control purposes.

Critical mass is large.

Over-all fuel inventory is large for small burnup (except for molten fuel).

Burnup of fuel severely limited by radiation damage (except for molten fuel).

Operates only on enriched fuels.

Removal of heat is serious problem, both during and after operation.

Choice of coolants is limited.

Fuel diluent may be required.

Specific power is often low.

Circulation of fuel is impractical because of high concentration of fuel required.

Neutron flux is high.

Fuel turnover is rapid.

Separations costs may be exorbitant.

Refabrication of fuel elements in breeding cycle must be carried out under remote control.

Has short prompt-neutron lifetime.

Geometry

Heterogeneous

Permits removal of each fuel piece after optimum irradiation, also rearrangement of fuel to increase burnup.

Reprocessing plant may be located where convenient.

Permits use of any enrichment of fuel in many cases.

Permits variation of concentration of fuel within elements or from element to element, as required for desired flux distribution.

Fission products are contained within fuel.

Coolant may be nonradioactive.

Direct maintenance of heat exchangers and pumps is often possible.

Fuel arrangement can be tailored to specific mechanical requirements.

Eliminates fuel-circulation problems.

Appreciable fast effect is achievable.

Complex and costly fuel shapes must often be fabricated.

Requires external control devices.

Large excess reactivity is required in thermal types.

Radiation damage may limit burnout.

Fission products may build up to high concentrations.

Fuel reprocessing is complicated and time-consuming.

Appreciable structural material (including cladding) is required.

Top coolant temperature is often limited by phase changes in fuel and temperature difference through cladding and film.

Homogeneous

Simplifies preparation and eliminates some steps in reprocessing of fuel.

Simplifies loading and unloading of reactor core.

Relatively high conversion or breeding ratio is possible.

Is inherently self-regulating.

Over-all heat-removal efficiency is high.

Mechanical design of core is usually uncomplicated.

Eliminates radiation damage to fuel.

Nonaqueous homogeneous systems are capable of high temperatures.

Fission and corrosion products may be removed and fuel replenished continuously.

Fission gases can be extracted from fuel in separate system.

Burnup is essentially unlimited if reprocessing is continuous.

Parasitic absorption of neutrons can be minimized.

Corrosion and mass transfer are major problems.

Coolant is highly radioactive and emits delayed neutrons.

Large quantities of radiolytic gases must be handled in aqueous types.

Requires separate fuel and blanket processes in many cases.

May require "on-site" reprocessing plant.

Fuel must be removed for reprocessing on an "average" basis.

Only enriched fuels can be used.

Primary circuit must be absolutely leakproof.

Maintenance costs for heat exchangers and pumps may be high.

Special precautions must be taken to prevent precipitation or settling out of fuel.

Temperature is often limited by phase changes in fuel solutions.

External system may be complicated and difficult to maintain.

Fuel inventory is high because of holdup in external system.

Fuel

Uranium 235

Occurs naturally—0.7 per cent of natural uranium.

Has greater delayed-neutron fraction than do U^{233} and Pu^{239}.

Can be concentrated at present only in government-owned gaseous diffusion plants.

Incapable of sustaining breeding in thermal reactors.

Becomes appreciably contaminated with U^{236} if recycled.

Uranium 233

Is produced from thorium, which is four times as plentiful in the earth's crust as uranium.

Can sustain breeding in either thermal or fast reactors.

Becomes highly contaminated with U^{232} (an alpha emitter) in course of production and therefore can be handled only by remote or other special methods.

Plutonium 239

Is produced from U^{238}, which consitutues most of natural uranium.

Yields high breeding gain in fast reactors.

Can sustain breeding in fast reactors only.

Becomes contaminated with Pu^{240}.

Radiative capture losses are high in thermal spectrum.

Is extremely toxic and can be handled only by remote or other special methods.

Fuel regeneration

Burner-upper (nonregenerative)

Minimizes core size.

Eliminates blanket and associated equipment.

Simplifies reactor design.

For a given power rating it provides the highest neutron flux.

Eliminates fertile and bred material from reprocessing cycle.

Usually requires reflector for efficient operation.

Employs only pure fissionable materials as fuel.

Useful only in certain applications where value of fuel consumed is outweighed by unusual advantages of nuclear power.

Limits nuclear-fuel reserves (on basis of current estimates) to one-sixth to one-seventh conventional fuel reserves.

Burnup limited to 1000 Mw-days/ton (of original natural U) per cycle.

Converter

Permits widest use of reactor designs and materials.

Probably will prove to be more economic than breeder type in smaller sizes.

Operates on all varieties of fuel.

Can be designed so that significant portion of new fuel is burned *in situ*, thus increasing life of core.

Often complicates chemical processing.

Breeder

Creates more fuel than is consumed.

Makes nuclear power industry independent of gaseous diffusion plants.

Can be designed so that significant portion of new fuel is burned *in situ*, thus increasing life of core.

Multiplies natural fissionable-material reserves by factor of 140 multiplied by relative abundance of thorium plus 1.

Limited in reactor designs and materials.

Economical only in large sizes.

Coolant

Gas (e.g., CO_2 or He)

Has negligible macroscopic capture cross section for neutrons and therefore does not cause increase in reactivity with loss of coolant.

In case of accident, negligible chemical activity between coolant and fuel will ensue.

Is noncorrosive to uranium, thorium, and structural materials, except possibly at very high temperatures.

Ordinary losses can be tolerated.

Eliminates danger of explosion or sudden vaporization of coolant.

Does not limit temperature.

CO_2 is usually cheap and available.

Helium diffuses through welds, flaws, and gaskets.

Must be highly pressurized to remove even moderate quantities of heat per unit volume of core.

Pumping power requirements are high if reasonably high heat-transfer rates are required.

Thermal conductivity is low.

Heat capacity on volume basis is low.

Water (H_2O)

Technology is well developed and generally accepted by electric-utility industry.

Is cheap and available.

Conventional equipment and materials may be used.

Can be employed as combination moderator and coolant.

Has high heat capacity and relatively good conductivity.

Provides negative temperature coefficient in most reactors.

Induced radioactivity is short-lived.

Provides transparent shielding for fuel-element handling.

May be used for decay-cooling of spent fuel elements.

Losses are not important.

Requires considerable pressurization for use at power-plant temperatures.

Is quite corrosive.

Reacts strongly with metallic nuclear fuels.

May react at high temperatures with structural metals.

Capture cross section for neutrons is appreciable.

Poison effect may be greater than moderating effect in some reactors, creating danger of excursion upon loss of coolant.

Flashes to steam (in part) if primary system ruptures.

Heavy water (D_2O)

Capture cross section for neutrons is negligible.

May permit breeding in a thermal reactor system.

Conventional equipment and materials may be used.

Can be employed as combination moderator and coolant.

Has high heat capacity and relatively good conductivity.

Permits use of both natural and enriched uranium as fuel.

Provides negative temperature coefficient.

Induced radioactivity is short-lived.

Requires considerable pressurization for use at power-plant temperatures.

Is quite corrosive.

Costs $28 per lb—adds considerably to investment charges.

Reacts strongly with metallic nuclear fuels.

May react at high temperatures with structural metals.

Fuel-element handling is complicated.

Equipment containing it must be leakproof.

Flashes to steam (in part) if primary system ruptures.

Liquid metal (e.g., Na or NaK)

Permits operation at high temperatures without pressurization.

Does not corrode certain common structural materials.

Does not react with uranium and thorium.

Heat capacity is reasonably high.

Thermal conductivity is excellent.

Electrical conductivity is excellent.

Electromagnetic pumps may be used.

Reacts violently with water or oxygen.

Is highly activated by neutron irradiation.

Secondary coolant system may be required to isolate radioactive coolant from water cycle.

May cause appreciable mass transfer of structural materials.

Special pumps, heat exchangers, etc., required.

Equipment containing it must be extremely leakproof.

Protective atmosphere of inert gas may be required around system.

Must be protected from freezing.

Moderators

H_2O

Greatest slowing-down power.

Reactors employing it will have small dimensions.

Is cheap and available.

May also serve as coolant.

Usually provides negative temperature coefficient.

Neutron-capture cross section is appreciable.

Has low boiling point.

Flashes to steam (in part) if rupture occurs in pressurized system.

Requires enriched fuels.

Permits an appreciable percentage of fast fissions.

D_2O

Highest moderating ratio of slowing-down power to neutron-capture cross section.

Permits use of either natural or enriched uranium as fuel.

Neutron-capture cross section is very small.

Requires a larger core for a given power than does H_2O.

May also serve as coolant.

Usually provides negative temperature coefficient.

Has low boiling point.

Is expensive.

Flashes to steam (in part) if rupture occurs in pressurized system.

Graphite

Has second-highest moderating ratio.

Permits use of either natural or enriched uranium as fuel.

Neutron-capture cross section is very small.

May also serve as structural material.

Will withstand high temperatures and is impervious to chemical attack in absence of oxygen.

Slowing-down power is marginal.

Reactors employing it will have large dimensions.

Creates control problems if used with H_2O.

Must be machined to close tolerances in some usages.

Be (BeO)

May also serve as structural material.

Will not volatilize.

Neutron-capture cross section is small.

Slowing-down power is best of all metals.

Is very expensive.

Is toxic.

Must be machined to close tolerances in some usages.

Corrosion characteristics unknown.

Reacts with water at high temperatures.

It is emphasized that the foregoing is not intended to be a complete compendium of the characteristics of the various broad classes of reactors. Every reactor designer or evaluator should make up his own list. The main value of these lists is not that they solve definitively particular design or evaluation problems but that they will gradually lead to a greater understanding of the reactor field by a greater number of people and thus to more intelligent and effective use of nuclear power for peaceful purposes.

14-5. Tabulation of Advantages and Disadvantages. Probably the best way to summarize what has been said in this chapter regarding nuclear-reactor types and their evaluation is to tabulate the significant advantages and disadvantages of some of the categories listed in Sec. 14-2. No attempt will be made to attach relative weights to the various items entered in these tables. *Also, the reader should bear in mind that this summary is being made at a stage in nuclear-reactor development when every day brings new ideas and concepts to the fore and experimental results are rapidly modifying existing knowledge. It is highly unlikely that these tables will be accurate or applicable after a few years.*

TABLE 14-2. POWER-REACTOR SUMMARY*

Reactor	Electric power output, kw	Date of operation	Notes
1. Pressurized H₂O cooled:			
a. Pressurized H₂O moderated:			
STR-I Submarine Thermal Reactor I............	....	1953	
STR-II Submarine Thermal Reactor II		1955	SSN 571, Nautilus
PWR Pressurized Water Reactor, Shippingport..	60,000	1957	
SFR Submarine Fleet Reactor............	....	?	5 will be built
LSR Large Ship Reactor....	....	?	
SRS Submarine Reactor Small............	....	?	
SAR Submarine Advanced Reactor............	....	?	
Albacore Hull Submarines........	....	....	7 will be built
Guided Missile Cruiser............	....	....	2 reactors
Nuclear Powered Aircraft Carrier..	....	....	8 reactors
Nuclear Merchant Ship Edisonvolta, Milan, Italy........	134,000		
Dominican Republic.............	12,000		
APPR-1 Army Package Power Reactor...........	2,000	1957	
APPR-1a Army Package Power Reactor...........	....	....	For Alaska
BTR Belgian Thermal Reactor.............	11,500	?	
CETR Consolidated Edison Thorium Reactor (license)	236,000	1960	96 Mw from oil-fired super-heater
YAER Yankee Atomic Electric Reactor (PDRP)†..........	134,000	1960	
University of Florida Reactor (PDRP)	2,000	1959	
b. D₂O moderated:			
Pacific Gas & Electric–Bechtel Corporation.................	100,000	Study	
c. Graphite moderated:			
APS-1 Atomic Power Station 1, Russian............	5,000	1954	
APS-2, 3 Atomic Power Station 2, 3, Russian........	100,000	1957	
Hanford Atomic Products Operation Group...................	223,000	Study	

TABLE 14-2. POWER-REACTOR SUMMARY* (*Continued*)

Reactor	Electric power output, kw	Date of operation	Notes
2. Pressurized D$_2$O cooled:			
a. Pressurized D$_2$O moderated:			
Commonwealth Edison–Public Service of Northern Illinois........	225,000	Study	
NPDR Nuclear Power Demonstration Reactor, Canada.................	20,000	1958	
Norway—Netherlands Reactor....	?	?	
Switzerland Reactor.............	10,000	?	
DuPont Study.....................	....	....	Natural uranium
3. Boiling H$_2$O cooled:			
a. Boiling H$_2$O moderated:			
BER-III Boiling Experimental Reactor, Borax III	2,000	1955	Being modified to Borax IV
EBWR Experimental Boiling Water Reactor......	5,000	1957	
SBR Small Boiling Reactor, Rural Cooperative Power Association	22,000	1960	4 Mw from coal-fired superheater
GEEBR General Electric Experimental Boiling Reactor (license)	5,000	1957	Electrical generating facility being furnished by Pacific Gas and Electric Co.
CEBR Commonwealth Edison Boiling Reactor (license).............	180,000	1960	
ALPR Argonne Low Power Reactor (U.S. Army)	200	(Early)	
American & Foreign Power.......	12,500	....	2 reactors, for 2 South American countries
4. Boiling D$_2$O cooled:			
a. Boiling D$_2$O moderated:			
Norwegian Power Boiling Reactor	5,000	1958	
Argonne National Laboratory.....	250,000	Study	
Kaiser Engineers................	100,000	Study	
5. CO$_2$ (or N$_2$) cooled:			
a. Graphite moderated:			
G-2 French Production Reactor..	50,000	1957	
British Power Reactors, Calder Hall	92,000	1957	70,000 kw net
6. He cooled:			
a. Graphite moderated:			

TABLE 14-2. POWER-REACTOR SUMMARY* (*Continued*)

Reactor	Electric power output, kw	Date of operation	Notes
Commonwealth Edison–Public Service of Northern Illinois.....	60,000	Study	
City of Holyoke (PDRP).........	15,000	?	May use nitrogen coolant
Air cooled‡			
a. Graphite moderated:			
BEPO British Experimental Pile O	None	1948	First use of reactor energy, 1.2 Mw, to heat building
G-1 French Production Reactor.................	5,000	1956	
GCRE Gas Cooled Reactor Experiment...............	....	1958	Aerojet General
7. Na cooled:			
a. Graphite moderated:			
Monsanto Chemical–Union Electric	220,000	Study	
SRE Sodium Reactor Experiment	7,500	1957	Electrical generating facility being furnished by Southern California Edison Co.
SGR Sodium Graphite Reactor, Consumers Public Power District (PDRP)........	75,000	1959	
b. Be or BeO moderated:			
SIR-A Submarine Intermediate Reactor A............	....	1955	
SIR-B Submarine Intermediate Reactor B		?	SSN 575, Seawolf
c. D$_2$O moderated:			
Chugach Electric, Alaska (PDRP)	10,000	1962	
d. Fast:			
British Fast Power Breeder, Dounreay.......................	60,000	1957	
EBR-II Experimental Breeder, Reactor II..........	15,000	1959	
FBR Fast Breeder Reactor, Power Reactor Development Company (PDRP)..	100,000	1960	

Table 14-2. Power-reactor Summary* (*Continued*)

Reactor	Electric power output, kw	Date of operation	Notes
LAMPRE-1 Los Alamos Molten Plutonium Reactor Experiment 1.........	None	1958	1000 kw heat
LAMPRE-2 Los Alamos Molten Plutonium Reactor Experiment 2........	5,000	1960	
8. NaK cooled:			
a. Fast:			
EBR-I Experimental Breeder Reactor I (CP-4)....	200	1951	
Zephyr British Fast Breeder Prototype..........	10	1954	
9. Diphenyl cooled:			
a. Diphenyl moderated:			
OMRE Organic Moderated Reactor Experiment.....	None	1957	16,000 kw heat
City of Piqua (PDRP)...........	12,500	1960	
American & Foreign Power... ...	12,500	....	For South America
10. Bismuth cooled:			
a. Graphite moderated:			
FPR Florida Power Reactor (license)—Florida Power & Light Co., Florida Power Corp., Tampa Electric Co.....................	200,000	Study	
11. Pressurized H_2O-UO_2SO_4 solution or slurry cooled:			
a. Pressurized H_2O moderated:			
HRE Homogeneous Reactor Experiment..............	150	1952	
Nuclear Power Group...	100,000	Study	
12. Pressurized D_2O-UO_2SO_4 solution or slurry cooled:			
a. Pressurized D_2O moderated:			
HRT Homogeneous Reactor Test	2,000	1956	
Wolverine Electric Cooperative (PDRP).............	10,000	1959	
Netherlands, FOM-KEMA Team	None	?	UO_2 suspended in D_2O; 250 kw heat

TABLE 14-2. POWER-REACTOR SUMMARY* (*Continued*)

Reactor	Electric power output, kw	Date of operation	Notes
13. Pressurized D_2O-UO_3-PuO_2 slurry cooled:			
a. Pressurized D_2O moderated......	....	....	Conceptual designs being developed at ORNL
14. Pressurized D_2O-UO_3-ThO_2 slurry cooled:			
a. Pressurized D_2O moderated:			
PAR Pennsylvania Advanced Reactor (license)—PP&L	150,000	1962	
15. Pressurized H_2O–uranium phosphate solution cooled:			
a. Pressurized H_2O moderated:			
LAPRE-1 Los Alamos Power Reactor Experiment 1	None	1956	2000 kw heat
LAPRE-2 Los Alamos Power Reactor Experiment 2	None	1956	1300 kw heat
16. Molten U-Bi alloy cooled:			
a. Graphite moderated:			
LMFRE Liquid Metal Fuel Reactor Experiment....	None	1959	5000 kw heat
LMFR Liquid Metal Fuel Reactor, Washington State Power Commission...............	30,000	?	
City of Orlando, Florida (PDRP)..	25,000–40,000	?	
17. Fused uranium compound cooled:			
a. Graphite moderated..............	....	....	Conceptual designs being developed at ORNL
b. Fast......................	....	....	Conceptual designs being developed at ORNL

* As of March, 1957.
† Power Demonstration Reactor Program.
‡ Added for historical completeness.

TABLE 14-3. PRESSURIZED H$_2$O REACTOR

Advantages	*Disadvantages*
Water technology well known	Water must be highly pressurized to achieve even reasonably high temperatures without boiling
Water is cheap	
Water is very effective attenuator of neutron energy—core is compact	Fuel-element fabrication expensive
Water has high heat capacity—permits high power density	Temperature limited in metallic fuel elements
Wide range of fuel enrichment may be used	Fission-product activity in core builds up to high level
Specific power high	Pure hot water is highly corrosive—requires special material for primary loop
Small critical mass possible	
Safe, stable system	
Negative temperature coefficient	Appreciable structural material required
Many safeguards against reactor runaway	Neutron economy poor unless Al or Zr is used
Ordinary leakage can be tolerated	Fuel must be at least slightly enriched
Fission products are contained, not circulated	Burnup usually limited by loss of reactivity of core
Radioactivity of coolant is short-lived if kept pure	Fuel suffers radiation damage
Incorporation of fertile material in core makes reasonable burnout and good long-term reactivity characteristics possible	Heat exchangers and control rods required
	Large excess reactivity at operating temperature
Conversion ratio may be high	Heat transfer only moderately efficient
Superheating steam in separately fired superheater is possible	Fuel handling necessitates complex equipment
Appreciable fast-fission effect attainable	Fuel reprocessing a difficult task
	Reactor must be shut down to unload and reload core
	Water would flash to steam (in part) in case of rupture of primary loop
	Water reacts with uranium, thorium, and structural metals under certain conditions
	Low volume ratio of moderator to fuel makes fuel-element design and insertion of control rods difficult
	Low thermal heads make heat exchangers, pumps, and piping large
	Hot-channel factors are significant

TABLE 14-4. PRESSURIZED D$_2$O REACTOR

Advantages	*Disadvantages*
Properties of heavy water well known	Heavy water expensive
Any fuel, including natural uranium, can be used	Special precautions must be taken to make primary loop leakproof and to prevent loss of heavy water during refueling operations
Heavy water is excellent moderator—core can be compact, and neutron economy is good if Al or Zr structural material is used	Primary loop must be highly pressurized to achieve even reasonably high temperatures without boiling
Heavy water has high heat capacity—permits high power density	Fuel-element fabrication expensive
Specific power high	Temperature limited in metallic fuel elements
Small critical mass possible	Fission-product reactivity in core builds up to high level
Safe, stable system	Pure hot heavy water is highly corrosive—requires special material for primary loop
Negative temperature coefficient	Appreciable structural material required
Many safeguards against reactor runaway	Burnup usually limited by loss of reactivity of core
Fission products are contained, not circulated	Fuel suffers radiation damage
Radioactivity of coolant is short-lived if kept pure	Heat exchangers and control rods required
Incorporation of fertile material in core makes reasonable burnout and good long-term reactivity characteristics possible	Large excess reactivity at operating temperature
May be designed to breed with U^{233}	Heat transfer only moderately efficient
	Fuel handling necessitates complex equipment
	Fuel reprocessing a difficult task
	Reactor must be shut down to unload and reload core
	Heavy water would flash to steam (in part) in case of rupture of primary loop
	Heavy water reacts with uranium, thorium, and structural metals under certain conditions
	Production of tritium may be a safety problem
	Low thermal heads make heat exchangers, pumps, and piping large
	Hot-channel factors are significant

TABLE 14-5. BOILING H₂O REACTOR

Advantages	Disadvantages
Some intermediate heat-exchange equipment eliminated	Boiling limits power density
Pressure is lower for given steam output conditions than in pressurized-water reactor	Changes in load on turbine reflected back to reactor as pressure changes
Metal surface temperatures are lower for given steam output conditions than in pressurized-water reactor	Radioactivity builds up in turbine
	Separately fired superheater cannot conveniently be employed
Power excursions quickly damped by formation of steam	System must be designed to overcome tendency to react negatively to load increases
Over-all thermal efficiency quite high	Corrosive effects of boiling water on cladding materials not well known
Conventional technology generally applicable	Temperature limited in metallic fuel elements
Water is cheap	Fission-product activity in core builds up to high level
Core is compact if void coefficient is low	Appreciable structural material required
Wide range of fuel enrichment may be used	Neutron economy poor unless Al or Zr is used
Specific power high, critical mass relatively small	Fuel must be at least slightly enriched
Negative temperature coefficient	Burnup usually limited by loss of reactivity of core
Ordinary leakage can be tolerated	Fuel suffers radiation damage
Fission products are contained, not circulated	Control rods required
Radioactivity of coolant is short-lived if kept pure	Large excess reactivity at operating temperature
Incorporation of fertile material in core makes reasonable burnup and good long-term reactivity characteristics possible	Fuel-element fabrication expensive
	Fuel handling necessitates complex equipment
Conversion ratio may be high	Fuel reprocessing a difficult task
Heat may be taken from circulating water, increasing power output	Reactor must be shut down to unload and reload core
	Water would flash to steam (in part) in case of rupture of primary loop
	Water reacts with uranium, thorium, and structural metals under certain conditions
	Condenser leaks may cause serious trouble

TABLE 14-6. BOILING D$_2$O REACTOR

Advantages	Disadvantages
Properties of heavy water well known	Heavy water expensive
Any fuel, including natural uranium, can be used	Special precautions must be taken to make entire system leakproof, including turbine, and to prevent loss of heavy water during refueling operations
Heavy water is excellent moderator—core can be compact if void coefficient is low, and neutron economy is good if Al or Zr structural material is used	Boiling limits power density
Specific power high, critical mass relatively small	Changes in load on turbine reflected back to reactor as pressure changes
May be designed to breed with U^{233}	Radioactivity builds up in turbine
Some intermediate heat-exchange equipment eliminated	Separately fired superheater cannot conveniently be employed
Pressure is lower for given steam output conditions than in pressurized-heavy-water reactor	System tends to react negatively to load increases
Metal surface temperatures are lower for given steam output conditions than in pressurized-heavy-water reactor	Control equipment more complicated than in pressurized-heavy-water reactor
Power excursions quickly damped by formation of steam	Corrosive effects of boiling heavy water on cladding materials not so well known
Over-all thermal efficiency quite high	Temperature limited in metallic fuel elements
Conventional technology generally applicable	Fission-product activity in core builds up to high level
Specific power high, critical mass relatively small	Appreciable structural material required
Negative temperature coefficient	Burnup usually limited by loss of reactivity of core
Fission products are contained, not circulated	Fuel suffers radiation damage
Radioactivity of coolant is short-lived if kept pure	Control rods required
Incorporation of fertile material in core makes reasonable burnout and good long-term reactivity characteristics possible	Large excess reactivity at operating temperature
Flashing tanks may be used to increase power output	Fuel-element fabrication expensive
	Fuel handling necessitates complex equipment
	Fuel reprocessing a difficult task.
	Reactor must be shut down to unload and reload core
	Heavy water would flash to steam (in part) in case of rupture of primary loop
	Heavy water reacts with uranium, thorium, and structural metals under certain conditions
	Production of tritium may be a safety problem
	Condenser leaks may cause serious trouble

Table 14-7. CO₂ Graphite Reactor

Advantages	*Disadvantages*
Corrosion by coolant is negligible	Reactor vessel and heat exchangers large and expensive
Coolant does not react with fuel or other core materials except possibly at very high temperatures	Core large, requires great number of carefully machined, absolutely clean, pure graphite pieces
Coolant will not flash into vapor if primary system is ruptured	Fuel-loading system is elaborate and costly
Coolant has very low capture cross section for neutrons	Temperature limited in metallic fuel elements
System can be designated with negative temperature coefficient and made exceptionally safe	Fission-product activity in core builds up to high level
Any fuel, including natural uranium, can be used	Heat-transfer efficiency is low
Coolant is cheap	Coolant must be pressurized
Fission products are contained, not circulated	Coolant circulation requires large fraction of power output of station
Ordinary leakage can be tolerated	Power density is low
Little structural material as such required in core	Critical mass is large
Incorporation of fertile material in core makes reasonable burnout and good long-term reactivity characteristics possible	Fuel suffers radiation damage
	Control rods required
	Large excess reactivity at operating temperature
Conversion ratio may be high	Flux pattern usually bad
Readily adaptable to ceramic fuel elements	Fuel-element fabrication expensive
	Fuel reprocessing a difficult task
Gas turbine may be employed	Reactor must be shut down to replenish fuel
Graphite is stable under irradiation at high temperatures	CO_2 dissociates above 300°C
	Elaborate filter system required to keep coolant free of radioactive dust

Table 14-8. Helium Graphite Reactor

Essentially the same as Table 14-7, except that helium has better heat-transfer characteristics and readily diffuses through solids

TABLE 14-9. SODIUM GRAPHITE REACTOR

Advantages	*Disadvantages*
High temperatures achievable without pressurization	Sodium reacts violently with water and actively with air
Corrosion not a serious problem	Sodium invades pore volume of graphite
Heat-removal characteristics excellent	Core is larger than equivalent water core; power density is relatively low
No reaction with uranium or thorium metal in case of fuel-tube failure	Appreciable structural material required in core
Wide range of fuel enrichment may be used	Critical mass is large
Electromagnetic pumps can be used with fair efficiency	Graphite must be canned in zirconium or stainless steel
Fission products are contained, not circulated	Sodium is strongly activated by neutron bombardment
Incorporation of fertile material in core makes reasonable burnup and good long-term reactivity characteristics possible	Intermediate cooling system required to separate active sodium from water-steam
Conversion ratio may be high	Special precautions must be taken to contain sodium which may leak out of primary or secondary loop and to prevent its contacting atmosphere or water
Fast effect can be appreciable	Special pumps, valves, etc., and double-walled tubing in steam generator required if Na or NaK is used in intermediate loop
Graphite stable under irradiation at high temperatures	Sodium must be kept free of oxygen
Fuel can be bonded to container with liquid metal	Temperature limited in metallic fuel elements
	Fission-product activity in core builds up to high level
	Flux pattern usually bad
	Maximum temperature may be limited by interaction between uranium and jacket materials
	Large excess reactivity at operating temperature
	Fuel suffers radiation damage
	Fuel handling extremely difficult
	Reactor must be shut down to replenish fuel
	Provision must be made to prevent coolant from freezing
	Thermal stresses make steam-generator and reactor-vessel design a difficult problem

Table 14-10. Sodium Fast Reactor

Advantages	Disadvantages
No moderator required	Specific power is not so high as in thermal reactor
Core is small; power density is very high	Large inventory of fuel required
High breeding gain possible	Enriched fuel must be used
Operation not limited by Xe poisoning	Sodium reacts violently with water and actively with air
Parasitic absorption of neutrons by structural materials and fission products is low	Fertile material must often be confined to external blanket
High burnup achievable	Fast-acting control system may be required
Excess reactivity at operating temperature can be small	Radiation damage a serious problem except with molten fuel
Full advantage is taken of excellent heat-removal characteristics of sodium	Coolant passages in core numerous and small
Can be operated within delayed-neutron fraction*	Good contact between fuel and coolant imperative
High temperatures achievable without pressurization	Sodium must be kept free of oxygen
Many cheap structural materials corrosion-resistant to sodium	High thermal stresses complicate reactor-vessel and steam-generator design
Sodium does not react with uranium and thorium	Fuel must be reprocessed frequently
Crude chemical reprocessing can be tolerated	High-atomic-weight materials must be used in core
Fuel can be bonded to container with liquid metal	Internal-conversion ratio is usually low
Electromagnetic pumps can be used with fair efficiency	Fuel handling probably difficult
Fission products are contained, not circulated	Neutron flux rises at center of core
	Hot-channel factors are significant
	Sodium is strongly activated by neutron bombardment
	Intermediate cooling system required to separate active sodium from water-steam
	Special precautions must be taken to contain sodium which may leak out of primary or secondary loop and to prevent its contacting atmosphere or water
	Special pumps, valves, etc., and double-walled tubing in steam generator required
	Temperature limited in solid-metal uranium fuel elements
	Fission-product activity in core builds up to high level
	Maximum temperature may be limited by interaction between uranium metal and jacket materials
	Reactor must be shut down to replenish fuel
	Provision must be made to heat coolant if it "freezes"

* This is not a real advantage actually.

TABLE 14-11. NaK FAST REACTOR

Advantages	Disadvantages
NaK has lower melting point than Na	NaK is more corrosive than Na because of higher solubility of oxide at a given temperature
	NaK absorbs appreciably more neutrons in fast spectrum than Na
	NaK costs more than Na
	NaK is a poorer heat-transfer medium than sodium

Otherwise same as Table 14-10

TABLE 14-12. DIPHENYL REACTOR

Advantages	Disadvantages
Boiling point of diphenyl high enough so that reasonably high steam temperatures can be achieved without pressurizing primary loop	Radiation damage to diphenyl an unknown
Diphenyl extremely good moderator—core will therefore be compact	High temperatures tend to polymerize diphenyl
Low fuel enrichment can be used	Decomposition products of diphenyl may be deposited on fuel elements and other heat-transfer surfaces
Fission products are contained, not circulated	Fission-product concentration will limit burnup
Negative temperature coefficient	Fuel suffers radiation damage
High specific power and power density achievable	Fuel reprocessing a difficult task
Critical mass small	Separate heat-exchange circuit and control rods required
Components and equipment for handling diphenyl well developed	Large excess reactivity at operating temperature
Neutron economy good if structural material with low capture cross section is used	Heat transfer not so efficient as in water system
Coolant would not vaporize to any great extent in case of rupture of primary loop	Must be shut down to replenish fuel
Incorporation of fertile material in core makes reasonable burnout and good long-term reactivity characteristics possible	Temperature limited in metallic fuel elements
Conversion ratio may be high	Fission-product activity in core builds up to high level
Superheating steam in separately fired superheater is possible	Appreciable structural material required in core
Coolant does not become radioactive	Fuel-element fabrication expensive
Ordinary leakage can be tolerated	Fuel handling necessitates complex equipment
System can be designed to be extremely safe	Reaction potentials of diphenyl with various metals still under study
Corrosion of metal surfaces is negligible	Elaborate coolant-purification equipment may be required
	Hydrogen gas is produced

TABLE 14-13. BISMUTH GRAPHITE REACTOR

Advantages	Disadvantages
Bismuth does not react strongly with either water or air	Melting point of bismuth is high, necessitating auxiliary heating equipment
High temperatures achievable without pressurization	Density of bismuth is so great that pumping power required is appreciable fraction of output of plant
Heat-removal characteristics excellent	Bismuth corrodes most metals
Wide range of fuel enrichment may be used	Polonium is produced in bismuth by neutron bombardment
Fission products are contained, not circulated	Leakage between bismuth and steam loops must be prevented
Bismuth compatible with graphite	Erosion and mass transfer major unknowns
Incorporation of fertile material in core makes reasonable burnup and good long-term reactivity characteristics possible	Core is larger than equivalent water core; power density is relatively low
Conversion ratio may be high	Critical mass is large
Fast effect can be appreciable	Bismuth must be kept relatively free of oxygen
Graphite stable under irradiation at high temperatures	Temperature limited in metallic fuel elements
Fuel slugs can be bonded to container with liquid metal	Fission-product activity in core builds up to high level
	Flux peaks markedly at center of core
	Maximum temperature may be limited by interaction between uranium and jacket materials
	Large excess reactivity at operating temperature
	Fast-acting control rods required
	Reactor must be shut down to replenish fuel
	Thermal stresses make heat-exchanger design a difficult problem
	Cost of bismuth is high, and availability is limited
	Fuel suffers radiation damage

TABLE 14-14. PRESSURIZED H_2O-UO_2SO_4 REACTOR

Advantages	Disadvantages
Fuel preparation, handling, and reprocessing simplified	Fuel solution is quite corrosive, especially at high concentrations and high temperatures
System has good over-all heat-transfer characteristics; heat exchanger can be optimized without affecting operation of core	Fuel solution is circulated outside core, necessitating absolute leaktightness throughout primary loop
Core has large negative temperature coefficient and is inherently self-regulating	Fuel and blanket materials require separate chemical reprocessing systems
No control rods required	Large quantities of H_2 and O_2 (radiolytic gases) given off by core during start-

TABLE 14-14. PRESSURIZED H_2O-UO_2SO_4 REACTOR (*Continued*)

Advantages

Excess reactivity at operating temperature is small if fission gases (primarily xenon) are continuously removed

Fission gases tend to be stripped out of core automatically by radiolytic gases (H_2 and O_2)

Solid fission products may also be continuously removed

Fuel suffers no radiation damage

Fuel can be continuously replenished, making unlimited burnup possible

System can be fueled conveniently with "hot" U^{233} or Pu

High specific power

Small critical mass

Fertile material may be incorporated in blanket, and higher conversion ratio may be attained

Appreciable fraction of power output can be generated in blanket

Solvent is cheap

Mechanical design of core is simple

Parasitic absorption by structural material is held to a minimum if liner between core and blanket is Zr or some other good neutron "window"

Iron and copper compounds inhibit decomposition of water by irradiation

Temperature control effected by regulating concentration of fuel in core

Power output governed solely by rate of heat removal from core

Steam may be superheated by conventional equipment

Disadvantages

up and shutdown, requiring complex off-gas system

Excess of oxygen must be provided to stabilize fuel solution and hold down corrosion

Blanket slurries not yet proved stable

System must be pressurized

Separation of fuel solution into two phases (one of which is highly corrosive) limits operating temperature to about 300°C

Enriched fuel required

Large fuel inventory in external circuit

Fuel has tendency to precipitate under certain conditions

Power density limited by fuel circulation rate and temperature difference allowable

Maintenance costs will probably be high

Fuel must be removed for reprocessing on an "average" basis

Delayed neutrons can produce appreciable flux in the external circuit, thereby creating additional structural damage and shielding problems

"On-site" reprocessing plant may be required

Response to load changes will be slow

Low thermal head makes heat exchanger, pumps, and piping large

TABLE 14-15. PRESSURIZED D_2O-UO_2SO_4 REACTOR

Differs from Table 14-14 in same way that Table 14-4 differs from Table 14-3—breeding possible with U^{233} fuel and thorium blanket, heavy water expensive, etc.

TABLE 14-16. PRESSURIZED H_2O–URANIUM–PHOSPHORIC ACID REACTOR

Advantages

Phosphate solution may be more stable under irradiation than is sulfate solution

Phosphate solution undergoes no phase separation at high temperatures

Radiolytic gases are automatically recombined internally

Disadvantages

Fuel solution extremely corrosive—primary loop must be lined with gold or platinum

Otherwise same general advantages and disadvantages as for pressurized H_2O-UO_2SO_4 reactor

TABLE 14-17. U-Bi GRAPHITE REACTOR

Advantages	Disadvantages
Fuel preparation, handling, and reprocessing simplified	Fuel solution is circulated outside core, necessitating absolute leaktightness throughout primary loop
High temperatures attainable without pressurization	Solubility of uranium in bismuth limited; enriched uranium must be used
Core has large negative temperature coefficient and is inherently self-regulating, but not quite so much as aqueous homogeneous reactor	Corrosion and erosion effects of U-Bi under irradiation unknown
Heat-removal characteristics are excellent —no transfer problems in core	Few materials of construction capable of handling fuel alloy
Core design is simple and inexpensive	Large inventory of expensive bismuth required
Burnup essentially unlimited because fuel undergoes no radiation damage and may be continuously replenished	Fuel and blanket materials require separate chemical reprocessing systems
Specific power and power density high	Thorium bismuthide-in-bismuth slurry process not developed
Excess reactivity at operating temperature is small if fission gases (primarily xenon) are continuously removed	Pumping losses are high owing to high density of bismuth
Fission gases tend to be expelled by bismuth—separation in degasser very simple	Auxiliary heating equipment required
	Intermediate heat-transfer loop necessary
Most fission and corrosion products may be continuously removed	System must be oxygen-free
System can be fueled conveniently with "hot" U^{233} or Pu	Polonium is produced in bismuth by neutron bombardment
Fertile material may be incorporated in blanket, and high conversion ratio or breeding may be attained	Inventory of uranium outside core is appreciable
	Mass transfer is major problem
Appreciable fraction of power output can be generated in blanket	Thermal stresses make heat-exchanger design difficult
Parasitic absorption of structural material may be held to a minimum if liner between core and blanket is good neutron "window"	Supply of bismuth is limited
	Delayed neutrons produce considerable flux in external circuit, thereby creating additional structural damage and shielding problems
Temperature control effected by regulating concentration of fuel in core	Bismuth expands upon freezing
Power output governed primarily by rate of heat removal from core	Control of core complicated by loss of delayed neutrons
Adaptable to pyrometallurgical reprocessing	Maintenance is a difficult problem
Bismuth does not react strongly with either water or air	
Bismuth compatible with graphite	
Graphite stable under irradiation at high temperatures	

REFERENCES

1. Kallman, D.: Electric Power from Nuclear Energy, "Standard Handbook for Electrical Engineers," 9th ed., McGraw-Hill Book Company, Inc., New York, in preparation.
2. Lane, James A.: "The Status of Reactors and Developments in Reactor Technology," 2d Annual Meeting, American Nuclear Society, Chicago, June 6–8, 1956.
3. Reactor News—Details on OMRE, *Nucleonics*, **14**(5): 22 (May, 1956).
4. Kolflat, Alf: "Special Engineering Features in Design of the Experimental Boiling Water Reactor Plant," American Power Conference, Chicago, Mar. 21–23, 1956.
5. Kallman, D.: Uses of Atomic Energy for Process Heat, "Proceedings of the Atomic Industrial Forum Meeting on Public Relations for the Atomic Industry," pp. 21–26, New York, Mar. 19, 1956.
6. Daniels, F.: Small Gas-cycle Reactor Offers Economic Promise, *Nucleonics*, **14**(3): 34–41 (March, 1956).
7. Atomic Industrial Forum, Inc.: U.S. Power Reactor Program, *Forum Memo*, New York, March, 1956.
8. Edlund, M. C.: "Engineering Problems in Reactor Technology," Society of Professional Engineers of Virginia, Lynchburg Chapter, Lynchburg, Va., Feb. 22, 1956.
9. Kallman, D.: "Economic Aspects of Nuclear Fuels," American Institute of Mining & Metallurgical Engineers, New York, Feb. 2, 1956.
10. Sykes, J. H. M.: British Reveal Details of A-Plant, *Elec. World*, **145**(1): 31–32 (Jan. 2, 1956).
11. Davis, W. Kenneth: "Industrial Applications of Atomic Power," American Association for the Advancement of Science, Atlanta, Dec. 29, 1955.
12. Schomer, R. T., J. A. Klapper, D. Mars, and R. W. Carlson: "Extracting Heat from Liquid Metal Fuel," Preprint 48, Nuclear Engineering and Science Congress, Cleveland, Dec. 13, 1955.
13. Katzin, L. I., and B. I. Spinrad: "A U-233 Breeder-U-235 Converter Reactor," Preprint 35, Nuclear Engineering and Science Congress, Cleveland, Dec. 12–16, 1955.
14. Koch, L. J.: "The Engineering Design of EBR-II," Preprint 280, Nuclear Engineering and Science Congress, Cleveland, Dec. 12–16, 1955.
15. Proceedings of Press Seminar on Civilian Power Reactor Development, U.S. Atomic Energy Commission, Dec. 6, 1955.
16. Rhode, G. K.: "Reactor Principles and Characteristics," a report to Gibbs & Cox, Inc., New York, Dec. 5, 1955.
17. Rowley, L. N., and B. G. A. Skrotzki: Nuclear Energy Today, *Power*, **99**(12): 73–96 (December, 1955).
18. Nuclear Power Group, Raytheon Manufacturing Company: "Nuclear Reactor Data," Waltham, Mass., Nov. 15, 1955.
19. Menke, J. R.: Optimum Reactor Designs for Commercial Power; H. A. Smith: The Preliminary Design of NPD; P. A. Herreng: Nuclear Power Reactors; and B. R. Prentice: The Dual Cycle Boiling Reactor for Electric Power, "Proceedings of the Fourth Annual Conference on Atomic Energy in Industry," pp. 134–164, National Industrial Conference Board, New York, Oct. 26, 1955.
20. Ford, G. W. K.: Power Reactor Projects throughout the World, *Engineering*, **180**(4680): 490–500 (Oct. 7, 1955).
21. Charpie, R. A.: Geneva: Reactors, *Sci. American*, **193**(4): 56–68 (October, 1955).
22. Hopping, R. L., and A. F. Nehrenz: Design and Economic Aspects of Packaged

Power Reactors, "Proceedings of the Atomic Industrial Forum Meeting on Commercial and International Developments in Atomic Energy," pp. 234–239, Washington, Sept. 28, 1955.

23. Amorosi, A., A. P. Donnell, and H. A. Wagner: A Developmental Fast Neutron Breeder Reactor, "Proceedings of the Atomic Industrial Forum Meeting on the Commercial and International Developments in Atomic Energy," pp. 34–45, Washington, September, 1955.

24. Okrent, D., R. Avery, H. H. Hummel, et al.: A Survey of the Theoretical and Experimental Aspects of Fast Reactor Physics, "Proceedings of the International Conference on the Peaceful Uses of Atomic Energy," vol. V, pp. 347–363, Geneva, Aug. 19, 1955.

25. Siegel, S., R. L. Carter, F. E. Bowman, and B. R. Hayward: Basic Technology of the Sodium Graphite Reactor, "Proceedings of the International Conference on the Peaceful Uses of Atomic Energy," vol. IX, pp. 321–330, Geneva, Aug. 18, 1955.

26. Lichtenberger, H. V., F. W. Thalgott, W. Y. Kato, M. Novick, et al.: Operating Experience and Experimental Results Obtained from a NaK-cooled Fast Reactor, "Proceedings of the International Conference on the Peaceful Uses of Atomic Energy," vol. III, pp. 345–360, Geneva, Aug. 5, 1955.

27. Hinton, C.: The Graphite-moderated Gas-cooled Pile—Its Place in Power Production, "Proceedings of the International Conference on the Peaceful Uses of Atomic Energy," vol. III, pp. 322–329, Geneva, Aug. 15, 1955.

28. Simpson, J. W., M. Shaw, et al.: Description of the Water Reactor (PWR) Power Plant at Shippingport, Pa., "Proceedings of the International Conference on the Peaceful Uses of Atomic Energy," vol. III, pp. 211–242, Geneva, Aug. 15, 1955.

29. Barnes, A. H., L. J. Koch, H. O. Monson, and F. A. Smith: The Engineering Design of EBR-II, a Prototype Fast Neutron Reactor Power Plant, "Proceedings of the International Conference on the Peaceful Uses of Atomic Energy," vol. III, pp. 330–344, Geneva, Aug. 15, 1955.

30. Parkins, W. E.: The Sodium Reactor Experiment, "Proceedings of the International Conference on the Peaceful Uses of Atomic Energy," vol. III, pp. 295–321, Geneva, Aug. 15, 1955.

31. Briggs, R. B., and J. A. Swartout: Aqueous Homogeneous Power Reactors, "Proceedings of the International Conference on the Peaceful Uses of Atomic Energy," vol. III, pp. 175–187, Geneva, Aug. 13, 1955.

32. Zinn, W. H.: Review of Fast Power Reactors, "Proceedings of the International Conference on the Peaceful Uses of Atomic Energy," vol. III, pp. 198–204, Geneva, Aug. 13, 1955.

33. Weinberg, A. M.: Survey of Fuel Cycles and Reactor Types, "Proceedings of the International Conference on the Peaceful Uses of Atomic Energy," vol. III, pp. 19–25, Geneva, Aug. 12, 1955.

34. McCullough, C. R., M. M. Mills, and E. Teller: The Safety of Nuclear Reactors, "Proceedings of the International Conference on the Peaceful Uses of Atomic Energy," vol. XIII, pp. 79–87, Geneva, Aug. 10, 1955.

35. Dietrich, J. R., H. V. Lichtenberger, W. H. Zinn, et al.: Design and Operating Experience of a Prototype Boiling Water Power Reactor, "Proceedings of the International Conference on the Peaceful Uses of Atomic Energy," vol. III, pp. 56–68, Geneva, Aug. 9, 1955.

36. Lane, J. A.: Economics of Nuclear Power, "Proceedings of the International Conference on the Peaceful Uses of Atomic Energy," vol. I, pp. 309–321, Geneva, Aug. 9, 1955.

37. Blokhintsev, D. I., and N. A. Nikolayev: The First Atomic Power Station of the USSR and the Prospects of Atomic Power Development, "Proceedings of the International Conference on the Peaceful Uses of Atomic Energy," vol. III, pp. 36–55, Geneva, Aug. 9, 1955.

38. Anderson, A. N., and J. W. Landis: "Consolidated Edison Company's Indian Point Nuclear Steam-Electric Generating Station," National Association of Power Engineers, New York, June 14, 1955.

39. Landis, J. W.: "The Small Power Reactor," Power Reactor Conference, University of Florida, Gainesville, Fla., May 27, 1955.

40. Landis, J. W.: Power from Atomic Energy, "Proceedings of the American Power Conference," vol. XVII, pp. 98–102, Chicago, Mar. 30, 31, and Apr. 1, 1955.

41. Landis, J. W.: "Power Plant Applications of Reactors," N.Y. Section, AIEE Course in Atomic Energy and Its Application, Jan. 6, 1955; Bulletin 3-605, The Babcock & Wilcox Company, New York.

42. Lane, J. A.: Reducing Nuclear Power Costs, *Nucleonics*, **13**(1): 24–27 (January, 1955).

43. Weinberg, A. M.: Power Reactors, *Sci. American*, **191**(6): 33–39 (Dec. 1, 1954).

44. Williams, C., and F. T. Miles: Liquid Metal Fuel Reactor Systems, *Nucleonics*, **12**(7): 11–42 (July, 1954).

45. McLain, S.: Materials for Nuclear Reactors, *Chem. Eng. Prog.*, **50**(5): 240–244 (May, 1954).

46. Abbott, W. E.: Power Reactor Design Fundamentals, *Chem. Eng. Progr.*, **50**(5): 245–248 (May, 1954).

47. Landis, J. W.: Status of Nuclear Power Reactor Technology, *Mech. Eng.*, **76**(2): 143–146 (February, 1954).

CHAPTER 15

LEGAL ASPECTS OF NUCLEAR POWER

By John Gorham Palfrey

15-1. Introduction.   The purpose of this chapter is to provide a general background of the provisions of the law in atomic energy which have a particular bearing on work in the field of nuclear engineering.   In 1954 a new law, the Atomic Energy Act of 1954 [1], was enacted, superseding the original (McMahon) Act of 1946 [2].   This chapter will consist primarily of a summary of the highlights of the new law.*   Significant provisions of the Atomic Energy Commission's regulations, issued pursuant to the Act, are noted in the text or in footnotes [4].

The provisions of the Atomic Energy Act of 1946 set up a structure of government monopoly and control over facilities, materials, patents, and information relating to the production and use of nuclear energy.   These controls embraced both military and nonmilitary development and tightly circumscribed the field of nuclear engineering and reactor development. The Act of 1954 substantially relaxed this monopoly structure, in each of the above categories, as it related to atomic-energy development for nonmilitary purposes.   However, the area of permissible nongovernmental activity, while enlarged, remains subject to control by extensive government regulation.

15-2. The Role of the Government.   The essential design of the new law, as it affects the nuclear engineering field, is to provide, first, for a continuation of the system of government ownership and government-directed operation of nuclear-material production and reactor development in the attainment of national defense objectives.   Second, it is to provide for a new system of government regulation of private activities in the industrial field by means of an elaborate licensing mechanism. Third, the law emphasizes the encouragement of research and development programs under both government and private auspices in the commercial applications of atomic energy.

This combination of functions of the Commission as an operating agency, as a regulatory agency, and as a promotional agency is reflected

* For more detailed analyses of the new statute, see Ref. 3.

in many sections of the new law. At times the distinctions in function are obscured because the structure of the Act, with sections on research, materials, facilities, commercial licensing, patents, information, etc., remains largely as before.

In this chapter, principal attention is given to the features of the law concerned with the regulation and control of those private activities which are newly authorized by the Act. The heart of the control scheme consists of the licensing provisions affecting facilities and materials; these are discussed first. Problems of government promotion, subsidy, and power policy will be outlined next, to be followed by a summary of the changes in the controls over information, international activities, military affairs, and patents.

15-3. The Regulatory Scheme. *The Legal Basis for Private Activity.* The private development and use of atomic energy for nonmilitary purposes are made possible by two major changes in the law. The first is to permit private ownership of facilities, producing and utilizing fissionable material, pursuant to licenses issued by the Commission [Secs. 41a(2) and 101*]. The second is to authorize the Commission to distribute and to license the possession of fissionable material, owned by the government, for the conduct of research and development and for the operation of production and utilization facilities (Secs. 52, 53).

The Commission's Licensing Authority. The private activity thus made possible is subject to Commission regulation through a system of licensing. The scope of the Commission's licensing authority, in general terms, embraces the area of work concerned with nuclear reactors, nuclear fuels and reactor products. It applies to production and utilization facilities; it applies to "special nuclear material" (replacing the term "fissionable material"), source material, and by-product material. The terminal processes in the commercial application of atomic energy, involving the generation, transmission and sale of electric power, are subject to the authority of the appropriate Federal, state or local agency, as is the case with electric power generally. Those licensees of the Commission, with facilities for generating electric power on a commercial basis, who transmit it or sell it wholesale in interstate commerce are subject to the regulatory provisions of the Federal Power Act (Secs. 271–272).

Licensing of Facilities. The Commission's licensing authority over facilities is derived from the requirement of a Commission license "for any person within the United States to transfer or receive in interstate commerce, manufacture, produce, transfer, acquire, possess, import, or export any utilization or production facility." These facilities are defined to include any equipment or device (including its specially

* In this chapter, "sections" refer to the Atomic Energy Act of 1954, not to sections in this book.

designed, important component parts), capable of producing or utilizing special nuclear material in such quantity as to affect significantly the national security and public health and safety (Secs. 11*p*, 11*v*).

THE LICENSING PROCESS.   The licensing process is actually a two-step operation.   Applicants proposing to construct or modify a production or utilization facility must first receive a construction permit (Sec. 185).   If the application is "otherwise acceptable," the Commission is to grant such a permit, setting forth the earliest and latest dates for the completion of the construction.   The Commission will thereafter issue a license if construction is completed within the time limit and the facility is found to operate in conformity with the license application and the Commission's regulations.

TYPES OF LICENSES.   The licenses issued by the Commission for utilization or production facilities are of two basic types: commercial licenses (Class 103 licenses), and licenses for medical therapy and for research and development (Class 104 licenses).   The guiding considerations in the issuance of both types of licenses are the public health and safety and the common defense and security.

1. *Commercial licenses* (Sec. 103).   The provisions for commercial licenses look to the future.   They may not be issued until after the Commission has made a finding in writing that the particular type of utilization or production facility "has been sufficiently developed to be of practical value for industrial or commercial purposes" (Sec. 102).   No such finding has yet been made and none is expected for the immediate future.

a. Issuance.   The Commission is to issue these licenses on a non-exclusive basis to those persons who are prepared to meet the safety standards set by the Commission, who agree to make available information on licensed activities to the Commission which it determines is needed to promote security and protect public safety, and whose activities would serve a useful purpose "proportionate to the quantities of special nuclear or source material to be utilized" (Sec. 103*b*).   Licenses may be issued for specified periods, not to exceed forty years, and are subject to renewal.

b. Public bodies.   No license for a commercial power facility will be issued by the Commission until notice has been given to the appropriate regulatory agency with jurisdiction over the proposed activity's rates and services, as well as to public and private bodies within transmission distance who are authorized to engage in the distribution of electric energy.   The Commission, in issuing such licenses, is required to give preferred consideration to facilities located in high-cost power areas and to public or cooperative bodies, whenever there are conflicting applications for a limited number of licenses.

2. *Research and development licenses* (Sec. 104). *a. Medical.* In issuing licenses for utilization facilities for use in medical therapy, the Commission is affirmatively directed to permit as much effective medical therapy as is possible with the amount of special material available for such use and to impose the minimum amount of regulation consistent with its security and safety obligations under the Act.

b. Research and development. With regard to research and development licenses, a distinction is made between experimental commercial power facilities and other research and development facilities (Secs. 104*b*, *c*). In both cases, there is to be imposed minimum regulation consistent with the Commission's security and safety obligations. But in the case of the experimental power facilities, it is also necessary that the regulations and terms of the license be "compatible" with those which would apply in the event that a commercial license was later issued for that type of facility.* In issuing these Class 104 licenses, priority is to be given to those activities which the Commission believes will lead to major advances in the commercial application of atomic energy.

3. *Conditions and terms of Section 103 and 104 licenses.* In addition to the requirement of construction permits, certain other requirements apply to both commercial and noncommercial licenses. Operators of nuclear reactors must be licensed by the Commission (Sec. 107). No license may be issued to a corporation which the Commission has reason to believe is subject to foreign control or domination (Sec. 104*d*).

The Commission may group facilities into classes on the basis of their operating and technical characteristics. It may define the activities to be carried out in each class and designate the amount of special material for use (Sec. 106). While no time limit on the terms of Sec. 104 licenses was included in the Act, the Commission's regulations apply the 40-year limit of commercial licenses to Sec. 104 licenses as well.†

Each application for a license submitted to the Commission must

---

* This "compatability" requirement is amplified somewhat by the AEC in its regulations [4*d*]. In a note to Sec. 50.41, on the standards for Class 104 licenses, it is stated that the Commission has determined that the existing regulations and terms of the Class 104 licenses for experimental power facilities are "compatible" with those of a Class 103 license, should it later be issued to apply to the facility in question.

Unanswered questions remain, however. There are antitrust and preference requirements of a commercial license that do not apply to a noncommercial (Class 104) license. Responsive to these factors, the Commission indicated in 1956 that a licensee would not be required to convert a noncommercial license into a commercial license after the facility has been constructed and operations begun. For other problems of AEC licensing and regulation, see Trowbridge, Licensing and Regulation of Private Atomic Energy Activities, 34 *Texas L. Rev.* 842 (1956).

† Section 50.51 of the regulations provides that each license will be issued for a fixed period of time to be specified in the license, but in no case to exceed 40 years from the date of issuance.

contain information which the Commission determines by regulation is necessary for the appraisal of technical, financial, and other qualifications (Sec. 182). The technical specifications should include information on the amount, kind, and source of special nuclear material required, the facility's location and its specific characteristics, and other information the Commission may require to determine that the operation would be in accord with the national security and public health and safety.* These specifications would be a part of the license issued.

The technical part of the license application, concerning the evaluation of the radioactive hazards, is known as the "hazards summary report." According to a Commission spokesman [5], it can be filed on a step-by-step basis. A preliminary license application, with data based on an initial hazard evaluation, may be submitted to and acted upon by the Commission. In the course of consultations, a Commission statement would then be developed, delineating areas where additional information is needed. Similarly, when a construction permit is issued, its final terms need not be fully worked out at the outset. Further supplemental details can be submitted to the Commission and reviewed by it "until all reservations have been removed, and upon completion of the reactor, the construction permit may be converted into a license."

In the summer of 1956, the Commission issued a "conditional" construction permit to the Power Reactor Development Company for its fast-breeder reactor. The conditions related to the demonstration of financial responsibility and the resolving of areas of uncertainty regarding the hazard potential of this type of reactor.†

4. *Processing and fabricating facilities.* Although not clearly specified in the Act, a license must be obtained for the construction and operation of chemical facilities to process irradiated fuels, according to Commission regulation,‡ which defined the term "production facility" to include these processing plants.

The Commission determined, however, not to include facilities for the fabrication of reactor fuel elements in its list of production facilities.§ While no construction permit and facilities license are required for fabrication plants, a materials license for the use and possession of special nuclear material is required, as discussed below.

5. *Component parts.* The definition of production and utilization facilities includes any important component part especially designed for the facility "as determined by the Commission." Under the McMahon

---

* Further details on the contents of applications for licenses are set forth in the Commission's regulations on the licensing of facilities [4, Secs. 50.33 to 50.36].

† AEC Press Release, Aug. 3, 1956.

‡ AEC Regulation on the Licensing of Facilities [4, Sec. 50.2a(3)].

§ AEC Press Release, Jan. 18, 1956.

Act, production facilities had been determined by the AEC to include components such as radiation-detection instruments and mass spectrometers. They could be manufactured and sold domestically under a general AEC license, but they required specific licenses for export. Licensing on this basis is authorized by the new law to apply to those component parts determined to be production or utilization facilities, if such licensing is thought not to constitute an unreasonable risk to the national security (Sec. 109).

The Commission in its regulations did not list any component parts as production or utilization facilities, although it reserved the right to do so at a later date.* The proposed regulation would thus remove AEC licensing controls over component parts but would not relieve anyone from compliance with export restrictions of other agencies, such as the Department of Commerce, under the general export control laws.

6. *Antitrust provisions* (*Sec.* 105). Existing antitrust laws are declared to be applicable to the atomic-energy program, and if, in the course of a licensed activity, a court or government agency finds that these laws have been violated, the Commission may suspend or revoke the license. The Commission is also to report to the Attorney General information on the use of special nuclear material which appears to the Commission to violate or to lead toward the violation of the antitrust laws or to restrict free competition. Commercial license applications are also subject to a preliminary referral to the Attorney General for advice as to whether their issuance would tend to create or maintain a situation inconsistent with the antitrust laws.

Licensing of Materials. SPECIAL NUCLEAR MATERIAL. The term "fissionable material" is replaced in the new Act by a broader term "special nuclear material," in order to make clear the inclusion of material essential to fusion processes. It is defined to mean plutonium, uranium 233 and 235, or any other material capable of releasing substantial quantities of atomic energy, which the Commission so denominates in the interests of the common defense and security, when agreed to by the President and placed before the Joint Committee on Atomic Energy for 30 days (Secs. 11t, 51).

The new Act, while permitting the private ownership of facilities, prohibits private ownership of the special material itself (Sec. 52), as the old Act did with regard to fissionable material. This prohibition, however, now applies to such material whether produced in government-owned or privately owned facilities.

---

* A note to Sec. 50.2 of the regulations on the licensing of facilities [4] states that the Commission may from time to time add to the foregoing definitions and may also include as a facility an "important component part especially designed for a facility, but has not at this time included any component parts in the definitions."

This preservation of government monopoly of special material is accompanied by new licensing provisions providing both for the distribution of government-produced special material to private persons and for the purchase by the government at a "fair price" of privately produced special material.

1. *Government distribution.*   The Commission is authorized (Sec. 53) to license qualified applicants and distribute to them special nuclear material (*a*) for the conduct of research and development activities of the types described in the chapter on research (discussed below), (*b*) for research and development activities and medical therapy conducted pursuant to a Commission license (Sec. 104), and (*c*) for commercial operations conducted pursuant to a commercial license (Sec. 103).

2. *Price.*   In the case of a commercial license, the Commission is directed (Sec. 53*c*) to make a "reasonable charge" on a nondiscriminatory basis for the material licensed and distributed.   In the case of research arrangements and noncommercial licenses, the imposition of a charge is a matter of Commission discretion, on the basis of criteria which the Commission establishes, taking into account the nature of the institution and the purposes for which the material is used.

In determining what the charge should be, the Act sets forth specific considerations to guide the Commission (Sec. 53*d*).   They include the cost to the Commission of producing the special material, the use to be made of the material, the extent to which its use will advance the development of peaceful uses of atomic energy, and the energy value of the material in the particular use in question.   Price schedules have been established by the AEC.*

If the material is for use pursuant to a commercial license, the AEC is directed to make a further charge for material consumed in the operation of the reactor.   This charge is to be based either on the estimated cost of the material to the Commission, or on the average fair price which the Commission pays for the private production of such material, depending on which is the lower amount.

3. *Conditions in the license.*   Several conditions apply to licensees for special material (Sec. 53*e*): the material is subject to the government's reserved right of recapture or control; possession of the material is subject to such terms that do not "permit" the user to construct an atomic weapon; and the material must be handled in accordance with safety standards established by the Commission.

* The unclassified price schedules established by the AEC for the sale of natural and enriched uranium and other materials are given in App. K, page 816.   For further description of the schedule of prices for special material and of charges for material and for services leased and sold, see Appendix 11 of the Twenty-first Semiannual Report of the AEC, January, 1957.

4. *Quantity.* The President determines annually (Sec. 41*b*) the quantity of special material to be available for private distribution. Within this limitation and to the "maximum extent practicable" (Sec. 53*f*), the Commission is directed to distribute enough material to permit widespread independent research and development activities.

If the demand exceeds supply, preference is to be given "to those activities which are most likely, in the opinion of the Commission, to contribute to basic research, to the development of peacetime uses of atomic energy or to the economic and military strength of the nation" (Sec. 53*f*).

5. *Long-term commitments.* The Commission is authorized to make special material available to licensees for the period of the license (Sec. 53*a*). Despite this language, the extent of the government commitment that is possible under the Act to supply special material on a long-term basis is as yet uncertain. Contributing to the uncertainty is the fact that special nuclear material is specifically excluded in the grant of general authority to the Commission to make long-term commitments to supply materials and services (Sec. 161*m*).

The Commission has evidently taken the position that the President can make allocations of enriched material on a long-term basis and that the Commission can undertake to supply reactor fuel on this basis. In early 1956 the Commission obtained an allocation from the President of 20,000 kg of enriched uranium to support the future operation of private reactors. A schedule of estimated quantities of material required would provide the basis for an allocation of material by the AEC at the time a construction permit is issued. A long-term materials license would later be issued at the time that a license to operate the facility was issued by the Commission.*

6. *Government purchase.* Section 56 provides that any person who lawfully produces special material, other than under government contract, shall be paid a "fair price" for it. In determining this price, the Commission takes into account the value of the material for its intended use by the United States and may give such weight to the actual production cost of the material as it finds to be equitable (Sec. 56). The determined fair price is to apply to all licensed producers of the same material at any one time, and the Commission may establish guaranteed prices for periods up to seven years. This schedule of "buy-back" prices is also provided in App. K.

SOURCE MATERIAL. 1. *Controls.* Source material is defined to mean uranium, thorium, or ores containing these materials in such concentration as the Commission determines by regulation, and other materials

---

* Licensing of Materials [4*f*, Sec. 70.31*b*(1)].

which the Commission determines, with the approval of the President, are essential to the production of special material (Secs. 11s, 61).

The Commission's control over source material is derived, as before, not from government ownership but from the requirement of an AEC license for the transfer, delivery, and possession of source material (Sec. 62). Further provision is made for the reporting of ownership, extraction, and shipment of source material, for its acquisition by the Commission, and for exploration.*

The new law deletes the reservation to the United States of source material in public lands, because of legal doubts as to the validity of mining claims based on the discovery of a source material alone. The effect of the deletion is to place source material on the same footing as other minerals under the general mining laws [8]. The Commission is now expressly authorized to lease public lands for mining and prospecting operations.

2. *Price.* The uses for which source material may be distributed and the Commission's discretion in imposing charges (Sec. 63) parallel the provisions on special-material distribution. The Commission may, in addition, enter into agreements with licensees to "sell, loan, or otherwise make available" source material as needed for the licensed activity. The Commission is to establish prices, on a nondiscriminatory basis, which it considers will provide the government reasonable compensation and "will not discourage sources of supply independent of the Commission" (Sec. 161m).

BY-PRODUCT MATERIAL. The new law preserves the Commission's authority to produce and distribute by-product material (Secs. 81–82). It is defined as radioactive material (other than special nuclear material) yielded in or made radioactive by exposure to radiation "incident to the process" of producing or using special material (Sec. 11e). Licensing requirements supersede the very similar system of authorizations set up by the Commission under the original Act [4c].

The law newly authorizes the distribution of by-products produced in private reactors. Domestic and foreign distribution of isotopes by the Commission and by private persons is authorized. The principal distinction in the requirements affecting foreign and domestic traffic is that both general and special licenses may be issued for domestic distribution, whereas by-products may be exported only upon an express finding of the Commission that each such distribution will not be inimical to the national interest. The Commission is also authorized to require reports from a foreign user of isotopes.

---

* The AEC has continued in effect its licensing regulations for source materials issued under the Act of 1946 [19 Fed. Reg. 5628 (1954), 10 C.F.R. pt. 40 (Supp. 1955)].

15-4. Government Assistance and Government Operation. Hand in hand with the government program of regulating private activity in the research and development of atomic power, the law also envisions a parallel research and development program under government auspices [9]. One element of government activity involves the continuation of the Commission's programmatic operations in the field of reactor development in pursuance of national defense objectives. Another element involves developmental programs undertaken by the Commission to further the peacetime applications of atomic energy.

AEC Research and Development Program. Authority for government-sponsored research activities is primarily to be found in the chapter on research (Secs. 31–33). While the new research provisions are much like the old, two additions point up (1) the conduct of government-directed research and development, and (2) government promotion of private development in the field of reactor engineering.

As before, the Commission is authorized and directed to make arrangements for the conduct of research and development in its own facilities or in the facilities of others in five broad areas,* in order to ensure the continued conduct of such activities by private and public bodies and "to assist in the acquisition of an ever-expanding fund of theoretical and practical knowledge."

However, the fourth area of industrial uses newly includes activities relating to "the demonstration of the practical value of utilization or production facilities for industrial or commercial purposes" [Sec. 31a(4)]. This provision gives specific statutory support for the $200 million program, previously launched, which involves the construction, under AEC contract, of five separate types of experimental nuclear power reactors. It is contemplated that in this government-sponsored program the major part of the cost will be borne by the government, although the commitment of an increasing proportion of private funds in future reactor development activities is a governmental objective [10].

---

* These areas are the following:

1. Nuclear processes

2. The theory and production of atomic energy, including processes, materials, and devices related to such production

3. Utilization of special nuclear material and radioactive material for medical, biological, agricultural, health, or military purposes

4. Utilization of special nuclear material, atomic energy, and radioactive material and processes entailed in the utilization or production of atomic energy or such material for all other purposes, including industrial uses, the generation of usable energy, and the demonstration of the practical value of utilization or production facilities for industrial or commercial purposes

5. The protection of health and the promotion of safety during research and production activities

Research and Development Services. The second addition to the chapter on research amplifies the services which the government is authorized to perform in the promotion of private development. If the Commission finds private facilities inadequate for the purpose, it may conduct for others, through its own facilities, such activities and studies as it deems appropriate to the development of atomic energy. The charges to be made for these services are a matter of Commission discretion (Sec. 33).

The Commission announced in 1956 that private groups would be permitted to make use of government facilities for their own purposes, provided that the user had appropriate security clearance, that the work did not interfere with the AEC program, and that private facilities or equipment was not reasonably available. Priorities for such permitted activities might be established if the AEC found them necessary. The arrangements would include an assessment of charges, based on the cost of the services to the AEC or else on going commercial rates [4*l*, p. 27].

Relationship with the Licensing Program. A major area of uncertainty in the new statute is the relationship between research and development arrangements, authorized in this chapter on research, and research and development (Class 104) licenses authorized in the chapter on licenses. This issue is rendered acute by the "no-subsidy" provision in the Commission's grant of general authority. Section 169 states that no Commission funds may be used in the construction or operation of facilities pursuant to a commercial or a research and development license, except under a contract or other arrangement entered into pursuant to the research section of the Act.

The distinction implicit in this provision is that which customarily exists between the promotional and regulatory functions of the government. Confusion arises here because the two roles are so closely intertwined under the Atomic Energy Act.

GOVERNMENT ASSISTANCE FOR LICENSED ACTIVITIES. Despite the "no-subsidy" language, applicable to Sec. 104 licenses, it is evident that various forms of government assistance without charge or below cost are authorized under the Act. Under Sec. 63*c*, source materials may be furnished without charge, in some instances. Under Sec. 33, research services—testing fuel elements, etc.—may be done with or without charge in government facilities. Under Sec. 53*c*, the Commission may waive charges for special material consumed in Sec. 104 reactor facilities. Under Sec. 161*m*, the Commission may enter into agreements with commercial and noncommercial licensees, in which the government may provide for the processing, fabricating, refining and separating of materials in government facilities, which are needed in reactor construction and operation. The AEC may also agree under this section to sell, lease, or otherwise make available material, other than special material, necessary for the conduct of the licensed activity. Charges under this section are those which will provide the government reasonable compensation and will not discourage the development of sources independent of the Commission.

POWER DEMONSTRATION REACTOR PROGRAM. In administering the law, the Commission has taken a further step in providing government assistance for licensed activities. It launched a Power Demonstration Reactor Program involving the participation of firms that qualify as Class 104 licensees on the basis of the reactor construction proposals submitted [11]. The AEC then enters into research and development contracts with these licensees, which involve payment of part of the development and operation expenses of experimental reactors [12]. Despite some uncertainty about the Commission's legal authority to subsidize its licensees, its action has found support, as a matter of policy, on the ground that such measurable assistance to licensees is preferable to the payment of hidden subsidy through premium prices that might otherwise be established for the special material produced in the private reactors of Sec. 104 licensees.

Independent Industrial Program. Other industrial groups, such as Consolidated Edison Company of New York and Commonwealth Edison's Nuclear Power Group at Dresden, Ill., have been issued construction permits, as Class 104 licensees, for reactors to be built with private funds. No accompanying contracts for AEC research-and-development assistance will be involved.

Distribution of Public Power. The Commission's authority to engage in the production and sale of atomic power was the outcome of extended controversy. Under Sec. 44, the Commission is not authorized to engage in the sale or distribution of energy for commercial use except for that energy which the Commission produces "incident to the operation" of its research and development and its production facilities. The energy that is produced at Commission production facilities or in its "experimental utilization facilities" may be used by the Commission, transferred to other government agencies, or sold to publicly, cooperatively, or privately owned utilities at reasonable, nondiscriminatory prices.

If the by-product is electric energy, the price is subject to regulation by the appropriate agency having jurisdiction. In determining who would receive this energy, the Act requires the Commission to "give preference to and priority to public bodies and cooperatives or to those privately owned utilities who are providing power to high cost areas not being served by public bodies or cooperatives."

15-5. Information. In the development of atomic energy by private institutions, a critical problem has been posed by the secrecy barriers closing off from the outside much information which is needed in order to engage in the field. In the new chapter on information (Secs. 141–146), the Commission is given authority that has resulted in the declassification of a greater amount of reactor technology information. The Commission is also authorized to provide wider access to information which is still classified for those interested in entering the atomic field.

Declassification. The policy basis for greater declassification to further industrial development is indicated by the inclusion of "industrial" as well as scientific progress as a basis for the disclosure of technical information (Secs. 3b, 141b). The Act maintains the special and comprehensive category of "restricted data" for classified atomic information (Sec. 11r). However, the test for declassifying such data has been changed from one of deciding whether its publication would "adversely effect the common defense and security" to one of whether the data could be published "without undue risk to the common defense and security" (Sec. 142a). This change formalizes the approach to declassification as a problem of balancing the dangers against the advantages of disclosure, rather than as a problem to be decided on the basis of possible adverse effects alone.

Further impetus to disclosure is provided by the directive that the Commission maintain a continuous review of restricted data and of classification guides to determine what information could be declassified "without undue risk" to security (Sec. 142b). The effectiveness of this directive may depend largely on the role of the Joint Committee in seeing to it that the continuous review by the Commission does in fact occur.

Personnel Security Clearance. The second important change affecting private activity in this chapter is the relaxation of clearance requirements for access to restricted data.

CONTRACTORS AND LICENSEES. Under the Act, the requirement of personnel security clearance for those with access to classified atomic information applies equally to employees working under a Commission contract and to licensees of the Commission. There is to be an investigation of character, association, and loyalty, by the Civil Service Commission or the Federal Bureau of Investigation, and a determination by the AEC that clearance would not endanger the common defense and security (Sec. 145a).

A new provision, however, authorizes the Commission to establish standards as to the scope and extent of background investigations (Sec. 145f). These standards are to be based on the location and kind of work to be done and are to take into account the degree of security importance of the restricted data to which access will be permitted. This provision makes it possible for the Commission to relate the degree of security clearance required to the security importance of the information involved.

L-CLEARANCE AND ACCESS PERMITS. On the basis of this authority, the Commission has established a category of information and a corresponding type of clearance designed to give industry greater access to classified information in the reactor field. The Commission is in the process of down-grading, from "secret" to "confidential," reactor technology information which it is not prepared to declassify altogether [13]. For access to such information a limited, or "L" clearance is required, which involves only a file check of government records, rather than a full field investigation of character, association, and loyalty, as in the case of "Q" clearance. The latter is still required for those with access to secret information and for direct employees of the AEC.

In April, 1955, the Commission established a new program for access to reactor technology information, thereafter formalized and amplified by Commission regulation [4b]. Under this program, the Commission provides access to "confidential" restricted data relating to nonmilitary uses of atomic energy to anyone who can demonstrate that he has a potential use or application for such data in his business, trade, or profession. The applicant must then obtain an L clearance and agree in writing to conform to AEC security regulations and must submit proposed procedures for safeguarding information for Commission approval. These permit holders are required to reimburse the Commission for cost of clearance and for the publication and other services furnished by the Commission under the permit.

Access to "secret" restricted data requires detailed evidence of actual need for the information, a listing of the specified categories of information required, and a statement of its proposed use in specific projects. A full AEC Q clearance is thereafter required.

In both types of access permits, the permitees may obtain access to information, within the scope of their permit, from private persons as well as from the AEC and its contractors. The confidential permit holder may permit any cleared individual in his own organization—or outside—to have access to such information. Secret data, on the other hand, may be made available only to those cleared individuals within the organization who require access to it in the performance of their duties. Each access permit is issued for a 2-year term and is subject to renewal.

Criminal Provisions. The criminal provisions for the unlawful handling of restricted data (Secs. 221–228) are largely unchanged in the new Act, providing punishments ranging from $10,000 or 10 years' imprisonment to life imprisonment or death, depending on the nature of the act or the intent with which the act was performed. The provisions on tampering with restricted data with intent to injure the United States or benefit a foreign nation have been extended. They now include data used in connection with the production of special nuclear material, or with research and development financed in whole or in part by Federal funds "or conducted with the aid of special nuclear material" (Sec. 226).

Section 227 newly provides for a fine of $2500 for anyone authorized to have restricted data who, having reason to believe the information is classified, communicates it or conspires to communicate it to an unauthorized person, or receives it from him, knowing or having reason to believe that the person is not authorized to have access to such information.

Other important changes in the information section, relating to international activities and to AEC–Defense Department relationships, will be dealt with below in the context of the respective topics.

15-6. International Activities. Under the McMahon Act, private international activities in nuclear engineering were largely prohibited. Production facilities could not be exported, except for certain component parts under a specific AEC license. Export of fissionable material was forbidden, and no one could engage directly or indirectly in the production of fissionable material abroad. Except for the limited governmental exchange of information permitted by an amendment in 1951, the old Act prohibited the international exchange of classified information in the absence of congressionally approved international controls over atomic weapons.

Agreements for Cooperation. The new Act extends the area of possible international cooperation by authorizing governmental "agreements for cooperation" with foreign nations (Secs. 144a, 123). They must be recommended by the AEC and approved by the President with a determination that the performance of the agreement would promote, and would not constitute an unreasonable risk to, the common defense and security. The agreements must then be submitted to the Joint Committee on Atomic Energy for 30 days, while Congress is in session. These agreements must include a guarantee that security safeguards will be maintained, that no material transferred will be used for military purposes, and that no restricted data will be transmitted to unauthorized persons or transferred beyond the cooperating party's jurisdiction.

Under such agreements, the Commission may communicate restricted data that do not relate to weapon design or fabrication.* The Commission may further distribute special nuclear material and production and utilization facilities.

Private Activities. The Commission has said that it expects that industry will play a major role in implementing these agreements by

---

* Restricted data that may be transmitted include information on refining, purification and subsequent treatment of source material, reactor development, production of special material, health and safety, and industrial applications of atomic energy (Sec. 144a).

supplying equipment, facilities, services, and materials (other than special nuclear material) [14]. However, if the transactions involve the communication of restricted data, they must be approved by the Commission.

Apart from agreements for cooperation, private export of component parts of production and utilization facilities and of source materials and of radioisotopes may be authorized pursuant to a Commission license. Section 57a(3) of the new Act modifies the absolute prohibition on engaging directly or indirectly in the production of special nuclear material abroad, by making it possible for the Commission to authorize those activities which it determines would not be inimical to the national interest.*

The Commission issued a regulation in 1956 [4j, Sec. 110.7] which included a Commission determination that private activities in the atomic-energy field abroad would not be inimical to the national interest if they were limited to noncommunist countries and did not involve the communication of restricted data or classified defense information. Thus, unclassified sales and promotional activities abroad are authorized. Specified activities, however, require a report to the Commission within 30 days of their occurrence. These activities include design, construction or operation of reactors, processing and fabricating facilities, or assistance in such design, etc. Transactions involving restricted data, special nuclear material, and completed facilities require a governmental agreement.

International Arrangements. Agreements for cooperation, in the new law, are to be distinguished from multilateral arrangements, such as the international atomic agency. These are labeled "international arrangements" and are defined to include treaties (ratified by the Senate) and other international agreements approved by both houses of Congress (Secs. 11k, 124).

15-7. AEC–Defense Department Relationships. *Weapons.* The Commission maintains its authority to conduct experiments and to do research and development work in the military application of atomic

---

* The role of industry in agreements for cooperation was further described in the proposed agreement between the United States and Canada, approved by the President and signed by representatives of Canada on June 15, 1955 [15]. Article V of the proposed agreement states that it is contemplated that private persons in each country may deal directly with each other. They may make arrangements "to transfer and export materials, including equipment and devices, to, and perform services for, the other government" and for persons authorized by the other government to receive the materials and use the services. Excluded from such arrangements are materials, equipment, or devices deemed by the AEC to be primarily of military significance. Private activities are subject to applicable laws of the respective governments, and governmental approval is required when the materials or services are classified or when providing them requires the communication of classified information.

energy, and to engage in the production of atomic weapons to the extent that the President may annually direct (Sec. 91a). "Atomic weapon" is newly defined to mean any device utilizing atomic energy as a weapon, weapon prototype, or weapon test device, exclusive of the separable or divisible means for transporting or propelling the device (Sec. 11d). According to the Committee report [16], the definition "specifically excludes airplanes, submarines, or rockets which may carry the weapons, unless the propulsive power is an integral part of the weapon itself."

The President may direct the Commission to transfer special material or atomic weapons to the Department of Defense. He may also authorize the latter to manufacture or acquire any atomic weapon or utilization facility for military purposes, but not to produce special material except as it is incidental to the utilization facility's operations (Sec. 91b).

Military Agreements for Cooperation. The information chapter of the Act authorizes the Department of Defense to enter into agreements for cooperation (Sec. 144b) under procedures similar to the cooperative agreements initiated by the AEC. The Defense Department agreements may be with foreign nations or with regional defense organizations. They may involve transmission of restricted data necessary for the development of defense plans, the training of personnel in using and defending against atomic weapons, and the evaluation of the atomic-weapon capabilities of potential enemies.

Defense Information. The information chapter of the Act also makes provision for the joint handling of atomic military information by the AEC and the Department of Defense. Section 142c authorizes the declassification of data jointly determined by the two agencies to relate primarily to the military utilization of atomic weapons, which they jointly decide can be published without undue risk to the national security. Furthermore, provision is made for the transfer of classified information within this military utilization area from the "restricted data" category to the category of "defense" information, upon a joint determination that it can be adequately safeguarded as such (Sec. 142d).

Supplementing these provisions on the handling of information, the Act authorizes the Commission to provide access to restricted data for authorized Department of Defense personnel who have been cleared by the customary security procedures of the Defense Department, which are found by the Secretary of Defense to be in reasonable conformity with Commission standards (Sec. 143).

15-8. Patents. *McMahon Act.* The McMahon Act eliminated patent rights in atomic-energy inventions to the extent that the inventions were useful in the production of atomic weapons or the production of fissionable material. It provided a system of just compensation and awards, applicable to patents revoked or forbidden. Patents were permitted for inventions relating to the utilization of atomic energy to the extent that they were not used in fissionable material or weapon production. However, these inventions could be declared affected with the public interest and a system of compulsory licensing imposed by the Commission.

In operation, patent rights of Commission contractors were determined less by the Act than by patent clauses written into the Commission contracts.* These clauses gave the government extensive rights in almost all inventions resulting from work under the contract. The outcome has been the issuance by the AEC of several hundred royalty-free nonexclusive patent licenses covering patents for unclassified inventions.

Inventions Newly Patentable. The new law continues to forbid patents on inventions in the weapons field but permits them on inventions useful in the production, as well as the utilization, of special material (Secs. 151a–c).

Government Rights in Inventions. Section 153 provides for a new system of compulsory licensing, during the next five years, applicable to the enlarged area now subject to private patents. The patent chapter also contains a provision asserting the government's rights in inventions made under government arrangements, and seemingly extends these rights to a vast area of private activity. Section 152 states that unless the Commission waives its claim, any invention in the atomic field "made or conceived under any contract, subcontract, arrangement, or other relationship with the Commission, regardless of whether the contract or arrangement involved the expenditure of funds by the Commission, shall be deemed to have been made or conceived by the Commission."

PURPOSE. Section 152 was initially proposed as a substitute to the compulsory licensing provision, but in the end both sections were included. Their common purpose was to prevent those already engaged in atomic work from gaining an unfair advantage and a controlling patent position [18].

APPLICATION. The requirements for compulsory licensing are stringent, and the authority is regarded as a reserve power to be invoked only in exceptional circumstances [19]. Section 152, however, has potential application to a broad area of arrangements with the AEC.

The Commission, however, has recently issued waivers of rights under this section for (1) inventions made or conceived as a result of access to restricted data, (2) licenses issued by the AEC, and (3) inventions made as a result of the use of materials sold, distributed, leased, or otherwise made available by the AEC.†

---

* Described by the AEC at the 1955 Hearings [17].

† See p. 105 of the Twenty-first Semiannual Report of the AEC, January, 1957.

15-9. Organization. The basic organizational structure of the Commission and of the various committees is preserved under the new law with a number of changes in detail (Secs. 21–28). In providing for a civilian commission of five members as before, the Act newly describes the role of the chairman, appointed by the President, as one of presiding at all Commission meetings, acting as official spokesman of the Commission, and, on the Commission's behalf, seeing to the faithful execution of its policies and decisions. Each member of the Commission, however, is to have equal responsibility and authority in all Commission decisions and acts.

Eleven program divisions are permitted by law, including a Division of Military Application, headed by an active member of the armed forces. The Commission established a new Division of Civilian Application, responsible for the administration of the licensing provisions of the Act. A statutory Inspection Division is provided for, responsible for gathering information as to whether contractors, licensees, and Commission employees are complying with the provisions of the Act and with the Commission's rules and regulations. Other divisions of importance in the reactor field are the programmatic Divisions of Research and of Reactor Development.

The General Advisory Committee of nine members continues in effect to advise the Commission on scientific and technical matters. The organization and operation of the Military Liaison Committee is somewhat altered to point up the responsibilities of the Defense Department itself, acting through its representatives on the MLC.

The role of the Joint Committee on Atomic Energy is amplified (Secs. 201–207). During the first 30 days of each session, the Committee is directed to conduct hearings on the state of the atomic-energy industry. It is to be kept fully and currently informed by the Commission on all its activities, and also by the Department of Defense on all its activities relating to atomic energy.

REFERENCES

1. 68 Stat. 919, 42 U.S.C. §§2011–2218 (Supp. 11, 1955).
2. 60 Stat. 755, 42 U.S.C. §§1801–1819 (1952).
3. The most extensive and authoritative analysis of the Act of 1954 is H. S. Marks and G. F. Trowbridge, "Framework for Atomic Industry," Bureau of National Affairs, 1955. Symposia on various legal aspects of atomic energy are presented in 21 *Law & Contemp. Prob.* (winter, 1956) and in 34 *Texas L. Rev.* (June, 1956).
4. The following regulations have been published in final form:
   a. Rules of Practice, 21 Fed. Reg. 804 (1956), adding 10 C.F.R. pt. 25.
   b. Access to Restricted Data, 21 Fed. Reg. 810 (1956), adding 10 C.F.R. pt. 30.
   c. Licensing of Byproduct Material, 21 Fed. Reg. 213 (1956), amending 10 C.F.R. pt. 30 (Supp. 1955).

d. Licensing of Production and Utilization Facilities, 21 Fed. Reg. 355 (1956), amending 10 C.F.R. pt. 50 (Supp. 1955).

e. Operators' Licenses, 21 Fed. Reg. 6 (1956), adding 10 C.F.R. pt. 55.

f. Special Nuclear Material Regulations, 21 Fed. Reg. 764 (1956), amending 10 C.F.R. pt. 70 (1949).

g. Procedure on Applications for Determination of Reasonable Royalty Fee or Other Compensation for Patents, Inventions or Discoveries, 10 C.F.R. pt. 80 (Supp. 1955).

h. Standard Specifications for the Granting of Patent Licenses, 21 Fed. Reg. 606 (1956), adding 10 C.F.R. pt. 81.

i. Safeguarding of Restricted Data, 21 Fed. Reg. 718 (1956), adding 10 C.F.R. pt. 95.

j. Unclassified Activities in Foreign Atomic Energy Programs, 21 Fed. Reg. 418 (1956), adding 10 C.F.R. pt. 110.

k. Standards for Protection against Radiation, 20 Fed. Reg. 5101 (1955), adding 10 C.F.R. pt. 20.

A compilation of the Commission's regulations, published in final form, is contained in the Twentieth Semiannual Report of the Atomic Energy Commission, pp. 169–252, July, 1956.

5. Price, Harold L. (Deputy General Counsel and Director of the Division of Licensing): speech to the Atomic Industrial Forum, AEC Release, May 24, 1955. See also AEC regulation on the licensing of facilities [4d, Sec. 50.34].

6. AEC Press Release 590, Jan. 10, 1955.

7. "Hearings before the Joint Committee on Atomic Energy, Development, Growth and State of the Atomic Energy Industry," 84th Cong., 1st Sess., 1955, pt. 1.

8. "Report of the Joint Committee on Atomic Energy" (H. R. Rep. 2181 and Sen. Rep. 1699), 83d Cong., 2d Sess., 1954, p. 18.

9. Ref. 8, p. 9.

10. Ref. 7, pp. 89ff.

11. Ref. 7, p. 6.

12. Ref. 7, pp. 152–153.

13. Ref. 7, p. 5.

14. Ref. 7, p. 208.

15. Department of State and AEC Release, June 20, 1955.

16. Ref. 8, p. 11.

17. Ref. 7, p. 193.

18. Ref. 8, p. 95.

19. Ref. 7, p. 194.

APPENDIXES

APPENDIX A

PHYSICAL PROPERTIES OF PRINCIPAL NONMETALLIC HEAT-TRANSFER LIQUIDS OF SEVERAL TYPES

Fluid	Boiling point, °C	Melting point or pour point, °C	Temperature, °C	Density, g/cm³	Viscosity, centipoises	Specific heat, cal/(g)(°C)	Thermal conductivity, Btu/(hr)(ft)(°F)	Prandtl No.	Vapor pressure, psia	Thermal-neutron absorption cross section, barns	Approximate cost for drum lots, $/lb	Corrosiveness	Stability
Anisole	153.8	-37.3	0	1.009	1.530	0.44 at 28°C	......		Negligible	......	0.80	Noncorrosive to mild steel	Stable to heat to at least 250°C
			100	0.913	0.419			~10⁶	2.72				
Aroclor 1248 (chlorinated biphenyl and polyphenyl)	340–375	-7*	0	1.48	~10⁵	0.270	0.0671		Negligible	......	0.19	Noncorrosive to metals; plastics have poor resistance	Stable to 300°C, resistant to acid and alkali
			100	1.37	4.2	0.295	0.0695	43.2	Negligible				
			200	1.28		0.328			0.348				
			300	1.17		0.360			7.17				
Aroclor 1268 (chlorinated biphenyl and polyphenyl)	435–450	......	0	1.807 at 25°C					Negligible	......	0.19	Noncorrosive to metals; plastics have poor resistance	Resistant to high temperature, acid, and alkali
			100	......									
1,2,4-Trichlorobenzene	205–235	10 (max)	0	1.463	0.746	0.20 at 25°C	......		Negligible	96.99	0.125	Noncorrosive to iron	Stable to 300°C
			100	1.31					0.445				
Dowtherm A. 73.5% $(C_6H_5)_2O$, 26.5% $(C_6H_5)_2$	......	12.0	0	1.31		0.45				3.26	......	No effect on common metals; steel recommended	Thermally stable to 372°C
			100	0.998	1.00	0.56	0.074	1800	0.1				
			200	0.908	0.45	0.66	0.079	863	3.54				
			300	0.809	0.35	0.68		7.6	34.9				
			400	0.689	0.30			6.2	161				
Perfluorocyclic ether, $C_8F_{16}O$	101	-100*	0	1.825	2.41	0.26 at 25°C	0.119	12.7		0.198	......	Noncorrosive to metals and plastics	Chemically inert to 500°C
			100	1.55	0.465				14.7				
Heptacosafluorotri-butylamine, $(C_4F_9)_3N$	177	-50*	0	1.922	10.6	0.27 at 25°C	......		Negligible	2.02	......	Will not attack common metals to several hundred °C; inert to plastics	Chemically stable to 500°C
			100	1.701	0.918				0.92				
			200	1.48	0.252								

Material	BP (°C)	Pour point (°C)	Temp (°C)	Density	Viscosity (cstk)	Sp. heat	Thermal cond.		Vapor pressure		Cost, $/lb	Corrosion	Thermal stability
HB-40 (partially hydrogenated mixture of isomeric terphenyls)	362	−25*	0	1.003 at 25°C	490	…	0.07	…	Negligible	…	0.180 in 400-lb drums	…	Thermally stable at 372°C in the absence of air
			100	0.96	3.9	…	0.07	…	Negligible 15 mm				
			200	0.89	0.9	…	0.07	…	230 mm				
			300	…	0.4	…	…	…	…				
Circo XX (Sun) naphthenic petroleum	…	−12 (max)*	100	0.895	14	0.495	0.068	247	(220°C flash point)		…	Noncorrosive	Copper catalyzes oxidation
			200	0.840	2	0.58	0.064	44					
			300	0.785	1	0.662	0.060	27					
Circo XXX naphthenic petroleum	…	−4 (max)*	100	0.897	30	0.50	0.068	534	(270°C flash point)		…	Noncorrosive	Copper catalyzes oxidation
			200	0.847	2.7	0.59	0.064	58					
			300	0.790	1	0.687	0.060	28					
Meproline (refined mineral oil)	…	−1*	0	0.925	41.1	0.46 (37°C)	…	…	…		0.104	Noncorrosive	Decomposition temperature above 315°C
			100	0.865	4.8	…	…	…	…				
			200	0.752	…	…	…	…	…				
GE Silicone SF-96 (40)	…	−54*	0	0.984	88.5	0.70 (316°C)	0.0893	…	Negligible		3.77	Noncorrosive to common metals up to 200°C	Unstable above 200°C
			100	0.889	14.7	0.347 at 27°C	0.0895	…					
			200	0.795	5.4	…	…	…					
DC-710 Silicone	…	−22	0	1.10 at 25°C	522–577 at 25°C	…	…	…	Negligible		…	Noncorrosive to metals	Stable at 250°C in air; will not jell or decompose
			100	…	…	…	…	…	Negligible				
			200	…	…	…	…	…					
Ethylene glycol	197.2	−12	0	1.1152	61.2 cstk	0.575 at 20°C	0.1461	583	Negligible	1.93	0.20	Common metals have fair resistance	
			100	…	2.15 cstk		0.1440		0.292				
Ucon LB-300-X (polyalkylene glycol and derivatives)†	…	−40*	0	1.008	700	0.42	0.085	…	Negligible		0.29	Noncorrosive to metals	Stable to 260°C in closed vented systems
			100	0.933	9.6	0.50	0.084	136	Negligible				
			200	…	2.4	0.59	…	…					
Hitec (HTS) 40% NaNO₂, 7% NaNO₃, 53% KNO₃ by weight	…	143	200	1.93	8	0.340	0.251	25.8	Negligible	3.31	0.13	Low below 450°C	Thermally stable to 427°C; slow decomposition above this
			300	1.86	3.15	0.340	0.226	11.2	Negligible				
			400	1.78	1.87	0.340	0.190	7.9	Negligible				
			500	1.72	1.34	…	0.154	7.0	Negligible				
Sodium hydroxide	1390	318	350	1.771	4.0	0.554	0.58	9.2	Negligible		0.04	Extremely corrosive; nickel acceptable	Thermally stable to at least 650°C
			400	1.746	2.8	0.524	0.58	6.1	Negligible				
			450	1.722	2.2	0.494	0.59	4.5	Negligible				
			500	1.698	1.8	0.464	0.71	2.9	Negligible				
KCl-LiCl eutectic (55.1 wt % KCl)	…	352	400	1.69	4.57	0.321	…	…	Negligible		0.60		
			500	1.63	2.36	0.316	…	…	Negligible				

* Pour point.

† LB series available in uninhibited and oxidation-inhibited types in nine different grades from 65 to 1800 Saybolt seconds at 100°F.

PHYSICAL PROPERTIES OF MOLTEN METALS

Metal	Boiling point, °C	Freezing point, °C	Temperature, °C	Density, g/cm³	Viscosity, centipoises	Specific heat, Btu/(lb)(°F)	Thermal conductivity, Btu/(hr)(ft)(°F)	Prandtl No.	Thermal-neutron absorption cross section, barns	Approximate cost, $/lb	Recommended containers
Bismuth..........	1477	271	300	10.03	1.665	0.0343	9.93	0.01395	0.032	2.25	Mo, W, Ta, Be
			400	9.91	1.378	0.0354	8.96	0.01320			
			600	9.66	0.996	0.0376	8.96	0.01015			
Lead.............	1737	327.4	400	10.51	......	0.037	9.20	....		0.13	Mo, Ta, Cb,
			500	10.39	1.88	0.037	8.96	0.0188	0.17		beryllia,
			700	10.15	1.350	....	8.72				quartz
Pb-Bi eutectic	1670	125	200	10.46	......	0.035	5.57	....	....	1.31	Ti-4% Cr
(44.5 wt. % Pb)			300	10.32	......	0.035	6.29	....	0.087		alloy; Be;
			400	10.19	1.48	0.035	7.0				high-Cr steels
Lithium..........	1317	179	200	0.507	0.5644	1.00	26.6	0.0513	67	9.00	Mo, Cb, Ta,
			600	0.474	......	1.00					Armco iron,
			1000	0.441	......	1.00					beryllia
Mercury..........	357	−38.87	100	13.352	1.21	0.03279	5.41	0.01775	360	2.63	Cr, Si, and Ti
			200	13.115	1.01	0.03245	6.19	0.01281			alloy steels
			300	12.881	0.92	0.03234	6.80	0.01059			
Potassium........	760	63.7	200	0.795	0.290	0.1887	26.0	0.00509	2.0	3.66	Stainless
			400	0.747	0.191	0.1826	23.1	0.00365			steels, Ni
			600	0.700	0.150	0.1825	20.5	0.00323			and Ni alloys, Zr
Sodium..........	883	97.8	200	0.903	0.440	0.3166	47.1	0.00715	0.49	0.18	Stainless
			400	0.854	0.269	0.3031	41.2	0.00479			steels, Ni
			600	0.805	0.202	0.2982	36	0.00405			and Ni alloys
Sodium-potassium	784	−11	100	0.847	0.475	0.227	14.1	0.0185	1.27	2.90	Stainless
alloy, 78 wt. %			500	0.751	0.180	0.2095	15.7	0.0058			steels, Ni
potassium			700	0.703	0.146	0.211					and Ni alloys
Sodium-potassium	825	19	100	0.867	0.50	0.255	13.1	0.0235	0.96	1.71	Stainless
alloy, 44 wt. %			500	0.768	0.18	0.235	15.9	0.0064			steels, Ni
potassium			700	0.727	0.15	0.236	15.7	0.0055			and Ni alloys
Tin.............	2270	231.9	400	6.841	1.38	0.061	19.1	0.0107	0.6	0.80	Be, Ti, Cr,
			600	6.709	1.05	0.065	18.4	0.0090			graphite

APPENDIX C

PHYSICAL PROPERTIES OF SATURATED WATER AND STEAM

Subscripts f and g represent liquid and vapor, respectively

| Temperature | | V_f, ft³/lb[a] | μ_f, centipoises[b] | k_f, Btu/(hr)(ft)(°F)[c] | Pr_f | P_{sat}, psia[a] | h_f, Btu/lb[a] | h_{fg}, Btu/lb[a] | V_g, ft³/lb[a] | μ_g, centipoises[b,d,e] | k_g, Btu/(hr)(ft)(°F)[f,g] | Pr_g |
°F	°C											
32	0	0.01602	2.35	0.320	17.966	0.08854	0.00	1075.8	3306	0.0088	0.00914	
50	10	0.01603	1.42	0.334	10.319	0.17811	18.07	1065.6	1703.2	0.0092	0.00958	
68	20	0.01605	1.01	0.347	7.047	0.3390	36.04	1055.5	925.9	0.0096	0.0100	
86	30	0.01609	0.774	0.357	5.246	0.6152	54.00	1045.2	527.3	0.0100	0.0104	
104	40	0.01615	0.625	0.365	4.145	1.0695	71.96	1034.9	313.1	0.0104	0.0109	
122	50	0.01621	0.523	0.373	3.394	1.7888	89.92	1024.6	192.95	0.0108	0.0113	
140	60	0.01629	0.448	0.380	2.857	2.8886	107.89	1014.1	123.01	0.0112	0.0118	
158	70	0.01638	0.391	0.385	2.465	4.519	125.89	1003.5	80.84	0.0116	0.0123	
176	80	0.01648	0.346	0.389	2.162	6.868	143.91	992.7	54.61	0.0120	0.0128	1.089
194	90	0.01659	0.310	0.391	1.931	10.168	161.97	981.6	37.83	0.0124	0.0133	1.083
212	100	0.01672	0.281	0.393	1.746	14.696	180.07	970.3	26.80	0.0128	0.0138	1.077
230	110	0.01684	0.256	0.395	1.588	20.780	198.23	958.8	19.382	0.0132	0.0144	1.087
248	120	0.01699	0.236	0.396	1.466	28.797	216.45	946.8	14.282	0.0136	0.0150	1.097
266	130	0.01714	0.218	0.396	1.360	39.182	234.76	934.5	10.704	0.0140	0.0156	1.108
284	140	0.01730	0.202	0.396	1.267	52.418	253.15	921.8	8.146	0.0144	0.0162	1.119
302	150	0.01747	0.189	0.395	1.195	69.046	271.66	908.6	6.287	0.0148	0.0168	1.151
320	160	0.01765	0.177	0.393	1.131	89.66	290.28	894.9	4.914	0.0151	0.0175	1.190
338	170	0.01785	0.166	0.391	1.076	114.89	309.04	880.6	3.886	0.0160	0.0182	1.255
356	180	0.01806	0.157	0.389	1.032	145.45	327.96	865.6	3.105	0.0164	0.0190	1.295
374	190	0.01829	0.149	0.386	0.999	182.07	347.04	850.0	2.504	0.0169	0.0198	1.343
392	200	0.01853	0.141	0.383	0.959	225.56	366.33	833.6	2.0372	0.0174	0.0206	1.410
410	210	0.01878	0.134	0.379	0.932	276.75	385.83	816.3	1.6700	0.0178	0.0216	1.436
428	220	0.01907	0.128	0.374	0.914	336.60	405.58	798.1	1.3790	0.0183	0.0226	1.509
446	230	0.0194	0.122	0.369	0.896	405.9	425.6	778.9	1.1451	0.0188	0.0241	1.529
464	240	0.0197	0.117	0.363	0.887	485.6	446.0	758.6	0.9559	0.0194	0.0254	1.608
482	250	0.0200	0.112	0.356	0.883	577.0	466.7	736.9	0.8018	0.0200	0.0268	1.680
500	260	0.0204	0.107	0.349	0.882	680.8	487.8	713.9	0.6749	0.0205	0.0275	1.804
518	270	0.0209	0.103	0.340	......	798.6	509.5	689.2	0.5701	0.0211	0.0290	1.919
536	280	0.0213	0.0991	0.330	......	931.1	531.6	662.8	0.4824	0.0217	0.0319	1.959
554	290	0.0219	0.0955	0.320	......	1079.8	554.4	634.2	0.4088	0.0223	0.0326	2.185
572	300	0.0225	0.0922	0.309	......	1246.1	578.1	603.1	0.3464	0.0230	0.0347	2.390
590	310	0.0232	0.0890	0.298	......	1431.2	602.8	569.0	0.2931	0.0237	0.0370	2.635
608	320	0.0240	0.0861	.....	......	1637.2	628.7	531.1	0.2474	0.0245	0.0396	2.950
626	330	0.0251	0.0833	.....	......	1865.5	656.0	488.9	0.2074	0.0255	0.0422	3.451
644	340	0.0263	0.0807	.....	......	2118.3	685.3	440.6	0.1724	0.0266	0.0430	4.431
662	350	0.0280	0.0782	.....	......	2398.2	718.2	383.1	0.1409	0.0280	0.0463	5.708
680	360	0.0305	0.0759	.....	......	2708.1	757.3	309.9	0.1115	0.0297	0.0513	7.832
698	370	0.0359	0.0737	.....	......	3053.4	814.5	191.5	0.0802	0.0306[d]	0.0678	
705.4	374.1	0.0503	0.0728	.....	......	3206.2	902.7	0	0.0503	0.0503	0.0497[e]	0.0827

[a] J. H. Keenan and F. G. Keyes, "Thermodynamic Properties of Steam," John Wiley & Sons, Inc., New York, 1936.
[b] G. A. Hawkins, W. L. Sibbitt, and H. L. Solberg, Dynamic Viscosity of Water and Superheated Steam, *Trans. ASME*, **70:**19–23 (1948).
[c] E. F. M. Van der Held, and F. G. Van Drunen, A Method of Measuring the Thermal Conductivity of Liquids, *Physica*, vol. 15, no. 10, October, 1949.
[d] N. E. Dorsey, "Properties of Ordinary Water-Substance," Reinhold Publishing Corporation, New York, 1940.
[e] F. G. Keyes, The Viscosity and Heat Conductivity of Steam, *J. Am. Chem. Soc.*, **72:**433 (1950).
[f] B. W. Gamson, A Generalized Thermal Conductivity Correlation for Gas State, *Chem. Eng. Progr.*, February, 1949, p. 154.
[g] F. G. Keyes, "Summary of Measurements of Heat Conductivity," Massachusetts Institute of Technology, Division of Industrial Cooperation, October, 1949.

EXTRAPOLATED WATER DATA AT 5000 PSI*

Temperature, °F	Viscosity, lb/(hr)(ft)	Thermal conductivity, Btu/(hr)(ft)(°F)	Enthalpy, Btu/lb	Density, lb/ft³	Prandtl No.
400	0.350	0.400	381	55.26	0.860
450	0.303	0.3865	435	53.14	0.840
500	0.273	0.3705	489	50.95	0.820
550	0.249	0.3494	546	48.47	0.834
600	0.227	0.3232	605	45.65	0.875
650	0.205	0.2856	670	42.02	0.995
700	0.181	0.2356	748	37.26	1.331
720	0.169	0.2080	785	34.76	1.653
740	0.155	0.1730	830	31.53	2.33
760	0.135	0.1365	887	27.16	3.35
770	0.1215	0.1237	924	24.31	3.88
780	0.109	0.1140	964	21.60	4.10
790	0.0975	0.1064	1008	19.04	3.71
800	0.0900	0.1000	1047	16.95	3.16
820	0.0820	0.0904	1109	14.07	2.40
840	0.0810	0.0833	1156	12.33	2.06
860	0.0812	0.0775	1194	11.22	1.879
880	0.0818	0.0723	1226	10.39	1.760
900	0.0823	0.0680	1255	10.01	1.719
950	0.0844	0.0606	1319	8.475	1.542
1000	0.0869	0.0556	1369	7.685	1.489
1100	0.0925	0.0505	1455	6.625	1.444
1200	0.0982	0.0488	1529	5.896	1.438
1300	0.1041	0.0488	1599	5.369	1.432
1400	0.1100	0.0497	1665	4.950	1.440
1500	0.1158	0.0514	1730	4.587	1.455
1600	0.1216	0.0536	1794	4.290	1.470

* Courtesy of K. Goldmann, Nuclear Development Corporation of America.

PHYSICAL PROPERTIES OF HEAVY WATER, D₂O

Molecular weight, 20.03; melting point, 3.81°C; boiling point, 101.43°C

Temperature, °C	Density, g/cm³	Viscosity, centipoises	Specific heat, cal/(g)(°C)	Thermal conductivity, Btu/(hr)(ft)(°F)	Prandtl No.	Vapor pressure, psia	Latent heat of vaporization, Btu/lb
10	1.1060	1.685	1.010	0.327	12.59	0.151	989.0
20	1.1054	1.260	1.006	0.338	9.08	0.294	
30	1.1032	0.972	1.005	0.346	6.83	0.542	
40	1.0999	. . . . .	1.004	0.353	. . . . .	0.954	956.7
50	1.0957	. . . . .	1.004	0.359	. . . . .	1.617	
60	. . . . . .	. . . . .	. . . . .	0.363	. . . . .	2.64	935.8
80	. . . . . .	. . . . .	. . . . .	0.370	. . . . .	6.41	914.4
100	. . . . . .	. . . . .	. . . . .	0.374	. . . . .	13.97	892.2
120	. . . . . .	. . . . .	. . . . .	. . . . .	. . . . .	27.8	869.0
160	. . . . . .	. . . . .	. . . . .	. . . . .	. . . . .	88.1	818.5
200	. . . . . .	. . . . .	. . . . .	. . . . .	. . . . .	224.3	760.9
240	. . . . .	. . . . .	. . . . .	. . . . .	. . . . .	487.6	
371.5*	0.363	. . . . .	. . . . .*	. . . . .	. . . . .	3213	0

* Critical temperature.

THERMAL PROPERTIES OF SOLIDS

Substance	Melting point, °F	Thermal conductivity, Btu/(hr)(ft)(°F)		Density, lb/ft³	Linear thermal expansion (68–212°F), (°F)⁻¹ × 10⁻⁶	Specific heat (32–212°F)	Approximate base price of tubes, $/lb	Thermal-neutron macroscopic capture cross section, cm⁻¹
		212°F	932°F					
Metals:								
Admiralty metal....	1715	63	......	532	11.2	0.09	1.175	0.26
Aluminum (Alcoa 2S, 99%+)......	1195–1210	106	......	169	13.1	0.22	0.56	0.018
Beryllium..........	2400–2460	85	......	115	6.9	0.5	175.00	0.0012
Boron.............	3600+	......	......	146	4.6	0.36	55.00^{a,b}	104
Cadmium...........	610	53	......	540	17.5	0.056	1.70^a	111
Carbon (AGOT graphite; blocks 4.37 × 4.37 × 51 in.)	5500	50–120	60 (new), 30 (aged)	103	0.8 (lengthwise) 1.5 (transverse)	0.39	0.44^a	0.00045
Copper, 99.9%.....	1980	218	207	556	9.1	0.092	0.577	0.36
Hastelloy "B".....	2410–2460	6.5	......	577	5.6	0.091	4.62	0.35
Inconel (wrought)..	2540–2600	8.7	13.8	530	6.4	0.11	1.525	0.36
Magnesium (Dow-metal J-1)........	1145	46	......	112	16	0.25	....	0.018
Monel (wrought)...	2370–2460	15	......	551	7.8	0.13	1.28	0.38
Nickel (pure), 99.99%........	2650	48	36	556	7.2	0.112	1.40	0.40
Phosphor bronze, 5%	1920	47	......	553	9.9	0.09	....	0.32
Sodium............	208	(73)	......	59	35	0.30	0.18^c	0.012
Steel, carbon, SAE 1020.............	2760	29.5	22.6	491	6.5	0.116	0.15	0.22
Steel, chrome, 5%..	2700–2780	15.4	15.4	501	7.1	0.12	0.40	
Steel, stainless, type 304 (austenitic) 18-8S...........	2550–2650	9.4	12.4	501	8.0	0.12	1.37	0.24
Steel, stainless, type 440 (martensitic)	2700–2790	13.3	15.3	479	5.1	0.15	1.15	0.23
Steel, stainless, type 430 (ferritic).....	2650–2700	12.8	13.7	475	5.1	0.15	1.20	0.23
Tantalum, 99.9%...	5391	31	......	1037	3.57	0.033	65.00	1.1
Titanium..........	3020	8.8	7.9	283	4.7	0.126	5.00^d	0.29
Tungsten..........	6152	96	......	1224	2.2	0.034	....	1.21
Zirconium (also Zircaloy)^e........	3353	8.3	6±	406	3.3	0.072	10–15^d	0.0076
Oxides and carbides:								
Aluminum oxide....	3720	16.2	5.8	250	4.3	0.20	0.04	0.012
Magnesium oxide...	5070	19.7	7.5	228	6	0.23	0.36	0.0032
Beryllium oxide....	4350–4550	121	36.3	175	......	0.26	....	0.00060
Boron carbide......	4450	3.6	6	153	2.5	0.31	8.00	70
Concrete (1–5 mix with sand and marble)..........	......	0.78	......	143	......	0.16		
Silicon carbide......	4000	......	7.5	205	2.5	0.20	0.15	
Fuels:								
Plutonium.........	1179							
α (below 243°F)..	......	......	2.8^f	1230	30			
β (243–392°F)....	......	......	6.1^f	1110	19			
γ (392–572°F)....	......	......	7.4^f	1070	20			
δ (572–887°F)....	......	......	9.4^f	1000	−12			
ε (887–1179°F)...	1179	......	9.2^f	1020	14			
Thorium...........	3074		......	724	6.2	0.028	....	0.17
α (below 2515°F)	......	22	26					
β (2515–3074°F)..	3074							
Uranium...........	2064			....	−0.8 to +13	......	35.00^c	0.35
α (80°F)........	......	......	(14.9)	....	7	0.0277		
α (440°F).......	......	......	(17.6)	1190	....	0.0322		
α (800°F).......	......	......	(19.1)		....	0.0379		
α (1160°F)......	......	......	(23)		....	0.0470		
β (1224–1417°F)..	......	......	(26)	1130	....	0.0430		
γ (1417–2064°F)..	2064	......	(26)	1126	10	0.0388		
Uranium alloys (with Zr, Al, etc.)^g								

^{a} Ingots, bars, etc. ^{b} Insoluble calcium borate, $50/ton; soluble borax and boric acid, $100/ton. ^{c} Extruded metal. ^{d} Sponge. ^{e} Zircaloy-2 contains 1.5% Sn, 0.05% Ni, 0.12% Fe, 0.10% Cr, and balance Zr. It is stronger and more corrosion-resistant than pure Zr. ^{f} By Eq. (9-20). ^{g} Physical properties may be roughly estimated by interpolation.

VAPOR PRESSURE OF METALS*

Constants in the equation: $\log P = B - A/T$, for P in millimeters of mercury and T in degrees Kelvin

Element	A	B
W.............	49,300	11.8
Ta.............	45,100	11.3
Nb.............	39,200	11.0
C.............	37,700	10.7
Os.............	37,000	10.7
Ir.............	35,400	10.4
Mo.............	34,700	10.3
Ru.............	33,300	10.2
Th.............	29,400	10.0
Rh.............	29,100	9.9
Pt.............	27,900	9.7
Zr.............	26,300	9.5
U.............	25,000	9.34
V.............	25,000	9.34
Co.............	21,400	9.15
Pd.............	20,000	9.12
Ti.............	19,700	9.00
Ni.............	19,700	9.00
Au.............	18,900	8.82
Fe.............	18,700	8.80
Sc.............	18,200	8.70
Si.............	17,900	8.60
B.............	17,200	8.60
Cu.............	16,600	8.50
Ge.............	16,300	8.45
Be.............	16,000	8.40
Cr.............	15,700	8.37
Sn.............	14,400	8.28
Ga.............	14,000	8.26
Ag.............	13,700	8.25
Al.............	13,300	8.23
Mn.............	12,900	8.20
In.............	12,600	8.18
Pb.............	10,000	8.11
Ba.............	9,900	8.00
Bi.............	9,600	7.95
Sb.............	9,500	7.95
Ca.............	8,650	7.80
Tl.............	8,450	7.75
Sr.............	8,100	7.67
Li.............	7,750	7.65
Mg.............	6,680	7.50
Zn.............	5,640	7.38
Na.............	5,220	7.40
K.............	4,205	6.98
Rb.............	4,250	7.28
Hg.............	2,760	7.20

* Mainly from R. L. Loftness, AEC Report NAA-SR-132, July 10, 1952. These constants are for the metals in liquid form but also hold reasonably well somewhat below the melting point. Most of the graphs of $\log P$ vs. $1/T$ converge at $P = 10^7$ mm and $1/T = 0$.

APPENDIX H

APPROXIMATE THERMAL EMISSIVITY OF NUCLEAR-REACTOR CONSTITUENTS

Constituent	Thermal emissivity		
Polished aluminum	0.04–0.1		
Commercial aluminum and alloys	0.1–0.2		
Oxidized aluminum	0.2–0.3		
Carbon steels	0.4–0.6		
Oxidized carbon steel	0.6–0.9		
Silicon carbide	0.88–0.95		
Stainless steels	0.5–0.7		
Graphite	0.7–0.75		
Molten coolant metals	0.1		
Liquid water	0.95 +		
	At PL, atm-ft,* of		
	0.01	0.1	1
Water vapor:			
1000°F	0.013	0.073	0.28
1500°F	0.008	0.054	0.25
2000°F	0.006	0.040	0.21
Gaseous CO_2:			
1000°F	0.030	0.080	0.14
1500°F	0.027	0.076	0.15
2000°F	0.021	0.063	0.14

* $L = D$ for gas in a pipe, $D_0 - D_i$ for gas in an annulus. Emissivity is negligible for air, H_2, He.

APPENDIX I

AVAILABILITY OF AEC DOCUMENTS

The following depository libraries in the United States receive essentially all non-classified documents published by the AEC, and guides to AEC-developed information. Each library agrees to provide reference and publication services to requesters. Similar libraries are available in many foreign countries.

California
  Berkeley, University of California General Library
  Los Angeles, University of California Library
Colorado
  Denver, Denver Public Library
Connecticut
  New Haven, Yale University Library
District of Columbia
  Washington, Library of Congress
Georgia
  Atlanta, Georgia Institute of Technology Library
Illinois
  Chicago, John Crerar Library
  Chicago, University of Chicago Library
  Urbana, University of Illinois Library
Indiana
  Lafayette, Purdue University Library
Iowa
  Ames, Iowa State College Library
Kentucky
  Lexington, University of Kentucky Library
Louisiana
  Baton Rouge, Louisiana State University Library
Massachusetts
  Cambridge, Harvard University Library
  Cambridge, Massachusetts Institute of Technology Library
Michigan
  Ann Arbor, University of Michigan Library
  Detroit, Detroit Public Library
Minnesota
  Minneapolis, University of Minnesota Library
Missouri
  Kansas City, Linda Hall Library
  St. Louis, Washington University Library
New Jersey
  Princeton, Princeton University Library
New Mexico
  Albuquerque, University of New Mexico Library

813

New York
  Buffalo, Lockwood Memorial Library
  Ithaca, Cornell University Library
  New York, Columbia University Library
  New York, New York Public Library
  Troy, Rensselaer Polytechnic Institute Library
North Carolina
  Durham, Duke University Library
  Raleigh, North Carolina State College Library
Ohio
  Cincinnati, University of Cincinnati Library
  Cleveland, Cleveland Public Library
  Columbus, Ohio State University Library
Oklahoma
  Stillwater, Oklahoma Agricultural and Mechanical College Library
Oregon
  Corvallis, Oregon State College Library
Pennsylvania
  Philadelphia, University of Pennsylvania Library
  Pittsburgh, Carnegie Library of Pittsburgh
Tennessee
  Knoxville, University of Tennessee Library
  Nashville, Joint University Libraries
Texas
  Austin, University of Texas Library
Utah
  Salt Lake City, University of Utah Library
Washington
  Seattle, University of Washington Library
Wisconsin
  Madison, University of Wisconsin Library

Nuclear Science Abstracts is available in most libraries and provides information on availability and prices of AEC reports. Many AEC reports are for sale at OTS; inquiries for official use should go to TIS (see App. J).

APPENDIX J

INITIALS OF REPORTS, LABORATORIES, ETC.

AEC	Atomic Energy Commission, Washington 25, D.C.
AECD	AEC declassified report
AECU	AEC unclassified report
AERE	Atomic Energy Research Establishment, Harwell, Berks, England
ANL	Argonne National Laboratory, Lemont (Chicago), Ill. (operated by the University of Chicago)
ANP	Aircraft Nuclear Propulsion, General Electric Company, Evendale, Ohio
BMI	Battelle Memorial Institute, Columbus, Ohio
BNL	Brookhaven National Laboratory, Upton, Long Island, N.Y. (operated by Associated Universities, Inc.)
HW	Hanford Works, Richland, Wash. (operated by the General Electric Company)
ISC	Iowa State College, Ames, Iowa
K	Gaseous Diffusion (K-25) Plant, Carbide and Carbon Chemicals Corporation, Oak Ridge, Tenn. (operated by Union Carbide Nuclear Company)
KAPL	Knolls Atomic Power Laboratory, Schenectady, N.Y. (operated by the General Electric Company)
LA, LADC	Los Alamos Scientific Laboratory, Los Alamos, N. Mex. (operated by the University of California)
NAA	North American Aviation, Downey, Calif.
NACA	National Advisory Committee for Aeronautics, Washington 25, D.C.
NEPA	Nuclear Energy for the Propulsion of Aircraft project (Oak Ridge, Tenn.)
NNES	National Nuclear Energy Series, McGraw-Hill Book Company, Inc., New York, N.Y.
NP	AEC file designation for non-AEC reports
NRB	Naval Reactors Branch, Department of the Navy, Washington 25, D.C.
NRL	U.S. Naval Research Laboratory, Washington 25, D.C.
NRTS	National Reactor Testing Station, Arco, Idaho (supervised by AEC)
NYO(O)	New York Operations Office (AEC), New York, N.Y.
ORNL	Oak Ridge National Laboratory, Oak Ridge, Tenn. (operated by Union Carbide Nuclear Company)
ORO	Oak Ridge Operations Office (AEC), Oak Ridge, Tenn.
OTS	Office of Technical Services, Department of Commerce, Washington 25, D.C.
TID	Technical Information Division of AEC, Washington 25, D.C.
TIS	Technical Information Service of AEC, Oak Ridge, Tenn.
UCRL	University of California Radiation Laboratory, Berkeley, Calif.
WAPD	Westinghouse Atomic Power Division, Bettis Field, Pittsburgh, Pa.

COSTS OF NUCLEAR MATERIALS AND PROCESSES

The AEC established classified prices effective July 1, 1955, at which it guaranteed to buy and sell nuclear materials for the next seven years, subject to changes in the Bureau of Labor Statistics index.   In August, 1955, it declassified the following prices:

D_2O.. \$28/lb
Natural U_3O_8, purified concentrates (1962+)........ \$8/lb
Natural U metal (0.7% U^{225})..................... \$40/kg
Thorium metal................................. \$43/kg

In November, 1956, and February, 1957, lower prices were announced for enriched U as UF_6:

Wt. % U^{235}	0.72	1.0	1.5	2	3	5	7
\$/total kg	40.50	75.75	145.50	220.00	375.50	698.25	1028.00
\$/g U^{235}	5.62	7.58	9.70	11.00	12.52	13.96	14.68

Wt. % U^{235}	10	20	30	50	80	90	95
\$/total kg	1529.00	3223.00	4931.00	8379.00	13,596.00	15,361.00	16,258.00
\$/g U^{235}	15.29	16.12	16.44	16.76	17.00	17.07	17.13

The AEC charges 4 per cent annual interest on the fuel inventory of any commercial nuclear enterprise.

Other prices:

Zr (average AEC purchases)... \$6.36/lb
Be (average AEC purchases).. \$47/lb
Pu^{239} (for 1962–1963, based on U^{235} price and relative fuel value)........ \$12/g
U^{233} (for 1962–1963, as nitrate)..................................... \$15/g

Fabrication and recovery costs may be taken from unofficial, foreign, and unclassified rough estimates, which are roughly correct within 25 per cent:

Fabrication costs:
Graphite... \$0.25–1.00/lb
Natural or slightly enriched U*....................... \$5/lb
Pu*.. \$10/lb
U^{235} or Pu^{239} in Al-clad rods*........................... \$1/g
U^{235} or Pu^{239} in Zr-clad rods*........................... \$2/g
MTR-type fuel elements, each........................ \$500
Recovery costs:
Pu and U from fuel elements (until 1967, as nitrates)..... \$7.65/lb†

* S. McLain, ANL-5424, 1955.
† AEC Press Release 999, Mar. 7, 1957.

SERIES EXPANSIONS

$$\sin x = x - \frac{x^3}{3!} + \frac{x^5}{5!} - \frac{x^7}{7!} + \cdots \qquad\qquad -\infty < x < +\infty$$

$$\cos x = 1 - \frac{x^2}{2!} + \frac{x^4}{4!} - \frac{x^6}{6!} + \cdots \qquad\qquad -\infty < x < +\infty$$

$$\sinh x = x + \frac{x^3}{3!} + \frac{x^5}{5!} + \frac{x^7}{7!} + \cdots \qquad\qquad -\infty < x < +\infty$$

$$\cosh x = 1 + \frac{x^2}{2!} + \frac{x^4}{4!} + \frac{x^6}{6!} + \cdots \qquad\qquad -\infty < x < +\infty$$

$$e^x = 1 + x + \frac{x^2}{2!} + \frac{x^3}{3!} + \cdots \qquad\qquad -\infty < x < +\infty$$

$$J_0(x) = 1 - \frac{x^2}{2^2} + \frac{x^4}{2^2 4^2} - \frac{x^6}{2^2 4^2 6^2} + \cdots \qquad\qquad -\infty < x < +\infty$$

$$I_0(x) = 1 + \frac{x^2}{2^2} + \frac{x^4}{2^4 (2!)^2} + \frac{x^6}{2^6 (3!)^2} + \cdots \qquad\qquad -\infty < x < +\infty$$

$$K_0(x) = (0.116 - \ln x) + (1.116 - \ln x)\frac{x^2}{2^2}$$

$$+ \frac{1.616 - \ln x}{(2!)^2}\frac{x^4}{2^4} + \frac{1.949 - \ln x}{(3!)^2}\frac{x^6}{2^6} + \cdots \qquad 0 < x < +\infty$$

CROSS SECTIONS OF FUEL ELEMENTS

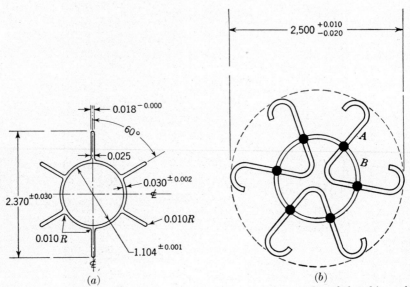

FIG. M-1. (*a*) Cross section of Brookhaven reactor fuel element finned sheathing tube (reduced size). Material is 2S aluminum, extruded, then cold-drawn 15 per cent to size. An anodized layer 0.0008 in. thick is applied internally to prevent U-Al alloying. Natural-uranium slugs 1.0995 ± 0.0005 in. OD are inserted and the sheath collapsed hydrostatically. Helium fills the residual crack. (*D. H. Gurinsky et al., "The Fabrication of Fuel Elements for the BNL Reactor," Paper 828, UN Conference on Atomic Energy, Geneva, 1955.*) (*b*) Cross section of new BNL reactor fuel elements, employing highly enriched U^{235}. Each element is 24.125 in. long and contains three sandwiches *A* bent as shown. Each sandwich contains 5.1 ± 0.5 per cent U as a 2S-Al alloy 0.02 in. ± 10 per cent thick, 2.855 ± 0.01 in. wide, and 22.5 in. long. Cladding is 2S-Al, 0.02 in. ± 10 per cent on each side. Spacing end rings *B* are anodized 61 ST Al, 0.5 in. long and 1.25 in. OD. Thermal power is controlled by the cooling system and remains at 30 Mw, but neutron flux is raised.

FIG. M-2. Cross section of 18-sandwich fuel-element assembly as used in MTR, etc. (slightly enlarged; actual size = 3 × 3 in.). See App. N-2.

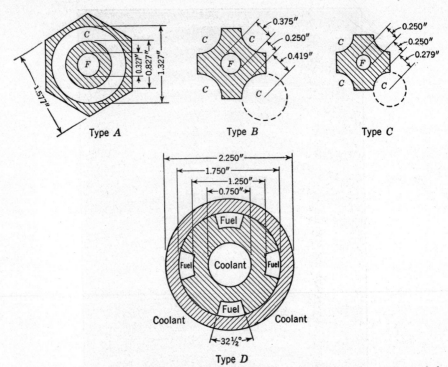

FIG. M-3. Cross sections of four proposed core-element designs for internally cooled LMFR (reduced size). The solid is graphite. Bi-U fuel solution is in channels F and coolant in channels C. (*C. Williams, F. T. Miles, and O. E. Dwyer, Nucleonics, July, 1954.*)

FIG. M-4. Cross section of a "seed" assembly for PWR (see App. N-1). The Zr-U[235] alloy sheets are clad with Zircaloy-2. Sandwich thickness and spacing = 0.080 in.

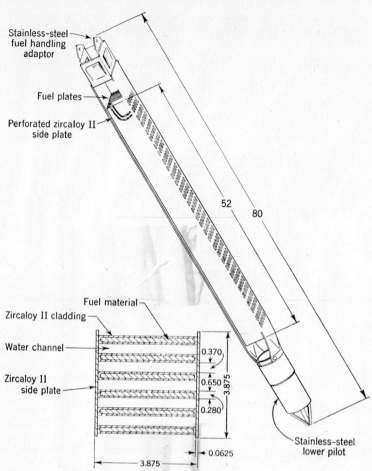

FIG. M-5. Cross section and outside view of an EBWR fuel assembly (see App. N-2).

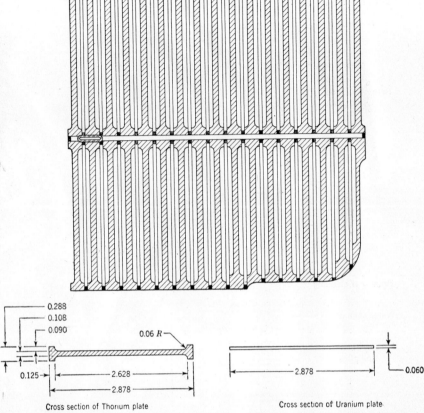

0.288
0.108
0.090
0.06 R
0.125
2.628
2.878
2.878
0.060

Cross section of Thorium plate   Cross section of Uranium plate

FIG. M-6. One fuel assembly considered by Babcock & Wilcox for Consolidated Edison converter and power reactor (dimensions in inches). Ninety assemblies in initial loading. Length of fuel plates is 6 ft 1 in. Maximum heat velocity for a new element is approximately 167,000 Btu/(hr)(ft^2) at the U and 0 at the Th. For a spent element approximately one-third will remain at the U and two-thirds will shift to the Th. The cooling water in the channels is at 1500 psi and 18 fps, entering at 480°F and leaving at 510°F. Total rated flow = 120,000 gpm. Total thermal power of reactor = 1.8 × 10^9 Btu/hr. (In 1957 a new core design was adopted because of radiation-damage data on the above Th and U-Zr plates. The new core employs 97 per cent Th and 3 per cent U^{235} oxides in Zircaloy-2 tubes. The new electrical power attributable to the core is 163 Mw.)

TABULATION OF REACTORS AND PRINCIPAL DATA*

1. PRINCIPAL REACTORS IN OPERATION BEFORE 1956
KEY

1   Thermal power and average thermal flux (unless otherwise designated)
2   Fuel, moderator, and fuel core arrangement
3   Reflector
4   Shielding
5   Over-all size
6   Controls
7   Coolant
8   Remarks (start-up date)

Thermal, Heterogeneous, Graphite-moderated Natural-uranium Reactors

CP-1, Chicago pile, uranium-graphite pile, West Stands, Chicago. **(1)** 200 watts (maximum). **(2)** 6 tons of natural uranium, 40 tons of natural uranium oxide (UO_2 and U_3O_8) (see CP-2). **(3)** 1 ft of graphite. **(4)** None. **(5)** $24\frac{1}{2}$ by $24\frac{1}{2}$ by 19 ft high; 385 tons of graphite. **(6)** 10 rods. **(7)** None. **(8)** World's first reactor (Dec. 2, 1942). After initial operation, assembly dismantled to form basis for CP-2. $1,500,000.

CP-2, Palos Park, Ill. **(1)** 200 watts (up to 2 kw, 3×10^8). **(2)** 3200 uranium metal lumps, $2\frac{1}{4}$ in. in diameter; 14,500 uranium oxide lumps, $2\frac{1}{4}$ in. in diameter; about 50 tons of uranium; lumps spaced on $8\frac{1}{4}$-in.-square lattice; fuel core 18 ft wide, 20 ft long, 19 ft high. **(3)** 1 ft of graphite. **(4)** 5-ft concrete walls; 6 in. of lead; and 4 ft of wood on top. **(5)** 30 ft wide, 32 ft long, $24\frac{1}{2}$ ft high; 472 tons of graphite; total weight about 1400 tons. **(6)** Bronze strips covered with Cd; 2 regulating, 1 shim, 3 safety. **(7)** None. **(8)** (Mar. 20, 1943.) In service until May, 1954. $2 million.

GLEEP, Graphite Low-Energy Experimental Pile, Harwell, England. **(1)** 100 kw; 3×10^{10}. **(2)** 12 tons of uranium bars, 0.9 in. in diameter, 12 in. long, sprayed with 0.003 in. of Al; 21 tons of uranium oxide in Al cans 1.6 in. in diameter, 12 in. long; line lattice with $7\frac{1}{4}$-in. pitch; fuel core 5.24 m long, 2.86 m radius; metal in central portions up to $1\frac{3}{4}$ m radius. **(3)** 2 ft. of graphite in shape of octagon. **(4)** Cubical concrete shield with 5-ft-thick walls. **(5)** Large cube containing 505 long tons of graphite. **(6)** Rods containing Cd; 4 coarse control, 1 fine control, and 6 safety. **(7)** Air, at least 5000 ft³/min at subatmospheric pressure. **(8)** $k_{ex} = 0.002$; maximum uranium cartridge temperature, 60°C (1947).

Oak Ridge, ORNL Reactor, X-10 Pile. **(1)** Increased from 1000 to 3800 kw; 10^{12}. **(2)** Uranium fuel slugs in 0.035-in.-thick 2S Al jackets, 1.1 in. in diameter, 4 in. long, 2.57 lb per slug; 39 to 54 slugs per channel; 1248 fuel channels on 8-in.

* H. S. Isbin, "Nuclear Reactor Catalog," UN Conference on Atomic Energy, Geneva, 1955; 19th and 20th Semiannual Reports of the AEC; etc. See also Chaps. 13 and 14.

rectangular lattice, 821 used (30 tons required for criticality, 54 tons used); diamond-shaped channel, $1\frac{3}{4}$ in. square; fuel core about 20-ft cube. (**3**) 2 ft of graphite. (**4**) 7 ft of concrete. (**5**) 47 ft long, 38 ft wide, 32 ft high; 620 tons of graphite. (**6**) 4 safety (each $0.003k$), 2 shim (each $0.007k$), 2 regulating (each $0.005k$); backup safety consists of boron steel shot; vertical rods $3\frac{1}{2}$ in. square, 8 ft long, contain layers of Cd encased in steel; horizontal steel rods $1\frac{3}{4}$ in. square, 200 in. long, contain $1\frac{1}{2}$ per cent boron. (**7**) Air, 120,000 ft³/min drawn through reactor, filtered, and discharged through 200-ft stack. (**8**) Average highest fuel temperature, ~270°C; average moderator temperature, 125°C (1943). An Al-silicon eutectic is used to bond Al jacket to uranium metal.

BEPO, British Experimental Pile, Harwell, England. (**1**) Increased from 4 to 6 Mw; central thermal flux $\sim 1.4 \times 10^{12}$. (**2**) 0.9-in.-diameter, 12-in.-long uranium bars, encased in Al; 28 tons for criticality, 40 tons for fuel load; 900 central channels used, 20 bars per channel, $7\frac{1}{4}$ in. between channels; cross section of empty channel is $3\frac{1}{2}$ in.²; fuel core 10 ft radius, 20 ft. long. (**3**) About 3 ft of graphite. (**4**) 6-in. cast-iron plate, 6 ft of concrete (barytes aggregate, $\rho = 3.5$) on sides and $7\frac{1}{2}$ ft. on top; 600 tons of steel, 3000 tons of concrete. (**5**) Cubical core of 8 tons of graphite. (**6**) 4 horizontal and 10 vertical boron-carbide-filled hollow steel rods, 2 in. in diameter. (**7**) Air, 260,000 ft³/min, filtered, and discharged from 200-ft stack; stopping power of filters down to 5 μ. (**8**) Fuel elements operate at temperature of 220°C; exit air temperature, 80°C; about 1.2 Mw of heat recovered in a hot-water system for heating a building (first practical utilization of heat) (1948).

Brookhaven, BNL. (**1**) 30 Mw; 4×10^{12}. (**2**) Al-clad uranium metal rods placed on 8-in. centers; 60 tons of uranium; 1369 circular fuel channels, circular cross section of 36 cm²; graphite moderator in two right rectangular prisms, $12\frac{1}{2}$ by 25 by 25 ft, separated by vertical opening for cooling air (see Fig. M-1); normally about 870 channels loaded; being changed to U²³⁵ strip fuel elements. (**3**) 4.5 ft of graphite. (**4**) 6-in. iron plate, $4\frac{1}{4}$ ft of heavy concrete, 3-in. iron plate. (**5**) 38 by 55 by 30 ft high; 20,000 tons (including foundation), 700 tons of graphite. (**6**) Horizontal rods containing boron, entering from two adjacent corners in two arrays of 8 rods each; back-up safeties consisting of boron-steel shot to 4 tubes in graphite, trichlorobenzene to eleven tubes. (**7**) Air, 300,000 ft³/min at subatmospheric pressure; each half of moderator complex cooled separately, filters on inlet and exit air. (**8**) Temperature ranges from 125 to 175°C at experimental openings. Complicated mechanical-electrical system for channel-leak detection; He in fuel element serves to indicate leak by pressure drop, provides an inert atmosphere here for the uranium, and improves heat transfer (1950). Plan to replace natural-uranium fuel with MTR-type fuel plate. $20 million.

Hanford, Hanford, Washington. (**1**) Classified. (**2**) Natural uranium, graphite moderator. (**3**) Graphite. (**7**) Water. (**8**) Three Pu-production reactors in 1945, and five additional units have been added; secondary recirculating glycol system to furnish heat to building. $1 billion.

Hanford 305 Test Reactor, Hanford, Washington. (**1**) 6 watts. (**2**) Uncooled graphite-moderated natural-uranium reactor. (**5**) About an 18- to 20-ft cube. (**6**) Control rod, shim rod, safety rod; BF₃ control system tested. (**7**) None. (**8**) Used to measure reactivity change due to a given amount of sample. Not sealed from atmosphere, and ambient pressure affects reactivity.

Windscale, Sellafield, England. (**1**) Classified. (**2**) Similar to BEPO; natural uranium, graphite-moderated. (**3**) Graphite. (**4**) Biological concrete shield, steel-plate thermal shield.   (**6**) 24 boron-steel control rods, horizontally at right angles to fuel channels.   (**7**) Air, inlet and exit filters, discharge through 400-ft stack.   (**8**) Two Pu-production reactors.

Thermal, Heterogeneous, Heavy-water-moderated Uranium Reactors

CP-3, ANL.   (**1**) 300 kw; 10^{12}.   (**2**) 120 (maximum 136) uranium metal rods 1.1 in. in diameter, 0.035-in. Al jacket, 6 ft long; 3 tons of natural uranium metal; square lattice, $5\frac{3}{8}$ in. centerline to centerline; $6\frac{1}{2}$ tons of D_2O contained in 72-in.-diameter, 105-in.-high Al tank.   (**3**) 2 ft of graphite.   (**4**) 4 in. of Pb-Cd, and 7-ft 8-in. octagonal concrete wall 13 ft high; top shield contains Cd, 1 ft of lead, and 4 ft of wood and steel.   (**5**) Octagonal, 26 ft across, 14 ft high.   (**6**) Rods containing Cd metal; 2 control, 3 shim, and 2 safety.   (**7**) 200 gpm of D_2O.   (**8**) World's first heavy-water reactor (May 15, 1944).   Helium sweep for dissociated D_2O.   Improved version CP-3' with enriched uranium alloyed with Al. $2 million.

CP-3', ANL.   (**1**) 300 kw; 3 to 4 $\times$ 10^{12}.   (**2**) 4.2 kg of U^{235}, about 122 fuel rods 2.16 cm in diameter, 1.68 m long; highly enriched in U^{235}, Al alloy of 98 per cent Al and 2 per cent U.   (**3–7**) See CP-3.   (**8**) Improved design of CP-3 (1950). Replaced by CP-5 in 1954. $2 million.

ZEEP, Zero Energy Experimental Pile, Chalk River, Canada.   (**1**) 3.5 watts; 6 $\times$ 10^6 (30 watts maximum).   (**2**) Natural-uranium metal in form of slugs, 1.285 in. in diameter, 6 in. long, jacketed with Al "stockings," 1.295 $\pm$ 0.005 in. ID, 0.040 $\pm$ 0.003 in. wall, and 9 ft $6\frac{3}{4}$ in. long, holding 9 slugs; total of 148 rods; square lattice spacing of 6 in.; 10 tons of D_2O contained in steel tank $6\frac{3}{4}$ ft in diameter, $8\frac{1}{2}$ ft high.   (**3**) Graphite, $2\frac{1}{2}$ ft thick under tank, 3 ft around.   (**4**) Water in tanks, 3 ft thick.   (**6**) 4 plates, 8 rods, Cd-coated stainless steel.   (**7**) None.   (**8**) Migration area M^2 = 237 cm²; slowing down length, $\sim \sqrt{118.5}$ cm; buckling factor B^2 = 794 $\times$ 10^{-6} cm⁻²; mean life of neutron 0.86 $\times$ 10^{-3} sec (1945).

ZOE, Chatillon, France.   (**1**) Increased from 5 to 150 kw, 2 to 3 $\times$ 10^{10}.   (**2**) Originally, UO_2 in tablets 3 cm high stacked in Al tubes of 66 mm ID and 180 cm effective height; effective density of oxide is 8.3; D_2O (at 40 $\pm$ 6°C) contained in cylindrical Al vat, 181 cm ID, 235.5 cm high; maximum number of fuel bars is 69 (3.55 tons of UO_2), set hexagonally (sides of hexagon, 18.6 cm); uranium oxide replaced by metal rods.   (**3**) Graphite (diffusion length 45 cm), 15 mm of D_2O at bottom.   (**4**) Concrete 150 cm thick.   (**6**) Two sets of 2 types of Cd safety rods; Cd regulating bars.   (**7**) D_2O circulated through an external circulation system.   (**8**) Laplacian = 5.8 m⁻² (1948).

Heavy-water Research Reactor, Russia.   (**1**) 500 kw.   (**2**) Uranium fuel rods suspended in D_2O contained in 175-cm-diameter tank, 195 cm high; critical experiments performed include use of 2.2- and 2.8-cm-diameter rods with a 0.1-cm Al envelope; square lattice spacing from 63 to 162.6 cm; number of rods from 86 to 292; critical D_2O level from 120 to 181.6 cm; rod length $\sim$160 cm.   (**3**) 100-cm-thick graphite reflector on sides and bottom.   (**4**) Concrete side shields 2.5 m thick.   (**6**) 4 Cd control rods.   (**7**) D_2O circulated.   (**8**) He atmosphere above D_2O (1949).

P-2, Saclay, France.   (**1**) 1500 kw, 7 $\times$ 10^{12} (maximum).   (**2**) 136 uranium rods, 1.1 in. in diameter, 7 ft long, triangular lattice with rod spacing of 5.93 in., fuel element consists of 4 concentric Al cylinders with innermost one as protective sheath for U rod, insulating space between second and third cylinders, 3.3 tons of uranium; fuel immersed in Al tank, $8\frac{2}{3}$ ft high and $6\frac{2}{3}$ ft in diameter, partly filled with D_2O (6.3 tons); with heavy water at 1794 mm, critical height $H_c = 224 \pm 3$ cm, critical radius $R_c = 132 \pm 2$ cm; $M^2 = 238$ cm^2, $B^2 = (5.30 \pm 0.15) \times 10^{-4}$ cm^{-2}.   (3) Graphite 3 ft thick outside tank and on sides and bottom.   (4) Cast-iron thermal shield about 8 in. thick, concrete shield about 7 ft.   (6) Cd plate moving between tank and reflector, 2 Cd rods in tank.   (7) Originally, recirculation of nitrogen at 10 atm, flows down space between outer two Al cylinders and up space between first and second cylinders; system now uses CO_2 under 7 atm.   (8) (1952).

JEEP, Kjeller, Norway.   (1) Designed for 100 kw, increased to 350 kw, 10^{12}.   (2) 2200 kg of uranium slugs in Al tubes; 35.5 kg per rod; rods 25.4 mm in diameter, 300 mm long, placed on 180-mm centers; fuel core 1.9 m long, 7 tons of D_2O contained in 2-m-diameter tank; 65 to 76 fuel elements.   (3) Graphite, 700 mm thick.   (4) Octagonal concrete shield, 2 m thick on sides, lined with Cd.   (6) 4 Cd plates 1300 mm long, 350 mm wide, 1.7 mm thick, held between Al plates, placed between tank wall and reflector.   (7) 4 liters/sec of D_2O (inlet 20°C, outlet 40°C).   (8) Average lifetime of neutron 2×10^{-3} sec (1951).

NRX, Chalk River, Canada.   (1) 30 Mw, $>5.8 \times 10^{13}$; rebuilt reactor, 40 Mw.   (2) 176 uranium rods, Al-clad, surrounded by two concentric tubes; uranium rod 1.36 in. in diameter; Al sheath thickness 0.079 in.; outer Al sheath ID 1.66 in., thickness 0.040 in.; uranium assemblies hang vertically; D_2O contained in cylindrical vessel 10 ft high, 8 ft in diameter; core contains about 10.5 tons of natural uranium and about 20 tons of D_2O; each fuel rod is 10 ft long; coolant annulus reduced from 0.100 to 0.071 in.   (3) Graphite.   (4) 8 ft concrete, steel thermal shield.   (5) 34 ft in diameter, 34 ft high.   (6) Varying D_2O level, Cd and boron rods; 18 shutoff rods, 1 control.   (7) Light water flows downward in inner annulus (air flows upward in outer annulus); D_2O circulated to external heat exchanger; air cooling of graphite; 95 per cent of heat removed by water flow in fuel channel.   (8) Helium purge system.   Shut down in December, 1952, following accident and severe contamination; operation restored in February, 1954.   Plutonium and U^{233} are produced (1947). $10 million.

CP-5, ANL.   (1) 1000 kw, 3×10^{13} (1×10^{13} epithermal, 1×10^{12} virgin); design is 4 Mw.   (2) Fuel elements are curved, sandwich plates, 0.05 cm fuel Al-U^{235} alloy clad with 0.05 cm 72S-Al; 12 plates mounted in a 3- by 3- by 24-in. fuel box; normal assembly has 12 fuel boxes and 16 required at 4 Mw; D_2O contained in Al tank 6 ft in diameter, $7\frac{1}{2}$ ft high; fuel core about 2 ft in diameter by 2 ft high; assembly contains about 1680 g of U^{235} and 7 tons of D_2O.   (3) 2 ft of D_2O and 2 ft of graphite around and below tank; zone is 10 ft high and 10 ft in diameter.   (4) $\frac{1}{4}$ in. boral, $3\frac{1}{2}$ in. lead, 4 ft 8 in. limonite-iron concrete ($\rho = 4.4$ g/cm^3).   (5) Octagonal, 20 ft across and $13\frac{1}{2}$ ft high.   (6) 4 shim safety rods (Al-clad Cd), signal-arm type, operating between parallel rows of fuel assemblies; one regulating rod moving up and down outside the fuel assemblies; quick-opening valve to drain D_2O.   (7) About 1200 gpm of D_2O flows up through fuel assemblies [about 1 m/sec in central section; maximum heat flux about 4.5 cal/(cm^2)(sec)] and down in tank with a 5°C rise; D_2O circulated through external, light-water heat exchanger and cooled from 126 to 118°F; chilled-water system to limit temperature rise and to increase reactivity; thermal shield cooled by light water (total shield heat load, including graphite, about 2 to 3 per cent of total); helium purge system graphite reflector cooled with helium.   (8) Reactor housed in an almost leakproof building (1954). $3 million.

SLEEP, Swedish Low-energy Experimental Pile, Stockholm, Sweden.   (1) 300 kw, 3×10^{11} (maximum).   (2) 126 uranium fuel rods, arranged in a hexagonal lattice, spacing 145 mm, 29 mm in diameter, canned in Reflectal (very pure Al alloyed with 0.5 per cent Mg) by forcing through a die, reducing wall thickness of tube from 1.25 to 1.10 mm; D_2O contained in 1.85-m-diameter and 2.54-m-high Reflectal tank. (3) 900-mm-thick graphite surrounding bottom and sides of tank. (4) Biological shield of poured concrete with iron ore, $\rho = 3.8$, 1.8 m thick, lined with a Cd-Al sandwich. (6) Two security rods equivalent to about 2 per cent in reactivity, two regulating plates between reactor tank and graphite reflector control about 0.6 per cent in reactivity. (7) D_2O recirculated, 1000 liters/min, to air-cooled heat exchanger; about 0.001 of circulation passes through mixed-bed ion exchange; air used for cooling, passed upward between reflector and tank and downward between graphite and shield. (8) Reactor built underground. Helium atmosphere above D_2O and circulated through palladium catalyst recombination unit (1954). Effective mean life of thermal neutrons 0.71 ± 0.03 msec.

Thermal, Homogeneous, Light-water-moderated Enriched-uranium Reactors; Water Boilers and Homogeneous Reactor Experiment

LOPO, Low Power Water Boiler, Los Alamos. (1) $\frac{1}{20}$ watt. (2) Enriched-uranium–sulfate solution, 580 g of U^{235}, 3378 g of U^{238}, 534 g of S, 14,068 g of O, 1573 g of H; $\rho = 1.348$ g/cm^3 at 39°C; solution contained in 1-ft-diameter, $\frac{1}{32}$-in.-thick, stainless-steel sphere (15 liters). (3) 3- by 3- by 6-in. bricks of BeO, $\rho = 2.7$; graphite on bottom. (4) None. (6) Cd cylinder, $\frac{3}{4}$ in. in diameter, 34 in. long, Cd safety curtain. (7) Water; maximum temperature, 39°C. (8) World's first water-boiler reactor (1944); replaced with HYPO.

HYPO, High Power Water Boiler, Los Alamos. (1) 6 kw, 3×10^{11}. (2) Enriched-uranium–nitrate solution, 896.6 g of U^{235}, 5,341 g of U^{238}, 731 g of N, 13,780 g of O, 1,312 g of H; 13.65 liters of $UO_2(NO_3)_2 \cdot 6H_2O$ in H_2O, $\rho = 1.615$, contained in 1-ft-diameter, $\frac{1}{16}$-in.-thick stainless-steel sphere. (3) 24 by 24 by 27 in. of BeO surrounded by graphite to form a 60- by 48- by 60-in. rectangular parallelepiped. (4) 4 in. of lead, $\frac{1}{32}$ in. of Cd, 5 ft of concrete. (6) Rods containing Cd; 1 shim, 2 control, 1 safety. (7) 50 g/hr water through 6-turn, $\frac{1}{2}$-in. ID, 157-in.-long coil; maximum solution temperature, 185°F. (8) About 50 cm^3/sec of air used to sweep out H_2 and O_2 from decomposition of H_2O (1944). Replaced by SUPO. $500,000.

SUPO, Super Power Water Boiler, Los Alamos. (1) 45 kw, 1.7×10^{12} (maximum) (estimated maximum intermediate flux, 2.8×10^{12}; fast flux, 1.9×10^{12}). (2) Enriched-uranium–nitrate solution, 88.7 per cent of U as U^{235}, 777 g of U^{235} for critical mass, 870 g of U^{235} used; $\rho = 1.10$, 12,700 cm^3 of solution contained in 1-ft-diameter stainless-steel sphere. (3) About a 55-in. cube of graphite. (4) $\frac{1}{2}$ in. of B_4C in paraffin; 2 in. of steel; 4 in. of lead; 5 ft of concrete. (5) About 15 by 15 by 11 ft. (6) See HYPO; two additional control rods move in reactor core in reentrant thimbles. (7) Water circulated through three 20-ft-long, $\frac{1}{4}$-in. OD, $\frac{3}{16}$-in. ID stainless-steel tubes. (8) 100 liters/min air circulation, recombiner for H_2 and O_2; rate of radiolytic hydrogen evolution about 0.55 moles/kwhr (1951). $500,000.

WBNS, Water Boiler Neutron Source, NAA, Downey, Calif. (1) 1 watt, 5×10^7. (2) 1.5 lb of uranium fuel in form of U^{235}-enriched uranyl nitrate in light-water solution; solution contained in a 1-ft-diameter stainless-steel sphere $\frac{1}{16}$ in. thick. (3) Graphite, 5 ft in diameter and 6 ft high. (4) 2 ft of concrete blocks. (6) Two safety rods (coarse control, fine control) move in reflector. (7) None. (8) Calculated $k_\infty = 1.561$ (leakage 0.360) (1951).

Livermore Water Boiler, Livermore, Calif. (1) 100 watts, 10^9 (increased to 500 watts, 2×10^{10} maximum thermal flux). (2) 14.524 liters of light-water solution of UO_2SO_4 containing 694.2 g of U^{235}; 798 g-moles of hydrogen; 420, oxygen; 5.64, sulfur; and 2.95, uranium; solution contained in stainless-steel sphere, $12\frac{1}{2}$ in. OD, 0.06-in. wall thickness. (3) Right cylinder of graphite, 5 ft in diameter by 5 ft high (Brookhaven graphite with thermal diffusion length of 51.6 cm). (4) Graphite in cylindrical steel tank, 0.030 in. of Cd shot and 5 in. of lead, 3-ft concrete blocks. (5) 13 by 26 by 9 ft high. (6) Two safety and two control rods; rods are flattened stainless-steel tubing packed with 1 lb (control rods) or 2.2 lb (safety rods) of boron carbide. (7) Cooling coil in sphere, six-turn helix of $\frac{5}{16}$-in. ID tubing, 160 in. long, distilled water used as coolant, and refrigeration unit used to reduce solution temperature. (8) Closed gas-handling system (1953).

NCSC, North Carolina State College Reactor, Raleigh, N.C. (1) 10 kw, 5×10^{11}. (2) 14 liters of light-water solution of UO_2SO_4 containing 790 g of U^{235} of 90 per cent isotopic enrichment, $\rho = 1.08$ g/cm³; solution contained in stainless-steel cylinder, $\frac{1}{16}$ in. thick, 11 in. high, $10\frac{3}{4}$ in. in diameter. (3) 20 in. of graphite; 105 ft³, 5.4 tons; $\rho = 1.65$ g/cm³. (4) 6 ft of concrete, barytes ore as coarse aggregate and colemanite ore as fine aggregate, $\rho = 3.4$ g/cm³; 4 to 6 in. of lead around graphite. (5) Octagon, 17 ft across, 12 ft high. (6) Two control rods, stainless-steel tubes of 2.5 g/cm³ sintered B_4C powder, in reentrant sheaths; two shim rods, 4-in.-wide Cd strips, periphery of reactor cylinder. (7) Light water (refrigerated city water), 1 gpm flow through each of 4 helical coils of $\frac{1}{4}$-in. stainless-steel tubing, 7 ft immersion length. (8) Maximum reactor solution temperature, 80°C (1953). $130,000 for the reactor and $500,000 for the building and laboratories.

HRE-1, Homogeneous Reactor Experiment, ORNL. (1) 1000 kw. (2) Enriched uranyl sulfate (enrichment > 90%) dissolved in distilled water; solution contained in an 18-in.-diameter 347 stainless-steel sphere with $\frac{3}{16}$-in. walls; boiling of fuel solution suppressed by maintaining a pressure of 1000 psi (in pressurizer, solution heated to 545°F); maximum fuel temperature, 482°F; fuel consumption 0.002 lb/day at 1000 kw. (3) 10-in. layer of D_2O, pressurized with helium to within ± 100 psi of fuel solution, temperature regulated to 350°F. (4) Reflector and core contained in an outer pressure vessel of forged steel, 39-in. ID with 3-in.-thick wall; 7-ft-thick concrete walls (barium sulfate ore used as aggregate to increase ρ to 3.5). (6) Reactor is self-stabilizing as a result of its large negative temperature coefficient (and mechanical safety devices found to be unnecessary); safety measures include: 2 safety plates fall in 0.01 sec and are completely effective within 0.2 sec, reduce reactor temperature from 250 to 243°C; reflector may be dumped within $6\frac{1}{2}$ sec, drops operating temperature to 160°C; dilution of fuel requires 15 min and is normal shutdown procedure; reduction in power demand and draining fuel solution. (7) Reactor solution circulation, 100 gpm; outlet temperature 482°F and cooled to 410°F in a U-tube heat exchanger, generating steam in the shell side at 3000 lb/hr at 200 psia; generated 150 kw of electricity; canned rotor centrifugal pump for circulating fuel solution; about 50 kw removed by recirculating D_2O reflector through boiler feed-water preheater. (8) Gas evolution ($2H_2 + O_2$) at rate of 10 cfm (STP) and 20 cm³/day of fission-product gases. About 40 kw liberated in flame recombiner, followed by a catalytic recombiner. Radioactive gases absorbed in activated carbon beds. Catalytic recombiner for D_2 and O_2 gases. Radioactivity of fuel solution about 30 curies/cm³. Reactor has been dismantled and to be replaced by HRE-2 (1952).

Thermal, Heterogeneous, Light-water-cooled (and -moderated) Enriched-uranium Reactors

BSR, BSF, Bulk Shielding Facility (typical of swimming-pool types), ORNL. (1) 100 kw, 5×10^{11} (10^{12} maximum available thermal flux, 3×10^{12} maximum available epithermal flux). (2) MTR-type fuel elements: 18 convex plates, each 3 in. wide, 24 in. long and 0.06 in. thick and 0.117 in. apart, comprise a fuel box; plates of enriched uranium ($>90\%$ U^{235}) encased in 2S Al sandwiches, clad with 72S (the latter corrodes preferentially); about 140 g of U^{235} per fuel box; ratio of Al (including structural members) to H_2O is 0.7; fuel boxes held in an Al grid 5 in. deep; critical mass about 3 kg of U^{235} in fuel core 12 by 12 by 24 in.; with allowances for other factors including beam holes, burnup, 3.5 kg; with BeO reflector, 2.4 kg of U^{235} in fuel core 9 by 12 by 24 in. (3) 10 cm of BeO on 4 sides of active lattice; or light water. (4) 16 ft of water above reactor; water and high-density concrete below reactor. (5) Pool 40 ft long, 20 ft wide, and 20 ft deep. (6) Two Cd-Pb safety rods (each equivalent to 3 to 4 per cent reactivity) and a Cd-Pb regulating rod (0.8 per cent reactivity); cross sections of $1\frac{1}{4}$ by $2\frac{1}{2}$ in., and travel inside special fuel boxes; other rods used are two B-Pb shim safety rods (mixture of Pb and boral) and one control rod. (7) Light water, convective flow. (8) Addition of 50 ppm Na_2CrO_4 to inhibit corrosion of Al (1950). \$250,000.

MTR, Materials Testing Reactor, National Reactor Testing Station, Arco, Idaho. (1) 30 Mw, 4×10^{14} (in reflector), 1×10^{14} fast flux. (2) Fuel assembly is a 3 by 3 by 24 in. box containing 18 curved, vertical, fuel plates; plates consist of uranium-Al alloy, 0.5 mm thick, sandwiched between 0.5 mm Al; 3.0-mm spacing between plates; U^{235} per box increased from 140 to 200 g; lattice configuration can be varied from a 3 by 9 (27 fuel assemblies) to a 5 by 9 array; initially about 4 kg of U^{235}; core of fuel assemblies held between 2 grids, and is contained in a 54-in.-diameter Al tank extended to form a well 30 ft deep filled with H_2O; Al-H_2O volume ratio 0.73 (see Fig. M-2). (3) Primary reflector, Be 3 ft high, 54 in. in diameter, in water zone; secondary reflectors, graphite in two zones outside water tank; pebble zone of 700,000 1-in.-diameter balls to form 7-ft 4-in. square by 9 ft high; outer stack of graphite to 12 by 14 ft, 9 ft 4 in. high. (4) Two 4-in. layers of steel thermal shield, 9 ft of barytes concrete, water 15 ft above core and 5 ft below. (5) About a 34-ft cube. (6) Up to 8 vertical shim safety rods, upper section containing Cd steel and lower containing fuel plates; Be-Cd shim rods; 2 vertical regulating rods in Be reflector. (7) Light water recirculated, cooled by flash vaporization, 20,000 gpm, enters at 100°F at top and leaves at bottom at 111°F; about one-sixth of heat generated in graphite and thermal shield, air-cooled; 4 per cent of heat generated in Be reflector. (8) Reactor used for study of materials exposed to intense radiations (1952). About 100 g of U^{235} equivalent to 1 per cent reactivity. Fuel plates are curved laterally on a 5.5-in. radius to strengthen them and to define direction of thermal strain. \$18 million.

RMF, Reactivity Measurement Facility, National Reactor Testing Station, Arco, Idaho. (1) Zero power. (2) 30 MTR fuel assemblies installed in MTR canal. (8) Cold clean reactor used for precise measurements in reactivity effects.

LITR, Low-Intensity Test Reactor, ORNL. (1) Increased from 500 to 3000 kw, 2×10^{13}. (2) 3.4 kg of U^{235}; MTR fuel assemblies (16 plates per fuel assembly). (3) Loosely stacked Be blocks, minimum thickness 8 in. (4) Unmortared concrete blocks, except for outermost 1-ft layer; minimum thickness $10\frac{1}{2}$ ft.

(**6**) Three shim safety rods, containing a cadmium section and a fuel section. (**7**) Light-water recirculation system with water-air heat exchanger to dissipate heat; cellulose filters used remove turbidity of hydroxides (Al and carbon-steel piping); pH maintained between 5.5 and 6.5; bypass line to demineralizer. (**8**) Installation served as mock-up for MTR; converted to a training and research reactor (1950). $1 million.

RFT, Experimental Nuclear Reactor, Russia. (**1**) 300 kw, 2×10^{12}. (**2**) Active core approximated by a cylinder, 30 cm in diameter, 50 cm high; core consists of 32 units with 24 units each containing 16 fuel elements (remaining units for control rods); cylindrical fuel elements, 9 mm OD, 50 cm high, square lattice with ~18-mm spacing; fuel elements contain 10 per cent enriched U^{235}; 3.5 kg of U^{235} in core; light-water moderator, reflector, and coolant; core in 500-mm-diameter, water-filled Al tank, and fuel units fixed at top and bottom in guide grids; Al and Al alloy used as structural materials in active core and for canning fuel elements. (**3**) Water. (**4**) Cast iron and water used as shielding. (**6**) Three borax carbide safety rods, one steel rod. (**7**) 240 m^3/hr water flow through core, enter at 30°C, 1°C rise; maximum fuel surface temperature, 70°C. (**8**) A 2-Mw thermal-power nuclear reactor for research purposes has been designed with 52 units; fuel elements 10 mm OD, fuel loading 4.5 kg of U^{235}; water flow 900 m^3/hr; 8 boron carbide rods (3 safety, 5 shim), one steel regulating rod.

RPT, Reactor for Physical and Technical Investigations, Russia. (**1**) 10 Mw. (**2**) Central core consists of 37 cylindrical ducts, 54 mm in diameter with 14-cm spacing, piercing graphite layers, core 1 m in diameter, 1 m high; tubes of Al alloy including fuel elements are inserted in ducts; fuel elements are hollow cylinders containing enriched uranium, covered with an Al casing; spacer ribs on Al end caps and on Al belt in the medium section; both graphite and water act as moderator. (**3**) Core and graphite reflector fill a cylinder 240 cm high, 260 cm in diameter, 80 cm of graphite on sides, 60 cm on bottom; graphite $\rho = 1.8$ g/cm³, thermal diffusion length ~52 cm. (**4**) Side shielding, iron layer-frame 2.5 cm thick; 320 cm of concrete, $\rho = 2.4$ g/cm³; top, 150 cm of graphite, 40 cm of Pb, 20-cm-thick iron slab. (**6**) 2 automatic control rods, 3 manual, 3 slowly moving automatic rods of boron carbide, 2 systems of safety rods. (**7**) Distilled water enters annulus between fuel tube and duct and flows out through internal duct of fuel element; circulated in closed system; 20 to 30°C inlet, 55 to 65°C outlet; individual flows to ducts measured; 6 m^3/hr per duct. (**8**) Use of He for heat transfer in graphite and to dry graphite if water should leak, in. Calculated k with and without water is 1.59 ± 0.03 (1952). (Replica of APS.)

APS, Atomic Power Station, Russia. (**1**) 30 Mw (heat), 5×10^{13} (5 Mw electrical power). (**2**) Enriched uranium (5 per cent U^{235}) graphite- and water-moderated; 128 fuel channels pierce central part of graphite brickwork, forming a core 150 cm in dia. and 170 cm high; thin-walled steel used for fuel channel, and hollow fuel elements placed in channel (see PRT); 60 uranium rods required for criticality; total uranium charge is 550 kg. (**3**) Graphite encased in a hermetical steel jacket, clearance allowed in graphite brickwork, and jacket filled with He or N_2; maximum graphite temperature, 650 to 700°C. (**4**) Side water shield 100 cm thick, concrete wall 300 cm thick. (**6**) 18 boron carbide rods. (**7**) Distilled water at 100 atm pumped down through the tubes and return upflow over surface of the uranium fuel elements; recirculation with make-up system; flow and temperature of each channel recorded; inlet temperature, 190°C, outlet temperature, 260–270°C; heat exchange to generate steam at 1.25 atm, 255 to 260°C; separate cooling facility for graphite. (**8**) U^{235} burnup about 15 to 20 per cent (1954).

Geneva Reactor Exhibit, Geneva, Switzerland. (1) 10 kw (nominal) to 100 kw; 10^{11} (at center). (2) Swimming-pool type, about 4.5 kg of U^{235}, enriched to 20 per cent, 23 MTR-type fuel assemblies with 18 plates per assembly; Al-H_2O volume ratio 0.65; active lattice 15 by 15 by 24 in. (3) Light water. (4) Light water plus earth. (5) Pool size, 10 ft in diameter, 21 ft deep, 13,000 gal capacity of demineralized water. (6) Three safety and control rods, boron carbide, each about 2 per cent. (7) Light water, natural convection. (8) Built by United States for the International Conference on Peaceful Uses of Atomic Energy (1955). $350,000.

STR, Submarine Thermal Reactor (STR–Mark I, prototype, National Reactor Testing Station, Arco, Idaho; STR–Mark II, U.S.S. *Nautilus*). (7) Light water. (8) World's first mobile reactor unit; first major use of zirconium (1953). Also designated S1W, S2W.

TTR, Low Power Thermal Test Reactor, KAPL. (1) 100 watts, 3.4×10^9 (Improved version up to 10 kw). (2) 20 fuel slug tubes, each 2 in. in diameter, 24 in. long, containing U-Al alloy disks with polyethylene spacers strung on $\frac{3}{16}$-in.-diameter rod; disks (slugs) immersed in light paraffin base oil contained in slug tube; tubes in cylindrical array in an Al tank of 12-in. ID and 18-in. OD; tank filled with water; 2.7 kg of U^{235} required for criticality; extra 0.1 kg provides about 0.6 per cent excess reactivity. (3) 30-in. cylindrical ring of graphite (water, hydrocarbon). (4) Reactor located in a room with 6-ft-thick concrete walls. (5) About a 5-ft cube. (6) Coarse control, 6 Fe-clad Cd sheets, 4 in. wide, 18 in. long, at periphery of tank; fine control, 2 Cd rods, 18 in. long, $\frac{1}{2}$ in. in diameter, move between fuel slug tubes; safety, 4 Cd rods, $\frac{1}{2}$ in. in diameter. (7) For 10-kw reactor, deionized water circulated through lattice. (8) Internal thermal column is a graphite cylinder 12 in. in diameter, 18 in. high. Test hole in center of column. Several reactors of this type have been built. (16 fuel assemblies, 3-in.-diameter disks, Al alloy containing 90 per cent enriched U^{235}, clad with 2S Al. 4 Cd safety sheets, 4 Cd safety rods, 3 control rods, and neutron source.) Also designated NTR.

Fast and Intermediate Reactors

"Clementine," Los Alamos Fast Reactor. (1) 25 kw, 10^{13} fast flux. (2) Rods of pure plutonium metal clad with steel, in 6-in. array, contained in a 6-in.-diameter mild steel pot; rods of clad natural uranium interspersed with the fuel rods. (3) A round fuel pot, 6-in.-thick reflector of natural uranium built out of silver-plated blocks, 6 in. of steel followed by 4 in. of lead to form 38-in. cube. (4) Alternating layers of 3 in. of iron and masonite, and later iron and boron-impregnated plastic, total thickness of 30 in. on 3 sides; 18 in. heavy aggregate concrete. (5) 11 by 15 by 9 ft high. (6) Two safety and two regulating rods in uranium reflector, natural uranium in lower section and B^{10} in upper section, reactivity equivalent 0.004; safety block, large section of reflector, dropped to produce a reactivity change of 0.013. (7) Mercury; water used to remove heat from uranium reflector. (8) World's first fast reactor (1946). Dismantled 1954 after failure of fuel element.

EBR-1, Experimental Breeder Reactor, National Reactor Testing Station, Arco, Idaho. (1) 1400 kw, fast flux, center of core, 1.1×10^{14}. (2) Close-packed array of rods each containing a U^{235} center section 7.5 in. long with a top and bottom natural-uranium blanket section; fuel slugs (2 per rod), enriched to >90 per cent U^{235}, $4\frac{1}{4}$ in. long, 0.384 in. in diameter, jacketed in stainless steel; provisions for 217 rods (rods not containing U^{235} are loaded with uranium);

48.2 kg of U^{235} for criticality, 52 kg used; double-walled reactor tank with gas space used for leak detection and insulation. **(3)** Natural-uranium blanket in two sections; first section tightly packed array of rods $1\frac{5}{16}$ in. in diameter, $20\frac{1}{4}$ in. long, clad with 0.020-in. stainless-steel jackets; core and inner blanket contained in a $15\frac{7}{8}$-in.-diameter stainless-steel tank; outer blanket consists of an array of keystone-shaped bricks forming a cylinder around sides (and bottom) of reactor tank; bricks are natural uranium, jacketed in 0.020 in. of stainless steel; outer blanket and shield mounted on hydraulic elevator; 18 in. of graphite for reflector. **(4)** Thermal neutron shield, 6 in. of iron (air-cooled), 9 ft of ordinary concrete. **(6)** 12 control rods of natural uranium in outer blanket, 8 for shutdown (0.2 per cent k), one bottom safety (0.07 per cent k), $4\frac{1}{4}$ in. blanket travel, 0.89 per cent k; complete removal of outer blanket, 8.9 per cent k. **(7)** 292 gpm NaK, temperature rise 88°C, 228°C in, flow upward in core and downward in first blanket; outer blanket air-cooled, 6000 ft³/hr maintains bricks at 200°C; 72 per cent heat generated in core, 14 per cent inner blanket, 14 per cent outer blanket; coolant adds a reactivity equivalent to 2 kg of U^{235}. **(8)** World's first production of electrical power from a nuclear reactor (1951). 200 kw produced. Conversion ratio determined by two methods as 1.00 ± 0.04 and 1.01 ± 0.05. Second uranium core to be replaced by plutonium core. (First core produced over 3×10^6 kwhr of heat.) Fission gases collected in top section above fuel slugs.

SIR, Submarine Intermediate Reactor (SIR–Mark A, prototype, West Milton, N.Y.; SIR–Mark B for U.S.S. *Sea Wolf*). **(7)** Liquid sodium. **(8)** Prototype reactor enclosed in 225-ft-diameter 1-in.-thick steel sphere, gastight. Also designated S1G, S2G. Abandoned after test run because of leaks in superheaters.

ZEPHYR, Zero Energy Fast Reactor, Harwell, England. **(1)** 4 watts. **(2)** Cylindrical core with height = diameter $\approx$ 15 cm, consisting of natural uranium and plutonium. **(3)** Uranium. **(4)** No shielding, but reactor in small concrete room. **(6)** Control and safety rods consist of uranium rods moving vertically in channels around core; uranium safety block can be pulled out. **(7)** None. **(8)** (1954).

2. REACTOR DEVELOPMENTS
United States

Borax-1.* Operated successfully at NRTS under large power excursions, but was destroyed in 1954 in a "runaway" simulated test with a power rise over 10^6 kw in 0.1 sec. MTR-type reactor core in reactor tank, 4 ft in diameter, 13 ft high. Reactor tank filled with water to within 3 to $4\frac{1}{2}$ ft from top.

Borax-2.* Operating at 6000-kw heat under 300 psi (ten fuel plates per assembly).

Borax-3.* Produced 2 Mw of electric power in 1955–1956. First reactor to light a whole city (July 17, 1955).

Borax-4.* Employs uranium oxide fuel elements. GE and Pacific Gas and Electric Company, San José, Calif. Experimental, 3 Mw, electrical.

EBWR,* Experimental Boiling Water Reactor, ANL. Began operation on schedule in 1957 with output of 5000 kw electricity, 20 Mw heat. Light-water coolant and moderator. Reactor and power-generating equipment housed in gastight steel shell with a 400,000-ft³ volume. 77 fuel elements are natural uranium, with 35 more Zr-U^{235} elements necessary for attaining criticality. Each element contains 6 plates 0.28 by 3.75 by 48 in. and 0.37 in. apart (see Fig. M-5). Loading is 4535 kg of natural U plus 19 kg of U^{235}. Natural and enriched elements

---

* Boiling-water reactors. "Borax" is misderived from Boiling Reactor Experiment.

have the same heat capacity and generation. Nine control rods give 12.5 per cent reactivity. Operates by natural convection at 600 psig with 2.56 per cent exit quality. $4 million plus fuel. Forced convection and D_2O to be tested later.

SRE, Sodium Reactor Experiment, Atomics International, Santa Susana, Calif. To produce 20 Mw (heat). Sodium-cooled, graphite-moderated, uranium enriched to 2.8 per cent U^{235}. 6-in.-long, 0.750-in.-diameter uranium slugs stacked in 6-ft-long stainless-steel tubes, 0.010-in. walls, with NaK in 0.010-in. annulus. He in gas space above slugs. Clusters of 7 tubes form the fuel elements. Core region 6 ft high, 6 ft in diameter, containing about 31 fuel clusters. Hexagonal graphite prisms, 10 ft long, are clad with zirconium. The core tank is stainless steel, 19 ft deep, 11 ft diameter $5\frac{1}{2}$ in. steel thermal shield surrounds tank, concrete biological shield. Sodium coolant inlet temperature, 500°F; average outlet temperature 960°C. Secondary sodium coolant employed. $10 million. Criticality tests, 1957.

Tower Shielding Facility, ORNL. Reactor operated at an altitude of 200 ft. Used for experiments requiring minimal scattering from the ground and adjacent structures.

EBR-2. Operated at NRTS by ANL, probably to 1959. To produce 62.5 Mw heat, 17,500 kw electricity. Cost of reactor, $15,300,000 plus fuel. To be eventually fueled with a uranium-plutonium alloy with a blanket of natural or depleted uranium. Can also use 150 kg of 45 per cent U^{235} or 90 kg of 24 per cent Pu^{239}. Coolant is sodium. Steam is at 1800 psig and 900°F. Zero-power fast critical assembly, ZPR-3, also at NRTS, Arco, Idaho, will furnish data for critical mass, breeding ratio, and power distribution in the core of EBR-2 with U^{235}, also Pu.

HRE-2 (improved version of HRE-1). At ORNL. Initial operation to be with D_2O blanket system; fuel solution to be dilute solution of uranyl sulfate (about 90 per cent U^{235}) in D_2O, 9.6 g of U^{235} per kg D_2O, power level of 10 Mw. Later operation will use thorium oxide suspension in D_2O in blanket system, and U^{233} to be substituted for U^{235}. Design bases: inlet fuel solution temperature, 256°C; outlet, 300°C; fuel circulation rate 400 gpm; control by variable solution concentration, negative temperature coefficient (-2×10^{-3} per °C); core ID 32 in., 2000 psia; 14-in. blanket, blanket ID 60 in., 2000 psia, wall thickness 4.4 in.; core material Zircaloy-2; system, 347 stainless steel. Saturated 520-psia steam is produced. Cost of reactor, $1,800,000 plus fuel. Completed in 1956. Has associated chemical processing plant.

PWR, Pressurized Water Reactor, Shippingport, Pa. To produce 230 Mw heat, 60 Mw electricity (full-scale power plant). Light-water cooled and moderated, enriched uranium in clad plates and natural uranium in tubes. Zirconium-uranium alloy plates clad with zircaloy-2, 0.080-in. plate thickness, 0.080-in. spacing for water coolant. UO_2 pellets in Zircaloy-2 tubing, about 10-in. long, 0.413-in. OD (0.028-in. wall); 100 tubes assembled in a bundle, 7 bundles fastened in a stack to form a total height of 6 ft. Reactor core is a 6-ft-diameter right cylinder, 6-ft high, containing both highly enriched "seed" and natural uranium blanket assemblies. Seed assemblies consist of the plates, and blanket assemblies contain the tube bundles. 52 kg of enriched U^{235} in seed assemblies, and 12 tons of natural uranium in form of UO_2 in blanket. Reactor vessel over-all height 33 ft, 9 ft in diameter, nominal wall thickness $8\frac{1}{2}$ in., 2000 psia. Water flow 45,000 gpm, inlet 508°F, outlet 542°F; 600 psia saturated steam produced. 24 hafnium control rods (see Fig. M-4). $37,500,000. Designed by WAPD, operated by Duquesne Light Company. Uniformly enriched second core designed for 100 eMw. Final cost may be $55 million.

WTR, Westinghouse Testing Reactor. To operate at 20 Mw and to be used for material and reactor-component testing. U^{235}-Al fuel, H_2O moderated. Maximum thermal flux, 10^{14} (1957).

HPRR, High Performance Research Reactor, ORNL. To be of MTR type and to operate at 10 Mw.

LMFR, Liquid-Metal-Fuel Reactor, BNL. Feasibility studies recommend development of reactor systems using a solution or dispersion containing U^{233} in molten bismuth (see Figs. 13-29 and M-3). LMFRE will have 5 to 10 thermal megawatts. Operated by Babcock & Wilcox Co. (1959).

Swimming Pool, Water Boiler, and CP-5-type Reactors. More than eight reactors of these types are planned for the national laboratories, universities, and industrial groups. More than 25 universities, alone, are considering reactors for their programs in nuclear-engineering education.

SR, Savannah River Reactors. Operated by Du Pont. Five production reactors utilizing heavy water as moderator and coolant. $1 billion.

OWR, Omega West Reactor, Los Alamos. 1- to 4-Mw reactor of MTR type.

ETR, Engineering Test Reactor, NRTS. 175-Mw reactor of MTR type. Through holes up to 9 in. square. Completion in 1957. Cost $15 million.

LAPRE I.* 2000 thermal kilowatts. Used H_2O–phosphoric acid fuel solution (8.4 kg of 90 per cent U^{235}) in forced convection at 3900 psig, heated to 805°F. Steam at 3600 psia. Reactor cost, $700,000 plus fuel. Completed in 1956, then abandoned owing to corrosion of the gold lining.

LAPRE II.* 1300 thermal kilowatts. Uses H_2O–phosphoric acid fuel solution (7.7 kg of 90 per cent U^{235}) in natural convection at 800 psig, heated to 600°F. Steam is saturated, at 600 psia. Reactor cost, $100,000 plus fuel. 1956 completion.

OMRE, Organic Moderated Reactor Experiment, NRTS (Atomics International). Completion in 1957. 15 thermal megawatts. Diphenyl moderator and coolant, outlet at 530°F and 300 psig. 20 kg of 90 per cent U^{235}. Reactor cost, $875,000 plus fuel (total $1,800,000).

Experimental Plutonium Fuel Reactor. Program at HW (GE), including a small experimental reactor to utilize solid Pu, or Pu and U^{235}, in power reactors.

Gas-cooled Reactors. Study contracts with Sperry Rand Corporation and Studebaker-Packard Corporation on reactors driving gas turbines; many industrial bids for the reactor.

LAMPRE, Molten Plutonium Fast Reactor Experiment, Los Alamos. 1 thermal megawatt. NaK coolant. 1959 completion.

FIR, Food Irradiation Reactor. Design studies for an Army reactor to be optimized for γ-ray production.

IRL, Industrial Reactor Laboratory, Plainsboro, N.J. 5 thermal megawatts. BSR (AMF). First cooperative industrial reactor (1958).

Foreign

Calder Hall Piles, Britain's first commercial-scale power reactors. Produce 92 Mw electricity. Natural-uranium fuel, graphite-moderated reactor, with CO_2 under pressure as coolant; heat transferred in 70-ft-high 18-ft-diameter towers to a double-pressure steam cycle; reactor contained in 40-ft-diameter by 60-ft-high pressure shells, several inches thick.

Dounreay Reactor, Britain's first breeder reactor. To be built at Dounreay in Northern Scotland; reactor housed in steel sphere 125 ft in diameter. To produce

* Homogeneous power-reactor experiments at Los Alamos. Completion in 1956.

60 Mw (heat); core 2 ft in diameter, 2 ft long, sodium-cooled. Neutron shield 4-ft-thick graphite containing boron.

Dimple, Deuterium Moderated Pile Low Energy, Harwell, England. Britain's first heavy-water reactor (1954).

E. 443, Higher Power, Heavy-Water Reactor, Harwell, England. Designed to remove 10 Mw from core, 1 Mw from experimental facilities. 10^{14} thermal flux. Core may contain as much as 2.5 kg of U^{235} in form of uranium-Al alloy plates arranged in boxes, forming a cylinder 60 cm long, equivalent diameter 86 cm. Core placed in center of 2-m-diameter Al tank, 2 m high. Forced upward flow of D_2O. Graphite reflector outside Al tank, 60 cm thick on sides and bottom, contained in steel tank in He atmosphere. Biological concrete shield.

NRU, Canada's third reactor. To operate in 1956; to produce plutonium as well as testing components for the prototype power reactor. 275 Mw.

Belgian Reactor. Heterogeneous, natural-uranium fuel, graphite-moderated reactor under construction; United States–fabricated Al-clad fuel slugs; 18-cm lattice spacing, air-cooled, 3.5 Mw (heat).

Argentina. Plans for a natural-uranium-fueled, beryllium-oxide-moderated, heavy-water-cooled reactor, patterned partially after Daniel's Oak Ridge design concept, to produce 17,000 kw.

Netherlands. Two suspension reactors are under study using natural or slightly enriched U for power only: NUPOP (Natural Uranium Power Only Pile) uses a D_2O slurry of UO_2 at 90 atm and 300°C, developing 25 kw/kg of U. The UO_2 particles are 10 μ in size for fission products to escape to the moderator. SUSPOP (Suspension POP) employs dry UO_2 particles, fluidized and circulated through the reactor by a gas stream, below the sintering temperature of 1500°F.

3. PLANS FOR POWER-PRODUCING REACTORS
United States

AEC Five-year Reactor-development Program

See Table 13-15 and App. N-2.

Industrial (or Power-demonstration) Reactor Program

	Heat output, Mw	Electrical power output, Mw	Estimated completion date	Notes
Yankee Atomic Electric Co. (Massachusetts)	480	134	1960	Pressurized light-water moderator and coolant; $41,000,000[a]
Nuclear Power Group (Commonwealth Edison, Chicago) and General Electric Company	682	180	1960	Boiling-water type; $45,000,000 total cost[b]
Atomic Power Development Associates (Detroit Edison)	300	100	1960	Fast breeders; $45,000,000[c]
Consumers Public Power District of Nebraska and Atomics International	250	75	1959	Sodium-graphite reactor; $45,000,000[d]
Consolidated Edison (New York) and Babcock & Wilcox Co.	500	250[e]	1960	Pressurized water, uranium-thorium converter; $55,000,000 total cost[f]
Pennsylvania Power and Light Co. ...	...	150	1962	Homogeneous; 600 psia steam

[a] Coolant averages 535°F, 2000 psig. Steam is saturated at 600 psia. 28,800 kg of 2.7 per cent U^{235}.

[b] Coolant and steam average 480°F, 600 psig. 68,000 kg of 1.1 per cent U^{235}.

[c] Sodium coolant averages 800°F, 150 psig. Steam is 730°F at 600 psia. 2100 kg of 20 per cent U^{235}.

[d] Sodium coolant averages 925°F, 300 psig. Steam is 825°F at 800 psig. 24,600 kg of 1.8 per cent U^{235}.

[e] 140 Mw from reactor, 110 from oil-fired superheater (1957 design, 163 + 112 Mw).

[f] Coolant averages 500°F, 1500 psig. Steam is 420 psia saturated. 275 kg of 90 per cent U^{235} and 8100 kg of Th (1957 design estimate, $70 million; see Fig. M-6).

Smaller (or Second-round) Power-demonstration Reactor Program

The AEC invited bids on nuclear power plants of 5 to 10 Mw, 10 to 20 Mw, and 20 to 40 Mw of electrical power. Seven plants of different types were submitted [Chap. 9, Ref. 182].

American & Foreign Power Co. has ordered from Ebasco Services three 10,000-electrical-kilowatt small commercial and training reactors for Latin America, for 1958 operation (for Brazil, Cuba, and Mexico).

Mobile Reactors

Submarines........ See STR (WAPD), SIR (KAPL); SAR (Submarine Advanced Reactor) (KAPL, also designated S3G, two to be used in a radar picket submarine); SFR (Submarine Fleet Reactor); SRS (Submarine Reactor Small) (Combustion Engineering), also designated S1C (SIR later abandoned)

Ships............. LSR (Large Ship Reactor, Pressurized-water type, WAPD), also designated S1W; NROE (Naval Reactor Organic Experiment, KAPL), for light vessels; MarAd tanker (B & W)

Aircraft........... ARE (Aircraft Reactor Experiment), low-power prototype of high-temperature power-producing reactor, etc.

Portable.......... APPR (Army Package Power Reactor), pressurized-water type to produce 2 Mw electricity by 1957 (at Fort Belvoir, Va., by Alco Products, Inc.); ALPR (Argonne Low Power Reactor), boiling-water heterogeneous-reactor power plant, 200 kw (at NRTS, by ANL, for the Army)

Locomotive........ Studies by Walter Kidde Nuclear Laboratories, Inc., for Baldwin Locomotive Works

Britain's Ten-year Program*

Type of reactor	Number of reactors	Electrical output, Mw	Estimated completion date
Gas-cooled, Calder Hall type (core = $14,000,000)..........	6	300	1958–1959
Gas-cooled, improved............	4	400–800	1963
Higher-power...................	4	1000	1963–1964
Liquid-metal-cooled............	4	1000	1965
Production and power generation	6	Addition to above program	

* In March, 1957, an upward revision was announced, to yield 5000 to 6000 Mw by 1966 at a plant cost of $2,077,000,000 and initial fuel cost of $496,000,000, totaling some $450 per kilowatt.

French Five-year Program

Two plutonium-producing reactors, Marcoule, France.

G-1. Similar to Brookhaven reactor, contains 100 tons of natural uranium, elements 26 mm in diameter, 100 mm long, sheathed in Mg. 1200 tons of graphite. Air-cooled at atmospheric pressure. Under construction, to produce about 40 Mw heat and 5 Mw electricity in 1956.

G-2. Graphite-moderated; 100 tons of natural uranium, elements 26 mm in diameter, 300 mm long, sheathed in Mg. CO_2-cooled in a pressurized closed circuit. Under construction, to produce 100 to 150 Mw heat, and 30 Mw electricity.

Other Foreign Power-reactor Plans

Dutch-Norwegian Program. Ship reactor-prototype for pressurized heavy-water reactor to be built in the Netherlands.

Australia. 10-Mw power reactor planned.

Russia. One reactor with 5-Mw electrical output. Second reactor being built with 100-Mw electrical capacity. Boiling homogeneous nuclear reactor for power.

Sweden. Plans for 20-Mw reactor by 1959 to be built south of Stockholm. 100-Mw reactor by 1965 to be built in western province of Bohuslan.

Switzerland. 10-Mw (electricity) experimental natural-uranium and heavy-water reactor to be built at Würenlingen.

Canada. Atomic Energy of Canada, Ltd., and Hydro Electric Power Commission of Ontario to build 20-Mw (electricity) prototype heavy-water, natural-uranium power reactor; to be in operation 1958.

EURATOM (France, Germany, Italy, Belgium, Luxemburg, and Holland). 15,000 Mw by 1968.

APPENDIX O

ACCUMULATION AND REQUIRED DILUTION OF LONG-LIVED FISSION PRODUCTS

Basis: steady production of 1 ton of fission products per day—about 700,000 thermal Mw or 200,000 electrical Mw

Isotope (and daughter)	Fission yield, %	Half-life	Radiation, Mev β	Radiation, Mev γ	Cumulative activity 10 years	Cumulative activity ∞ years	Permissible μc/ml* (continuous nonoccupational) Air	Permissible μc/ml* (continuous nonoccupational) Water	Volume to dilute to tolerance (∞ years), mi³ Air	Volume to dilute to tolerance (∞ years), mi³ Water
Zr^{95}	6.4	65 d	$0.39(98\%)$, $1.0(2\%)$	$0.73(93\%)$, $0.23(93\%)$, $0.92(7\%)$	4.3×10^{10}	4.3×10^{10}	4×10^{-8} (as Cb)	4×10^{-4} (as Cb)	2.6×10^{8}	2.6×10^{4}
(Cb^{95})		35 d	0.15	0.76						
Ce^{144}	5.3	280 d	0.35	0	3.7×10^{10}	3.7×10^{10}	7×10^{-10}	3.6×10^{-3}	1.3×10^{10}	2.5×10^{3}
(Pr^{144})		17.5 m	3.0	0.2, 1.2						
Ru^{106}	0.5	1 y	0.03	0	3.3×10^{10}	3.3×10^{10}	2.6×10^{-9}	1.3×10^{-2}	3.1×10^{9}	620
(Rh^{106})		30 s	$3.5(82\%)$, $2.3(18\%)$	$0.51(17\%)$, $0.73(17\%)$, $1.2(1\%)$						
Pm^{147}	2.6	4.4 y	0.22	0	1.7×10^{10}	1.7×10^{10}	2×10^{-8}	0.1	2×10^{8}	40
Sr^{90}	5.3	19.9 y	0.61		1.1×10^{10}	3.5×10^{10}	2×10^{-10}	8×10^{-7}	4×10^{10}	10^{7}
(Y^{90})		62 h	2.3	0						
Cs^{137}	6.2	33 y	$0.5(95\%)$, $1.19(5\%)$		7.8×10^{9}	4.5×10^{10}	2×10^{-8}	1.5×10^{-4}	5.4×10^{8}	7.2×10^{4}
(Ba^{137})		2.6 m	0	0.66						
Tc^{99}	6.2	210,000 y	0.3	0	1.3×10^{6}	4.5×10^{10}	3×10^{-7}	3×10^{-3}	3.6×10^{7}	3.6×10^{3}
Pu^{239} (sol.)	(0.1% loss in processing)	24,000 y	(5.1α)	0.05	2.1×10^{6}	1.5×10^{9}	2×10^{-13}	1.5×10^{-7}	9×10^{11}	1.2×10^{6}
Total......	38.1/200				1.5×10^{11}	2.5×10^{11}			9.6×10^{11}	1.1×10^{7}

* From AEC, Proposed Standards for Protection against Radiation, released July 11, 1955.

FISSION CROSS SECTION OF U²³⁵

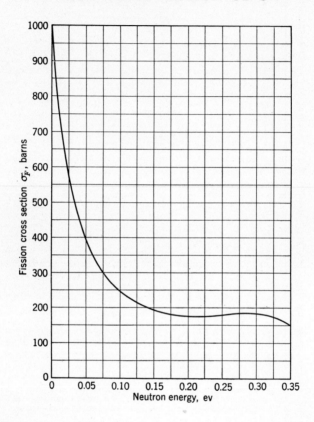

Appendix Q

NEUTRON CROSS SECTIONS OF SELECTED MATERIALS*

Element	Thermal σ_a, barns	Thermal σ_s, barns	Half-life	Radiation	Energy, Mev	σ, barns	Resonances, ev <1000 ev
$_2\text{He}^4$	0	0	....	None			
$_3\text{Li}^6$	945 (n, γ)	1.4	....	None	....	945 (n, γ)	None
$_3\text{Li}^7$	0	1.4	....	None	....	0	None
$_4\text{Be}^9$	0.010	7.5	2.7 $\times$ 10^6 years	β^-	0.560	0.009	None
$_{29}\text{Cu}^{63}$	4.3	7.2	12.8 hr	β^-	0.571		600
				β^+	0.66	3.9	
				γ	1.34, 1.20		
$_{29}\text{Cu}^{65}$	2.11	7.2	5.14 min	β^-	2.9	1.8	280
				γ	1.32		
$_{45}\text{Rh}^{103}$	150 $\pm$ 7	5 $\pm$ 1	4.5 min	β	2.0	1.26	155
				γ	0.052		
$_{47}\text{Ag}^{107}$	30 $\pm$ 2	....	2.3 min	$\text{Ag}^{108}\beta^-$	1.8	5.12	
				γ	0.60		
$_{47}\text{Ag}^{109}$	84 $\pm$ 7	....	270 days	$\text{Ag}^{110}\beta^-$	2.9		
				γ	0.7–2.0		
$_{47}\text{Ag}$	62 $\pm$ 2	6 $\pm$ 1	....	......	....	....	67
$_{49}\text{In}^{113}$	56 $\pm$ 12	....	49 days	$\text{In}^{114}\beta^-$	2.0		
				γ	0.2–0.8		
$_{49}\text{In}^{115}$	145 $\pm$ 15	....	54 min	$\text{In}^{116}\beta^-$	3.3	1.46	
				γ	1.8–2.1		
$_{49}\text{In}$	190 $\pm$ 10	2.2 $\pm$ 0.5	....	......	....	....	192
$_{53}\text{I}^{127}$	6.7 $\pm$ 0.6	3.6 $\pm$ 0.5	25 min	β^-	2.0	3.8	10.3
				γ	0.4		
$_{79}\text{Au}^{197}$	98 $\pm$ 1	9.3 $\pm$ 1	2.7 days	β^-	0.97	4.91	107
				γ	0.4		
$_{90}\text{Th}^{232}$	7.0	12.6	23.3 min	β^-	1.2	7.7	Many
$_{90}\text{Th}^{233}$	....	....	24.1 days	β^-	0.11, 0.20	1400	?
				γ	0.092		
$_{91}\text{Pa}^{233}$	....	....	1.18 min	β^-	1.52, 1.32	40	
				γ	0.8, 0.4		
			6.7 hr	β^-	0.56, 1.55	26	
				γ	0.70		
$_{92}\text{U}^{238}$	2.75	10	23.5 min	β^-	1.20, 2.06	2.8	Many
				γ	0.076, 0.92		
$_{92}\text{U}^{239}$	....	....	17 hr	β^-	....	22	

* See also Table 3-4, p. 86.

INDEX